Lecture Notes in Networks and Systems

Volume 917

The series “Lecture Notes in Networks and Systems” publishes the latest developments in Networks and Systems—quickly, informally and with high quality. Original research reported in proceedings and post-proceedings represents the core of LNNS.

Volumes published in LNNS embrace all aspects and subfields of, as well as new challenges in, Networks and Systems.

The series contains proceedings and edited volumes in systems and networks, spanning the areas of Cyber-Physical Systems, Autonomous Systems, Sensor Networks, Control Systems, Energy Systems, Automotive Systems, Biological Systems, Vehicular Networking and Connected Vehicles, Aerospace Systems, Automation, Manufacturing, Smart Grids, Nonlinear Systems, Power Systems, Robotics, Social Systems, Economic Systems and other. Of particular value to both the contributors and the readership are the short publication timeframe and the world-wide distribution and exposure which enable both a wide and rapid dissemination of research output.

The series covers the theory, applications, and perspectives on the state of the art and future developments relevant to systems and networks, decision making, control, complex processes and related areas, as embedded in the fields of interdisciplinary and applied sciences, engineering, computer science, physics, economics, social, and life sciences, as well as the paradigms and methodologies behind them.

Indexed by SCOPUS, INSPEC, WTI Frankfurt eG, zbMATH, SCImago.

All books published in the series are submitted for consideration in Web of Science.

For proposals from Asia please contact Aninda Bose (aninda.bose@springer.com).

Giancarlo Fortino · Akshi Kumar ·
Abhishek Swaroop · Pancham Shukla
Editors

Proceedings of Third International Conference on Computing and Communication Networks

ICCCN 2023, Volume 1

Springer

Editors
Giancarlo Fortino
University of Calabria
Arcavacata, Rende CS, Italy

Akshi Kumar
Department of Computing and Mathematics
Manchester Metropolitan University
Manchester, UK

Abhishek Swaroop
Department of Information Technology
Bhagwan Parshuram Institute
of Technology
New Delhi, India

Pancham Shukla
Department of Computing
Faculty of Engineering
Imperial College London
London, UK

ISSN 2367-3370 ISSN 2367-3389 (electronic)
Lecture Notes in Networks and Systems
ISBN 978-981-97-0891-8 ISBN 978-981-97-0892-5 (eBook)
https://doi.org/10.1007/978-981-97-0892-5

This Springer imprint is published by the registered company Springer Nature Singapore Pte Ltd.
The registered company address is: 152 Beach Road, #21-01/04 Gateway East, Singapore 189721, Singapore

ICCCNet-2023 Advisory Committee and Technical Program Committee

Prof. Dr. Vincenzo Piuri, University of Milan, Italy
Prof. Dr. Oscar Castillo, Tijuana Institute Technology, Tijuana, Mexico
Prof. Dr. Valentina Emilia Balas, Aurel Vlaicu University of Arad, Romania
Prof. Dr. Aboul Ella Hassanien, Cairo University, Egypt
Prof. Dr. Joel J. P. C. Rodrigues, Federal university of Piaui, Brazil
Prof. Dr. Vineet Kansal, ProVC, AKTU; Director, IET, Lucknow, India
Prof. Dr. Giancarlo Fortino, Università della Calabria, Italy
Prof. Dr. João Manuel R. S. Tavares, Universidade do Porto (FEUP), Portugal
Prof. Dr. Victor Hugo C. de Albuquerque, Federal University of Ceará, Brazil
Prof. Dr. Yu-Dong Zhang, University of Leicester, LE1 7RH, UK
Prof. Dr. Neeraj Kumar, Thapar Institute of Engineering and Technology, Patiala, Punjab, India
Prof. Dr. Zdzislaw Polkowski, Jan Wyzykowski University, Poland
Prof. Dr. George A. Tsihrintzis, University of Piraeus, Greece
Prof. Dr. Arun Sharma, Indira Gandhi Delhi Technical University for Women, India
Prof. Dr. Abhishek Swaroop, Bhagwan Parshuram Institute of Technology, India
Prof. Dr. Giorgos Karagiannidis, Aristotle University of Thessaloniki, Greece
Prof. Dr. Sheng-Lung Peng, National Dong Swa University, Taiwan
Dr. Dijana Oreski, University of Zagreb, Varazdin, Croatia
Dr. Ahmed Zobaa, Brunel University London, UK
Dr. Vicente García Díaz, University of Oviedo, Spain
Prof. Dr. A. K. Singh, National Institute of Technology Kurukshetra, India
Prof. Dr. Anil Kumar Ahlawat, KIET Group of Institutes, India
Dr. Jafar A. Alzubi, Al-Balqa Applied University, Salt—Jordan
Prof. Dr. Alex Norta, Tallinn University of Technology, Estonia
Dr. Utku Kose, Suleyman Demirel University, Isparta, Turkey
Prof. Dr. Isabel De La Torre Díez, University of Valladolid, Spain
Dr. Oana Geman, Universitatea Stefan cel Mare din Suceava, 720229 Romania
Dr. Keping Yu, Waseda University, Tokyo, Japan
Dr. Varun G. Menon, SCMS School of Engineering and Technology, Kochi, India
Dr. Mohammad Shojafar, University of Surrey, UK

Dr. Muhammad Habib ur Rehman, King's College London, UK
Dr. Irfan Mehmood, University of Bradford, UK
Dr. Muddesar Iqbal, London South Bank University, UK
Dr. Shahid Mumtaz, institute de telecommunication, Portugal
Dr. Waleed Ejaz, Lakehead University, Canada
Dr. Anish Jindal, University of Essex, UK
Dr. Suresh Chavhan, Vellore Institute of Technology, Vellore, India
Dr. D. Jude Hemanth, Karunya University, Coimbatore, India
Dr. GaganGeet Singh Aujla, Durham University, UK
Dr. Sachin Kumar, South Ural State University, Chelyabinsk, Russian Federation
Dr. Prayag Tiwari, Aalto University, Finland
Dr. Pradeep Malik, KIIT University, India
Dr. Akash Kumar Bhoi, Sikkim Manipal Institute of Technology, India
Dr. Hamid Reza Boveiri, Sama College, IAU, Shoushtar Branch, Shoushtar, Iran
Dr. Sahil Garg, École de technologie supérieure, Université du Québec, Montreal, Canada
Dr. Gulshan Shrivastava, Sharda University, Gr. Noida, India
Dr. Gabriella Casalino, University of Bari, Italy
Dr. Aditya Khamparia, Babasaheb Bhimrao Ambedkar University, India
Dr. Amit Kumar Jaiswal, University of Leeds, United Kingdom
Dr. Qianqian Xie, University of Manchester, Manchester, UK
Dr. Yousaf Bin Zikria, Yeungnam University, South Korea
Dr. Francesco Piccialli, University of Naples Federico II, Italy
Dr. Ashiq Anjum, University of Leicester, UK
Prof. Dr. Nuno M. Garcia, University of Beira Interior, Covilhã, Portugal
Prof. Dr. Kashif Saleem, Universiti Teknologi Malaysia, Riyadh, Saudi Arabia
Dr. Le Hoang Son, University of Dannang, Vietnam
Dr. Jaafar Alghazo, Virginia Military Institute, Lexington, VA
Dr. Jalil Piran, Sejong University, South Korea
Dr. Hari Mohan Pandey, Edge Hill University, UK
Prof. Dr. P. Sanjeevikumar, Aarhus University, Herning, Denmark
Dr. Kemal Polat, Abant Izzet Baysal University, Turkey
Dr. Juhriyansyah Dalle, Universitas Lambung Mangkurat, Indonesia
Dr. Ahmed Elngar, Beni—Suef university, Egypt
Dr. Prajoy Podder, Institute of Information and Communication Technology, BUET, Dhaka
Dr. M. Rubaiyat Hossain Mondal, Institute of Information and Communication Technology, BUET, Dhaka
Prof. Dr. Arun Kumar Sangaiah, VIT University, India
Prof. Dr. Robert Hsu, National Chung Cheng University, Taiwan
Dr. Sarada Prasad Gochhayat, Old Dominion University, USA
Dr. Daniel Nogueira, University of Minho, Brazil
Prof. Dr. Gwanggil Jeon, Incheon National University, South Korea
Dr. Khan Muhammad, Sejong University, South Korea
Dr. Yenumula B. Reddy, Grambling State University, Louisiana

ICCCNet-2023 Steering Committee

General Chair(s)

Prof. Dr. Omer Rana, Cardiff University, UK
Prof. Yang Xiao, The University of Alabama, USA

Honorary Chairs

Prof. Dr. Vincenzo Piuri, University of Milan, Italy
Prof. Dr. Janusz Kacprzyk, FIEEE, Polish Academy of Sciences, Poland
Prof. Dr. Manu Malek, EiC Computer and Electrical Engineering, Stevens Institute of Technology, USA
Dr. Jon G. Hall, EiC Expert Systems (WILEY), The Open University, UK

Conference Chair

Dr. Ali Kashif Bashir, Manchester Metropolitan University, UK
Prof. Dr. Darren Dancey, Manchester Metropolitan University, UK
Prof. Abhishek Swaroop, Bhagwan Parshuram Institute of Technology, India

Technical Program Chair

Prof. Dr. Joel J. P. C. Rodrigues, Senac Faculty of Ceará, Fortaleza—CE, Brazil
Prof. Dr. Giancarlo Fortino, Università della Calabria, Italy
Dr. Rajkumar Rathore, Cardiff Metropolitan University, UK

Convener

Dr. Akshi Kumar, Godsmith University of London, UK
Dr. Pancham Shukla, London Metropolitan University, London, UK
Dr. Ashish Khanna, Maharaja Agrasen Institute of Technology, India
Dr. Deepak Gupta, Maharaja Agrasen Institute of Technology, India
Dr. Utku Kose, Suleyman Demirel University, Isparta, Turkey

Publicity Chair

Prof. Dr. Valentina Emilia Balas, Aurel Vlaicu University of Arad, Romania
Prof. Dr. Zdzislaw Polokoswski, Jan Wyzykowski University, Polkowice
Prof. Dr. Vijay Singh Rathore, IIS (Deemed to be University), Jaipur
Dr. Hamid Reza Boveiri, Sama College, Islamic Azad University, Shoushtar Branch, Iran
Dr. Syed Bilal Shah, Darul Hekma University, Saudi Arabia
Dr. Umesh Gupta, Bennett University, India

Co-convener

Mr. Moolchand Sharma, Maharaja Agrasen Institute of Technology, India

Preface

We hereby are delighted to announce that Manchester Metropolitan University, Manchester, United Kingdom, has hosted the eagerly awaited and much coveted International Conference on Computing and Communication Networks (ICCCNet-2023) in Hybrid Mode during 17–18 November 2023. The conference has attracted many high-quality submissions and stimulated cutting-edge research discussions among many academic pioneering researchers, scientists, industrial engineers, and students with the reception of papers including more than 1,800 authors from different parts of the world. The committee of professionals dedicated towards the conference is striving to achieve a high-quality technical program with tracks on Networks & Computing Technologies, Advances in Artificial Intelligence & Machine Learning, Security & Privacy, Emerging Topics in 5G/6G Communication Systems, Cyber-Physical Systems, Emerging Trends in Data Analytics, Cybersecurity for Industry 4.0, and Smart and Sustainable Environmental Systems. All the tracks chosen in the conference are interrelated and are very famous among present-day research community. Therefore, a lot of research is happening in the above-mentioned tracks and their related sub-areas. More than 600 full-length papers have been received, among which the contributions are focused on theoretical, computer simulation-based research, and laboratory-scale experiments. Among these manuscripts, 121 papers have been included in the Springer proceedings after a thorough two-stage review and editing process. All the manuscripts submitted to the ICCCNet-2023 were peer-reviewed by at least two independent reviewers, who were provided with a detailed review proforma. The comments from the reviewers were communicated to the authors, who incorporated the suggestions in their revised manuscripts. The recommendations from two reviewers were taken into consideration while selecting a manuscript for inclusion in the proceedings. The exhaustiveness of the review process is evident, given the large number of articles received addressing a wide range of research areas. The stringent review process ensured that each published manuscript met the rigorous academic and scientific standards. It is an exalting experience to finally see these elite contributions materialize into two volumes as ICCCNet-2023 proceedings by Springer entitled "International Conference on Computing and Communication Networks".

ICCCNet-2023 invited two keynote speakers, who are eminent researchers in the field of computer science and engineering, from different parts of the world. In addition to the plenary sessions on each day of the conference, 12 (2 in physical mode and 10 in online mode) concurrent technical sessions are held every day to assure the oral presentation of around 121 accepted papers. Keynote speakers and session chair(s) for each of the concurrent sessions have been leading researchers from the thematic area of the session. A technical exhibition is held during all the 2 days of the conference, which has put on display the latest technologies, expositions, ideas, and presentations. The research part of the conference was organized in a total of eight special sessions. These special sessions provided the opportunity for researchers conducting research in specific areas to present their results in a more focused environment.

An international conference of such magnitude and release of the ICCCNet-2023 proceedings by Springer has been the remarkable outcome of the untiring efforts of the entire organizing team. The success of an event undoubtedly involves the painstaking efforts of several contributors at different stages, dictated by their devotion and sincerity. Fortunately, since the beginning of its journey, ICCCNet-2023 has received support and contributions from every corner. We thank them all who have wished the best for ICCCNet-2023 and contributed by any means towards its success. The edited proceedings volume by Springer would not have been possible without the perseverance of all the steering, advisory, and technical program committee members.

All the contributing authors owe thanks from the organizers of ICCCNet-2023 for their interest and exceptional articles. We would also like to thank the authors of the papers for adhering to the time schedule and for incorporating the review comments. We wish to extend my heartfelt acknowledgment to the authors, peer-reviewers, committee members, and production staff whose diligent work put shape to the ICCCNet-2023 proceedings. We especially want to thank our dedicated team of peer reviewers who volunteered for the arduous and tedious step of quality checking and critiquing the submitted manuscripts. The management, faculties, administrative, and support staff of the university have always been extending their services whenever needed, for which we remain thankful to them.

Lastly, we would like to thank Springer for accepting our proposal for publishing the ICCCNet-2023 conference proceedings. The help received from Mr. Aninda Bose, the acquisition senior editor, in the process has been very useful.

Arcavacata, Italy — Giancarlo Fortino
Manchester, UK — Akshi Kumar
New Delhi, India — Abhishek Swaroop
London, UK — Pancham Shukla

Contents

Editors and Contributors

About the Editors

Giancarlo Fortino (SM'12) is Full Professor of Computer Engineering at the Dept. of Informatics, Modeling, Electronics, and Systems of the University of Calabria (Unical), Italy. He received a Ph.D. in Computer Engineering from Unical in 2000. He is also Distinguished Professor at Wuhan University of Technology and Huazhong Agricultural University (China), High-End Expert at HUST (China), Senior Research Fellow at the Italian ICAR-CNR Institute, CAS PIFI Visiting Scientist at SIAT—Shenzhen, and Distinguished Lecturer for IEEE Sensors Council. He is Web of Science Highly Cited Researcher 2020. His research interests include wearable computing systems, e-Health, Internet of Things, and agent-based computing. He is currently the scientific responsible of the Digital Health group of the Italian CINI National Laboratory at Unical. He is Author of 500+ papers in int'l journals, conferences, and books.

Dr. Akshi Kumar is Senior Lecturer-Data Science and AI in the Department of Computing & Mathematics, Faculty of Science & Engineering, Manchester Metropolitan University, UK. She is Post-doc from Federal Institute of Education, Science and Technology of Ceará, Fortaleza, Brazil and a Ph.D. from Faculty of Technology, University of Delhi, India. She has worked as Associate Professor at the Netaji Subhas University of Technology (NSUT), New Delhi, India, and as Assistant Professor at the Delhi Technological University (DTU), New Delhi, India, and has a total teaching and research experience of 16 years. She has published more than 85 peer-reviewed journal papers including 50 SCIE publications, 65+ conference papers with 4 best paper awards and 2 patents with Indian Patent Office. She has successfully guided 7 doctorates, 33 Master thesis candidates. She has been serving as Associate Editor and Guest Editor in various high impact journals with reputed publishers.

Prof. Dr. Abhishek Swaroop completed his B.Tech. (CSE) from GBP University of Agriculture & Technology, M.Tech. from Punjabi University Patiala, and Ph.D. from NIT Kurukshetra. He has 28 years of teaching and Industrial experience. He has served in reputed educational institutions such as Jaypee Institute of Information Technology, Noida, Sharda University Greater Noida and Galgotias University, Greater Noida. He is actively engaged in research. He has guided 12 M.Tech. Dissertations also. He has authored 3 books and 5 book chapters. His 10 papers are indexed in DBLP and 16 papers are SCI. He had been part of the organizing committee of three IEEE conferences (ICCCA-2015, ICCCA-2016, ICCCA-2017), one Springer conference (ICICC-2018) as Technical Program Chair. He is Member of various professional societies like CSI and ACM and editorial board of various reputed journals.

Dr. Pancham Shukla is Principal Teaching Fellow in the Faculty of Engineering at Imperial College London. He holds more than 25 years of teaching and research experience at various Universities in UK and India. He holds Ph.D. in Electrical and Electronic Engineering from Imperial College London. He has supervised a broad range of projects and published papers in signal/image processing, electronic, and communications systems. He has organized conferences and has served as Technical Session Chair along with reviewing many international conference and journal papers. He received Vice Chancellor's award for outstanding contribution to Learning and Teaching as well as Student Union's award as an outstanding member of academic staff while working at London Metropolitan University.

Contributors

Laith Abualigah Computer Science Department, Al al-Bayt University, Mafraq, Jordan;
MEU Research Unit, Middle East University, Amman, Jordan;
Applied Science Research Center, Applied Science Private University, Amman, Jordan;
Jadara Research Center, Jadara University, Irbid, Jordan

Vaibhav Agrohi Department of Computer Science and Engineering Department, Arya College of Engineering, Jaipur, Rajasthan, India

Oindrila Ajha Kalinga Institute of Industrial Technology, Deemed to Be University, Bhubaneswar, India

Abiola Akinnubi University of Arkansas, Little Rock, Arkansas, USA

Md. Akkas Ali School of Computing Science and Engineering, Galgotias University, Greater Noida, UP, India

Mohd Mohsin Ali Bennett University, Greater Noida, India

Ericson O. Alieto Western Mindanao State University, Zamboanga City, Philippines

Ahmed Alkhayyat College of Technical Engineering, Islamic University, Najaf, Iraq;
College of Technical Engineering, Islamic University of Al Diwaniyah, Al Diwaniyah, Iraq;
College of Technical Engineering, Islamic University of Babylon, Baghdad, Iraq

Heba Al-Mamouri Software Department, College of Information Technology, University of Babylon, Babylon, Iraq

Jafar A. Alzubi Faculty of Engineering, Al-Balqa Applied University, As-Salt, Jordan

Ayan Ambesh Vivekananda School of Information Technology, VIPS-TC, Delhi, India

Ridwan Amure University of Arkansas, Little Rock, Arkansas, USA

Harshit Anand Kalinga Institute of Industrial Technology, Deemed to Be University, Bhubaneswar, India

R. J. Anandhi New Horizon College of Engineering, Bengaluru, India

Greeshma Andrew Cochin University of Science and Technology, Kochi, India

K Antony Kumar Department of Computer Science and Engineering, Vel Tech Rangarajan Dr. Sagunthala R&D Institute of Science and Technology, Chennai, Tamilnadu, India

Jayachandran Arumugam Presidency University, Bangalore, India

Mohammad Atique Department of Computer Science and Engineering, Sant Gadge Baba Amravati University, Amravati, Maharashtra, India

Wadhah R. Baiee Software Department, College of Information Technology, University of Babylon, Babylon, Iraq

Tamoghno Bakshi Kalinga Institute of Industrial Technology, Deemed to be University, Bhubaneswar, India

Avishek Banerjee Department of CSBS, Asansol Engineering College, Asansol, India

Aniruddha Bardekar Department of Computer Science and Engineering, Sant Gadge Baba Amravati University, Amravati, Maharashtra, India

Noorulden Basil Department of Electrical Engineering, College of Engineering, Mustansiriyah University, Baghdad, Iraq

Delphine Battistelli Paris Nanterre University, Nanterre, France

Malika Bendechache University of Galway, Galway, Ireland

Ajay Kumar Bharti Babu Banarsi Das University, Lucknow, India

Anuj Kumar Bharti Bennett University, Greater Noida, India

Swastika Bishnoi Kalinga Institute of Industrial Technology, Deemed to Be University, Bhubaneswar, India

Spardha Bisht School of Management, Graphic Era Hill University, Bhimtal Campus, Dehradun, Uttarakhand, India

Rob Brennan University College Dublin, Dublin, Ireland

Declan Byrne Saint James's Hospital, Dublin, Ireland

M. J. Carmel Mary Belinda Saveetha School of Engineering, SIMATS, Chennai, Tamilnadu, India

Princess Krizzle M. Casiano Western Mindanao State University, Zamboanga City, Philippines

Subhajit Chakrabarty Guwahati, India

Partha Chakraborty Department of Computer Science and Engineering, Comilla University, Cumilla, Bangladesh

E. Chandralekha Vel Tech Rangarajan Dr. Sagunthala R&D Institute of Science and Technology, Chennai, Tamilnadu, India

Soumia Chandran Government Engineering College, Thrissur, Kerala, India

Vaibhav Chaudhary Thapar Institute of Engineering and Technology, Patiala, Punjab, India

Anand Chauhan 1SCSET, Bennett University, Greater Noida, India

Priyanka Chhabra Indira Gandhi Delhi Technical University for Women, Kashmere Gate, New Delhi, India

Shailja Chhillar Indira Gandhi Delhi Technical University for Women, Kashmere Gate, New Delhi, India

Sibashish Choudhury Kalinga Institute of Industrial Technology, Deemed to be University, Bhubaneswar, India

Donnacha Creagh Saint James's Hospital, Dublin, Ireland

Nahidah T. Darweesh College of Computer Science and Information Technology, University of Anbar, Ramadi, Iraq

Anupam Das Guwahati, India

Soham Das Kalinga Institute of Industrial Technology, Deemed to Be University, Bhubaneswar, India

Souryadipta Das Kalinga Institute of Industrial Technology, Deemed to Be University, Bhubaneswar, India

S. R. Deepak Department of Computer Science and Engineering, Rajalakshmi Engineering College, Chennai, Tamilnadu, India

M. P. Deepika Cochin University of Science and Technology, Kochi, India

O. Rama Devi Department of Artificial Intelligence and Data Science, Lakireddy Bali Reddy College of Engineering (Autonomous), Mylavaram, AP, India

Rajesh Kumar Dhanaraj Symbiosis Institute of Computer Studies and Research (SICSR), Symbiosis International (Deemed University), Pune, India

Zouina Doukha RIIMA Laboratory, Computer Science Faculty, USTHB, Bab Ezzouar, Algiers, Algeria

Valentina Dragos ONERA-The French Aerospace Lab, Palaiseau, France

Ashit Kumar Dutta College of Applied Sciences, Al Maarefa University, Riyadh, Saudi Arabia

Tiyasha Dutta Kalinga Institute of Industrial Technology, Deemed to Be University, Bhubaneswar, India

Malick Ebiele University College Dublin, Dublin, Ireland

Damodar Reddy Edla National Institute of Technology, Farmagudi, Ponda, Goa, India

Bashar Bahaa Qas Elias Department of Information and Communication Engineering, College of Information Engineering, Al-Nahrain University, Baghdad, Iraq

Bernadeth A. Encarnacion Western Mindanao State University, Zamboanga City, Philippines

Venkateswara Prasad Evani GVP College of Engineering for Women, Visakhapatnam, India

Amit Kumar Garg Department of Electronics and Communications, Deenbandhu Chhotu Ram University of Science and Technology, Murthal, India

Ankit Garg AIT-CSE, Chandigarh University, University Centre for Research and Development (UCRD), Mohali, Punjab, India

Una Geary Saint James's Hospital, Dublin, Ireland

Deepa Gupta GL Bajaj Institute of Management, Greater Noida, India

Deepak Gupta Department of Computer Science and Engineering, National Institute of Technology Arunachal Pradesh, Jote, Arunachal Pradesh, India; MAIT, Delhi, India

Mukul Gupta GL Bajaj Institute of Management, Greater Noida, India

Pradeep Gupta Ajay Kumar Garg Engineering College, Ghaziabad, Uttar Pradesh, India

Sonam Gupta Ajay Kumar Garg Engineering College, Ghaziabad, Uttar Pradesh, India

Sudhanshu Gupta 1SCSET, Bennett University, Greater Noida, India

Umesh Gupta School of CSET (SCSET), Bennett University, Greater Noida, India

Ameer Hamza Department of Computer Science, HITEC University, Taxila, Pakistan

Afshan Hashmi Department of Computer Science & Engineering, Jamia Hamdard, New Delhi, India

Yaqot Mohsin Hazzaa College of Computer Science and Information Technology, University of Kirkuk, Kirkuk, Iraq;
Ministry of Youth and Sport, Baghdad, Iraq

Haider K. Hoomod College of Education, Computer Science Department, Al-Mustansiriya University, Baghdad, Iraq

Imran Hussain Department of Computer Science & Engineering, Jamia Hamdard, New Delhi, India

Gaurav Indra Indira Gandhi Delhi Technical University for Women, Kashmere Gate, New Delhi, India

Al-Amin-Ul Islam Department of Information and Communication Technology, Comilla University, Cumilla, Bangladesh

Shania H. Jaafar Mindanao State University-Sulu, Sulu, Philippines

Kiran Jabeen Department of Computer Science, HITEC University, Taxila, Pakistan

Akshat Jain Bennett University, Greater Noida, India

Vasu Jain 1SCSET, Bennett University, Greater Noida, India

Srikanth Jannu Vaagdevi Engineering College, Warangal, Telangana, India

T. Jayaganeshan Department of Artificial Intelligence and Machine Learning, Rajalakshmi Engineering College, Chennai, Tamil Nadu, India

Tuka Kareem Jebur Department of Accounting, Al-Mustansiriyah University, Baghdad, Iraq

S. Kabilesh Department of Artificial Intelligence and Machine Learning, Rajalakshmi Engineering College, Chennai, Tamil Nadu, India

Aqeel A. Kadhim Computer Techniques Engineering Department, Imam AlKadhim University College, Baghdad, Iraq

Sahil Kadiyan Thapar Institute of Engineering and Technology, Patiala, India

Gauri Kalnoor Department of Computer Science and Engineering, BMSCE, Bangalore, India

Deepali Kamthania Vivekananda School of Information Technology, VIPS-TC, Delhi, India

Shailesh Kasande Suryadatta Group of Institutes, Pune, India

Harkiran Kaur Thapar Institute of Engineering and Technology, Patiala, India

Raminder Kaur Department of Computer Science and Engineering Department, JECRC University, Jaipur, India

Vandana Dixit Kaushik Department of Computer Science & Engineering, Harcourt Butler Technical University Kanpur, Kanpur, Uttar Pradesh, India

Balasubramanian Prabhu Kavin Department of Data Science and Business Systems, College of Engineering and Technology, SRM Institute of Science and Technology, Kattankulathur, Chengalpattu District, Tamil Nadu, India

Tamanna Kewal University School of Business, Chandigarh University, Mohali, Punjab, India

Thura Jabbar Khaleel Iraqi Commission for Computers and Information/ Informatics Institute for Postgraduate Studies, Baghdad, Iraq

Muhammad Attique Khan Department of Computer Science, HITEC University, Taxila, Pakistan

Ibtihal Ali Khanjar Education College, Mustansiriyah University, Baghdad, Iraq

Leila Kheroua RIIMA Laboratory, Computer Science Faculty, USTHB, Bab Ezzouar, Algiers, Algeria

Oyindamola Koleoso University of Arkansas, Little Rock, Arkansas, USA

S. Siva Rama Krishna Department of Artificial Intelligence and Data Science, Lakireddy Bali Reddy College of Engineering (Autonomous), Mylavaram, AP, India

Alok Kumar Department of Computer Science and Engineering, School of Engineering and Technology (Formerly Known As UIET), Chhatrapati Shahu Ji Maharaj University, Kalyanpur, Kanpur, Uttar Pradesh, India

Ashutosh Kumar Kalinga Institute of Industrial Technology, Deemed to be University, Bhubaneswar, India

Bijendra Kumar Department of Computer Science and Engineering, NSUT, Delhi, India

Mahesh Kumar Department of Computer Science and Engineering Department, Graphics Era Deemed to Be University, Dehradun, India

Manish Kumar Bisht Graphic Era Hill University, Haldwani Campus, Uttarakhand, India

Waleed Hadi Madhloom Kurdi Altoosi University College, Najaf, Iraq

Monica Lamba Department of Computer Science and Engineering Department, Arya College of Engineering, Jaipur, Rajasthan, India

Angeline Lydia Department of Computer Science and Engineering, Vel Tech Rangarajan Dr. Sagunthala R&D Institute of Science and Technology, Chennai, Tamil Nadu, India

S. Madhesh Department of Artificial Intelligence and Machine Learning, Rajalakshmi Engineering College, Chennai, Tamil Nadu, India

Kaushik Majumdar Department of CSE, MAKAUT, Kolkata, India

J. Alamelu Mangai School of CSE & IS, Presidency University, Bengaluru, India

Eashaan Manohar Department of Computer Science and Engineering, Rajalakshmi Engineering College, Chennai, Tamilnadu, India

Hamzah M. Marhoon Department of Systems Engineering, College of Information Engineering, Al-Nahrain University, Jadriya, Baghdad, Iraq

Fozilatunnesa Masuma Department of Computer Science and Engineering, Comilla University, Cumilla, Bangladesh

Md Tabrez Nafis Department of Computer Science & Engineering, Jamia Hamdard, New Delhi, India

Jade Mekki Paris Nanterre University, Nanterre, France

P. S. Metkewar Symbiosis Institute of Computer Studies and Research (SICSR), Symbiosis International (Deemed University), Pune, Maharashtra, India; School of Computer Science and Engineering, Dr Vishwanath Karad MIT World Peace University, Kothrud, Pune, India

Prakash B. Metre Department of Computer Science and Engineering, Presidency University, Bangalore, India

Momina Mir Department of Computer Science, HITEC University, Taxila, Pakistan

Sushruta Mishra Kalinga Institute of Industrial Technology, Deemed to be University, Bhubaneswar, India

F. Mohammed Abraar Khan Department of Artificial Intelligence and Machine Learning, Rajalakshmi Engineering College, Chennai, Tamil Nadu, India

Md. Mohi Uddin Department of Computer Science and Engineering, The people's University of Bangladesh, Dhaka, Bangladesh

Mohd. Mohsin 1SCSET, Bennett University, Greater Noida, India

K. Moorthi Department of Computational Intelligence, College of Engineering and Technology, SRM Institute of Science and Technology, Kattankulathur, Chengalpattu District, Tamil Nadu, India

Samira Moussaoui RIIMA Laboratory, Computer Science Faculty, USTHB, Bab Ezzouar, Algiers, Algeria

Sameena Naaz Department of Computer Science, University of Roehampton, London, UK

Jolan R. Naif Informatics Institute for Postgraduate Studies, Iraqi Commission for Computers and Informatics, Baghdad, Iraq

Harsh Pal Vivekananda School of Information Technology, VIPS-TC, Delhi, India

Sourojit Pal Department of IT, Jadavpur University, Kolkata, India

Parnani Panda Kalinga Institute of Industrial Technology, Deemed to be University, Bhubaneswar, India

Ayushman Pranav Bennett University, Greater Noida, India

P. C. Rafeeque Government College of Engineering, Kannur, India;
APJ Abdul Kalam Technological University, Thiruvananthapuram, Kerala, India

Priti Rai GL Bajaj Institute of Management, Greater Noida, India

Saloni Rai Department of Electronics and Communications, Deenbandhu Chhotu Ram University of Science and Technology, Murthal, India

Manish Raj 1SCSET, Bennett University, Greater Noida, India

B. Rajalakshmi New Horizon College of Engineering, Bengaluru, India

T. S. Rajarajeswari Department of Artificial Intelligence and Data Science, Lakireddy Bali Reddy College of Engineering (Autonomous), Mylavaram, AP, India

Saurabh Rana Department of Mathematics (SCSET), Bennett University, Greater Noida, India

S Ravikumar Department of Computer Science and Engineering, Vel Tech Rangarajan Dr. Sagunthala R&D Institute of Science and Technology, Chennai, Tamil Nadu, India

Vivian Rawade Kalinga Institute of Industrial Technology, Deemed to Be University, Bhubaneswar, India

Shailendra Kumar Rawat Amity Institute of Information Technology, Lucknow, India

P. C. Reghu Raj Government Engineering College, Kozhikkode, India;
APJ Abdul Kalam Technological University, Thiruvananthapuram, Kerala, India

K. Reji Rahmath Government Engineering College, Palakkad, India;
APJ Abdul Kalam Technological University, Thiruvananthapuram, Kerala, India

Joel J. P. C. Rodrigues College of Computer Science and Technology, China University of Petroleum (East China), Qingdao, China;
Instituto de Telecomunicações, Covilhã, Portugal

Proshanta Roy Department of Computer Science and Engineering, Comilla University, Cumilla, Bangladesh

Sandip Roy VMASC, Old Dominion University, Suffolk, VA, USA

Ali Makki Sagheer College of Computer Science and Information Technology, University of Anbar, Ramadi, Iraq

Chayti Saha Department of Computer Science and Engineering, Comilla University, Cumilla, Bangladesh

Abhipsa Sahoo Kalinga Institute of Industrial Technology, Deemed to Be University, Bhubaneswar, India

Shubham Sahoo Kalinga Institute of Industrial Technology, Deemed to Be University, Bhubaneswar, India

B. Saiganesh Department of Electronics and Communication Engineering, Rajalakshmi Engineering College, Chennai, Tamilnadu, India

Jatinderkumar R. Saini Symbiosis Institute of Computer Studies and Research, Symbiosis International (Deemed University), Pune, India

Muna M. Salman Information Technology Directorate, Ministry of Science and Technology, Baghdad, Iraq

Ankita Samal Kalinga Institute of Industrial Technology, Deemed to Be University, Bhubaneswar, India

Kaipa Sandhya Presidency University, Bangalore, India

S. Sanjai Department of Computer Science and Engineering, Rajalakshmi Engineering College, Chennai, Tamilnadu, India

Riya Sanjesh School of CSE & IS, Presidency University, Bengaluru, India

Shweta Saraswat Department of Artificial Intelligence and Data Science Department, Arya College of Engineering, Jaipur, Rajasthan, India

Charu Saxena University School of Business, Chandigarh University, Mohali, Punjab, India

U. Surya Venkata Sekhar Department of Artificial Intelligence and Data Science, Lakireddy Bali Reddy College of Engineering (Autonomous), Mylavaram, AP, India

Santoshi Sengupta School of Management, Graphic Era Hill University, Bhimtal Campus, Dehradun, Uttarakhand, India

Alpna Sharma Vivekananda School of Information Technology, VIPS-TC, Delhi, India

Anupam Kumar Sharma School of Computing Science and Engineering, Galgotias University, Greater Noida, UP, India

Pritika Sharma Thapar Institute of Engineering and Technology, Patiala, India

Shilpi Sharma Department of Computer Science and Engineering, NSUT, Delhi, India

Soumya Sharma The NorthCap University, Gurugram, India

Srishti Sharma The NorthCap University, Gurugram, India

Sachin Shetty VMASC, Old Dominion University, Suffolk, VA, USA

Nadia Adnan Shiltagh Computer Engineering Department, College of Engineering, University of Baghdad, Baghdad, Iraq

Shivendra Shivani Thapar Institute of Engineering and Technology, Patiala, Punjab, India

Akansha Singh Bennett University, Greater Noida, India

Akshat Singh Vivekananda School of Information Technology, VIPS-TC, Delhi, India

Divya Singh Ajay Kumar Garg Engineering College, Ghaziabad, Uttar Pradesh, India

Milind Singh Kalinga Institute of Industrial Technology, Deemed to Be University, Bhubaneswar, India

Sanjay Kumar Singh Amity Institute of Information Technology, Lucknow, India

Shubham Singh Vivekananda School of Information Technology, VIPS-TC, Delhi, India

Vineeta Singh Department of Computer Science and Engineering, School of Engineering and Technology (Formerly Known As UIET), Chhatrapati Shahu Ji Maharaj University, Kalyanpur, Kanpur, Uttar Pradesh, India

Abhigyan Sinha Kalinga Institute of Industrial Technology, Deemed to be University, Bhubaneswar, India

K. R. Sowmia Department of Artificial Intelligence and Machine Learning, Rajalakshmi Engineering College, Chennai, Tamil Nadu, India

Divya Srivastava Bennett University, Greater Noida, Uttar Pradesh, India

Himanshu Srivastava Bennett University, Greater Noida, India

Snahil Subhra Kalinga Institute of Industrial Technology, Deemed to Be University, Bhubaneswar, India

Shrishti Swaraj Kalinga Institute of Industrial Technology, Deemed to be University, Bhubaneswar, India

Prince Chandra Talukder Department of Computer Science and Engineering, Comilla University, Cumilla, Bangladesh

Riya Tanwar Indira Gandhi Delhi Technical University for Women, Kashmere Gate, New Delhi, India

Chaitanya Thuppari National Institute of Technology, Farmagudi, Ponda, Goa, India

Md. Tofazzal Hosen Department of Computer Science and Engineering, The people's University of Bangladesh, Dhaka, Bangladesh

Venkata Bhanu Prasad Tolety Scholar, Jawaharlal Nehru Technological University, Kakinada, India

Muhammad Sami Ullah Department of Computer Science, HITEC University, Taxila, Pakistan

Shahla U. Umar College of Computer Science and Information Technology, University of Kirkuk, Kirkuk, Iraq

Bhawna Upadhayay Ajay Kumar Garg Engineering College, Ghaziabad, Uttar Pradesh, India

Shraddha Vaidya Symbiosis Institute of Computer Studies and Research, Symbiosis International (Deemed University), Pune, India

Michele Della Ventura Department of Music Technology, Music Academy "Studio Musica", Treviso, Italy

Deepak Kumar Verma Department of Computer Science and Engineering, School of Engineering and Technology (Formerly Known As UIET), Chhatrapati Shahu Ji Maharaj University, Kalyanpur, Kanpur, Uttar Pradesh, India

Muskan Verma Indira Gandhi Delhi Technical University for Women, Kashmere Gate, New Delhi, India

Sanchit Verma Kalinga Institute of Industrial Technology, Deemed to Be University, Bhubaneswar, India

K. Vijay Department of Computer Science and Engineering, Rajalakshmi Engineering College, Chennai, Tamilnadu, India

Marie Ward Saint James's Hospital, Dublin, Ireland

Rohit Yadav Department of Computer Science and Engineering, Vel Tech Rangarajan Dr. Sagunthala R&D Institute of Science and Technology, Chennai, Tamil Nadu, India

Intisar Abd Yousif Education College, Mustansiriyah University, Baghdad, Iraq

Mohammad Abu Yousuf Institute of Information Technology, Jahangirnagar University, Savar, Dhaka, Bangladesh

Multiscale Feature Pyramid Network-Enabled Deep Learning and IoT-Based Pest Detection System Using Sound Analytics in Large Agricultural Field

Md. Akkas Ali, Anupam Kumar Sharma, and Rajesh Kumar Dhanaraj

Abstract Modern farming techniques can now be implemented globally and at a reasonable cost. In the annals of agriculture, this is a significant turning point moment. The widespread dissemination of information and the advent of cutting-edge technologies have facilitated this revolution. However, pests cause substantial harm to farmland, which has economic, ecological, and societal consequences. As a result, there is a need to employ sophisticated computational methods to eradicate pests before they cause irreparable harm. Recently, ML-based studies have focused on agricultural issues. The study's primary objective is to develop a workable method for identifying pests in big agricultural fields using dependable pest analytics made possible by the Internet of Things. The proposed method incorporates several different sound pre-processing techniques drawn from the field of sound analytics. BPF, Triangular window, Kaiser window, FFT method, DFT, and PLP are all examples. A Multiscale Feature Pyramid Network (MFP-Net) model was trained, tested, and validated using data from an analysis of 650 pest sounds. Compared to the existing MS-ALN, YOLOv5, Faster RCNN, and ResNet-50 models, the recommended MFP-Net model correctly identified the pest with a 99.76% accuracy, 99.03% recall, and 99.06% F1-score. The capability of this research to detect pests in large agricultural fields early on is the primary reason for its significance. As a direct result, the number of crops produced will increase, improving the economic growth of the farmers, the nation, and the world.

Keywords Sound analytics · ACF · GFA · MFP-Net · IoT · Pest detection

Md. Akkas Ali · A. K. Sharma (✉)
School of Computing Science and Engineering, Galgotias University, Greater Noida, UP, India
e-mail: anupam.sharma@galgotiasuniversity.edu.in

R. K. Dhanaraj
Symbiosis Institute of Computer Studies and Research (SICSR), Symbiosis International (Deemed University), Pune, India

G. Fortino et al. (eds.), *Proceedings of Third International Conference on Computing and Communication Networks*, Lecture Notes in Networks and Systems 917,
https://doi.org/10.1007/978-981-97-0892-5_1

1 Introduction

Agricultural pests can damage the economy, environment, and society. These effects may hurt agriculture and nutrition. Pests and diseases cost farmers money. This lowers revenue and farming profitability. Pesticides and other equipment may cost farmers more. If these expenses lower profit margins, agriculture may suffer. Quarantine and trade restrictions may follow agricultural pests. Crop pests would reduce export markets and farmers' income. Pests impact other organisms, disrupting ecosystems. High pesticide dosages may impact the local environment. Birds, pollinators, and beneficial insects may be harmed. Root-eating insects can damage soil fertility and agriculture. Pesticide runoff, water contamination, aquatic habitat loss, and health concerns are further issues. Pests diminish crop yield, affecting global food security. Meeting global requirements is getting harder. Rats and mosquitoes carry dangerous infections. Malaria, dengue, lyme, and hantavirus. Pests can ruin the agricultural economy. Farm revenue declines may generate rural migration, urbanization, unemployment, and poverty. Pests may adapt to new settings, reproduce quickly, and withstand insecticides. Climate change may alter pest populations and distribution, complicating agriculture. IPM reduces pests and their environmental impacts through preventative approaches, biological controls, and appropriate pesticide use. This plan promotes sustainable farming, biodiversity conservation, and non-chemical pest management. Priorities are long term. Without agriculture, many nations would fail. Pests that carry agricultural diseases harm plant health and output. Exterminating many pests requires vigilance. Solutions have been suggested for traps, new parts, chopping and breaking, and collected feces. Manual crop inspection is time-consuming and inaccurate. Farmers poison enormous areas with chemicals. Experts meticulously count and identify insect species [1]. This approach approximates agricultural pest density. Agriculture workers must detect and classify the many pests that might damage crops. Early insect pest detection and classification are essential for chemical and biological management. Insect identification is tedious and error-prone. Agricultural research challenges prompted a machine learning-based automated photo processing system [2–4]. This method classifies insects. An intelligent agricultural system needs computer vision to progress agricultural technology quickly, accurately, and reliably [5]. The automated agriculture support system is promising despite crop monitoring and pest tracking. ML and DL can now forecast insects and other pests from low-quality pictures. This is much better. DL replicates neural networks to emulate brain processing. DL features updated hardware and software. Pictorial categorization is traditional. This paper examines insect classification [6]. 75,222 IP102 pictures test the categorization algorithm. Pests include 102 species. Experiment with different models and integrate the results with a projected ensemble model technique that maximizes output using many hyperparameters and an ensemble of voting classifiers. [7–9] Research-based reliability. Manual bug classification is tedious. AI and computer science classify pests. Despite stagnation, new AI technologies applying learning algorithms identify pests as ready-made solutions. Handmade parts are inefficient. Some insects are good for motion and postural variation statistical sequences. Detection needs tracking, pest identification, and population estimates [10]. SSD and YOLO coexist. They use vector layers to provide default margins for different sizes. High-level SSD performance maps reveal concerns. SSD identifies grain bugs immediately. Several criteria quickly and accurately detect grain pests. Empty containers signify a white background and data-gathering grains. Nine thousand nine hundred ninety pictures have been uploaded. All 1998 old photos were used [11]. This study investigates the MFP-Net model for an Internet of Things-enabled pest detection system that uses reliable pest analytics in critical agricultural regions. MFP-Net-trained insects. We statistically examined 650 pest-related noises from 13 species using a PLP coefficient matrix. The study supports human-free field monitoring. The following is the list of the objectives, specific goals, and contributions of this research:

- To evaluate the literature, locate unmet needs in the rapidly expanding Internet of Things (IoT)-based agricultural sector, specifically concerning pest identification. This literature evaluation provided direction for future study. This involves boosting the accuracy of IoT-enabled networks in large agricultural areas for identifying pest sounds and using cutting-edge analytics to boost GDP and crop yields.
- To implement the Internet of Things and multi-module DL technologies into a pest detection system that is both highly effective and inexpensive.
- To be optimal to train, validate, and test the MFP-Net model in a setting with fewer distractions, since this would lead to better results in the long run.
- To achieve optimum agricultural production requires enhancing the efficiency with which pests may be detected.

2 Literature Review

Qian et al. [12] created a unique automatic data augmentation method that uses adaptive methods to pick the appropriate augmentation strategy to increase data modeling productivity. Res2Net injects FPNs to understand tiny annoyances. It helped identify minute pests. TTA shows pest images at different scales. They tested our technology on 4000 photos of four pests. Pest detection mAP is 81.0% using the technique. Chu et al. [13] enlarged seven typical granary pests photographed. Microscope, smartphone, DSR camera, and internet scanner took 5230 photos. These photographs illustrate many barn pests. They developed a BiFPN with intersecting join distances to detect objects. This model averages 98.2% across seven pests. Ablation proved it. YOLOv5s network model and attention mechanism averaged 96.9% accuracy. BiFPN replaced PANet with 97.2% accuracy. Visual representation studies were conducted. The model captures essential pest features. The model can detect grain storage facility pests at various scales. Anwar et al. [14] suggested creating an ensemble-based model with transfer learning. The experimental configuration chooses the optimum solution using pre-trained models like VGG16, VGG19, and ResNetv50. The Ensemble Voting Classifier learns and predicts the input sample.

The IP102 reference dataset contains 75,000 samples from 102 classes for model training and evaluation. It detects pests with 82.5% accuracy. Chithambarathanu and Jeyakumar [15] said the plan maximizes crop protection and yields. This cropping method maximizes plant protection and yield. It detects and classifies citrus, rice, and cotton pests. It reviews pest detection methods for agricultural field monitoring. These technologies automate big domain monitoring, saving time and preventing errors. Umamaheswari et al. [16] tried to detect and identify numerous plant diseases before. Crops, fruits, and vegetables are in short supply due to the late twentieth century population increase. Pests and diseases cut yield by 30% yearly. Pests destroy crops and spread disease. Experts can see sickness or infestation. This ailment may require expensive doctor appointments. Pests must be eradicated immediately due to the harm. AI and ML can detect early. Recurrent and convolutional neural networks classify plant diseases. Object detection and picture recognition models identify and classify plant diseases early. This model currently forecasts regional pest-caused plant diseases better. This research involves seeking a better way to stop pests from causing substantial damage to crops, vegetables, and fruits. Insect infestations cause crop failures and food shortages [17]. Early pest identification helps farmers enhance crop quality and yield with cost-effective preventative measures. Because bugs look alike, hand-identifying them takes time. Expert counsel is also scarce. Deep learning automated bug identification and classification to prevent manual inspection issues. CornerNet is a lightweight drone-based DenseNet-100 version. The three-part structure follows. A DenseNet-100-based customized CornerNet follows. CornerNet categorizes pests. DenseNet integrates feature maps from all layers to help CornerNet recognize insect pests as critical point pairs. The IP102 pest recognition reference dataset was chosen to evaluate the proposed model since it comprises pests of diverse sizes, colors, orientations, categories, chromaticity, and lighting conditions. It outperforms current field insect detection methods. YOLOv5m helped Dai et al. [18] identify plant insects. SWin Transformer and Transformer techniques improve YOLOv5m network global function capture and receptivity. ResSPP captures spine data. Finally, WConcat must be added to the aggregation function to improve network function aggregation. YOLOv5m achieved 97% accuracy. YOLO wins. YOLOv5m's enhanced pest-detecting accuracy. Mallick et al. [19] developed mung crop damage detection. Since mung bean crop pictures are scarce, our model was trained via transfer learning. This approach could discover dangerous pathogens faster. Using seasonal leaf samples, the model identified six mung bean illnesses and four pests. 93.65% accuracy. Yang et al. [20] provide Maize-YOLO, a real-time, high-precision corn pest monitoring technique. CSPResNeXt-50 and VoVGSCSP modules extend YOLOv7's backbone. The model discovers new networks faster and more accurately with less computing complexity. IP102 tested Maize-YOLO. 4533 photos of 13 common corn-damaging insects were tested and taught. 77.3% mAP and 76.3% recollection increase experiment findings. This approach identifies maize field pests in real time. Zhu et al. [21] showed how Raspberry Pi can fight crop pests and diseases. Lychee and Longan tree risks were compiled first. Bacteria and insects were captured by the Raspberry Pi controlling the camera. The algorithm interpreted the images. Bluetooth speakers announced the final tally. Data mining, extraction, merging, and storage created a knowledge graph. The Raspberry Pi-trained VGG-16 model has a 94.9% detection rate.

A literature review shows that pest-sound analytics have not been sufficiently developed. Analyzing pest sound data and building IoT-based networks will help close this knowledge gap and improve pest detection across large agricultural areas. Automation and AI increase pest identification using strong analytics. Insect detection research has increased. Modern sound analytics classifies insects and other pests. However, thorough analytical methods can help identify problems early and reduce contamination. Pest detection training, validation, and testing should use MFP-Net models.

3 Dataset and Data Pre-processing

The pest sound library used in this study can be found in [22]. There are 13 kinds of pests, and 650 sounds were analyzed during this research, shown in "Table 1." To characterize and analyze the pest audio data before fusing it with a proposed deep learning system model, the proposed system made use of the BPF for denoising the data, the Triangular for reducing the amount of audio spectrum leakage, the Kaiser window for transforming overlapping frames into non-overlapping, the FFT, DFT, STFT, and the feature extraction via PLP. There was a 70% utilization rate throughout the training phase, 20% during testing, and 10% during validation. The deep multi-branch fusion residual network uses a 16-kilohertz audio sample rate, although this can be altered. Adjusting the sampling rate makes this possible. Python's resample feature and other applications can help. Labeling data helps classify insect species and their noises. This innovation aids HDL-Net model training and IoT-based pest identification. The proposed method developed a template by computing pest sound mean, median, variance, and standard deviation. Machine learning could improve pest identification models with this template. This form identifies pests. Background noise makes pest detection difficult in most agricultural settings. Noise, background noise, and anomalies impair pest detection accuracy. Filtering and denoising remove noise and other unwanted data. Data quality improves. Spectrum subtraction, Blackman–Harris window, Kaiser window. Bandpass filters improved insect noises in the suggested research. In large agricultural areas, auditory cues may be missed. Because of this, sound transmissions must be divided into tiny fragments that deliver only the sound the unwelcome bug enjoys. Energy, motion, or abstraction segmentation. The hamming window could reduce spectral leakage. Pitch shifting, time stretching, and noise injection can augment training set data for system stability. Normalizing data standardizes size and features. We can normalize data by removing the mean and dividing the variance by the standard deviation. The suggested approach calculated each pest sound's LPC feature matrix, mean, median, variance, and standard deviation to train pest identification models. Deep learning models use acoustic data in numerous steps. Feature extraction. A deep learning model requires structured audio signals. MFCC may help. Labeled data creates test, validation, and training

Table 1 List of Pests used for analysis

SN	Name of Pest	Samples
1	Greyback children's grubs	50
2	Galleria mellonella	50
3	White grubs	50
4	Rice weevils	50
5	Cock roach	50
6	Plodia interpunctella moth	50
7	Noctuid moth	50
8	Scapteriscus vicinus (Crickets)	50
9	Prostephanus truncatus	50
10	Weevils	50
11	Rodents	50
12	Geotrupes egeriei	50
13	Grasshopper	50

sets. Deep multibranch fusion residual network training sets teach new tricks. Data validating sets update model hyperparameters, while the test set evaluates system efficacy.

4 Proposed Methodology

4.1 MFP-Net System Model

The Multiscale Feature Pyramid Network (MFP-Net) uses deep learning and IoT-oriented sound analytics to automatically identify and monitor pests in large agricultural areas. This revolutionary technology combines deep learning, Internet of Things sensors, and advanced analytics to help farmers mitigate pest infestation and maximize agricultural harvests. The feature pyramid uses the rising feature maps from input data steps 4, 8, 16, and 32. These maps are labeled c_1, c_2, c_3 and c_4. Feature pyramid networks start with feature maps p_1, p_2, p_3 and p_4. Functions will store data in contexts apart from themselves. Set C to set P will lose universal semantic information. Acquiring multiscale feature map spatial information is the most common method. We think functional channel-to-world information exchange should be equally important. ResNet-50 was the system's backbone.

4.1.1 Adaptive Channel Fusion Phase

The research extensively uses the relationships between the several feature maps to merge features successfully. Let c_i Stand for the ith c in the sequence $[c_2, c_3, c_4, c_5]$. . To save space, we condense the spatial information into channel descriptors. In particular, a convolution kernel with dimensions of 1 by 1 for maintaining the same quantity of input channels, which can be written as (1), where H is the horizontal, W is the vertical resolution, g_c is the compressed channel response descriptor, and $F_c(x, y)$ is the corresponding frequency component, $F_c(x, y)$ is a point on the response map that denotes the corresponding frequency component. To locate associated channels for a specific channel descriptor $g_c \in R^{H*W*C}$, , we use the adaptive global averaging (AGAP) and adaptive maximum pooling (AMP) algorithms:

$$g_c = \frac{1}{H * W} \sum_{x=1}^{H} \sum_{y=1}^{W} F_c(x, y) \tag{1}$$

where σ denotes the sigmoid activation function. At the outset, we split the process in half: the first part focuses on generating a feature plane $g_c \in R^{H*W*C}$ with an adaptive gradient averaging pooling (AGAP) $F_1 \in R^{H*W*C}$ and the second half on extracting a high-level feature map from the feature plane $g_c \in R^{H*W*C}$ with maximum adaptive pooling $F_2 \in R^{H*W*C}$. . The obtained channel descriptor is then subjected to 1*1 operation kernels in the following phase. The high-level feature map is being used here. $\overline{F_1} = f_{1D}(F_1) \in R^{H*W*C/8}$ and $\overline{F_2} = f_{1D}(F_2) \in R^{H*W*C/8}$ are the resulting channel descriptors. The channel ID obtained is used to construct these two identifiers. Finally, the activation function ReLU and the Conv1Doperations are used to calculate the HWC values for the two-channel descriptors $\overline{F_1}$ and $\overline{F_2}$ as follows: $\bar{\bar{F}} = \overline{f_{1D}}(\text{ReLU}(\overline{F_1})) \in R^{H*W*C}$ and $\bar{\bar{F}}_2 = \overline{f_{1D}}(\text{ReLU}(\overline{F_2})) \in R^{H*W*C}$. Inputting data is the third stage. Next, we combine the channel descriptions for the $\bar{\bar{F}}_1$ and $\bar{\bar{F}}_2$ bands to create the $A_c \in R^{H*W*C}$ channel descriptions. After activating the A_c with a sigmoid function, we apply the Hadamard product operation on the initial high-level feature map, which produces the $M_c \in R^{H*W*C}$. The steps involved in computing are briefly described below:

$$A_1 = \overline{f_{1D}}(\text{ReLU}(f_{1D}(\text{AvgPool}(y_c)))) \tag{2}$$

$$A_2 = \overline{f_{1D}}(\text{ReLU}(f_{1D}(\text{MaxPool}(y_c)))) \tag{3}$$

$$A_c = \sigma(A_1(g_c) + A_2(g_2)) \tag{4}$$

$$M_c = \text{Had}(A_c g_c) \tag{5}$$

where AvgPool(y_c) and MaxPool(y_c) are stands for the AGAP and AMP procedures, respectively, for each supplied feature map. The notation f_{1D} and $\overline{f_{1D}}$ is used to refer to one-dimensional convolution processes with a kernel size of one. These actions have the potential to either decrease or raise the total number of channels, depending on which direction they are taken. A rectified linear unit's activation function is denoted by the function "ReLU," which stands for "reversed linear unit." "Had" refers to the procedure implemented during the Hadamard product's production.

4.1.2 Global Feature Adaptation Phase

Adapting global feature data can improve the efficiency and accuracy of discovery efforts. As a result, the MFP-Net includes a global features module for added support. Figure 1 shows how this module was integrated after ACFM was applied. Here, we use the GAP operation to gather aggregate data, and then we employ 1 * 1 convolution to combine data from several sources. We then include the output characteristics into the primary data stream.

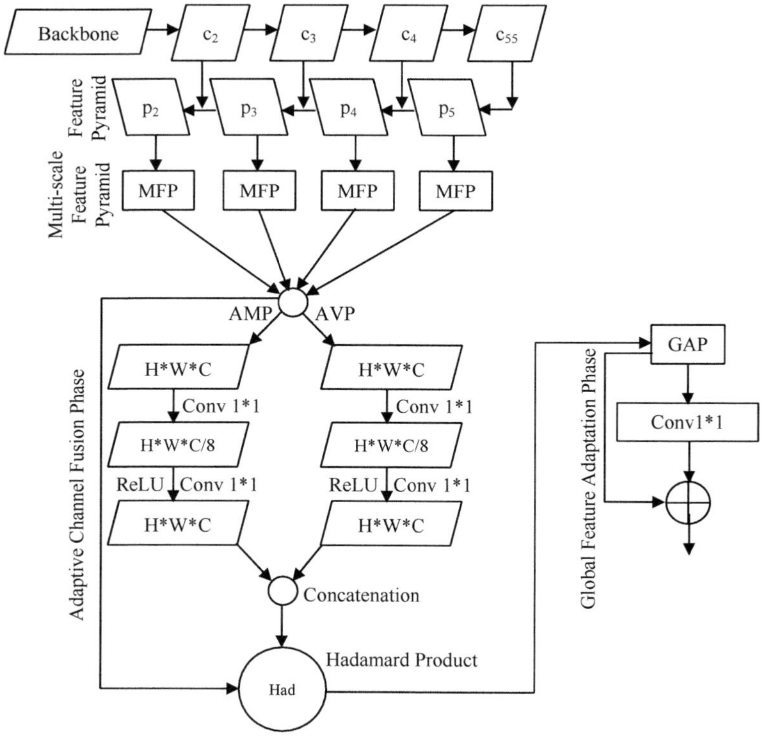

Fig. 1 The architecture of the proposed methodology

5 Experimentation, Results, and Analysis

The proposed method incorporates a structure for carrying out experiments and analyzing the results of such tests. This structure is divided into three stages: the examination of the experimental setup, the discussion of performance, and the comparison of performance.

5.1 Experimental Setup

For the research, we made use of a personal computer that was configured with Windows 10, an additional effects processor (an Intel(R) Core (TM) i5-9300H operating at 2.40 GHz), 64 gigabytes of random-access memory (RAM), and a video card with an additional effect (either a 2 GB Nvidia GeForce GTX 970 M or a 1 GB Nvidia GeForce GTX 1650). The research was written in Python version 3.8.10 for ease, and PyTorch version 1.9.0 was utilized to assist in constructing the network model. In addition, the amount of available computing power for computer graphics is increased when CUDA 10.2 is used for GPU acceleration.

5.2 Performance Parameters

All of its facets were examined before comparing the alternative plan and the proposed technique, which comprised several steps to accomplish this objective. The terms "true positive," "true negative," "false positive," and "false" are interchangeable and correspond to the notions "truep," "truen," "falsep," and "falsen," respectively. In addition, the suggested system measured the MSE and PE Optimization to arrive at the most accurate model design. Adam, when we first started the investigation, the subspace had many limitations. The proposed approach circumvented these constraints by decreasing the subspace, and the results of numerous preliminary tests showed that the model's performance had improved.

$$\text{Accuracy} = \frac{(\text{Truen} + \text{truep})}{(\text{Truen} + \text{truep} + \text{falsen} + \text{falsep})} \tag{6}$$

$$\text{F1-score} = \frac{\text{Two} * \text{precision} - \text{recall}}{\text{Precision} + \text{recall}} \tag{7}$$

$$\text{Recall} = \frac{\text{TrueP}}{(\text{FalseN} + \text{TrueP})} \tag{8}$$

Table 2 Performance parameters measurement for proposed and our previous works

Methods		Accuracy	Recall	F1score	MSE	PE
Our previous works	DenseNet	99.64	98.91	98.86	0.0002	0.36
	DQN	99.68	98.93	98.94	0.0001	0.32
	CNN-Bi-LSTM	98.91	98.54	98.58	0.0023	1.09
Proposed	**MFP-Net**	**99.76**	**99.03**	**99.06**	**0.00008**	**0.24**

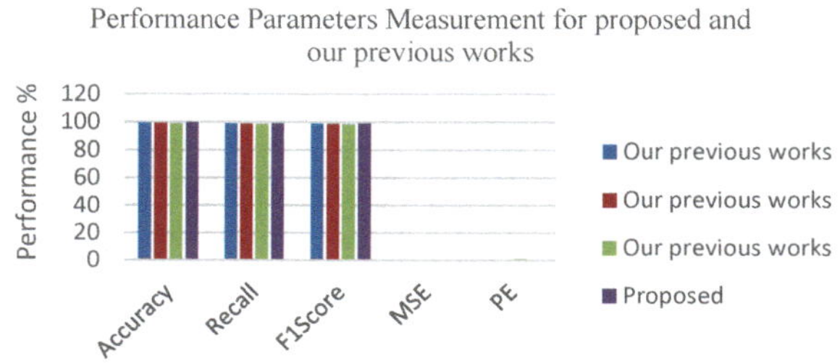

Fig. 2 Performance parameters measurement for proposed and our previous works

$$\text{MSE} = (1/n) * \Sigma\,(\text{actual}-\text{forecast})^2 \tag{9}$$

$$\text{PE} = (\text{actual}-\text{forecast}) * 100. \tag{10}$$

5.3 Results

The MFP-Net model was used to get the accuracy scores, recall, F1-scores, MSE scores, and PE values summarized in "Table 2" and displayed in "Fig. 2" The model was examined in three stages: training, validation, and testing. The training was the first stage.

5.4 Performance Comparisons

"Table 3," which follows and is presented in "Fig. 3" in a graphical perspective, is a summary of the results of a comparative study between these current planned research efforts and the state-of-the-art MS-ALN [23], YOLOv5 [13], Faster RCNN [24], and ResNet-50 [25] techniques. The findings are presented in a graphical format for better understanding.

Table 3 Performance comparison of proposed work with state-of-the-art works

Methods	Accuracy (%)	Year
MS-ALN [23]	74.61	2022
YOLOv5 [13]	96.90	2023
Faster RCNN [24]	92.66	2022
ResNet-50 [25]	99.33	2023
MFP-Net [Proposed]	**99.76**	**2023**

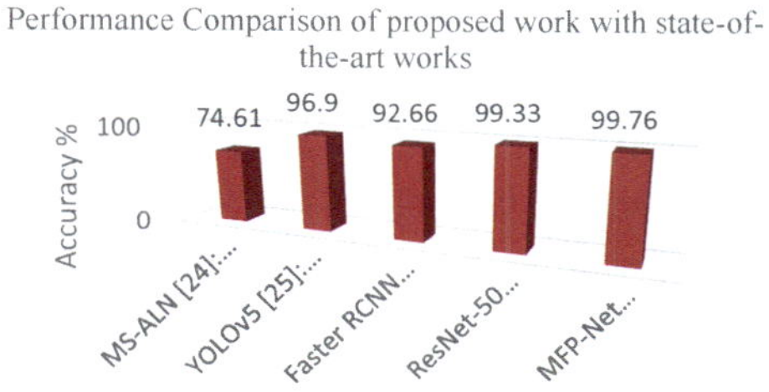

Fig. 3 Performance comparison of proposed work with state-of-the-art works

5.5 *Analysis*

Trustworthy analytics have improved pest detection as automation technologies increase. Identifying audio signals is being investigated increasingly. Robust analytics classifies invasive species. Anything imaginable. It fits both, making it a viable option. Advanced pest management analytics are being developed. This project would use real-time Internet of Things analytics to manage pests across a large agricultural area. It separates beetle noises. MFP-Net learns, validates, and tests invasive species. BPFs, Triangular windows, Kaiser windows, FFTs, and DFTs reduced background noise and spectrum leakage. The proposed system completed both processes. Combining these PLP parts' benefits. Pre-processing, carefully optimized, doesn't degrade the signal. Pest noises create PLP coefficient matrices. The new MFP-Net model diagnoses pests with 99.76% accuracy, 99.03% recall, and 99.06% F1-score. It outperforms MS-ALN, YOLOv5, Faster RCNN, and ResNet-50.

6 Limitations and Future Recommendation

The fact that the proposed research has an accuracy rate lower than one hundred percent is a restriction. So, we will keep working on this research until we reach an accuracy rate of one hundred percent in our detection methods. We can identify, prevent, and ultimately control key agricultural pests using trustworthy analysis. Imagine a government agency or an industrial development company interested in contributing to funding. We intend to submit applications for this job in real time to vast agricultural areas nationwide. If at all possible, we would appreciate it.

7 Conclusion

Modern farming is scientific, technological, and data-driven. IoT enabled "smart farming" in agriculture. This disaster transformed farming and opened new doors. Farmer issues lower crop output. Pests cost farmers a lot of money. Farmers use pesticides to control weeds, plant diseases, and other pests. Pesticides harm ecosystems, health, and the economy. MFP-Net will develop an IoT-based infrastructure for autonomous, real-time insect tracking. IoT and acoustic analytics reduce pesticides. Field-level audio data confirms pests. PLP produced coefficients from beetle noise. MFP-Net was trained, verified, and tested using the PLP coefficient matrix and statistical data from 650 distinct pest noises. MFP-Net outperforms MS-ALN [23], YOLOv5 [13], Faster RCNN [24], and ResNet-50 [25] with an F1-score of 99.06%, recall of 99.03%, and accuracy of 99.76%.

References

1. Kong, J.L., Wang, H.X., Yang, C.C., Jin, X.B., Zuo, M., Zhang, X.: Fine-grained pests & diseases recognition via Spatial Feature-enhanced attention architecture with high-order pooling representation for precision agriculture practice. Agriculture, 1592804 (2022)
2. Deng, F., Mao, W., Zeng, Z., Zeng, H., Wei, B.: Multiple diseases and pest detection based on federated learning and improved faster R-CNN. IEEE Trans. Instrum. Meas. **71**, 1–11 (2022)
3. Chen, C.-J., Huang, Y.-Y., Li, Y.-S., Chen, Y.-C., Chang, C.-Y., Huang, Y.-M.: Identification of fruit tree pests with deep learning on embedded drone to achieve accurate pesticide spraying. IEEE Access **9**, 21986–21997 (2021)
4. Hadi, M.K., Kassim, M.S.M., Wayayok, A.: Development of an automated multidirectional pest sampling detection system using motorized sticky traps. IEEE Access **9**, 67391–67404 (2021)
5. Martin, J., Ansuategi, A., Maurtua, I., Gutierrez, A., Obregón, D., Casquero, O., Marcos, M.: Using a mobile manipulator, a generic ROS-based control architecture for pest inspection and treatment in greenhouses. IEEE Access **9**, 94981–94995 (2021)
6. Li, S., Wang, H., Zhang, C., Liu, J.: A self-attention feature fusion model for rice pest detection. IEEE Access **10**, 84063–84077 (2022)
7. Albanese, A., Nardello, M., Brunelli, D.: Automated pest detection with DNN on edge for precision agriculture. IEEE J. Emerg. Sel. Top. Circ. Syst. **11**(3), 458–467 (2021)
8. George Cosmin, S., Anghelescu, P., Mazăre, A.G., Ionescu, L.M., Şerban, G.: Complex electronic system for monitoring, warning, and preventing water stress and pest detection in crops. In: 2022 14th International Conference on Electronics, Computers and Artificial Intelligence (ECAI), pp. 1–4. IEEE (2022)
9. Chen, C.J., Huang, Y.Y., Li, Y.S., Chen, Y.C., Chang, C.Y., Huang, Y.M.: Identifying fruit tree pests with deep learning on embedded drones to achieve accurate pesticide spraying. IEEE Access **9**, 21986–21997 (2021)
10. Lyu, Z., Jin, H., Zhen, T., Sun, F., Xu, H.: Small object recognition algorithm of grain pests based on SSD feature fusion. IEEE Access **9**, 43202–43213 (2021)
11. Mamdouh, N., Khattab, A.: YOLO-based deep learning framework for olive fruit fly detection and counting. IEEE Access **9**, 84252–84262 (2021)
12. Qian, S., Du, J., Zhou, J., Xie, C., Jiao, L., Li, R.: An effective pest detection method with automatic data augmentation strategy in the agricultural field. SIViP **17**(2), 563–571 (2023)

13. Chu, J., Li, Y., Feng, H., Weng, X., Ruan, Y.: Research on multiscale pest detection and identification method in granary based on improved YOLOv5. Agriculture **13**(2), 364 (2023)
14. Anwar, Z., Masood, S.: Exploring deep ensemble model for insect and pest detection from images. Procedia Comput. Sci. **218**, 2328–2337 (2023)
15. Chithambarathanu, M., Jeyakumar, M.K.: Survey on crop pest detection using deep learning and machine learning approaches. Multimedia Tools Appl. 1–34 (2023)
16. Umamaheswari, S., Vishal, N.R., Pragadesh, N.R., Lavanya, S.: Performance analysis of resnet50 architecture-based pest detection system. In: 2023 9th International Conference on Advanced Computing and Communication Systems (ICACCS), vol. 1, pp. 578–583. IEEE (2023)
17. Albattah, W., Masood, M., Javed, A., Nawaz, M., Albahli, S.: Custom CornerNet: a drone-based improved deep learning technique for large-scale multiclass pest localization and classification. Comp. Intell. Syst. **9**(2), 1299–1316 (2023)
18. Dai, M., Dorjoy, M.M.H., Miao, H., Zhang, S.: A new pest detection method based on improved YOLOv5m. Insects **14**(1), 54 (2023)
19. Mallick, M.T., Biswas, S., Das, A.K., Saha, H.N., Chakrabarti, A., Deb, N.: Deep learning-based automated disease detection and pest classification in Indian mung bean. Multimedia Tools Appl. **82**(8), 12017–12041 (2023)
20. Yang, S., Xing, Z., Wang, H., Dong, X., Gao, X., Liu, Z., ... Zhao, Y.: Maize-YOLO: a new high-precision and real-time method for maize pest detection. Insects **14**(3), 278 (2023)
21. Zhu, D., Xie, L., Chen, B., Tan, J., Deng, R., Zheng, Y., ... Andrew, W.H.: Knowledge graph and deep learning-based pest detection and identification system for fruit quality. Internet of Things **21**, 100649 (2023)
22. https://data.nal.usda.gov/dataset/bug-bytes-sound-library-stored-product-insect-pest-sounds
23. Feng, F., Dong, H., Zhang, Y., Zhang, Y., Li, B.: MS-ALN: multiscale attention learning network for pest recognition. IEEE Access **10**, 40888–40898 (2022)
24. Butera, L., Ferrante, A., Jermini, M., Prevostini, M., Alippi, C.: Precise agriculture: effective deep learning strategies to detect pest insects. IEEE/CAA J. Automatica Sinica **9**(2), 246–258 (2021)
25. Hadipour-Rokni, R., Asli-Ardeh, E.A., Jahanbakhshi, A., Sabzi, S.: Intelligent detection of citrus fruit pests using a machine vision system and convolutional neural network through transfer learning technique. Comput. Biol. Med. **155**, 106611 (2023)

Detection of Wormhole Attacks Using the DCNNBiLSTM Model to Secure the MANET

B. Rajalakshmi, R. J. Anandhi, K. Moorthi, Balasubramanian Prabhu Kavin, and Rajesh Kumar Dhanaraj

Abstract The obvious response to an anytime, anywhere network in the current era of heterogeneous networks is MANET. It is a collection of wirelessly communicative mobile gadgets that operate on their own. In order to stop the damage that rogue network nodes could do, security in MANET is a crucial duty. A category of denial-of-service (DoS) assault, the wormhole attack bombards the network with numerous phoney packets and messages in an effort to exhaust its resources. Wormhole attacks are a particular type of network layer attack that imitates routing algorithms. Because of deep learning's exceptional performance in many detection and identification tasks, this research presents an approach for intelligent and effective attack detection in MANET. Our goal is to mimic a wormhole attack in a network environment with many wormhole tunnels. The CNN layer is used by the DCNNBiLSTM architecture to extract features from input data, and BiLSTMs are used to predict sequences. Five performance indicators were used to compare CNN, LSTM and DCNNBiLSTM under wormhole attack.

Keywords Bidirectional long short-term memory · Nodes · Deep learning · Ad hoc networks · Wormhole attacks · Deep convolutional neural network

---

B. Rajalakshmi · R. J. Anandhi
New Horizon College of Engineering, Ring Road, Bellandur Post, Bengaluru 560103, India

K. Moorthi
Department of Computational Intelligence, College of Engineering and Technology, SRM Institute of Science and Technology, Kattankulathur, Chengalpattu District 603203, Tamil Nadu, India
e-mail: moorthik@srmist.edu.in

B. P. Kavin (✉)
Department of Data Science and Business Systems, College of Engineering and Technology, SRM Institute of Science and Technology, Kattankulathur, Chengalpattu District 603203, Tamil Nadu, India
e-mail: ceaserkavin@gmail.com

R. K. Dhanaraj
Symbiosis Institute of Computer Studies and Research (SICSR), Symbiosis International (Deemed University), Pune, India

G. Fortino et al. (eds.), *Proceedings of Third International Conference on Computing and Communication Networks*, Lecture Notes in Networks and Systems 917,
https://doi.org/10.1007/978-981-97-0892-5_2

1 Introduction

Ad hoc networks have taken on a variety of shapes as a result of the wireless mobile computing technology's quick development and growth in popularity, which has sparked a revolution in the computer industry. In order to create provisional networks deprived of pre-existing substructure base stations, MANETs are made up of mobile nodes that connect with others directly within their radio coverage [1]. There is no central node in the MANET that can monitor the transit of data between the nodes [2]. It functions by transmitting messages. Nodes take part in routing by passing packets to the node after them until they reach their target [3]. Ad hoc networks are susceptible to assaults for a diversity of reasons, including a lack of secure borders, risks from rogue nodes already present, a lack of centralized management tools, limited power availability, scalability, etc. [4]. MANET can be simply deployed everywhere where it would be challenging to set up an infrastructure because it doesn't require one [5].

Denial of Service (DoS) is a threat to ad hoc networks that can result from attacks against MANET's security. In a DoS attack, malicious nodes insert themselves into the network and pretend to be the quickest path between sources and destinations Malicious nodes either create fake messages or choose which messages to discard in order to completely shut down the network. Wormhole assaults are one type of DoS attack used in MANET [6]. The well-known wormhole assault is one of the gravest dangers to the security of MANET. Damage can also be sustained by various MANET routing methods, including AODV), (DSR) and others [7]. Wormhole attacks are a key MANETS tactic. In this attack, a node snoops on network packets coming from one point and 'tunnels' them to node further away, which copies the packets locally [8]. To create the tunnel, a variety of techniques can be utilized, including encapsulation, high-power transmission, etc. [9]. Compared to packets delivered over standard multi-hop paths, the tunnelled packets arrive either rapidly or with fewer hop counts [10]. Even though the network offers secrecy, this kind of assault is still possible [11]. A wormhole attack could impede the discovery of any routes besides the wormhole when it is employed against an on-demand routing protocol like AODV/DSR [12]. Since the wormhole assault needs off-channel data delivery, the MANET nodes have a very difficult time detecting it. Due to the off-channel data transfer required for the wormhole assault, the MANET nodes have a very difficult time detecting it [13]. There have been many approaches proposed to identify wormhole assaults [14].

To address the drawbacks of conventional techniques, many scholars have proposed ML-based solutions [4]. However, deep learning (DL), a development of neural networks, does not always require features since it can mechanically extract the best features from the raw data. Numerous cyber security challenges, such as virus detection, botnet identification and intrusion detection, have been successfully resolved in the present day thanks to DL design [15]. Deep learning (DL) developments have had a positive impact on the fields of image processing, and computer vision [16].

Below are listed some of this work's main contributions:

(1) With a small number of malicious and helpful nodes, this has set up a wormhole attack in an ad hoc network setting.
(2) Follow the usual output files, which have the complete network and node deployment information. A method for choosing useful features and data collection criteria depending on those attributes.
(3) The CNN and BiLSTM networks should be combined in a model with a covert layer. To progress the classification accuracy of the model, clean the data before including the attention mechanism. The CNN and BiLSTM networks should be combined in a model with a covert layer. To improve the model's classification accuracy, clean the data before adding the attention mechanism.
(4) The method's detection accuracy is increased by manually eliminating characteristics from the design element and using them to train the deep learning to perfect the properties of high-dimensional data.

1.1 Organization of Paper

The remaining paper is arranged as shadows: After Sect. 2 summarizes the related works, Sect. 3 brings a brief conversation of the suggested model, Sect. 4 exemplifies the consequences of the validation investigation and Sect. 5 offers a summary, a conclusion is formed.

2 Related Works

Actual data delivery in the network is also impacted by the assaulter's harmful activities, which are predicted to use link bandwidth. This study by Rathod and Saxena [17] aims to quantify the malicious Remapping attack's performance within the MANET and identify the node or nodes that are impacted by the attack's detrimental effects on routing effectiveness. The recommended security theme is to look at the precedence sender node in the network that must connect, and the only offender is one node that floods the network with reserve packets. Furthermore, neither the sender nor the recipient is the attacker node. The suggested security measure against remapping is to prevent the offender's actionsPerformance metrics are used to assess a network's effectiveness, including as PDF and routing overhead, and attacker node credentials.

A brand-new security paradigm for wormhole mitigation is outlined in this study by Rajkumar and Mercy Shalinie [18]. In order to identify an ongoing wormhole attack in MANET, our security system makes use of the attack model. The most important part of our approach is that an admin of trustworthy nodes. A boomerang packet is periodically sent across the network by the admin node. The packet intended to be returned to the admin node itself is known as a boomerang packet. Admin determines whether or not the packet was a victim of a wormhole attack based on the condition of the packet when the boomerang packet returns. Precautions are taken to limit the harm in the case of a wormhole attack. In the NS-3.26 simulator, we performed extensive trials, and it compared the outcomes with cutting-edge solutions. With a 97% accuracy rate, our method was able to find the wormhole nodes. When compared to other cutting-edge solutions during the wormhole attack, the packet loss rate was also lower. This was due to MANET's accurate and timely discovery of the wormhole attack.

The suggested remedy by Bhatti et al. [19] deals with an attack tactic that is less frequently used by the research community: knocking out an attacker. Furthermore, it is compatible with the current network stack and doesn't call for any changes to or additions to the hardware. The idea is simulated in ns2.30, and it is found that the suggested solution's detection rate is typically between 98 and 99%. Theoretically, the search time for a wormhole node is between 0.07 and 0.71 s. The simulation, on the other hand, indicates that the typical time for 0.67 s. Regarding packet loss, the suggested method has about a 22% overhead. Both static and mobile situations favour its performance; nevertheless, movement conditions result in greater frame losses than static ones. The solution has a computed complexity of O. (n). The method is advised as a result of simulation findings, which show that it is efficient in terms of memory, computing, bandwidth and energy use. Utilizing statistical measures like accuracy and Matthew's correlation coefficient, the solution is verified.

In order to provide a system for MANET (EHGSO), this study employed a deep network based on Exponential-Henry Gas Solubility Optimization by Ninu et al. [20]. In the first stage of the secure routing process, the newly designed EHGSO algorithm is used to choose the optimal paths. As fitness indicators, the method makes use of energy, distance, local quality and connection quality. The projected EHGSO (EWMA) combines the (EWMA) and the Henry Gas Solubility Optimization (HGSO). To begin the intrusion prevention phase, the base station must first modify transmitted data packets and extract knowledge discovery in database (KDD) properties, which constitutes the second phase. After KDD features have been retrieved, the data is augmented. The provided EHGSO method may be used to train the deep neuro-fuzzy network, which can then be used for intrusion detection. When compared to existing technologies, the proposed approach performs exceptionally well. Without any assaults, the suggested method yields the following results: (in order) 0.342 J, 134,975 kbps, 4.123, 0.086, 95.877%, 0.950 and 0.924. Multiple metrics, including power consumption, throughput, packet loss, jitter and recall, were used to arrive at these estimates.

The presented algorithms, Cascading Back (CBPNN) and feedforward neural network (FNN), are used by Abbood et al. [21] in this paper's MANET to identify complex patterns while focusing on security standards through efficiency determination, identifying malicious nodes and mitigating network attacks (FFNN). These fundamental types of deep neural network building designs, in addition to Convolutional Neural Networks (CNNs), are frequently employed to enhance the functionality of intrusion detection systems (IDS) and the use of IDS in conjunction with machine learning (ML). More so than their statistical and logical counterparts, Machine Learning (ML) approaches enable MANET network learning and promote adaptation to various contexts. In terms of Average Receiving Packet (ARP) and

End-to-End (E2E) performance, the suggested model performs better than another present model. Results from CBP, FFNN and CNN were used to compile, in order, 74, 82 and 85%, respectively, of the data.

In order to provide beneficial qualities to IDS for green smart mobility on MANET, this paper by Singh and Vigila [22] described ML approaches. When learning about the MANET context, users can determine incursion with the aid of the performance of ML-IDSs and a study of their suitability in MANETs. KDD IDS with improved ML. In this IDS method, ensemble learning awarded controlled packet anomaly scores. This ML method is our answer to the real MANET dataset shortages. An IDS for green smart transportation that is more resilient has been produced using ML approaches, simulation and a working prototype. ML-enhanced IDS found and curbed dangerous MANET activity. This study boosted IDS knowledge using ubiquitous learning. The proposed PCA-FELM is put into practice using the MAT LAB simulator. A factor of 99.08% separates the proposed PCA-FELM from competing methods like DBN-IDS, GOA-SVM and SDAE-ELM. Studies on the KDD Cup99 dataset for the Content Discovery and Data Gathering Tools Competition show that the suggested PCA-FELM model performs better than alternative methods.

The Whale Optimization Network (DNN) is used to optimize the preprocessed data to construct an effective and efficient system for classifying and predicting unexpected cyber-attacks. This study by Edwin Singh and Celestin Vigila [23] suggested that WOA-DNN be used to detect and categorize incursions in MANET. This prevents intrusion assaults by providing secure data transfer to other nodes. The ML-IDS and WOA-DNN (Machine Learning) approaches are used to locate the invaders. Using Principal Component Analysis (PCA), the data is de-dimensionalized, increasing the outputs' accuracy. To assess whether a result is good or bad, forward propagation uses a classifier. To evaluate the effectiveness of the conventional and proposed models, classification accuracy, attack detection rate, precision rate, F-measure and recall are all used. Higher evaluation metrics and a 99.1% accuracy rate are features of the proposed WOA-DNN model. There are less false alarms thanks to WOA-Superior DNN's attack detection rate compared to other systems. The suggested WOA-DNN model has a 99.1% classification accuracy rate.

The Vijayan et al. proposed model [24] performs intrusion detection using fuzzy clustering and fuzzy Naive Bayes models and employs an ensemble optimization approach to address the security challenges in MANET. Here, the Grasshopper and Chimp optimization methods are the algorithms that are taken into consideration. The Chimp Optimization method is used to identify the cluster head that will provide the optimal route for MANET's system efficiency. The following factors are used to select the best routes: bandwidth. This Grasshopper Optimization Procedure is being used to improve global optimization by reducing mistakes. The effectiveness of the aforementioned techniques is assessed using a variety of assaults, such as floods, blackholes and selective packet drops. It is discovered that ensembled optimization performs better than individual algorithms.

3 Proposed Work

3.1 Problem Statement

Since wormhole attacks do not rely on the Medium Access Control (MAC) layer protocol and are not immune to cryptographic techniques, they pose a greater security risk to an ad hoc network. It links two nodes together to enable the transmission of data or packets across the network. When network communication ensures authenticity and anonymity, there is a chance for an assault. Without addressing themselves, an attacker can make a wormhole for packets. Attackers rebroadcast packets that they receive by tunnelling them through a second end. Using an ad hoc network routing technique when a destination node is out of broadcast range is dangerous because some nodes may believe they are in the range of the destination node. Each ROUTE REQUEST packet in a communication network can be tunnelled or forged by malicious nodes without changing the packet's header or internal data.

Direct tunnelling of packets occurs at one point, followed by replaying at another. Both the transmitter and the receiver in this wormhole attack believe they are right next to each other. A hidden attack known as a wormhole occurs when an evil node fails to properly update the packet header. A different type of wormhole assault is an exposed attack, in which the attacker influences the packet header rather than the actual packet content to change the routing. When they locate or are close to the initiator of route discovery, malicious nodes prevent any other path save a wormhole. Data packets can be discarded rather than forwarded by an attacker. Wormhole tunnels are fast; therefore, if attackers tunnel without damaging data packets, the network service will be more dependable and helpful.

Upon receiving a ROUTE REQUEST from the source node (S) and transmitting it to the destination node (D), a malicious node will simply pass the request to the next node without altering the hop count or neighbour node information in the packet header. When two malicious nodes are nearby, they build a tunnel that speeds up packet transmission compared to a regular network. Source and destination nodes are supposed to be neighbours because of packet header information at the destination node (D), however, they are not.

When two malicious nodes are close neighbours, you should build a tunnel so that other nodes can send Route Requests without having to change the essential packet header information. Both graphs show that when a device is out of transmission range, there are many false neighbours by utilizing packet header information. Wormhole attacks can be used to snare network traffic, deliberately drop packets and launch a variety of assaults that harm data packets and network traffic protocols. The primary goal of this work is to identify and isolate harmful activity nodes. Particularly in a wired network, intelligent IDS is frequently employed for attack detection. Machine learning systems maintain high detection rates for attacks, compromised routes and traffic skewness. Our research of these models served as a starting point for designing a classification model for wormhole identification in wireless ad hoc networks. There has been less categorization work reported for wireless ad hoc networks. Due to the classification model's application to diverse traffic situations, including network mobility, the presence of malicious nodes, sampling frequency and attack choice, the task is more difficult. In a fixed environment, conventional methods only offer security against wormhole attacks. These methods of detection are ineffective when an attack alters its characteristics or the network environment. Then, this developed a framework for wormhole attack finding in a wireless ad hoc network.

3.2 *Ad Hoc Network*

Ad hoc networks' peer nodes connect with one another devoid of any infrastructure. These nodes have the ability to communicate wirelessly using multi-hop technology. The ROUTE REQUEST is initiated and sent to neighbours who are within communication range by nodes using their self-organizing property. Nodes deliver messages directly to each other when they are within communication range; otherwise, a path is established. These nodes typically utilize the same frequency range, hopping pattern, spreading code and communication medium. Coordinating medium access between nearby nodes and managing the wireless link resources are duties of the data link layer. Medium access control is the key technique that permits nodes to share a broadcast channel in an ad hoc network. The property of the network layer is responsible for preserving a multi-hop communication path. A node may be a neighbour of any other node due to the mobility or volatility of an ad hoc network. Mobility, self-organization, wireless communication and a lack of infrastructure are the main characteristics of an ad hoc network that increase its susceptibility to attacks.

3.3 *DCNNBiLSTM*

The CNN layer, used in the DCNNBiLSTM architecture, extracts features from the input data, the BiLSTM forecast sequences to produce the necessary outputs and the DNN enhances errors and loss optimization. The Long-term Recurrent Convolutional Network (LRCN) network is another name for the CNN BiLSTM design, which was developed using the CNN BiLSTM design to produce a hybrid CNN and BiLSTM model. A Deep CNN-BiLSTM can be created by first layering the front end using CNN, then layering the BiLS levels, a DNN layer and finally the output layer. The algorithm is demonstrated by Algorithm 1 in action.

Algorithm 1 DCNNBiLSTM model

Require: Input df = dataset
Set ***Features*** = $f1, \ldots, fn$ from df set as Subset, where $\boldsymbol{n} = \textbf{\textit{no. of selected features}}$
Split $\mathbf{df}_{\textbf{\textit{Features}}}$ into training, and testing
Procedure model ()
Add Conv1D Layer to model, input shape = (76, 1), $kernelsize = n$
Set Activation functions: ReLU
Add BatchNormalisation to model
Add Bidirectional LSTM Layer to model, $Neuron = 64$
Add Reshape Layer, $input\_shape = 128$
Add BatchNormalisation to model
Add Bidirectional LSTM Layer to model, $Neuron = 128$
Add Dropout Layer
Add DNN Layer o model, $Neuron = 64, 32, 16$
Set Activation functions: ReLU
Add Output Layer with $Dense = 3$
Set Activation functions: Softmax

Equation (1) provides the formula for activating the ReLU-equipped CNN layer. Equation (2) is used to construct the batch normalization. The output layer is given the softmax function in Eq. (3).

$$\text{Relu}(x) = \max(0, x) \tag{1}$$

$$\text{BatchNormalization}(x) = \frac{X - \text{Min}}{\text{Max} - \text{Min}} \tag{2}$$

$$\text{softmax}(x_i) = \frac{\exp(x_i)}{\sum_j \exp(x_j)} \tag{3}$$

In essence, an LSTM is the result of combining two separate LSTMs. With the help of this architecture, the networks can get information about the forward and backward sequences for each input sequence.

The parameters used to represent the DCNNBiLSTM neural network are the 0.0001 learning rate parameters, Adam optimization function and ReLU activation function. Different numerical techniques also use different loss functions. This essay employs an original numerical strategy. The loss function should therefore be categorical cross-entropy.

4 Result and Discussion

4.1 Data Collection

We have accumulated 6,37,862 different samples, including 1,52,144 normal and 4,85,718 harmful samples. It creates a dataset with 20 carefully chosen features that have been labelled. The value −1 fills in a dataset field that is blank. For the purpose of wormhole attack detection, a high-volume dataset is created in an ad hoc network environment.

4.2 Performance Metrics

By employing a wormhole node network attack, we were able to simulate and test the proposed DCNNBiLSTM using NS-2.35. By employing a wormhole node network attack, we were able to simulate and test the proposed DCNNBiLSTM using NS-2.35. The NS-2.35 program known as the CMU tool, which creates a file with random node placement and movement over a predetermined amount of time, was used to create the scenarios. Wormhole assault interval was set to 0.06 s. Table 1 demonstrates the simulation's environment settings.

Five performance indicators were used to compare CNN, LSTM and DCNNBiLSTM under wormhole attack: throughput, end-to-end delay and load are all metrics (NRL).

Table 1 Environment limits for simulation

Environment parameters for simulation	
Node placement	Random
Transport protocol	UDP
Simulation	Simulation time 200 s
Node number	200, 400, 600, & 800
Data rate	0.5 Mbps
Pause time	5 s
Routing protocol	DCNNBiLSTM
Wormhole attack interval	0.06 (16 requests per sec)
Coordination	850 * 850 m
Speed	Maximum 15 mps
MAC protocol	802.11
Radio type	802.11b radio
Connection	Constant item size 512(byte)
Transmission range	150 m

Packet Delivery Ratio

The PDR is the proportion of the sum of data packets positively sent from the source node to those successfully received by the destination (PDR). PDR can be determined by using Formula (4).

$$\text{PDR} = \frac{R_{\text{packets}}}{S_{\text{packets}}} \tag{4}$$

The terms R_{packets} and S_{packets} stand for the quantity of packets transmitted and received, respectively.

Throughput

Throughput, or how quickly packets arrive over time, from the source node, is displayed. Throughput is able to be determined using a Formula (5).

$$\text{PDR} = \frac{R_{\text{packets}}}{C_{\text{time}}} * \frac{8}{1024} \tag{5}$$

R_{packets} is the sum of received packets, together with C-time. is how long the nodes have been connected.

End-to-End Delay

End-to-end delay represents the average sum of time required for a packet to move from its source node to its terminus node across the network. A formula can be used to determine end-to-end delay (6).

$$\text{Avg}_{EtE} = \frac{\sum_{i=1}^{N} Rt_i - St_i}{N} \tag{6}$$

N is the total sum of nodes in the network, Rt_i, St_i are the conventional and directed times of the ith packet.

Average Residual Energy

The ARE metric computes the average amount of energy that is still present in each network node. Formula (7) can be used to determine ARE:

$$\text{ARE} = \frac{R_E}{N} \tag{7}$$

R_E is the remaining energy and N is the network's total sum of nodes.

Table 2 Packet delivery ratio with sum of nodes

Number of nodes	CNN	LSTM	DCNNBiLSTM
200	0.58	0.74	0.89
400	0.73	0.97	0.99
600	0.92	0.99	1.05
800	0.99	0.99	1.09

Normalized Routing Load

The 'Normalized Routing Load' is the proportion of received routing packets to those received at the (NRL). Formula (8) can be used to calculate NRL:

$$\text{NRL} = \frac{Rt_{\text{packets}}}{R_{\text{packets}}} \tag{8}$$

Rt_{packets}, or the sum of received nodes, is equal to Rt_{packets}, or the number of routing packets.

Table 2 shows the numeric consequences of assessment among CNN, LSTM and DCNNBiLSTM in terms of PDR.

According to Fig. 1, The recommended model clearly outperforms CNN, LSTM and DCNNBiLSTM when it comes to counting, as demonstrated by a number of rounds and (PDR). Using the provided methods, we were able to improve PDR results by getting effective packets at the destination. Comparing the two methods reveals that the proposed model frequently leads to the sending of additional packets, with a 6.25% increase over LSTM and a 9.67% increase over CNN.

Table 3 compares the numerical results of throughput between CNN, DCNNBiLSTM and LSTM.

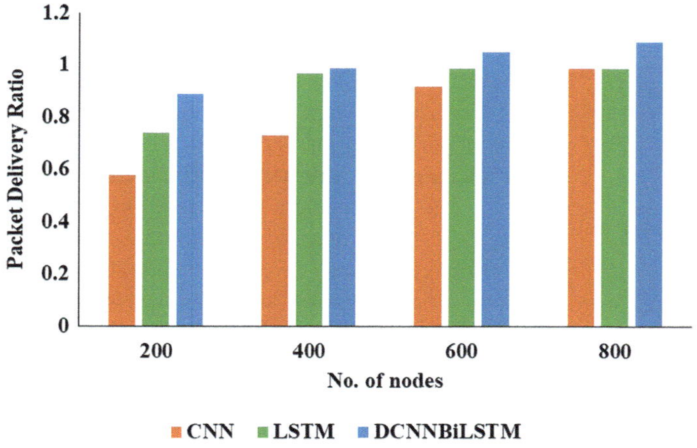

Fig. 1 Packet delivery ratio analysis

Table 3 Throughput with sum of nodes

Number of nodes	CNN	LSTM	DCNNBiLSTM
200	148.8	174.4	190
400	86.45	186.4	192.5
600	92.31	194.1	212.8
800	92.74	195.2	244.2

Figure 2 illustrates how, in the event of a wormhole attack, due to the rebroadcasting of erroneous requests, CNN's throughput drops, while its node count increases. Because of network congestion brought on by the wormhole attack, regular packets will be dropped or delayed, which will have an impact on the throughput and PDR. The DCNNBiLSTM result displays a greater throughput than both CNN and LSTM.

The comparison of CNN, DCNNBiLSTM and LSTM in terms is shown in Table 4 in numerical form.

Figure 3 shows how, during a wormhole attack, as the sum of nodes increases due to the congestion brought on by the wormhole node, the end-to-end delay in CNN increases. The latency will increase overall as a result of regular packets being lost or delayed. When there isn't a wormhole attack on the network, DCNNBiLSTM's

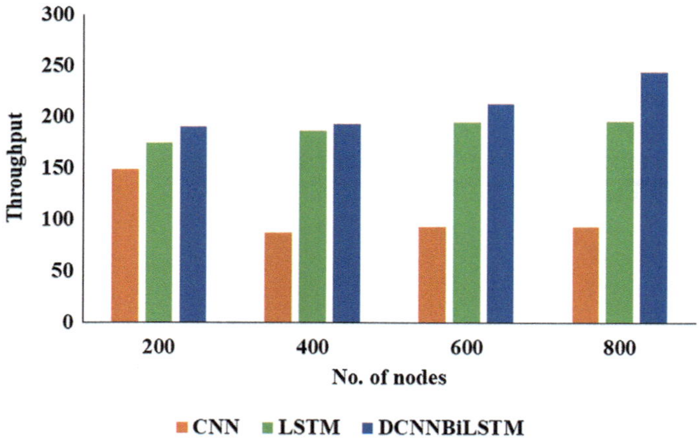

Fig. 2 Throughput analysis

Table 4 Average of end-to-end delay with sum of nodes

Number of nodes	CNN	LSTM	DCNNBiLSTM
200	0.7	0.99	0.48
400	1.02	1.16	0.79
600	1.38	1.26	0.69
800	1.73	1.51	0.99

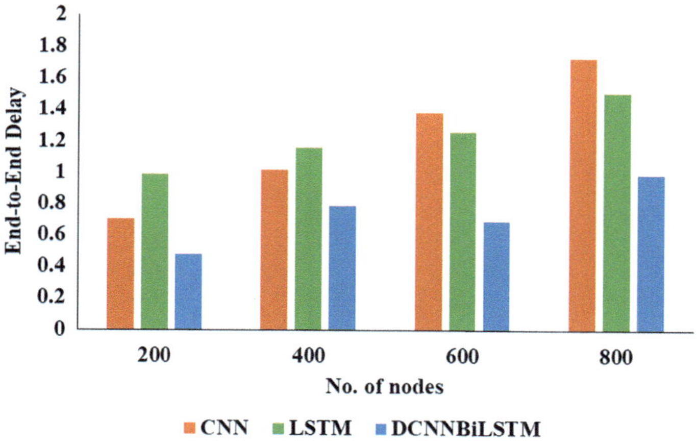

Fig. 3 Avg of end-to-end delay analysis

end-to-end delay is at its shortest. The outcome of a wormhole attack using DCNNBiLSTM has a smaller end-to-end delay than LSTM since it recognizes and isolates the network. The end-to-end latency of the suggested technique is reduced by 17.03% compared to CNN and by 39.81% compared to LSTM, according to the results of the imitation experiments.

In terms of ARE analysis, Table 5 compares the numerical results of CNN, DCNNBiLSTM and LSTM.

Figure 4 demonstrates that a wormhole assault is when the DCNNBiLSTM ARE outcome is at its lowest. The wormhole attack consumes up nodes' energy by keeping them busy sending incorrect requests around the network, which is especially true as the sum of nodes grows. When the network is not being attacked by wormholes, CNN performs the best in terms of ARE. DCNNBiLSTM thus has a little lower ARE than CNN and LSTM. The proposed model, LSTM, reduces ARE by 11.6 and 19.79%, respectively, in comparison to CNN.

In terms of normalized routing load analysis, Table 6 compares the numerical findings of CNN, DCNNBiLSTM and LSTM.

Figure 5 illustrates how the result of NRL in CNN during a wormhole attack is growing as the number of nodes upsurges because the wormhole node phoney requests and the nodes keep rebroadcasting these fake requests, which volume of routing packets. Rebroadcasting of fake requests will become more common as

Table 5 Average residual energy with sum of nodes

Number of nodes	CNN	LSTM	DCNNBiLSTM
200	0.97	0.85	0.67
400	0.99	0.98	0.88
600	1.32	1.21	1.09
800	1.43	1.35	0.99

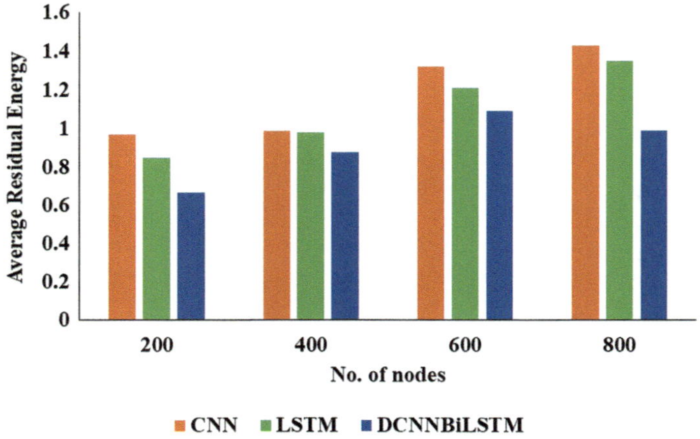

Fig. 4 Average residual energy analysis

Table 6 Normalized routing load with sum of nodes

Number of nodes	CNN	LSTM	DCNNBiLSTM
200	0.96	0.75	0.77
400	0.99	0.88	0.68
600	1.22	1.11	0.59
800	1.33	1.25	0.49

nodes multiply. The DCNNBiLSTM result has a lower NRL than CNN and LSTM. The suggested model uses 42.09% less energy than CNN, while the proposed model uses 35.53% less energy than LSTM.

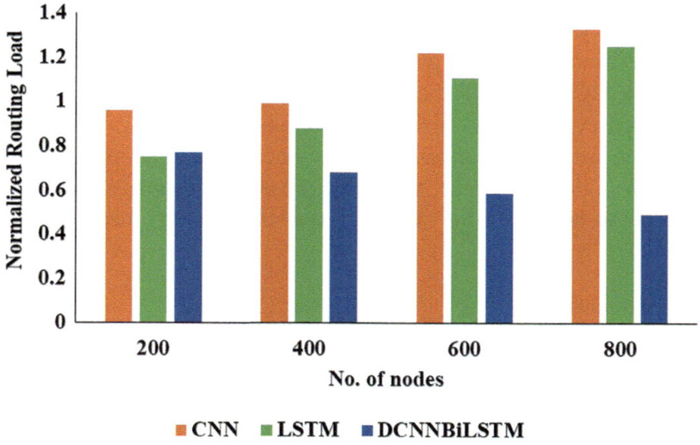

Fig. 5 Normalized routing load analysis

5 Conclusion

The wormhole assault is one sort of (DoS) attack that depletes the network's resources. Different performance parameters on the network are impacted by a flooding assault. To minimize the impact on the network of a wormhole node, it is crucial to prevent and detect it. Deep learning is one of the innovative methods that wormhole attack detection systems have recently used extensively to improve their effectiveness in protecting hosts and computer networks. This article focuses on wormhole attack detection in MANET using deep learning. Use of CNN and LSTM advantages are combined in the DCNNBiLSTM topology. For system administrators and network businesses to enhance wormhole attack detection, the suggested DCNNBiLSTM architecture is ideal. A high-volume labelled dataset was created utilizing data that was gathered using chosen features. In terms of PDR, ARE and NRL, the simulation results demonstrate that the suggested wormhole assault for MANET is reasonably comparable to the DCNNBiLSTM. DCNNBiLSTM demonstrated promise for reducing the effects of wormhole assaults, particularly in low-mobility environments. In the upcoming work, intend to combine the existing technique with packet loss as a wormhole detection criterion to increase the detection ratio.

Ad-hoc network comprises mobile nodes with each node having the movement throughout the network. Communication among these mobile nodes happens through the wireless links by direct or intermediate nodes without possessing any fixed infrastructure because of their mobile nature. The node works as host as well as router for other nodes in network for transmission of data. The nodes need to cooperate with each other during communication. However, in ad-hoc network, nodes may not cooperate with each other to reduce the power consumption and bandwidth requirements. These non-cooperative nodes, which tend to drop as well as alter the data, without forwarding them, are referred to as malicious or selfish.

References

1. Tahboush, M., Agoyi, M.: A hybrid wormhole attack detection in mobile ad-hoc network (MANET). IEEE Access **9**, 11872–11883 (2021)
2. Gupta, N., Singh, S.N.: Wormhole attacks in MANET. In: 2016 6th International Conference-Cloud System and Big Data Engineering (Confluence), pp. 236–239. IEEE (2016)
3. Biswas, J., Gupta, A., Singh, D.: WADP: a wormhole attack detection and prevention technique in MANET using modified AODV routing protocol. In: 2014 9th International Conference on Industrial and Information Systems (ICIIS), pp. 1–6. IEEE (2014)
4. Singh, Y., Khatkar, A., Rani, P., Barak, D.D.: Wormhole attack avoidance technique in mobile adhoc networks. In: 2013 Third International Conference on Advanced Computing and Communication Technologies (ACCT), pp. 283–287. IEEE (2013)
5. Anju, J., Sminesh, C.N.: An improved clustering-based approach for wormhole attack detection in MANET. In: 2014 3rd International Conference on Eco-friendly Computing and Communication Systems, pp. 149–154. IEEE (2014)

6. Shin, S.Y., Halim, E.H.: Wormhole attacks detection in MANETs using routes redundancy and time-based hop calculation. In: 2012 International Conference on ICT Convergence (ICTC), pp. 781–786. IEEE (2012)
7. Imran, M., Khan, F.A., Jamal, T., Durad, M.H.: Analysis of detection features for wormhole attacks in MANETs. Procedia Comput. Sci. **56**, 384–390 (2015)
8. Jamali, S., Fotohi, R.: Defending against wormhole attack in MANET using an artificial immune system. New Rev. Inform. Netw. **21**(2), 79–100 (2016)
9. Rmayti, M., Begriche, Y., Khatoun, R., Khoukhi, L., Mammeri, A.: Graph-based wormhole attack detection in mobile ad hoc networks (MANETs). In: 2018 Fourth International Conference on Mobile and Secure Services (MobiSecServ), pp. 1–6. IEEE (2018)
10. Jamali, S., Fotohi, R.: DAWA: Ddfending against wormhole attack in MANETs by using fuzzy logic and artificial immune system. J. Supercomput. **73**(12), 5173–5196 (2017)
11. Alam, M.R., Chan, K.S.: RTT-TC: a topological comparison based method to detect wormhole attacks in MANET. In: 2010 IEEE 12th International Conference on Communication Technology, pp. 991–994. IEEE (2010)
12. Bashir, M., Tahir, S., Almufareh, M.F., Hamid, B., Qamar, F.: Wormhole attack detection technques in MANET. In: 2023 International Conference on Business Analytics for Technology and Security (ICBATS), pp. 1–7. IEEE (2023)
13. Gorlatova, M.A., Mason, P.C., Wang, M., Lamont, L., Liscano, R.: Detecting wormhole attacks in mobile ad hoc networks through protocol breaking and packet timing analysis. In: MILCOM 2006–2006 IEEE Military Communications Conference, pp. 1–7. IEEE (2006)
14. Nouri, M., Aghdam, S.A., Aghdam, S.A.: Collaborative techniques for detecting wormhole attack in MANETs. In: 2011 International Conference on Research and Innovation in Information Systems, pp. 1–6. IEEE (2011)
15. Diro, A.A., Chilamkurti, N.: Distributed attack detection scheme using deep learning approach for Internet of Things. Futur. Gener. Comput. Syst. **82**, 761–768 (2018)
16. Sriram, S., Vinayakumar, R., Alazab, M., Soman, K.P.: Network flow based IoT botnet attack detection using deep learning. In: IEEE INFOCOM 2020-IEEE Conference on Computer Communications Workshops (INFOCOM WKSHPS), pp. 189–194. IEEE (2020)
17. Rathod, I., Saxena, A.: Remapping attack detection and prevention for reliable data service in MANET (2023)
18. Rajkumar, K., Mercy Shalinie, S.: Boomerang packet testing to mitigate wormhole attack in MANET (2023)
19. Bhatti, D.S., Saleem, S., Imran, A., Kim, H., Kim, K.I.: DAIWN: detection and isolation of wormhole nodes in wireless ad hoc networks (2023)
20. Ninu, S.B.: An intrusion detection system using exponential henry gas solubility optimization based deep neuro fuzzy network in MANET. Eng. Appl. Artif. Intell. **123**, 105969 (2023)
21. Abbood, Z.A., Atilla, D.Ç., Aydin, Ç.: Intrusion detection system through deep learning in routing MANET networks. Intell. Autom. Soft Comput. **37**(1) (2023)
22. Singh, C.E., Vigila, S.M.C.: An investigation of machine learning-based intrusion detection system in mobile ad hoc network. Int. J. Intell. Eng. Inform. **11**(1), 54–70 (2023)
23. Edwin Singh, C., Celestin Vigila, S.M.: WOA-DNN for intelligent intrusion detection and classification in MANET services. Intell. Autom. Soft Comput. **35**(2) (2023)
24. Vijayan, P., Anbalagan, P., Selvakumar, S.: An ensembled optimization algorithm for secured and energy efficient low latency MANET with intrusion detection (2023)

Enhancing Security in Wireless Sensor Networks: A Broadcast/Multicast Authentication Framework with Identity-Based Signature Schemes

Shilpi Sharma and Bijendra Kumar

Abstract WSN applications are feasible for real-time access to the data. Real-time access to the data may become the cause of security breaches. The proposed authentication scheme incorporates cryptographic credentials, secure communication channels, and biometric authentication to ensure robust user and node authentication. Formal security analysis is conducted using the ROR model, and simulations are performed using AVISPA. Our analysis considers a one-way hash function execution time (Th) based on meticulous examination of experimental results. We approximate Th to be around 0.0005 s, establishing its validity for comparative analysis. The results demonstrate the scheme's efficacy, ensuring confidentiality, integrity, and availability of critical data. Our proposed approach strikes a balance between security demands and WSN limitations, providing a secure framework for seamless access to real-time data. The empirical results reinforce the scheme's effectiveness, practicality, and suitability for WSN applications.

Keywords WSN · Wireless sensor network · Security · Authentication · Sensor node

1 Introduction

The resource constraints in WSN nodes necessitate a careful balance between security requirements and limited resources. Authentication algorithms should be designed to minimize computational complexity and energy consumption while still providing the necessary security levels. WSN user access control mechanism is essential

---

S. Sharma (✉) · B. Kumar
Department of Computer Science and Engineering, NSUT, Delhi, India
e-mail: shilpi.sharma.cs19@nsut.ac.in

B. Kumar
e-mail: bizender@nsut.ac.in

G. Fortino et al. (eds.), *Proceedings of Third International Conference on Computing and Communication Networks*, Lecture Notes in Networks and Systems 917,
https://doi.org/10.1007/978-981-97-0892-5_3

for authorization and authentication [1]. Two crucial constraints that need to be considered are computational limitations and energy consumption constraints.

1. Computational Limitations: WSN nodes often have limited processing power, memory, and computational capabilities. This restricts the complexity of authentication algorithms that can be implemented on these nodes.
2. Energy Consumption Constraints: WSN nodes are typically powered by batteries, which have limited energy capacity. Authentication algorithms that consume excessive energy can deplete the node's power quickly, leading to reduced network lifetime and frequent battery replacements. Thus, energy-efficient authentication algorithms are crucial for prolonging the network's operational duration.

1.1 Security Challenges for WSN

1. Identity and authentication: In IoT applications utilizing WSNs, the presence of diverse entities such as computers, servers, sensors, actuators, and RFID tags requires the distinction of three categories of identifiers: object identifiers, communication identifiers, and application identifiers. This differentiation is essential to effectively manage and identify the various entities involved in the system [17]. Traditional user-password authentication [19] may not be suitable, and there are cases where users delegate credentials to virtual entities known as digital shadows.
2. Protocol and network security: Symmetric Key Cryptography (SKC) with pre-shared keys [4] works optimally when entities come from a specific local area. However, key distribution becomes challenging when entities connect with unknown entities at a specific time.
3. Privacy: In order to ensure data security and privacy, each entity within a wireless sensor network (WSN) should determine the level of granularity for generated and shared data, and enforce suitable access control policies. Providing users with usable interfaces to define granularity and access control policies [14] can be achieved through advanced technologies like privacy-preserving distributed data mining algorithms, AI, metadata, and applications.

1.2 Our Contribution

The contributions of the research paper are listed below:

- Introduction of a novel authentication scheme tailored for Wireless Sensor Networks (WSNs).
- Incorporation of cryptographic credentials, secure communication channels, and biometric authentication for robust user and node authentication.
- Comprehensive evaluation through simulations and empirical analysis.

- Comparative analysis with existing schemes, considering computational and communication costs.
- Formal security analysis using the ROR model to assess the scheme's resilience against security threats.

1.3 Related Work

The paper [6] highlights the need for secure and efficient authentication schemes in wireless sensor networks (WSNs). The proposed three-factor authentication scheme claims to address these vulnerabilities using elliptic curve cryptography (ECC). It includes features including smart card revocation, local password changing, dynamic sensor node addition, and password guessing protection. In [7] as a two-factor scheme, the combination of password and smartcard is used, and as a three-factor scheme, biometric is also used along with password and smartcard. The system enables users to register on several gateway nodes, lowering the risk of a single point of failure and enhancing the effectiveness of authentication. In [11], the energy saving protocol is proposed to encrypt the message and claimed to achieve authentication and confidentiality. A secure multifactor authentication approach [9] for IoT-enabled WSNs to address security threats in data transmission is proposed. It introduces a physically unclonable function-based authentication method and establishes secure sessions to ensure confidentiality. Formal and informal security analyses are conducted to evaluate the proposed approach. The study [3] surveys data aggregation protocols in WSN and IoT networks, addressing optimization and requirements trade-offs. It highlights existing gaps and suggests research solutions. A transitive device authentication protocol [8] for large-scale IoT systems addresses the challenge of authenticating devices connected to intermediate devices rather than directly to gateways. The protocol utilizes session keys established in mutual authentication and undergoes security analysis and verification. Formal analyses and simulations demonstrate its resistance to critical attacks, while experimental comparisons show improved performance compared to existing approaches. The proposed protocol [5] combines fuzzy verifier and honeyword security enhancement techniques with discrete elliptic curve encryption for forward security and robustness against sensor capture attacks. The protocol's efficacy in terms of security and efficiency has been shown through formal security analysis and comparisons.

The issues of authentication and key management in IoT are discussed in the article [12], especially in vital infrastructures like health care, intelligent transportation, and home automation. In this survey article, we look at key management and authentication in IoT for critical infrastructures. It divides the development into four components that handle privacy and security concerns. Paper addresses the security concerns in wireless sensor networks (WSNs), a subset of the Internet of Things (IoT). It proposes a key management protocol based on pool-hash to enhance information security and ensure connectivity and resilience. The experimental results validate the suitability of the proposed scheme for WSN applications. The research [2] focuses on security services in the Internet of Everything (IoE) and the challenge of key exchange in insecure networks. Existing symmetric key exchange techniques lack authentication and are therefore unauthenticated, leaving them open to man-in-the-middle attacks. The technique for symmetric key exchange that is suggested in this research is compact and reliable, making it ideal for smart devices with little computational power. The work [13] focuses on integrating artificial intelligence (AI) and IoT for efficient agriculture management in smart cities. Bi-level authentication in [16], which can utilize polynomial key share and can provide node-to-gateway authentication through mapping verification of nodes. Instead of deploying one security mechanism hybrid security mechanism can be a better option. To improve resource management and decision-making, the suggested approach makes use of metaheuristic optimization. Using secure IoT devices and dependable connection, it enables reliable data collecting. Simulations demonstrate the model's importance in reducing overheads on IoT devices and improving resource monitoring in agriculture.

2 Proposed Authentication Scheme

The specific cryptographic credentials used in the proposed authentication scheme play a vital role in ensuring data authenticity and integrity in WSN applications. These credentials include:

1. Digital Signatures: Digital signatures are used to verify the authenticity of data requests and responses exchanged between nodes and users.
2. Public Key Infrastructure (PKI): PKI is employed to manage the distribution and verification of digital certificates. Digital certificates bind public keys to the identity of the entities, such as WSN nodes and users, involved in the authentication process.
3. Secure Communication Channels: The data transfer is encrypted by the authentication mechanism to prevent eavesdropping and manipulation and is protected by secure communication channels like Transport Layer Security (TLS) or Secure Socket Layer (SSL).
4. Biometric Authentication: The identification of users accessing the WSN data is confirmed using biometric authentication technologies like fingerprint or iris recognition.

Secure communication channels are developed in WSNs to safeguard data transfer by using a variety of encryption mechanisms and techniques. The Transport Layer Security (TLS)/safe Socket Layer (SSL) protocols, which employ asymmetric encryption algorithms like RSA for safe key exchange and symmetric encryption methods like Advanced Encryption Standard (AES) for data encryption, are one frequently used method. Additionally used are the shared secret key encryption and decryption protocols Data Encryption Standard (DES) and Triple Data Encryption

Standard (3DES). AES, a widely adopted symmetric encryption algorithm, operates on fixed-size data blocks and employs a substitution–permutation network. To address resource constraints, lightweight cryptographic algorithms like PRESENT, LEA, or SIMON are designed specifically for WSN applications, optimizing encryption and decryption operations. Key exchange protocols such as Diffie–Hellman (DH) or Elliptic Curve Diffie–Hellman (ECDH) enable secure key agreements without transmitting the secret keys over the network.

In the proposed scheme, biometric data, such as fingerprints or iris patterns, is collected from users or nodes using specialized biometric sensors or devices. This data is then securely stored in a dedicated storage module, either locally or in a centralized server, ensuring the confidentiality and integrity of the biometric information.

By incorporating biometric authentication methods, the proposed scheme strengthens the authentication process in the resource-constrained WSN environment, providing an additional layer of security and enhancing the overall robustness of the system.

While developing a novel authentication scheme following assumptions are considered:

- Cryptographic credentials which are employed in the scheme are computationally secure and lightweight.
- Messages which are sent by base stations may not be secured or can be modified.
- There is a communication channel between sensor nodes, i.e., WSN is connected.
- Due to the mobility of nodes, as well as due to the revocation of malicious nodes, the network topology is likely to change frequently.
- The battery life, storage space, and processing power of every sensor node in the WSN are all limited (Table 1).

2.1 *Phases of the Proposed Scheme*

Phase 1: Sensor Node Registration

This step involves the registration of the sensor node with the gateway node (GW) to access WSN services. The registration process involves using a pairwise key along with authentication at the base station. The pairwise key is generated through a series of steps:

- First, the base station generates a large number of random functions, which are stored in a symmetric square matrix with unique IDs.
- This matrix can be decomposed into a triangular form. When a node is assigned to the matrix, it is placed in the lower part of the row and the column from the upper part of the same index.
- During communication, the nodes exchange their columns. To establish communication between two nodes, they must have a common key. This key is generated by combining the row of one node with the column of the other node in the matrix.

Table 1 Numerous symbols\notations used in the proposed scheme

Notation	Definition
Udi	User's mobile device i
Ui	User (i)
Idi	Identity (Ui)
PWi	Password (Ui)
Ni	Secret random number is known to the user (Ui)
GW	Gateway node/Base station
H	Cryptography Hash function of Password
G	General information (Username & Password)
Ai, Bi, Ci	Biometric smart template
Sn	Sensor node
Xs	Pre-installed secret key
Tid	Temporary ID
Ts	Timestamp of registering a Gateway
Xa	A secret is known to the base station
K	Base Station's symmetric key
xCHj	A secret that is shared between CHj and the Base Station
(.)	One-way cryptographic hash function
$\oplus$	XOR operation
||	Operator for concatenating strings bitwise
Q	The sequence of elements in the underlying finite field Fq
E	An Elliptic curve selected appropriately from Fq
P	A base point among the generator points on the Elliptic curve
N	The prime order of the point P
O	The point at infinity, where multiplying the point P by a scalar n results in the identity point O, and P is not equal to O

- Each node has a unique row and column associated with it, and the combination of these two elements generates a pairwise key. This method of generating pairwise keys ensures that each pair of nodes has a unique key, and the keys are not shared between nodes.

To access WSN services, the sensor node registers itself with the gateway node (GW) by using pairwise key along with authentication at the base station. The pairwise key can be generated through the following steps:

- The base station is capable of generating a large number of random functions, and these functions are stored in a symmetric square matrix with unique IDs.
- The matrix can be decomposed in triangular form.

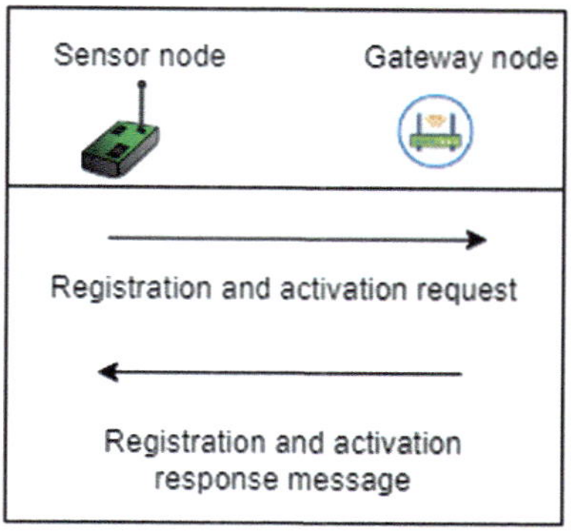

Fig. 1 Sensor node registration phase

- While a node is assigned in the matrix, it is with the lower part of the row and the column from the upper part of the same index. While communication is taking place, the nodes exchange their columns.
- While communication between two nodes, they must have a common key for communication (Fig. 1).
- The user's mobile device (*UDi*) has a registration interface, and a user inputs his/her ID (*IDi*) and a chosen password *PWi.* Although we assume that the connection between the UD and GW is secure, it is not feasible to send the user password to *GW* in plain text to avoid a stolen verifier attack. Therefore, the user computes the Hash of *PWi,* and 512-bit output is stored as *vpwi.*

$$UD\text{: Compute } vpwi = H(PWi).$$

- UD provides its identity *UDi* and *vpw* to the GW node in a secure way.

$$UD \Rightarrow GW\text{: } UDi, vpw$$

- The GW node has the pre-installed secret key *Xs* stored in it, and only an authorized system administrator has the permission to access the *GW* data. The *GW* generates a random nonce *N*0 and computes *Tid* and *Xs*. Every time a user registers to *GW*, a fresh nonce ensures a fresh Temporary ID (*Tid*). The secret key *Xs* computed by *GW* is also refreshed. Then GW stores (*Tid, vpw, Xs, and TS*) in the user database. *TS* represents the timestamp that the gateway recorded when a user was doing the registration. A one-bit flag is associated with each *Tid* to keep track of login.

$$\begin{aligned}
&GW\text{: Compute } g = H(UDi)\\
&\text{Compute } Tid = g \oplus N0\\
&\text{Compute} = H(Tid||Xs)\\
&\text{Store } Tid, vpw, Xs, TS
\end{aligned}$$

- GW responds to the user for a registration successful with *N0* and

$$Xs.GW \Rightarrow UD\text{: } Succ\_Reg\,(N0,\ Xs)$$

- Upon receiving $N0$ and Xs from GW, UDi computes Tid and stores Tid and Xs for future use.

$$UDi\text{: Compute } g = H(UI)$$
$$\text{Compute } Tid = g \oplus N0$$
$$\text{Store } Tid, Xs$$

- GW provides (Tid, Xs, TS) to sensor devices to provide login interface to the user $GW \Rightarrow SNs$: Tid, Xs, TS.
- The data is maintained in the sensor device to store (Tid, Xs, TS) for user.

Phase 2: User Registration

The whole process involves the following steps:

- User with a user identity and secret security code sends a request through a secure channel to register to GW Node.
- Upon receipt, the request for registration from the user GW node verifies the user's ID in the existing database. If it already exists, then the request is rejected; otherwise, the session number is initiated, and registration data is recorded (Fig. 2).

When a user (Ui) who is new to the system wishes to retrieve information from a particular sensor node (UDi), they are obligated to initiate a registration request to the gateway node (GW). In response to the registration request, the GW node creates a biometric card specifically for Ui, serving as a means of authentication and authorization for the user's future access to the system.

For registering a user to sensor nodes following steps are executed:

- Selection of user identity Ui, selection of passwords PWi.

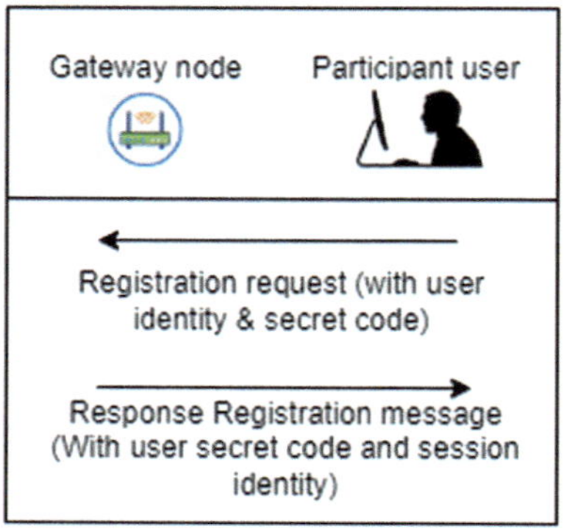

Fig. 2 User registration phase

- Select the random number Ni to compute hashed password $PWi = h\ (PWi\ ||Ni)$.
- If the request for registration is accepted, the *GW* node computes biometric template $A_i = h\ (UDi\ ||xa)\ ||MPW\ i)$, $B_i\ = h\ (UDi\ ||xa)\ MPWi$ and $C_i = h(xa||K)$.
- Send UDi, PWi, and PW1 to the GW node as a registration request message via a secure channel. The GW node verifies if the user's UDi is available in the user database after receiving the registration request message from Ui. Then *GW* issues a smart biometric template which is tamper-proof with A_i, B_i, C_i, and $h(.)$ stored in it.
- After receiving the biometric templates, the user enters *Ni* into the biometric template.

Phase 3: Login and Authentication Phase

- User requests for login to access the services from WSN.
- The GW node conducts a search within the database of the user to retrieve the pair of IDs and biometric information. In this instance, various cases are possible,
  - Case 1: IDs not matched; In this scenario, the GW node terminates the login and authentication session and denies the login request.
  - Case 2: IDs matched; in this case, GW updates the session number, computes a new secret key, and executes the next steps (Figs. 3 and 4).

When GW receives a login request, it will perform the following steps to authenticate Ui:

- Verify current timestamp Ts, if $(T' - Ts) > \Delta T$, authentication process terminated, ΔT is the time interval, while transmission delay of the WSN. If $(T' - Ts) \leq \Delta T$, the next steps will be proceeding.
- *GW* deciphers the user's identity $IDi = IDi \oplus h\ (X\ ||\ Ts)$ and $Mi' = h\ ((IDi \oplus h(S)\ ||\ X\ ||\ T)$.
- If $Mi' = Mi$, *GW* authenticates for login; otherwise rejects.
- GW forms message *M: GW –> U {Sn}*, where *Sn* is the sensor that will respond to the *Ui.*

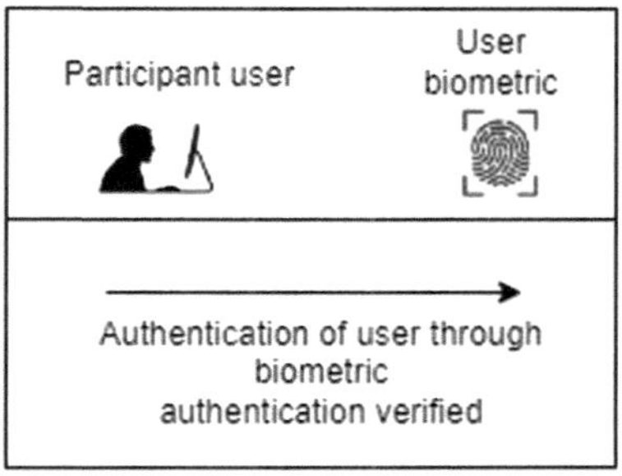

Fig. 3 User login phase

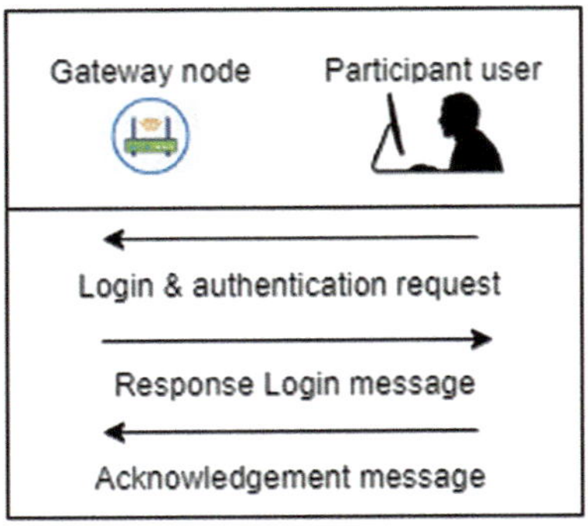

Fig. 4 User authentication phase (Sensor node and GW)

Phase 4: Password Update and Sensor Node Addition Phase

Due to security purposes, user Ui can change his/her password. The following steps are executed while the user wishes to update the password PWi.

- The first user (Ui) gives the inputs of biometric credential, immediately reader terminal asks the user to insert *IDi,* current password *PWi* (current). Then terminal computes $ai = Bi \oplus MPWi$ (current).
- If the computed password matches with the credentials stored in the biometric template, the user proceeds to the next steps. If the password entered does not match, the user is intimated as the wrong password *PWi.* The process of taking an old password to verify and entering a new password performs only up to a certain number of times to prevent password-guessing attacks by malicious users.
- After taking input for the new password, i.e., *PWi* (new), the masked value of it computed $MPWi = h$ (*PWi* (new)) and $Ai' = h$ *(ai* || *MPWi* (new).
- Finally, *PWi* (old) is replaced with *PWi* (new).

3 Cryptanalysis of the Authentication Scheme Proposed in Xie et al. [19]

One limitation of the existing work was that the user's biometric information was stored in a smart card. The smart card calculated α = Rep (BIOi, β) and verified whether V1 = h(UIDi||UPWDi||α). However, if the smart card was stolen or lost, the user would be unable to enter the registration phase, and adversaries could easily access all the sensitive information. This is not the case in the proposed work, where biometric templates are used for user authorization.

Offline password guessing attacks: In Xie et al. [19], the claim that an adversary cannot speculate the true password even after having the information kept in the smart card and all messages communicated in a public channel seems to be an illusion.

In Xie et al. [19], the session key, SK = h(abP||UIDi||SIDj||T3), is generated through an elliptic curve cryptosystem and the fuzzy extractor methods impose a heavy calculation for resource-constrained devices and WSN. In our proposed scheme, the session key is generated with a random number and the biometric template, which avoids the heavy calculation.

Anonymity of sensor node: The proposed scheme has not kept real identity of the user and sensor is transmitted through a public channel in any of the phases of communication. Even if during the authentication and verification phase a malicious user eavesdrops on the message IDi = IDi ⊕ h (X || Ts) and tries to get the identity of the sensor node, adversary cannot obtain IDi without h (X || Ts).

Replay attack: Replay attack can be prevented by adding timestamps and random numbers. The proposed scheme uses timestamps and secret random numbers. Timestamps for each session are different. Hence any session cannot be replayed.

Man-in-the-middle attack: The proposed protocol verifies that Mi′ is equal to Mi, when the gateway receives a login request from a user. As a result, the user is authenticated based on the value of Mi′. If an adversary attempts to impersonate a legitimate user, they would need to know the pre-shared secret key parameters Ts and Xs. However, obtaining these parameters is not feasible. Therefore, the proposed scheme effectively mitigates man-in-the-middle attacks.

4 Security Analysis of the Proposed Protocol

4.1 Informal Security Analysis

The proposed scheme demonstrates resilience against offline password-guessing attacks.

Proof Let's there is an attacker A who possesses a genuine user Ui's smart card SCi. Through power analysis attacks, A can extract all the information stored in SCi, including TIDi, Kpub, etc. In our scenario, we make the assumption that A has acquired the table of user identity verifier from the database of base station GW. It is of utmost importance to recognize that merely obtaining the virtual identity of Ui does not furnish A with sufficient information to make an educated guess about the password PWi of Ui. There exist compelling rationales behind this assertion:

- Guessing PWi would require A to have simultaneous knowledge of secret credentials IDi. However, obtaining such simultaneous knowledge is computationally infeasible and extremely difficult to achieve.
- Additionally, for A to successfully guess PWi and perform a login, they would need to acquire both the password and biometric credentials simultaneously. This poses a computationally challenging problem that is highly unlikely to be achieved in practice.

Therefore, the proposed scheme is not susceptible to these types of attacks.

4.2 Security Analysis Using BAN Logic

The BAN logic involves various types of elements, including principals, encryption and decryption keys, and formulas. The typical symbols used to represent principals are A, B, P, Q, and S. Principals A, B, and P's public keys are represented by Ka, Kb, and Ks, respectively the following elements are utilized in constructing a BAN logic proof:

- $P \mid \equiv X$: Principal P is either entitled to believe that assertion X or P believes it.
- $\neq(X)$: Formula X is new, or X has just been created.
- $P \mid \Rightarrow (X)$: P has authority over X.
- $P \lhd (X)$: P sees X
- $P \mid \backsim X$: P once said X
- $P \overset{k}{\leftrightarrow} Q$: P and Q can safely interact with one another using the shared key K.

Rule I: The Significance of the Message

$$\text{A.} \quad \frac{P| \equiv P \rightleftharpoons Q, P \lhd (X)Y}{P|\equiv Q| \backsim X}$$

The BAN logic can be applied in scenarios where principal P is more likely to trust principal Q if Q has previously mentioned X. In addition, if P believes that Y is a shared secret with Q and P observes the message (X)Y, this can be used as a basis for logical deduction. This can be particularly useful in situations where communicated messages involve shared secrets.

$$\text{B.} \quad \frac{P| \equiv P \leftrightarrow Q, P \lhd \{X\}k}{P|\equiv Q| \sim X}$$

If it is assumed that Q has K, a shared key, and X is encrypted using K, then when P observes this, P can conclude that Q has previously made the statement X. This principle can be helpful in making logical deductions when encrypted messages are transmitted using a shared key K.

Rule II (Nonce-Verification Rule)

$$\frac{P| \equiv \neq (X), P| \equiv Q| \backsim X}{P| \equiv Q| \equiv X}$$

P infers that Q shares P's belief in X if P thinks that the statement X can only be uttered now and Q has previously stated X.

Rule III (Jurisdiction Rule)

$$\frac{P| \equiv Q_1 \Rightarrow (X), P| \equiv Q| \equiv X}{P| \equiv X}$$

If P believes that Q has control or expertise over X, then P considers Q to be truthful about X.

Rule IV (Freshness-Conjunct Rule)

$$\frac{P| \equiv\neq (X)}{P| \equiv\neq (X, Y)}$$

If X is considered current at present, then any message containing X is also considered current because of the current status of X.

Rule V (Other Rules)

$$\frac{P| \equiv X, P| \equiv Y}{P| \equiv (X, Y)}, \frac{P| \equiv (X, Y)}{P| \equiv X}, \text{ and } \frac{P| \equiv Q| \equiv (X, Y)}{P| \equiv Q| \equiv X}.$$

The feature required for the belief operator |≡ is that P will believe a group of statements only if P infers each statement individually. The recommended method enables a strong mutual authentication between a legitimate user (Ui) and the reachable sensor node (Sn) during the authentication and key negotiation phase. BAN logic may be used to verify the protocol's security and integrity, which is an ideal technique to demonstrate this authentication.

Proof The following test goals, which are customized particularly to the needs of the authentication and key agreement phase, must be met in order to evaluate the success of the proposed system.

Goals

$$\text{Goal G1: } Sn \mid \equiv Ui \overset{Xs}{\longleftrightarrow} Ui$$

$$\text{Goal G2: } Ui \mid \equiv \; Ui \Rightarrow Ui \overset{Xs}{\longleftrightarrow} Sn.$$

Generic Form

The subsequent messages are employed in the generic formulation of the proposed scheme for the login, authentication, and key agreement phases:

Msg1: Ui $\rightarrow$ GW: (Xs, Tid, IDi)
Msg2: GW $\rightarrow$ Sn: (UDi, Ni, xa)
Msg3: Sn $\rightarrow$ GW: (UDi, Ni, K)
Msg4: GW $\rightarrow$ Ui: (UDi, K, Ni, Xs)

The following assumptions are made:

$$A_1\text{: } GW| \equiv \text{Ui} \overset{IDi,Ai,Bi,Ci}{\longleftrightarrow} GW$$

$$A_2\text{: } \text{Ui} \mid \equiv \text{ Ui} \overset{Xs}{\longleftrightarrow} GW$$

$$A_3\text{: } \text{Sn} \mid \equiv \text{ Sn} \overset{Xa,K}{\longleftrightarrow} \text{Ui}$$

$$A_4\text{: } \text{Ui} \mid \equiv GW \Rightarrow \text{Ui} \overset{K}{\longleftrightarrow} GW$$

Proof The following stages are the main proofs utilizing the BAN logic's rules and presumptions:

- From Msg1, we have

$$\text{S1: Ui} \overset{Tid}{\longrightarrow} GW\text{: } (\text{IDi}, (\text{Ui} \overset{Xs}{\longrightarrow} GW))$$

- We obtain S2 from Rules 1, 4, and 5 as well as from A1, A2, and S1

$$\text{S2 : } \text{GW}| \equiv \text{Ui}| \equiv (\text{Ui} \overset{Xs}{\longleftrightarrow} GW)$$

- From Rule 3, A2, and S2, we have

$$\text{S3: } \text{GW}| \equiv (\text{Ui} \overset{Xs}{\longrightarrow} GW)$$

- From Msg2, Rule 2 and A3, we obtain

$$\text{S4: } \text{Sn}| \equiv Sn \rightarrow GW\text{: } (\text{UDi, Ni, K})$$

- From Msg3, Rule 2,3 and S4, we obtain Goal G1

$$\text{S5: } \text{Sn} \mid \equiv \text{Ui} \overset{Xs}{\longrightarrow} \text{Ui} \qquad (\textbf{Goal G1})$$

- From Rules 1, 2, 3 & 4, A4 and S5, we get

$$\text{S6: } \text{Ui}| \equiv GW \Rightarrow \text{Ui} \overset{K}{\longleftrightarrow} Sn$$

- From Rules 4 & 5, A3, A4 and S6, we get

$$\text{S7: } \text{Ui}| \equiv \text{Ui} \Rightarrow \text{Ui} \overset{Xs}{\longleftrightarrow} Sn \qquad (\textbf{Goal G2})$$

Based on objectives G1 and G2, it can be concluded that the anticipated scheme successfully accomplishes mutual authentication between Ui and Sn.

4.3 *Formal Security Verification Using ProVerif Simulation*

ProVerif is a protocol verification tool, used to assess and verify the robustness of the protocol against all possible active attacks. Table 2 displays variables and the whole user and sensor authentication procedure. Figure 5 depicts the simulation result, which demonstrates that the suggested approach complies with the authentication and security criteria.

Table 2 Security analysis of the proposed scheme

Security properties	Xie et al. [19]	Masud et al. [11]	Shivanna and Venkatesiah [16]	Proposed
Password Guessing	√	×	×	×
Man-in-the-middle attack	√	√	√	×
Mutual authentication	×	×	×	×
Insider attack	√	√	×	×
User anonymity	×	×	×	×
Impersonation attack	×	×	√	×
Modification attack	×	√	×	×

```
-- Query inj-event(endSGparam(x)) ==> inj-event(beginSGparam(x)) in process 1
Translating the process into Horn clauses...
Completing...
ok, secrecy assumption verified: fact unreachable attacker(Ki[])
ok, secrecy assumption verified: fact unreachable attacker(NC[])
Starting query inj-event(endSGparam(x)) ==> inj-event(beginSGparam(x))
RESULT inj-event(endSGparam(x)) ==> inj-event(beginSGparam(x)) is true.
-- Query not attacker(secretA[]); not attacker(secretB[]); not attacker(secretC[]);
Translating the process into Horn clauses...
Completing...
ok, secrecy assumption verified: fact unreachable attacker(Ki[])
ok, secrecy assumption verified: fact unreachable attacker(NC[])
Starting query not attacker(secretA[])
RESULT not attacker(secretA[]) is true.
Starting query not attacker(secretB[])
RESULT not attacker(secretB[]) is true.
Starting query not attacker(secretC[])
RESULT not attacker(secretC[]) is true.
Starting query not attacker(secretD[])
RESULT not attacker(secretD[]) is true.
```

Fig. 5 Simulation result

4.4 Formal Security Analysis Under the ROR Model Through AVISPA Simulation

In the context of the proposed scheme, the ROR model is employed to conduct a formal security analysis. It verifies the correctness and security properties of the authentication protocol. By considering the protocol in the ROR, potential vulnerabilities and threats can be identified and evaluated.

The protocol involves three participants: user Pt1 (Ui), gateway Pt2 (GW), and sensor Pt3 (Sn). Each participant is represented as the t-th instance during the execution. For clarity, Pt1 (Ui), Pt2 (GW), and Pt3 (Sn) correspond to the t-1th instance of the user, the t-2th instance of the gateway, and the t-3th instance of the sensor, respectively. Furthermore, we consider the presence of an adversary A with the capability to modify, remove, insert, or gain knowledge of transmitted messages during the communication process.

The simulations using AVISPA involve the consideration of various attack scenarios within the ROR (real or random) model. The model includes several queries designed to replicate actual attacks and assess the protocol's security robustness. The specific queries used in the simulations are as follows:

- Execute (Pt1(Ui), Pt2(GW), Pt3(Sn)): This query allows adversary A to intercept the messages exchanged between the wireless entities Ui, GW, and Sn over public channels.
- Corrupt SC: This query enables the adversary to extract all sensitive parameters smartcard stored in the user's.
- Reveal (Pt): This query facilitates the disclosure of the session key K between Pt and its corresponding partner in the active session.
- Send (Pt, Msg): This query represents an active assault, where the adversary has the ability to transmit a message Msg to Pt and react accordingly.
- Test (Pt): This query assesses the robustness of the session key shared between Ui, GW, and Sn according to the ROR model. A coin is flipped in an unbiased manner before the simulation begins.

The simulation setup involves the execution of the proposed protocol using the back-ends OFMC and CL-AtSe. The goal of the simulation is to analyze the protocol's behavior and verify its executability. The results demonstrate that the proposed scheme has been accurately translated into HLPSL specifications and fulfills the design objectives by guaranteeing executability. The protocol is tested for execution and model checking, with a specific focus on a limited number of sessions.

By conducting these simulations, the protocol's ability to withstand potential attacks and vulnerabilities is assessed, ensuring its security robustness and effectiveness in real-world scenarios (Figs. 6, 7, 8 and 9).

The user, Gateway, and sensor roles—which make up the fundamental roles in our suggested scheme—have their HLPSL specifications given. Simulations of the suggested procedure were performed utilizing the well-known back-ends OFMC and CL-AtSe. The fact that the technique uses bitwise XOR operations is notable.

```
// Phase 2: User Registration and Authentication
process phase2 =
 ( process userRegistration =
  ( input(U, ID(U), PWi(U), Bi(U)) ->
   send(GW, ID(U), PWi(U), Bi(U)) ->
   recv(GW, Tid(U)) -> send(U, Tid(U)) -> STOP
  )
 )
 |||
 ( process userAuthentication =
  ( recv(U, Tid(U)) -> input(GW, Tid(U)) ->
   send(SN, Tid(U)) -> input(SN, authenticate(U, GW, SN)) ->
   send(U, authenticate(U, GW, SN)) -> STOP
  )
 )
```

Fig. 6 Role of user

```
role gateway(GW)
const
 H: protocol,
 ID: protocol,
 Xs: protocol,
 K: protocol

 ( process gatewayRegistration =
  ( recv(U, pairKey(U, GW)) -> send(GW, pairKey(U, GW)) -> STOP )
 )
```

Fig. 7 Role of gateway

```
role sensorNode(SN, GW)
const
 H: protocol,
 ID: protocol,
 Xs: protocol,
 K: protocol
// Phase 1: Sensor Node Registration
process phase1 =
 ( process sensorNodeRegistration =
  ( recv(GW, pairKey(U, GW)) -> send(SN, pairKey(U, GW)) -> STOP
  )
 )
```

Fig. 8 Role of sensor

It is assured that the protocol can reach a state where a potential attack could occur throughout its execution by testing the protocol's executability. The results unequivocally demonstrate that the proposed scheme has been accurately translated into HLPSL specifications, effectively fulfilling the design objectives by guaranteeing executability.

Protocol Analysis Summary: - Backend: OFMC - Version: 2006/02/13 - Analysis Result: Safe - Details: Bounded number of sessions - Protocol: C:\progra~1\SPAN\testsuite - Goal: As specified - Comments: None - Statistics: - Parse Time: 0.00 seconds - Search Time: 0.25 seconds - Visited Nodes: 160 nodes - Depth: 7 plies	Protocol Analysis Summary: - Backend: CL-AtSe - Analysis Result: Safe - Details: Bounded number of sessions, Typed model - Protocol: C:\progra~1\SPAN\testsuite - Goal: As specified - Statistics: - Analyzed: 4 states - Reachable: 0 states - Translation Time: 0.39 seconds - Computation Time: 0.00 seconds

Fig. 9 Result of simulation

Scheme	User	Sensor node	Gateway node	Total cost	Communication cost
Amin, R. and Biswas, G.P., 2016	7Th	5^{Th}	8Th	20^{Th}	408 bytes
Xu, L. and Wu, F., 2019	9Th	4^{Th}	11Th	24^{Th}	560 bytes
Our proposed	12Th	4^{Th}	7Th	22Th	580 bytes

Fig. 10 Computational and communication costs

5 Evaluation of Security and Efficiency Characteristics

5.1 Computational and Communication Costs Comparison

In Table Fig. 10, we present a comparison and assessment of the computational overheads and communication cost between our proposed scheme and other relevant schemes [20, 21] during the login and authentication phases. We consider the following time value: Th for executing a one-way hash function. To determine the time value, we refer to the experimental results provided in Amin and Biswas [20] for the cryptographic primitives. Therefore, we approximate the value as follows: Th is approximately 0.0005 s.

5.2 Security Features Comparison

The following section will concentrate on evaluating the efficiency of the newly proposed approach in comparison to other related approaches The evaluation will

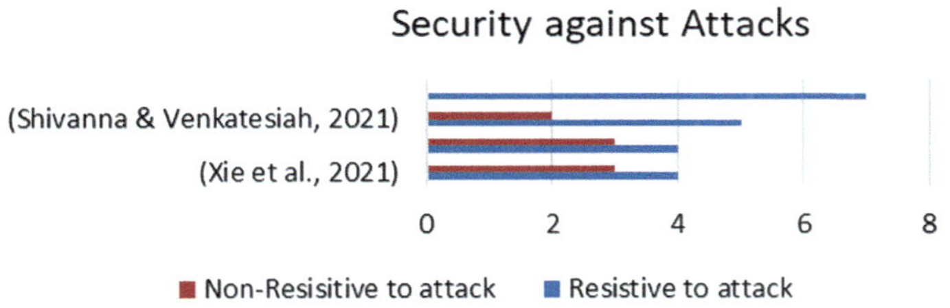

Fig. 11 Security comparison of the proposed scheme

be restricted to the authentication stages since users usually do not modify their passwords and biometric data frequently, and the registration stage of users and sensors occurs only once.

To assess the security attributes of the proposed approach, a comparative analysis is conducted with alternative methods of a similar nature. The comparison focuses on security aspects such as password guessing, man-in-the-middle attack, mutual authentication, insider attack, user anonymity, and impersonation attack (Fig. 11).

5.3 *Limitations*

While the proposed authentication scheme demonstrates effectiveness in ensuring the security of wireless sensor networks (WSNs), there are certain trade-offs and limitations that need to be considered.

1. Computational Overhead: The scheme's reliance on cryptographic operations and biometric authentication introduces additional computational overhead, which can impact the responsiveness of the WSN nodes, particularly those with limited processing power and resources.
2. Energy Consumption: The use of encryption algorithms and secure communication channels can increase the energy consumption of WSN nodes, which can be a concern in resource-constrained environments.
3. Biometric Challenges: While biometric authentication adds an extra layer of security, it also introduces certain challenges. The collection, storage, and processing of biometric data require careful consideration to ensure privacy and protection against unauthorized access. Robust mechanisms for secure biometric data transmission, storage, and authentication are necessary.

To enhance the proposed authentication scheme, future research can focus on addressing these limitations and challenges.

6 Conclusion

In this study, we analyzed a recently proposed authentication scheme by Xie et al. [19] and demonstrated that it generates the session key, using computationally intensive methods like elliptic curve cryptosystem and fuzzy extractors. These methods impose a significant computational burden on resource-constrained devices and wireless sensor networks. However, in the context of wireless sensor networks, it is crucial to develop security solutions with minimal computational overhead. To address this issue, our proposed scheme employs a combination of biometric authentication and pre-shared keys, which significantly reduces the computation time required for user authentication and session key establishment. Additionally, our scheme eliminates the use of smart cards, mitigating security risks associated with smart card loss. Hence, we can say our proposed scheme offers promising security solutions for wireless sensor network applications by minimizing computational overhead and communication requirements. It provides a balance between security and resource limitations, making it suitable for practical implementation in wireless sensor networks.

References

1. Ardiansyah, F., Budi, A.S., Primananda, R.: Hybrid cryptography for data security in wireless sensor network. ACM Int. Conf. Proc. Ser. **256**, 221–225 (2021). https://doi.org/10.1145/3479645.3479694
2. Ashraf, Z., Sohail, A., Yousaf, M.: Robust and lightweight symmetric key exchange algorithm for next-generation IoE. Internet of Things (Netherlands), **22**(November 2022), 100703 (2023). https://doi.org/10.1016/j.iot.2023.100703
3. Begum, B.A., Nandury, S.V.: Data aggregation protocols for WSN and IoT applications—a comprehensive survey. J. King Saud Univ.—Comput. Inform. Sci. **35**(2), 651–681 (2023). https://doi.org/10.1016/j.jksuci.2023.01.008
4. Bettoumi, B., Bouallegue, R.: Efficient reduction of the transmission delay of the authentication based elliptic curve cryptography in 6LoWPAN wireless sensor networks in the Internet of Things. In: 2021 International Wireless Communications and Mobile Computing, IWCMC 2021, pp. 1471–1476. https://doi.org/10.1109/IWCMC51323.2021.9498578
5. Chen, C., Guo, H., Wu, Y., Gao, Y., Liu, J.: A novel two-factor multi-gateway authentication protocol for WSNs. Ad Hoc Netw. **141**(December 2022), 103089 (2023). https://doi.org/10.1016/j.adhoc.2023.103089
6. Dai, C., Xu, Z.: A secure three-factor authentication scheme for multi-gateway wireless sensor networks based on elliptic curve cryptography. Ad Hoc Netw. **127**(December 2021), 102768 (2022). https://doi.org/10.1016/j.adhoc.2021.102768
7. Das, P., Vashisth, A., Chadha, D., Ananda Kumar, S., Banerjee, A., Shiaeles, S.: LIMAP: a lightweight multilayer authentication protocol for WBAN. Wireless Pers. Commun. **121**(4), 2857–2884 (2021). https://doi.org/10.1007/s11277-021-08853-x
8. Krishnasrija, R., Mandal, A. K., Cortesi, A.: A lightweight mutual and transitive authentication mechanism for IoT network. Ad Hoc Netw. **138**(September 2022), 103003 (2023). https://doi.org/10.1016/j.adhoc.2022.103003
9. Kumar, R., Singh, S., Singh, P.K.: A secure and efficient computation based multifactor authentication scheme for Intelligent IoT-enabled WSNs. Comput. Electr. Eng. **105**(June 2022), 108495 (2023). https://doi.org/10.1016/j.compeleceng.2022.108495

10. Majeed, A. A., Mahmood, B. A., Shakir, A.C.: A secure and energy saving protocol for wireless sensor networks. Bull. Electr. Eng. Inform. **10**(6), 3353–3360 (2021). https://doi.org/10.11591/eei.v10i6.2796
11. Masud, M., Gaba, G.S., Alqahtani, S., Muhammad, G., Gupta, B.B., Kumar, P., Ghoneim, A.: A lightweight and robust secure key establishment protocol for Internet of Medical Things in COVID-19 patients care. IEEE Internet Things J. **8**(21), 15694–15703 (2021). https://doi.org/10.1109/JIOT.2020.3047662
12. Rao, P.M., Deebak, B.D.: A comprehensive survey on authentication and secure key management in internet of things: challenges, countermeasures, and future directions. Ad Hoc Netw. **146**(March), 103159 (2023). https://doi.org/10.1016/j.adhoc.2023.103159
13. Rehman, A., Abunadi, I., Haseeb, K., Saba, T., Lloret, J.: Intelligent and trusted metaheuristic optimization model for reliable agricultural network. Comput. Stand. Interfaces **87**(May 2023), 103768 (2024). https://doi.org/10.1016/j.csi.2023.103768
14. Rezaeibagha, F., Mu, Y., Huang, K., Zhang, L., Huang, X.: Secure and privacy-preserved data collection for IoT wireless sensors. IEEE Internet Things J. **8**(24), 17669–17677 (2021). https://doi.org/10.1109/JIOT.2021.3082150
15. Sasirega, L., & Shanthi, C.: Bi-level authentication and mannequin routing for improving security features of WSN-IoT. Sci. Tech. J. Inform. Technol. Mech. Opt. **21**(6), 929–935 (2021). https://doi.org/10.17586/2226-1494-2021-21-6-929-935
16. Shivanna, P., Venkatesiah, S.S.: Secure multimodal authentication scheme for wireless sensor networks. Int. J. Saf. Sec. Eng. **11**(6), 653–661 (2021). https://doi.org/10.18280/ijsse.110605
17. Thahniyath, G., Mishra, P., Moorthy, S.R.: Two phase secure data collection technique for wireless sensor networks. Indonesian J. Electr. Eng. Comput. Sci. **26**(1), 512 (2022). https://doi.org/10.11591/ijeecs.v26.i1.pp512-520
18. Wan, T., Wang, L., Liao, W., Yue, S.: A lightweight continuous authentication scheme for medical wireless body area networks. Peer-to-Peer Netw. Appl. (2021). https://doi.org/10.1007/s12083-021-01190-7
19. Xie, Q., Li, K., Tan, X., Han, L., Tang, W., Hu, B.: A secure and privacy-preserving authentication protocol for wireless sensor networks in smart city. Eurasip J. Wirel. Commun. Netw. **2021**(1) (2021). https://doi.org/10.1186/s13638-021-02000-7
20. Amin, R., Biswas, G.P.: A secure light weight scheme for user authentication and key agreement in multi-gateway based wireless sensor networks. Ad Hoc Netw. **36**, 58–80 (2016)
21. Xu, L., Wu, F.: A lightweight authentication scheme for multi-gateway wireless sensor networks under IoT conception. Arab. J. Sci. Eng. **44**(4), 3977–3993 (2019)

TSFTO: A Two-Stage Fuzzy-Based Tasks Orchestration Algorithm for Edge and Fog Computing Environments

Leila Kheroua, Zouina Doukha, and Samira Moussaoui

Abstract One of the main issues addressed nowadays in IOT environments is reducing data latency and ensuring an efficient workload balancing. The classical cloud-computing paradigm can no more face the huge amount of data generated by thousands of nearby connected objects. Consequently, the new emerging edge and fog paradigms can help reducing latency, preserving network bandwidth, and improving service quality. Following this trend, we investigate the use of fuzzy logic to improve tasks orchestration. Our proposed TSFTO algorithm is deployed on a multi-layer architecture challenging various device types and important data volume. It also allows managing workload balancing in a cloud, fog, edge, and IoT infrastructure. This process takes into account the unloaded tasks features and the computational resources states. In addition, we have used a list of new criteria as fuzzy entries having a major impact on improving the orchestration method. To validate our proposed TSFTO protocol, we have conducted several simulation tests using the PureEdgeSim simulator. On average, results show that TSFTO reduces the task execution response time by 46.66%, enhances task failure rate by 3%, and saves 10% energy gain.

Keywords Internet of Things · Cloud · Fog and edge computing · Service placement · Orchestrator · Fuzzy logic · Latency · Offloading tasks · PureEdgeSim

L. Kheroua (✉) · Z. Doukha · S. Moussaoui
RIIMA Laboratory, Computer Science Faculty, USTHB, Bab Ezzouar, Algiers, Algeria
e-mail: lkheroua@usthb.dz

Z. Doukha
e-mail: zdoukha@usthb.dz

S. Moussaoui
e-mail: smoussaoui@usthb.dz

G. Fortino et al. (eds.), *Proceedings of Third International Conference on Computing and Communication Networks*, Lecture Notes in Networks and Systems 917,
https://doi.org/10.1007/978-981-97-0892-5_4

1 Introduction

Internet of Things (IoT) is commonly recognized as a worldwide network of interconnected objects, each with its own unique identity, that communicate through various networks in order to exchange data and transform it into valuable information [1]. Nowadays, there has been a significant rise in the number of objects connected to the Internet in home automation and smart cities, with the need to handle large volumes of data transmitted by IoT. As the cloud is not suitable for addressing latency and transmission delay issues, companies are now turning towards promising paradigms that can avoid network bottlenecks, such as fog and edge computing. As per the scientific community, fog and edge computing pertain to the use of virtualized environments to manage and process data for IoT devices [2, 2]. In this paper, we use a network with a four-layer architecture [4, 5] (IoT, edge, fog, and cloud). The purpose is to use close-range devices that possess adequate computing power to place latency-sensitive and bandwidth-intensive applications at the data source layer itself if possible. Otherwise, we try to leverage other resources located at the fog layer in order to minimize latency and network usage. High computational-intensive applications will be placed directly on the cloud or at the fog layer if they are latency-sensitive. However, it is necessary to define where the service should be supported in order to achieve maximum network efficiency. This raises an issue related to the Service Placement Problem (SPP). In SPP, data must be sorted and placed at the appropriate resource level capable of handling them. In fact, this decision is a critical task, so to overcome these challenges, we propose an intelligent service placement mechanism based on a fuzzy logic orchestrator. The decision to offload tasks must take into account many parameters. Thus, in a distributed and heterogeneous edge environment, fuzzy logic presents a good alternative to manage the imprecision of the real world.

The intent of this work is basically to enhance the task orchestration process by reducing task execution time, energy consumption of IoT devices, and task failure rate. The main contributions of the proposed TSFTO algorithm are as follows:

1. Integration of cloud, fog, edge, and IoT: TSFTO works on a four-layer architecture to take advantage of the computation capacity of the cloud while reducing task's response-time execution by processing data at the network edge.
2. Adopting a fuzzy-based new orchestration method at two stages that refine the task offloading: in the first stage, the best layer among cloud, fog, and edge is selected to offload the task. If the edge layer is selected, then a second selection process is executed to choose the best edge device.
3. Introducing new metrics that consider the heterogeneous and dynamic IoT devices characteristics: Mobility (Mob) to avoid disconnection between devices; (FogUsage and DestCPU) to ensure the more available devices for task processing.

The rest of this paper is organized as follows: Sect. 2 presents a review of related work; Sect. 3 presents the proposed Two-Stage Fuzzy-Based Tasks Orchestration

(TSFTO) algorithm steps; Sect. 4 deals with performance evaluation of the proposed TSFTO algorithm along with comparison results. Finally, Sect. 5 concludes the paper and discusses some future works.

2 Related Work

In literature, some studies [3, 5–8, 14] have addressed Service Placement Problem (SPP) in an edge–fog–cloud environment. These studies have been motivated by several objectives such as minimizing latency, reducing energy consumption, and improving quality of service. In [3], Olena Skarlat et al. addressed the Fog Service Placement Problem (FSPP). The proposed architecture implements a fog node that handles orchestration control in each Fog colony. The FSPP has been solved using several approaches, namely: genetic algorithm, greedy first fit heuristic, and an exact optimization method. Results indicate that genetic algorithm and exact optimization method does not exceed the application execution deadline, and the optimization method produces a more efficient service placement plan ensuring more QoS criterions. In addition, in [6, 7] Naas et al. study the SPP in a cloud and fog infrastructure. Their main goal was minimizing global latency when processing data in fog devices. For that purpose, three strategies have been proposed, including one exact approach and two heuristics. An orchestrator has been implemented for each strategy in order to reduce resolution time. The proposed scenario is centralized, online, and static, and does not consider the mobility constraint. The problem has been addressed as a Generalized Assignment Problem, (GAP) using a Linear Integer Programming (ILP). Some other works propose a fuzzy workload orchestration to deal with SPP. For example, in [8], Cagatay et al. proposed a fuzzy-based orchestration algorithm that outperforms other solutions but still needs enhancements considering the same mobility speed of end devices. Another interesting work was proposed in [5], where Mechalikh et al. propose an orchestrator designed for a realistic environment. The IoT layer involves more heterogeneous and constrainted devices. The task orchestration algorithm is based on fuzzy decision tree. The reinforcement learning approach adopted ensures a better adaptability to different workloads with less service time. However, task failure due to mobility can still be improved by proposing a more rational model governing the mobility of end devices. Recently, in Hamzei et al. [11], the authors proposed a fuzzy hybrid algorithm to address service composition challenges in IoT context with specific QoS requirements. Results validate that the proposed method improved reliability, availability, and cost-effectiveness compared to alternative approaches. However, it should consider various and dynamic QoS parameters [12]. Also, Fereira et al. [13] is a recent work that proposes an energy-efficient framework based on Integer Linear Programming (ILP) to select nodes for task offloading while meeting other task requests.

3 The Two-Stage Fuzzy-Based Tasks Orchestration Algorithm (TSFTO)

When receiving a request, an orchestrator has to decide on which resource tasks should be executed. The request provides elementary data such as device status information and the application requirements to find the best offloading destination. In our proposed TSFTO algorithm, we have placed an orchestrator on each edge device adopting a distributed strategy. The orchestration process is executed twice at the edge layer. Hence, edge layer devices have to communicate their state periodically via Wi-Fi access technologies. The main TSFTO steps are as follows.

3.1 The First Task Orchestration Algorithm Stage

The first stage consists of selecting the best resource to offload the task. Several criteria come into play to determine the best target resource. In addition to the three criteria already used in Sonmez et al. [8]: WAN Bandwidth (*WB*), Task Length (*TL*), and the task's sensitivity to Latency (*L*), we propose using two more criteria: device mobility (*Mob*), and Fog resource Usage (*FU*). Device mobility will help reducing task execution failure rate by privileging static devices, whereas the fog resource usage will help avoiding the fog layer resources which are generally favored due to their low latency, especially in hotspots. Offloading is then directed to more performant cloud servers. To this end, the five sets of membership functions: μWB, μTL, μL, μMob, and μFU will, respectively, define the membership function of the WAN bandwidth, the task length, the task sensitivity to delay, the mobility of the device, and finally the utilization rate of the fog resource. In the case where the edge layer is selected, a second stage is carried out to select the best edge device.

3.2 The Second Task Orchestration Algorithm Stage

Given the number, diversity, and heterogeneity of devices in the edge layer, the second TSFTO algorithm step involves four criteria: Task Length measured in Giga Instructions Per Second (*GIPS*), CPU usage by edge nodes (*DestCPU*), remaining energy of the potential offloading destination (*DestRem*), and finally the mobility of the potential edge node (*Mobility*). Therefore, four sets of membership functions: μLT, $\mu DestCPU$, $\mu DestRem$, and μM define the previous criteria, respectively.

3.3 The Fuzzy Inference System

We have adopted the Mamdani method [9] where fuzzy logic process includes the fuzzification, the inference engine, the rule aggregation, and finally, the defuzzification.

- **Fuzzification**: Each one of the membership functions defined previously will be evaluated according to the three linguistic variables: Low (L), Medium (M), and High (H). The membership functions of the algorithm's first and second stages are presented as follows:
  - First stage
    $\mu_{\mathrm{WB}}(x) = [\mu_{\mathbf{L}\mathrm{WB}}(x), \mu_{\mathbf{M}\mathrm{WB}}(x), \mu_{\mathbf{H}\mathrm{WB}}(x)]$
    $\mu_{\mathrm{TL}}(x) = [\mu_{\mathbf{L}\mathrm{TL}}(x), \mu_{\mathbf{M}\mathrm{TL}}(x), \mu_{\mathbf{H}\mathrm{TL}}(x)]$
    $\mu_{\mathrm{L}}(x) = [\mu_{\mathbf{L}\mathrm{L}}(x), \mu_{\mathbf{M}\mathrm{L}}(x), \mu_{\mathbf{H}\mathrm{L}}(x)]$
    $\mu_{\mathrm{Mob}}(x) = [\mu_{\mathrm{LMob}}(x), \mu_{\mathbf{M}\mathrm{Mob}}(x), \mu_{\mathbf{H}\mathrm{Mob}}(x)]$
    $\mu_{\mathrm{FU}}(x) = [\mu_{\mathbf{L}\mathrm{FU}}(x), \mu_{\mathbf{M}\mathrm{FU}}(x), \mu_{\mathbf{H}\mathrm{FU}}(\mathrm{x})]$
  - Second stage
    $\mathrm{M}_{\mathrm{LT}}(x) = [\mu_{\mathrm{lLT}}(x), \mu_{\mathbf{m}\mathrm{LT}}(x), \mu_{\mathbf{h}\mathrm{LT}}(x)]$
    $\mu_{\mathrm{DestCPU}}(x) = \left[\mu_{\mathbf{L}\mathrm{destCpu}}(x), \mu_{\mathbf{M}\mathrm{destCpu}}(x), \mu_{\mathbf{H}\mathrm{destCpu}}(x)\right]$
    $\mu_{\mathrm{DestRem}}(x) = [\mu_{\mathbf{L}\mathrm{destRem}}(x), \mu_{\mathbf{M}\mathrm{destRem}}(x), \mu_{\mathbf{HdestRem}}(x)]$
    $\mu_{\mathrm{M}}(x) = [\mu_{\mathrm{M}}(x), \mu_{\mathbf{m}\mathrm{M}}(x), \mu_{\mathbf{h}\mathrm{M}}(x)]$
- **Inference engine**: Our inference system is based on the minimum function. Following Formulas (1) and (2) are applied for each rule of, respectively, TSFTO algorithm's first and second stages.

$$\mu\ \mathrm{offload} = \min(\mu WB(x), \mu TL(x), \mu L(x), \mu Mob(x), \mu FU(x)) \quad (1)$$

$$\mu(\mathrm{edge}) = \min(\mu TL(x), \mu DestCPU(x), \mu M(x), \mu DestRem(x)) \quad (2)$$

- **Aggregation**: This step involves searching the resulting global membership function by combining the already evaluated inference rules. According to Zadeh [9], the calculation of the activation degree of each rule uses the MIN method, while the rule composition phase) uses the MAX one. Formulas (3) and (4) are applied to each one of the rules of the TSFTO steps 1 and 2, respectively.

$$\mu\ \mathrm{offload} = \max(\mu \mathrm{R1offload}\ \ldots, \mu Rn\ \mathrm{offload}) \quad (3)$$

$$\mu(\mathrm{edge}) = \max(\mu \mathrm{R1edge}, \ldots, \mu Rn\ \mathrm{edge}) \quad (4)$$

- **Defuzzification**: The global fuzzy output set constructed by aggregation is used in this step to obtain a unique numerical value as an interpretable output for the user. In our work, we use the center of gravity method [9] to determine the value of the output membership function using the following Formula (5).

$$Cg = \frac{\sum x\mu(x)}{\sum \mu(x)}. \tag{5}$$

4 Performance Evaluation and Results

The performance evaluation of TSFTO algorithm is conducted using a simulation scenario where three different types of applications are executed including an augmented reality application handling images of about 1 Mbytes, an e-health application, and a request for infotainment. Four types of devices are deployed in our scenario as depicted in Table 1 to include the factor of the heterogeneity of edge devices. All simulation tests are performed using PureEdgeSim [10] according to the parameters shown in Table 2. The cloud will be represented by a data center with a total of 2000 GIPS, distributed over 16 virtual machines, while the fog will have a similar data center, but with a lower computing capacity of 100 GIPS, which will be sufficient given the small simulation area. To prove the effectiveness of our proposed algorithm, we evaluated it compared to the closest algorithm in terms of architecture and proposed criteria [8] using the following metrics: the average service time, the energy consumption of battery-powered devices, the decision accuracy, and the task failure rate. The average service time is an important performance metric. Its reduction is one of the objectives targeted by our proposed algorithm.

Table 1 Edge devices configuration

Edge devices	Laptop	Smartphone	Gateway	Sensor
Mobility	No	Yes	No	No
Battery-powered	Yes	Yes	No	No
Generate tasks	No	Yes	No	yes
Power percentage devices (%)	20	30	10	40
Battery power (Wh)	56.2	18.75	–	–
Energy consumption rate (Idle state) (Wh)	1.7	0.078	1.6	0.036
Energy consumption rate (maximum) (Wh)	3.96	3.96	3.96	3.96
CPU (GIPS)	70	25	16	–
CPU cores	8	8	4	-
RAM (Gbytes)	8	4	4	4
Stockage (Gbytes)	1024	128	32	–

Table 2 Simulation parameters

Parameter	Value
Simulation duration	5 (min)
Task generation rate	1 task/device/second
Number of edge devices	100–800
Mobile devices velocity	1.4 m/s
Simulation area size	200 × 200 (m)
Network update interval	10 s
WAN bandwidth	300 (Mbits/s)
WLAN bandwidth	300 (Mbits/s)
WAN propagation delay	0.2 s
Orchestrator deployment	Edge
Architecture	Edge, fog, and cloud

The average response time includes both the task processing time and the network delay. We compared the average response time of our proposed TSFTO algorithm and the algorithm proposed in Sonmez et al. [8] FUZZY LOGIC (FL). The graph shown in Fig. 1 shows that below 300 devices, both algorithms favor offloading tasks to fog and edge because they take into consideration the criterion of task sensitivity to delay when selecting the resource. Additionally, the resource utilization rate is low, which reduces the processing time. However, beyond 300 devices, the 2 algorithms diverge and the FL one generates a longer time response of about more than 46.66% compared to the TSFTO algorithm. This is because the FL algorithm favors offloading tasks to the fog due to its low latency, which generates a long queue at the layer, and therefore it will take more time to perform the offloaded tasks. Meanwhile, the TSFTO algorithm provides better results and remains stable with a response time of 0.3 s. This is guaranteed by adding the FogUsage and DestCPU criteria. Figure 2 displays the energy usage of devices using both TSFTO and FL algorithms. Similarly, beyond 300 devices, the energy rate appears to be more constant, with around a 10% energy gain margin for the TSFTO algorithm that avoids using battery-powered and mobile devices (if possible) and relies more on the cloud for non-latency-sensitive tasks. The failed task ratio is due to mobility and high delays. However, the task failure ratio is low for both algorithms in the scenario with low edge device density. As the density increases, the network workload also increases. Since the FL algorithm relies solely on cloud and fog layers for task offloading, an overload is expected at these layers, which will increase the failure rate. This failure may also be caused if users change their location or if the edge device battery discharges before receiving a response to previously requested tasks. Nevertheless, the TSFTO algorithm provides better results because:

- TSFTO algorithm mobility criterion ensures avoiding offloading tasks to mobile devices that may leave their coverage areas taking the task with them.

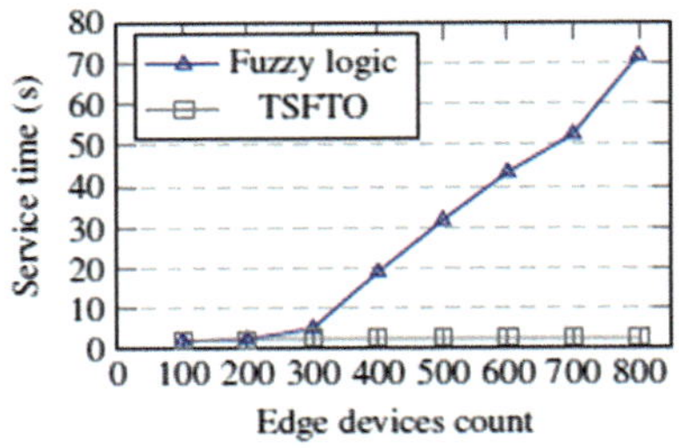

Fig. 1 The average service time

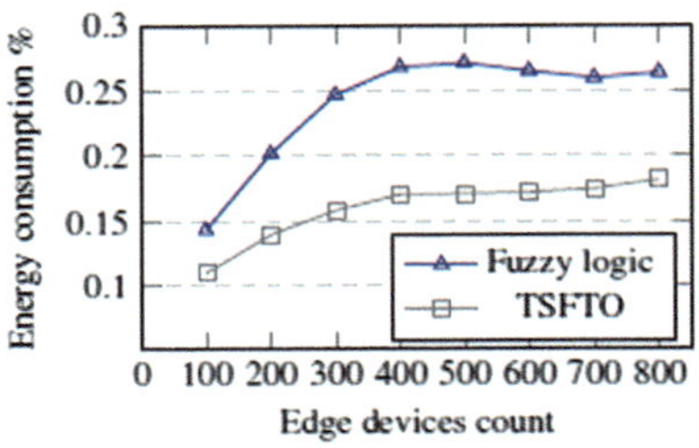

Fig. 2 Energy consumption of battery devices

- The remaining energy of the edge node allows avoiding offloading tasks into edge devices with low battery charge, risking device shutdown at the time of task execution.
- As shown in Fig. 3, the edge device layer has allowed a better tasks-workload across three layers instead of two, thereby contributing to the improvement of the task success rate.

The task accuracy defines the success rate of their execution as well as the judicious choice of the appropriate layer for their offloading. The graph in Fig. 4 shows that the success rate of the TSFTO algorithm is between 98.95 and 100% for all the architecture three layers. This can be explained by the integration of multiple criteria, including mobility (Mob), which allowed for better decision-making. Although the tests on the proposed algorithm have achieved the set objectives in terms of workload, latency, and successful tasks, the current study has some limitations that need to be addressed: 1. It is essential to meet the dynamic IoT devices characteristics by

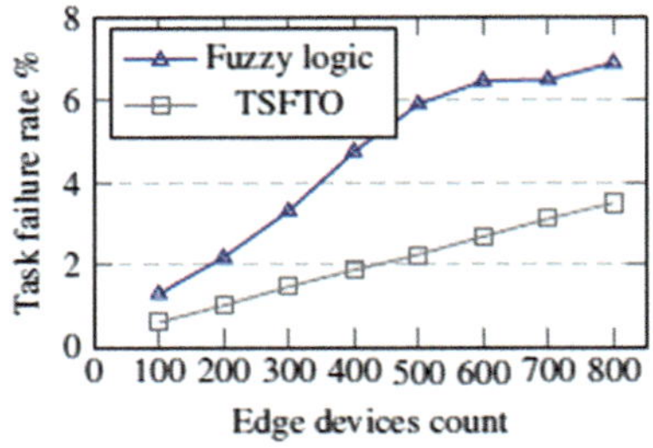

Fig. 3 Tasks failure rate

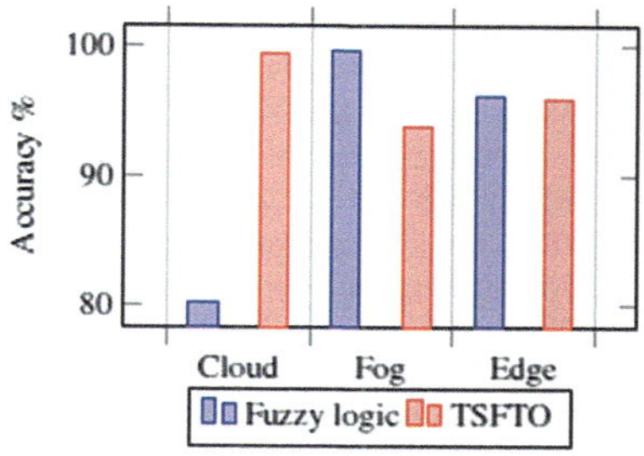

Fig. 4 Tasks accuracy

integrating some real-world constraints, for example, a random speed movement of mobile edge devices and considering the fog MAN bandwidth that may affect the network delay. 2. To better face scalability constraints, it is crucial to consider hierarchical methods for orchestrator's management at the edge layer.

5 Conclusion and Perspective

In this work, we propose a fuzzy logic-based tasks orchestration algorithm designed for a multi-layer architecture, with a distributed management. The algorithm execution includes two steps: the first step consists of selecting the best resource among cloud, fog, and edge layers for offloading the task, taking into consideration five criteria: task sensitivity to delay, WAN bandwidth, task length, device mobility, and fog resource utilization rate. The second step consists on selecting the best edge device according to four criteria: task length, processor utilization rate, remaining energy, and mobility. Indeed, the above criteria's selections have a significant impact on improving our orchestration algorithm performances in terms of task failure ratio and response time. However, it would be interesting to measure the energy and the network bandwidth consumption of the algorithm implementation. Also, as future work, we propose following recommendations: Evaluating various QoS criteria to enhance the orchestrator performances; improving the placement of the orchestrator by deploying a strategy based on clustering; and creating bidirectional communication links between fog nodes. Finally, we can improve the mobility criterion in both steps by considering device speed and direction.

References

1. Kavre, M., Gadekar, A., Gadhade, Y.: Internet of Things (IoT): a survey. In: International Conference (PuneCon), IEEE Pune Section, pp. 1–6 (2019)
2. Elmroth, E., Leitner, P., Schulte, S., Venugopal, S.: Connecting fog and cloud computing. IEEE Cloud Comput. **4**(2), 22–25 (2017)

3. Skarlat, O., Nardelli, M., Schulte, S., Borkowski, M., Leitner, P.: Optimized IoT service placement in the fog. Serv. Oriented Comput. Appl. **11**(4), 427–443 (2017)
4. Muneeb, M., Ko, K.-M., Park, Y.-H.: A fog computing architecture with multi-layer for computing-intensive IoT applications. Appl. Sci. **11**(24), 11585 (2021)
5. Mechalikh, C., Taktak, H., Moussa, F.: A fuzzy decision tree based tasks orchestration algorithm for edge computing environments. In: International Conference on Advanced Information Networking and Applications, pp. 193–203. Springer (2020)
6. Naas, M.I., Raipin, P., Lemarchand, L., Boukhobza, J.: iFogStor: an iot data placement strategy for fog infrastructure. In: IEEE 1st International Conference on Fog and Edge Computing (ICFEC), pp. 97–104 (2017)
7. Ait-Oucheggou, L., Naas, M.I., Hadjadj-Aoul, Y., Boukhobza, J.: When IoT data meets streaming in the fog. In: IEEE 6th International Conference on Fog and Edge Computing (ICFEC), Messina, Italy, pp. 50–57 (2022)
8. Sonmez, C., Ozgovde, A., Ersoy, C.: Fuzzy workload orchestration for edge computing. IEEE Trans. Netw. Serv. Manag. **16**(2), 769–782 (2019)
9. Zadeh, L.: A Mamdani-style fuzzy controller. IEEE Trans. Syst. Man Cybern. **7**(2), 153–162 (1975)
10. Mechalikh, C., Taktak, H., Moussa, F.: PureEdgeSim: a simulation toolkit for performance evaluation of cloud, fog, and pure edge computing environments. In: International Conference on High Performance Computing and Simulation (HPCS), pp. 700–707, Dublin, Ireland (2019)
11. Hamzei, M., Khandagh, S., Jafari Navimipour, N.: A: quality-of-service-aware service composition method in the Internet of Things using a multi-objective fuzzy-based hybrid algorithm. Sensors **23**, 7233 (2023)
12. Chai, Z.-Y., Du, M.-M., Song, G.: Z: A fast energy-centered and QoS-aware service composition approach for Internet of Things. Appl. Soft Comput. **100**, 106914 (2020)
13. Fereira, R., Ranaweera, C., Lee, K., Schneider, J.-G.: Energy efficient node selection in edge-fog-cloud layered IoT architecture. Sensors **23**(13), 6039 (2023)
14. Tavousi, F., Azizi, S., Ghaderzadeh, A: A fuzzy approach for optimal placement of IoT applications in fog-cloud computing. Cluster Comput. **25** (2023)

Tracking Climatic Variations Through Smart IoT-Driven Approach: An Exploratory Analysis

Ashit Kumar Dutta, Abhipsa Sahoo, Ankita Samal, Swastika Bishnoi, Sanchit Verma, Sushruta Mishra, and Joel J. P. C. Rodrigues

Abstract Climate crisis refers to the impact of climate change and global warming. We connect sensors and gadgets to Internet servers so that they can produce accurate reports regarding the state of the environment. It is very crucial to choose the right IoT sensors for each domain for climate monitoring. It would be beneficial for the environment if the materials used in constructing the sensors are non-toxic and biodegradable so that after they are used, there won't be adverse effects of the sensors and contribute to the local region pollution. Due to various IoT devices operating, a notable amount of external power is required, thus increasing the carbon footprint. To lower the carbon footprint, sensors with the capability of harvesting energy from the surrounding environment such as air pressure, solar light, temperature, humidity, etc., are put into use. In this research study, we would be studying the major changes in climate experienced within the past decades, and how IoT has been working for the betterment of the environment, and with the help of two case studies, we will learn about the functioning of IoT components in the monitoring process of the environment as a whole, and the intervention of IoT in groundwater management.

Keywords Climate · Environment · IoT · Sensors · Wireless sensor networks · Raspberry Pi · Temperature · Carbon emissions

A. K. Dutta
College of Applied Sciences, Al Maarefa University, Ad Diriyah, Riyadh 13713, Saudi Arabia
e-mail: adotta@mcst.edu.sa

A. Sahoo · A. Samal · S. Bishnoi · S. Verma · S. Mishra (✉)
Kalinga Institute of Industrial Technology, Deemed to Be University, Bhubaneswar, India
e-mail: sushruta.mishrafcs@kiit.ac.in

J. J. P. C. Rodrigues
College of Computer Science and Technology, China University of Petroleum (East China), Qingdao 266555, China

Instituto de Telecomunicações, Covilhã, Portugal

J. J. P. C. Rodrigues
e-mail: joeljr@ieee.org

G. Fortino et al. (eds.), *Proceedings of Third International Conference on Computing and Communication Networks*, Lecture Notes in Networks and Systems 917,
https://doi.org/10.1007/978-981-97-0892-5_5

1 Introduction

The term "Climate crisis" relates to global warming, climatic change, and its effects. Climate crisis and "climate emergency" have been defining the possible threats of sudden temperature change to the environment and mankind. It also impulses hostile mitigation of change in the climate. Those who "think it conjures the gravity of the threats the earth confronts from prolonged greenhouse gas emissions and can help generate the kind of political willpower that has long been lacking from climate campaigning" use these terms. They think that declaring climate change a catastrophe could have an even more significant impact, although "global warming" garnered hysterical support, also exerted more than "climate change" did.

The climatic changes that the world has experienced, within the past few decades, are as follows:

- **Increase of CO_2 by 10% due to fossil fuels**: According to research from the Global Carbon Project, "The world CO_2 emissions from cement and fossil fuels are expected to reach a record high in 2019," the increase in global emissions slowed between 2014 and 16 was just temporary. Fossil fuel emissions increased by 1.5% in 2017, 2.1% in 2018, and 0.6% in 2019. According to research, even though countries implement the current climatic commitments entirely, we would still witness a gap of 15 $GtCO_2e$ for 1.5 °C by the year 2030, i.e., more than two times the size of the combined emissions of India and China. The overall carbon emission year-wise is shown in Fig. 1.
- **Increase in the average temperature, globally**: According to an assessment from the International Panel on Climate Change, a mere 1.5° of global warming will have a substantial impact on the climate, and 2° will worsen the situation

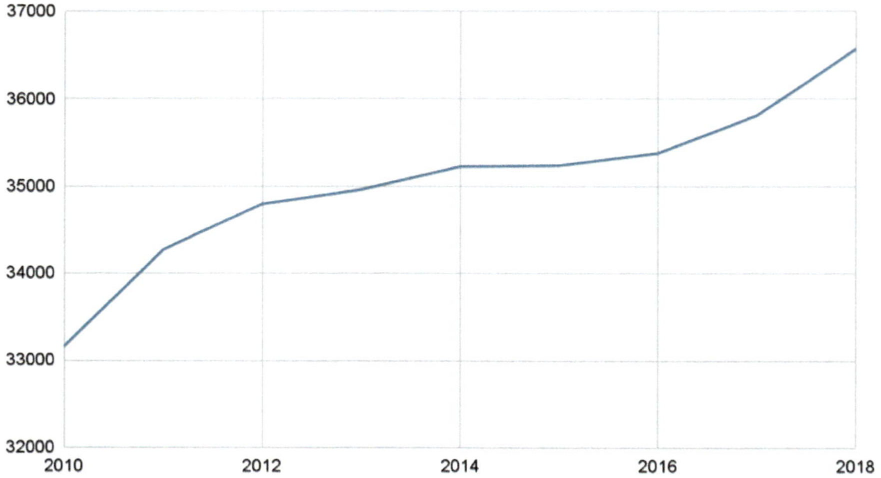

Fig. 1 Graph representing the carbon emissions per year from 2010 to 2018

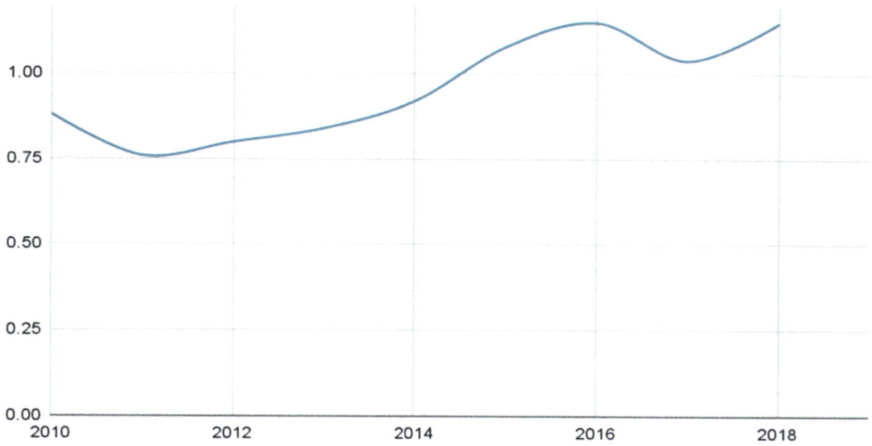

Fig. 2 Graph representing the average temperature on a global scale per year from 2010 to 2018

significantly. It is expected that the Arctic would have only one summer without sea ice every century under 1.5 °C of global warming, and at 2 °C, the possibility increases to at least once every decade. Figure 2 shows the mean temperature on a global level from 2010 to 2018.

- **Rise in sea level:** Between the year 1993 and the past year, the average global sea level rose by about 3.3 mm each year. Over the past decade, the sea level rose by around 2 inches throughout the time course of the last decade. In 2018, the world witnessed the highest mean sea level recorded by satellites. It is recorded that out of the top ten largest cities of the world, eight are situated around weak coastal regions. Even a few inches of rise in sea level would accelerate the changes of high tide flood, causing exertion on the inland, leading to storms, and threatening the infrastructure and habitat. The mean global rise in sea level is shown in Fig. 3.

2 IoT and Its Impact on Climate Monitoring

IoTs are not restricted to boosting the comfort level of smart wearable devices, creating efficient business, logistics strategies, etc. It has started assisting in disjointing the trajectory of disastrous climate change. Innovators across the globe are participating in the climate protection move by incorporating IoT devices with international standards. The ideal scenarios as outlined by IPCC for the substantial reduction in carbon emissions at the global level range from 40 to 70% lesser than that of in the year 2010. The organization predicts that the average temperature might keep on increasing by 2 °C until the year 2050.

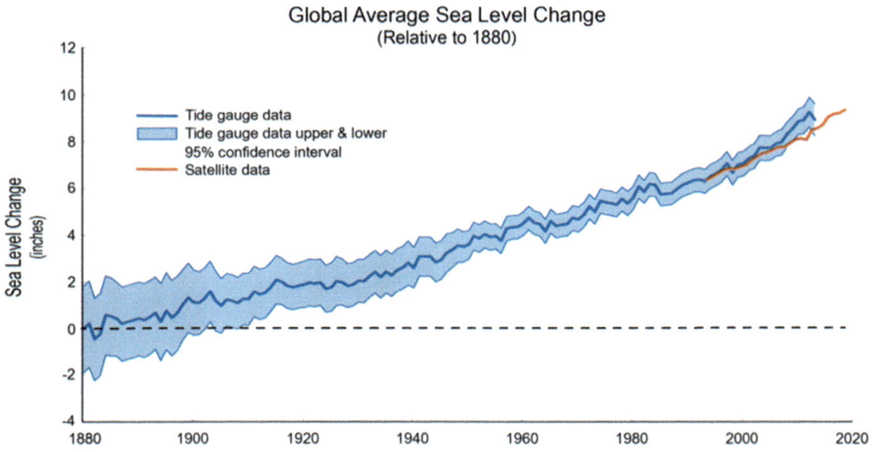

Fig. 3 Graph representing average sea level rise at global sea level, produced by Environmental and Energy Study Institute

- **Potential to decrease carbon output**: A 3% reduction in the amount of carbon emission can be witnessed by IoT smart devices intervening for mending the efficiency and dependency on disposable elements in both public and commercial sectors. Ericsson Research has stated that intelligent agriculture can help in reducing more than 3%. Early execution of IoT monitoring and shift for cargo transport via ocean, has revealed that it is possible to decrease fuel consumption by 15%.
- **Sensors for the Environment as a Service**: Given the IoT techniques existent currently, the day isn't too far when transitioning to a connected and more responsive network would become convenient to everyone. The first step towards it is to standardize environmental sensors that are powered by IoT. IoT intends to offer useful climate information. Waste can be reduced by increasing the citizen and energy flow, goods, and data flow.

3 Factors Affecting Climate Variations

Earth constantly undergoes changes driven by internal and external forces. By creating temperature gradients in the atmosphere and oceans, the Sun nourishes life on Earth and regulates weather and climate. Deforestation, urbanization, and auto pollution are examples of human influences. Because of human activities, the amount of greenhouse gases in the atmosphere has increased, which also contributes to climate change.

- **Temperature**: The average temperature on our planet is 15 °C or 59 °F. It measures how hot or cold something is. Temperature affects climate when it warms or cools up the earth's surface. Being able to absorb the heat from the sun, the surface of the Earth will cause global warming when it becomes hotter. However, ice on top of the ground indicates a colder environment, which means less heat will be absorbed by the earth, protecting us from climate change.
- **Humidity**: The climate is affected by humidity. High air humidity can reduce both the amount of solar energy that reaches the earth and the amount that escapes spacing directly, especially when combined with stagnant atmospheric layers. This causes more warming due to greater absorption of sunlight's heat by the earth's surface and the atmosphere. Because moist air retains more heat than dry air, the amount of moisture in the air increases with humidity, which can have an impact on how much energy is required to heat an object or place.
- **Precipitation**: A measurement of the amount of water that falls from the sky is known as precipitation. It can be in various forms including rain, snow, hail, and other liquid and solid forms of precipitation. This affects the climate by cooling the soil and bringing more water. Evaporation and condensation also affect the water cycle, which ultimately affects other things like plants. It can also cause natural disasters such as floods and landslides.
- **Solar radiation**: The sun's heating properties make solar radiation a significant contributor to climate change. It may result in a positive feedback loop whereby increased atmospheric CO_2 concentrations warm the environment, which increases the absorption of solar radiation and raises atmospheric CO_2 concentrations even further. Infrared radiation from the heat that travels through Earth's atmosphere and then reflects off the planet's surface and back into space is what creates the greenhouse effect.

4 Related Works

In Kimothi et al. [1], the research scholars have provided a dummy implementation of a glacier monitoring system in areas of glaciers where the parameters for glaciers such as surface area and volume of each glacier were monitored. The paper has listed down a few of the possible methods using IoT such as intelligent ecosystems around glaciers, edge gadgets powered ML, less costing IoT devices with smart energy algorithms, and rescue robots with resisting power against glaciers. In Achar [2], the author has mentioned that in order to enhance the safety of the environment, IoT can prove to be beneficial. In Kumar [3], the author has made an attempt to incorporate the effects of smart cities and their components on the mitigation and adaptation cycle of climate, and the requirement of correct decision frameworks within the city. In Okafor et al. [4], the research scholars have used selection methods like forward selection, backward elimination, and EFS, to determine the factors affecting cairclipO3 and cairclipNO2 sensors for measuring O_3 and NO_2, respectively. The result of their study showcases the relevance of feature selection in constructing accurate calibration models working on multiple parameters and the effective work efficiency of SLR, ANN, and MLR in calibrating the less costing sensors for the monitoring networks made for environmental monitoring. In Piromalis et al. [5], the author has made an attempt to create an evaluation of technologies and utilization of IoT in the facility agriculture sector. The study aims to spot the challenges and tendencies of deploying IoT techniques w.r.t. the idea of a smart climate for sustainability. In Salam [6], the author has presented a framework for the creation, explanation, and application of projections due to climate change, and continuous innovation by stakeholders. An IoT infrastructure is also mentioned to support continuous innovation in the climate and environmental science field. In Fang et al. [7], the research scholars have prepared a paper where an IIS prototype for monitoring and managing the environment is prepared. With this, a case study on climate change in regional areas along with their ecological responses is drawn. The outcome of this study shows that the availability of water resources is the deciding factor for the initial production of the terrestrial ecosystem. In Ming et al. [8], the research scholars have proposed an effective solution for the monitoring of CO_2 using IoT and cloud computing technologies. A monitoring structure to generate, collect, store, and visualize the content of CO_2 using an MQ135 CO_2 sensor, WiFi module, firebase storage service, and carbon insight android mobile application for visualization, is built. In Shete and Agrawal, [9], the research scholars have proposed a system that is low-cost and compact and monitors the data remotely. It aids the growth of the city and helps in the improvement of the lifestyle of the netizens. In Ting et al. [10], the research scholars have designed an agricultural system that monitors the important parameters of climate such as temperature and humidity. It provides an efficient and safe decision system. It can notify about the running status of the water motor. It collects real-time data from the sensors deployed and gives suggestions on the pretext of precautions and decisions. In Salam [6], the author has drawn a critical review of different purpose-driven environment monitoring systems. The main focus of their study is water and air quality monitoring. It aims to study other factors of the environment like pollution and disasters. In Mabrouki et al. [11], the research scholars have proposed an auto-weather monitoring system. It allows dynamic and real-time data associated to the regional climatic state. The project works with the objective of sensing the climatic criterion according to the sensors. In Arshad [12], the research scholars envisaged included a third party Raspberry Pi-based smart embedded system, IoT analytics, and sensors working with climatic factors. It aims to benefit the sustainability, productivity, and profit, of the greenhouse farming infrastructure. In Mitro et al. [13], the research scholars have developed smart gadgets that monitor the climatic factors in the jurisdiction of the heritage sites. In Karunarathne et al. [14], the research scholars have built a technological infrastructure with three divisions; acquire data, curate data, and present data.

5 IoT-Based Case Studies in Monitoring Climatic Conditions

In this section, we will be talking about the involvement of IoT techniques in two different fields, in the form of two case studies. The first case study talks about the usage of WSNs in monitoring environmental factors. The second case study talks about the role of IoT in groundwater management.

5.1 Smart Climate Tracker with Application of IoT-Based Radio Sensor Grid

5.1.1 Overview

This project aims to create a system that can be used at the global level at any size to monitor the variables in any environment. Web servers display and retrieve real-time data collected by Raspberry Pi and sensors attached to record data from the environment. With the use of the internet, data becomes handy and can be worked upon irrespective of the location of the user [15]. The ZigBee protocol avails Raspberry Pi as the base station, that links distributed sensor nodes. WSN gathers data about the physical phenomenon in habitat monitoring. IoT can be considered as the connecting brick for cellphones, smart TVs, sensors, and actuators with the network. It allows a new type of interhuman interaction. In WSN, the photoelectric node with the help of sensors, senses the information and gathers it from the end points, which delivers the information to router which further sends it to the coordinator and then provides services to various clients. The collected data is then stored at the base station and sent to the clients via the cloud. Some examples of sensors are temperature vibration, pressure, light, and moisture.

5.1.2 Methodology

The Raspberry Pi monitors all sensor nodes. The design of a Raspberry Pi-based environment monitoring WSN system comprises sensor interfaces such as humidity, temperature, and CO_2. The web server retrieves instantaneous data gathered from sensory nodes. The base station acts as an entry node of WSN that incorporates the web and database server on a single platform, thus reducing the cost and complexity of deploying the model. The role of the sensor node is to recognize the sensor data which then gets its end tag, and to send the information extracted from the neighboring router within the coverage area with a valid end tag, to the next router or the coordinator. The work of the coordinator is to stay connected to the base station. Then the base station sends all the information to the database server or Ethernet. Thus, a WSN is set up with a coordination node, various sensor nodes, a workstation, and a database.

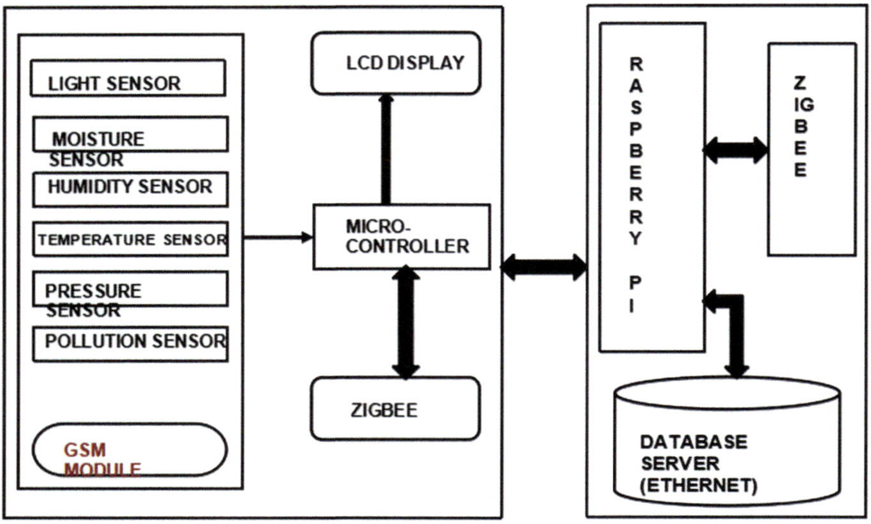

Fig. 4 Overall structure of the environment monitoring WSN system

Raspberry Pi has incorporated XBee as the coordinator.USB connection and UART serial communication connect the XBee and Raspberry Pi instantly. Here, the base station serves as a gateway. The base station analyzes the data that the sensory nodes have gathered, and the Raspberry Pi modifies its database with the newfound information. The Raspberry Pi communicates with the clients through a network outside the system via ZigBee. The case study uses Digi International XBee series module S2 for wireless transmission of information and a multi-hop protocol for networking purposes. Figure 4 illustrates the entire methodology.

5.1.3 Outcomes

The router, end tags, and coordinator are the three different hardware components used in WSN. The open-source data platform for IoT gives users access to a wide range of embedded gadgets and web services. As an outcome, XBee is set up in the instance that acts as a coordinator and is connected to the Raspberry Pi via an interface called UART. The sensor node acts as a router and an end tag, to establish communication with the neighboring router to send real-time data. During the entire networking process, there exists only one coordinator which stays connected with the base station or Raspberry Pi.

5.2 An Overview of IoT-Based Systems in Groundwater Management Models

5.2.1 Overview

Groundwater is crucial to the sustainability of ecosystems and aids in human adaptation to climatic fluctuation and change. Groundwater resources, however, are dynamic and frequently vary as a result of alterations in water abstraction, land use, and climate changes [16]. To solve these problems, many conventional solutions have been proposed for groundwater modeling, such as certain numerical techniques. The management of groundwater resources now collects more data than before thanks to IoT, which is expanding globally. Also, changes and availability of groundwater resources have a significant impact on how well groundwater resources are managed using data. According to various studies in the literature, groundwater managers do not currently have an effective groundwater management system that can gather the required data in real time [17, 18]. In addition, it is clear from the literature that existing data collection methods lack the necessary efficiency to achieve the requirements of calculation models and management objectives. This research intends to highlight contemporary groundwater management models and the role of IOT, in contrast to prior studies that exclusively focused on certain groundwater challenges linked to simulation and optimization approaches.

5.2.2 Methodology

The IoT-based groundwater-level system architecture is composed of three levels, the presentation layer, the service layer, and the physical layer. The communication infrastructure and a large number of sensor nodes are incorporated into the physical layer to collect the necessary data on the groundwater level and aquifer. The service layer receives the raw data that have been measured and gathered. At the service layer, a variety of technologies are available for data analytics. The information that was obtained from the physical layer is also kept in this layer. This layer serves as the implementation layer for both the application and business logic, making it a crucial component of the design [19]. While viewing information on a screen monitor, users are permitted to interact with the presentation layer, which is the visualization layer. Figure 5 displays a high-level view and functional depiction of the IoT-based groundwater level monitoring system's process. This demonstrates how the end-user platform, nodes, and well-deployed sensors are all connected. The direction of the data flow is shown by the arrows. The direction of the data received is shown by the solid line, while the direction of the data request is shown by the broken lines. In this system, the IoT and GIS are coupled to guarantee that groundwater level data is uploaded in real time, which enhances the database's quality and offers geographic information about the system. The sensors positioned in the observation wells measure the groundwater level parameters, and the related values

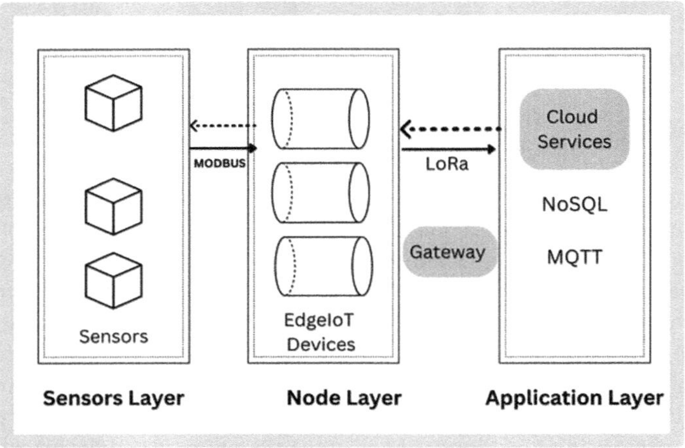

Fig. 5 Design of an IoT-based groundwater resource monitoring system

are transmitted to their nodes using the Modbus protocol. With the Long-Range (LoRa) wireless network, the information is sent from the nodes to the gateway [20]. The data is kept on the MQTT server platform, which is connected to a NoSQL database or another GPU platform for real-time display. Instances of Google Maps API, ArcGIS, ThingSpeak, and WebGL are among these web-based client GPUs.

5.2.3 Outcomes

Historically, most groundwater resource management models in use today have been integrated with simulation and optimization approaches that use the necessary mathematical programming to provide answers to problems within the aquifer. The total limitations, however, include the uncertainties resulting from the system modeling's input parameters. Moreover, there is little research data on the creation of scalable and effective computer models for managing groundwater resources in real time, despite the high level of research focus given to tackling such challenges [21]. The use of IoT-based solutions for managing groundwater resources is a highly useful tool for data collecting, monitoring, and modification. This technique has great potential for water management when combined with GIS. IoT data processing, however, often takes place via a computer resource located in a distant data center utilizing either the internet or cloud computing. As a result, there is now a risk to consumers' privacy, low latency, and scalability issues. As a significant portion of the IoT's enormous data volumes is uploaded into the cloud, an effective and scalable IoT platform is needed to extract useful data in real time for managing groundwater resources.

5.3 Comparative Study of the Discussed Case Studies

The major differences between the above-mentioned case studies in Table 1 are as follows:

Table 1 A comparison of the above-identified case studies on different parameters

Parameters	Case study—I	Case study—II
Domain of research	Monitoring the environmental factors using IoT-based WSN	Sustainability of groundwater on a global level
Methodology	The architecture includes various sensors; *light sensors, moisture sensors, humidity sensors, temperature sensors, pressure*, etc., GSM module, ZigBee, Raspberry Pi, Ethernet, microcontroller, and LCD	The sensors make the sensory layer, that is linked with the node layer with Modbus. The node layer consists of EdgeIoT devices which are connected to the application layer through LoRa. The application layer has cloud services, a NoSQL database, and an MQTT communicational protocol
Outcomes	The open-source data platform for IoT gives users access to a wide range of embedded gadgets and web services. As an outcome of this, XBee is set up in the instance that acts as a coordinator and is connected to the Raspberry Pi via UART. The sensor node acts as a router and an end tag, to establish communication with the neighboring router to send real-time data	This assessment provided a broader view of IoT-enabled management than previous research, which provided a more focused analysis that incorporated simulation and optimization management models. This is because resource data measurement is a key component of effective groundwater management. There are uncertainties in the input of the model in parameters, nevertheless modeling forms a part of the overall constraints
Pros	– The proposed methodology is cost-effective, as it uses WSN that is easy to install, maintain, and operate – It is highly scalable, as it allows for the addition of new sensors with minimal effort – It uses low-power WSN, which consumes lesser energy compared to traditional wired sensors	– It makes use of algorithms that examine large datasets and identify patterns. This improves the accuracy of rainfall predictions compared to traditional methods – It uses automated algorithms that analyze large data quickly, reducing the time required to generate predictions – It is highly scalable, as it can be applied on larger datasets, allowing comprehensive analysis
Cons	– The wireless communication makes it vulnerable to security breaches – Wireless sensors have a limited range of coverage	– The accuracy depends majorly on the quality of the input data – The method is designed primarily for the prediction of rainfall patterns

6 Future Challenges of IoT in Climate Monitoring

- **Privacy and Security**: As IoT becomes a significant part of the future Internet, it must be employed for large-scale, partially mission-critical systems in order to manage trust and security chores effectively. Shared information models should provide confidence and quality of the information to facilitate usage across many applications [22].
- **Cost versus Usability**: IoT uses technology to link physical objects to the Internet [23]. If the usage of IOT is to grow in the next years, it is necessary that the cost of the components needed to allow capabilities like sensing, tracking, and control mechanisms be reasonably cheap.
- **Data management**: When considering a world of networked objects that are constantly exchanging various types of information, it becomes necessary to consider the volume of data generated and the processes involved in processing that data.
- **Device-level energy issues**: Understanding that communication uses the majority of energy on devices and figuring out how to connect "things" in an interoperable way while taking energy limitations into account is one of the main challenges in the Internet of Things.

7 Conclusion

The Internet of Things is a brand-new Internet revolution and a key research topic for researchers in embedded computer science and information technology because of its broad variety of applications and the heterogeneous blend of many embedded and communications technologies in its architecture. IoT tools assist in assessing various environmental variables in accordance with plant needs. All sensors and equipment offer reliable information about environmental conditions since they are all connected to Internet servers. For the purpose of monitoring the climate, it's crucial to choose the right kind of IoT sensor for each area. It is important to choose non-toxic, biodegradable materials when creating sensors so that, once they have served their purpose, they do not harm the environment or add to local pollution. Also, a huge quantity of external power is needed to run such a vast number of IoT devices, which raises their carbon footprint.

References

1. Kimothi, S., et al.: Intelligent energy and ecosystem for real-time monitoring of glaciers. Comput. Electr. Eng. **102**, 108163 (2022). https://doi.org/10.1016/j.compeleceng.2022.108163
2. Achar, S.: Influence of IOT technology on environmental monitoring. Asia Pacif. J. Energy Environ. **7**(2), 87–92 (2020). https://doi.org/10.18034/apjee.v7i2.649

3. Kumar, S.: An Overview of Smart City Approaches for Addressing Global Climate Crisis, vol. 10, no. 4, pp. 1–12 (2020)
4. Okafor, N.U., et al.: Improving data quality of low-cost IOT sensors in environmental monitoring networks using data fusion and machine learning approach. ICT Exp. **6**(3), 220–228 (2020). https://doi.org/10.1016/j.icte.2020.06.004
5. Piromalis, D. D., et al.: Current trends and challenges in the deployment of IOT technologies for climate smart facility agriculture. Int. J. Sustain. Agricult. Manag. Inform. **5**(2/3), 181 (2019). https://doi.org/10.1504/ijsami.2019.10023178
6. Salam, A.: Internet of Things for environmental sustainability and climate change. Internet of Things, 33–69 (2019). https://doi.org/10.1007/978-3-030-35291-2_2
7. Fang, S., et al.: An integrated system for regional environmental monitoring and management based on Internet of Things. IEEE Trans. Indus. Inform. **10**(2), 1596–1605 (2014). https://doi.org/10.1109/tii.2014.2302638
8. Ming, F.X., et al.: Real-time carbon dioxide monitoring based on IOT & cloud technologies. In: Proceedings of the 2019 8th International Conference on Software and Computer Applications (2019).https://doi.org/10.1145/3316615.3316622
9. Shete, R., Agrawal, S.: IOT based urban climate monitoring using Raspberry Pi. In: 2016 International Conference on Communication and Signal Processing (ICCSP) (2016). https://doi.org/10.1109/iccsp.2016.7754526
10. Ting, L., et al.: A secure framework for IOT-based smart climate agriculture system: toward blockchain and edge computing. J. Intell. Syst. **31**(1), 221–236 (2022). https://doi.org/10.1515/jisys-2022-0012
11. Mabrouki, J., et al.: IOT-based data logger for weather monitoring using arduino-based wireless sensor networks with remote graphical application and alerts. Big Data Min. Anal. **4**(1), 25–32 (2021). https://doi.org/10.26599/bdma.2020.9020018
12. Arshad, J.: Intelligent greenhouse monitoring and control scheme: an arrangement of sensors, Raspberry Pi based embedded system and IOT platform. Indian J. Sci. Technol. **13**(27), 2811–2822 (2020). https://doi.org/10.17485/ijst/v13i27.311
13. Mitro, N., et al.: Smart tags: IOT sensors for monitoring the micro-climate of cultural heritage monuments. Appl. Sci. **12**(5), 2315 (2022). https://doi.org/10.3390/app12052315
14. Karunarathne, S.M., et al.: A technological framework for data-driven IOT systems: application on landslide monitoring. Comput. Commun. **154**, 298–312 (2020). https://doi.org/10.1016/j.comcom.2020.02.076
15. Jaldi, A. R., et al.: Environmental monitoring using wireless sensor networks (WSN) based on IoT, **4**(1), 1–8 (2017)
16. Aderemi, B., et al.: A review of groundwater management models with a focus on IOT-based systems (2021). https://doi.org/10.20944/preprints202107.0227.v3
17. Patel, K.K., Patel, S.M.: Internet of Things-IOT: definition, characteristics, architecture, enabling technologies, application & future challenges (2016)
18. Subhra, S., Mishra, S., Alkhayyat, A., Sharma, V., Kukreja, V.: Climatic temperature forecasting with regression approach. In: 2023 4th International Conference on Intelligent Engineering and Management (ICIEM), London, United Kingdom, pp. 1–5 (2023). https://doi.org/10.1109/ICIEM59379.2023.10166883
19. Mohanty, S., Behera, A., Mishra, S., Alkhayyat, A., Gupta, D., Sharma, V.: Resumate: a prototype to enhance recruitment process with NLP based resume parsing. In: 2023 4th International Conference on Intelligent Engineering and Management (ICIEM), London, United Kingdom, pp. 1–6 (2023). https://doi.org/10.1109/ICIEM59379.2023.10166169
20. Soumyadeep, C., Rohan, K., Aditya, K.S., Sushruta, M.: 14 explainable intelligence enabled smart healthcare for rural communities. In: Explainable Artificial Intelligence for Biomedical Applications, pp. 289–308. River Publishers (2023)
21. Adrija, M., Yash, A., Sushruta, M.: 8 pragmatic study of IoT in healthcare security with an explainable AI perspective. In: Explainable Artificial Intelligence for Biomedical Applications, pp. 145–166. River Publishers (2023)

22. Bhavya, M., Pranshu, S., Sushruta, M., Sibanjan, D.: 17 comparative analysis of breast cancer diagnosis driven by the smart IoT-based approach. In: Explainable Artificial Intelligence for Biomedical Applications, pp. 353–374. River Publishers (2023)
23. Swain, T., Mishra, S.: Evolution of machine learning algorithms for enhancement of self-driving vehicles security. In: 2022 International Conference on Advancements in Smart, Secure and Intelligent Computing (ASSIC), Bhubaneswar, India, pp. 1–5 (2022). https://doi.org/10.1109/ASSIC55218.2022.10088396

Design and Manufacturing of an Efficient Low-Cost Fire Surveillance System Based on GSM Communications and IoT Technology

Hamzah M. Marhoon, Noorulden Basil, and Bashar Bahaa Qas Elias

Abstract In addition to recent technological advancements, a noticeable trend has emerged toward the development of smart cities and homes. By integrating intelligent home control components with existing household appliances, it aims to significantly reduce the use of human effort. This integration has led to decreased electricity consumption, enhanced home security against accidents and theft, and an overall improvement in residents' comfort levels. This paper proposes the utilization of Internet of Things (IoT) technology for the design of an intelligent home fire alarm system. The suggested system is comprised of an Arduino microcontroller, a NodeMCU microcontroller, a GSM shield, a flame sensor, a buzzer, and a servo motor. Given the limitations and relatively low accuracy of flame sensors, the LM35 temperature sensor is employed in conjunction with it. Upon detecting a flame and a subsequent temperature rise beyond 50 °C, the Arduino Uno will promptly dispatch an SMS warning to the homeowner. Simultaneously, a logical "1" signal will be transmitted to the NodeMCU microcontroller. This microcontroller will then utilize IoT technology to exhibit the warning message on the pre-programmed graphical user interface (GUI). To conclude the alert sequence, the Arduino Uno microcontroller will activate the servo motor. This activation serves the purpose of initiating the opening of the exit door and triggering the alarm. Rigorous practical testing has been conducted on the implemented system, yielding consistently positive outcomes and successful results.

H. M. Marhoon (✉)
Department of Systems Engineering, College of Information Engineering, Al-Nahrain University, Jadriya, Baghdad, Iraq
e-mail: hamzah.marhoon@nahrainuniv.edu.iq

N. Basil
Department of Electrical Engineering, College of Engineering, Mustansiriyah University, Baghdad, Iraq
e-mail: noorulden@uomustansiriyah.edu.iq

B. B. Q. Elias
Department of Information and Communication Engineering, College of Information Engineering, Al-Nahrain University, Jadriya, Baghdad, Iraq
e-mail: bashar.bahaa@nahrainuniv.edu.iq

G. Fortino et al. (eds.), *Proceedings of Third International Conference on Computing and Communication Networks*, Lecture Notes in Networks and Systems 917,
https://doi.org/10.1007/978-981-97-0892-5_6

Keywords IoT · Fire alarm system · GSM · Arduino · NodeMCU

1 Introduction

In recent years, there has been notable progress in the field of smart homes, mostly propelled by technological breakthroughs and the growing incorporation of Internet of Things (IoT) devices. Smart houses are fitted with sophisticated systems that provide automation and remote control of numerous processes, hence boosting convenience, efficiency, and security for homeowners. These residences employ interconnected gadgets and sensors to oversee and control lighting, temperature, security systems, and household appliances. Smart homes provide customized experiences and energy-conserving functionalities by means of seamless connectivity and clever algorithms, thereby converting conventional living spaces into interconnected and intelligent ecosystems. The ongoing advancement of technology has led to the emergence of smart homes, which have significant potential for enhancing our everyday experiences and transforming our interactions within domestic environments [1, 2].

The IoT framework comprises a network of interconnected devices, sensors, and systems that engage in continuous communication and data sharing. The IoT facilitates the connectivity and interaction of objects, devices, and humans via the Internet. The interconnectivity of systems enables the efficient gathering and dissemination of extensive data, hence enabling improved surveillance, regulation, and automation across diverse fields. The structure of the IoT commonly consists of a blend of hardware elements, including sensors, actuators, and gateways, in addition to software applications and cloud-based platforms for the purpose of data processing and analysis, as depicted in Fig. 1. The IoT framework possesses the capacity to connect the physical and digital realms, hence holding the promise of transforming various sectors and fields, including healthcare, transportation, agriculture, and smart cities. The IoT framework establishes the groundwork for a future characterized by enhanced connectivity and technological advancement through the facilitation of intelligent decision-making, optimisation of resource utilization, and improvement of overall efficiency [3–6].

The fire alarm system serves a vital function in protecting both human lives and valuable assets from the destructive consequences of fires. Ensuring building safety and facilitating early detection and prompt notifications are crucial aspects of fire prevention and mitigation. These systems are comprised of strategically positioned smoke detectors, heat sensors, and audible sirens that are specifically designed to identify and notify individuals of the existence of smoke or increased temperatures. Upon the detection of a fire, the alarm system is triggered, resulting in the activation of loud sirens or bells. This serves the purpose of promptly notifying the occupants to escape the premises. In order to optimize operational efficiency and efficacy, fire alarm systems undergo ongoing advancements through the incorporation of cutting-edge technology, such as wireless communication and intelligent integration. These

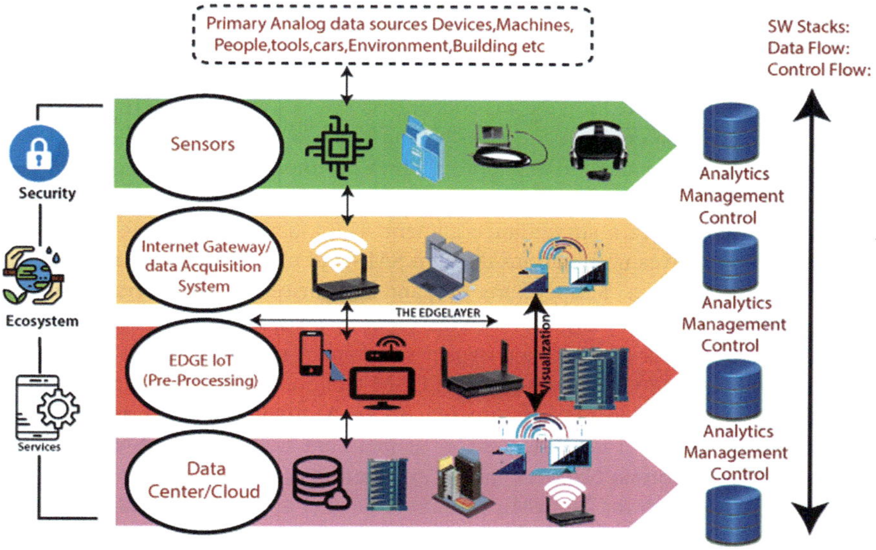

Fig. 1 Functional diagram of IoT layers

devices offer crucial protection and assurance in home and commercial environments by consistently monitoring potential fire dangers [7–9].

The use of IoT technology has brought about a significant transformation in fire alarm systems, resulting in improved efficacy and heightened responsiveness. IoT-enabled fire alarm systems employ interconnected sensors, detectors, and control units to offer real-time monitoring and intelligent identification of fire occurrences. Whereas, these systems have the capability to effectively identify smoke, heat, and flames, so guaranteeing timely detection and expeditious warning to homeowners or relevant authorities. Moreover, the integration of IoT connectivity facilitates the capability of remotely monitoring and controlling various devices. This functionality empowers homeowners to promptly get real-time notifications and undertake essential measures, even in situations where they are physically distant from their residences. The integration of IoT technology with fire alarm systems enhances the overall safety of residential dwellings. The implementation of this technology facilitates prompt and effective reaction in the event of a fire, hence potentially preserving human lives and mitigating harm to physical assets. The aforementioned fire protection technique showcases the potential of IoT in revolutionizing the creation of secure and interconnected living spaces [10].

Several studies have been made in the literature regarding the design of fire alarm systems. The authors in [11] presented a novel of a complex fire alarm system that utilizes a fuzzy neural network to achieve perfect intelligence. By leveraging sensor data, the system derives a fuzzy neural network model tailored to the specific characteristics of fire detection signals, thereby enabling self-learning and adaptability. The authors in [12] have developed a wireless intelligent home alarm system that incorporates anti-theft, anti-fire, and anti-harmful gas leak features. This system utilizes the microcomputer AT89C51 and the voice chip ISD1420. The authors have utilized two microcomputers to exhibit the gas concentration and serve as the alarm host, with wireless transmission employed to transmit the alarm signals. In smoke detection, a voice message is dispatched to the appropriate department. The authors suggested a novel home safety system, employing Zigbee and Arduino to oversee crucial areas within a household via a surveillance camera. Upon activation due to fire or gas leaks, the user receives notifications through SMS and multimedia message service. The utilized sensors (i.e., temperature, fire, and gas sensors) were integrated into the system's mainboard using Zigbee modules, creating a wireless sensor network [13].

This research paper proposes a compact-sized smart system with low-cost applications for home safety based on IoT and GSM technology. The system comprises a flame sensor, an LM35 temperature sensor, an Arduino Uno, and a NodeMCU microcontroller board. The flame sensor, temperature sensor, and GSM shield are connected to the Arduino Nano. A special connection is utilized in this work to link the Arduino Uno with the NodeMCU microcontroller using the digital input pins. The buzzer is activated once the flame is detected, the temperature increases above 50 °C, and an SMS is sent to the homeowner. Finally, the Arduino Uno sends a logic "1" to the NodeMCU, which, in turn, displays a warning message on the GUI accessed via IoT.

This paper is organized as follows: Sect. 2 exhibits the utilized sensors and shields such as Arduino Uno, NodeMCU, SIM900 GSM shield, flame sensor, temperature sensor, and servo motor. Section 3 exhibits the design procedure for the introduced work. Section 4 exhibits the obtained outcomes after applying the practical test. Finally, Sect. 5 exhibits the conclusion from the designed work.

2 Proposed System Hardware

This section of the paper presents the fundamental sensors and modules employed to build the system, such as Arduino Uno, NodeMCU, flame sensor, servo motor, and Lm35 temperature sensor.

2.1 Arduino Uno

The Arduino Uno stands out as a highly popular microcontroller board, finding extensive employment across a spectrum of electronic projects and prototype endeavors. Central to its design is the Atmega328P microcontroller, a stalwart component that underpins both novices and proficient creators. The Arduino Uno boasts a repertoire of 14 pins, meticulously allocated to cater to digital input/output demands. Among these, six pins embrace the capacity for analog output through the ingenious PWM

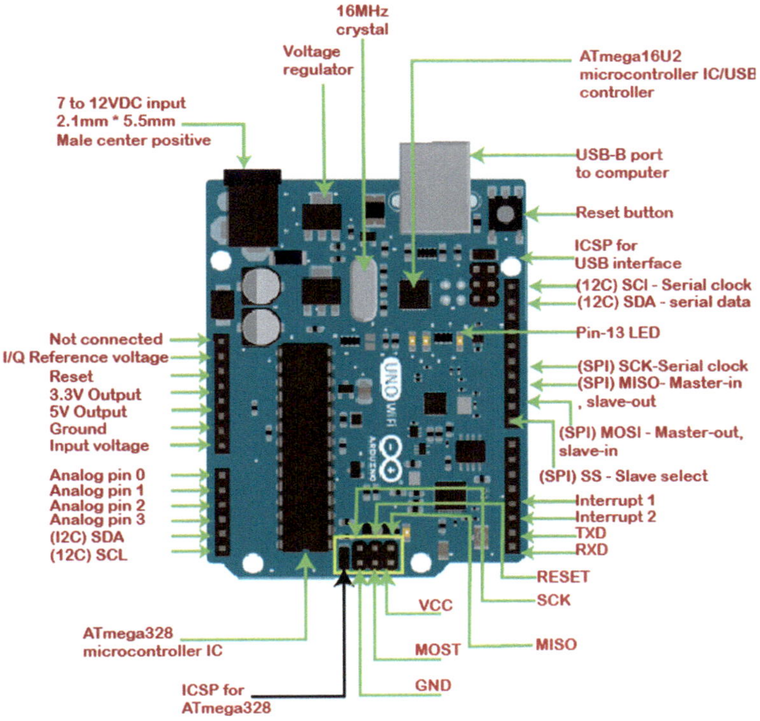

Fig. 2 Arduino Uno pins

methodology. Additionally, the board integrates six inputs for analog functionality, a 16 MHz oscillator, a USB connector, a power jack for the requisite energy supply, an ICSP header, and a conveniently sized reset button. The intricate configuration of the Arduino Uno's pins is visually represented in Fig. 2 [14, 15].

2.2 *NodeMCU Microcontroller*

NodeMCU is a highly versatile microcontroller based on the built-in ESP8266 Wi-Fi module, which offers a wide range of functionalities, especially for IoT projects. It integrates an ESP-12E module and a Lua interpreter firmware, making developing and deploying IoT applications easy. NodeMCU can be programmed using the Arduino IDE by integrating the ESP community library with the exact instructions as Arduino that lead to facilitating the programming. NodeMCU has 15 input/output pins with a single analog input pin and a 3.3 V VCC pin [16]. Figure 3 exhibits the NodeMCU microcontroller with pins configuration.

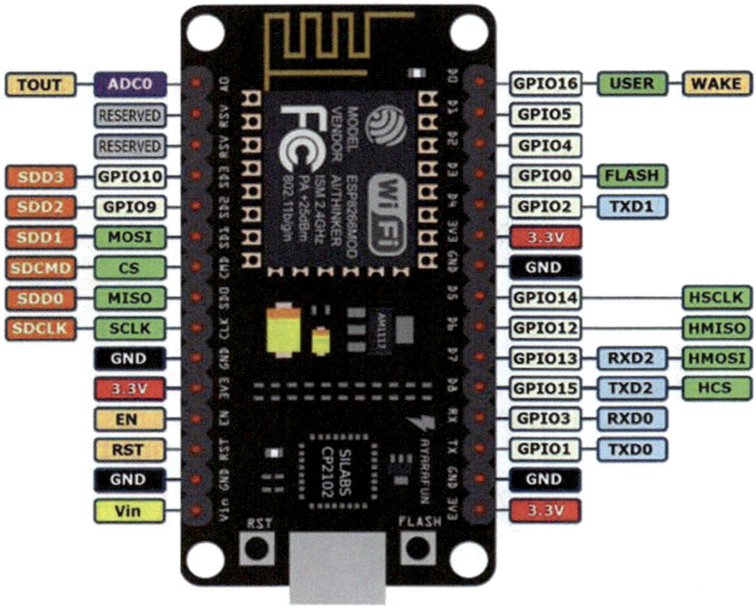

Fig. 3 NodeMCU pins

2.3 *KY-026 Flame Sensor*

The KY-026 Flame Sensor module is designed to perceive infrared light emitted by flames. It features digital and analog outputs and a potentiometer for fine-tuning the sensitivity. This module is frequently employed in the fire detection systems [17]. This sensor has four fundamental pins for analog output, digital output, VCC, and GND, as shown in Fig. 4.

2.4 *Servo Motor*

Servo motors are DC motor types that incorporate position feedback to enhance their functionality. This means that with the help of programming, the microcontroller can control the servo motor to move to a specific position by adjusting the pulse time or pulse length, as illustrated in Fig. 5.

Conventional servo motors have a range of motion from 0 to 180°, while another type can rotate from 0 to 360°. What sets the servo motor apart from a regular DC motor is its impressive torque, superior performance, and user-friendly operation, eliminating the need for an extra driving circuit. Additionally, it exhibits a consistent speed even when the load varies. The servo motor consists of three terminals, with

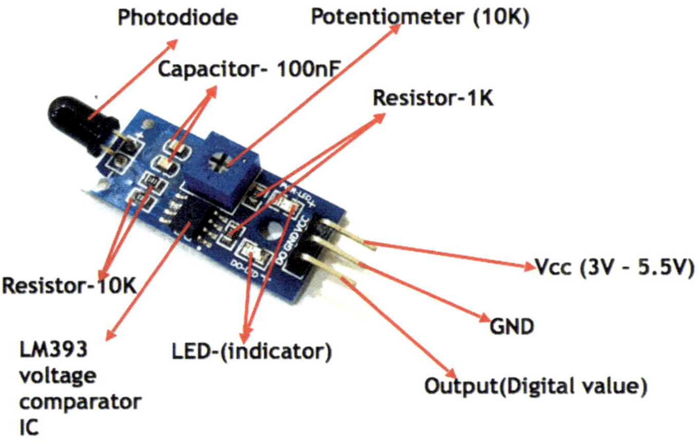

Fig. 4 KY-026 flame sensor structure

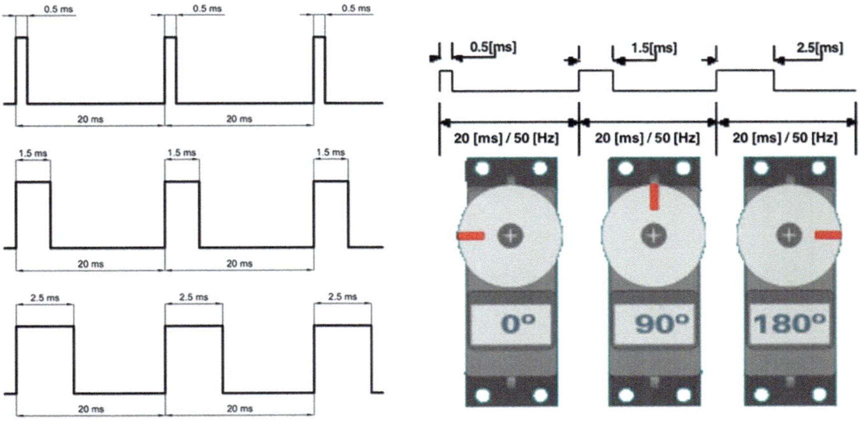

Fig. 5 Relationship between the pulse length and the associated angle of servo rotation

two dedicated to a power supply and the remaining terminal directly connected to the Arduino microcontroller, as depicted in Fig. 6 [18].

2.5 *SIM900 GSM Shield*

The SIM900 GSM/GPRS shield is a sophisticated communication module that offers uninterrupted connectivity for GSM and GPRS networks. The aforementioned shield is notable for its commendable dependability, adaptability, and effective operational capabilities. This technology facilitates the seamless integration of users' projects

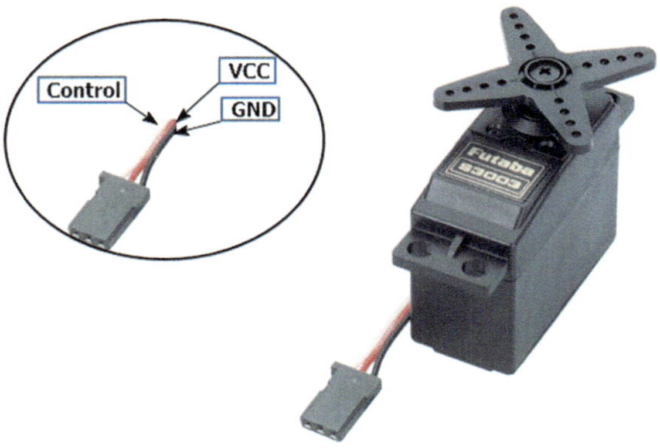

Fig. 6 Servo motor pins

with cellular connection capabilities, hence enabling remote monitoring, control, and data transmission. The SIM900 shield presents a practical solution for a range of applications including IoT devices, home automation, security systems, and others, owing to its compact size and user-friendly interface. The broad array of capabilities and its compatibility with various microcontrollers render it a favored option among developers and enthusiasts. The SIM900 GSM/GPRS shield is designed to facilitate reliable and efficient communication through various means such as sending SMS messages, making voice conversations, and establishing a GPRS data connection. Figure 7 shows the structure of the SIM900 GSM shield [19].

2.6 Lm35 Temperature Sensor

The LM35 Temperature Sensor is widely recognized for its exceptional precision and dependability in temperature measurement. The notable attributes of its simplicity and precision render it a widely favored option across diverse applications. The sensor functions by adhering to the notion of a linear alteration in voltage that is directly proportional to the temperature. The device has an analog output that exhibits a direct correlation with the observed temperature. The LM35 sensor's wide operating temperature range and low power consumption make it ideal for a variety of commercial and domestic uses [20]. As seen in Fig. 8, the Lm35 sensor has three pins: VCC, analog output, and GND.

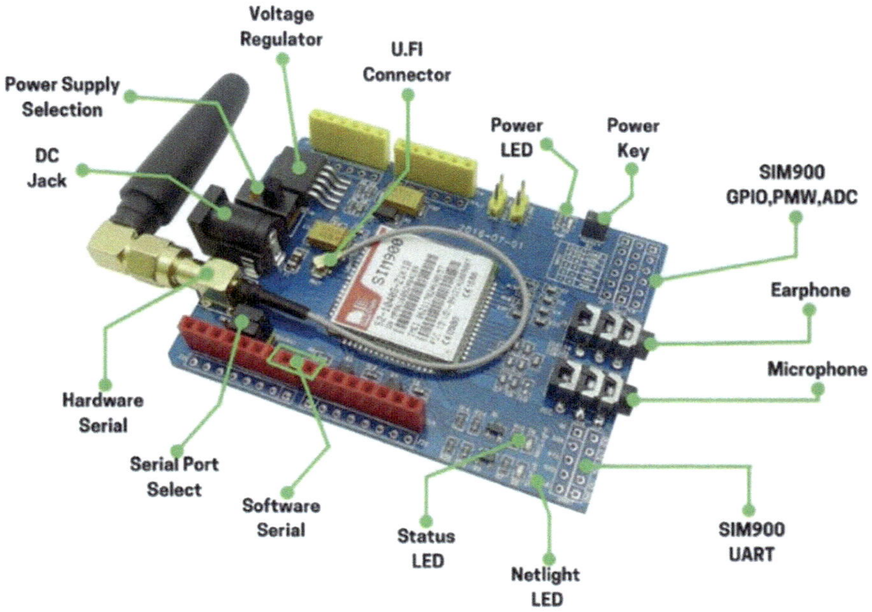

Fig. 7 Configuration for the SIM900 module

Fig. 8 Lm35 temperature sensor

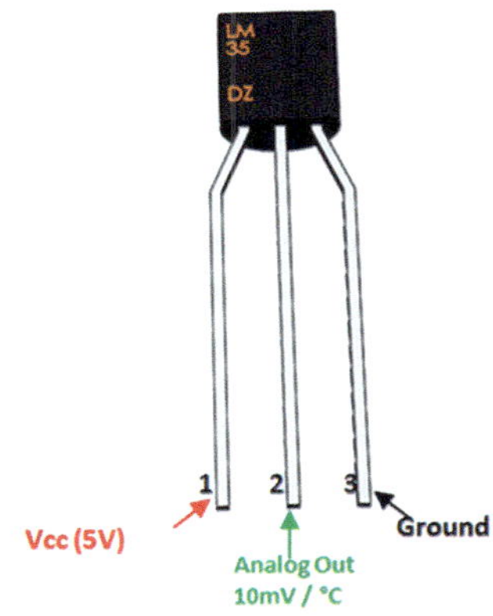

3 System Implementation

In this section, we will discuss the essential measures that must be taken in order to put into effect the proposed system to protect homes from fire hazards. In addition, we will provide succinct explanations for each stage, making it easy to comprehend how the system works. Results from real-world testing of the implemented system will also be given to demonstrate its efficacy and dependability. By doing so, we hope

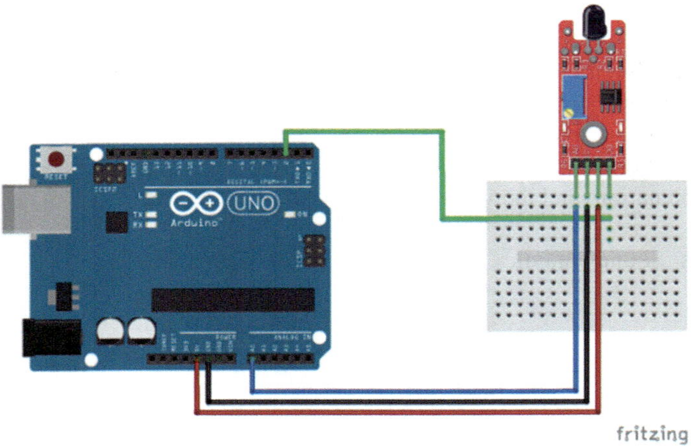

Fig. 9 Connection diagram for KY-026 flame sensor

to shed light on the feasibility and efficacy of the suggested fire protection system for habitations or other institutions.

3.1 System Design

In this project, a smart fire alarm system is designed and implemented using Arduino, a flame sensor, a NodeMCU microcontroller, a SIM900 GSM module, a small buzzer, and a 180° servo motor. The objective of this fire detector, utilizing Arduino, is to sense the surroundings for the occurrence of fire with the help of the flame sensor. Figure 9 illustrates the circuit diagram depicting the connection between the KY-026 flame sensor module and the Arduino Uno microcontroller.

After establishing the connection of the flame sensor, the GSM module was directly connected to the Arduino Uno microcontroller board. This connection allows for transmitting warning messages to the homeowner's mobile phone in the event of fire detection. It is important to note that GSM shields of this nature require a steady voltage with a stable current. Therefore, a 9 V power supply with a capacity of 2A is utilized to ensure optimal performance and prevent any disconnection of the shield from the service tower. Figure 10 exhibits the connection of the SIM900 shield with the Arduino Uno board.

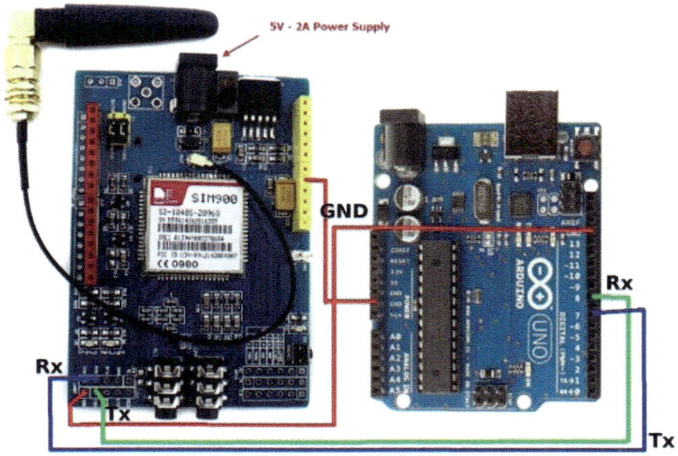

Fig. 10 Connection for SIM900 shield with the Arduino Uno board

Finally, the NodeMCU microcontroller is connected to the Arduino microcontroller board through a digital pin, utilizing the built-in pull-up resistor. Once a fire is detected, the Arduino microcontroller will send a logic "1" signal to the NodeMCU microcontroller, which, in turn, displays a warning message on the GUI by leveraging IoT technology. In addition, a small white LED is integrated with the NodeMCU and the programmed GUI to be employed during emergencies. Figure 11 depicts the flowchart of the proposed smart fire alarm system.

Figure 12 shows cases of the implemented prototype for the system, and Fig. 13 illustrates the graphical block diagram to describe the working principle for the proposed system.

4 Results and Discussion

This section presents the results obtained after conducting the practical test for the proposed system. A flame source, such as a lighter, was positioned near the flame sensor to evaluate the implemented system. Subsequently, the sensor relayed a logic "1" signal to the Arduino microcontroller, transmitting a warning message to the homeowner's mobile phone via the GSM module, as depicted in Fig. 14.

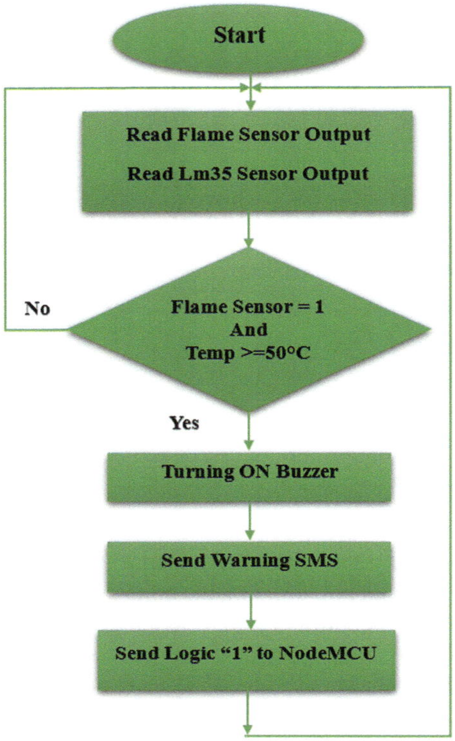

Fig. 11 Proposed system flowchart

Then the Arduino Uno also sends a logic "1" signal to the NodeMCU microcontroller, enabling the display of an additional warning message on the programmed GUI through the IoT technology, as previously mentioned. This functionality is illustrated in Fig. 15. Additionally, the programmed GUI incorporates a button for controlling the emergency light. In this work, we aimed to use two different types of microcontroller boards to improve the system's reliability. In the event of a failure in one of the microcontrollers, the other part of the system continues to operate normally, ensuring uninterrupted functionality. To showcase the effective performance of the implemented system alongside the traditional system, Table 1 has been compiled.

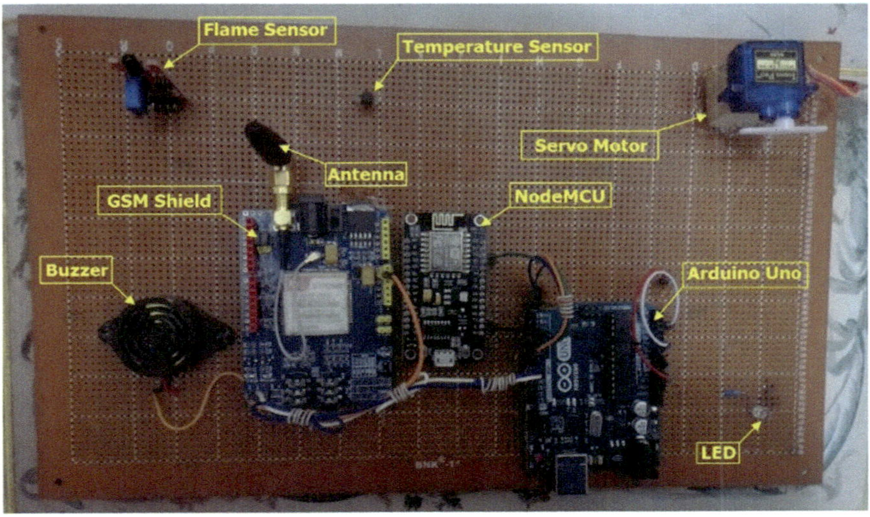

Fig. 12 Implemented fire alarming system

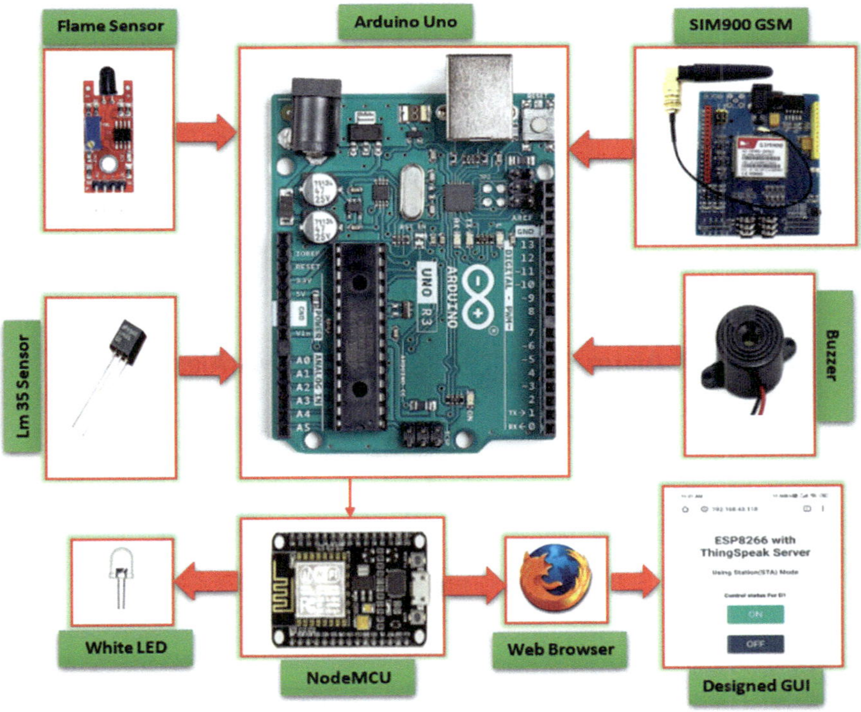

Fig. 13 Graphical block diagram for the proposed work

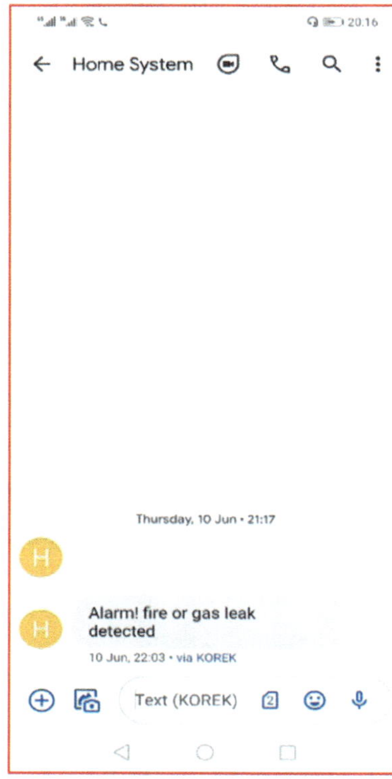

Fig. 14 Received warning message

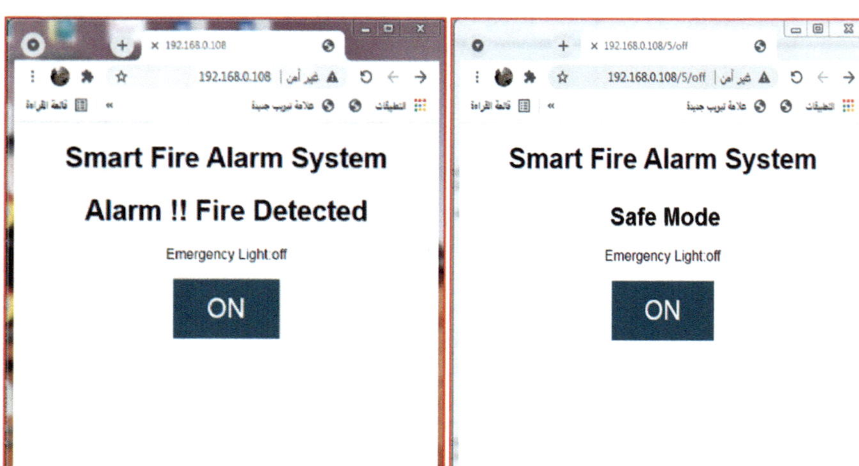

Fig. 15 Obtained results from IoT segment

Table 1 Comparison between the proposed system and the traditional systems

Aspect	Proposed system	Traditional systems
Technology utilized	IoT, GSM, flame sensor, LM35 temperature sensor, and buzzer	Wired alarms, smoke detectors
Detection speed	Rapid flame detection and temperature sensing	Depends on sensor responsiveness
Alert mechanism	SMS alerts, IoT-based GUI notifications	Audible alarms, no remote alerts
Remote monitoring	Yes	No
Human intervention	Reduced need for human intervention	Often requires manual action
Real-time updates	Real-time alerts and updates	Delayed response in some cases
Automation	Automatic door opening and alarm triggering	Limited automation capabilities
Installation	Wireless setup requires IoT/GSM components	Wired installation with detectors
Maintenance	Requires occasional software updates	Regular sensor maintenance
Cost	Bundling costs are higher	Generally lower upfront costs
Expandability	Easily expandable with additional devices	Limited expandability
Home automation integration	Seamless integration with smart home tech	Limited integration potential
Accessibility during power outages	GSM backup power ensures continued alerts	Limited functionality without power
Response time in emergency situations	Swift response due to real-time alerts	Potentially slower response time

5 Conclusion

This work has been undertaken to assist homeowners in addressing the issue of fire outbreaks when they are not present in the building. Critical situations or unpredictable events often occur in residential areas or buildings without the residents' knowledge. Moreover, the system built is cost-effective compared to other alarm systems available in the market, and it can be easily implemented in houses and other establishments. The LM35 temperature sensor has been integrated into the system to enhance fire detection capabilities. When a flame is detected, and the temperature exceeds 50 °C, the alarm is activated, a warning message is sent to the homeowner via GSM, and a notification is displayed on the GUI through IoT technology. Based on the obtained results, it is evident that the home alert system is feasible and effective in protecting houses from the danger of fires.

References

1. Venkatraman, S., Overmars, A., Thong, M.: Smart home automation—use cases of a secure and integrated voice-control system. Systems **9**(4), 77 (2021). https://doi.org/10.3390/systems9040077
2. Almusaed, A., Yitmen, I., Almssad, A.: Enhancing smart home design with AI models: a case study of living spaces implementation review. Energies **16**(6), 2636 (2023). https://doi.org/10.3390/en16062636
3. Murad, M., Bayat, O., Marhoon, H.M.: Design and implementation of a smart home system with two levels of security based on IOT technology. Indonesian J. Electr. Eng. Comput. Sci. **21**(1), 546 (2021). https://doi.org/10.11591/ijeecs.v21.i1.pp546-557
4. Saputro, J.S., et al.: Energy monitoring and control of automatic transfer switch between grid and solar panel for home system. Int. J. Robot. Control Syst. **3**(1) (2023). https://doi.org/10.31763/ijrcs.v3i1.843
5. Al-Kadhimi, A.M., Marhoon, H.M., Karam, Z.A.: Implementation of cell phone detection mobile robot for restricted areas using NodeMCU. Iraqi J. ICT **1**(1), 27–35 (2018). https://doi.org/10.31987/ijict.1.1.6
6. Baqir, N.E., Al-Nouman, M.I.: RFID-based users' localization in the crowded areas. Iraqi J. Inform. Communi. Technol. **4**(3), 1–11 (2021). https://doi.org/10.31987/ijict.4.3.145
7. He, X., Feng, Y., Xu, F., Chen, F.-F., Yu, Y.: Smart fire alarm systems for rapid early fire warning: advances and challenges. Chem. Eng. J. **450**, 137927 (2022). https://doi.org/10.1016/j.cej.2022.137927
8. Sharkawy, A.-N., Hasanin, M., Sharf, M., Mohamed, M., Elsheikh, A.: Development of smart home applications based on Arduino and Android platforms: an experimental work. Automation **3**(4), 579–595 (2022). https://doi.org/10.3390/automation3040029
9. Taha, I.A., Marhoon, H.M.: Implementation of controlled robot for fire detection and extinguish to closed areas based on Arduino. Telkomnika (Telecommun. Comput. Electron. Control) **16**(2), 654–664 (2018). https://doi.org/10.12928/telkomnika.v16i2.8197
10. Ehsan, I., et al.: Internet of things-based fire alarm navigation system: a fire-rescue department perspective. Mob. Inf. Syst. **2022**, 1–15 (2022). https://doi.org/10.1155/2022/3830372
11. Yu, Q., Zheng, D., Fu, Y., Dong, A.: Intelligent fire alarm system based on fuzzy neural network. In: 2009 International Workshop on Intelligent Systems and Applications, Wuhan, China, pp. 1–4 (2018). https://doi.org/10.1109/IWISA.2009.5072924
12. Shunxia, C., Yanda, C.: Design of wireless intelligent home alarm system. In: 2012 International Conference on Industrial Control and Electronics Engineering, Xi'an, China, pp. 1511–1513 (2019). https://doi.org/10.1109/ICICEE.2012.397
13. Hou, J., Wu, C., Yuan, Z., Tan, J., Wang, Q., Zhou, Y.: Research of intelligent home security surveillance system based on ZigBee. In: 2008 International Symposium on Intelligent Information Technology Application Workshops, Shanghai, China, pp. 554–557 (2018). https://doi.org/10.1109/IITA.Workshops.2008.223
14. Murad, A., Bayat, O., Marhoon, H.M.: Implementation of rover tank firefighting robot for closed areas based on Arduino microcontroller. Indonesian J. Electr. Eng. Comput. Sci. **21**(1), 56–63 (2021). https://doi.org/10.11591/ijeecs.v21.i1.pp56-63
15. Maharmi, B., Widyastomo, B., Palaha, F.: Water flow measurement-based data acquisition using Arduino Microcontroller and PLX-DAQ software. Jurnal Ilmiah Teknik Elektro Komputer dan Informatika **8**(1), 107–118 (2022). https://doi.org/10.26555/jiteki.v8i1.23637
16. Mukalaf, A.H., Marhoon, H.M., Suwarno, I., Ma'arif, A.: Design and manufacturing of smart car security system with IoT-based real-time tracking. Int. J. Intell. Syst. Appl. Eng. **11**(6s), 745–752 (2023)
17. Marhoon, H.M., Mahdi, M.I., Hussein, E.D., Ibrahim, A.R.: Designing and implementing applications of smart home appliances. Mod. Appl. Sci. **12**(12), 8–17 (2018). https://doi.org/10.5539/mas.v12n12p8

18. Singh, A., Adhav, S., Dalvi, A., Chippa, A., Rane, M.: Arduino based dual axis solar tracker. In: 2022 Second International Conference on Artificial Intelligence and Smart Energy (ICAIS), Coimbatore, India, pp. 1789–1793 (2022). https://doi.org/10.1109/ICAIS53314.2022.9742876
19. Hossain, M.A., Jahura, F., Rahman, M.A., Ahmed, T.: IoT-based drunk driving detection and engine locking system with real-time location in app. In: TENCON 2022—2022 IEEE Region 10 Conference (TENCON), Hong Kong, Hong Kong, pp. 1–5 (2022). https://doi.org/10.1109/TENCON55691.2022.9978102
20. Sahrab, A.A., Marhoon, H.M.: Design and fabrication of a low-cost system for smart home applications. J. Robot. Control (JRC) **3**(4), 409–414 (2022). https://doi.org/10.18196/jrc.v3i4.15413

A Comparative Analysis of Optimized Routing Protocols for High-Performance Mobile Ad Hoc Networks

Ayushman Pranav, Akshat Jain, Mohd Mohsin Ali, Manish Raj, and Umesh Gupta

Abstract Mobile ad hoc networks (MANETs) let wireless devices interact without a fixed infrastructure, yet routing algorithms are still crucial. MANETs are used in disaster relief, military operations, and smart cities. This study compares MANET Bellman-Ford, Dijkstra, and Genetic routing algorithms. These methods use hop count and connection cost to find the fastest paths between nodes. The report evaluates routing algorithms using six criteria (bandwidth, mobility, transmission power, battery capacity, packet loss, and health). We run numerous simulations with different parameter settings and statistically analyze the results. The paper offers a battery capacity and transmission power-based network node health measure. Many MANET routing methods are tailored for specific performance metrics and simulated conditions. Simulations show that the recommended measure is particularly beneficial for analyzing MANET network node status. The study also finds that certain routing methods perform better in low packet loss and high battery capacity. These findings underscore the need to consider performance indicators and simulated scenarios when choosing a MANET routing algorithm. This research provides insights into enhancing MANET routing algorithms with 90% accuracy and performance dependent on performance measurements and network factors. The findings show the merits and weaknesses of each algorithm, helping choose MANET routing solutions.

Keywords Battery capacity · Bellman-Ford · Dijkstra · Genetic algorithm · High bandwidth · Low mobility · Mobile Ad Hoc Networks (MANETs) · Network health · NS-2 simulation · Packet loss · Performance metrics · Routing algorithms · Transmission power

A. Pranav · A. Jain · M. M. Ali · M. Raj · U. Gupta (✉)
Bennett University, Greater Noida, India
e-mail: er.umeshgupta@gmail.com

M. Raj
e-mail: manish.raj@bennett.edu.in

G. Fortino et al. (eds.), *Proceedings of Third International Conference on Computing and Communication Networks*, Lecture Notes in Networks and Systems 917,
https://doi.org/10.1007/978-981-97-0892-5_7

1 Introduction

Mobile ad hoc networks (MANETs) are studied in this research. The network is nomadic and has no physical connections [5, 13]. Nodes may move. When required, mobile nodes may be connected instantly. A dynamic topology can build a MANET. Functional network nodes might be hosts or routers. Few performance measurements assess the two reactive methods other than overhead transmission. The findings demonstrate that DSR is more trustworthy and stable than AODV. Classic, evolutionary, and swarm intelligence-based multicast routing methods are examined in [16]. Introducing an innovative virus classification architecture, Ali et al. (2022) achieve high accuracy, with prospects for synergy with MANETT algorithms [21]. Multicast routing is becoming more prevalent in high-speed networks due to its bandwidth efficiency, resource reduction, and cost-effective communication. Multicast networks' dynamic and extensive nature makes finding the best routes with guaranteed QoS difficult. Their swarm-based and multicast routing technique is also explored. Based on [7, 10, 19], Ant Colony Optimization is preferable for MANETs and other adaptive networks. Biological ants follow a predefined itinerary and leave pheromones. Active routing protocols are optimized link state routing systems. The essay proposes revising the Ad hoc On-demand Distance Vector routing technology to improve MANET QoS [8]. Proposing a novel deep learning technique for skin disorder classification, Ali et al. (2022) lay the foundation for potential synergies with MANETT algorithms in mobile ad hoc networks [20]. The suggested technique uses a unique "Link Stability Factor" to assess node link quality (LSF). Ali et al. (2022) pioneer a deep learning approach for precise fruit freshness assessment, offering potential integration with MANETT algorithms for broader deployment [21]. A revolutionary protocol feature, "Route Error Recovery" (RER), helps quickly resume normal operation following a route interruption. A new academic work [3, 4, 6] suggests using a genetic algorithm to enhance MANET routing. Genetic algorithms use evolution and natural selection to optimize. This study improves fuzzy-based routing's membership functions and starting rule sets via an algorithm. Genetic method-based routing finds the shortest route faster than traditional routing protocols. Simulations reveal that the Fuzzy-GA technique has a 21.1% higher throughput than the conventional AODV protocol. In MANETs, routing allows data packets to be transported from their source to their destination while reducing travel time and overhead. Its unstable basis and changeability make routing difficult [1, 2]. The contributions of the papers are

1. Bellman-Ford, Dijkstra, and the Genetic Algorithm are three well-known routing algorithms for Mobile Ad Hoc Networks (MANETs), and here we compare and evaluate them.
2. Quantity of Service (QoS) metrics improvement via the creation of adapted variants of these algorithms to enhance such qualities as bandwidth, mobility, transmission power, battery life, packet loss, and health.
3. Extensive simulation studies under different network situations evaluated the effectiveness of the algorithms.

4. Testing the algorithms' performance in various settings, such as those with high mobility and heavy network traffic.
5. Studying how these results could affect network administration and end-user experiences from a social and managerial point of view.

2 Literature Reviews

Citation	Summary	Research gaps
[3]	Studies cross-layer optimization for MANETs to improve QoS in IoT applications, proposing a new optimization framework	Explore the performance of the proposed framework under various network topologies and traffic patterns. Evaluate the framework using different QoS metrics to enhance service quality in MANETs
[4]	Review the use of MANET for routing protocols, discussing challenges and proposed solutions	Include recent research on MANET routing protocols to provide a comprehensive overview. Address challenges and solutions for MANET routing in various application domains
[5]	Review different MANET protocols, discussing their pros, cons, and selection factors	Extend the review to incorporate recent research on MANET protocols for a more up-to-date perspective. Discuss selection factors, pros, and cons across diverse application domains
[6]	Provides an overview of MANET routing protocols, focusing on zone routing protocol's advantages, disadvantages, and applications	Add more detailed information about the zone routing protocol. Explore the applications and drawbacks of the zone routing protocol in various network scenarios
[7]	Evaluates MANET routing protocols under different network scenarios, comparing performance metrics	Expand the evaluation to include a broader range of routing protocols and network scenarios. Study the influence of mobility, traffic load, and node density on routing protocol performance
[8]	Assesses MANET routing protocol performance, comparing metrics like packet delivery ratio, delay, and overhead	Increase the number of evaluated routing protocols and performance metrics. Analyze how mobility, traffic, and node density impact routing protocol performance
[9]	Compares MANET routing protocols based on features, advantages, and protocol selection considerations	Extend the comparison to include more routing protocols and their features. Provide more profound insights into protocol selection factors for various scenarios
[10]	Studies mobility model impact on MANET performance, comparing metrics such as packet delivery, delay, and overhead	Expand the study to include additional mobility models and performance metrics. Investigate how factors like traffic load and node density affect mobility model impact

(continued)

(continued)

Citation	Summary	Research gaps
[11]	Introducing a genetic algorithm (GA)-driven multicast routing strategy for MANETs	The GA-driven approach needs more adaptability to evolving network topology, overlooks node energy consumption, and neglects security aspects in multicast transmissions
[12]	The paper compares integrating mobile ad hoc networks (MANETs) with satellite and cellular networks	Advancing routing protocols for synergistic resource use in satellite and cellular networks, enhancing security measures for data transmission, and optimizing power consumption in integrated MANETs with satellite and cellular networks

3 Proposed Methodology

The importance of routing algorithms in MANETs and their justification. Non-fixed infrastructure, restricted bandwidth, energy limitations, and dynamic topology have sparked interest in Bellman-Ford, Dijkstra, and Genetic routing algorithms. Figure 1 aims to uncover insights that may optimize bandwidth usage, power consumption, and mobility and develop more efficient and adaptive wireless networks.

The dataset was created using several routing techniques in a synthetic network. The study included six critical performance parameters [9, 11]: bandwidth, packet loss, mobility, transmission power, battery capacity, and health measures to increase performance. The goal was to optimize the method by 90%.

Dataset Representation

The following Table 1 represents a subset of the generated dataset, providing insight into the attributes of the first few nodes.

Methodology and Proposed Solution

This research examines and assesses three MANET routing methods under different network situations. A wireless network of N nodes is simulated in the study.

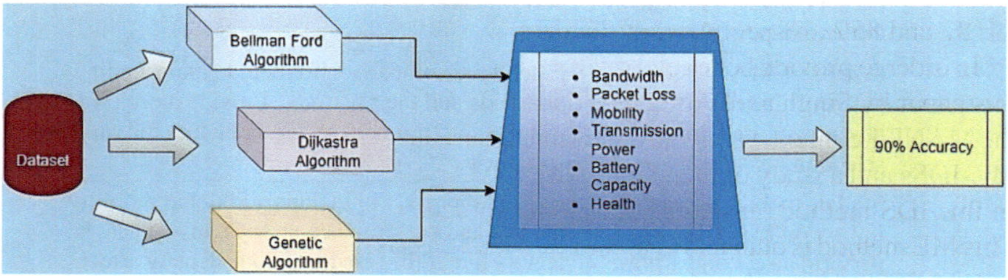

Fig. 1 Research design of modified research protocol

Table 1 Table representing the sample-generated dataset

Node	Position	Bandwidth	Mobility	Transmission_power	Packet loss	Battery capacity
0	(26.91, 67.54)	5	0.99	11.82	0.08	98.36
1	(70.49, 41.30)	5	0.79	7.92	0.19	30.96
2	(54.05, 62.51)	3	0.76	11.14	0.5	43.21
3	(39.70, 18.89)	5	0.61	5.65	0.3	71.34
4	(16.59, 66.17)	10	0.39	13.64	0.88	56.79

Randomly selecting source–destination pairs and transmitting packets creates traffic. Bellman-Ford, Dijkstra, and Genetic algorithms identify the fastest and optimal node pathways [4]. The algorithms' efficiency is measured by bandwidth, mobility, transmission power, battery capacity, packet loss, and health. Data is gathered and evaluated from many simulations with different parameter values. The network environment is created and simulated using NS-2. Bellman-Ford-based Mobile Ad Hoc Algorithm for Shortest Path Routing with Inverse Bandwidth Edge Weights and Negative-Weight Cycles.

The Bellman-Ford method [14, 15, 17] finds the shortest route between two nodes in a network with negative-weight edges. It assigns 0 to the source node and infinite to all other nodes as their distance from it. If it discovers a shorter route, it examines all graph edges and changes each node's distance and predecessor. It also finds graph cycles with negative total weight, indicating no shortest route. Finally, it finds the shortest route by following each node's predecessor from destination to source. The Bellman-Ford method takes O (X * Y) time, where X is the number of vertices and Y is the number of edges. Thus, spatial complexity is O (Y) [18].

Optimization of Modified Bellman-Ford Algorithm for High Bandwidth, Low Mobility, High Transmission Power, High Battery Capacity, and Low Packet Loss (Fig. 2).

Dijkstra's algorithm finds the shortest path between nodes in a network with non-negative edge weights. It initializes distances with 0 for the source node and infinity for others. The algorithm explores the graph, updating distances based on cumulative edge weights. A priority queue helps pick the next node efficiently. On finding a shorter path, distances and predecessors are updated. This continues until all nodes are visited or the destination's shortest path is known. Not for negative weights or cycles. The shortest path is traced back from the destination to the source using stored predecessors. Time complexity is $O((X + Y) * \log(X))$, with spatial complexity $O(X + Y)$ (Fig. 3).

Using Bandwidth as a Fitness Metric in the Genetic Algorithm Shortest Path Routing Protocol for MANETsGenetic algorithms [16] have been used to solve optimization issues, including shortest path problems. Genetic algorithms are used in MANETs for several diverse types of work, including channel assignment, resource distribution, and routing.

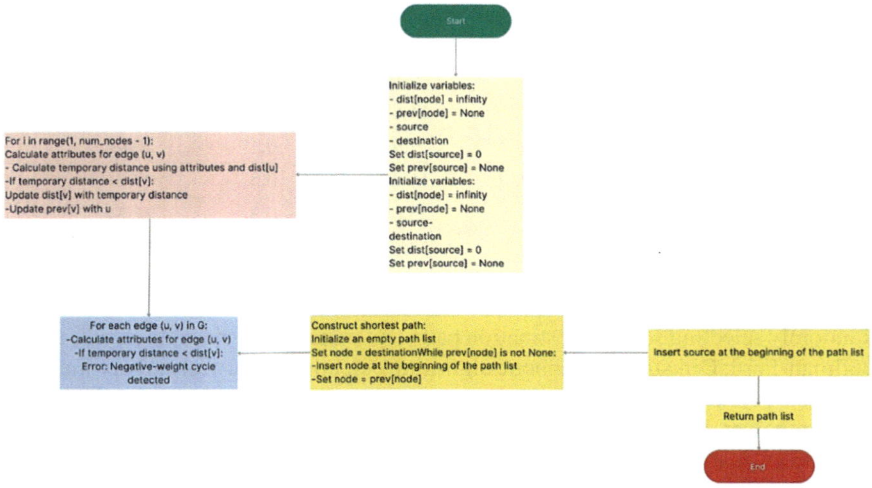

Fig. 2 Modified Bellman-Ford algorithm flowchart

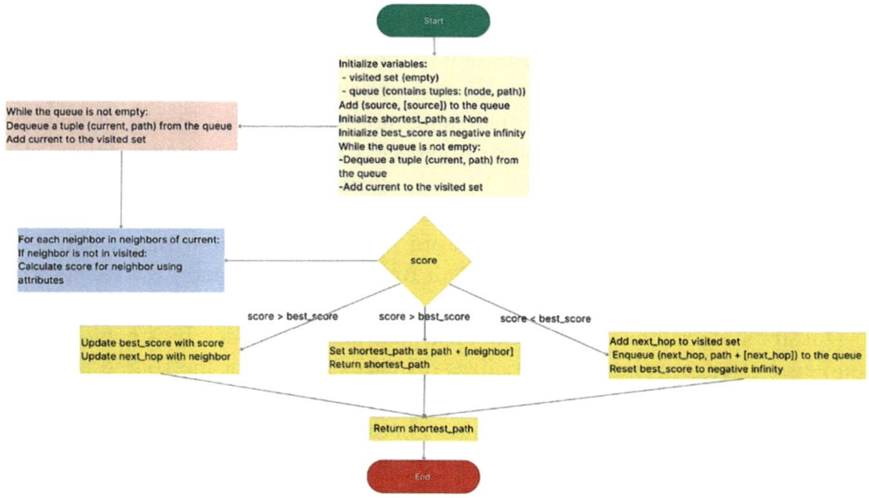

Fig. 3 Modified Dijkstra algorithm flowchart

Initialization: Establish the community of people, where each person stands for a path that joins the Source node to the Destination node.

Fitness Function: Defining fitness function that considers various node properties such as bandwidth, mobility, transmission power, battery capacity, and packet loss to calculate the fitness of an individual in the population (Fig. 4).

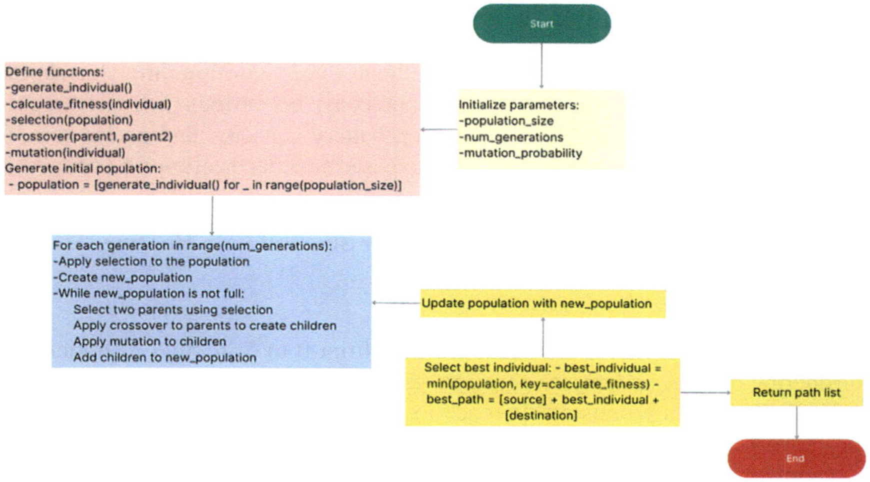

Fig. 4 Modified Genetic algorithm flowchart

4 Results

Bellman-Ford-Based Mobile Ad Hoc Algorithm for Shortest Path Routing

Using Inverse Bandwidth as Edge Weights and Detecting Negative-Weight Cycles

In the Bellman-Ford algorithm, the nodes traverse from 0 to 5 through a shorter path, considering positive and negative weights edge cycles as in Fig. 5.

The basic algorithm inhibits the performance of QoS factors, due to which the custom-modified algorithm of Bellman-Ford is used to improve the QoS performance

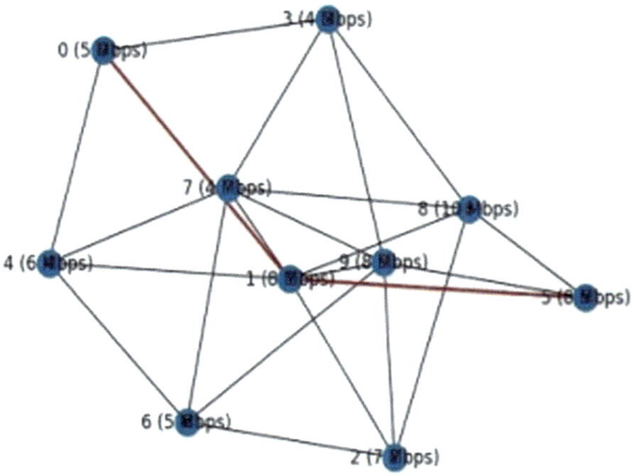

Fig. 5 Bandwidth Distribution along the Shortest Path from Node 0 to Node 5 using the Bellman-Ford Algorithm

metrics- Bandwidth, Mobility, Transmission Power, Battery Capacity, Packet Loss, and Health.

The nodes traverse from 0 to 5 in the Bellman-Ford algorithm through a shorter path. Still, the Bellman-Ford custom algorithm considers various QoS factors, such as bandwidth, mobility, transmission power, battery capacity, and packet loss, to traverse the nodes from 1 to 9 with the increased transfer health of 57.564, as in Fig. 6 and Table 2.

Dijkstra-Based Mobile Ad Hoc Algorithm for Shortest Path Routing

Using Inverse Bandwidth as Edge Weights

In the Dijkstra Algorithm, the nodes traverse from 0 to 5 through a shorter path considering positive and negative weights edge cycles as in Fig. 7.

The Dijkstra is a promising approach to finding the shortest path in wireless ad hoc networks, especially in scenarios where network parameters change rapidly. The nodes traverse from 0 to 5 in the Dijkstra Algorithm through a shorter route. Still, the Dijkstra custom algorithm considers various QoS factors, such as bandwidth, mobility, transmission power, battery capacity, and packet loss, to traverse the nodes from 1 to 6 with the increased transfer health of 40.1146, as in Fig. 8 and Table 3.

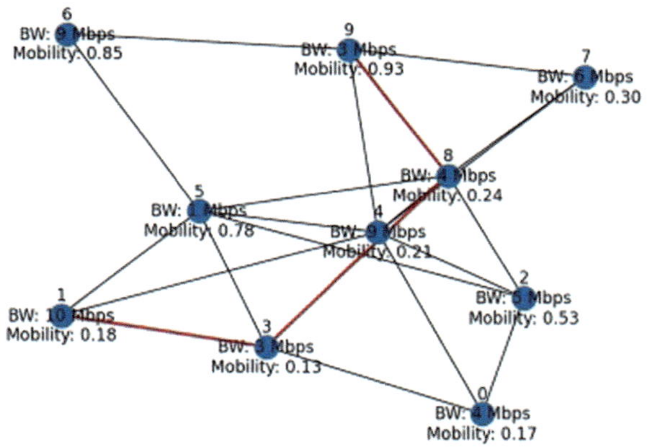

Fig. 6 Bandwidth and Mobility along the Improved Path from Node 1 to 9 using Modified Bellman-Ford Algorithm

Table 2 Enhanced routing protocol for high-performance MANETs: a modified Bellman-Ford algorithm

Node_a	Node_b	Bandwidth	Mobility	Transmission_ power	Battery_ capacity	Packet_loss
1	3	3	0.180358	8.351630	18.422820	0.725788
3	8	5	0.242995	8.644484	18.422820	0.725788
8	9	3	0.934960	10.903120	17.431564	0.725042

Average health of transfer 57.564

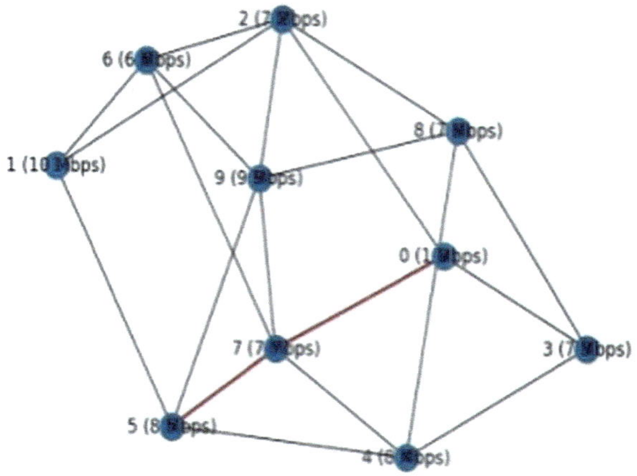

Fig. 7 Bandwidth Distribution along the Shortest Path from Node 0 to Node 5 using Dijkstra Algorithm

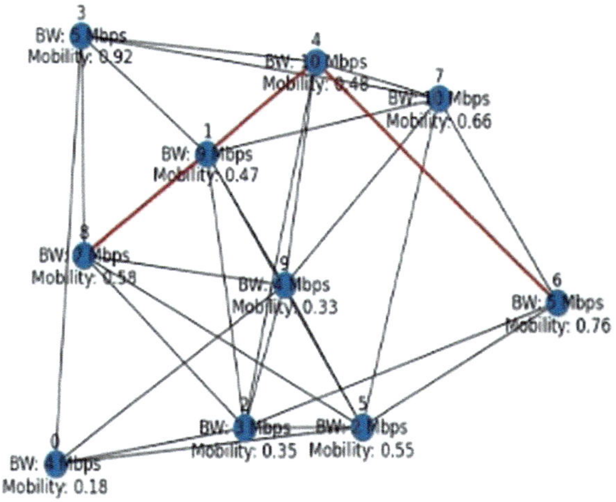

Fig. 8 Bandwidth and Mobility along the Improved Path from Node 1 to Node 6 using Modified Dijkstra Algorithm

Table 3 Enhanced routing protocol for high-performance MANETs: a modified Dijkstra algorithm

Node_a	Node_b	Bandwidth	Mobility	Transmission_ power	Battery_ capacity	Packet_loss
1	8	7	0.582725	17.246334	39.221364	0.848823
8	4	7	0.582725	8.676017	39.221364	0.146374
4	6	5	0.759435	8.676017	53.431171	0.748964

Average health of transfer 40.1146

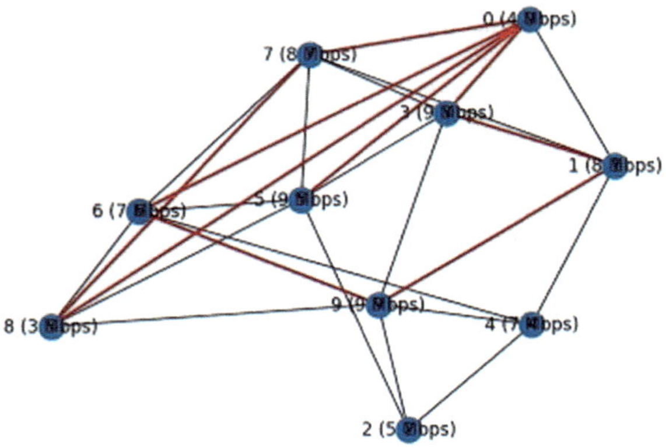

Fig. 9 Bandwidth Distribution along the Shortest Path from Node 0 to Node 5 using Genetic Algorithm

Genetic Algorithm-Based Mobile Ad Hoc Algorithm for Shortest Path Routing

Using Bandwidth as a Fitness Metric in the Genetic Algorithm Shortest Path Routing Protocol for MANETs

The Genetic algorithm follows the fitness-based metric approach and uses a population-based approach to find the shortest path between a source and destination node, as in Fig. 9.

Modified Genetic Algorithm Shortest Path Routing Protocol for MANETs using High Bandwidth, Low Mobility, High Transmission Power, High Battery Capacity, and Low Packet Loss Optimized Routing

The modified-custom Genetic Algorithm involves initializing a population, evaluating their fitness, and improving the QoS metrics for the traversal of nodes from 2 to 5, thus selecting individuals for reproduction and applying crossover, mutation, and repetition to create improved solutions, thus improving the average health of transfer to 61.968 as in Fig. 10 and Table 4.

Metric	Bellman-Ford algorithm	Dijkstra algorithm	Genetic algorithm
Bandwidth	Optimizes bandwidth usage	Optimizes bandwidth usage	Optimizes bandwidth usage
Mobility	Considers mobility changes	Adapts to varying mobility scenarios	Adapts to varying mobility scenarios
Transmission Power	Optimizes transmission power	Optimizes transmission power	Optimizes transmission power
Battery capacity	Considers battery capacity	Considers battery capacity	Considers battery capacity

(continued)

(continued)

Metric	Bellman-Ford algorithm	Dijkstra algorithm	Genetic algorithm
Packet loss	Minimizes packet loss	Enhances route quality with packet loss	Optimizes routes for reduced packet loss
Health	Focuses on maintaining network health	Enhances network stability and path quality	Optimizes for network health and performance

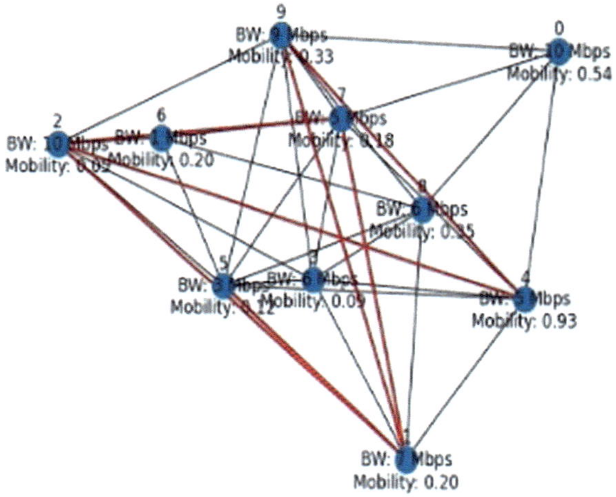

Fig. 10 Bandwidth and Mobility along the Improved Path from Node 2 to Node 5 using Modified Genetic Algorithm

Table 4 Enhanced routing protocol for high-performance manets: a modified Genetic algorithm

Node_a	Node_b	Bandwidth	Mobility	Transmission_ power	Battery_ capacity	Packet_loss
2	7	5	0.182529	10.162777	12.631048	0.252171
7	1	5	0.182529	10.162777	12.631048	0.252171
1	1	5	0.182529	10.162777	12.631048	0.252171
1	1	5	0.182529	10.162777	12.631048	0.252171
1	1	5	0.182529	10.162777	12.631048	0.252171
1	9	5	0.182529	10.162777	12.631048	0.252171
9	4	5	0.182529	10.162777	12.631048	0.252171
4	2	5	0.182529	10.162777	12.631048	0.252171
2	1	5	0.182529	10.162777	12.631048	0.252171
1	5	3	0.195732	14.417211	61.683832	0.518757

Average health of transfer 61.968

5 Conclusion

Simulations and experimental data evaluated three MANET routing approaches for each algorithm—Bellman-Ford, Dijkstra, and Genetic Algorithm—in terms of throughput, mobility, transmission power, and energy consumption. The study subjected the algorithms to several network setups and tuning factors. In the study, we showed the algorithms' performance curves' average and standard deviation. The Dijkstra algorithm utilized bandwidth effectively with low power regardless of network design. The Bellman-Ford technique enhanced energy efficiency and mobility, whereas the Traveling Salesman algorithm worked best in smaller networks. Genetic and Swarm algorithms excelled in mobile and congested networks. Their limits on more extensive networks and higher traffic volumes have been highlighted. In conclusion, our study illuminates MANET routing strategy advantages and downsides. Researchers and network designers may pick the best algorithm to maximize throughput and routing based on their networks. Our findings suggest swarm algorithms may not be ideal for large-scale networks with significant traffic. These algorithms can function in highly mobile and congested environments, but they may require aid to maximize efficiency across massive networks. Thus, researchers and designers must examine their networks and consider alternative routing algorithms to handle the challenges of more extensive networks and higher traffic volumes. In addition to technical contributions, this study's findings have ramifications for Mobile Ad Hoc Network management and user experiences (MANETs). These changes will influence network administrators and end users administratively and socially. This study may assist network administrators and designers pick routing methods for diverse network architectures. Knowing the pros and cons of different algorithms may boost network performance and resource utilization.

6 Limitations

While our research presents valuable insights into the performance of Genetic Algorithm (GA), Bellman-Ford, and Dijkstra routing algorithms for Mobile Ad Hoc Networks (MANETs), certain limitations must be acknowledged. Firstly, our study's findings are based on simulations and experimental data, which may not fully capture the complexities of real-world MANET scenarios. Furthermore, selecting specific QoS metrics such as bandwidth, mobility, transmission power, battery capacity, packet loss, and health may only cover some possible metrics relevant to different MANET applications. Lastly, our research focuses on a comparative analysis of these three algorithms, and other emerging routing algorithms or techniques might need to be covered in this study. Therefore, while our findings provide valuable insights, they should be considered within these limitations, and further research is encouraged to explore a broader spectrum of routing algorithms and network scenarios.

References

1. Gupta, U., Gupta, D.: Least squares structural twin bounded support vector machine on class scatter. Appl. Intell. **53**, 15321–15351 (2023). https://doi.org/10.1007/s10489-022-04237-1
2. Hazarika, B.B., Gupta, D., Gupta, U.: Intuitionistic fuzzy kernel random vector functional link classifier. In: Sisodia, D.S., Garg, L., Pachori, R.B., Tanveer, M. (eds.) Machine Intelligence Techniques for Data Analysis and Signal Processing. Lecture Notes in Electrical Engineering, vol. 997. Springer, Singapore (2023). https://doi.org/10.1007/978-981-99-0085-5_72
3. Hasan, N., Mishra, A., Ray, A.K.: Cross-layer optimization aspects of MANETs for QoS-sensitive IoT applications. In: Dhar, S., Mukhopadhyay, S.C., Sur, S.N., Liu, C.M. (eds.) Advances in Communication, Devices and Networking. Lecture Notes in Electrical Engineering, vol. 776. Springer, Singapore (2022). https://doi.org/10.1007/978-981-16-2911-2_45
4. Ramphull, D., Mungur, A., Armoogum, S., Pudaruth, S.: A review of mobile ad hoc NETwork (MANET) protocols and their applications. In: 2021 5th International Conference on Intelligent Computing and Control Systems (ICICCS), Madurai, India, pp. 204–211 (2021). https://doi.org/10.1109/ICICCS51141.2021.9432258
5. Kurode, E., Vora, N., Patil, S., Attar, V.: MANET routing protocols with emphasis on zone routing protocol—an overview. In: 2021 IEEE Region 10 Symposium (TENSYMP), Jeju, Republic of Korea, pp. 1–6 (2021). https://doi.org/10.1109/TENSYMP52854.2021.9550879
6. Mahamune, A.A., Chandane, M.M.: Evaluating routing protocols for mobile ad hoc networks under varying network scenarios. In: 2021 Third International Conference on Intelligent Communication Technologies and Virtual Mobile Networks (ICICV), Tirunelveli, India, pp. 220–225 (2021).https://doi.org/10.1109/ICICV50876.2021.9388574
7. Saudi, N.A.M., Arshad, M.A., Buja, A.G., Fadzil, A.F.A., Saidi, R.M.: Mobile Ad-Hoc Network (MANET) routing protocols: a performance assessment. In: Kor, L.K., Ahmad, A.R., Idrus, Z., Mansor, K. (eds.) Proceedings of the Third International Conference on Computing, Mathematics and Statistics (iCMS2017). Springer, Singapore (2019). https://doi.org/10.1007/978-981-13-7279-7_7
8. Mishra, A., Singh, S., Tripathi, A.K.: Comparison of MANET routing protocols. Int. J. Comput. Sci. Mob. Comput. **8**(March), 67–74 (2019). https://www.researchgate.net/profile/Shweta-Singh-35/publication/331980713_Compari-son_of_Manet_Routing_Protocols/links/5c98a669299bf11169457ef3/Comparison-of-Ma-net-Routing-Protocols.pdf
9. Appiah, M.: The impact of mobility models on the performance of mobile ad hoc network (MANET). In: 2016 International Conference on Advances in Computing and Communication Engineering (ICACCE), Durban, South Africa, pp. 208–213 (2016). https://doi.org/10.1109/ICACCE.2016.8073749
10. Rajan, C., Shanthi, N.: Genetic based optimization for multicast routing algorithm for MANET. Sadhana **40**, 2341–2352 (2015). https://doi.org/10.1007/s12046-015-0437-8
11. Miao, Y., Sun, Z., Wang, N., Cruickshank, H.: Comparison studies of MANET-satellite and MANET-cellular networks integrations. In: 2015 International Conference on Wireless Communications & Signal Processing (WCSP), Nanjing, China, pp. 1–5 (2015). https://doi.org/10.1109/WCSP.2015.7341337
12. Puri, S., Arora, V.: Routing protocols in MANET: a survey. Int. J. Comput. Appl. **96**, 7–12 (2014). https://doi.org/10.5120/16852-6718
13. Natarajan, D., Rajendran, A.P.: AOLSR: hybrid ad hoc routing protocol based on a modified Dijkstra's algorithm. J Wirel. Commun. Netw. **2014**, 90 (2014). https://doi.org/10.1186/1687-1499-2014-90
14. Ismail, Z., Hassan, R.: A performance study of various mobility speeds on AODV routing protocol in homogeneous and heterogeneous MANET. In: The 17th Asia Pacific Conference on Communications, pp. 637–642 (2011)
15. Zonghua, M., Xiaojing, M.: A modified AODV routing protocol based on route stability in MANET. In: IET Conference Proceedings, pp. 63–67 (2011). https://doi.org/10.1049/cp.2011.0959. IET Digital Library, https://digital-library.theiet.org/content/conferences/10.1049/cp.2011.0959

16. Mo, Y., Huang, J., Huang, B.: Manet node based mobile gateway with unspecific Manet routing protocol. In: 2006 International Symposium on Communications and Information Technologies, Bangkok, Thailand, pp. 886–889 (2006). https://doi.org/10.1109/ISCIT.2006.339864
17. Rai, A.: A study on Bellman Ford algorithm for shortest path detection in global positioning system. Int. J. Res. Appl. Sci. Eng. Technol. (2006)
18. Schmidt, R., Trentin, M.: MANETs routing protocols evaluation in a scenario with high mobility MANET routing protocols performance and behavior, pp. 883–886 (2008). https://doi.org/10.1109/NOMS.2008.4575238
19. Ahuja, R.K., Magnanti, T.L., Orlin, J.B.: Network Flows, Theory, Algorithms, and Applications. Prentice-Hall (1993)
20. Mohsin Ali, M., Chandra Joshi, R., Kishore Dutta, M.: An automated and efficient deep learning based classification of multiple skin disorders from skin lesion images. In: 2022 International Conference on Edge Computing and Applications (ICECAA), Tamilnadu, India, pp. 1156–1161 (2022). https://doi.org/10.1109/ICECAA55415.2022.9936097
21. Ali, M.M., Joshi, R.C., Dutta, M.K., Burget, R., Mezina, A.: Deep learning-based classification of viruses using transmission electron microscopy images. In: 2022 45th International Conference on Telecommunications and Signal Processing (TSP), Prague, Czech Republic, pp. 174–178 (2022). https://doi.org/10.1109/TSP55681.2022.9851305

Integrating Sandpiper Optimization Algorithm and Secure Aware Routing Protocol for Efficient Cluster Head Selection in Wireless Sensor Networks

Tuka Kareem Jebur

Abstract Wireless sensor networks (WSNs) play a crucial role in various applications, ranging from environmental monitoring to industrial automation. secure and efficient data routing in WSNs remains a significant challenge. This paper proposes an approach for opting optimal CH in WSNs using the Sandpiper Optimization Algorithm (SOA) and Chaos-Secure Aware Routing Protocol (SARP). The methodology also enhances network lifetime, energy efficiency, packet delivery ratio, and coverage while maintaining robustness, scalability, and resource efficiency. A simulation system using Matlab R2020a and the CICIDS dataset has been used to test and evaluate the performance in terms of attack detection and mitigation mechanisms including blackhole, gray hole, and wormhole attacks the results demonstrate that the proposed methodology outperforms existing algorithms in terms of network lifetime, energy consumption 3500 J, packet delivery ratio 95%, and throughput1.2 Mbps, while effectively mitigating routing attacks. This research contributes to the advancement of secure and efficient WSNs.

Keywords Wireless sensor networks · Cluster head · Sandpiper optimization algorithm · Chaos · Secure aware routing protocol

1 Introduction

Wireless Sensor Networks (WSNs) have gained considerable attention as a highly promising technological solution across diverse domains, such as environmental monitoring, surveillance, and healthcare [1]. Within a WSN, a multitude of sensor nodes work collaboratively to acquire and transmit data to a centralized base station. Ensuring efficient network management and communication is pivotal to optimize performance and prolong the network's operational lifespan. Among the critical considerations in WSNs is the judicious selection of cluster heads, which play a

T. K. Jebur (✉)
Department of Accounting, Al-Mustansiriyah University, Baghdad, Iraq
e-mail: tukakareem@uomustansiriyah.edu.iq

G. Fortino et al. (eds.), *Proceedings of Third International Conference on Computing and Communication Networks*, Lecture Notes in Networks and Systems 917,
https://doi.org/10.1007/978-981-97-0892-5_8

crucial role as intermediaries between the sensor nodes and the base station [2]. Clustering represents a highly advantageous option for the transmission of data in wireless networks by virtue of its capacity to conserve energy. Through the capability of data aggregation, clustering affords the opportunity to effectively curtail the transmission of redundant data between cluster members and the network sink [3]. The appropriate designation of cluster heads can exert a substantial influence on vital aspects such as energy consumption, data routing, and overall network performance [4]. Blackhole attacks in wireless sensor networks are a major security issue given their capacity to interfere with network traffic and undermine data integrity. This type of attack entails generating malevolent nodes that lure network traffic by presenting themselves as the most direct route to the base station. Sinkhole attacks, in particular, represent a grave danger, as they have the potential to manipulate, alter, and impair the network structure [5]. Grayhole attacks in (WSNs) represent a major security apprehension. The malevolent nodes in these attacks selectively forward or drop packets, leading to a disturbance in network communication [6]. The utilization of chaos provides a heightened level of precision and efficacy in detecting abnormal network activity within (WSN), ultimately leading to a decrease in false positives that can arise from subjective factors [7]. A method centered on chaos has been proposed for the authentication of identities in wireless sensor networks, with the aim of guaranteeing security and minimizing resource usage [8].

The rest sections of this paper are partitioned as follows: Sect. 2 provides an overview of existing research on CH selection algorithms and secure routing that utilize optimization algorithms. Section 3 general introduction, Sect. 4 expounds upon the proposed CH selection process, which is premised on SOA and chaos Secure Aware Routing. Section 5 provides a detailed analysis and discussion of the results.

2 Related Works

The past decade has seen much attention to study WSN. The purpose of this section is the aim of this particular section is to conduct a comprehensive examination of a multitude of methodologies utilized for the discovery of CH and routing. It is worth noting that a myriad of techniques has been put in previous research endeavors [9]. In this paper a new clustering algorithm is proposed for WSNs to reduce energy consumption and thus prolong the life of the WSN, which gives a new perspective to the selection of CH while creating a more efficient WSN than the popular clustering algorithms LEACH and PEGASIS, the scalability of the proposed algorithm, which could be a potential limitation [10]. presents a secure-aware multipath routing protocol designed for wireless sensor networks (WSN) by leveraging the Rider-ASO algorithm. The protocol employs a trust-based approach to select cluster heads and effectively discovers multiple secured paths from the source to the destination, thereby outperforming other conventional techniques with notably low latency and significantly enhanced throughput rate [11]. The paper proposes a protocol for selecting cluster heads (CH) in wireless sensor networks (WSNs) using the firefly algorithm (FA) and hesitant fuzzy. The proposed protocol improves energy saving and increases the network lifetime, limitations such as the proposed protocol not be suitable for all types of WSNs and not always reflect real-world performance [12]. suggested a new energy-aware cluster head selection framework for wireless sensor networks using a hybrid optimization algorithm. The algorithm combines the concepts of Sea Lion Optimization and Particle Swarm Optimization to select the optimal cluster head based on energy, distance, delay, and Quality of Service, not be scalability for large-scale wireless sensor networks. It also overlooks security and potential hardware/software components, potentially increasing system cost and complexity [13]. The paper proposes a new algorithm called Sandpiper Optimization Algorithm (SOA) for selecting Cluster Head (CH) in the Internet of Things (IoT) network. The proposed algorithm extends the network lifespan and increases the throughput while reducing energy consumption among the nodes in the network, More convergence time and Time-consuming computation of the fitness function.

3 Methodology

The proposed methodology involves the integration of the Sandpiper Optimization Algorithm (SOA) to opt for optimal CH in the network based on some parameter such as distance, connectivity, and energy of node then applying chaos to determine any network abnormal activity then applying Secure Aware Routing Protocol (SARP) using for security transfer data in (WSN). The cluster head selection process is a crucial component in wireless sensor networks (WSNs) that involves the identification and assignment of cluster heads from the sensor nodes in the network. Cluster heads act as intermediaries between the individual sensor nodes and the base station, performing tasks such as data aggregation, routing, and network management the following algorithm to opt for optimal CH in WSN Fig. 1 showed the proposed method.

3.1 Secured Aware Routing Protocol (SARP)

This algorithm can be defined as an algorithm inspired by collective behavior, in addition to the characteristics of coastal birds, especially the bird species known as the “gull”. This bird species exhibits interesting behavior, such as its method of gathering, searching for food, and navigating. This behavior has been utilized to solve problems, including complex problems [14]. In search operations, this algorithm imitates the behavior of sandpipers in order to find the best solutions by repeating different methods and patterns of sandpiper movement, enabling them to make better decisions. This type of algorithm adheres to the principles of exploration and exploitation, which are crucial for the success of sandpipers’ food search

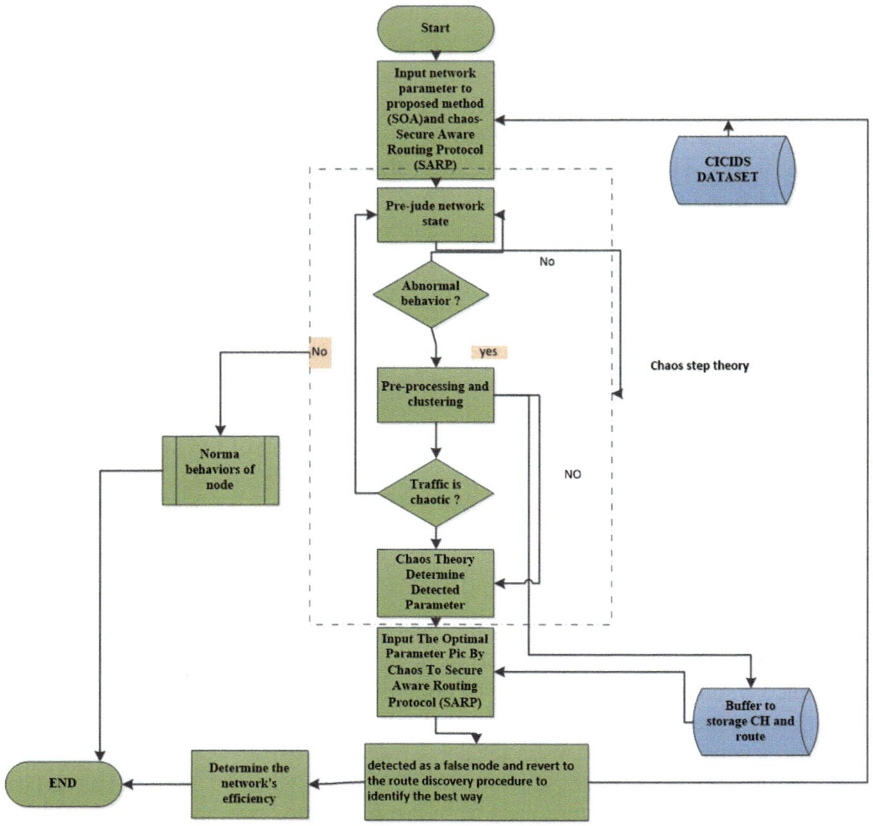

Fig. 1 Diagram for the suggested method

techniques. The Sandpiper Optimization Algorithm (SOA) is a simple and scalable architectural engineering that integrates the fundamental principles of initialization, evaluation, movement, location update, exploration, exploitation, communication, and termination. Sandpipers strike a balance between exploration and exploitation, learning from each other's experiences and fine-tuning their search techniques. The Sandpiper Optimization Algorithm provides a fresh perspective for solving optimization problems and has shown promising results in various fields [14]. For the purpose of selecting the best and most suitable Cluster Head (CH) in wireless sensor networks, this type of algorithm based on the Symmetric Optimization Approach (SOA) has been used. This algorithm provided the best CH selection process based on several factors, including the amount of available energy in the node, the efficiency of communication between nodes, and the distribution of nodes in the network. Through simulation, the application of this algorithm achieved effective balance by reducing energy consumption and improving network performance, resulting in improved energy efficiency and increased network lifetime [15].

3.2 Support Vector Machine (SVM)

SVM is a routing protocol designed for wireless sensor networks (WSNs) that prioritizes security considerations while establishing and maintaining communication paths between sensor nodes, Chaos-SARP aims to ensure secure and reliable data transmission within the network, protecting against various security threats such as data tampering, eavesdropping, and node compromise [16–18].

Algorithm 1: Pseudo-code for Sandpiper Optimization Algorithm (SOA) to opt for optimal CH:

- Initialize the population of sandpipers
- Set the maximum number of iterations
- Set the exploration rate
- Set the attractiveness coefficient
- while (not reached maximum iterations) do
- for each sandpiper in the population do
- Select a random sandpiper from the neighborhood
- Calculate the distance between the sandpipers
- Calculate the attractiveness of the current sandpiper
- Calculate the exploration probability
- if (random number < exploration probability) then
  - Perform exploration (random movement)
- else
  - Perform exploitation (move toward the selected sandpiper)
- end if
- Update the position of the current sandpiper
- end for
- Update the exploration rate based on the iteration count
- Update the attractiveness coefficient based on the iteration count
- end while
- Select the best sandpiper from the population as the solution.

Algorithm 2: Pseudo-code representation of integrating the (SOA) and Chaos-Secure Aware Routing Protocol for a WSN:

- Initialize the wireless sensor network
- Initialize the Sandpiper Optimization Algorithm (SOA) parameters
- Initialize the Chaos-Secure Aware Routing Protocol (SARP) parameters
  while (not reached termination condition) do
- // Cluster Head Selection using SOA
- Perform the Sandpiper Optimization Algorithm (SOA) to select cluster heads
- // chaos -Secure Aware Routing Protocol (SARP)
- function findSecureOptimalRoute (node, clusterHead)

- // Find the secure optimal route from a node to its cluster head, Adjust the route discovery algorithm based on the network model and security requirements
- // Consider energy efficiency, distance, and security metrics for route selection
- // Implement appropriate routing algorithm considering chaos-based security
- // Return the secure optimal route from the node to the cluster head
- **Abnormal Node Detection**:
  - Define the abnormal Node Detection (network) function that takes the network as input and detects abnormal nodes using chaos-based metrics.
  - For each node in the network:

    Collect relevant sensor data from the node.
    Calculate a chaos-based metric (e.g., Lyapunov exponent) to determine the level of chaos in the node's data.
    Compare the chaos-based metric with a predefined threshold to identify abnormal nodes.
    Mark the abnormal nodes for further action or isolation.
- **Cluster Formation**
- Initialize an empty set to store eligible cluster head nodes (cluster heads). Perform the Sandpiper Optimization Algorithm (SOA) to select cluster heads.
- **Route Generation**:
  - For each cluster head in the cluster heads set:

    Generate optimal routes for nodes within the cluster using a routing
- **Optimal Path Selection**:
  - Define the find secure optimal path (source, destination, network) function that takes the source node, destination node, and the network as inputs.
  - Implement a secure-aware routing algorithm to find the most secure and energy-efficient path from the source to the destination.
  - The secure-aware routing algorithm should consider energy efficiency, security metrics, distance metrics, and the presence of abnormal nodes for route selection.
- **Data Transmission**:
  - Initialize the network and other necessary variables.
  - Call the abnormalNodeDetection function to detect and mark abnormal nodes.
  - Select the source node and the destination node for data transmission.
  - Call the findSecureOptimalPath function to get the secure optimal path from the source to the destination, avoiding abnormal nodes.
- Modified all network factors
- end while
- Terminate the system.

4 Result Analysis and Discussion

Here and in this part of the research, a comprehensive analytical explanation and presentation of the results are provided, where the obtained results are compared based on a variety of criteria. The results of the current study demonstrated the effectiveness of the proposed method in efficiently selecting the best and optimal cluster heads, reducing energy consumption, and securely transmitting data in wireless sensor networks. The SOA algorithm showed better and faster performance in the process of selecting CH compared to other traditional methods. By using this method, the selection of CH was improved based on several factors, including the amount of energy in the node, the distance between the node and the base station, and the communication between them. Additionally, the use of the SARP-Chaos protocol was employed to find a secure path for data transmission and detect malicious nodes, which clearly resulted in improving network lifetime and energy efficiency.

4.1 Performance Matrices

Within the framework of the proposed CH selection using the SOA algorithm, and the selection of the best and optimal path (Chaos-SARP) is applied to WSNs networks, the network performance is evaluated by examining several elements of network functionality and efficiency. The following are some of the basic items that were applied during the examination:

1. **Lifetime of Network**: Is the time during which the duration of the first node in the network is measured until its energy is exhausted.
2. **Consumption of Energy**: Is the amount of total power consumption of the network during operation and data transfer.
3. **Efficiency of Routing**: To determine the efficiency of data transmission, calculate the packet delivery ratio (PDR).
4. **Scalability**: It is the process of evaluating the performance of network efficiency compared to the increase of nodes in it.
5. **Security**: Estimate the security features that providing by the Chaos-SARP.
6. **Coverage of Network**: Calculate the percentage of network coverage providing by the installed nodes.
7. **Stability**: It is a measurement of the efficiency of the network in terms of the failure of one of the nodes or an increase in the size of the network.
8. **Quality of the Service**: Assess the network's ability to satisfy QoS criteria such as latency, bandwidth, and reliability.

4.2 Testing Dataset

In this study dataset CICIDS was used to measurements the efficiency of the suggested method.

4.3 Result Discussion

The proposed method in this study was implemented using Matlab R2020a. An evaluation was conducted to measure the efficiency of the WSN performance using multiple variables. This method enabled researchers to gain insights into the effectiveness and efficiency of the proposed method compared to algorithms or other methods presented in Table 2 and Fig. 2. The evaluation results showed the availability of valuable information for understanding the strengths, limitations, and other possible approaches to improve the selection of the proposed CH using the SOA algorithm and the application of secure data routing through the Chaos-SARP application on wireless sensor networks. Thus, this method provided accurate information, as shown in Table 3 and Fig. 3, about the comparison between AODV, LEACH, and the proposed method based on different metrics. According to the results obtained, it was observed that the proposed network coverage is 90%, and the virtual network lifespan has become 8.5 days, while the packet delivery rate is 95%. Additionally, when a malicious node is detected, its contents are delivered and the node's location is alerted through the application of chaos theory. The data transmission process is considered incorrect due to its spread in the network, leading to disconnection or incorrect delivery of data to the relevant interface. Therefore, as a result of this, the proposed method provides a high degree of accuracy in detecting malicious nodes, reducing power consumption, and increasing the network lifespan by selecting the best channel and choosing a secure path for data transmission.

Table 1 Simulation value in the proposed method

Parameter	Value
Network area	200×200 m^2
Sink location	(100, 100)
Number of sensor nodes	100, 200, and 500
Percentage of CH	5%
Control packet size	200 bits
Data packet size	4000 bits

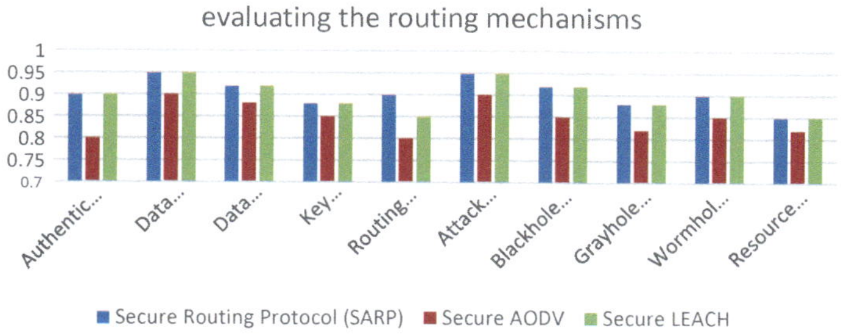

Fig. 2 Evaluating the routing mechanisms employed in the Secure Aware Routing Protocol (SARP)

Table 2 Evaluating the routing mechanisms

Routing mechanism	Secure routing protocol (chaos-SARP)	Secure AODV	Secure LEACH
Authentication	0.9	0.8	0.9
Data confidentiality	0.95	0.9	0.95
Data integrity	0.92	0.88	0.92
Key management	0.88	0.85	0.88
Routing protocol type	0.9	0.8	0.85
Attack detection	0.95	0.9	0.95
Blackhole attack mitigation	0.92	0.85	0.92
Grayhole attack mitigation	0.88	0.82	0.88
Wormhole attack mitigation	0.9	0.85	0.9
Resource efficiency	0.85	0.82	0.85

Table 3 and Fig. 3 display a thorough comparison of the Secure AODV, Secure LEACH, and the proposed method in terms of various criteria such as Authentication, Data Confidentiality, Data Integrity, and Key Management. Upon analyzing the results, it is evident that the proposed method has demonstrated exceptional performance with a high degree of accuracy.

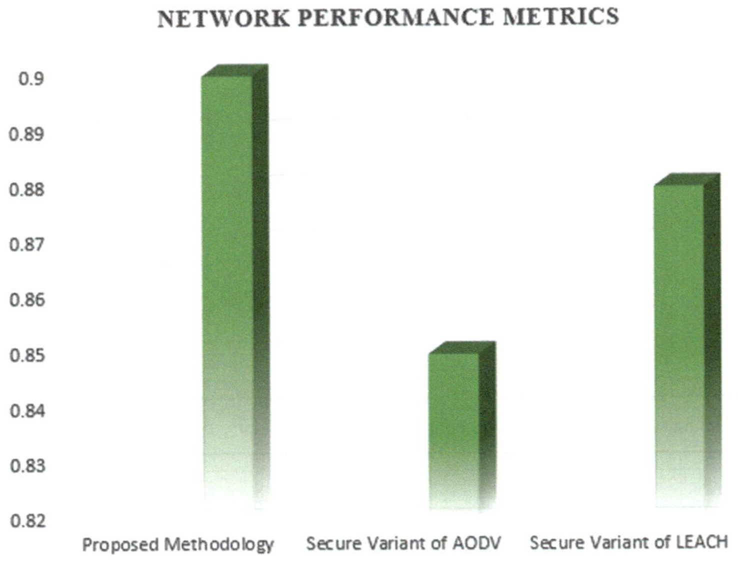

Fig. 3 The evaluation of network performance metrics for the different methods

Table 3 The evaluation of network performance metrics for the different methods

Performance metric	Proposed methodology	Secure variant of AODV	Secure variant of LEACH
Network lifetime	8.5 days	6.2 days	7.8 days
Energy consumption	3500 J	4100 J	3800 J
Packet delivery ratio	95%	88%	91%
End-to-End delay	50 ms	75 ms	65 ms
Throughput	1.2 Mbps	0.9 Mbps	1.1 Mbps
Network scalability	Good	Moderate	Moderate
Security effectiveness	High	Medium	Low
Network coverage	90%	85%	88%
Stability	Resilient	Susceptible	Moderate
Quality of the service	Reliable	Unreliable	Unreliable

5 Conclusion

The proposed CH Selection using the SOA and CHAOS-SARP offers a promising approach to enhance the performance and security of wireless sensor networks. The integration of the SOA algorithm for optimal CH opt that reduces energy consumption and reduces reddened of data enhanced network layer therefore prolonging network lifetime and then using in the second step chaos-SARP protocol for secure routing contributes to improved network lifetime, energy efficiency, packet delivery ratio, and network coverage. While the proposed methodology has shown promising results, there are still challenges and avenues for future research. The computational complexity of the SOA algorithm, real-world implementation. Overall, the proposed methodology holds great potential for improving the performance, efficiency, and security of wireless sensor networks, opening up opportunities for various applications in diverse domains. The future work areas will further strengthen the proposed methodology and pave the way for its practical implementation in real-world WSN deployments, ensuring improved network performance and enhanced security for various applications.

References

1. Keerthika, M., Shanmugapriya, D.: Wireless sensor networks: active and passive attacks—vulnerabilities and countermeasures. Glob. Transit. Proc. **2**(2), 362–367 (2021). https://doi.org/10.1016/j.gltp.2021.08.045
2. Subramani, N., Mohan, P., Alotaibi, Y., Alghamdi, S.: An efficient metaheuristic-based clustering with routing (2022). https://doi.org/10.3390/s22020415
3. Hoomod, H K., Jebur, T.K.: Applying self-organizing map and modified radial based neural network for clustering and routing optimal path in wireless network. J. Phys. Conf. Ser. **1003**(1) (2018). https://doi.org/10.1088/1742-6596/1003/1/012040
4. Evangelakos, E.A., Kandris, D., Rountos, D., Tselikis, G., Anastasiadis, E.: Energy sustainability in wireless sensor networks: an analytical survey. J. Low Power Electron. Appl. **12**(4) (2022). https://doi.org/10.3390/jlpea12040065
5. Mehta, A., Sandhu, J.K., Pundir, M., Kaur, R., Sapra, L.: Sinkhole attack detection in wireless sensor networks BT. In: Proceedings of Data Analytics and Management, pp. 85–94 (2022). https://doi.org/10.1007/978-981-16-6285-0_8
6. Rehman, A., Rehman, S.U., Raheem, H.: Sinkhole attacks in wireless sensor networks: a survey. Wirel. Pers. Commun. **106**(4), 2291–2313 (2019). https://doi.org/10.1007/s11277-018-6040-7
7. Jebur, T.K.: Proposed hybrid secured method to protect against DDOS in n Vehicular Adhoc Network (VANET). Int. J. Interact. Mob. Technol. **17**(11), 141–154 (2023). https://doi.org/10.3991/ijim.v17i11.38907
8. Meshram, C., Lee, C.C., Meshram, S.G., Meshram, A.: OOS-SSS: an efficient online/offline subtree-based short signature scheme using Chebyshev chaotic maps for wireless sensor network. IEEE Access **8**, 80063–80073 (2020). https://doi.org/10.1109/ACCESS.2020.2991348
9. Lakshmanna, K., Subramani, N., Alotaibi, Y., Alghamdi, S., Khalafand, O.I., Nanda, A.K.: Improved metaheuristic-driven energy-aware cluster-based routing scheme for IoT-assisted wireless sensor networks. Sustainability **14**(13) (2022). https://doi.org/10.3390/su14137712
10. Shyjith, M.B., Maheswaran, C.P., Reshma, V.K.: Secure-aware multipath routing using atom search rider optimization algorithm in wireless sensor networks. Int. J. Mob. Comput. Multimed. Commun. **12**(2), 36–55 (2021). https://doi.org/10.4018/IJMCMC.2021040103
11. Rayenizadeh, M., Kuchaki Rafsanjani, M., Borumand Saeid, A.: Cluster head selection using hesitant fuzzy and firefly algorithm in wireless sensor networks. Evol. Syst. **13**(1), 65–84 (2022). https://doi.org/10.1007/s12530-021-09405-1
12. Yadav, R.K., Mahapatra, R.P.: Hybrid metaheuristic algorithm for optimal cluster head selection in wireless sensor network. Pervasive Mob. Comput. **79**, 101504 (2022). https://doi.org/10.1016/j.pmcj.2021.101504

13. Sankar, S., et al.: Cluster head selection for the internet of things using a sandpiper optimization algorithm (SOA). J. Sens. **2023** (2023). https://doi.org/10.1155/2023/3507600
14. Goel, S.: Sandpiper optimization algorithm: a novel approach for solving real-life engineering problems (2020). https://doi.org/10.1007/s10489-019-01507-3
15. Rajalakshmi, S., Kanmani, S.: A comprehensive review on recent intelligent metaheuristic algorithms. Int. J. Swarm Intell. **7**(2), 182–205 (2022)
16. Abdel-Basset, M., El-Shahat, D., Elhoseny, M., Song, H.: Energy-aware metaheuristic algorithm for Industrial-Internet-of-Things task scheduling problems in fog computing applications. IEEE Internet Things J. **8**(16), 12638–12649 (2021). https://doi.org/10.1109/JIOT.2020.3012617
17. Singh, D., Gupta, M., Kumar, R., Obaid, A.J.: A systematic review of human fall detection algorithms using machine learning. In: Rani, A., Kumar, B., Shrivastava, V., Bansal, R.C. (eds.) Signals, Machines and Automation. SIGMA 2022. Lecture Notes in Electrical Engineering, vol. 1023. Springer, Singapore (2023). https://doi.org/10.1007/978-981-99-0969-8_60s
18. Iqbal, M.O., Obaid, A.J., Agarwal, P., Mufti, T., Hassan, A.R.: Blockchain technology and decentralized applications using blockchain. In: Tuba, M., Akashe, S., Joshi, A. (eds.) ICT Infrastructure and Computing. Lecture Notes in Networks and Systems, vol. 520. Springer, Singapore (2023). https://doi.org/10.1007/978-981-19-5331-6_57

The Effects of Weather Conditions on Data Transmission in Free Space for Wireless Sensor Networks

Aqeel A. Kadhim, Jolan R. Naif, Muna M. Salman, and Haider K. Hoomod

Abstract It is necessary to collect data measured in open lands, whether for example for monitoring purposes or for border protection. The process of data transmission from active nodes and its next level to the cluster head (CH), followed by follow-up to arrive at the base station (BS), with safety and acceptability, are the main objectives of wireless sensor networks (WSNs). The key benefits of wireless systems are low-cost system installation and loss-free data transfer. Weather conditions in the spacious area affect wireless data transmission that is done between sensor nodes in free space. This study attempted to ascertain the influence of atmospheric and meteorological conditions on the loss of data when utilizing the wireless data transmission systems in a wireless sensor network. For this reason, the volume of data transfer losses that happen in free space because of three types of weather conditions is the main discussion point. The meteorological criteria that can be considered in this research are temperature, clear weather, dust, and rain as the weather elements that affect the Iraqi climate. The correlation analysis's outcomes show that the amount of precipitation and the size of raindrops have a main effect on the losses of data transmission. From a practical point of view, there is no confirmed statistical evidence of the amount of loss in the volume of transmitted data as a result of the amount of rainfall with the use of different transmission frequencies. In this research, we designed an objective routing protocol called Fuzzy Clustering with Ant Colony System FCACS, which has been utilized as a simulation program using the MATLAB 2018b application. The practical results of the proposed protocol for data transmission in free space and

A. A. Kadhim (✉)
Computer Techniques Engineering Department, Imam AlKadhim University College, Baghdad, Iraq
e-mail: kadhimaqeel@gmail.com

J. R. Naif
Informatics Institute for Postgraduate Studies, Iraqi Commission for Computers and Informatics, Baghdad, Iraq

M. M. Salman
Information Technology Directorate, Ministry of Science and Technology, Baghdad, Iraq

H. K. Hoomod
College of Education, Computer Science Department, Al-Mustansiriya University, Baghdad, Iraq

G. Fortino et al. (eds.), *Proceedings of Third International Conference on Computing and Communication Networks*, Lecture Notes in Networks and Systems 917,
https://doi.org/10.1007/978-981-97-0892-5_9

under different weather conditions were examined, and the simulation results were compared with those of two well-known routing protocols, which are DEEC and Z-SEP. In terms of the quantity of data received from the CH or BS and for the various weather circumstances that were suggested in the simulation, the proposed protocol demonstrated a high level of performance. Through the practical results linked to the energy dissipation for the three routing protocols, it is obvious that sending data via WSN consumes high energy in the case of rainy weather.

Keywords FCACS · Attenuations · Weather conditions · Data sending

1 Introduction

Both indoor and outdoor settings are suitable for the deployment of WSNs. The transmission of sense data by active nodes in a network is impacted by several factors. Climate factors in free space settings can alter the quality and quantity of data transmission, whereas metal-dense surroundings in enclosed spaces have an impact on wireless transmission. On an open field, wireless sensors may be installed everywhere and in any number of locations. One of the great benefits of wireless sensor networks is the transmission of data over long distances in the open air. However, data transmission may also be impacted by the variance in *atmospheric* conditions from one place to another. In outdoor settings, wireless data transmission may have an influence on both the amount and quality of data [1].

This study tries to locate the impact of some *atmospheric* variables or *weather* conditions that have an important impact on data transmission in free space in a country like Iraq.

The most significant attenuating elements for free space communications (especially in Iraq) include air temperature, rain, and dust, among other weather conditions. This encourages researchers to think about the most significant impacts of weather conditions on radio communications used in long-lasting, earthly applications of WSN [2–4].

One of the major contributions that has been adopted in this study is the selection of more than one algorithm and technique for disseminating and grouping sensor nodes, cluster heads HD election, and finding the perfect and short way for data transmission to base station BS. The goal of this study is to develop a new routing protocol with good performance in the proposed environmental conditions while delivering data from the sensor nodes to the base station in an efficient manner. The proposed protocol has been called Fuzzy Clustering with Ant Colony System (FCACS). Besides that, the suggested protocol presents an efficient strategy for managing energy for wireless sensor networks (WSN), by utilizing a data-driven model that depends on the techniques for collecting data from active nodes, sending it to cluster heads, and routing the data to the base station.

A comparison study between the FCACS protocol and the well-known energy-efficient routing protocols for heterogeneous wireless sensor networks, which are

Z-SEP and DEEC protocols, in terms of the volume of data packets that can be sent in free space from sensor nodes to their CH and then sent by one-hop or multi-hop to the base station with different weather conditions The size of the dissipated energy of the entire network and its impact on the network's lifetime are also studied. This research also shows that there is a significant improvement in the FCACS protocol in extending the life and stability of the network. In addition, the average energy consumed for all weather conditions is better than the rest of the protocols that were compared with and taken into consideration in this research.

2 Related Works

In this and the previous decade, several studies related to the study of data transmission in free space have appeared using fiber optics for data transmission, and there are also many studies on the effect of weather conditions on data transmission. However, there is no study dedicated to the transmission of data in the free space for sensor networks or the effect of weather conditions on data transmission, especially in Middle Eastern countries like Iraq. Nassir Abuhamoud et al. [5] presented an analysis of the effect of dust storms on the transmission of radio signals. What is the effect of dust storm characteristics in terms of visibility and size of mobile particles on data transmission in free space? Also, the size of dust particles has a significant impact on the attenuation and obstruction of data transmission. The practical results show that the effect of attenuation will increase with increasing frequencies and air humidity and decrease with visibility Vasilis Christofilakis et al. [6]. Through this study, which presents a literature survey of previous studies, especially in the past decade, on the issue of the effect of rain on data transmission in free space, and also what the research challenges and future directions are based on the literature, the use of microwave path attenuation for rain rate estimation is considered a novel and feasible method. Taking into account many important points for the purpose of achieving the highest possible accuracy of any model showing the relationship between attenuation and rainfall rate Skirelis et al. [2]. This article reviewed the vulnerability of wireless sensors to various climatic factors. The study also presented practical research on the effect of different weather conditions such as snow, rain, fog, and clouds, in addition to the effect of wind on wireless signals. Especially weather effects on wireless sensor networks using free space optical or radio.

3 Methodology

3.1 Weather Conditions Effect on the Transmitted Signal

Weather refers to the condition of the atmosphere, including how warm or chilly, quiet or windy, clear or dusty it is. The term "weather" often refers to the daily temperature and precipitation patterns. Water vapor condensation that falls from clouds or is deposited from the atmosphere onto the earth is referred to as precipitation. Fog, dribble, and rain are the basic types of precipitation. Recent studies have shown that various weather conditions have an array of effects on transmitting data over wireless networks. The studies show that the propagation, refraction, and absorption of microwaves by the gases in the atmosphere are prime factors affecting the transmission signal power in wireless communications. Dust and rain are examples of environmental conditions that might interfere with electromagnetic communications. Wireless signal integrity can be significantly impacted by weather conditions. Any obstacle that faces the radio signal between the transmitter device and the receiver will reduce the strength of the signal through attenuation. These obstacles include liquid objects such as fog or rain in addition to solid objects such as trees or walls [2, 7].

The transmission of wireless signals from one antenna device to another will expose them to some losses in power. For the purpose of overcoming the signal losses that may occur in the propagation route, systems that have been properly built employ the appropriate antennas, signal frequencies, and transmission power. The design of a wireless network will take into consideration the requirements of careful consideration for radio interference. Weather conditions (like temperature, rain, and dust) are some examples of environmental cases that might interfere with electromagnetic communications, especially in the climatic conditions in Iraq. All weather conditions are contributing to attenuating the signal's rout. One of the most important climatic phenomena that have been taken into account in their impact on the strength of the transmitted signal is the amount of rain and the size of the raindrop. There is a direct proportion between the size of the raindrop and the quantum of attenuation in the transmitted signal; the larger the raindrop, the greater the amount of attenuation in the signal [4, 7].

3.2 Some Local Weather Conditions that Affecting on Data Transmission

3.2.1 Rain

The amount of wireless signal scattering increases exponentially with the size and quantity of the droplets. More than any other type of weather, heavy rains can commonly degrade wireless signals. The radio signal is randomly weakened by rain as it travels through the atmosphere; the attenuation effect of rain on wireless networks cannot be disregarded, particularly in certain regions where rain is more common. Raindrops grow in size until they are large enough to generate the processes of reflection and refraction. Raindrops that are big enough to generate reflection processes are those that are larger than normal. The growing droplets create scattering that is not dependent on wavelength [2]. One of the main significant issues with long-range propagation is the absorption of the transmitted signal due to precipitation. When there is heavy rain, the presence of droplets can have a considerable negative impact on the communication channel's performance and dependability. The rain impact on the transmission of millimeter-range waves is rather complex to address since it is important to determine the size and form of a droplet, how frequently they fall, how much precipitation is falling, etc. The most often used formula for calculating absorption because of rain is [2]

$$A = aR^b \tag{1}$$

where the absorption is expressed in decibels for each kilometer (dB/km) and the intensity of rain is expressed in millimeters for each hour (m/h), the factors a and b rely on the frequency and the droplet distribution size. The absorption is dependent on R, the rain's intensity, for certain values of a and b. The biggest issue is with the times when rain intensity will reach a particular limit. Everything is based on the climatic zone, where there are 15 climate zones on earth that are separated based on the precipitation [2, 8].

It was discovered that the attenuation that results grows linearly with the pace of precipitation. In addition, the average size of the raindrops, which is in the range of a few millimeters, was also observed to increase with the rate of precipitation. The particular attenuation of a wireless connection is written as follows, where R is the rain rate given in mm/h [2]:

$$A_{spec} = 1.076 \cdot R^{0.67} \text{ [dB/km]} \tag{2}$$

Rain attenuation significantly reduces the RF communications performance in the GHz region, which limits the usage of GHz frequencies for direct visual communication links. The primary attenuating factor at frequencies >10 GHz is rain [2], despite the fact that fog, clouds, and dust particles all have a significant impact on the propagation of signals. The following equation will give the link between the rate of rain and a given attenuation

$$\text{Att}_{\text{rain}} = k_1 R^{k_2} \tag{3}$$

where

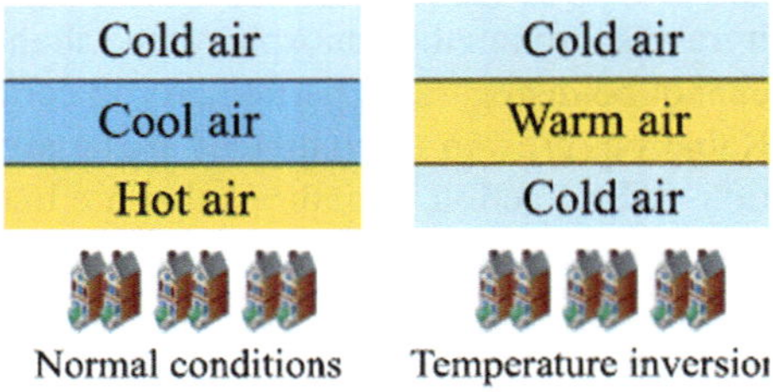

Fig. 1 A pictorial illustration of stream-like conditions in the atmosphere

R = Represents rain rate in mm/h. k_1 and k_2 = Are model parameters that depend upon raindrop size and rain temperature. The values of the rain attenuation prediction model for free space communication recommended are $k_1 = 1.074$ and $k_2 = 0.67$, respectively [8, 15].

3.2.2 Air Temperature

Any association between signal intensity and air temperature that has been found is almost always directly related to variations in humidity. Temperature changes alone shouldn't have an effect on the signal's intensity as it approaches the receiving equipment. It's also critical to realize that ultra-high frequencies (UHFs) move along lines of sight, in contrast to very low frequencies that rebound between the ground and the layer of ionosphere and hence cover very vast distances. The range of a line-of-sight cellular transmission cannot be increased under atmospheric circumstances that may allow a sloppy radio operator to bounce the belonging signal halfway across the world. Temperature inversions, which take place if a heated air layer is trapped above a cool air layer (which can be seen in Fig. 1), provide circumstances related to an atmospheric outlet where radio waves can bounce off [2, 14].

3.2.3 Dust and Sand

Dust and sand also cause signal attenuation, but really heavy dust and sand are the main reasons that make satellite links unavailable. Depending on the local weather conditions, different things contribute more to attenuating the propagation signal. Early studies had focused on the homogeneous distribution or distinctive geometric shape of the dust and sand attenuation. This approximation provided the designers with acceptable results in either high or low visibility, but it will no longer do so in low visibility. In Iraq, dust storms are a common meteorological occurrence. In some parts of the world, dust storms may be the most serious environmental issues. The dust storms sometimes traveled up to 4000 km from their source. One of the known effects of dust storms is the decrease in solar radiation, which has an influence on the effectiveness of solar devices. Radio waves are attenuated by the sand and dust particles' energy absorption and scattering as they travel through an environment that contains deposition-like dust particles [6, 8–10].

3.3 *Signal Attenuation Due to Dust Storms*

One of the main issues with using radio waves for earthly and free space communication, especially in arid and semi-desert areas, is the signal attenuation induced by dust storms. During a dust storm, the dust particles climb high enough to be in the radio wave link's line of sight, reducing the signal's strength and creating a service outage. Consistent with the distribution of vertical dust particle size, the particle radius decreases as height increases, increasing visibility with height and resulting in reduced attenuation. Therefore, precise measurement of the dust storm height results in a more accurate measurement of the channel attenuation, which ultimately reduces the need for additional transmission power and enhances QoS. These storms are the main factor in satellite communication channels' attenuation [6, 11, 12].

3.3.1 Dust Storm Attenuation

Due to the fact that each of the dust and sand particle groupings is a discrete random medium, the general model for scattering in these populations is very similar to that of the hydrometeor population. As a result, the techniques for anticipating the effects of sign attenuation because of rain may be used for dirt typhoons. By establishing the forward scattering function of the amplitude of one particle, the signal attenuation caused by a dust storm is commonly predicted. To complete the solution, it may be done by using the Mie solutions or Rayleigh approximations. The number of particles and their radius have a big impact on the procedure. electromagnetic radiation's attenuation (AT's) along a channel with length L due to particles precipitating, which may be written as [13]:

$$A_T(db) = \int_0^L A_p(dx) \tag{4}$$

where A_p (db/km) is the particular attenuation defining for the precipitate particles.

4 Proposed Protocol Simulation Environments and Results

This section will present the simulation results for the FCACS protocol in the proposed domain of WSN and the comparison results with well-known energy-efficient routing protocols for heterogeneous wireless sensor networks (Z-SEP and DEEC). The simulation is also implemented with member nodes that are deployed in a random manner in the monitoring area, and its default area is equal to (100 × 100) m^2. Table 1 lists the network scenarios and simulation settings for the suggested schema.

Simulation and analysis were performed using the Graphical User Interface GUI for the MATLAB-18b software package, which is implemented on a machine running the Windows 10 × 64 operating system. In this simulation, a network of 100 sensor nodes that are dispersed across a landscape of 100 m by 100 m is modeled. Nodes will be static and randomly distributed in the region of interest. Each simulation takes approximately 300 s to complete. An omnidirectional antenna is employed with a 2 Mbit/s channel capacity. The data streams have constant bit rates (CBR), and each data packet has a size of 4000 bits. Only after the node's energy is depleted is it deemed to have died.The FCACS proposed protocol was utilized together with two well-known routing protocol modules (the Z-SEP and DEEC protocols). The FCACS protocol is simulated, and the practical results are compared with the selected protocols from the point of view of transmitted data packets to the cluster head and base station. The energy dissipated for each protocol was also computed. Through the use of the simulation program, which introduced the weather conditions in Iraq, that had an important impact on the data transmission in free spaces. Two locally influencing weather factors were taken into account, which are rain and dust, in addition to clear weather condition. From the practical process, useful and important results have been

Table 1 The proposed parameters

Parameter	Value
Network size	100 m × 100 m
Monitoring area nature	Flat
Sensor nodes	100
Sink nodes	1
Nodes behavior	Static
Position of sink node	[50,100]
Number of clusters	5
Termination condition	0.00001
CH node buffer size	35 packet
Sink node buffer size	100 packet
E = Initial energy for node	2 J
E_{CH} = Initial energy for CH node	2 J
E_{elec}	50 nJ/bit

extracted regarding the data transmission density within the sensor network. Taking into consideration the importance of network stability and the network lifetime that makes the network operate properly and continue data flow for a longer time, the volume of energy dissipation during the simulation process for the three protocols is considered one of the important elements for the sustainability of the network. Therefore, the dissipation energy for the three protocols has been calculated during the simulation process. Also, through the practical results of data transmission under the selected weather conditions, the data obtained from the simulation program was recorded for each protocol separately, as were the practical results of the combined protocols, in order to give us an adequate perception of the influences of weather conditions on data transmission in free space.

With the suggested protocol, more data packets may be appropriately sent from active sensors to corresponding CH nodes and from CH to BS with better residual energy and network stability. The practical result in all simulation tests will take into account the effective weather conditions affecting data transmission in free space.

4.1 Experimental Results and Discussion

This study's objective is to assess and examine the effectiveness of three existing heterogeneous routing protocols, FCASC, Z-SEP, and DEEC, under various environmental and weather intensity conditions. What was extracted from the simulation environments of the selected protocols is the outcome of a practical result, which is to construct a database, from which we obtained a number of figures that reflect the practical results and also reflect the behavior of the network during the simulation process. The following points can be summarized from the experimental results:

- Figure 2 shows the volume of data packets sent to CH with clear weather for all proposed protocols. The volume of data transferred to the cluster heads during the life of the network and for all the proposed protocols has been computed. According to the practical results, the amount of data transferred when using the FCACS protocol is the highest, followed by the Z-SEP protocol. The data transfer volume for the FCACS protocol at the last round amounted to 63,661 packets.
- Also, if the amount of data transmitted over the network to the BS is computed, the performance of the FCACS protocol is the best as compared to other protocols. The volume of data transferred until the last round of the network's life reached 49,641. Figure 3 shows the volume of data packets that are sent to BS with clear weather for all proposed protocols.
- In Fig. 4, a simulation of the system is shown using rainy weather. The computing of the data transmitted over the network to the CHs was less than in the case of clear weather. Also in Fig. 5, the volume of data transmission to the BS is less than clear weather. This is obvious because of the effect of raindrops on the volume of data transmitted over the network. However, through practical results, it became clear that the performance of the protocol is the best compared to other protocols.

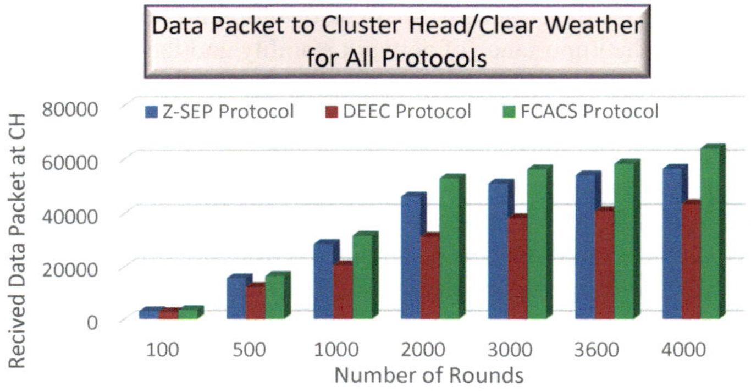

Fig. 2 A bar chart shows the volume of data packets that send to CH with clear weather for all proposed protocols

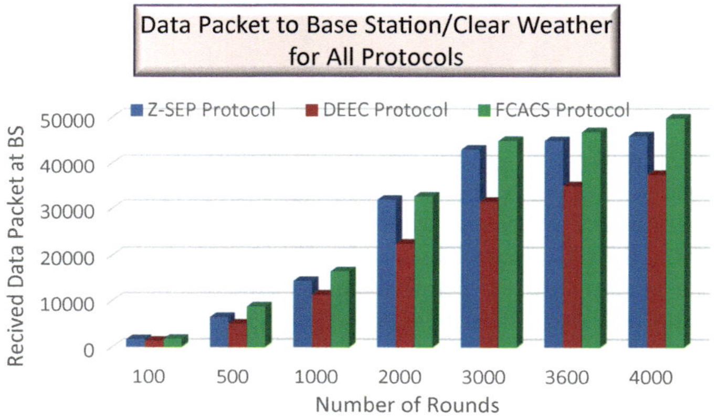

Fig. 3 A bar chart shows the volume of data packets that send to BS with clear weather for all proposed protocols

- To view the volume of data transmitted during the system simulation using dusty weather. The volume of data transferred over the entire network to CHs, as shown in Fig. 6, is less than in the case of clear weather but greater than that resulting from rainy weather. Figure 7 shows the volume of data sent to the BS in dusty weather. The volume of transmitting data was logical and simulated reality. Through the practical results, it became clear that the performance of the FCACS protocol is also the best compared to other protocols.
- In order to study the amount of energy dissipation during network lifetime, a simulation of the system was carried out to show the amount of energy consumed for the proposed protocol and for all weather conditions. Through the practical results, it was found that the energy expended up to 400 rounds is approximately the same for all weather conditions; after that, the consumed energy begins to increase

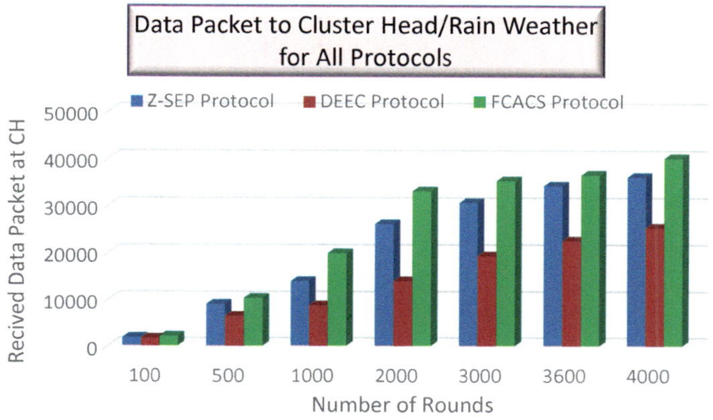

Fig. 4 A bar chart presents the data packets quantity that send to CH with rain weather for all proposed protocols

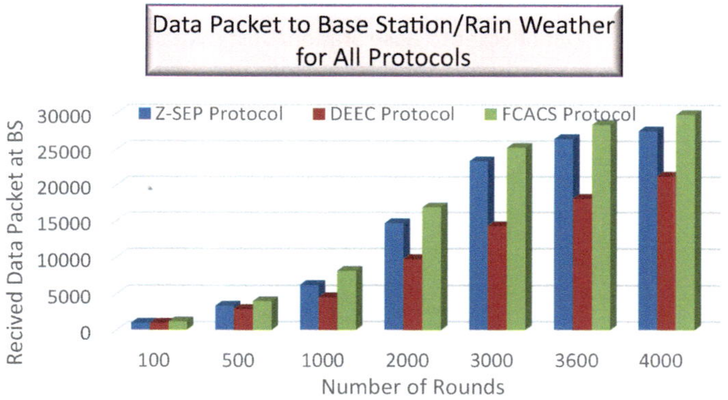

Fig. 5 A bar chart presents the data packets quantity that send to BS with rain weather for all proposed protocols

in rainy weather. In the case of dusty weather, the energy dissipation begins to increase by 1000 rounds. This indicates the correctness of the data extracted from the simulation results in the previous paragraphs, as the volume of transmitted and correct data is affected by weather factors, and this is reflected negatively on the volume of energy spent for the entire network. Figure 8 presents the dissipation energy that is used in all weather conditions for the FCACS protocol.

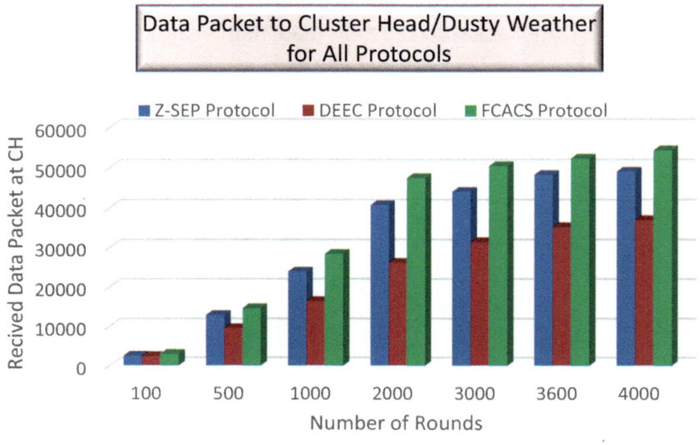

Fig. 6 A bar chart shows the data packets quantity that send to CH with dusty weather for all proposed protocols

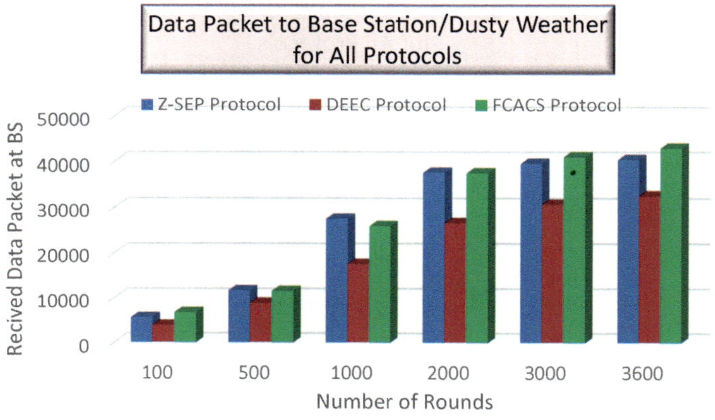

Fig. 7 A bar chart shows the data packets quantity that send to BS with dusty weather for all proposed protocols

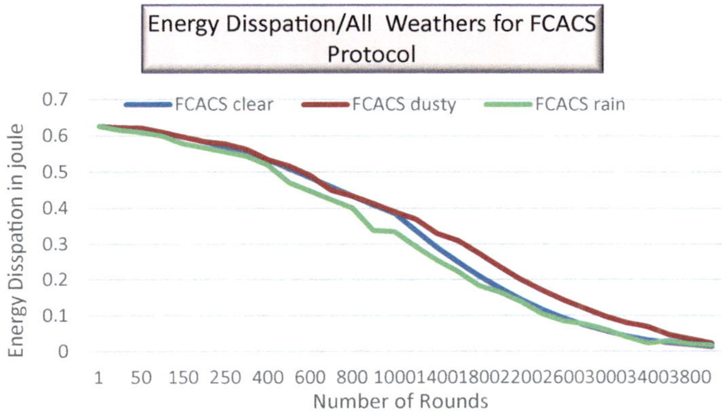

Fig. 8 The dissipation energy that used in all weather conditions for FCACS protocol

5 Conclusion

It is necessary to correctly predict weather changes, on the basis of which wireless sensor networks could be utilized. Through the simulation results, it has been established that selecting the best protocol in a given wireless network environment depends on the prevalent weather conditions.

The practical results proved that the performance of the proposed protocol FCACS is better than that of the DEEC and Z-SEP protocols in terms of the volume of data received from the CHs or BS and for the different weather conditions proposed in the simulation. Through the obtained practical results related to the energy dissipation for the three protocols, we can clearly notice that the energy consumed in WSN to send data in the case of rainy weather is the most. The explanation for this is that the transmitted signal in rainy weather is exposed to reflection and refraction, so the network requires re-transmission of the data from sensor nodes to their CHs or from CHs to BS, which increases rapidly the consumed energy. Also, in the case of dusty weather, the signal is exposed to absorption and scattering or dispersion due to particles of dust, and this aspect also leads to additional energy consumption, but it is less when compared to the amount of energy consumed in rainy weather. This also proves that the FCACS protocol has outstanding performance in reducing energy consumption for the entire network, which reflects positively on the network's stability and also keeps the network operating for a longer period.

References

1. Tezcan, C.S., Gündoğdu, K.S.: The effect of some meteorological parameters on wireless data transmission. J. Agric. Nat. 1127–1133 (2022). https://doi.org/10.18016/ksutarimdoga.vi.910496
2. Skirelis, J., et al.: Wireless sensor networks: towards resilience against weather-based disruptions. Electr. Control Commun. Eng. **15**(2), 79–87 (2019). https://doi.org/10.2478/ecce-2019-0011. https://content.sciendo.com
3. Nadeem, F., Chessa, S., Leitgeb, E., Zaman, S.: The effects of weather on the life time of wireless sensor networks using FSO/RF communication. Radio Eng. **19**(2), 262–270 (2010)
4. Navakauskas, D., Pupeikis, R.: On-line approach for fast convolution over sensor networks. TEM J. Technol. Educ. Manag. Inform. **7**(2), 243–249 (2018)
5. Abuhamoud, N., Fadil, S., Alsadi, S., Omar, B.: Attenuation of dust and sand storms of microwave signal at Sabha City. Int. J. Eng. Res. Dev. **17**(4), 14–21 (2021) e-ISSN: 2278-067X, p-ISSN: 2278-800X, www.ijerd.com
6. Christofilakis, V., et al.: Earth-to-earth microwave rain attenuation measurements: a survey on the recent literature. Symmetry **12**, 1440 (2020). www.mdpi.com/journal/symmetry
7. Rangarajan, J., Baskaran, K.: Evaluating the impact of weather condition on MANET routing protocols. Int. J. Electr. Eng. Inform. **7**(3) (2015)
8. Harb, K., Omair, B., Abdul-Jauwad, S., Al-Yami, A.: Systems adaptation for satellite signal under dust, sand and gaseous attenuations. J. Wirel. Netw. Commun. 39–49 (2013)
9. Nassar, Y., Aissa, K., Alsadi, S.: Air pollution sources in libya, research & reviews. J. Ecol. Environ. Sci. **6**, 63–79 (2018)
10. Hafez, A., Nassar, Y., Hammdan, M., Alsadi, S.: Technical and economic feasibility of utility-scale solar energy conversion systems in Saudi Arabia. Iranian J. Sci. Technol. Trans. Electr. Eng. **44**(1), 123–225 (2019)
11. Nassar, Y., Awiedat, M., Iessa, K., AlShareef, I., Ali, K., Fakher, M.: Estimation of CO_2 emission factor for the energy industry sector in Libya: a case study. Environ. Dev. Sustain. 1–29 (2021)
12. Abuhamoud, N.: Evaluation of L2 trigger impact on fast mobile IPv6 handover. Int. J. Innov. Technol. Explor. Eng. (IJITEE) **4** (2014)
13. Fadil, S., Abuhamoud, N.: Link analysis model for direct broadcast satellites case of study: Benghazi and Sebha. In: Proceedings of the 6th International Conference on Engineering & MIS 2020, pp. 1–7 (2020)
14. Alghurabi, K.A., Obaid, A.J., Alkhayyat, H.K., Abdulabbas, Y.M., Albermany, S.A.: Using symmetric group to generate dynamic S-box. In: Kumar, A., Fister Jr., I., Gupta, P.K., Debayle, J., Zhang, Z.J., Usman, M. (eds.) Artificial Intelligence and Data Science. ICAIDS 2021. Communications in Computer and Information Science, vol. 1673. Springer, Cham (2022). https://doi.org/10.1007/978-3-031-21385-4_41
15. Suresh, R., Kumar, K.S., Regan, M., Kumar, K.N., Devi, R.N., Obaid, A.J.: Dynamical properties of a modified chaotic Colpitts oscillator with triangular wave non-linearity (2023). https://doi.org/10.24425/acs.2023.145112

Using Wireless Sensor Network Performance and Optimizing in Underground Mines with Virtual MIMO Antenna

Shailendra Kumar Rawat, Sanjay Kumar Singh, and Ajay Kumar Bharti

Abstract Wireless sensor networks rely heavily on battery power for continuous operation, making energy efficiency a critical factor in their longevity. To address this challenge, a novel architectural system known as Virtual MIMO has been developed. Virtual MIMO operates in conjunction with traditional MIMO technology and holds the potential to establish an energy-efficient network. Leveraging the synergistic benefits of Virtual MIMO, which include enhanced system generalization and energy conservation, it becomes possible to employ MIMO technology within a single antenna system. This paper primarily centers on the utilization of Virtual MIMO to prolong the lifespan of wireless sensor networks while minimizing energy consumption. Alongside various S-parameters, the research reveals varying return loss values in decibels (dB) at different frequencies, such as S21, S22, S11, and S12. In summary, Virtual MIMO techniques play a pivotal role in the creation of efficient wireless sensor networks and offer a means to optimize their overall performance.

Keywords MIMO · SISO · WSN · RF · MSA

1 Introduction

Wireless sensor networks (WSNs) are formed by strategically placing sensors across diverse terrains to capture data regarding various environmental parameters, including temperature, sound levels, moisture, and wind speed, among others. This collected data is then wirelessly transmitted to a central hub through autonomous

S. K. Rawat (✉) · S. K. Singh
Amity Institute of Information Technology, Lucknow, India
e-mail: rawat.nokia@gmail.com

S. K. Singh
e-mail: sksingh1@amity.edu

A. K. Bharti
Babu Banarsi Das University, Lucknow, India

G. Fortino et al. (eds.), *Proceedings of Third International Conference on Computing and Communication Networks*, Lecture Notes in Networks and Systems 917,
https://doi.org/10.1007/978-981-97-0892-5_10

wireless ad hoc networks. WSNs are employed to comprehensively cover the area, and modern advancements facilitate the monitoring of sensor functionality. These networks consist of sensor nodes that establish connections with neighboring nodes, collectively forming the network infrastructure. Typically, each node is equipped with a radio transmitter, a microcontroller, electronic components, and a power source. The energy supply can either be batteries or an integrated mechanism for harnessing energy from the surroundings. Despite variations in sensor node complexity affecting costs, miniaturization to the scale of a dust grain remains a technological challenge. Resource limitations such as space and funding constraints complicate the realization of WSNs. Diverse WSN configurations, ranging from simple star topologies to intricate wireless mesh networks, can be configured based on specific needs. Many experts anticipate WSNs to hold substantial importance in the future, and the corporate sector is increasingly demonstrating interest in this domain. A key hurdle in WSN design revolves around minimizing energy consumption to enhance its operational lifespan.

1.1 Mimo

MIMO technology involves the utilization of a pair of antennas at both the transmitting and receiving points. MIMO, an abbreviation for "multiple input, multiple outputs," operates by employing two antennas at both ends of the communication setup. This strategy aims to enhance communication quality by concurrently utilizing multiple signal pathways, which leads to error reduction, data transfer acceleration, and an augmentation in radio transmission capacity. By enabling radio frequency (RF) systems to effectively manage a higher volume of information simultaneously, MIMO contributes to establishing connections that are more stable and less congested. The configuration of a MIMO antenna setup is depicted in Figs. 1 and 2.

Fig. 1 MIMO antenna

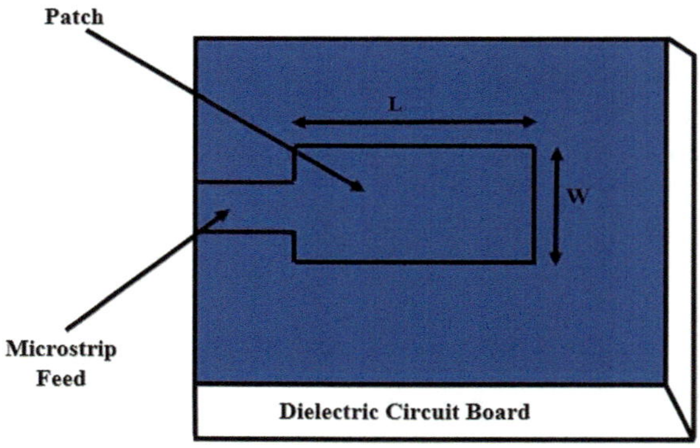

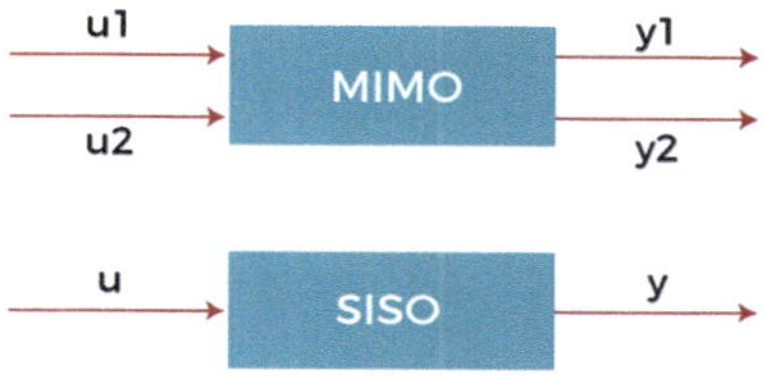

Fig. 2 Single input single output (SISO)

Input

The control system receives an external energy signal to generate the intended output. Within a control system, the input signifies the desired action that can trigger various responses within the system. There are two common types of inputs employed: SISO (Single Input Single Output) and MIMO (Multiple Input and Multiple Output). SISO implies that the system generates a single output in response to a single input, whereas MIMO generates multiple outputs in response to multiple inputs. In the context of a control system, the reference input, also known as the set point or desired value, holds significant importance. It acts as a target or objective for the system to attain, guiding its behavior and output in accordance with the goal. This reference input forms the foundation for error-controlled regulation, utilizing negative feedback to regulate and control errors within the system.

Output

In a control system, the output signifies the tangible outcome produced as a response to the input signal applied. Inputs are introduced to stimulate the system, and the outputs represent the processed outcomes of these inputs. The outputs mirror the behavior, performance, or sought-after results of the system, which can encompass the results of a portion of the process or the entirety of it.

2 Wireless Sensor Network (WSN)

In Fig. 3, the collaborative MIMO transmission, spanning from the source node S to the destination node D over a certain distance, comprises three distinct phases.

Local Data Exchange: In this initial phase, node S has the capability to collaborate with its neighboring nodes for the purpose of data exchange. The distance between cooperating nodes (denoted as "dm") is significantly smaller compared to the overall distance "D.“

1. **Cooperative MIMO Transmission**: Subsequently, the cooperative transmission stage takes place. Node S employs Space–Time Block Coding (STBC) techniques to encode and simultaneously transmit the data. This transmission can target either the destination node or multiple cooperative nodes, resembling traditional

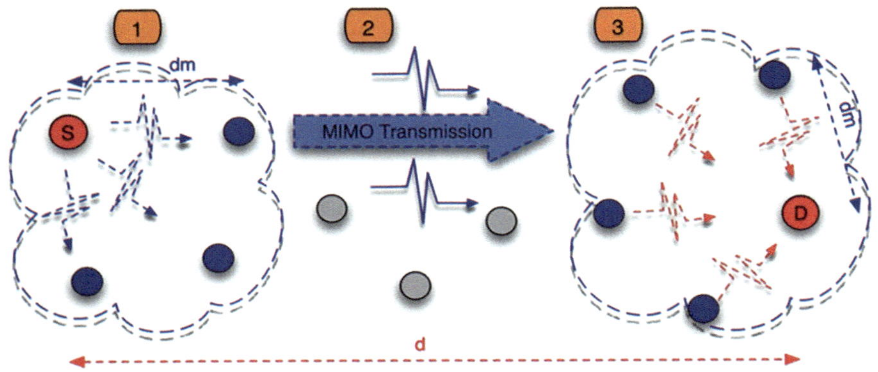

Fig. 3 MIMO transmission from source node S to destination node D

MIMO systems where each cooperative node effectively functions as one antenna within the MIMO system.

2. **Cooperative Reception**: On the reception end, the cooperative neighbors of the destination node D receive the MIMO-modulated symbols. They then sequentially retransmit these symbols to the destination node D, facilitating the joint combination of MIMO signal.

   However, in a cooperative MIMO system where nodes are physically separated, several challenges emerge. These include synchronization errors during cooperative transmission and the introduction of additional noise during cooperative reception. Consequently, the system's performance may suffer, leading to a higher Bit Error Rate (BER) compared to traditional MIMO systems with an equivalent number of transmission-reception antennas.
3. **Microstrip Antenna**: The dimensions of a microstrip antenna are typically chosen to achieve resonance at the operating frequency, resulting in a favorable real impedance that enhances the antenna's performance. To enable rectangular microstrip antennas to capture radio waves effectively. Additionally, the antenna's width plays a role in determining its impedance, as it affects the radiation produced by fringing fields that extend through its open sidewalls. This narrow bandwidth proves advantageous for various communication applications. It's worth noting that the frequency range of radiation may not always align with that of the antenna under examination. The configuration of the microstrip antenna is depicted in Fig. 4.

Wireless Sensor Networks (WSNs) comprise sensor nodes that communicate wirelessly and spontaneously among themselves. While WSNs have gained popularity, their security remains uncertain due to constraints such as low power, limited memory, computational capacity, battery life, and bandwidth. These node groups are employed for collecting environmental data, but if you intend to safeguard your data within a WSN, you may face various vulnerabilities. WSNs offer continuous monitoring capabilities for specific areas. When deployed in particular regions to oversee

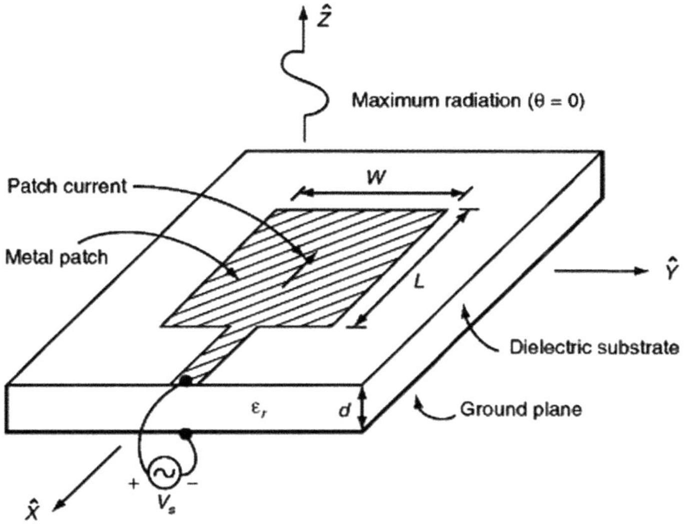

Fig. 4 Microstrip antenna

specific activities, it is referred to as area monitoring. For example, in military applications, sensors are utilized to detect the presence of adversaries nearby. Another instance of area monitoring is geo-fencing for monitoring gas or oil pipelines. Medical devices can be inserted inside the body, placed on the skin's surface, or positioned in close proximity to the user. These systems employ sensors found within the environment. The Wireless Anti-Terrorism System (WATS) can detect potential threats, including a nuclear "briefcase bomb" designed to detonate on the ground. WATS is equipped with wireless gamma and neutron detectors that communicate via a wired network [24]. Sensor-collected data undergoes a process known as "data fusion," a critical system component. MIMO allows simultaneous transmission and reception of multiple data signals over the same wireless channel, enabled by multipath propagation when orthogonal frequency division multiplexing encodes each channel, increasing memory capacity. It's important to note that while multipath is an interesting aspect, it's not the sole reason for MIMO's functionality, which aim to enhance single data signal performance [26]. In summary, MIMO should not be equated with smart antenna techniques. SDMA refers to a method for simultaneously communicating with individuals situated in different locations near the same base station. This technique employs directional antennas or smart antennas [29]. In such scenarios, the base station boasts "multiple receiving antennas" while "several individuals are situated at a distance (see Fig. 5).

Those utilizing MIMO can categorize their usage into three primary groups: Precoding, Spatial Multiplexing (SM), and Diversity Coding.

In the case of precoding, each antenna transmits a distinct signal with varying phases and gain weights to ensure optimal signal strength at the receiver. This method

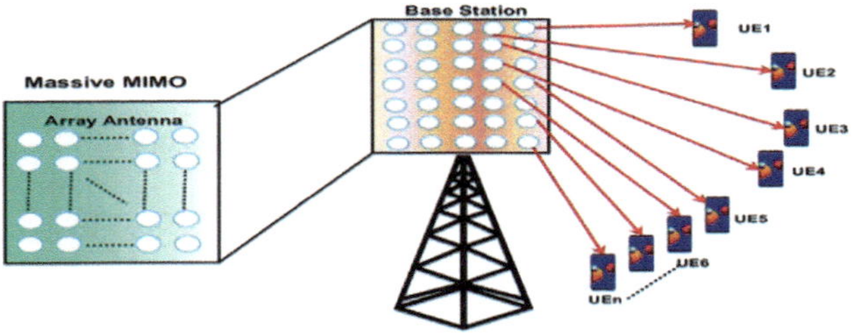

Fig. 5 MIMO Antenna Functions for WSN

is particularly useful in cellular networks with numerous paths rather than focused beams. Maintaining consistent signal strength across multiple antennas simultaneously can be challenging when the receiver possesses multiple antennas, making precoding with multiple streams a preferred choice.

1. Conversely, spatial multiplexing relies on significant spatial differences in signal appearance to divide streams into nearly parallel channels at the receiver.
2. Diversity coding comes into play when the transmitter lacks knowledge of the active channel. In this scenario, only a single stream is transmitted, and space–time coding is employed to enhance signal clarity. Most transmit antennas employ orthogonal coding, ensuring independence or near-independence for each of them. This results in a more robust signal, even when multiple antennas are transmitting and receiving signals, and each antenna may experience its own attenuation. However, determining precisely which specific input led to a particular output can be challenging due to system complexity and various influencing factors when channel knowledge is lacking [42].

3 Enhancing Network Bandwidth in Mining Operations

Enhancing network bandwidth within mining environments can be accomplished by strategically positioning an antenna, either at the central ceiling location or along one of the sidewalls. This placement optimizes the coupling efficiency of the antenna's radiated power. By fine-tuning the polarization and placement of the antenna, improvements can be achieved in both the near and far field communication ranges. The efficacy of the grating-based testing system was validated through systematic measurements conducted within an underground mining tunnel, yielding results that closely aligned with the initial assumptions. Additionally, the study assessed electromagnetic interference originating from the mining infrastructure, comparing it to signals received from antennas located at various positions. Notably, the study may not have accounted for certain cables, such as power and communication lines, due to their enduring and stable requirements.

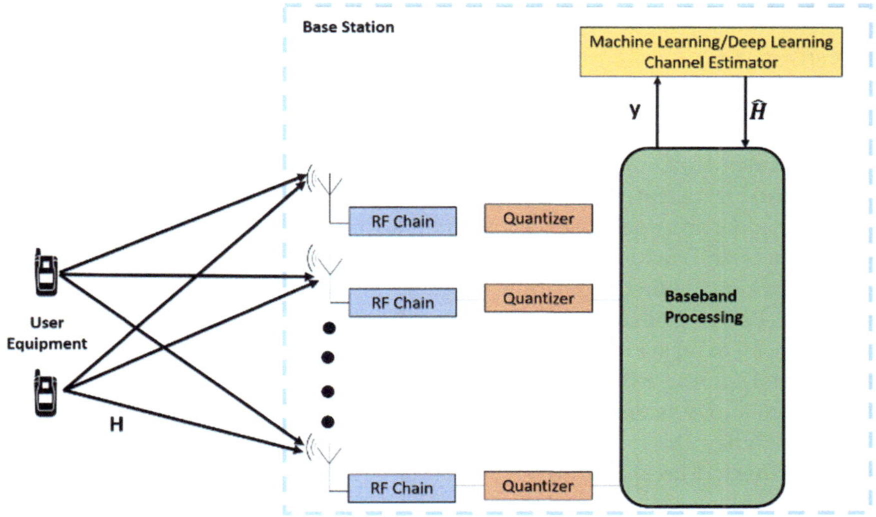

Fig. 6 MIMO and 5G massive systems

4 MIMO and 5G Massive Systems

See Fig. 6.

5 Ongoing Advancements in MIMO Technology

The evolution of MIMO technology remains a dynamic endeavor within the wireless industry, as it strives to accommodate an increasing number of antennas, networks, and devices in limited spatial confines. This continuous development has led to significant enhancements in MIMO technology over time. An illustrative example is the 5G network, which leverages an array of small antennas to deliver substantially greater bandwidth compared to its 3G and 4G predecessors and can support a significantly higher volume of connected devices.

5.1 Benefits of Massive MIMO for 5G Networks

Massive MIMO technology presents several advantages for 5G networks.

Enhanced Spectrum Efficiency: Massive MIMO achieves superior spectrum efficiency by directing focused, narrow beams to individual users. This can result in spectral efficiency more than ten times greater than the 4G/LTE MIMO systems currently in use.

- **Improved Energy Efficiency**: Due to its concentrated antenna array in a compact area, massive MIMO consumes less energy and requires less radiated power in large-scale MIMO systems.
- **Increased Data Rates**: Massive MIMO enhances data rates and network capacity by providing array gain and spatial multiplexing.
- **Precise User Tracking**: Massive MIMO improves the accuracy and reliability of user tracking by directing narrow signal beams toward users.
- **Lower Power Consumption**: Massive MIMO employs ultra-low power linear amplifiers, reducing power consumption substantially.
- **Reduced Fading Effects**: Massive MIMO exhibits resistance to fading, thanks to its numerous antennas at the receiver.
- **Low Latency**: Air interface latency is minimized with massive MIMO.
- **Robustness**: Massive MIMO systems are resilient to internal jamming and unintended interference, and they can withstand failures of individual antennas.
- **Enhanced Reliability**: A high number of antennas in massive MIMO provides increased diversity gain, enhancing connection reliability.
- **Improved Security**: The use of orthogonal mobile station channels in massive MIMO enhances physical security.
- **Simplified Linear Processing**: The abundance of base station antennas allows for simpler signal detection and precoding [42].

6 Microstrip Antennas

Microstrip antennas are typically fabricated using printed circuit boards and photolithography techniques, creating compact and visible antennas. These antennas consist of a metal foil layer with various patterns, referred to as "pattern antennas," on the surface, mounted on a base plate of metal foil. Often arranged in 2D patterns, these antennas employ microstrip transmission lines for radio wave transmission and reception, facilitating radio wave transmission. Their popularity arises from their slim and flat profile, making them suitable for various applications, including consumer products, aircraft, and missiles. Additionally, higher-power active antennas can be fashioned using active devices such as microwave integrated circuits. The patch antenna is the most prevalent type of microstrip antenna, and it can be grouped to form more complex antenna configurations. These antennas are crafted by etching antenna elements onto a metal track attached to an insulating dielectric substrate, with a metal layer added to the substrate's other side to create a ground plane. While microstrip antennas can assume various shapes, squares, and rectangles are common. Some patch antennas lack a dielectric base. Although microstrip antennas may be less rugged than other antenna types, they offer a broad frequency range and a low-profile design, making them suitable for external use on aircraft and spacecraft.

7 Advantages of Microstrip Antennas

Microstrip antennas offer several advantages, including their ease of design and manufacture due to their simple two-dimensional structure. They excel at higher frequencies, particularly in the UHF and above bands, thanks to their resonant frequency. Utilizing lithography techniques, multiple patches can be printed on a single substrate, enhancing efficiency and potential gain compared to single-patch designs (Figs. 7 and 8).

The process of ensuring that antennas meet or adhere to certain criteria is called "antenna engineering." This involves examining all antenna parameters, including S-parameters and return loss. Figure 9 shows the recorded frequency response of the S11 parameter of the MIMO antenna, which can also be seen in Figs. 10, 11, and 12 that show the return loss. S-parameter measurements are made to determine how

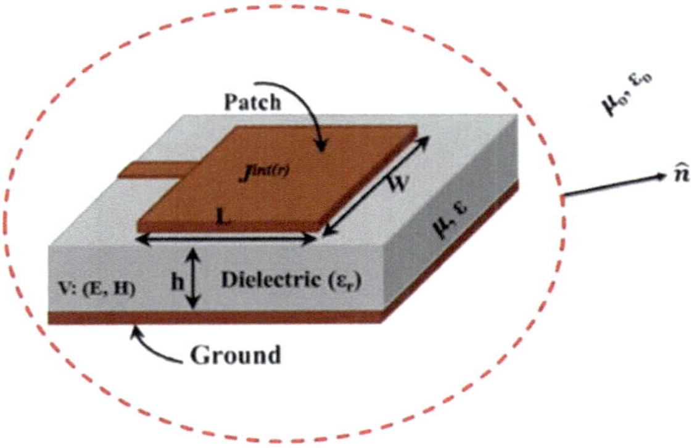

Fig. 7 Design and performance of microstrip antenna

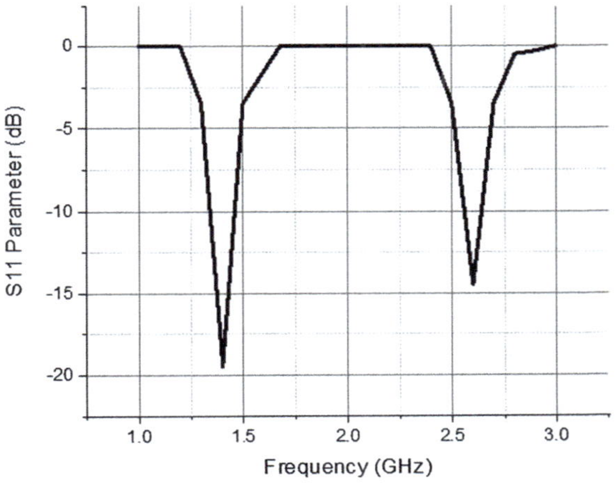

Fig. 8 Plotted frequency response of the MIMO antenna S11 parameter

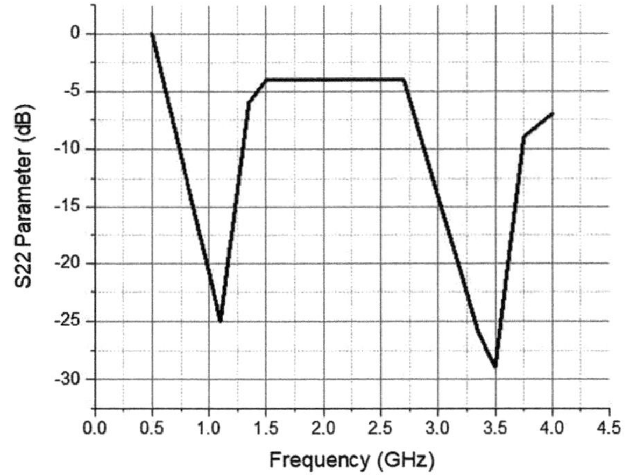

Fig. 9 Plotted MIMO antenna frequency S22 parameter

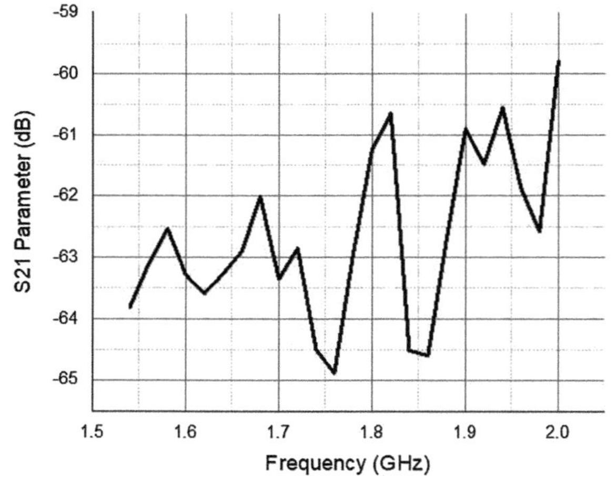

Fig. 10 Plotted frequency response of the MIMO Antenna S21 parameter

much power is lost and how isolated the prototype is. For MIMO antennas, the S-parameter can measure antenna performance. Lower return loss means the prototype is more efficient calculated at the frequency where the return loss is the lowest.

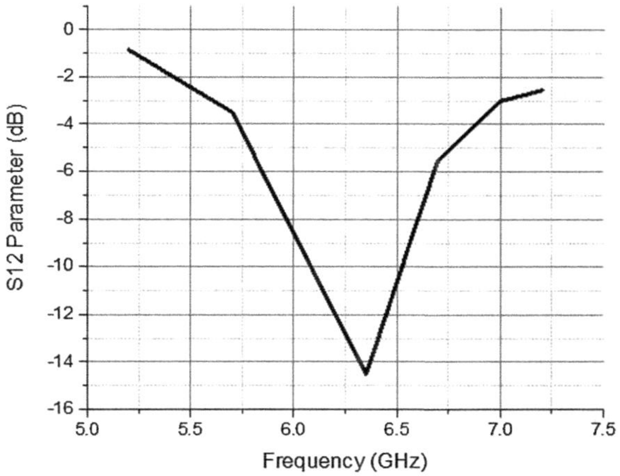

Fig. 11 Plotted MIMO antenna with frequency S12 parameter

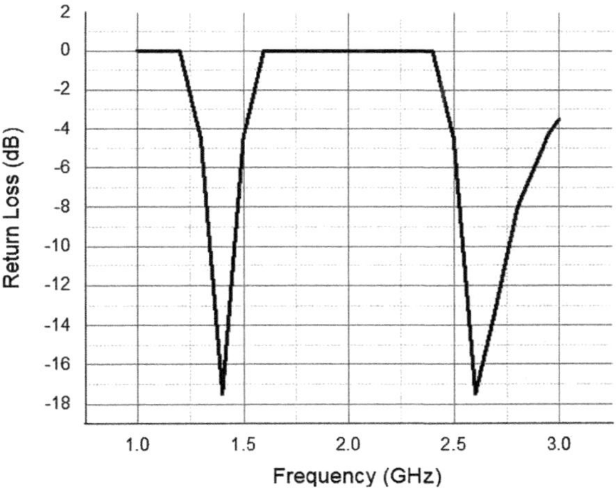

Fig. 12 Result

8 Conclusion

In the realm of wireless sensor networks, techniques known as "virtual MIMO" are harnessed to enhance communication efficiency, functioning in a manner akin to actual MIMO systems. Our research has delved into virtual MIMO for both fixed and variable data rates, revealing its potential to deliver energy and cost savings. The reliability of our wired testing system was validated through data measurements conducted within an underground mining tunnel, underscoring its suitability for application in coal mines. Wireless sensor networks have been gaining prominence due to their capacity to gather, process, and transmit data via numerous compact sensor nodes powered by small batteries. By implementing efficient modulation and transmission methods, these networks reduce the energy required for transmitting equivalent data volumes. If you seek a deeper understanding of MIMO and its energy dynamics, we recommend perusing this paper, which explores the energy consumption associated with various MIMO schemes and their distinctions. Among these schemes, Multiple Input and Multiple Output (MIMO) stands out as the most energy-efficient approach to conserve power.

References

1. Khurshid, A., Dong, J., Ahmad, M.S., Shi, R.: Optimized super-wideband MIMO antenna with high isolation for IoT applications. Micromachines (2022). https://www.ncbi.nlm.nih.gov
2. Mudda, S., Gayathri, K.M., Mallikarjun, M.: Wide-band frequency tunable antenna for 4G, 5G/Sub 6 GHz portable devices and MIMO applications. Progr. Electromagn. (2022). go.gale.com
3. Mohanty, A., Behera, B.R.: Characteristics mode analysis: a review of its concepts, recent trends, state-of-the-art developments, and its interpretation with a fractal UWB MIMO antenna. Progr. Electromagn. Res. (2021). https://www.researchgate.net
4. Ikram, M., Nguyen-Trong, N.: IEEE Antennas and …, (2020). https://espace.library.uq.edu.au
5. Kraus, J.D., Marhefka, R.J., Khan, A.S.: Antennas and Wave Propagation, 4th edn. McGraw Hill, New Delhi (2010)
6. Department of Communications and Electronics, School of Electrical and Computer Engineering, Shiraz University, Shiraz, Iran. Received 14 August 2020, Accepted 13 October 2020, Available online 18 October 2020, Version of Record 22 October 2020
7. Haupt, R.L.: Antenna Arrays: A Computational Approach. Wiley (2010)
8. See, C.H., Abd-Alhameed, R.A., Abidin, Z.Z.: Wideband printed MIMO/diversity monopole antenna for WiFi/WiMAX applications. IEEE Trans. Antennas Propag. (2012). https://cir.nii.ac.jp
9. Jobs, M.: Design and performance of diversity based wireless interfaces for sensor network nodes. Dissertation, Institutionen för teknikvetenskaper (2013)
10. Yoshida, S., Hasegawa, N., Kawasaki, S.: Experimental demonstration of microwave power transmission and wireless communication within a prototype reusable spacecraft. IEEE Microw. Wirel. Compon. Lett. **25**(8), 556–558 (2015)
11. Wang, R., Wang, B.Z., Huang, W.Y., Ding, X.: Compact reconfigurable antenna with an omnidirectional pattern and four directional patterns for wireless sensor systems. Sensors (2016). https://www.scienceopen.com
12. Sharawi, M.S.: Current misuses and future prospects for printed multiple-input, multiple-output antenna systems [wireless corner]. IEEE Antennas Propag. Mag. (2017). https://ui.adsabs.harvard.edu
13. Galajda, P., Svecova, M., Drutarovsky, M., Slovak, S., Pecovsky, M., Sokol, M., Kocur, D.: Wireless UWB sensor system for robot gripper monitoring in non-cooperative environments. Recent Adv. Intell. Eng. Springer (2020)
14. Janhunen, J., Mikhaylov, K., Petäjäjärvi, J., Sonkki, M.: Wireless energy transfer powered wireless sensor node for green IoT: design, implementation and evaluation. Sensors (Basel) (2019). https://www.ncbi.nlm.nih.gov
15. Goswami, P.K., Goswami, G.: Compact corner truncated fractal slot antenna for cognitive radio sensor network. IETE J. Res. (2023). https://www.ingentaconnect.com
16. Palanisamy, P., Subramani, M.: An integrated antenna for cognitive radio wireless sensor networks and HD video transmission applications. Int. J. RF Microw. Comput.-Aided Eng. (2021). https://search.ebscohost.com
17. Song, R., et al.: A graphene-assembled film-based MIMO antenna array with high isolation for 5G wireless communication. Appl. Sci. **11**(5), 2382 (2021)

18. Fu, Z., Shen, W.: Eight-element self-decoupled MIMO antenna design for 5G smartphones. Int. J. RF Microw. Comput.-Aided Eng. (2021). https://search.ebscohost.com
19. Khan, M.K., Feng, Q., Zheng, Z.: Experimental investigation and design of UWB MIMO antenna with enhanced isolation (2020). https://www.researchgate.net
20. Mohanty, A., Behera, B.R.: Design of a 3-port compact MIMO antenna based on characteristics mode analysis approach (2021). https://www.academia.edu
21. Gupta, A., Kansal, A., Chawla, P.: Design of a wearable MIMO antenna deployed with an inverted U-shaped ground stub for diversity performance enhancement. Int. J. Microw. Wirel. Technol. (2021). https://www.proquest.com
22. Sanmuga Sundaram, R., Natarajan, S., Rajkumar, R.: A compact MIMO Antenna with electromagnetic bandgap structure for isolation enhancement (2020). https://www.researchgate.net
23. Eslami, A., Nourinia, J., Ghobadi, C.: Four-element MIMO antenna for X-band applications. Int. J. Microw. Wirel. Technol. (2021). https://search.ebscohost.com

A Novel Logistic Regression-Based Fire Detection Model Using IoT in Underground Coal Mines

Chaitanya Thuppari, Srikanth Jannu, and Damodar Reddy Edla

Abstract The environment of underground coal mines (UCMs) is vulnerable to many environmental problems and consequential endangerment. Among those problems, mine fire is a liable threat that causes the loss of lives of mine workers and other valuable infrastructure and resources in the mines. Therefore, continuous monitoring is very important for early detection of mine fire in UCMs. Internet of Things (IoT), nowadays, is widely used for continuous monitoring of the environment of UCMs. For the purpose of monitoring, deploy sensor nodes in the region of UCMs to sense the local information of the environment and transfer that information to the sink for further process. As the gathered information is indistinct, it is necessary to consider the information for taking precautions. Therefore, we propose a method to predict the occurrences of mine fire in UCMs by using logistic regression model. Then, we applied active learning (AL) and semi-supervised learning (SSL) methods on the dataset sequentially to train the model by dividing the dataset into training and testing datasets. The training dataset is used to train a model. On the other hand, the testing dataset is used to evaluate the trained model. The overall process is implemented at the sink instead of control stations. In case of any hazardous situation, the sink takes immediate and necessary actions based on the sensed data rather than the control station. The trained model is simulated using WEKA tool for the data of continuous monitoring on different hazard conditions. This model is more reliable and works effectively when compared to offline monitoring system in any kind of hazards situations in UCMs. The simulated results show the accuracy of the proposed

C. Thuppari (✉) · D. R. Edla
National Institute of Technology, Farmagudi, Ponda 403 401, Goa, India
e-mail: chaitu786chaitanya@gmail.com

D. R. Edla
e-mail: dr.reddy@nitgoa.ac.in

S. Jannu
Vaagdevi Engineering College, Warangal 506 005, Telangana, India
e-mail: j.srikanth@live.com

G. Fortino et al. (eds.), *Proceedings of Third International Conference on Computing and Communication Networks*, Lecture Notes in Networks and Systems 917,
https://doi.org/10.1007/978-981-97-0892-5_11

method as 98% whereas the accuracy of linear regression and naive Bayes has been shown as 93% and 96%, respectively, that shows our proposed method outperforms over the existing techniques.

Keywords Coal mines · Logistic regression · Internet of things · Fire prediction

1 Introduction

1.1 Background

The hazardous environment of underground coal mines (UCMs) is very danger such as roof fall, explosions, and mine fires, which are a serious problem that harm people's lives, and damage to mining machinery and infrastructure [1–3]. Among these problems, mine fire has become the major concern for the coal miners as well as the researchers. There are many factors that cause mine fires like mine exploding, mine explosions, friction sparks during welding, drilling, cutting processes, electric sparks, and unpredictable combustion. However, the spontaneous combustion accounts for 75% to 90% of coal mine fires. The auto-oxidation process of coal and other carbon materials that results in the production of toxic and flammable gases such as carbon dioxide (CO_2), carbon monoxide (CO), and heat that leads to self-combustion. During combustion process, various hazardous gases are produced due to condensation of O_2 which increases heat in the UCM environment. As a result, there is a high chance of occurrence of fire in the UCM. However, prediction of fire in coal mines is difficult as it is monitored manually. Therefore, the early and regular maintenance of the combination can prevent the mines from burning and safe guard the miners. There are numerous works which have been developed on UCM monitoring for fire prediction or fire detection.

In [4], an interval type-2 fuzzy logic controller (IT2FLC) has been developed to control the environment. In this method, the sensor nodes collect the local data such as temperature, humidity, CO, CO_2, and other gases and send it to the sink. Then, based on the received data, IT2FLC takes the decision at the sink instead of sending it to the control center which saves time. Ashok et al. [5] proposed an SVM model to analyze the effect of water absorption on iron ore samples. However, the method has not addressed the fire prediction. Muduli et al. [6] have developed a data mining method to monitor and forecast the mine fire hazards in UCMs. In [7], the authors have stated that all dynamic circumstances cause damage to underground excavations that are known as rock busters. Mine fire due to rock busters in UCMs is complex as that cannot be predicted easily. The authors used machine learning algorithm models to analyze the risk of rock explosion. A uniform manifold approximation model has been proposed to forecast in UCMs [8].

1.2 Contributions

In this paper, an IoT-based fire detection/prediction model for UCM has been developed using a logistic regression model (LRM). Initially, a number of sensor nodes are deployed to form IoT network in the underground coal mine. Those sensor nodes in IoTUSN collect various local data like temperature, CO, CO_2, and O_2. Then, that will be sent to the sink for fire prediction [7]. Finally, the fire prediction model uses the above sensed data as input to predict mine fires. Our focal contributions are mentioned as follows:

- Random deployment of nodes to sense the local data form in the UCM.
- Design a coal mine fire early prediction model based on logistic regression algorithm.
- Integration of AL and SSL for the uncertainty and unlabeled samples.
- Analysis of simulation results of various performance metrics.

The organization of the remaining paper is given as follows: in Sect. 2, we discuss LRM. The detailed proposed work is explained in Sect. 3. Then the simulation results are analyzed in Sect. 4. Finally, Sect. 5 concludes the paper.

2 Logistic Regression Model

Here, we discuss about the logistic regression model (LRM) with respect to the UCM [9]. For example, the coal dataset contains N samples that include few labeled samples as NF (non-fire samples) and unlabeled samples F (fire samples) where $N = NF + F$. Moreover, assume that the data contains p attributes and β is the coefficient between NF and F that represents $\beta = (\beta_0, \beta_1, \ldots, \beta_p)$. Then, the individual sample is represented as $(y_i, c_i, x_i)_i^n$ where y_i is the data output and $x_i = (x_1, x_2 \ldots x_p)$ i is the representation of the sample data and the c_i denotes that whether the data is labeled or unlabeled. Then, the LRM can be defined as follows:

$$p(y = 1|x_i) = \frac{e^{(x_i\beta)}}{1 + e^{(x_i\beta)}} \tag{1}$$

The logarithmic likelihood of LRM can be represented as

$$\ln(\beta) = -\sum_{i=1}^{N} y_i \times log\left(\frac{e^{(x_i\beta)}}{1 + e^{(x_i\beta)}}\right) + (1 - y_i) \times log\left(\frac{e^{(x_i\beta)}}{1 + e^{(x_i\beta)}}\right) \tag{2}$$

2.1 Uncertainty Sampling

Here, we use uncertainty sampling to choose the samples from unlabeled dataset in active learning (AL). In the LRM, the probability of a sample that is close to the decision boundary can be considered as the most uncertain sample in AL [9]. Therefore, the modeled can be expressed as

$$min \sum_{i=1}^{NF} l(x_i^T, y_i, \beta) + \sum_{j=1}^{F} v_j \times l(x_j^T, y_j, \beta) + f_{AL}(v_j, \alpha) + \lambda \times P(\beta) \quad (3)$$

where v represents the weight of the samples, f_{AL} is the objective function to generate the weight value, and α is the control parameter. The $F_{AL(v,\alpha)}$ can be defined as

$$v_i = \begin{cases} 1 & (0.5 - \alpha) < l(x_j^T, y_i, \beta) < (0.5 + \alpha) \\ 0 & Otherwise \end{cases} \quad (4)$$

2.2 Self-training

In the LRM, the probability of a sample is close to either 0 or 1 can be assumed as the high-confidence sample in semi-supervised learning. Then, the self-training model is expressed as

$$min \sum_{i=1}^{NF} l(x_i^T, y_i, \beta) + \sum_{j=1}^{F} w_j \times l(x_j^T, y_j, \beta) + f_{SSL}(w_j, \gamma) + \lambda \times P(\beta) \quad (5)$$

where w describes the weight samples in SSL, $f_{SSL}(w, \gamma)$ is the objective function to select samples, and γ represents the control parameter. The $f_{SSL}(w, \gamma)$ can be represented as

$$w_j = \begin{cases} 0 & \gamma < l(x_j^T, y_i, \beta) < (1 - \gamma) \\ 1 & Otherwise \end{cases} \quad (6)$$

2.3 Integration of Semi-supervised Learning and Active Learning

In this paper, we integrate SSL and AL using logistic regression for the better classification of samples. Therefore, the SSL selects high-confidence unlabeled samples that restrict the classifier to be near-sighted. Moreover, AL selects the low-confidence data samples [9]. Therefore, the model is represented as

$$min \sum_{i=1}^{NF} wl(x_i^T, y_i, \beta) + \sum_{j=1}^{F} (w_j \oplus v_j) l(x_j^T, y_j, \beta) + f_{AL}(v_j, \alpha) + f_{SSL}(w_j, \gamma) + \lambda P(\beta) \quad (7)$$

From the above model, the uncertainty of classified samples, the incorrectly classified data samples must have a chance to be revised in further iterations. Therefore, the updation process works as follows:

- If the sample is incorrectly classified by SSL, then the sample will be placed in unlabeled dataset and it weighs to be re-classified.
- If it is incorrectly classified by AL, the label of the data sample can be changed. Moreover, it will be sent to the labeled dataset.

3 Proposed Work

In this section, we deploy few sensor nodes in the mining area such as walls and roofs of the random locations. These sensor nodes along with other IoT devices form IoTUSN to communicate each other. The sensor nodes sense the parameters of UCM such as temperature, carbon monoxide, carbon dioxide, and oxygen and send to the sink for further processing. The detailed LRM is described as follows:
Step 1: Initially, the labeled dataset of underground coal mine, i.e., $[T, CO, CO_2, O_2, CLASS]$ is used to train the LRM.
Step 2: The LRM is used to label the unlabeled data and high-confidence data based on Equation (7).
Step 3: Then, the sample data is placed in labeled dataset if it is selected correctly by SSL. Otherwise, it will be placed in unlabeled data. Whereas the sample is selected correctly by AL, it will be placed in labeled dataset. Otherwise, the label will be changed and send that sample directly to labeled dataset.
Step 4: Then, identify the unlabeled dataset based on Step 3.
Step 5: Run this process until all the unlabeled data classify correctly or run the process for certain number of rounds.

Then, the sink transmits this collected information to the monitoring stations. However, the sink endures the decisions and provides preventive measures in emergencies to save the time in emergency cases. Therefore, some judgments or estimates must be combined on the observation data to improve the accuracy and reliability of the IoTUSN.

During the combustion process, O_2 is consumed and heat is released and various flammable gases such as CO and CO_2 are produced that rise the ambient temperature (T) of the UCM. More than normal consumption of T, CO and CO_2 as well as O_2 production is a clear indication of ground fire. However, fire detection at the exit using the above fire detector is done offline, and it cannot solve the problem of pressure. Therefore, we propose a novel logistic regression-based fire prediction algorithm where we use feature matrices T, CO, CO_2, and O_2 as input parameters of UCM.

Algorithm 1 Algorithm for Logistic Regression Model

Require: The labeled Coalmine dataset NF and F
Ensure: Model coefficient β
Initialize the weights w, v, pseudo-labeled dataset PD
Set training dataset $TD = NF$
while (!Not convergent) **do**
  Construct Logistic classifier based on TD
  Label the unlabeled data as F
  Build the most confidence dataset as N_{SSL} and most uncertain dataset as N_{AL}
  if $PD \neq \phi$ **then**
    Evaluate the pseudo-labeled dataset as PD
    Update PD
  end if
  Update training data TD based on f_{SSL}, f_{AL} and PD
  Update unlabeled data F and pseudo-labeled data PD
  Initialize f_{SSL}, f_{AL}, w and v
end while
Return coefficient β

Algorithm 1 provides a detailed logistic regression-based fire prediction algorithm for UCMs.

4 Simulation Results

For the simulation purpose, the proposed algorithm is simulated using WEKA tool [10]. When entering the underground monitoring data in the coal mine into the WEKA tool, the temperature was used as input data CO, CO_2, and O_2. Fire, as shown in Table 1, is considered as no fire (NF) and fire (F) plans that is shown in confusion matrix. The dataset consists of 1065 instances and those are classified into training and testing datasets. Out of these 1065 instances, 1054 are classified correctly and the remaining are wrongly classified. From Table 1, we calculate the following performance measures:

Precision shows the number of instances positively classified from the dataset that are predicted positively and it is shown below:

$$Precision = \frac{TP}{TP + FP} \tag{8}$$

Recall describes the correct classification to particular class. In other words,

$$Recall = \frac{TP}{TP + FN} \tag{9}$$

Table 1 Confusion matrix

	False Positive (FP)	False Negative (FN)
True Positive (TP)	686	11
True Negative (TN)	8	368

Table 2 Comparison of reliability and accuracy over existing techniques

	Precision	Recall	F-measure	Accuracy
Linear regression	0.651	0.955	0.774	0.93
Naive Bayes	0.961	0.95	0.95	0.96
Proposed	0.985	0.97	0.983	0.98

F-Measure: F-Measure is a metric derived from the Precision and Recall. In other words,

$$F - Measure = \frac{2 \times Precision \times Recall}{Precision + Recall} \tag{10}$$

Accuracy: It is a measure of correct prediction of each individual class from the total classes. In other words,

$$Accuracy = \frac{TP + TN}{TP + TN + FP + FN} \tag{11}$$

Finally, these metrics are calculated from the confusion matrix that is mentioned in Table 1 and compared with the existing machine learning models that are presented in Table 2 and Fig. 1.

The simulation results indicate that the proposed algorithm is reliable and accurate than the existing machine learning models in terms of the above performance measures. However, the limitations of the proposed model are that the algorithm is restricted to use exclusively for early prediction of mine fire in the UCM.

5 Conclusion

In this study, a prediction technique for IoTUSN called a logistic regression model was developed at the sink. This model is the integration of semi-supervised learning and active learning that was used to test/evaluate the data and take the decision, respectively. The WEKA tool is used to simulate the proposed LRM-based fire detection algorithm. The results demonstrate that the fire detection model based on the LRM is accurate than the existing models. The accuracy of our model is 98% whereas the accuracies of existing methods, namely, naive Bayes and linear regression are 96% and 93%, respectively.

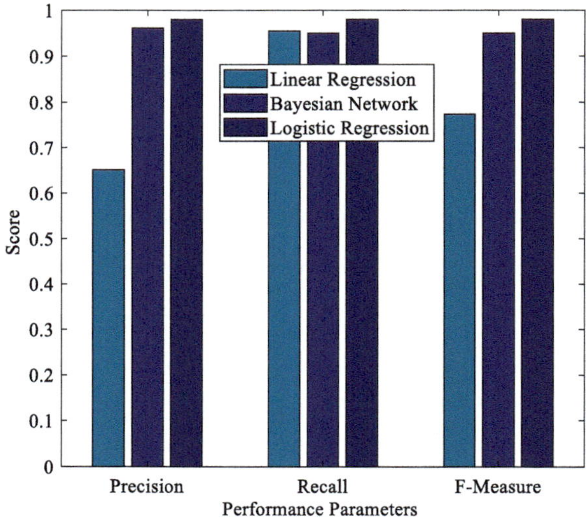

Fig. 1 Comparison of performance metrics over existing methods

References

1. Tripathy, D.P., Ala, C.K.: Identification of safety hazards in Indian underground coal mines. J. Sustain. Mining **17**(4), 175–183 (2018)
2. Wang, T., et al.: A spatio-temporal temperature-based thresholding algorithm for underground coal fire detection with satellite thermal infrared and radar remote sensing. Int. J. Appl. Earth Observat. Geoinf. **110**, 102805 (2022)
3. Wang, T., et al.: Research progress and visualization of underground coal fire detection methods. Environ. Sci. Pollut. Res. 1–20 (2023)
4. Basu, S., Pramanik, S., Dey, S., Panigrahi, G., Jana, D.K.: Fire monitoring in coal mines using wireless underground sensor network and interval type-2 fuzzy logic controller. Int. J. Coal Sci. Technol. **6**, 274–285 (2019)
5. Patel, A.K., Chatterjee, S., Gorai, A.K.: Effect on the performance of a support vector machine based machine vision system with dry and wet ore sample images in classification and grade prediction. Pattern Recognit Image Anal. **29**, 309–324 (2019)
6. Muduli, L., Mishra, D.P., Jana, P.K.: Wireless sensor network based underground coal mine environmental monitoring using machine learning approach. In: Proceedings of the 11th International Mine Ventilation Congress, pp. 776–786. Springer Singapore (2019)
7. Wojtecki, Ł, Iwaszenko, S., Apel, D.P., Bukowska, M., Makowka, J.: Use of machine learning algorithms to assess the state of rockburst hazard in underground coal mine openings. J. Rock Mech. Geotech. Engin. **14**(3), 703–713 (2022)
8. Kumari, K., et al.: UMAP and LSTM based fire status and explosibility prediction for sealed-off area in underground coal mine. Process Saf. Environ. Prot. **146**, 837–852 (2021)
9. Chai, H., Liang, Y., Wang, S., Shen, H.W.: A novel logistic regression model combining semi-supervised learning and active learning for disease classification. Sci. Rep. **8**(1), 13009 (2018)
10. Weka 3: Data mining with open source machine learning software. https://www.cs.waikato.ac.nz/ml/weka/

An MQTT IoT Intrusion Detection System Using Deep-Learning

Greeshma Andrew, M. P. Deepika, and Soumia Chandran

Abstract Nowadays the devices connected to an Internet of Things (IoT) network is enormously increasing day by day. IoT devices widely using Message Queuing Telemetry Transport (MQTT) protocol for their communication. Because of the heterogeneous nature and lack of security in IoT devices manufacturing, security systems against MQTT traffic are essential. In this paper, we proposed a zero biased Convolutional Neural Network (CNN) for the detection of intrusion. By removing the bias term, it reduces the computational complexity and it would be beneficial to deployed as Intrusion Detection System (IDS) for resource constrained IoT devices. The performance of binary classification of the model is compared with other Deep Neural Network (DNN) with the help of two abstract level features such as bidirectional flow and unidirectional flow from a publicly available dataset MQTT-IoT-IDS2020. The proposed model achieves superior results to the other model with the F1 score of 0.99 for bidirectional and unidirectional flow.

Keywords IoT · MQTT · Security

1 Introduction

The Internet of Things (IoT) is the connectivity of real-world objects or things with the help of the Internet without human intervention. For the efficient communication of devices in the IoT platform several protocols are designed [1]. Because of low bandwidth, low memory, and low packet loss, the Message Queuing Telemetry

G. Andrew (✉) · M. P. Deepika
Cochin University of Science and Technology, Kochi, India
e-mail: greeshmaandrew@cusat.ac.in

M. P. Deepika
e-mail: deepika.mp@cusat.ac.in

S. Chandran
Government Engineering College, Thrissur, Kerala, India
e-mail: soumiacm@gectcr.ac.in

G. Fortino et al. (eds.), *Proceedings of Third International Conference on Computing and Communication Networks*, Lecture Notes in Networks and Systems 917,
https://doi.org/10.1007/978-981-97-0892-5_12

Transport (MQTT) protocol is tremendously used in different domains such as smart homes, smart agriculture, smart healthcare, and industrial applications [2].

IoT platform produces an abundance of data that needs to be transmitted safely and securely. So IoT security has to improve to protect the data from security threats and attacks such as Denial of Service (DoS), Malware, Data privacy, and breaches [3].

Intrusion Detection System (IDS) based on Machine Learning (ML) can effectively find and classify anomalous behavior in IoT [4]. But ML does not perform well with large amounts of data, this issue emerges as a new trend in the research area. When dealing with large amounts of data, Deep Learning (DL) is the best choice to train the models [5]. Zero bias Deep Neural Network (DNN) has proved high accuracy for identification of IoT devices [6].

An MQTT IoT protocol Intrusion Detection System (IDS) is proposed in this paper and it is based on zero bias Convolutional Neural Network (CNN) that makes the system more robust. The model is trained by using two abstract level features such as bidirectional and unidirectional flow from MQTT-IoT-IDS2020 [7] public dataset. The performance metric of our proposed model shows 99.60% accuracy for bidirectional and 99.58% accuracy for unidirectional flow.

2 Related Works

In [8] paper, authors proposed a network IDS based on deep learning by using the MQTT-IoT-IDS2020 dataset. IDS performance is evaluated by standard performance metrics such as weighted average, accuracy, precision, and F1-score. accuracy was 97.09% and F1-score was 98.33% were obtained.

The paper [9] presents Generative Adversarial Network—Auto Encoder (GAN-AE) IDS for MQTT networks. In terms of accuracy and F1-score of 97% GAN-AE model performed better than Isolation Forest (IF), One-Class SVM (OCSVM), and AutoEncoder(AE) on their custom and public MQTT dataset.

In [10], the authors presented a DNN-based model for binary classification and multi-classification of attack and normal data based on public MQTT datasets. Their proposed DNN model showed the highest accuracy against other DL models. An ensemble learning (EL) based IDS is proposed in the paper [11]. They used bagging, boosting, and stacking as ensemble learning methods. To train the proposed model they generated a binary dataset that is balanced from the MQTT dataset. Their EL-based IDS showed an increase in accuracy as well as an F1-score equal to 95%.

3 Methodology

In this section, we discuss the steps and methods that we followed to propose our IDS based on deep neural networks.

Table 1 Before data preprocessing

	Bi-flow	Uni-flow
Normal	188,378	376,186
Attack	71,001	119,339

Table 2 After data preprocessing

	Bi-flow	Uni-flow
Normal	188,378	376,186
Attack	188,378	376,186

3.1 Dataset and Preprocessing

For the implementation and experimentation of IDS which we proposed, we used the dataset MQTT-IoT-IDS2020 [7]. This dataset contains three flow-level features such as Packet-flow, Unidirectional flow (Uni-flow), and Bidirectional flow (Bi-flow) features. Each flow level feature contains normal and four attack scenarios. In total five scenarios, such as a normal operation and four attacks include MQTT brute-force attack, Sparta SSH brute-force, UDP scan, and aggressive scan. In our experiment, we mainly focus on two abstract flow-level features: Uni-flow and bi-flow.

The above-mentioned five recorded scenarios of each flow level feature are combined into one CSV file with the help of Python script. For the attack/normal scenarios CSV file contains a binary label attribute as our classification is a binary classification. To test the efficiency of the proposed DNN technique we have gone through several data preprocessing steps. Data preprocessing involves data cleaning, data transformation, and feature scaling that improves the performance of the model. We removed the attributes protocol, source and destination IP address, source port, and destination port address for both flow-level features. Bi-flow feature contains 71,001 attack samples and 188,378 normal samples and in the case of uni-flow feature, it contains 119,339 attack samples and 376,186 normal samples. SMOTE technique [12] is used to balance this imbalanced dataset.

To normalize the data, in this work, we used Pythons standard scalar function. Tables 1 and 2 tell us the number of normal samples and attack samples before and after data preprocessing, respectively.

3.2 Proposed Model

Our proposed model is a zero-biased convolutional neural network (CNN). Zero-biased CNN is another form of the traditional convolutional neural network [6, 13], here the bias is set to zero. Bias is added along with weighted inputs to each neuron to shift the activation function. In the case of Internet of Things (IoT) security, it requires less memory and computational complexity. Avoiding bias terms reduces

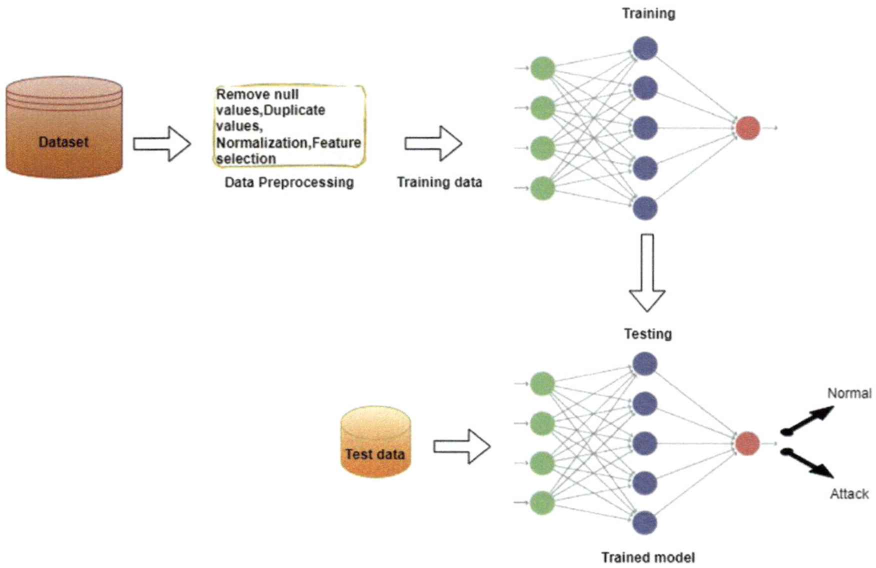

Fig. 1 Architecture of our model

the number of parameters thus leading to increase in the energy efficiency of IoT devices that often operate on energy-constrained resources.

The proposed model comprises a convolutional input layer, followed by a flattening operation, a hidden layer that is fully connected and an output layer. Figure 1 shows the architecture of our model. Rectified Linear Unit (ReLU) as activation function in input layers as well as in hidden layer, whereas output layer with sigmoid as activation function as this problem is a binary classification. Binary cross entropy is used as a loss function and Adaptive Moment Estimation (Adam) is used as an optimizer.

4 Experimental Setup

To implement this experiment we chose Google Colaboratory as an Integrated Development Environment (IDE) and Python 3 was our programming language. The other libraries used to implement this experiment were Numpy, Pandas, Keras, Tensorflow, and Scikit-learn.

4.1 Parameters Setup

There are many factors that affect the training process and the result of a DNN algorithm such as the selection of activation function, optimizer, and learning rate. We experimented with our model with all activation functions except for the output layer, which was sigmoid, ReLU, and tanh. Among them, ReLU showed better and faster training of the model. Like that, we experimented our model with Adam and Nesterov-accelerated Adaptive Moment Estimation (Nadam) optimizer. Adam showed better results within fewer epochs.

4.2 Evaluation Metrics

To evaluate how well the model is performed, some statistical ground truth values are required. In our work, we used metrics for evaluation such as precision, recall, F1-score, and accuracy.

5 Performance Analysis and Discussion

The results of the proposed deep learning method are going to be discussed in this section. The results of bidirectional flow and unidirectional flow show that our model performs better in detecting attacks. Table 3 shows the confusion matrix for bidirectional flow as well as the unidirectional flow of the proposed model. It obtained True Positive (TP) are 37,710, a False Positive (FP) are 5 and True Negative (TN) are 37,327 and False Negative (FN) are 310 for bidirectional flow (Bi-flow). And in the case of unidirectional flow (Uni-flow), it shows 75,289 TP, 14 FP, 74,557 TN, and 615 FN. Due to the stochastic nature, each time model learns differently. To determine the model's efficacy we used precision, recall, F1-score, and accuracy as metrics for evaluation.

Table 4 depicts the results of above-mentioned evaluation metrics.

We compare the performance of our work with a related work [10] which includes binary classification of bidirectional and unidirectional flow features. Tables 5 and 6 show the comparison of the bidirectional and unidirectional flow of binary classification. When comparing Table 5 metrics, the proposed model performs better with an accuracy of 99.84 and precision of 99.99% compared to 99.75 and 99.81% in the

Table 3 TP, TN, FP, AND FN for unidirectional and bidirectional flow

	TP	TN	FP	FN
Bi-flow	37,710	37,327	5	310
Uni-flow	75,289	74,557	14	615

Table 4 Results of unidirectional and bidirectional flow

	Accuracy	Precision	Recall	F1-score
Bi-flow	99.60	99.99	99.14	99.56
Uni-flow	99.58	99.94	99.18	99.57

Table 5 Performance comparison for bidirectional flow

Models	Accuracy	Precision	Recall	F1-score
[10]	99.75	99.81	88.69	93.95
Proposed	99.84	99.99	99.14	99.56

Table 6 Performance comparison for unidirectional flow

Models	Accuracy	Precision	Recall	F1-score
[10]	99.14	98.39	98.98	98.68
Proposed	99.58	99.94	99.18	99.57

model [10] respectively. On the other hand, when we consider recall (minimizing false negatives), our model performs better with a recall of 99.14% compared to 88.69% in the Proposed Model. If we consider overall performance, the F1-score provides a balanced measure between precision and recall. In this case, the Proposed Model has a higher F1-score of 99.56% compared to 93.95% in the Model. Therefore, based on the F1-score, the Proposed Model appears to perform better overall. When comparing the performance metrics of models in Table 6 the proposed model performs better for metrics such as accuracy and precision. The accuracy of our model is 99.58, precision 99.94, Recall 99.18, and F1-score 99.57.

6 Conclusion

The model that we proposed in this paper is a zero-biased CNN model for the detection of intrusion on MQTT IoT traffic for binary classification. To evaluate the model's performance we used the abstract level flow features, bidirectional and unidirectional from the MQTT-IoT-IDS2020 dataset. From this experiment, we have realized that the model proposed possesses high precision, recall, and F1-score. Overall the model we proposed performs better for intrusion detection systems. Our future work intends to develop an application for intrusion detection systems that can embedded on edge devices.

References

1. Al-Masri, E., Kalyanam, K.R., Batts, J., Kim, J., Singh, S., Vo, T., Yan, C.: Investigating messaging protocols for the Internet of Things (IoT). IEEE Access **8**, 94880–94911 (2020)
2. Soni, D., Makwana, A.: A survey on mqtt: a protocol of Internet of Things (IoT). In: International Conference on Telecommunication, Power Analysis and Computing Techniques (ICTPACT-2017), vol. 20, pp. 173–177 (2017)
3. Deogirikar, J., Vidhate, A.: Security attacks in IoT: a survey. In: 2017 International Conference on I-SMAC (IoT in Social, Mobile, Analytics and Cloud) (I-SMAC), pp. 32–37. IEEE (2017)
4. Da Costa, K.A., Papa, J.P., Lisboa, C.O., Munoz, R., de Albuquerque, V.H.C.: Internet of Things: a survey on machine learning-based intrusion detection approaches. Comput. Netw. **151**, 147–157 (2019)
5. Zhou, L., Pan, S., Wang, J., Vasilakos, A.V.: Machine learning on big data: opportunities and challenges. Neurocomputing **237**, 350–361 (2017)
6. Liu, Y., Wang, J., Li, J., Song, H., Yang, T., Niu, S., Ming, Z.: Zero-bias deep learning for accurate identification of Internet-of-Things (IoT) devices. IEEE Internet Things J. **8**(4), 2627–2634 (2020)
7. Hindy, H., Bayne, E., Bures, M., Atkinson, R., Tachtatzis, C., Bellekens, X.: Machine learning based IoT intrusion detection system: an MQTT case study (MQTT-IoT-IDS2020 dataset). In: International Networking Conference, pp. 73–84. Springer (2020)
8. Mosaiyebzadeh, F., Rodriguez, L.G.A., Batista, D.M., Hirata, R.: A network intrusion detection system using deep learning against MQTT attacks in IoT. In: 2021 IEEE Latin-American Conference on Communications (LATINCOM), pp. 1–6. IEEE (2021)
9. Boppana, T.K., Bagade, P.: GAN-AE: an unsupervised intrusion detection system for MQTT networks. Eng. Appl. Artif. Intell. **119**, 105805 (2023)
10. Khan, M.A., Khan, M.A., Jan, S.U., Ahmad, J., Jamal, S.S., Shah, A.A. Pitropakis, N., Buchanan, W.J.: A deep learning-based intrusion detection system for mqtt enabled IoT. Sensors **21**(21), 7016 (2021)
11. Zeghida, H., Boulaiche, M., Chikh, R.: Securing MQTT protocol for IoT environment using IDS based on ensemble learning. Int. J. Inform. Secur. 1–12 (2023)
12. Chawla, N.V., Bowyer, K.W., Hall, L.O., Kegelmeyer, W.P.: Smote: synthetic minority oversampling technique. J. Artif. Intell. Res. **16**, 321–357 (2002)
13. Alzubaidi, L., Zhang, J., Humaidi, A.J., Al-Dujaili, A., Duan, Y., Al-Shamma, O., Santamaría, J., Fadhel, M.A., Al-Amidie, M., Farhan, L.: Review of deep learning: concepts, CNN architectures, challenges, applications, future directions. J. Big Data **8**, 1–74 (2021)

Multi-objective Optimal Feature Selection for Cyber Security Integrated with Deep Learning

Anupam Das and Subhajit Chakrabarty

Abstract Cyber security in the context of big data has emerged as a critical issue and an upcoming challenge in the research community. The security problems in big data are handled using machine learning algorithms. In recent years, artificial intelligence has been an emerging technology that machine imitates human behaviors. The Intrusion Detection System (IDS) is the major component for the detection of malicious activities or cyber attacks. In intrusion detection, Artificial intelligence engages a major role widely, which is the better way of building and adapting IDS. Nowadays, Neural Network algorithms are promising techniques for framing the novel artificial intelligence, which is used for real-time problems. The proposed cyber security model uses a cyber defense dataset and a secured network is created based on multi-objective criteria. Here, the optimal feature selection is performed on the basis of a multi-objective function, which focuses on the correlation and accuracy of attack detection using a meta-heuristic algorithm called Beetle Swarm Optimization (BSO). Further, the deep learning model called Recurrent Neural Network (RNN) is used for detecting the attack. The multi-objective-based optimal feature selection is helpful for enhancing the performance of cyber security with reduced redundancy and complexity.

Keywords Artificial intelligence · Cyber security · Recurrent Neural Network · Optimal feature selection · Beetle Swarm Optimization · Intrusion Detection System

A. Das (✉) · S. Chakrabarty
Guwahati, India
e-mail: anupam.cotton@gmail.com

G. Fortino et al. (eds.), *Proceedings of Third International Conference on Computing and Communication Networks*, Lecture Notes in Networks and Systems 917,
https://doi.org/10.1007/978-981-97-0892-5_13

Nomenclature

Abbreviations	Descriptions
ANN	Artificial Neural Networks
BAS	Beetle Antennae Search
BSO	Beetle Swarm Optimization
CSA	Cuckoo Search Algorithm
DT	Decision Tree
EENS	Expected Energy Not Supplied
FDR	False Discovery Rate
FNR	False Negative Rate
FPR	False Positive Rate
GRU	Gated Recurrent Unit
IDS	Intrusion Detection System
ILP	Integer Linear Programming
LSTM	Long Short Term Memory
KNNK	Nearest Neighbor
MCC	Mathews Correlation Coefficient
NN	Neural Networks
NPV	Negative Predictive Value
RNN	Recurrent Neural Network
PSO	Particle Swam Optimization
ROC	Receiver Operator Characteristic
SSEP	Safety, Security, And Emergency Preparedness
SVM	Support Vector Machine

1 Introduction

In recent times, cyber attacks have increased rapidly and caused the complexity and risk to various organizations by constantly adapting management strategies [6]. An efficient cyber security is necessary for preventing interruption in the business for both private and public organizations and hence operational continuity is ensured. The modeling of better cyber security is a knowledge-concentrated task [8]. To provide this defense, various software and hardware components are mandatory and these are comparatively homogeneous and readily offered at free of cost or less cost. To deploy and combine such components, professional knowledge is needed for efficient organization defense, and flexible architecture systems are designed and implemented effectively. Therefore, cyber security is a complex ability, which is not readily formed by the buying of technical components. However, this is the expert knowledge of how to classify and organize such components, which produces the authentic defense [7].

Various intensity levels are developed for every cyber security control [10]. For example, the security control of patch management is developed based on various levels with rising intensities like (a) patch once; (b) patch monthly, (c) patch weekly, (d) patch daily, and (e) patch upon release. Likewise, a network firewall could be developed by deploying a security control using any one of the levels mentioned below (a) circuit-level gateway, (b) investigating firewall in its full state, (c) packets filter, and (d) firewall for application proxy with deep-packet-inspection. The selection of intensity and controls is based on the particular logical or physical type of an organization and it differs between them [9]. The choice of planning a cyber security is a multiple criteria involved decision problem. In an organization, the overall security state is improved based on the balancing of indirect and direct plan costs. Further, a decision problem with numerous criteria of diverse natures is generally approached by multi-objective optimization. The concept of the solutions is the 80/20 rule, and the 80/20 optimality, as the set of 80/20 optimal solutions [11].

Safety and security functions can influence each other and this can cause the degradation of unexpected performance and safety problems [13]. The cyber security controls are not degrading the reliability and performance of safety-grade digital models and these are not addressed in the literature dealing with cyber security problems. Fundamentally, the conventional cyber security guidelines have previously confirmed that several security controls, which contain the possible negative impacts on the SSEP functions, should be experimented on to ensure that there is no unfavorable effect [12]. Therefore, the security controls have been recognized and evaluated from a subjective view, and a systematic authentication procedure for security controls has not been suggested yet. The applications of cyber security control methods were securely managed by providing the required investigation and agreed upon by the appropriate quality guarantee technology like several inspection techniques and test models [15]. However, such quality guarantee technologies occasionally use a major part of the whole implementation attempt. As a result, it is required to implement some methods for handling the optimization issues in cyber security [14].

The major contributions of this paper are listed below.

- To select the optimal features concerning multi-objective function with parameters like correlation and accuracy for improving the performance of the proposed cyber security model.
- To classify the best optimal features with the deep learning model called RNN for ensuring cyber security.

This paper is organized as follows, Section 2 discusses the literate. Section 3 explains the cyber security model based on deep learning concept. Section 4 describes the multi-objective-based optimal feature selection for cyber security. Section 5 discusses the experimental results. Section 6 concludes this paper.

2 Literate Review

1. *Related works*

In 2019, Kanimozhi and Jacob [1] have developed a system for the detection of botnet attacks that created a severe risk to banking services and various domains financial industry. The proposed system was formed using artificial intelligence "on a realistic cyber defence dataset (CSE-CIC-IDS2018), the latest IDS Dataset in 2018 by Canadian Institute for Cybersecurity (CIC) on AWS (Amazon Web Services)". The proposed system based on ANN has established better accuracy and performance. The artificial intelligence-based IDS concerning the botnet attack was powerful and more accurate. This proposed model can be employed in traffic analysis methods like conventional network and real-time network traffic data analysis and cyber-physical systems.

In 2019, Liu et al. [2] have offered a structure for effectively solving a multi-objective optimization problem under cyber security defense. The general probabilistic security controls were controlled, which affected each vulnerability, more number of attack paths, and various vulnerabilities, thus the ILP conversion, linear problem relaxation, and dualization techniques were used. Moreover, min–max multi-objective optimization was used for the above-mentioned problems. An efficient mixed ILP was developed by converting the problem using the above-mentioned techniques.

In 2018, Yi et al. [3] have developed the configuration process of SVM for a bi-objective optimization complexity, where two objectivities like system difficulty and accuracy were considered. To solve the optimization problem, the hyper-heuristic structure was proposed. This structure was included with low-level heuristics for generating a novel SVM configuration and a high-level scheme for searching the performance to manage the selection process. Diverse rules were used by the low-level heuristics for the efficient exploration of the search space of SVM configuration. The proposed method has shown better effectiveness than conventional models.

In 2019, Illindala et al. [4] have proposed a new model called a tri-level optimization for formulating the coordinated attack state and recognized the optimal defending approach that was studied about the shielding resource allocation to evade adjacent to the corresponding attack. The interactions were formulated for the proposed model between operator, planner, and attacker in a time horizon of 24-h. Numerous operating rules were employed through stochastic programming. On the IEEE 14-bus system and IEEE 57-bus system, the proposed system experimented. Therefore, the results have shown the changes in the defending and attacking strategies regarding defend budgets, attack budgets, and durations of restoration. The attackers were launched the attacks, which was the reason for cascading failures. If the proposed model was used for the allocation of defending resources, then considerable unserved energy minimization was found.

In 2019, Vahidi [5] have implemented a new approach by introducing an efficient configuration to the cyber network model for lodging cyber protocols. This method has shown a better result in reducing EENS when it was differentiated from the previous models. Hence, two numerical concepts graph theory and teaching–learning-based optimization were proposed for considering the cyber protocols and for selecting the better cyber configuration along with the lowest EENS, respectively. At first, the better configuration was selected with each particular goal for n-number of devices for a cyber network model that was relevant by altering it into an adjacency matrix $n \times n$ and by the proposed teaching–learning-based optimization technique, which was combined with the proposed graph theory. The simulation was done by the popular Monte Carlo simulation method. Various kinds of cyber networks were analyzed to evaluate the proposed system.

B. *Problem statement*

As there are many features that help to resolve the optimization problem in cyber security, still there exist some challenges that must be considered in the coming days. Some of the main features and challenges are listed in Table 1. Artificial intelligence [1] is purely dependent on the CPU and the optimization can be tuned by Google's open-sourced Tensor-flow. However, it does not involve all the existing attacks and real-time attacks. ILP conversion and Linear programming relaxation [2] seizes multiple target nodes, recovery type protocols, and uncertainty in the parameters and it also attains validation with the help of complex real-world scenarios. Still, a methodical analysis of the effect of risk aversion is not considered. SVM [3] estimates the Pareto group of configurations and it is also very efficient when contrasted with the conventional approaches. Yet, it needs to update the new software for maintaining security up to date. The tri-level optimization model [4] prevents cascading failures and the defending and attacking approaches are altered through defend budgets, restoration durations, and attack budgets and it also causes lower unserved energy by corresponding attacks. But, it is not implemented in distribution systems and DC microgrids and it also does not assess and improve the strength of combined energy infrastructures like electricity-transportation system, gas-electricity systems, etc. and it also does not integrate the unreliability of dynamic issues, restoration and attacker process, and energy storage integration. Teaching–learning-based optimization and graph theory [5] minimizes the blackout and it also enhances the economic benefits and it is capable of every bulk power system. Still, it is complex to configure the firewalls in an accurate manner. These challenges are motivated to find a novel method to handle the optimization problem in cyber security.

3 Cybersecurity Model Based on Deep Learning Concept

A. *Proposed cyber security model*

The cyber security model is required for the protection of data, privacy, and human life. It is extremely dependent on context. The integrity of personal privacy is ensured by the cyber security and also ensures the protection of military threats and intellectual property. The protection of data against attacks or threats is an important

Table 1 Features and challenges of state-of-the-art optimization problems in cyber security

Author [citation]	Methodology	Features	Challenges
Kanimozhi and Jacob [1]	Artificial intelligence	• It is purely dependent on the CPU • The optimization can be tuned by Google's open-sourced Tensor-flow	• It does not involve all the existing attacks and real-time attacks
Liu et al. [2]	ILP conversion and Linear programming relaxation	• It seizes multiple target nodes, recovery type protocols, and uncertainty in the parameters • It attains validation with the help of complex real-world scenarios	• A methodical analysis of the effect of risk aversion is not considered
Yi et al. [3]	SVM	• It estimates the Pareto group of configurations • It is very efficient when contrasted with the conventional approaches	• It needs to update the new software for maintaining security up to date
Illindala et al. [4]	Tri-level optimization model	• Defending and attacking approaches alter through defend budgets, restoration durations, and attack budgets • It prevents the cascading failures • It causes lower unserved energy by corresponding attacks	• It is not implemented in distribution systems and DC microgrids • It does not assess and improve the strength of combined energy infrastructures like electricity-transportation systems, gas-electricity systems, etc • It does not integrate the unreliability of dynamic issues, restoration and attacker process, and energy storage integration
Vahidi [5]	Teaching–learning-based optimization and Graph theory	• It minimizes the blackout • It enhances the economic benefits • It is capable of every bulk power system	• It is complex to configure the firewalls in an accurate manner

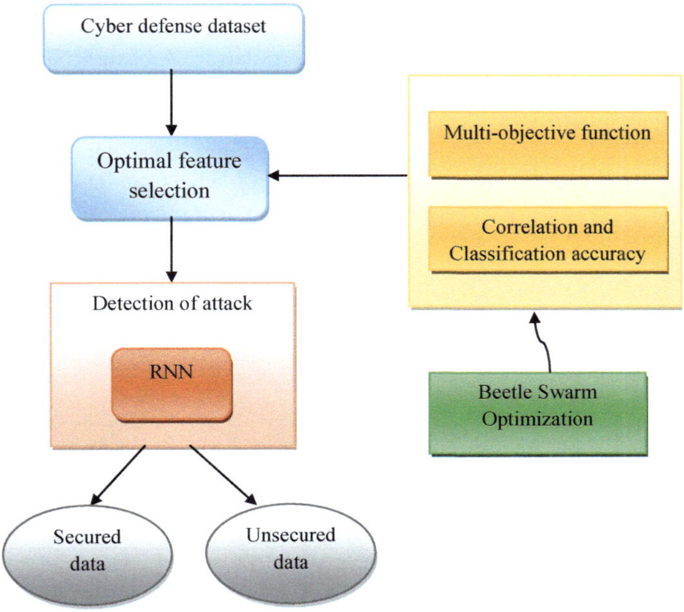

Fig. 1 Proposed cyber security model

challenge of cyber security. Machine learning is an emerging technique for the cyber security. By using machine learning, the cyber security systems can learn about the attacks and different types of attacks can be detected. In Fig. 1, the proposed cyber security model using the RNN is given.

The proposed cyber security model takes the input data from the cyber defense dataset. These data are subjected to optimal feature selection, in which the relevant data attributes are selected. The multi-objective function is considered for the optimal feature selection with parameters such as correlation and accuracy. The correlation between the two features should be attained minimum and the detection accuracy should be maximized. Here, the BSO algorithm is applied for the optimization of data attributes. The best-optimized features are classified using RNN, where the attack is detected.

B. *Recurrent neural network-based attack detection*

RNN [17] is one of the deep learning models. "RNN is a feed-forward artificial neural network". For solving the difficulty of RNN training, LSTM is developed. This model has "three gate units such as forget, input and output gates and memory cell unit". LSTM is utilized to remove unrelated information and to capture valuable information in sequence, which is obtained by updating three gate and memory units. GRU is one type of LSTM that makes the efficient RNN by removing the memory cell unit from LSTM. For the better enhancement of RNN structure, the GRU is used and the performance is also improved. The GRU is combined with the forgetting and output gates into a specific update gate represented as x, which is employed to get the output of the current state using linear interpolation. A basic RNN architecture can be formulated using input, output, and hidden sequence for each timestamp. It is represented as $(in_1, in_2, \ldots.., in_T)$, $(O_1, O_2, \ldots.., O_T)$, and $(hi_1, hi_2, \ldots.., hi_T)$, respectively. The input and output vector at timestamp T are denoted as in_T, O_T, respectively. The input is indicated as $\left(in_{if}, hi_{if-1}\right)$, where the input features of the *if*th image slice, r_g is the reset gate and the hidden state input features are represented as hi_{if}.

$$x_{if} = \delta\left(Wt^{ifin} in_{if} + Wt^{hix} hi_{if-1}\right) \tag{1}$$

$$r_{g_{if}} = \delta\left(Wt^{itr_g} x_{if} + Wt^{hir_g} hi_{if-1}\right) \tag{2}$$

Here, the term δ is the logistic sigmoid function WE_{RNN} is the weight metrics that are indicated as Wt^{ifin}, Wt^{hix}, Wt^{itr_g} and Wt^{hir_g}. In Eq. (3), the candidate state in the hidden unit is mathematically formulated.

$$\tilde{hi}_{if} = \tan\left(Wt^{ifhi} in_{if} + Wt^{hihi}\left(hi_{if-1} \otimes r_{g_{if}}\right)\right) \tag{3}$$

In Eq. (3), $\otimes$ is the element-wise multiplication. "When $r_{g_{im}}$ is close to 0, the reset gate r_g effectively makes the unit act as if it is reading the first symbol of an input sequence, allowing it to forget the previously computed state." The *if*th hidden activation state hi_{if} of GRU is a linear interpolation among h_{im-1} and h_{im-1} the previous state and the candidate state, respectively, which is given in Eq. (4).

$$\begin{aligned} hi_{if} &= \left(1 - x_{if}\right) \otimes \tilde{hi}_{if} + x_{if} \otimes hi_{if-1} hi_{if} \\ &= \left(1 - x_{if}\right) \otimes \tilde{hi}_{if} + x_{if} \otimes hi_{if-1} \end{aligned} \tag{4}$$

BGRU is used to capture the correlation features. At last, the learned features are classified using BGRUs that are combined with a soft-max layer and fully connected layer. RNN is trained using the new features.

4 Multi-objective-Based Optimal Feature Selection for Cyber Security

A. *Optimal feature selection*

The data from the dataset is subjected to an optimal feature selection process. Here, the relevant and best data features are selected. Here, the features are optimized using the BSO algorithm. The solution encoding of the attribute indexes is given in Fig. 2.

The attribute indexes are encoded using the BSO algorithm. The length of the features is equivalent to the length of the features. The bounding limit is fixed from 1

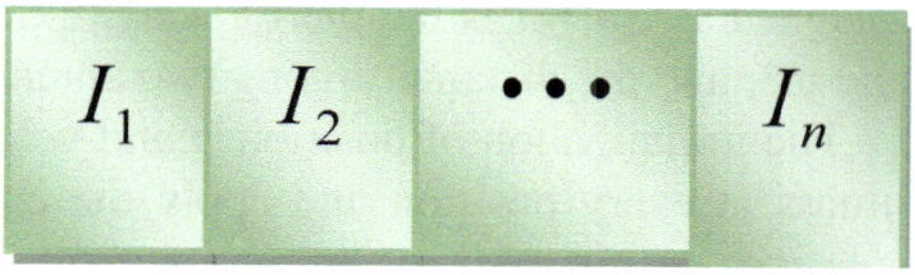

Fig. 2 Solution encoding of the attribute indexes

to the length of the features. The attribute index is termed as I_n, where $n = 1, 2, ..., N$. For the proposed cyber security model, a multi-objective function is considered with two major constraints. The first one is correlation and the second is accuracy. The correlation between the two features should be minimal and the accuracy should be maximized. By considering the multi-objective function, the fitness function is calculated based on Eq. (5).

$$Fi = \min\left(Cor + \frac{1}{Acc}\right) \tag{5}$$

Here, the correlation between the two features is formulated in Eq. (6).

$$Cor = \frac{nf \sum afbf - \sum af \sum bf}{\sqrt{\left(nf \sum af^2 - \left(\sum af\right)^2\right) - \left(nf \sum bf^2 - \left(\sum bf\right)^2\right)}} \tag{6}$$

The two features are considered as af and bf, and the number of feature pairs is represented as nf. Accuracy is described as "a ratio of the observation of exactly predicted to the whole observations". It is denoted in Eq. (7).

$$Acc = \frac{(tp + tn)}{(tp + tn + fp + fn)} \tag{7}$$

Here, the true positives are denoted as tp, fn is the false negatives, true negatives is referred to as tn, and fp is the false positives.

B. *Feature selection using beetle swarm optimization*

The cyber security model is proposed for the detection of attacks against threats using the BSO algorithm. The BSO [16] is inspired by the optimization algorithms like PSO and BSA. The better global searching is enabled by using the BSO than PSO. The multi-dimensional functions contain optimization problems that are effectively solved by BSO. The BSO algorithm is formulated based on the PSO algorithm. Here, in the PSO, each particle is considered as a beetle and searches. The initialization of preliminary position and velocity in the BSO algorithm is done like the PSO algorithm. In the BSO algorithm, the left and right individuals of the fitness values are compared at each iteration and evaluate the best values of the two that are employed for updating the beetle swarm position. Initially, the algorithm parameters like PSO size, learning factors, distance, and inertia weight are initialized. The size for PSO is assigned as Sp,Wt is the inertia weight, the learning factors are denoted as $f1$,$f2$, and $f3$, and the distance among two antennae of every beetle is represented as d.

The initialization of position p and velocity Ve is done randomly. Calculate the fitness at every position and apply the current position as the individual optimal solution p_{best} based on the comparison attains the recent global optimum g_{best}. Secondly, the beetle direction heads are randomized. The left and right individuals of the fitness values are compared at each iteration and evaluate the best values of the two that are employed for updating the beetle swarm position are represented as F_{left} and F_{right}, respectively. The position updating is given in Eq. (8).

$$Veb_i = -\alpha^c \cdot b \cdot sign\left(F\left(y_{pc}\right) - F\left(y_{qc}\right)\right) \tag{8}$$

Here, the update rate formed using the BSO is denoted as Veb_i, and $sign()$ is the symbolic function. The speed update rule is computed using the recent and global optimum solution by contrasting the position of the recent position of every beetle with p_{best} and g_{best}. The two-speed rules are combined to obtain the current update rules for the speed of each antenna using Eq. (9).

$$\begin{aligned} Ve_i^{j+1} &= Ve_i^{j+1} + f1 \cdot rand \cdot \left(pb_i^j - y_i^j\right) \\ &+ f2 \cdot rand \cdot \left(ps_i^j - y_i^j\right) + f3 \cdot rand \cdot Veb_1 \end{aligned} \tag{9}$$

Here, the speed Ve_i^{j+1} of the ith particle after jth iteration, the learning factor is termed as $f3$, the position is represented as y_i^{j+1} after jth iteration. The current position update rule is given in Eq. (10).

$$y_i^{j+1} = y_i^j + Ve_i^{j+1} \tag{10}$$

The learning factors and inertia weights are updated and also updates the optimal solutions like p_{best} and g_{best}. When the iteration is completed, the positions of g_{best} and $F(g_{best})$ are updated. The pseudo code of the BSO algorithm is given in Algorithm 1.

Algorithm 1: BSO [16]

Initialize swarm population
Initialize speed
Compute fitness of search agent
While $(j < J)$
    for each search agent
        Update speed using Eq. (9)
        Update position using Eq. (8)
    end for
    Compute fitness of each search agent $F(y)$
Update y^* if there is a best solution
end while
Return g_{best} and $F(g_{best})$

The BSO algorithm is selected for the optimization due to its features. The multi-dimensional functions contain optimization problems that are effectively solved by traditional BSO. It has better convergence speed and improves global optimization.

5 Experimental Results and Discussion

A. *Experimental setup*

The proposed cyber security model was developed in PYTHON, and the experimental analysis was carried out. Here, the conventional methods were analyzed with the proposed method through several performance measures. The cyber defense dataset was taken from the "https://archive.ics.uci.edu/ml/machine-learning-databases/00442/Provision_PT_737E_Security_Camera/: access date: 27–06-2020". The BSO-based optimal feature selection was compared with the optimization algorithms like PSO [18] and CSA [23] and machine learning algorithms such as SVM [19], KNN [20], DT [22], NN [21] and BSO-RNN [16, 17].

B. *Performance metrics*

Ten performance measures are considered for analyzing the performance and each performance is described below.

(a) Sensitivity: "the number of true positives, which are recognized exactly".

$$SE = \frac{tp}{tp + fn} \tag{11}$$

(b) Specificity: "the number of true negatives, which are determined precisely".

$$SP = \frac{tn}{fp} \tag{12}$$

(c) Precision: "the ratio of positive observations that are predicted exactly to the total number of observations that are positively predicted".

$$\text{Pr} = \frac{tp}{tp + tp} \tag{13}$$

(d) FPR: "the ratio of count of false positive predictions to the entire count of negative predictions".

$$FPR = \frac{fp}{fp + tn} \tag{14}$$

(e) FNR: "the proportion of positives which yield negative test outcomes with the test".

$$FNR = \frac{fn}{tn + tp} \tag{15}$$

(f) NPV: It is the "probability that subjects with a negative screening test truly don't have the disease".

$$NPV = \frac{fn}{fn + tn} \tag{16}$$

(g) FDR: It is "the number of false positives in all of the rejected hypotheses".

$$FDR = \frac{fp}{fp + tp} \tag{17}$$

(h) F1 score: It is defined as the "harmonic mean between precision and recall. It is used as a statistical measure to rate performance".

$$F1 \text{ score} = \frac{SE \cdot \text{Pr}}{\text{Pr} + SE} \tag{18}$$

(i) MCC: It is a "correlation coefficient computed by four values".

$$MCC = \frac{tp \times tn - fp \times fn}{\sqrt{(tp + fp)(tp + fn)(tn + fp)(tn + fp)}}. \tag{19}$$

C. *Performance analysis on optimal feature selection*

The BSO-based optimal feature selection is compared with the conventional algorithms for ensuring the performance of the proposed cyber security model by analyzing the various performance measures by considering various learning

percentages of RNN. In Fig. 3, the performance analysis for the BSO-RNN algorithm in terms of accuracy, precision, FPR, and FNR are given. The better accuracy and precision are observed in Fig. 3a and b, respectively. The accuracy of the BSO-RNN algorithm is 5.88% better than PSO-RNN and CSA-RNN algorithm at a learning percentage of 65. The precision of the BSO-RNN algorithm is 4.65 and 2.27% improved than PSO-RNN and CSA-RNN at learning percentage 65. The FNR and FPR are reduced for the BSO algorithm than the conventional algorithm based on Fig. 3c and d, respectively. At learning percentage 65, FPR for the BSO-RNN algorithm is 86% and 87% superior to PSO-RNN and CSA-RNN. Similarly, for the BSO-RNN algorithm, the minimum FNR is attained. It is 25% and 84% superior to PSO-RNN and CSA-RNN at a learning percentage of 65. In Table 2, the overall performance analysis of the proposed and conventional optimal feature selection is depicted. The BSO algorithm has shown better performance for all the performance measures than the conventional algorithms. The sensitivity of the BSO-RNN is 2 and 2.7% superior to PSO-RNN and CSA-RNN. The specificity of the BSO is 1.9% and 0.86% superior to PSO-RNN and CSA-RNN. The MCC of the BSO-RNN algorithm is 4.75% superior to PSO-RNN and CSA-RNN. Hence, when compared to the conventional algorithms, the proposed algorithm is outperformed.

D. *Classifier analysis*

The cyber attack detection using BSO-RNN over the conventional machine learning models at different learning percentages is shown in Fig. 4. The accuracy of the BSO-RNN is improved than the conventional algorithms. It is 35, 26, 20, and 15% improved than BSO-SVM, BSO-KNN, BSO-DT, and BSO-NN, respectively at a learning percentage of 65. The precision of the BSO-RNN is 18, 15, 13, and 10% improved than the BSO-SVM, BSO-KNN, BSO-DT, and BSO-NN, respectively for the learning percentage 55. The FNR and FPR are reduced for the BSO-RNN than the conventional algorithms. In Table 3, the overall classifier analysis for the BSO-RNN for various measures is represented. Therefore, the BSO-RNN algorithm has established better results than conventional methods for various performance measures.

6 Conclusion

The proposed cyber security model was focused on developing a secured model by considering multi-objective criteria with a deep learning mechanism. Here, based on the multi-objective function, the optimal feature selection was performed that focuses on the correlation and accuracy of attack detection. Further RNN was employed for detecting the attack. The accuracy of the BSO-RNN algorithm was 5.88% better than the PSO-RNN and CSA-RNN algorithms. Moreover, the accuracy of the BSO-RNN was 35, 26, 20, and 15% improved than BSO-SVM, BSO-KNN, BSO-DT, and BSO-NN, respectively. Therefore, the BSO-based optimal feature selection with RNN for

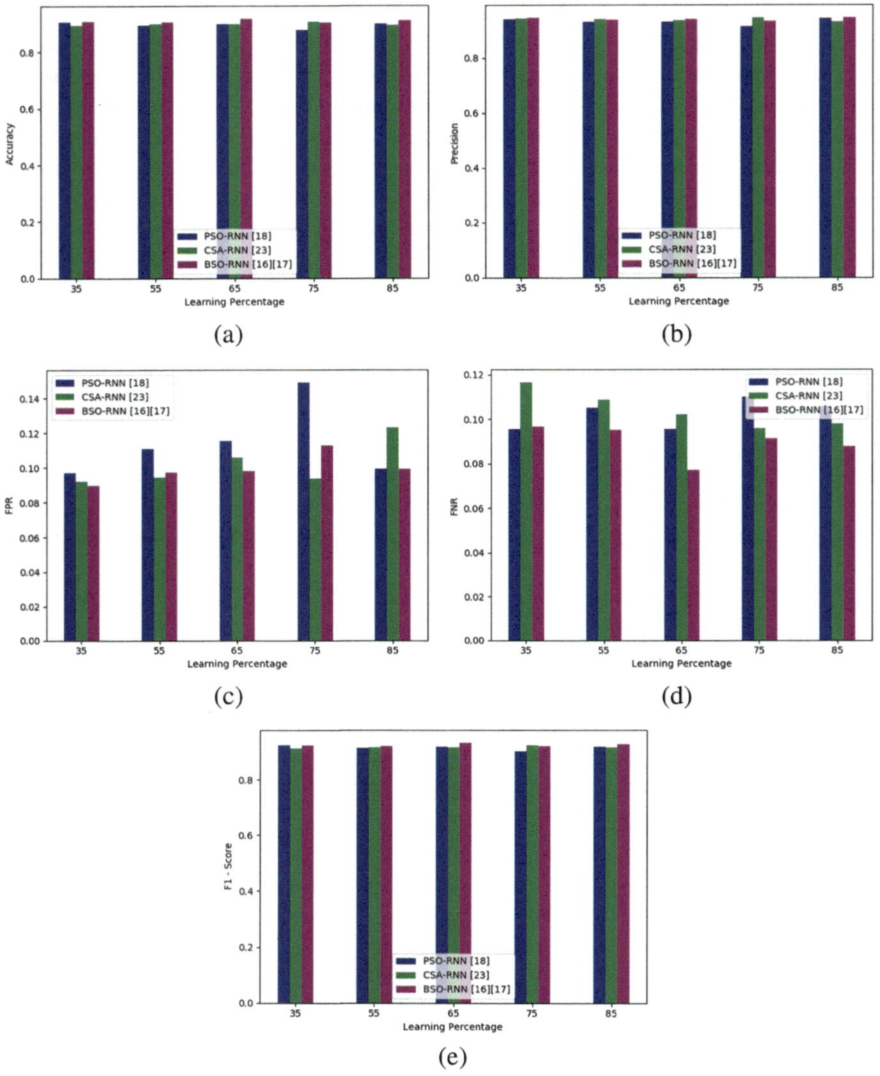

Fig. 3 Performance analysis of optimal feature selection with RNN-based classification by varying learning percentage. **a** Accuracy, **b** precision, **c** FPR, **d** FNR, and **e** F1-score

the proposed cyber security model has attained better results than the conventional methods.

Table 2 Overall performance analysis of optimal feature selection with RNN-based classification

Measures	Accuracy	Sensitivity	Specificity	Precision	FPR	FNR	NPV	FDR	F1-Score	MCC
PSO-RNN [18]	0.897069	0.904545	0.884393	0.929907	0.115607	0.095455	0.884393	0.070093	0.917051	0.782044
CSA-RNN [23]	0.896355	0.897727	0.894027	0.934911	0.105973	0.102273	0.894027	0.065089	0.915942	0.782046
BSO-RNN [16]	0.914939	0.922727	0.901734	0.940904	0.098266	0.077273	0.901734	0.059096	0.931727	0.819233

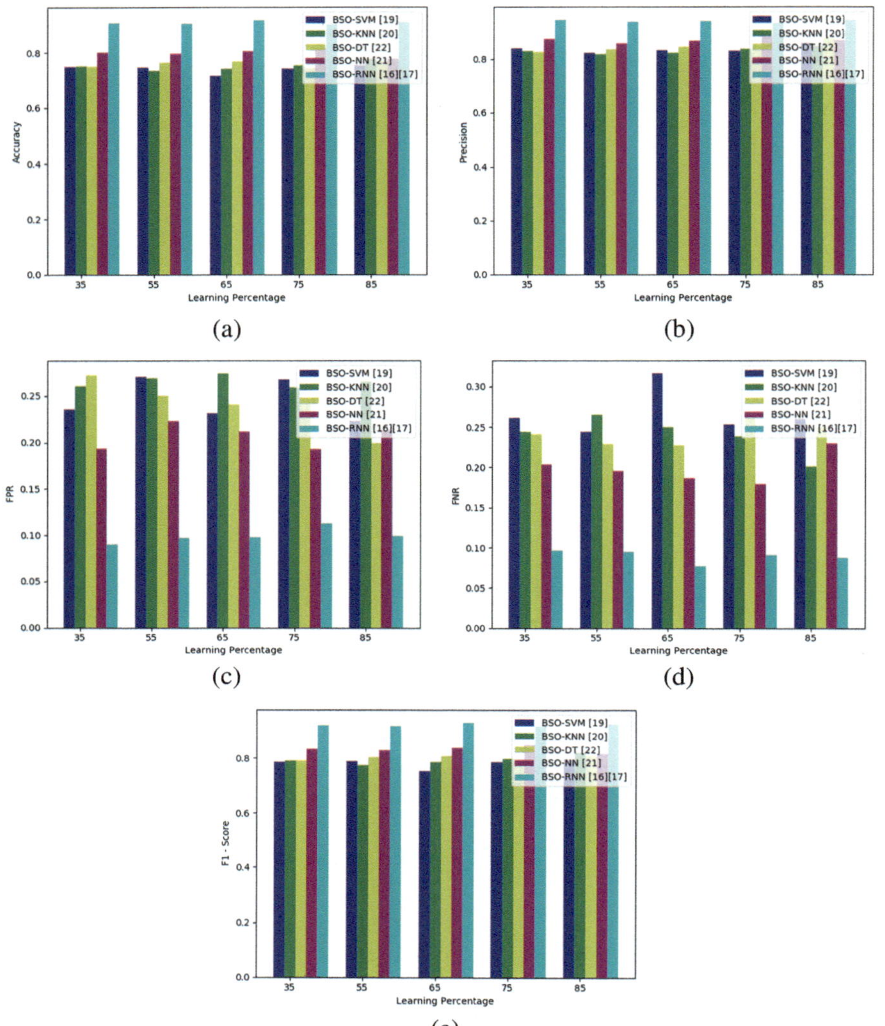

Fig. 4 Analysis on various machine learning algorithms for cyber security by varying learning percentage. **a** Accuracy, **b** precision, **c** FPR, **d** FNR, and **e** F1-score

Table 3 Overall performance analysis on various machine learning algorithms for cyber security by varying learning percentage

Measures	Accuracy	Sensitivity	Specificity	Precision	FPR	FNR	NPV	FDR	F1-Score	MCC
BSO-SVM [19]	0.715511	0.684091	0.768786	0.833795	0.231214	0.315909	0.768786	0.166205	0.751561	0.437766
BSO-KNN [20]	0.741244	0.75	0.726397	0.822943	0.273603	0.25	0.726397	0.177057	0.78478	0.465286
BSO-DT [22]	0.767691	0.772727	0.759152	0.84472	0.240848	0.227273	0.759152	0.15528	0.807122	0.519813
BSO-NN [21]	0.804146	0.813636	0.788054	0.866828	0.211946	0.186364	0.788054	0.133172	0.83939	0.591059
BSO-RNN [16] [17]	0.914939	0.922727	0.901734	0.940904	0.098266	0.077273	0.901734	0.059096	0.931727	0.819233

References

1. Kanimozhi, G.V., Prem Jacob, T.: Artificial Intelligence based network intrusion detection with hyper-parameter optimization tuning on the realistic cyber dataset CSE-CIC-IDS2018 using cloud computing. In: ICT Express, vol. 5, no. 3, pp. 211–214, September 2019
2. Khouzani, M.H.R., Liu, Z., Malacaria, P.: Scalable min-max multi-objective cyber-security optimisation over probabilistic attack graphs. Eur. J. Oper. Res. **278**, 894–903 (2019)
3. Sabar, N.R., Yi, X., Song, A.: A bi-objective hyper-heuristic support vector machines for big data cyber-security. IEEE Access **6**, 10421–10431 (2018)
4. Lai, K., Illindala, M., Subramaniam, K.: A tri-level optimization model to mitigate coordinated attacks on electric power systems in a cyber-physical environment. Appl. Energy **235**, 204–218 (2019)
5. Hamzeh, M., Vahidi, B., Nematollahi, A.F.: Optimizing configuration of cyber network considering graph theory structure and teaching–learning-based optimization (GT-TLBO). IEEE Trans. Indus. Inform. **15**(4), 2083–2090 (2019)
6. Almohri, H.M.J., Watson, L.T., Yao, D., Ou, X.: Security optimization of dynamic networks with probabilistic graph modelling and linear programming. IEEE Trans. Depend. Sec. Comput. **13**(4), 474–487 (2016)
7. Fielder, A., Panaousis, E., Malacaria, P., Hankin, C., Smeraldi, F.: Decision support approaches for cyber security investment. Decis. Support Syst. **86**, 13–23 (2016)
8. Homer, J., Ou, X.: SAT-solving approaches to context-aware enterprise network security management. IEEE J. Sel. Areas Commun. **27**(3), 315–322 (2009)
9. Nandi, A.K., Medal, H.R., Vadlamani, S.: Interdicting attack graphs to protect organizations from cyber attacks: a bi-level defender-attacker model. Comput. Oper. Res. **75**, 118–131 (2016)
10. Poolsappasit, N., Dewri, R., Ray, I.: Dynamic security risk management using Bayesian attack graphs. IEEE Trans. Depend. Sec. Comput. **9**(1), 61–74 (2012)
11. Sommestad, T., Ekstedt, M., Holm, H.: The cyber security modeling language: a tool for assessing the vulnerability of enterprise system architectures. IEEE Syst. J. **7**(3), 363–373 (2013)
12. Dezaki, H.H., Agah, M.M., Abyaneh, H.A., Haeri, H.: Sensitivity analysis of smart grids reliability due to indirect cyber-power interdependencies under various DG technologies, DG penetrations, and operation times. Energy Convers. Manag. **108**, 377–391 (2016)
13. Falahati, B., Fu, Y., Wu, L.: Reliability assessment of smart grid considering direct cyber-power interdependencies. IEEE Trans. Smart Grid **3**(3), 1515–1524 (2012)
14. Dezaki, H.H., Abyaneh, H.A., Haeri, H.: Impacts of direct cyber-power interdependencies on smart grid reliability under various penetration levels of microturbine/wind/solar distributed generations. IET Gener. Transm. Distrib. **10**, 928–937 (2016)
15. Facchinetti, T., Della Vedova, M.L.: Real-time modeling for direct load control in cyber-physical power systems. In: IEEE Transactions on Industrial Informatics, vol. 7, no. 4, pp. 689–698, November 2011
16. Chen, T. Zhu, Y., Teng, J.: Beetle swarm optimisation for solving investment portfolio problems. J. Eng. **2018**(16), 1600–1605 (2018)
17. Li, F., Liu, M: A hybrid convolutional and recurrent neural network for hippocampus analysis in Alzheimer's disease. J. Neurosci. Methods **323**, 108–118 (2019)
18. Bonyadi, M.R., Michalewicz, Z.: Analysis of stability, local convergence, and transformation sensitivity of a variant of the particle swarm optimization algorithm. IEEE Trans. Evol. Comput. **20**(3), 370–385 (2016)
19. Wu, J., Yang, H.: Linear regression-based efficient svm learning for large-scale classification. IEEE Trans. Neural Netw. Learn. Syst. **26**(10), 2357–2369 (2015)
20. Zhang, S., Li, X., Zong, M., Zhu, X., Wang, R.: Efficient kNN classification with different numbers of nearest neighbors. IEEE Trans. Neural Netw. Learn. Syst. **29**(5), 1774–1785 (2018)
21. Niu, H., Bhowmick, C., Jagannathan, S.: Attack detection and approximation in nonlinear networked control systems using neural networks. IEEE Trans. Neural Netw. Learn. Syst. **31**(1), 235–245 (2020)

22. Nancy, P. Muthurajkumar, S., Ganapathy, S., Santhosh Kumar, S.V.N., Selvi, M., Arputharaj, K.: Intrusion detection using dynamic feature selection and fuzzy temporal decision tree classification for wireless sensor networks. IET Commun. **14**(5), 888–895 (2020)
23. Joshi, A.S., Kulkarni, O., Kakandikar, G.M., Nandedkar, V.M.: Cuckoo search optimization—a review. Mater. Today: Proc. **4**(8), 7262–7269 (2017)

Application of Gradient Boosting Classifier-Based Computational Intelligence to Detect Drug Addiction Threat in Society

Ashutosh Kumar, Abhigyan Sinha, Tamoghno Bakshi, Sibashish Choudhury, Sushruta Mishra, and Laith Abualigah

Abstract Currently, drug and alcohol addiction has become a major menace to society's youth. As responsible citizens of this country, we must act now to keep these young brains from succumbing to this lethal addiction. In this research, a method based on intelligent analytics is designed for forecasting the threat of drug addiction using machine learning algorithms. Initially, we identify several reasons for drug addiction by speaking with health professionals, individuals addicted to drugs, and reading related studies and articles. Then we gather information from both addicts and abstainers. After preprocessing the data, we use Random Forest, Naive Bayes, Adaptive Boosting, and Gradient Boosting machine and assess the interpretation of each of these classifiers using certain well-known performance criteria. Outcome with Gradient Boosting showed promising with the highest accuracy rate of 95.4% and minimum error rate of 0.0854.

A. Kumar · A. Sinha · T. Bakshi · S. Choudhury · S. Mishra (✉)
Kalinga Institute of Industrial Technology, Deemed to be University, Bhubaneswar, India
e-mail: sushruta.mishrafcs@kiit.ac.in

A. Kumar
e-mail: 2005299@kiit.ac.in

A. Sinha
e-mail: 2005631@kiit.ac.in

T. Bakshi
e-mail: 2005624@kiit.ac.in

S. Choudhury
e-mail: 20051224@kiit.ac.in

L. Abualigah
Computer Science Department, Al al-Bayt University, Mafraq 25113, Jordan
e-mail: aligah.2020@gmail.com

MEU Research Unit, Middle East University, Amman 11831, Jordan

Applied Science Research Center, Applied Science Private University, Amman 11931, Jordan

Jadara Research Center, Jadara University, Irbid 21110, Jordan

G. Fortino et al. (eds.), *Proceedings of Third International Conference on Computing and Communication Networks*, Lecture Notes in Networks and Systems 917,
https://doi.org/10.1007/978-981-97-0892-5_14

Keywords Gradient boosting · Logistic regression · Machine learning · Prediction system · Addiction · Drug addiction

1 Introduction

Addiction to drugs is a widespread problem that every country should address, but each country has its unique way of dealing with it. The issue is related to social and familial norms, behaviors, and conduct and may have negative outcome on a person's health, as well as their public and private life. Our study investigated the correlation between these factors and drug addiction, using a dataset of 26 characteristics collected through extensive research. In contrast, another study analyzed referral sources and completion rates of prescription opined addiction treatment centers using an existing dataset, with a focus on drug usage frequency and specific substance categories, but they did not consider sociocultural aspects such as peer relationships and social behaviors. Another paper addressed the influence of peers on adolescent smoking, focusing on theoretical frameworks and empirical evidence. Our study will employ machine learning techniques to analyze a realistic dataset and report our findings.

The system shown in Fig. 1 employs an inference engine that consists of three main components: an expert interface, a prediction process, and a knowledge base. The expert interface serves as the user's main point of contact with the system. It is designed to provide a user-friendly interface that guides the user through the process of inputting the relevant information. The expert interface collects the data required for the prediction process and presenting it to the system in a format that can be easily processed. The prediction process is the central component of the inference engine. It takes in the data collected by the expert interface and applies a set of predefined rules and algorithms to make a prediction about the likelihood of the person being a drug addict. The prediction process is designed to be highly accurate, with a low error rate and to produce results quickly. The knowledge base is the final component of the inference engine. It contains all the data, rules, and algorithms that are required to make accurate predictions about the likelihood of a person being a drug addict. The knowledge base is designed to be constantly updated with new data, research, and insights into drug addiction, to ensure that the system remains up-to-date and accurate.

2 Related Works

Different recent works are done in the drug addiction detection domain by many researchers. Laura Acion et al. [1] used classical analytical strategies (such as logistic regression) to compare the estimates of treatment success generated using logistic regression and more modern techniques (like ANN). It also uses SL, an ensembling

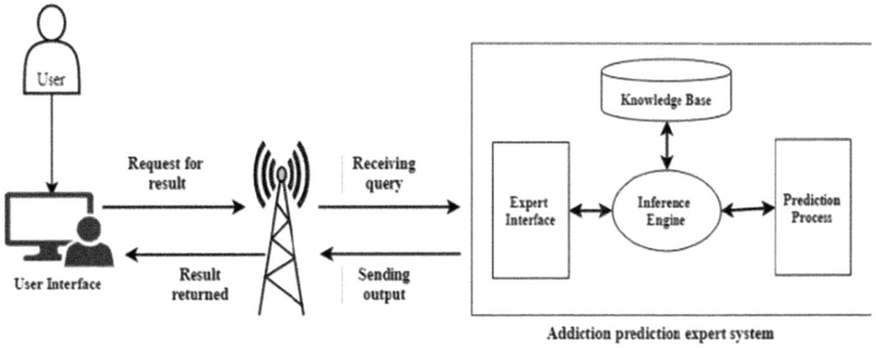

Fig. 1 System architecture of drug addiction prediction

learner technique that incorporates all of the other techniques in its module. Yankang Jing et al. [2] used k-nearest neighbors for imputation of missing data and then Random Forest algorithm for feature selection. Priyanka Nath et al. [3] applied a method that detects drug use and the time it occurs using machine learning. A tool for pattern identification is an ANN. The accuracy of identifying drug users is 81.1%, whereas that of identifying the time of usage is 71.9%. Md. Ariful Islam Arif et al. [4] implemented PCA (Principal Component Analysis) for processing the data. The algorithms used for the model are: k-NN, SVM, Logistic Regression, Naive Bayes, AdaBoost, and Perceptron. Hassanpour et al. [5] used social media to capture insights about substance use and addiction by using a deep learning system, which classifies individual risk of drug use based on their content of Instagram profiles. Banafsheh Rekabdar et al. [6] investigated, developed, and tested deep neural network and data analytics models to find drug and alcohol-dependent individuals in a Screening, Brief Intervention, and Referral to Treatment (SBIRT) program. It involves categorizing various forms of drug and alcohol misuse. Dae-Hee Han et al. [7] studied how well three different analytics algorithms predicted teen opioid addiction in the US. The predictive methods were created with methods: distributed ensemble methods, boosting technique, and artificial neural networks. Shahriar et al. [8] used a machine learning approach assisting in identifying some characteristics that distinguish the addicted and non-addicted people. This will assist people in determining whether or not a person will get addicted based on their health factors, family and social trends. Mohan S. et al. [9] used five distinct algorithms, including Logistic Regression, Decision Tree, Random Forest, Naive Bayes, and Support Vector Machine, were used to evaluate the factors that contribute to drug addiction as well as workable ways to reduce the drug addiction rate. In [10], authors demonstrated a comparative analysis using different ensemble classifiers for prediction analysis.

3 Proposed Model

For the given problem, we have done the data acquisition from different sources such as Kaggle. After getting our data, we convert it into a format which can be interpreted by Python. Now we import the given dataset and preprocess it using EDA, One-Hot Encoding, Feature Engineering, etc. After all this, we get a clean dataset to work upon. We split the dataset into training and testing dataset [11]. After all the exploration, we start to implement different ML models like KNN, SVM, Neural Networks, Random Forest, Linear Regression, Decision Tree with the concept of hyperparameter tuning. Since the model building part is over now, we calculate the F-Score, Precision, Recall, etc. for analyzing and comparing the different ML models used. The proposed model workflow is denoted in Fig. 2.

Training phase in predictive approach may be segregated into unsupervised and supervised types. In former category, a series of samples are utilized for training the system which are tagged to provide accurate outcome. But in the later type, no predefined labels with desired results are present, which makes the task difficult. Classification is the common technique under supervised category where past historical instances are used to design model, and it is applied to predict outcomes. In healthcare centers, massive patient-related data records are maintained [12]. Here, these prediction-based concepts are used to make decisions using past data.

In this work, a machine learning-based hybrid model for classifying the drug addiction has been proposed, having three stages.

- *Handling imbalance in the dataset stage:* The effects of imbalanced data are implicit, which means they do not cause an immediate error while building and executing the model, but the outcomes may be ambiguous [13, 14]. When the dataset is skewed for the major labeled category, a well-trained classification method could produce excellent mean rate of precision. Figure 3 shows the data skewing scenario in dataset.

One of the most widely used libraries is the imbalanced-learn package which provides functions for handling an imbalanced dataset by oversampling or undersampling it.

- Variable mapping phase: Variable mapping techniques are applied for mapping of non-separable data instances to separable instances [15, 16]. Methods like MAD

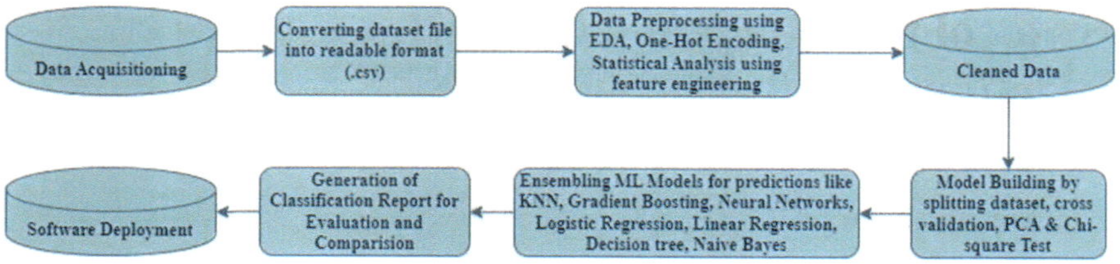

Fig. 2 Workflow model of proposed drug addiction detection methodology

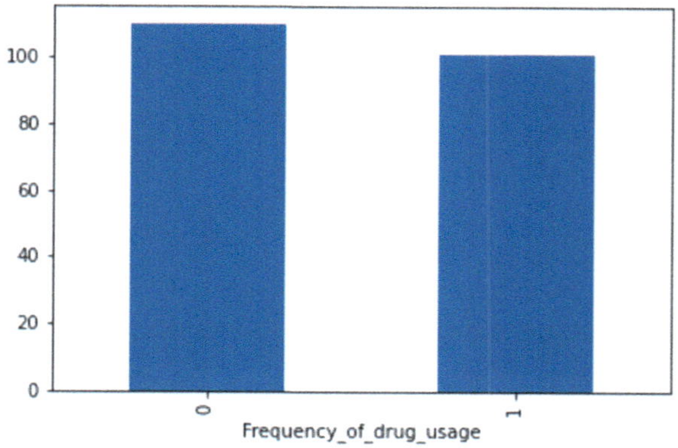

Fig. 3 A pivot table illustrating the absence of data imbalance

and correlation analysis use heatmap. Figure 4 shows the MAD values for the drug addiction dataset.

- Classification Stage: The developed model is also known as the Gradient Boosting Classifier as depicted in Fig. 5. In Gradient Boosting, every estimator aims to override its previous element by reducing errors. Gradient Boosting argues that in place of placing an estimator to samples in every round, it maps a new estimator to the resultant bugs developed by the previous forecast [17].

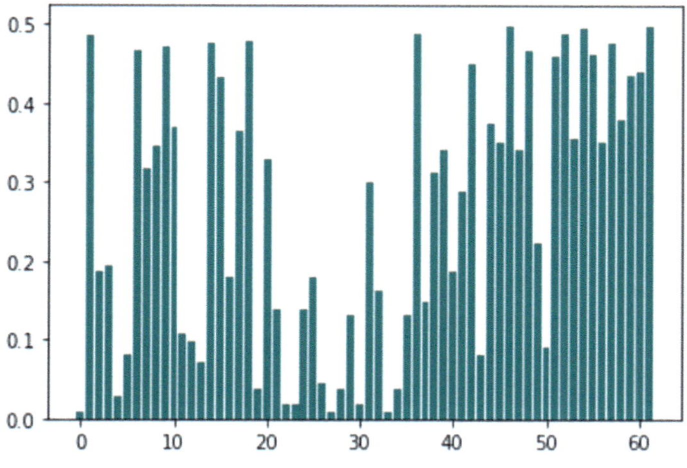

Fig. 4 MAD (mean absolute difference) of the dataset

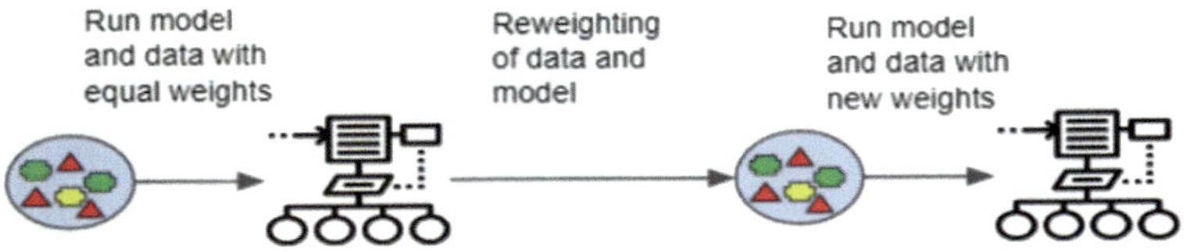

Fig. 5 Gradient Boosting classification using scikit-learn in Python

4 Result Analysis and Discussion

According to the existing model, logistic regression was used to determine addiction to drugs. Estimates derived from the model are based solely on family history, mental health, etc. but it assumes the linear relationship between dependent and independent variables. It also leads to over-fitting at times and it is very difficult to obtain complex relationships using this model and models like neural networks, Clustering algorithms and gradient boosters can outperform this model [18]. As per the model, the Gradient Boosting Classifier creates frameworks repeatedly, with every method aiming to reduce errors of the previous state. It is done on the basis of a novel prototype upon predefined module's error. It uses a Loss Function called log-likelihood. It has the role of estimating to which extent the model's predictions are correct. The potential benefits include that it is very easy to make an interpretation of its prediction. It also solves the problem of overfitting happening in Logistic Regression. Missing data present in the dataset can be handled by this algorithm. It also provides many choices for hyperparameter tuning and offers the advantage to optimize various loss functions.

Figure 6 demonstrates the comparative analysis of the proposed Gradient Boosting model with other computational approaches in context to accuracy rate analysis. It is observed that Gradient Boosting generates an optimal accuracy of 95.4% in comparison to other models. While Naive Bayes records the least accuracy rate.

The model efficiency is also determined by considering the RMSE metric to validate the errors noted as shown in Fig. 7. The Gradient Boosting method noted a least error rate of 0.0854 while Naive Bayes showed the maximum RMSE value in detecting drug addiction threats in society's youths.

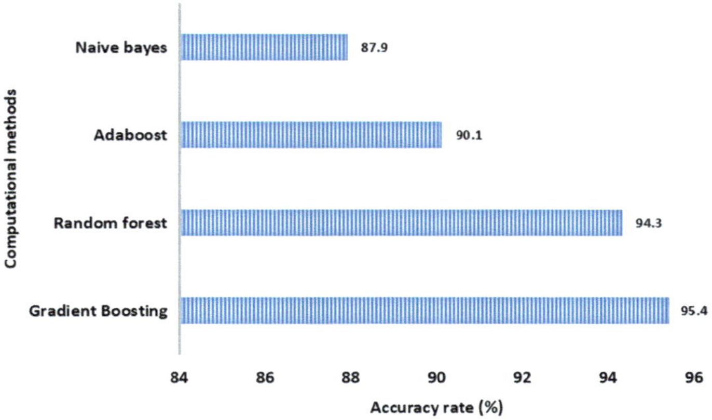

Fig. 6 Accuracy analysis of the proposed Gradient Boosting model with other models for the prediction of drug addiction

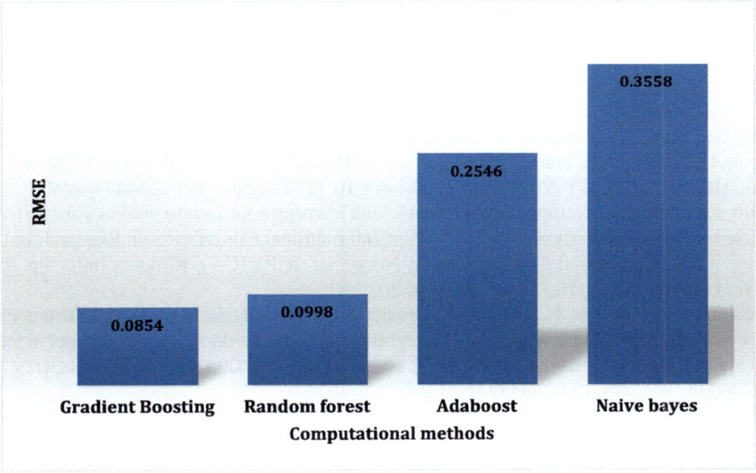

Fig. 7 RMSE analysis of proposed Gradient Boosting model with other models for prediction of drug addiction

5 Conclusion

The fight against drug and alcohol addiction is crucial and demands immediate action. Machine learning techniques can help by predicting the likelihood of addiction, identifying individuals who are susceptible, and offering necessary support before addiction becomes a severe issue. This research employed machine learning algorithms to anticipate addiction probability, and the initial outcomes were encouraging. Advanced deep learners like convolutional neural networks and recurrent neural networks show promise in improving the accuracy of predictive models. Personalized prediction models that consider a person's unique circumstances utilizing sophisticated machine learning techniques like reinforcement learning and causal inference are potential areas for future investigation. Clinical trials and longitudinal studies in real-life settings are necessary to verify the usefulness of the prediction models and applications, which can inform government decisions and guidelines on addiction prevention and treatment. These future studies could provide valuable insights into the practicability, precision, and effectiveness of these addiction prevention and treatment methods.

References

1. Acion, L., Kelmansky, D., van der Laan, M., Sahker, E., Jones, D., Arndt, S.: Use of a machine learning framework to predict substance use disorder treatment success. PLoS One **12**(4), e0175383 (2017). https://doi.org/10.1371/journal.pone.0175383.PMID:28394905; PMCID:PMC5386258
2. Jing, Y., Hu, Z., Fan, P., Xue, Y., Wang, L., Tarter, R.E., Kirisci, L., Wang, J., Vanyukov, M., Xie, X.Q.: Analysis of substance use and its outcomes by machine learning I. Childhood evaluation of liability to substance use disorder. Drug Alcohol Depend. 2020 January 1;**206**, 107605. https://doi.org/10.1016/j.drugalcdep.2019.107605. Epub 2019 Oct 22. PMID: 31839402; PMCID: PMC6980708
3. Arif, M., Sany, S.I., Sharmin, F., Rahman, S., Habib, T.: Prediction of addiction to drugs and alcohol using machine learning: a case study on Bangladeshi population. Int. J. Electr. Comput. Eng. (IJECE). **11**, 4471 (2021). https://doi.org/10.11591/ijece.v11i5.pp4471-4480
4. Nath, P., Kilam, S., Swetapadma, A.: A machine learning approach to predict volatile substance abuse for drug risk analysis. In: 2017 Third International Conference on Research in Computational Intelligence and Communication Networks (ICRCICN), Kolkata, India, pp. 255–258 (2017).https://doi.org/10.1109/ICRCICN.2017.8234516
5. Hassanpour, S., Tomita, N., DeLise, T., Crosier, B., Marsch, L.A.: Identifying substance use risk based on deep neural networks and Instagram social media data. Neuropsychopharmacology **44**(3), 487–494 (2019). https://doi.org/10.1038/s41386-018-0247-x. Epub 2018 Oct 24. PMID: 30356094; PMCID: PMC6333814
6. Rekabdar, B., Albright, D.L., McDaniel, J.T., Talafha, S., Jeong, H.: From machine learning to deep learning: a comprehensive study of alcohol and drug use disorder. Healthc. Anal. **2**, 100104 (2022). ISSN 2772-4425, https://doi.org/10.1016/j.health.2022.100104
7. Han, D.-H., Lee, S., Seo, D.-C.: Using machine learning to predict opioid misuse among U.S. adolescents. Prevent. Med. **130**, 105886 (2020). ISSN 0091-7435, https://doi.org/10.1016/j.ypmed.2019.105886
8. Shahriar, A., Faisal, F., Uddin Mahmud, S., Chakrabarty, A., Alam, Md. G.R.: A machine learning approach to predict vulnerability to drug addiction, pp. 1–7 (2019). https://doi.org/10.1109/ICCIT48885.2019.9038605
9. Mohan, S., Shailaja, L.: Prediction of drug addiction using supervised learning. Int. Res. J. Innov. Eng. Technol. (IRJIET) **6**(5), 213–216 (2022). ISSN (online): 2581-3048 https://doi.org/10.47001/IRJIET/2022.605030
10. Mishra, N., Mishra, S., Tripathy, H.K.: Rice yield estimation using deep learning. In: Innovations in Intelligent Computing and Communication: First International Conference, ICIICC 2022, Bhubaneswar, Odisha, India, December 16–17, 2022, Proceedings, pp. 379–388. Springer International Publishing, Cham (2023)
11. Chakraborty, S., Mishra, S., Tripathy, H.K.: COVID-19 outbreak estimation approach using hybrid time series modelling. In: Innovations in Intelligent Computing and Communication:

First International Conference, ICIICC 2022, Bhubaneswar, Odisha, India, December 16–17, 2022, Proceedings, pp. 249–260. Springer International Publishing, Cham (2023)
12. Verma, S., Mishra, S.: An exploration analysis of social media security. In: Predictive Data Security using AI: Insights and Issues of Blockchain, IoT, and DevOps, pp. 25–44. Springer Nature Singapore, Singapore (2022)
13. Singh, P., Mishra, S.: A comprehensive study of security aspects in blockchain. In: Predictive Data Security using AI: Insights and Issues of Blockchain, IoT, and DevOps, pp. 1–24. Springer Nature Singapore, Singapore (2022)
14. Swain, T., Mishra, S.: Evolution of machine learning algorithms for enhancement of self-driving vehicles security. In: 2022 International Conference on Advancements in Smart, Secure and Intelligent Computing (ASSIC), pp. 1–5. IEEE (2022)
15. Sahoo, S., Mishra, S.: A comparative analysis of PGGAN with other data augmentation technique for brain tumor classification. In: 2022 International Conference on Advancements in Smart, Secure and Intelligent Computing (ASSIC), pp. 1–7. IEEE (2022)
16. Mohapatra, S.K., Mishra, S., Tripathy, H.K.: Energy consumption prediction in electrical appliances of commercial buildings using LSTM-GRU model. In: 2022 International Conference on Advancements in Smart, Secure and Intelligent Computing (ASSIC), pp. 1–5. IEEE (2022)
17. Adrija, M., Yash, A., Sushruta, M.: 8 pragmatic study of IoT in healthcare security with an explainable AI perspective. In: Explainable Artificial Intelligence for Biomedical Applications, pp. 145–166. River Publishers (2023)
18. Bhavya, M., Pranshu, S., Sushruta, M., Sibanjan, D.: 17 comparative analysis of breast cancer diagnosis driven by the smart IoT-based approach. In: Explainable Artificial Intelligence for Biomedical Applications, pp. 353–374. River Publishers (2023)

Insurance 4.0: Smart Insurance Technologies and Challenges Ahead

Tamanna Kewal and Charu Saxena

Abstract The insurance value chain is being rapidly challenged by insurers as they integrate with technology. Quick customer service, convenience, and customization are three of InsurTech key selling points. The following are some of the most active and well-known InsurTech-related topics covered in this article: Mobile Applications, Smart Insurance Contracts, Blockchain/Distributed Ledger Technology, AI, IoT, Robot Advisors are all examples of digital insurance solutions. Based on a careful evaluation of the literature, this study provides cogent ideas under the umbrella of "InsurTech." The background of InsurTech and its numerous subfields are reviewed in this paper. Using innovative digital platforms in Insurance 4.0 can result in several advantages like helping with everyday commercial and administrative responsibilities, strategic decision-making, and process improvement and controlling. Blockchain, smart contracts, robot advisors, and mobile applications can all help to smooth the transition to Insurance 4.0. This paper demonstrates that, in order to execute Insurance 4.0, insurance procedures must be significantly digitized. Traditional work cultures, mindsets, methods, and monetary costs may provide problems in this regard.

Keywords InsurTech · Artificial intelligence · Big data · Blockchain · IoT · Telematics

1 Introduction

In a world of rising unpredictability and instability, the economic and societal relevance of being insured appears undeniable. InsurTech refers to the application of technological advancements to the present insurance sector system. InsurTech is

T. Kewal · C. Saxena (✉)
University School of Business, Chandigarh University, Mohali, Punjab, India
e-mail: charu.e8966@cumail.in

T. Kewal
e-mail: tamannakewal04@gmail.com

G. Fortino et al. (eds.), *Proceedings of Third International Conference on Computing and Communication Networks*, Lecture Notes in Networks and Systems 917,
https://doi.org/10.1007/978-981-97-0892-5_15

a term used to describe innovative solutions for the insurance sector [1]. InsurTech began in the insurance business with quick and easy pricing comparison tools but has now expanded to include e-marketing of insurance policies, P2P insurance selling, usage-based insurance, underwriting, cloud-based policies management, artificial intelligence-based customer support, and rapid claim resolution [2].

In our day and age, younger age groups have matured during a period of accelerated technological development. Consumer expectations, from shopping to managing financial and insurance needs, have reached a stage that traditional insurance cannot handle. The rapid introduction and adoption of technology in their daily lives has changed this generation's sense of expectation. The supply side of insurance value chain has been pressurized to acknowledge the gaps and fulfill these expectations by technology-enabled solutions. The recent growth of insurance technology and associated start-ups is really concealed and wrapped up in consumer expectations of service quality and performance. Insurance providers are already integrating IoT, AI, big data, telematics, and blockchain technologies swiftly and bringing substantial changes to conventional business procedures [3]. The sweeping transformation of the insurance sector is being fueled by these innovative technologies, which are changing how companies and customers interact, automating business operations, and designing customized offerings. Blockchain and smart contract technologies are growing as a platform with added value for banks and insurers as advancements in technology occur [4]. Blockchain use in the insurance sector could bring up new prospects for creating novel products [5]. Chatbots are now being used widely as an emerging communication alternative [6]. The Internet of Things (IoT) has long been recognized as a technology that combines with big data analytics and accumulates enormous volumes of data. Telematics is an IoT invention that is used to calculate auto insurance rates, gather driving behavior data and make it easier to spot fraud throughout the claim settlement process [7]. We must thus comprehend these topics covered by InsurTech and their ramifications. Main contributions of this study are:

- This paper firstly examines the existing literature on the development of insurance technology, outlining its background.
- A brief discussion of some of the most popular InsurTech topics, including AI, IoT, blockchain, insurance solutions offered through mobile applications, robot advisors, smart contracts, and so forth, is provided in the next section.
- The comparison of traditional insurance and InsurTech in the following section is discussed.
- Following part examines the security challenges surrounding InsurTech and why it is crucial to deal with these issues properly.
- The final section outlines the key research gaps in InsurTech and the direction of future study.

2 InsurTech—Background

InsurTech is a collection of technologies that combine information and insurance services [7]. For many years, insurance industry has been characterized as conservative and struggled with digital transformation but the pandemic has acted as a catalyst for innovation in this industry too like other industries. Insurers began making the shift to digital technology in quest of more customized and efficient services, and numerous technology companies launched new products to the market. Technological innovations like artificial intelligence, IoT, and blockchain have aided the rise of new insurance business models, ranging from peer-to-peer insurance and personalization via wearable devices to insurance plans suited to individual behavior, such as pay-as-you-live policies, which may result in economic rewards in the form of low premiums.

The term InsurTech was coined not long ago as one of the fintech sub-disciplines. InsurTech is a collection of technologies that combine information and insurance services [8]. For many years, insurance industry has been characterized as conservative and struggled with digital transformation but the pandemic has acted as a catalyst for innovation in this industry too like other industries [9–11]. Insurers began making the shift to digital technology in quest of more customized and efficient services, and numerous technology companies launched new products to the market [12]. Technological innovations like artificial intelligence, IoT, and blockchain have aided the rise of new insurance business models, ranging from peer-to-peer insurance and personalization via wearable devices to insurance plans suited to individual behavior, such as pay-as-you-live policies, which may result in economic rewards in the form of low premiums [11, 13–15]. Drones have been offered as a tool for measuring risk while approving insurance applications and calculating losses; Chatbots were introduced to communicate with clients, saving time for personnel and, eventually, money for insurance firms; Telematics encourages improved driving behaviors and also lowers insurance claims; Internet of things refers to a network of physical objects with sensors and connection that allows data to be transmitted and communicated hence those consumers that are ready to provide personal information can use the Internet of Things to automate most of their data sharing; Artificial intelligence is based on the idea that it is possible to create machines that can process data and learn without requiring human interaction, to automate the processes of loss settlement and general administration, removal of the human factor can also lead to faster data access and more accurate reporting [12, 16, 17].

3 Technologies Behind Insurance 4.0

3.1 *Big Data*

A revolution has taken place with the introduction of "big data analytics" in the insurance industry as it alters the way in which data were collected, managed, and used [13]. The five Vs of big data offer insurers with a beneficial foundation for transforming raw data into meaningful information [14–16]. These five Vs are: (a) Volume: the quantity of raw data collected; (b) Velocity: The speed at which that data is collected; (c) Variety: data collected comprised of structured and unstructured data; (d) Veracity: it considers the questions of whether the data is trustworthy and can companies rely on it; and (e) Value: it answers the question of worthiness of that data [15, 17, 18]. Big data converts human behavior into something numerical [19]. To completely analyze massive data, insurance businesses must apply big data analytics, including its subsets of explanatory, forecasting, and proactive analytics [20, 21]. In addition to specific algorithms and analytics, insurance companies must gather knowledge and hire personnel with the required abilities to apply the appropriate big data technologies and assess the quality of the data. The timeliness, correctness, and dependability of the underlying data also affect how useful big data is EIOPA [22]. As a result, gathering and preparing the data is an essential stage in extracting information from the data. A single data platform must be used to store all data in order to do comprehensive big data analysis [23].

3.2 *Telematics, Internet of Things, Wearables*

Big data has had an effect on the insurance industry through Internet of Things, wearable technology, and telematics [17]. These technologies may be used in health insurance, home insurance, and vehicle insurance. Data gathering using sensors is now possible in all these domains: wristbands or wearables that gather biological information [24, 25], home automation sensors for smoke, leakage, and moisture protection, and telematics devices that collect driver's whereabouts and speed in cars [13]. Telematics technology devices sometimes known as black boxes capture driving data, track behavior in hazardous zones, pace, patterns of acceleration, and braking [26]. As a result, these variables may be employed in a more precise and tailored pricing model, enabling previously pooled policyholders to transcend characteristics like age and position themselves as safe drivers who pose limited risk and are unlikely to get into a crash [17]. Wearables give the insurer a way to identify the genuine underlying risks of the insured policyholder in a manner like telematics. Although wearables are often thought of as wrist-worn gadgets, this technology is already producing vast amounts of data from a range of sensors implanted in devices. Similar to the use of telematics, these devices are becoming more accessible to the general population, and insurers are taking advantage of the influx of new information to improve their pricing models [17]. With telematics and wearables, IoT makes a variety of new data streams relevant to the insurance sector available. The IoT devices can exchange data without requiring human or computer intervention, commonly referred to smart devices connected through the internet e.g., wearables, other devices with sensors, etc. [12]. Applications range from environmental monitoring, such as assessing the quality of the air and water, to smart home devices, such as smoke detector, monitoring water pressure to locate potential leaks [17]. Activity trackers too use the IoT to motivate clients to exercise and then provide lower premiums when they achieve their health goals [1]. Insurers may more effectively focus on underwriting and adjust rates to meet customized requirements by incorporating big data [27].

3.3 Artificial Intelligence

Machine intelligence, such as problem-solving and reasoning, is referred to as artificial intelligence [28, 29]. Smart machines do tasks that are normally performed by people. The application of Artificial Intelligence in health and accident insurance is now being embraced by the insurance sector as well. Insurance companies now have access to useful information on the health of their customers because of innovations like sensors and AI-powered smartwatches [17]. Additionally, this AI system will advise and alert clients about bad lifestyle choices, possibly leading in lesser prices for both insurers and insured. AI assists in detecting fraudulent claims and managing the claim management process through apps [17, 30]. Insurance brokers made chatbot investments to offer instant insurance quote comparisons through online platforms [31]. Such chatbots serve as a virtual personal counselor. AI-powered chatbots facilitate quicker and easier user engagement with present and prospective clients than when a person is present [17]. When compared to traditional approaches, artificial intelligence allows insurers to improve customer segmentation through pattern recognition, for example, by removing human subjectivity, discovering underlying patterns, identifying customized micro-segments [29].

3.4 Blockchain

A distributed ledger is a decentralized database that uses encryption keys linked blocks to store information securely and completely on distributed general ledgers. The blockchain is the most popular type of distributed ledger technology and it stores transactions that cannot be altered by unauthorized person [29]. To authenticate a transaction in the database, strong mechanisms based on crypto-economic protocols are utilized. This is also the reason that blockchain is associated with "trust" and because transactions tied to it do not call for further intermediary permission [32]. Initially, blockchain was created to facilitate the purchase and sale of bitcoin. The core notion of blockchain is to make numerous copies of a record(s) that are spread over various systems and are public while also prohibiting them from being edited; when one copy is updated, all copies are updated simultaneously [2]. Blockchain is a system that uses blocks to create records of transactions and encrypts the data so that only authorized individuals can access it [33]. Blockchain technology can be extremely helpful in the insurance business, where quick processes are necessary for things like paying premiums, filing claims for damages, and payment of claims. By keeping the data in a separate database that can be quickly updated or deleted by giving approved access to the authorized persons involved with the transactions, encoding or decoding may be applied to safeguard the data and will cut down on time spent maintaining physical records as well [33].

3.5 Smart Contracts

Another use for the blockchain is smart contracts, which are computer protocols that make it easier for sellers and purchasers to carry out transactions without the need for intermediaries. This system enables insurers to stay clear of the expensive expenditures associated with contract preparation, agents' involvement, opportunistic behavior, and the misunderstandings that come with written language. To succeed where EDIs failed, smart contracts were created in a more technologically advanced manner. These programming languages provide greater scalability and a wider range of features. They can seamlessly interface with the operational and financial systems that are the backbone of modern businesses. Only very basic and underdeveloped digital environments might support EDIs since they need human intervention. The core characteristics of smart contracts are automation, decentralization, and anonymity, each of which has pros and cons. In unstable or unclear circumstances, creating smart contracts is quite expensive. They are responsive and adaptable at the same time, enabling the parties to integrate business standards in the contract.

3.6 Robot Advisors

Robot advisors are often built as an automated interface to imitate consumer interaction with actual human advisors. The procedure through which clients contact insurance intermediaries and obtain advice has been completely computerized. Yet, the technologies behind robot advisors might vary greatly. On the one hand, many insurance distribution robot advisors are driven by rule-based algorithms that are meant to offer predefined answers and perform specified actions. Insurance intermediaries can be relieved of recurring, monotonous tasks by rule-based virtual advisor. Virtual assistants and chatbots are used to enhance contact between insurance agents and clients. Consumers should anticipate a faster response to their issues even beyond business hours. As compared to rule-based systems, such robot advisors are revolutionary facilitators, capable of producing deductive reasoning advice and improving it over time. Insurance intermediaries have a greater grasp of their customers' risk profiles thanks to learning-based systems that capture behavioral data, allowing them to personalize insurance for consumers in accordance with data-driven algorithms. Consumers may also expect a quick and economical approach to personalized assistance without having to deal with insurance dealers for hours on end [34].

3.7 Mobile Applications

A greater audience may now be reached by numerous businesses thanks to the network effect of mobile phones and the creation of applications for these devices. Depending on the generation of mobile networks in use and the most popular phone kinds, mobile technology may be benefiting InsurTech in various ways. App-based advancements are made possible by smartphones and internet connectivity. Mobile networks that support big data transfers, prepaid mobile phones, and brief messages are required for this. This is especially important for developing countries with low insurance penetration rates and weak distribution networks [35].

3.8 RegTech

Insurance regulators are also impacted by the growing "InsurTech" involvement in the insurance sector, which puts additional pressure on them to keep up with technological advancements and requires them to swiftly adjust to the new reality. Regulators must strike a difficult balance between their traditional responsibilities for ensuring financial stability and consumer protection, on the one hand, and the need to promote free competition within the relevant market and avoid stifling innovation to meet consumers' and the market's constantly changing needs, on the other [18].

The operations of the insurance business are not the only ones affected by the InsurTech sector's rapid expansion; the activities of the relevant regulatory bodies are also severely hampered. When regulators exercise their supervisory competencies and powers, new InsurTech apps of a wide variety and with very diverse nature, understandings, and operations, new methods and channels of product distribution, new forms of cooperation between industry players, and even the entry of InsurTech start-ups add additional complexity.

4 Comparison Between Traditional Insurance Services and InsurTech

Traditional insurance	InsurTech
Conventional methods of providing insurance services and products without the aid of technology	Starts with providing modern insurance and products and adds data and technology to it
Begins as a purely financial arrangement	The customer relationship is intended to be improved by technology
Not technologically advanced or data-driven	Powered by data and technologically capable of specializing in insurance-related activities throughout the value chain
Not likely to cause disruption and likely to be replaced	Disruptive in nature
Lack of inventiveness in designing the policies and customization is difficult	Highly inventive, allowing for cheaper, faster, and easier access

5 Analyzing Gaps in Literature

There are few analytical studies that focus on the long- and medium-term changes in the structure of different insurance innovative products that occur as a result of InsurTech. Also, even though many studies examine InsurTech from the perspective of the client, in-depth research on how it is affecting businesses in terms of numbers is lacking. There is a huge need for more research on themes like fraud, identity theft, and financial crimes with reference to the former, as well as the evolving insurance purchasing platforms and client attitudes about them. Furthermore, there has not been much research done on the disparity between insurance technology and the rapidly expanding regulatory framework. RegTech and InsurTech are relatively recent concepts that sprang from FinTech; therefore their potential has not yet been completely realized [19]. Although there is literature that describes their characteristics, evolution, and influence, analytical works that elaborate their potential are hard to come across [20]. To keep up with the advances, various regulatory agencies at the national level have proposed and implemented special regulations. It seems that the whole financial industry is aware of FinTech, but InsurTech regulations are receiving less attention, and there are various issues to be addressed in this research domain. It will be necessary to develop the regulatory framework for the emerging InsurTech industry, considering its interaction with established financial services companies [18]. There will be a need for a competent regulator as more InsurTech services may become available in the financial industry. The geographical extent of the regulators' jurisdiction will need to be addressed in the context of new InsurTech services, which target a global audience. Another area with considerable study potential is this one.

Future research may be conducted using analytical approaches like thematic analysis to find and locate new ideas, definitions, and terminologies in the InsurTech domain as there have been very few studies entirely focused on evaluating the InsurTech literature.

6 Challenges Ahead

Not many clients can easily acquire it due to the standard insurance products' considerably higher costs, their narrower reach, and their stringent rules. IoT and InsurTech integration is currently acknowledged as a cutting-edge technology that has the potential to alter the way insurance solutions are provided and threaten the very viability of conventional insurers. Data and information about users must be safeguarded against breaches and other threats since InsurTech uses a significant quantity of data, much more than do traditional insurance companies. Due to cultural barrier, the insurance sector has been relatively hesitant to accept new technology. Yet, this resistance needs to be removed for the sector to stay competitive. The capacity of organizational structures to thrive will diminish if they are rigid and unable to take full use of the opportunities offered by digitalization and big data. Insurers adopting new technology often face constraints from poorly integrated IT systems and technical skills gaps. Cumbersome legacy technology limits their ability to rapidly launch new products and react to the change [35]. The creation of several distribution channels will probably increase expenses for insurers, at least temporarily. This reflects both the greater upfront expenses of, for instance, building a strong brand, which is sometimes necessary for direct, e-commerce sales, as well as the increased operational complexity of maintaining many more channels and potential points of client connection. This calls on insurers to prioritize increasing productivity across all channels of distribution. The effectiveness of both established and emerging digital channels may be more accurately monitored with the use of smart analytics. While the benefits that robot advisors provide, the hazards that they entail should not be underestimated. These intelligent algorithms' inaccurate recommendations might lead to some monetary costs or, in the worst-case situation, unjustly restrict certain consumers from obtaining coverage or force them to incur a substantially greater price for less benefits. The experimental study has found that when people are exposed to automated choices, they develop mistrust, a phenomenon known as algorithmic aversion. Regulatory safeguards must be put in place to guarantee that robot advisors are reliable [34].

7 Conclusion

Some of the most active fields of InsurTech research are highlighted with an overview of technologies including big data, Blockchain, IoT, etc. The InsurTech business model is based on these technologies and is supposed to overcome difficulties in the traditional value chain, including delays in resolving claims, processing papers, updating policies, and premature termination of the policy. These technologies will continue to have a revolutionary effect on the insurance sector. With the help of developments in AI, a wealth of data from fields like telematics, wearables, and IoT is enabling the creation of far more individualized products. These data sources are also beginning to reveal a totally different kind of customer connection, which enables the insurer to interact with policyholders more deeply. Big data will therefore probably continue to drastically change the sector in the coming years. The insurance industry as we now know it is beginning to be disrupted by blockchain, telematics, IoT, and AI. It is still unknown how much disruption will occur and how exactly it will occur. However, it appears that as the digital innovation in the insurance industry expands, InsurTech will play a larger role.

References

1. Yan, T.C., Schulte, P., Lee Kuo Chuen, D.: InsurTech and FinTech: Banking and Insurance Enablement. Elsevier Inc. (2018)
2. Nicoletti, B.: Insurance 4.0 benefits and challenges of digital technology (2021)
3. Łyskawa, K., Kędra, A., Klapkiv, L., Klapkiv, J.: Digitalization in insurance companies. In: International Scientific Conference "Contemporary Issues in Business, Management and Economics Engineering, May, pp. 9–10 (2019). https://doi.org/10.3846/cibmee.2019.086
4. Nam, S.O.: How much are insurance consumers willing to pay for blockchain and smart contracts? A contingent valuation study. Sustainability, **10**(11) (2018). https://doi.org/10.3390/su10114332
5. Heinonen, K., Strandvik, T.: Reframing service innovation: COVID-19 as a catalyst for imposed service innovation. J. Serv. Manag. **32**(1), 101–112 (2021). https://doi.org/10.1108/JOSM-05-2020-0161
6. Lanfranchi, D., Grassi, L.: Examining insurance companies' use of technology for innovation. Geneva Pap. Risk Insur. Issues Pract. (2021). https://doi.org/10.1057/s41288-021-00258-y
7. Volosovych, S., Zelenitsa, I., Kondratenko, D., Szymla, W., Mamchur, R.: Transformation of insurance technologies in the context of a pandemic. Insur. Mark. Co. **12**(1), 1–13 (2021). https://doi.org/10.21511/ins.12(1).2021.01
8. Stoeckli, E., Dremel, C., Uebernickel, F.: Exploring characteristics and transformational capabilities of InsurTech innovations to understand insurance value creation in a digital world. Electron. Mark. **28**(3), 287–305 (2018)
9. McFall, L.: Personalizing solidarity? The role of self-tracking in health insurance pricing. Econ. Soc. **48**(1), 52–76 (2019). https://doi.org/10.1080/03085147.2019.1570707
10. Dijksterhuis, C., Lewis-Evans, B., Jelijs, B., Tucha, O., de Waard, D., Brookhuis, K.: In-car usage-based insurance feedback strategies. A comparative driving simulator study. Ergonomics **59**(9), 1158–1170 (2016). https://doi.org/10.1080/00140139.2015.1127428
11. Wargin, J.: 11 insurance technology trends for 2021 (2020). https://www.duckcreek.com/blog/insurance-technology-trends/. Accessed 21 Jan 2022
12. Lynn, T., Mooney, J.G., Rosati, P., Cummins, M.: Disrupting finance: fintech and strategy in the 21st century (2018)
13. Barry, L., Charpentier, A.: Personalization as a promise: can big data change the practice of insurance? Big Data Soc. **7**(1) (2020). https://doi.org/10.1177/2053951720935143
14. De Mauro, A., Greco, M., Grimaldi, M.: A formal definition of big data based on its essential features. Libr. Rev. **65**(3), 122–135 (2016). https://doi.org/10.1108/LR-06-2015-0061/FULL/HTML

15. Ravi, V., Kamaruddin, S.: Big data analytics enabled smart financial services: opportunities and challenges. In: International Conference on Data Science and Big Data Analysis, pp. 15–39. Springer, Cham (2017). https://doi.org/10.1007/978-3-319-72413-3_2
16. Owadally, I., Zhou, F., Otunba, R., Lin, J., Wright, D.: Time series data mining with an application to the measurement of underwriting cycles. North Am. Actuar. J. **23**(3), 469–484 (2019). https://doi.org/10.1080/10920277.2019.1570468
17. Cortis, D., Debattista, J., Debono, J., Farrell, M.: Disrupting Finance, pp. 71–84
18. Dahiya, M., Sharma, S., Grima, S.: Big data analytics application in the indian insurance sector. Big Data Anal. Insur. Mark., 145–164 (2022). https://doi.org/10.1108/978-1-80262-637-720221008/FULL/HTML
19. Pentland, A.: Social physics: how good ideas spread-the lessons from a new science. Penguin (2014). https://doi.org/10.1007/s11138-014-0276-6
20. Pusala, M.K., Amini Salehi, M., Katukuri, J.R., Xie, Y., Raghavan, V.: Massive data analysis: tasks, tools, applications, and challenges. Big Data Anal. Methods Appl., 11–40 (2016). https://doi.org/10.1007/978-81-322-3628-3_2/COVER
21. Sivarajah, U., Kamal, M.M., Irani, Z., Weerakkody, V.: Critical analysis of big data challenges and analytical methods. J. Bus. Res. **70**, 263–286 (2017). Accessed 17 Dec. 2022. https://www.sciencedirect.com/science/article/pii/S014829631630488X
22. EIOPA: Big data analytics in motor and health insurance: a thematic review (2019)
23. BaFin: Big data meets artificial intelligence: challenges and implications for the supervision and regulation of financial services (2018). https://www.bafin.de/SharedDocs/Downloads/EN/dl_bdai_studie_en.html
24. Gilmore, J.N.: Everywear: the quantified self and wearable fitness technologies, **18**(11), 2524–2539 (2015). https://doi.org/10.1177/1461444815588768
25. Lupton, D.: The diverse domains of quantified selves: self-tracking modes and dataveillance. Econ. Soc. **45**(1), 101–122 (2016). https://doi.org/10.1080/03085147.2016.1143726
26. Azzopardi, M., Cortis, D.: Implementing automotive telematics for fleet insurance. J. Technol. Manag. Innov. **8**(4), 9–10 (2013). https://doi.org/10.4067/S0718-27242013000500005
27. Eling, M., Lehmann, M.: The impact of digitalization on the insurance value chain and the insurability of risks. Geneva Pap. Risk Insur. Pract. **43**(3), 359–396 (2018)
28. Kelley, K.H., Fontanetta, L.M., Heintzman, M., Pereira, N.: Artificial intelligence: implications for social inflation and insurance. Risk Manag. Insur. Rev. **21**(3), 373–387 (2018). https://doi.org/10.1111/RMIR.12111
29. Eckert, C., Osterrieder, K.: How digitalization affects insurance companies: overview and use cases of digital technologies. Zeitschrift für die gesamte Versicherungswiss. **109**(5), 333–360 (2020)
30. Eling, M., Lehmann, M.: The impact of digitalization on the insurance value chain and the insurability of risks. Geneva Pap. Risk Insur. Issues Pract. **43**(3), 359–396 (2018). https://doi.org/10.1057/s41288-017-0073-0
31. Riikkinen, M., Saarijärvi, H., Sarlin, P., Lähteenmäki, I.: Using artificial intelligence to create value in insurance. Int. J. Bank Mark. **36**(6), 1145–1168 (2018). https://doi.org/10.1108/IJBM-01-2017-0015/FULL/HTML
32. Davidson, S., De Filippi, P., Potts, J.: Blockchains and the economic institutions of capitalism. J. Inst. Econ. **14**(4):639–658 (2018). https://doi.org/10.1017/S1744137417000200
33. Shah, T., Shah, M.: Blockchain technology: a way ahead in insurance sector. Sambodhi **43**(4), 6–8 (2020). https://papers.ssrn.com/sol3/papers.cfm?abstract_id=3803773
34. Marano, P., Li, S.: Regulating robo-advisors in insurance distribution: lessons from the insurance distribution directive and the AI act. Risks **11**(1), 1–13 (2023). https://doi.org/10.3390/risks11010012
35. Cappiello, A.: Digital disruption and insurtech start-ups: risks and challenges. In: Technology and the Insurance Industry, pp. 29–50 (2018)

36. Kavuri, A.S., Milne, A.: FinTech and the future of financial services: What are the research gaps? (2019)
37. Gomber, P., Koch, J.A., Siering, M.: Digital finance and FinTech: current research and future research directions. J. Bus. Econ. **87**(5), 537–580 (2017). https://doi.org/10.1007/s11573-017-0852-x

Review on Vision Transformer for Satellite Image Classification

Himanshu Srivastava, Akansha Singh, and Anuj Kumar Bharti

Abstract Satellite image classification has been a topic of interest in the research community for the last two decades. Neural network researchers have given advanced models year by year for this problem. A lot of attention is given to the transformers from 2017 onwards and the vision transformer is a transformer-based model for computer vision problems. This paper reviews the application of vision transformers (and its variants) to the satellite image classification. The article provides a detailed working of the vision transformer and the history of its chronological development. The review prospects that the vision transformer is suitable for a sufficiently large dataset. For small datasets, convolutional network-based models perform well as compared to vision transformers, but pre-trained vision transformers beating convolutional models and transfer learning have produced better results. The review suggests, considering limited research in the field of vision transformer application for satellite data, due to the fairly new model, there are high possibilities for making this model producing good results. The article also explores the ongoing challenges and research opportunities in the vision transformer development for satellite image classification.

Keywords Self-attention mechanism · Vision transformer · Satellite images

H. Srivastava (✉) · A. Singh · A. K. Bharti
Bennett University, Greater Noida, India
e-mail: himanshu.shree@gmail.com

G. Fortino et al. (eds.), *Proceedings of Third International Conference on Computing and Communication Networks*, Lecture Notes in Networks and Systems 917,
https://doi.org/10.1007/978-981-97-0892-5_16

1 Introduction

1.1 Satellite Imagery

Satellite images are highly desirable for measuring and tracking human activities as well as natural activities on the earth. These are the images of earth taken by the satellites, using remote sensing technology. These satellites are operated by the government as well as private organizations for different purposes such as defense, environmental protection, natural disasters, urban planning, etc. Satellites have different sensors for capturing different information such as spatial, spectral, radiometric, geometric, and temporal. Each image can contain a number of bands and is known as visible image, infrared image, multi-spectral image, and hyper-spectral image. Today, satellite image resolution varies from centimeters to kilometers such as very high-resolution images (30 cm^2 per pixel), high-resolution images (3–5 m^2 per pixel), medium-resolution images (10–30 m^2 per pixel), and low-resolution images (60 m^2 per pixel). Each kind of satellite data is utilized in different applications. Satellite images are used in many computer vision tasks such as object detection, recognition, classification, and segmentation. Among these computer vision tasks, classification is more popular and the maximum time used by the computer vision researchers [21]. Since satellite images cover a larger area on the earth hence its database is very large and processing is very costly and challenging. Traditional methods of image classification suggested by Ojha et al. [19] were working on low level, global and local features of an image but the latest technology like deep learning is working on high level features of an image and getting better performance [12]. Nowadays, convolutional neural network (CNN)-based transfer learning is quite popular for the satellite image classification. But the latest technique is a vision transformer (ViT) that outperforms pre-trained convolutional neural networks for satellite image classification. Vision transformers are not only used in classification but also utilized in many computer vision problems, including image processing, object detection, segmentation, and video understanding.

1.2 Technological Background on Vision Transformer

Today, vision transformer is the latest technique while analyzing satellite images. Vision transform for computer vision tasks is highly inspired by transformers that come from natural language processing. It is necessary to dive into the neural machine translation concept before going through vision transformers. Neural machine translation is a very important application of natural language processing. Researchers proposed encoder-decoder based sequence to sequence models for neural machine translation such as Cho et al. [8] proposed recurrent neural network (RNN)-based encoder-decoder model, and Sutskever et al. [22] proposed long short-term memory

(LSTM)-based encoder-decoder model. In these models, the encoder took the variable length input and gave a fixed length vector representation of the input data. This fixed length vector representation is utilized by the decoder for creating variable length output sequences. These models were giving better performance as compared to previous models. Later on, it can be seen that encoder-decoder models were working well only for short sentences. It means these models were not capturing long sentence dependencies and were creating a bottleneck in terms of model performance. These issues are resolved by the attention mechanism inside the encoder-decoder model and given by Bahdanau et al. [3]. In the attention mechanism, all the hidden states produced by all the encoder blocks move towards the decoder stage. At the decoder side, it selects the hidden states based upon attention score for producing the output sequence. This approach frees up the model from generating fixed length context vectors and focuses only on relevant information for generating the output. This new approach achieves better translation performance as compared to state-of-the-art methods. This experiment shows that the BLEU score generated by this model is better for 30 and 50 word long sentences as compared to the older models and it is not degraded with the lengthy sentences. Afterwards, many researchers started doing modifications in the working of attention mechanisms and released new versions such as local attention mechanism and global attention mechanism given by Luong et al. [15], self-attention mechanism by Cheng et al. [7], and multi-head self-attention mechanism.

Attention mechanism is combined with RNN and LSTM and became the main architecture for doing any text-related task until 2017, after it, a new architecture, i.e. transformer came into light. Transformer is a deep learning model for natural language processing which adopts the concept of self-attention mechanism and was given by Vaswani et al. [2]. This is the first sequence model entirely based on self-attention mechanism where the recurrent layer in encoder-decoder is replaced by multi-head self-attention layers. This model is utilizing query, key, and value concepts for calculating self-attention score. Self-attention is the recent development in capturing long-range dependencies. It works by correlating different positions of the same sequence sentence and trying to create a representation of the given sequence. This model outperforms the other available models by getting a higher BLEU score with different tasks such as English to German and English to French translation. Subsequently, the transformer became a well-organized architecture for neural machine translation. Inspired by the success of transformers, researchers used the concept of transformers for image classification tasks. Some researchers combined self-attention with convolutional models. It was very challenging and complicated as self-attention required one pixel to attend every other pixel. Wang et al. [25] proposed the recurrent attention-based structure for image classification. This model selectively focuses on regions while creating high level features and processing them sequentially. Bello et al. [5] achieve improved classification results when some convolutional layers are replaced with self-attention layers on ImageNet but face some computational complexities. Ramachandran et al. [20] introduced a purely self-attention-based model without any convolutional layer for image classification and this was the first model which adopted self-attention as a stand-alone building block for image-related tasks. Wu et al. [26] used a transformer on top of the convolutional models. This model extracted features from convolutional models and fed them to the visual transformer and got the output. Chen et al. [6] used transformer architecture on low-resolution images for classification tasks. This model reshaped images such as long sequences of pixels.

1.3 Vision Transformer

Transformer has become popular for natural language processing and motivated by the success of this architecture. Dosovitskiy et al. [1] proposed a vision transformer for the computer vision community. Vision transformer shows that dependency of the computer vision community on convolutional neural networks no longer exists. A pure transformer can be applied to the sequence of image patches. Their experiment shows that this model outperforms convolutional networks when trained with larger dataset and with lesser computational resources. When the model is trained on a medium size dataset such as ImageNet-1 k, this model is not giving accuracy better than the CNN models. But this scenario changes when the model is trained with larger datasets such as ImageNet-21 k and JFT-300 M dataset. A comparison of the vision transformer model pre-trained on JFT-300 and the CNN model is shown in Table 1 and is given by Dosovitskiy et al. [1].

The architecture of the original vision transformer is shown in Fig. 1. It uses only encoders, no decoder is required. Input image is divided into equal size patches and these patches go through a linear layer for creating patch embedding. Model creates positional embedding for each patch and this positional embedding is added with patch embedding. The model utilized positional embedding to keep information about the position of each patch within an image. The model adds one extra learnable token (CLS token) at the front of the embedding sequence. Now, these sequences go to the encoder as input and the encoder produces output by the multi-layer perceptron (MLP) head. The MLP head is responsible for predicting the final class using the softmax function for a given image. Vision transformer architecture has an L number

Table 1 ViT outperforms state-of-the-art CNN models

Dataset/model	ViT-H/14 accuracy (%)	Previous SOTA accuracy (%)
ImageNet	88.55	87.54
ImageNet-ReaL	90.72	90.55
Cifar-10	99.50	99.37
Cifar-100	94.55	93.51
Pets	97.56	96.62
Flowers	99.68	99.63
VTAB	77.63	76.29

of identical encoders. Each encoder has two main blocks, (1) a multi-head self-attention (MHA) block and (2) a multi-layer perceptron block. Each block uses skip connections and is preceded by a normalization layer. First block is based on self-attention technique and extended to a number of heads. The self-attention for each element in the sequence is calculated by generating three vectors of the Q-Query, K-Key, and V-Value. These vectors are generated by multiplying each element with learnable query, key, and value matrices. Now, single-head self-attention is calculated by the given formula in Eq. (1).

$$\text{Attention\_Head}(Q, K, V) = \text{Softmax}\left(Q \cdot K^T/(d_k)^{1/2}\right) \cdot V \tag{1}$$

Where d_k represents the dimension of the key vector. Second block, multi-layer perceptron having two linear layers with Gaussian error linear unit (GELU) activation function [9].

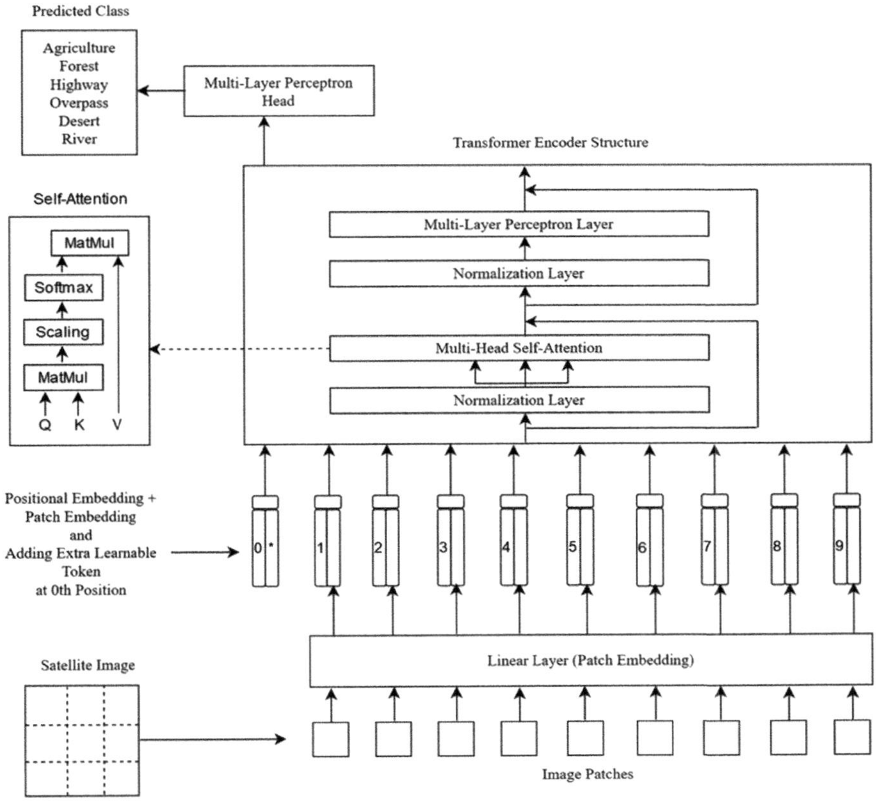

Fig. 1 Plain vision transformer architecture

Table 2 Vision transformer variants

Model	Layers	Heads	MLP count	Hidden count	Parameters (M)
ViT-base	12	12	3072	768	86
ViT-large	24	16	4096	1024	307
ViT-huge	32	16	5120	1280	632

Original vision transformer is available in three versions, i.e. ViT-Base, ViT-Large, and ViT-Huge. Each version has different hyperparameters in terms of number of layers or blocks, hidden size, MLP size, number of heads, and total trainable parameters. A detail of original vision transformer variants can be seen in Table 2 and is given by Dosovitskiy et al. [1].

After Dosovitskiy et al. [1], a number of researchers have published different versions of vision transformers for the image classification. Out of them, the Swin transformer proposed by Liu et al. [14] is a recent one that is gaining popularity.

There are various factors that play a role in the efficiency of a vision transformer. Hyperparameters are crucial in case of any neural networks. Hence, tuning of hyperparameters is a key factor for getting an efficient vision transformer. Hyperparameters include learning rate, batch size, weight decay, dropout, image resolution, optimization algorithm, patch size, training epochs, data augmentation, etc. The hyperparameters of each vision transformer model are optimized to give the best solution.

Vision transformer efficiency also depends upon transfer learning techniques. Here, transfer learning techniques play an important role when working with smaller datasets. Vision transformer models effectively work with larger datasets. Therefore, transfer learning technique is used for obtaining effective model performance when working with limited datasets.

Data augmentation technique is very important while working with deep learning models. Deep learning models such as vision transformers require a lot of data to train and this can be fulfilled by data augmentation techniques.

A vision transformer model is always treated as black box model because of their computational complexity. Interpretability helps the users to understand why the model is making a particular classification. Vision transformer explainability can be supportive where understanding the prediction is necessary like satellite image classification. Unfortunately, very little work is done for the explainability of the transformer.

2 Literature Review

Many researchers have introduced satellite image classification models using a vision transformer and its variants. Some researchers also combine convolutional neural networks and vision transformers for doing satellite image classification. Bazi et al.

[4] utilized a vision transformer model for satellite image classification. Experiment is performed on different datasets with the greater accuracies on UCM, AID, Optimal-31, and NWPU datasets. Lv et al. [16] modified the standard vision transformer for preserving spatial channel features and introduced two new blocks such as progressive aggregation (PA) and lightweight channel attention layer (LCA). PA layer is used to extract more local information by combining overlapped patches and LCA is utilized to give weightage to different channels in the classification token. This experiment used UCM, AID, and NWPU datasets and achieved better results. Mei et al. [17] incorporate grouped pixel embeddings (GPE) before encoder blocks inside vanilla vision transformers for spatial-spectral feature-based hyperspectral image classification. Transformer encoder with multi-head self-attention mechanism is used for creating long-range global dependencies, and GPE is used to make local relationships for spatial-spectral features of the hyperspectral images. This work is accomplished with satisfactory accuracies on SA, PU, WHU-LK, and HR-L datasets. Tang et al. [23] made an effort to combine the advantages of convolutional network and visual transformer and created four modules. Multi-layer feature extraction module extracts global visual information and multi-level convolutional features using ResNet34. Next, the contextual information extraction module extracts very rich contextual information from multi-level convolutional features. Thereafter, a cross-level attention module is used to aggregate multi-level features. Finally, the class score fusion module is used to combine the contribution of global and aggregated multi-level features for scene classification. This experiment utilized UCM, AID, and NWPU datasets and achieved better results. Wang et al. [24] proposed slightly changed vision transformer architecture and implemented on UCM, AID, and NWPU datasets with improved performance than plain vision transformers. This model replaced the original attention mechanism with rotated varied size window attention for extracting maximum information from generated windows and at the end global average pooling layer is used for the classification purpose. Kaselimi and Doulamis [11] proposed the ForestViT model for multilabel satellite image classification of deforestation and achieved better results as compared to other convolutional-based models. They have analyzed different benchmarking protocols on 12 different multilabel classes and utilized three accuracy metrics such as per class accuracy, overall accuracy, and multilabel accuracy. Liu et al. [13] introduced a model for extracting the high-resolution SAR image local feature using a fast convolutional neural network and global features by using a compact vision transformer and finally applying a fusion net to fuse global and local features. This experiment used Gaofen-3 SAR and F-SAR datasets, and this model outperformed the result for both datasets. Nabi et al. [18] attempted to incorporate knowledge distillation concept into vision transformer and convolutional neural network. This experiment utilized a student-based visual transformer and teacher-based convolutional neural network for single-label scene classification. The model got better accuracies on AID, optimal-31, and NWPU datasets. A new approach is presented by Horry et al. [10] where two speed networks such as convolutional neural network and vision transformer are combined and perform staggered training for satellite images. These ensemble models utilized

Table 3 Satellite Image Classification with Vision Transformer and Variants

Authors	Dataset with accuracy (%)
Bazi et al. [4]	UCM (98.49), AID (95.86), Optimal-31 (95.56), NWPU (93.83)
Lv et al. [16]	UCM (99.57), AID (96.98), NWPU (94.66)
Mei et al. [17]	SA (96.8), PU (97.87), WHU-LK (99.26), HR-L (82.85)
Tang et al. [23]	UCM (99.57), AID (96.41), NWPU (93.65)
Wang et al. [24]	UCM (99.70), AID (97.76), NWPU (94.81)
Liu et al. [13]	Gaofen-3 (91.10)
Horry et al. [10]	So2SaT (92.00)
Nabi et al. [18]	UCM (99.05), AID (96.74), Optimal-31 (96.9), NWPU (93.88)

Table 4 Available datasets for satellites image classification

Dataset name	Source	Classes	Images	Image resolution
WHU-RS19	Google Earth	19	1,005	600 × 600
UCM	USGS	21	2,100	256 × 256
NWPU-RESISC45	Google Earth	45	31,500	256 × 256
AID	Google Earth	30	10,000	600 × 600
Optimal-31	Google Earth	31	1,860	256 × 256
PatternNet	Google Earth	38	30,400	256 × 256
EuroSAT	Sentinel-2A	10	27,000	64 × 64

the So2Sat LCZ42 dataset that is divided into four intervals for staggered training and finally outperformed each of the individual models.

Latest work by the researchers for satellite image classification using different variants of vision transformers can be seen in Table 3.

Dataset plays an important role when analyzing variants of vision transformer models. There exist a number of datasets for satellite image classification as shown in Table 4. Among them, UCM, AID, and NWPU are very popular datasets.

3 Open Challenges and Future Scope

Till now, many variants of vision transformers have been discussed. These models are giving better performance as compared to the existing methods and chronologically performing well. When analyzing satellite images with vision transformer models, researchers have to face many challenges. Despite these challenges, the research study opens a path for future research. Some areas for future investigation include: (1) Huge data handling: Satellite images cover a larger area on the earth hence its size is very large. Thus, managing and processing such a huge amount of data is a challenging job. It can be handled with big data and cloud computing technologies.

(2) Image quality: Satellite images are taken by the satellites, therefore, environmental conditions can affect the image quality. There is a need for good models that can handle poor quality data. (3) Inherent characteristics: Vision transformer model works on supervised learning techniques. It requires labeled data but unfortunately less labeled datasets for satellite images are available. (4) Scalability: It is an issue in terms of model size, number of models, and computing power that fulfills computational demand. It can be done by powering the computer hardware and parallel programming concepts. (5) Privacy issues: Satellites are taking images of the property, hence there exist privacy issues because owners may not wish to share their property. (6) Lack of availability: Satellite images exist in public domain and private domain but private domain datasets are not available free of cost. Therefore, research is only limited to the public domain datasets for ordinary researchers. (7) Model interpretability: Deep learning is widely used but when deploying in real-time applications there exist security and reliability issues. Any loop hole either accidentally or intentionally could result in wrong prediction. (8) Slower training: Another limitation is the slow training process because models have to query against each pixel but it can be accelerated with optimization algorithms and transfer learning concepts. (9) Hybrid approach: Combining many deep learning models such as CNN and vision transformer can enhance satellite image classification model performance. Future research can inspect hybrid architectures for enhancing model computation. The growth of artificial intelligence is remarkable and solutions would be available in the near future.

4 Conclusion

Deep learning techniques are popular while working with satellite image classification. This paper presents the latest deep learning technique known as vision transformer for satellite image classification. A technological evolution of the transformer and its application in satellite images are introduced in this paper. Many researchers proved that the performance of standard visual transformers can be improved by slight modification in the architecture. Some researchers tried to enhance the performance of the model by combining the advantage of both the convolutional neural networks and vision transformers. Still, researchers have many challenges related to satellite image processing. These challenges can be resolved soon. Due to elevation in the sensor technology and launching a number of satellites, the importance of satellite images is boosted. There exist many applications of satellite images where deep learning methods such as convolutional neural networks and vision transformers can be benefitted. But, unfortunately, little literature exists, it means researchers need to concentrate on this highly valuable field, i.e. satellite image classification. Therefore, the analysis of satellite images with deep learning models such as vision transformers can be benefitted to our society and country.

References

1. Dosovitskiy, A., Beyer, L., Kolesnikov, A., Weissenborn, D., Zhai, X., Unterthiner, T., Dehghani, M., Minderer, M., Heigold, G., Gelly, S., Uszkoreit, J., Houlsby, N.: An image is worth 16 × 16 words: transformers for image recognition at scale (2020). https://doi.org/10.48550/arXiv.2010.11929
2. Vaswani, A., Shazeer, N., Parmar, N., Uszkoreit, J., Jones, L., Gomez, A.N., Kaiser, L., Polosukhin, I.: Attention is all you need (2017). arXiv:1706.03762
3. Bahdanau, D., Cho, K., Bengio, Y.: Neural machine translation by jointly learning to align and translate (2014). https://doi.org/10.48550/arXiv.1409.0473
4. Bazi, Y., Bashmal, L., Rahhal, M.M.A., Dayil, R.A., Ajlan, N.A.: Vision transformers for remote sensing image classification. Remote Sens. **13**(3), 516 (2021). https://doi.org/10.3390/rs13030516
5. Bello, I., Zoph, B., Vaswani, A., Shlens, J., Le, Q.V.: Attention augmented convolutional networks. In: Proceedings of the IEEE/CVF International Conference on Computer Vision, pp. 3286–3295 (2019). https://doi.org/10.48550/arXiv.1904.09925
6. Chen, M., Radford, A., Child, R., Wu, J., Jun, H., Luan, D., Sutskever, I.: Generative pretraining from pixels. In: International Conference on Machine Learning, pp. 1691–1703. PMLR (2020, November)
7. Cheng, J., Dong, L., Lapata, M.: Long short-term memory-networks for machine reading (2016). https://doi.org/10.48550/arXiv.1601.06733
8. Cho, K., Van Merriënboer, B., Gulcehre, C., Bahdanau, D., Bougares, F., Schwenk, H., Bengio, Y.: Learning phrase representations using RNN encoder-decoder for statistical machine translation (2014). https://doi.org/10.48550/arXiv.1406.1078
9. Hendrycks, D., Gimpel, K.: Gaussian error linear units (GELUs) (2016). https://doi.org/10.48550/arXiv.1606.08415
10. Horry, M.J., Chakraborty, S., Pradhan, B., Shulka, N., Almazroui, M.: Two-speed deep-learning ensemble for classification of incremental land-cover satellite image patches. Earth Syst. Environ. **7**(2), 525–540 (2023). https://doi.org/10.1007/s41748-023-00343-3
11. Kaselimi, M., Voulodimos, A., Daskalopoulos, I., Doulamis, N., Doulamis, A.: A vision transformer model for convolution-free multilabel classification of satellite imagery in deforestation monitoring. In: IEEE Transactions on Neural Networks and Learning Systems (2022). https://doi.org/10.1109/TNNLS.2022.3144791
12. Kumar, S., Gupta, S.K., Kaur, M., Gupta, U.: VI-NET: a hybrid deep convolutional neural network using VGG and inception V3 model for copy-move forgery classification. J. Vis. Commun. Image Represent. **89**, 103644 (2022). https://doi.org/10.1016/j.jvcir.2022.103644
13. Liu, X.., Wu, Y., Liang, W., Cao, Y., Li, M.: High resolution SAR image classification using global-local network structure based on vision transformer and CNN. In: IEEE Geoscience and Remote Sensing Letters, vol. 19. IEEE (2022). https://doi.org/10.1109/LGRS.2022.3151353
14. Liu, Z., Lin, Y., Cao, Y., Hu, H., Wei, Y., Zhang, Z., ... Guo, B.: Swin transformer: hierarchical vision transformer using shifted windows. In: Proceedings of the IEEE/CVF International Conference on Computer Vision, pp. 10012–10022 (2021). https://doi.org/10.1109/ICCV48922.2021.00986
15. Luong, M.T., Pham, H., Manning, C.D.: Effective approaches to attention-based neural machine translation (2015). https://doi.org/10.48550/arXiv.1508.04025
16. Lv, P., Wu, W., Zhong, Y., Du, F., Zhang, L.: SCViT: A spatial-channel feature preserving vision transformer for remote sensing image scene classification. IEEE Trans. Geosci. Remote Sens. **60**, 1–12 (2022). https://doi.org/10.1109/TGRS.2022.3157671
17. Mei, S., Song, C., Ma, M., Xu, F.: Hyperspectral image classification using group-aware hierarchical transformer. IEEE Trans. Geosci. Remote Sens. **60**, 1–14 (2022). https://doi.org/10.1109/TGRS.2022.3207933
18. Nabi, M., Maggiolo, L., Moser, G., & Serpico, S.B.: A CNN-transformer knowledge distillation for remote sensing scene classification. In: IGARSS 2022–2022 IEEE International Geoscience

and Remote Sensing Symposium, pp. 663–666. IEEE (2022). https://doi.org/10.3390/rs13204143
19. Ojha, N.K., Srivastava, H., Rakesh, N.: Evaluation of feature extractors for sign language recognition. In: 2021 10th IEEE International Conference on Communication Systems and Network Technologies (CSNT), pp. 381–386. IEEE (2021). https://doi.org/10.1109/CSNT51715.2021.9509588
20. Ramachandran, P., Parmar, N., Vaswani, A., Bello, I., Levskaya, A., Shlens, J.: Stand-alone self-attention in vision models. Adv. Neural Inform. Process. Syst. **32** (2019). https://doi.org/10.48550/arXiv.1906.05909
21. Singh, K.K., Sachan, V.K., Singh, A., Padmanaban, S. (eds.): Deep Learning in Visual Computing and Signal Processing. CRC Press (2022). https://doi.org/10.1201/9781003277224
22. Sutskever, I., Vinyals, O., Le, Q.V.: Sequence to sequence learning with neural networks. Adv. Neural Inform. Process. Syst. **27** (2014). https://doi.org/10.48550/arXiv.1409.3215
23. Tang, X., Li, M., Ma, J., Zhang, X., Liu, F., Jiao, L.: EMTCAL: Efficient multiscale transformer and cross-level attention learning for remote sensing scene classification. IEEE Trans. Geosci. Remote Sens. **60**, 1–15 (2022). https://doi.org/10.1109/TGRS.2022.3194505
24. Wang, D., Zhang, Q., Xu, Y., Zhang, J., Du, B., Tao, D., Zhang, L.: Advancing plain vision transformer toward remote sensing foundation model. IEEE Trans. Geosci. Remote Sens. **61**, 1–15 (2022). https://doi.org/10.48550/arXiv.2208.03987
25. Wang, Q., Liu, S., Chanussot, J., Li, X.: Scene classification with recurrent attention of VHR remote sensing images. IEEE Trans. Geosci. Remote Sens. **57**(2), 1155–1167 (2018). https://doi.org/10.1109/TGRS.2018.2864987
26. Wu, B., Xu, C., Dai, X., Wan, A., Zhang, P., Yan, Z., ... Vajda, P.: Visual transformers: token-based image representation and processing for computer vision (2020). https://doi.org/10.48550/arXiv.2006.03677

Primitive Roots and Euler Totient Function-Based Progressive Visual Cryptography

Vaibhav Chaudhary, Shivendra Shivani, and Divya Srivastava

Abstract Visual cryptography is a branch of cryptography that focuses on the encryption and decryption of images or visual data. The main objective of this paper is to introduce a visual cryptography scheme based on primitive root modulo 251 for encryption of a grayscale image, making it visually incomprehensible to unauthorized users. The mapping algorithm generated enhances security and efficiency of original image and subsequently shares are generated using bit-plane slicing. The generated shares possess high quality, ensuring accurate reconstruction of the original image. Moreover, for Inverse Mapping we took discrete logarithms and recovered image with no major contrast loss.

1 Introduction

Visual Cryptography (VC) can be defined as a cryptographic technique that utilizes visual images to conceal and reveal secret information. Traditional visual cryptography offers a unique and visually appealing approach to secure information sharing, where the decryption process does not require complex computations but relies on the human visual system's ability to perceive patterns and shapes. This technique finds applications in various domains, including secure communication, and is continuously evolving to address emerging security challenges in this digital age [1].

Visual cryptography finds versatile applications in the realm of security and privacy. It allows for secure image sharing, authentication, watermarking, and steganography while protecting digital signatures, medical records, and sensitive information.

V. Chaudhary (✉) · S. Shivani
Thapar Institute of Engineering and Technology, Patiala 147001, Punjab, India
e-mail: vaibhavr7796@gmail.com

S. Shivani
e-mail: shivendra.shivani@thapar.edu

D. Srivastava
Bennett University, Greater Noida 203206, Uttar Pradesh, India
e-mail: divya.srivastava@bennett.edu.in

G. Fortino et al. (eds.), *Proceedings of Third International Conference on Computing and Communication Networks*, Lecture Notes in Networks and Systems 917,
https://doi.org/10.1007/978-981-97-0892-5_17

VC can be used to strengthen voting systems and security of QR codes, safeguarding of copyrights, and to enhance CAPTCHA systems.

2 Related Works

Visual cryptography was first introduced in 1995 by Naor and Shamir in their seminal paper [2]. They proposed a method based on k-out-of-n visual cryptography scheme in which a secret image is divided into shares in such a way that each share discloses only partial information about the original image. Any k of those shares can reveal the secret image by stacking (OR operation) their transparencies, but any k - 1 of them provide no information about it.

Subsequently, **bit-plane slicing** was introduced to generate shares. The value of each pixel in an 8-bit quantized grayscale digital image is typically an integer between 0 and 255 and can be represented uniquely by an 8-bit binary sequence. Consequently, binary bit-plane slicing can convert a grayscale image into eight binary bit-planes [3, 4].

Given a number X in decimal notation, one can convert it into binary form by

$$X = \sum_{k=0}^{M-1} a_k \cdot 2^k, \quad \text{where } a_k \in \{0, 1\}$$

Later, **XOR-based visual cryptography** technique is proposed which uses the XOR operation to combine shares and reveal concealed data. The procedure involves dividing the confidential image into shares, and applying XOR operation between to recover the original image. The original hidden image is revealed visually without complex computational operations. The security of XOR-based Visual Cryptography is due to the fact that individual shares disclose no information about the secret image. The visual nature of the scheme makes it easy to confirm the accuracy of the reconstruction [5].

Researchers have used properties of prime numbers in cryptography for a considerable time. A grayscale image has pixel value from 0 to 255, and 251 is nearest odd prime to 255 [6]. Also, primitive root for the number 251 is an integer that, when multiplied by different powers modulo 251, yields all possible remainders from 1 to 250. Primitive root modulo 251 is significant because we can map each pixel of an image to another unique pixel (one–one mapping) [7].

3 Proposed Algorithm

Definition 1 For $m \geq 1$, the number of integers $\ngtr m$ and co-prime to m and which $\in \mathbb{Z}^+$ is denoted by $\phi(m)$. For example, $\phi(45) = 24$ [7]

The function ϕ is called the ***Euler totient function***.
Also, $\phi(m) = m - 1 \iff$ m is prime.

Theorem 1 *If* $r > 0$ *and* α *is prime , then*

$$\phi(\alpha^r) = \alpha^r - \alpha^{r-1} = \alpha^r \left(1 - \frac{1}{\alpha}\right) [7]$$

Proof For proof, refer [7]. □

Definition 2 Let $m > 1$ and gcd(β, m) = 1. The order of β modulo m is the smallest value of g for which $\beta^g \equiv 1 \pmod{m}$ exists. ($g \in \mathbb{Z}^+$) [7]

Definition 3 β is called the **primitive root** of an integer m if $\gcd(\beta, m) = 1$ and order of β is $\phi(m)$ modulo m [7].

For example, 3 is a primitive root of 7.

Primitive roots exist for every prime number. But, when m is not a prime, primitive root can exist. (For example, 2 is a primitive root of 9.)

Theorem 2 *Let* $\gcd(\beta, m) = 1$ *and* $[\beta_1, \beta_2, \ldots, \beta_{\phi(m)}] \in \mathbb{Z}^+$ *which are relatively prime to m and less than m. If* β *is a primitive root of m, then*

$$\beta, \beta^2, \beta^3, \ldots, \beta^{\phi(m)}$$

are congruent modulo m to $\beta_1, \beta_2, \ldots, \beta_{\phi(m)}$ *in some order [7].*

Proof Since β is co-prime to m, so are the exponents of β; therefore, each β^k is congruent modulo m to some β_i, and $\phi(m)$ numbers in the set $\{\beta, \beta^2, \ldots, \beta^{\phi(m)}\}$ are incongruent; thus, these exponents represent the integers $\beta_1, \beta_2, \ldots, \beta_{\phi(m)}$ not necessarily in same order of appearance. □

Remark The number of primitive roots if they exist of m are exactly $\phi(\phi(m))$ [7].

$\phi(251) = 251 - 1 = 250$

For example, $\phi(\phi(251)) = \phi(250) = \phi(2 \cdot 5^3) = 250 \cdot \left(1 - \frac{1}{2}\right) \cdot \left(1 - \frac{1}{5}\right) = 100$.

So, there are 100 primitive root modulo 251, which are given below:

6, 11, 14, 18, 19, 24, 26, 29, 30, 33, 34, 37, 42, 43, 44, 46, 53, 54, 55, 56, 57, 59, 61, 62, 70, 71, 72, 76, 77, 78, 82, 87, 90, 95, 96, 97, 98, 99, 104, 107, 109, 111, 116, 120, 127, 129, 130, 132, 133, 134, 136, 137, 139, 141, 143, 145, 146, 148, 150, 158, 159, 162, 163, 165, 166, 167, 168, 170, 172, 176, 177, 178, 183, 184, 185, 186, 191, 193, 199, 202, 203, 206, 210, 212, 213, 215, 216, 220, 223, 224, 228, 229, 230, 234, 236, 238, 239, 242, 244, 248.

Each of these numbers can be used as a *key* for mapping function.

Algorithm 1 Mapping algorithm

---

1. Consider any grayscale image of size $M \times N$ and convert it into an array.
2. Round-off all pixels greater than 250.

$$\begin{cases} \text{pixel}(\gamma) = 250, & \text{if } \gamma = 251 \text{ or } \gamma = 252 \text{ or } \gamma = 253 \text{ or } \gamma = 254 \text{ or } \gamma = 255 \\ \text{pixel}, & \text{otherwise} \end{cases}$$

3. Map the image pixels in the array using the function given below:

$$f(x) = (key)^x \mod 251$$

4. Now , use a key for mapping function out of primitive roots modulo 251.
5. Save the mapped pixels array as a new mapped image.

The function $f(x) = (key)^x \mod 251$ which is used in the algorithm takes primitive root modulo 251 as *key*. We know there are 100 primitive root modulo 251, listed above, which can be used as keys for mapping. Each pixel is mapped to a unique pixel which is a direct consequence of Theorem 2, resulting in a perpetually meaningless image as output (mapped image).

Algorithm 2 Bit-plane slicing and share generation

---

1. Use bit-plane slicing on **mapped image array** to get initial 8 share arrays.
2. Now, apply **Algorithm 1** on those arrays to get mapped share arrays.
3. Generate a random grayscale image of size $M \times N$(same as original image) using a **secret key** and convert it into an array.
4. Apply bitwise XOR of this image array with each and every mapped share array.
5. Save each received share array after XOR operation as an image to receive final 8 shares.

For mapping of shares, different keys can be used from the one used in original image mapping, out of possible 100 keys (Fig. 1).

The inverse mapping algorithm uses the **baby-step giant-step technique** to efficiently compute the discrete logarithm (DL). The algorithm divides the search space into two parts: the *baby-steps* and the *giant-steps*[8, 9]. By constructing a table of

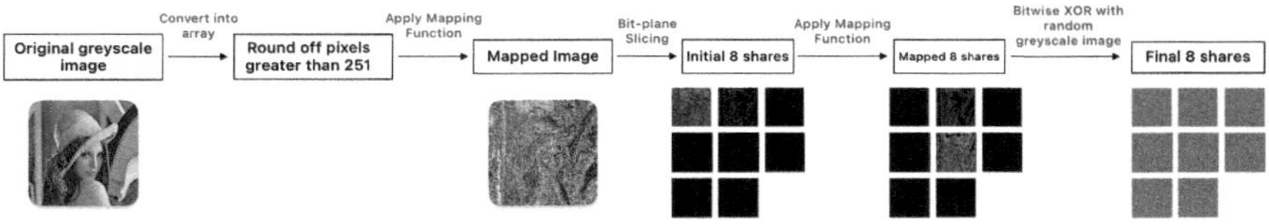

Fig. 1 Mapping and share generation

Algorithm 3 Inverse mapping algorithm

We have, $\eta = (\delta)^x (mod\, \lambda)$, where,
δ - the primitive root modulo 251 or *key*.
η - the element to be found (log base δ of η).
λ - modulus of odd prime (mod 251).

function SOLVEDISCRETELOGARITHM(δ, η, λ):
  $N \leftarrow \lceil\sqrt{\lambda - 1}\rceil$ ▷ Compute the value of N using the ceiling function
  Initialize an empty dictionary 'd' to store the baby steps.
  for i **from** 0 **to** $N - 1$ **do** ▷ Baby-step phase
    $d[\text{pow}(\delta, i, \lambda)] \leftarrow i$
  end for
  ▷ Giant-step phase using Fermat's Little Theorem
  $c \leftarrow \text{pow}(\delta, N \cdot (\lambda - 2), \lambda)$
  for j **from** 0 **to** $N - 1$ **do**
    $y \leftarrow (\eta \cdot \text{pow}(c, j, \lambda))$
    if *y exists in t* **then**
      Return $j \cdot N + d[y]$
    end if
  end for
  Return *None*
end function

Algorithm 4 Image recovery

1. Convert received shares into arrays and take bitwise XOR with random grayscale image generated using the same secret key used in share generation.
2. Apply inverse mapping function in **Algorithm 3** to inversely map pixels in arrays received after XOR operation.
3. Add these bit-plane arrays to obtain a single array and apply Inverse Mapping on this array.
4. Save this new array as recovered image.

---

the baby-steps and performing a search among the giant-steps, the algorithm can find the DL in a more efficient manner than a naive brute-force search.

The **discrete logarithm problem (DLP)** serves as the foundation for creating secure algorithms that ensure confidentiality and integrity of digital images. DL-based encryption methods often involve operations in finite fields. In public key cryptography, DLP is crucial. DLs are notoriously difficult to compute, and this fact forms the foundation of many of the most widely used cryptosystems. The higher the difficulty in computation, the more secure is the data transfer. Without the proper key

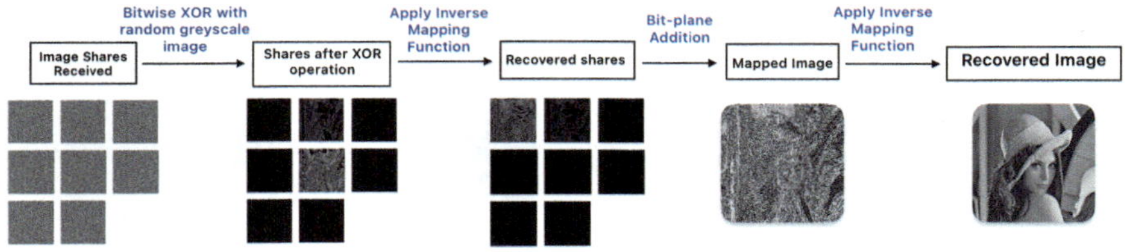

Fig. 2 Inverse mapping and image recovery

or knowledge of the DL solution, decrypting the image becomes extremely difficult, providing a high level of security against unauthorized access (Fig. 2).

4 Results

The **peak-signal-to-noise ratio** ($PSNR$) is used to compare the similarity between the original image I and the recovered image I' and also helps in evaluating the visual quality of I' [10]. A high $PSNR$ indicates excellent quality (i.e., has minimal distortion). In contrast, a low $PSNR$ value indicates poor visual quality (i.e., significant distortion). If $PSNR = \infty$, I' is identical to I, and there is **no contrast loss**. The PSNR is given by

$$PSNR = 10 \log_{10} \left(\frac{255^2}{MSE} \right) \text{ (dB)},$$

where $\log_{10}$ represents the base-10 logarithm, 255^2 is the maximum possible pixel value squared (for an 8-bit grayscale image), MSE is the **Mean Squared Error**, which measures the difference between the original and compressed images and is calculated as follows:

$$MSE = \frac{1}{L \times B} \sum_{i=1}^{L} \sum_{j=1}^{B} (I_{ij} - I'_{ij})^2,$$

where L is the height of the image, B is the width of the image, I_{ij} is the pixel value at position (i, j) in I, and I'_{ij} is the pixel value at position (i, j) in I'.

The **Structural Similarity** ($SSIM$) is a metric that is used to compare the similarity of two images. A higher $SSIM$ score indicates that the images are substantially similar to one another. However, a low $SSIM$ value indicates that the images are not very similar [10] (Fig. 3).

In Table 1, the $PSNR$ for six test images is exceptionally high which implies that image recovery is good. When $PSNR$ values are ∞, that implies, original and

Fig. 3 Test images

Table 1 Performances matrices for six test images

Test images	PSNR	SSIM	RMSE
Lena	77.7223	0.9997	1.3811
Peppers	∞	1.0	0
Baboon	∞	1.0	0
Elaine	86.2659	0.9999	0.5900
Boat	75.7937	0.9993	1.8296
Flower	∞	1.0	0

Table 2 PSNR values of recovered "LENA" image, with varying numbers of shares

Test image shares	PSNR
1 out of 8	28.1810
2 out of 8	27.8767
3 out of 8	27.9298
4 out of 8	28.8580
5 out of 8	29.4114
6 out of 8	29.5573
7 out of 8	32.3140
8 out of 8	77.7223

recovered image are identical. In addition, it can also be seen that the $SSIM$ values are nearly equal to 1, implying that the structure of the original and recovered images is identical or almost identical. Moreover, the $RMSE$ values for test images and the recovered image are nearly zero, implying that original and recovered images have no significant difference.

Performances in terms of $PSNR$ when k-out-of-n shares are used for Image Recovery of first test image Lena ($k \leq n$). Table 2 shows progressive nature of this VC scheme, as number of shares increases, more visual information about test images gets revealed (Fig. 4).

Fig. 4 Recovered images

5 Conclusion

In this research, we introduces a novel algorithm levergaing elementray number theory, proposing an effective. The results demonstrate that suggested approach provides a practical and aesthetically pleasing method for securely sharing sensitive 8-bit grayscale images. By mapping the original image into a mapped image and then generating shares of mapped image using bit-plane slicing and further mapping those generated shares, this technique ensures that none of the shares disclose any information about the original image. The decryption procedure progressively reveals the image, giving users control over the amount of information disclosed because it is critical to ensure the confidentiality and integrity of images during transmission.

References

1. Agarwal, S., Suri, J.S., Shivani, S.: In: Handbook of Image-Based Security Techniques, 1st edn., pp. 1–442. Chapman and Hall/CRC, New York (2018)
2. Naor, M., Shamir, A.: Visual cryptography. In: De Santis, A. (ed.) Advances in Cryptology - EUROCRYPT'94, pp. 1–12. Springer, Berlin (1995)
3. Lukac, R., Plataniotis, K.N.: Bit-level based secret sharing for image encryption. In: Pattern Recognition
4. Liu, Y.L., Liu, T., Yan, B., Pan, J.S., Yang, H.M.: Visual cryptography using computation-free bit-plane reconstruction. In: Security and Communication Networks, pp. 1–1168. Wiley, New Jersey (2022)
5. Kannojia, S.P., Kumar, J.: Xor-Based Unexpanded Meaningful Visual Secret Sharing Scheme, pp. 1–9. Inder- Science Publishers, Switzerland (2019)
6. Asmuth, C., Bloom, J.: A modular approach to key safeguarding **29**, 208–210 (1983)
7. David, M.: Burton. Elementary number theory. Wm. C, Brown (1989)
8. Coppersmith, D., Odlzyko, A.M., Schroeppel, R.: Discrete logarithms inGF(p). Algorithmica **1**, 1–15 (1986)
9. Buchmann, J., Weber, D.: Discrete logarithms: recent progress. In: Buchmann, J., Høholdt, T., Stichtenoth, H., Tapia-Recillas, H. (eds.) Coding Theory. Cryptography and Related Areas. Springer, Berlin (2000)
10. Setiadi, D.I.M.: PSNR versus SSIM: imperceptibility quality assessment for image steganography. Multimed Tools Appl **80**, 8423–8444 (2021)

Approach for Fire Detection Using Image Processing

Aniruddha Bardekar and Mohammad Atique

Abstract Huge devastation to humans, and the environment is caused by forest fires, which are one of the forces of nature that are hard to control. Forest ecosystems are destroyed because of this uncontrollable calamity. In this paper, a review is made on previous works done in the field of technology that could be beneficial to detect, control forest fires. Technologies such as GIS, image processing are main sources of this review, further study about these technologies is done in that light. This paper provides, focuses on the usage of the above-mentioned technologies in combating forest fire incidences.

Keywords Forest fires · Geographic information system · Geoprocessing · Satellite imagery · Image processing · Object-based image analysis · Spatial data

1 Introduction

Fires represent a hazardous danger to environmental systems, infrastructure, and human lives. Multiple instances of such dangerous fires were seen in the past. Huge devastation and damage to humans and the environment is caused by forest fires, which are one of the forces of nature that are hard to control and are hazardous for forest ecosystems. Wildfires simply damage the habitable and adaptable land for specific animal and plant species. Forest fire changes or causes the death of the plant life traits, which reinforce thousands of wildlife thereby forcing the animals out of the regions or even killing them. Certain measures are taken to prevent occurrences of forest fires, some of them are cutting of fire lines, patrolling forest areas by authorities during fire season, forest fire awareness drives that help common people to realize

A. Bardekar · M. Atique (✉)
Department of Computer Science and Engineering, Sant Gadge Baba Amravati University, Amravati, Maharashtra, India
e-mail: mohd.atique@gmail.com

A. Bardekar
e-mail: fgordon51@gmail.com

G. Fortino et al. (eds.), *Proceedings of Third International Conference on Computing and Communication Networks*, Lecture Notes in Networks and Systems 917,
https://doi.org/10.1007/978-981-97-0892-5_18

how dangerous forest fires can be to natural habitats and which in turn indirectly affects human beings. These measures are preventive and do not help detect forest fire prior to its occurrence. By the time authorities reach ground zero, the forest fire has already spread over large area and is uncontrollable for long period of time.

The proposed methodology mentioned here in this approach is an alternative preventive method that allows us to analyze forest fires after their occurrence and create information of such areas that are highly prone to forest fires by making the use of image processing. Once the forest fire-prone areas are known, it would be beneficial for authorities to patrol these areas properly and control the fire before it spreads over large area thereby minimizing the hazard. The images used here are high resolution, multi-feature satellite images, which are made readily available and are open source. These images contain spectral features such as infrared, near-infrared, shortwave infrared, thermal infrared, etc., which are not available in most commonly used cameras and can be utilized to map large fire areas. Satellites have a wide variety of sensors; therefore, it is beneficial to use these images for image processing. Another important aspect of satellite images which is usually overlooked is that these images are aerial and are of great use for areas that cannot be usually covered by normal cameras.

2 Background

The problem that we are trying to address is to prevent large forest fires that affect natural habitat as well as ecology. Certain measures are taken to prevent forest fires. Traditionally, awareness drives are undertaken to help people understand the drastic effects of forest fires. Watchtowers are another traditional way of monitoring of forest fires. Authorities patrol protected areas to detect forest fires. The problem with these traditional methods is that these are done at ground level, and it is important for the person monitoring forest for fire occurrences to be keen at observation. No matter how keen observer a person may be he ought to make a mistake. Also, patrolling large areas at same time, by a small team sound nearly impossible. The problem with watch towers is that they are not movable and only a certain range can be monitored. Even if forest fires are detected by these traditional methods, reaching the spot where fire occurred could be time-consuming if the fire is at long distance from patrolling unit or if the place is difficult to access due to location being geographically remote. Also, the problem with watchtowers is that the person monitoring from a watchtower has to wirelessly inform authorities patrolling the area or authorities present at the base station. The possibility of wireless communication in remote areas not occurring due to technical problems such as radio interference or the receiver not being in range cannot be overlooked. Misinterpretation of information conveyed over wireless communication is another issue that may occur. No matter how perfectly we try to implement these traditional methods, errors tend to occur.

To address such issues technological intervention comes into play. Technologies such as remote sensing satellites, geographic information system, satellite imageries can be utilized to monitor large areas. Modern satellites have capabilities to record geospatial imageries which are useful for a variety of applications ranging from military, disaster management, environment and what not. Utilizing this information for monitoring forest fires can be achieved by proper use of imagery along with GIS and exploiting the information present in spatial imagery for a specific use. The important features present in satellite imagery with respect to proposed method are red, near infrared, and thermal infrared spatial feature, which are helpful for us to determine the parameters required to spot large fires.

3 Previous Work Done

A workflow was presented to assess the extension of the burned areas due to wildfire events over green vegetated areas, in particular woods and forests, based on open access and high spatial resolution multispectral images from the Landsat-8 and Sentinel-2 satellites, together with open access high temporal resolution products (MODIS and VIIRS) [1]. The author in their article gave the precisely and timely extraction of burned areas after forest fire plays an important role in reducing disaster losses, maintaining ecological balance and protecting forest resources. In their article, they proposed a method of automatically extracting forest fire burned areas by using Gaofen-1 remote sensing images [2]. In their study, the authors presented recent fire history reconstruction using a series of United States Geological Survey (USGS) Landsat archived data from 1984 to 2011 in Attica region, Greece. Their results show that the fire scar perimeters were captured with high accuracy [3]. The discussion had shown that the use of data provided by geographic information systems can help precisely determine the areas affected by fires as well as the afflicted infrastructure [4]. In their research, authors aimed to highlight the capabilities of Synthetic Aperture Radar—SAR satellite sensors for investigating the environmental evolution of forest areas affected by Mediterranean fire events [5]. In their work, opinion was given addressing frequent forest fire incidents especially in reserved areas such as National parks is very significant to restore the natural habitat for wild life [6]. Authors have developed a semiautomatic method to map forest fires during large periods by means of remote sensing imagery [7]. In this research, authors during the demonstration phase, all the PREFER products have been, from the pre-operational point of view, demonstrated. All the products have been provided in time and with the planned frequency to the service infrastructure in order to be made available to the users. A part from the possible improvement in the wildfire management that could result from the adoption of the PREFER products in the decisional procedures of the institutions involved, in the paper, it has been shown how the information collected during the fire season can be exploited to extract, immediately after the summer season, statistical data on the fire incidence (number of fire, total burned areas, type of vegetation involved, level of damage, temporal distribution of the hazards, etc.) [8]. In the paper, authors showed how forest fires are detected using the Landsat-8 OLI spectral reflectance bands based on the fire detection algorithm developed by Schroeder et al. (2015). The results were compared with the MODIS TERRA and AQUA Active fire product (MOD14), which can be downloaded from the Fire Information for Resource Management System (FIRMS) website. In this study, forest fires were detected from the Landsat-8 OLI data for the Nainital district using the fire detection algorithm developed by Schroeder et al. (2016). Results were compared with the MODIS TERRA and AQUA thermal anomaly product. Results showed that OLI sensor can detect smaller fires whereas the MODIS sensor cannot [9].

4 Proposed Methodology

The method to analyze the fire area involves the usage of satellite imagery and performing certain operations on the imagery using GIS tool of some kind. The operations performed provide us with parameters that allow us to differentiate fire spots with other features of the imagery. The methodology to be followed is as follows.

4.1 *Gathering of Geospatial Data*

Primarily, for our analysis, geospatial data or satellite imagery is to be procured from online databases. The imagery features that we require for our analysis are red, near infrared, and thermal infrared. These features are called bands in technical terminology. The red feature of imagery is known as Band 4, near infrared feature is known as Band 5, and thermal infrared feature of imagery is known as Band 10. These bands are only compatible with GIS tools or to some extent in CAD software. Two different packages of these images are required—one before occurrence of fire and another after occurrence of fire. Once these images are available, we proceed further by importing geospatial data in GIS tools.

4.2 *Importing Geospatial Data in GIS Tools*

The downloaded imageries are to be imported in GIS tools. These tools enable us to manipulate the information content present inside the imagery. It helps us remove the parameters that are not desirable which helps to perform operations on relevant information. A variety of GIS tools are available in market, some of them are licensed tools while others are open source. Every tool is different from another and performs specific task. These tools usually utilize Digital Signal Processing to help manipulate data. Once the choice is made of a proper GIS tool, we import the images and prepare to manipulate the information. Figure 1 shows how imported bands appear in the GIS

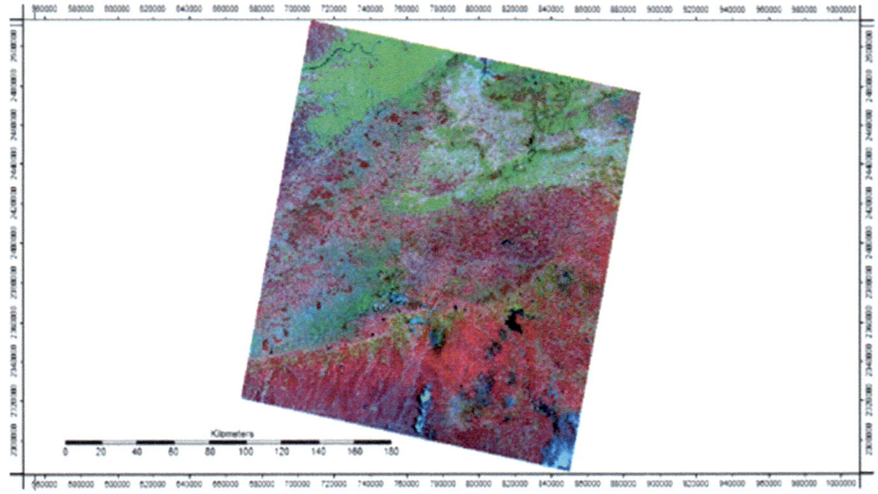

Fig. 1 Band 4, Band 5, and Band 10 Imported in GIS tool

tool. Here, we can see latitudes and longitudes shown along a rectangle which implies the geosynchronous characteristic of satellite bands and scale which helps understand the distances.

4.3 Manipulating Data

To manipulate the data, we begin with geo-processing of images. Three bands of satellite imagery before fire are imported viz. band 4, 5, 10 and create a package. Likewise, three bands of satellite imagery after fire are also imported and another package is made. This helps us differentiate between the imageries removing any confusion such as which band is analogous to before fire and which is analogous to after fire. Next, we clip area of interest, that is the area where the fire has occurred, ignoring all other area shown in the imagery. Both packages are clipped at the same extent. Once the area of interest is clipped, we create a difference image, which simply shows the spatial difference between bands before occurrence of fire and bands after occurrence of fire. This difference image is our main parameter on which we perform classification (Fig. 2).

Fig. 2 Clipped area showing "Fire Scar"

4.4 Classification of Burnt Area

The difference image shows the area that is burnt. The burnt area in the difference image is visible as dark red patches and obvious to human eye and cannot go unnoticed. To classify this burnt area properly, Object-Based Image Analysis of burnt area is done, which involves segmentation of area of interest, i.e. clipped area and classification of the segments of difference image. Segmentation in simple words is to individually generate separate parts of image based on spatial features of clipped area. Thereafter, these segments are classified according to similar spatial features of the difference image in clipped imagery. Those segments that cover "Burn Scar" of the difference image are merged, and other segments are removed from Object-Based Image Analysis to generate vector polygon of burnt area in the difference image. These vector polygons can be saved and used to identify fire-prone areas.

5 Result

In order to determine fire-prone areas, vector polygon samples were generated of area that were affected with forest fire using Object-Based Image Analysis. These polygons were imported in Google Earth application where fire-prone areas are visible as a closed boundary around the fire-affected area. To verify if the generated polygons correctly identify the forest fires, a comparison is made between the burnt

Table 1 Comparison of vector polygon burnt area and area calculated at ground level

S. No	Sample	Area according to vector polygon (ha.)	Area Measured at ground level (ha.)
1	Polygon 1	2.52	2.184
2	Polygon 2	3.162	3.053
3	Polygon 3	0.72	0.511
4	Polygon 4	3.783	3.4
5	Polygon 5	14.4	14.874

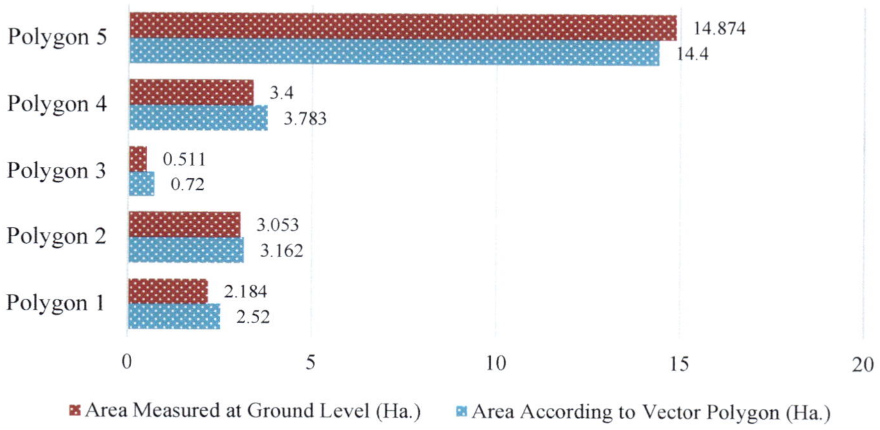

Fig. 3 Analysis of area measured at ground level and area according to vector polygon

area covered by vector polygon and the burnt area measured at ground level. The comparison is shown in Table 1.

Figure 3 depicts the performance of the methodology as compared to traditional methods. Apparently, the proposed methodology gives approximate measure of the burn extent.

6 Conclusion

A preventive measure for detection of fire by using satellite imagery features and processing this imagery to filter out the area that burnt due to fire can be implemented. Many such burnt areas can be found using Object-Based Image Analysis from a single imagery as it has capability to capture spatial imagery over wider areas. Polygons are generated by creating difference image and using Object-Based Image Analysis by segmentation indicating fire-prone area These polygons then can be collected and imported on maps and thereby areas having high occurrences of fire could be monitored.

7 Future Scope

The methodology is supervised as it requires human intervention to perform operations on satellite imagery. Unsupervised methodology could be developed that has minimum human intervention to determine fire scar using Object-Based Image Analysis. Further, a new method could be developed using the same principle to detect small fires as well as fires that occur in urban settlements.

References

1. Angelino, C.V., Cicala, L., Parrilli, S., Fiscante, N., Ullo, S.L.: Post-fire assessment of burned areas with landsat-8 and sentinel-2 imagery together with MODIS and VIIRS active fire products. In: 2020 International Geoscience and Remote Sensing Symposium, pp. 6770–6773 (2020).https://doi.org/10.1109/IGARSS39084.2020.9324512
2. Bin, W., Ming, L., Dan, J., Suju, L., Qiang, C., Chao, W., Yang, Z., Huan, Y., Jun, Z.: A method of automatically extracting forest fire burned areas using GF-1 remote sensing images. In: International Geoscience and Remote Sensing Symposium, pp. 9953–9955 (2019)
3. Pleniou, M., Xystrakis, F., Dimopoulos, P., & Koutsias, N.: Maps of fire occurrence—spatially explicit reconstruction of recent fire history using satellite remote sensing. J. Maps, 1–8 (2012).
4. Petkovic, M., Garvanov, I., Knežević, D., Aleksić, S.: Optimization of geographic information systems for forest fire risk assessment. In: IEEE Xplore (2020)
5. Ruiz-Ramos, J., Marino, A., Boardman, C.P.: Using Sentinel 1-SAR for monitoring long term Variation in burnt forest areas. In: International Geoscience and Remote Sensing Symposium, pp. 4901–4904 (2018)
6. Anath, S., Manjula, T. R., Niranjan, G. S., Kumar, S., Raghuveer, A., Raju, G.: Mapping of burnt area and burnt severity using Landsat 8 images: a case study of Bandipur forest fire region of Karnataka. In: IEEE Recent Advances in Geosciences and Remote Sensing: Technologies, Standards and Applications (2019).https://doi.org/10.1099/TENGARSS48957.2019.8976037.
7. Salvador, R., Díaz-Delgado, R., Valeriano, J., Pons, X.: Remote sensing of forest fires. In: Center for Ecological Research and Applied Forestry (2019)
8. Laneve, G., Díaz-Delgado, R., Valeriano, J., Pons, X.: Satellite based products for supporting forest fires prevention and recovery in Europe. In: IEEE Xplore (2016)
9. Suresh, B.K.V., Sai Krishna Vanama, V.: Fire detection in varying topography using Landsat-8 for Nainital Region, India. In: 3rd International Conference for Convergence Technology (2018).https://doi.org/10.1109/I2CT.2018.8529366

Review of Various Approaches for Authorship Identification in Digital Forensics

Riya Sanjesh and J. Alamelu Mangai

Abstract Authorship identification involves extracting and analyzing the author writing styles. Digital Forensics along with cyber investigations employ writing style to identify the author and their traits. Authorship identification was done on lengthy and short texts in English, Arabic, Chinese, and Greek. However, it emphasizes a unique and difficult scenario: identifying whether two writings published in distinct discourse types are produced by the same person. We provide sets of texts that use four different discourse styles, including essays, emails, text messages, and business notes, based on a new corpus of English texts. The cross-discourse-type ownership verification assignment is highly challenging due to the disparities in communication intent, target audience, and formality level. This paper evaluates various aspects of authorship identification and provides a thorough analysis of the assessment findings. This study also explores the language proficiency and problems in authorship identification tasks. A number of significant authorship identification domain studies were assessed for data, characteristics, techniques, and outcomes. After reviewing the research, we conclude that the outcomes of authorship identification task depend primarily on the specified stylometric characteristics and dataset used. The beneficial qualities also vary by the language type.

Keywords Stylometry · Machine learning · Authorship attribution · Identification · Verification · Obfuscation · Disclosure styles

R. Sanjesh (✉) · J. A. Mangai
School of CSE & IS, Presidency University, Bengaluru, India
e-mail: riya.sanjesh@presidencyuniversity.in

J. A. Mangai
e-mail: alamelu.jothidurai@presidencyuniversity.in

G. Fortino et al. (eds.), *Proceedings of Third International Conference on Computing and Communication Networks*, Lecture Notes in Networks and Systems 917,
https://doi.org/10.1007/978-981-97-0892-5_19

1 Introduction

Creator identifying (or authorship attribution) reveals the author of a text [1, 2]. Closed-set and open-set ownership attribution tasks mimic real-world conditions [3]. The first scenario works when just a limited list of people might be responsible for the creation of disputed texts, whereas the latter works when such lists are unavailable or unreliable. Verification of authorship by a single author is an example of open-set attribution in its most basic form [4]. Since each case may be broken down into authorship verification instances, authorship attribution is crucial to author identification. In order for a computer to determine whether or not two texts provided to it are by the same author, they must be written in the author's typical style [5, 6]. In a situation, where one authorship-known text is offered, one can identify the author of two works, usually as one recognized and one unknown [7].

Quantitative methods for authorship identification, verification, profiling, and author clustering assume that document writing styles may be defined, taught, and utilized to develop prediction models [8]. Modeling style-focused texts is a field goal. Any effective author identification system, whether for attribution or verification, must reliably identify authors across genres, themes, and target audiences. Characteristics like function words as well as common character-level n-grams are often favored since they have little to do with text content or genre. However, such features need extensive papers and provide sparse, less usable representations for brief documents [9, 10]. Thus, the field's key objective is to design algorithms that don't overfit training texts and adapt well across text types.

The purpose of authorship analysis is to enable applications in digital literature and text forensics by providing details on the creators of digital texts and to reveal such information to the public. The statistical evaluation of the style of a piece of writing is the focus of the subject of linguistics known as authorship analysis. In authorship analysis, textual elements are evaluated to determine who wrote a particular literary work. Stylometric features are those that may be used to identify an author based on their writing style. These features may be broken down into four categories: lexical, syntactic, basic, and content-specific. Since there is no attribute set that is sufficiently optimized to be relevant to all persons, all languages, and all domains [11], the stylometric collection of attributes is not standardized for every language or text source (social media vs. literature). Therefore, the inventory of characteristics differs between text sources.

Authorship verification efficiency depends on many aspects. Obviously, text length is important, and systems usually perform worse with short messages. Another tough type is writings by well-known and unnamed writers from other areas. Cross-domain authorship verification may reveal differences in subject (politics vs. athletics), genre (review vs. essay), and language (English vs. German). Various cross-topic and cross-genre authorship authentication and findings were quite low evaluated, especially for a Dutch cross-genre sample of articles and evaluations [12]. The profile-based paradigm, in which every sample of known authorship by a putative author is processed cumulatively, is followed by exactly one method [13, 14]. The instance-based paradigm in which every given piece of text may be used as recognized writers is shown in a distinct format. The need for instance-based approaches may be explained by the comparatively modest size of the prospective author set related to attribution issues and the equal split of training texts among the candidate authors. Fanfiction writings from diverse fandoms (writing inspired by particular massively famous works) were used. A huge training set with over 350,000 testing cases was created to apply powerful deep learning frameworks [14]. Unexpectedly, the best findings were high, suggesting that most fanfiction writers may keep their artistic preferences across fandoms. However, other variables may have falsely raised the results. Previous cross-domain attribution research explored cross-topic circumstances most often. General thematic groupings and pre-assignment of texts govern subject.

A constrained corpus where certain people write about a given subject is another method [15, 16]. Topic control is fine-grained. However, controlled corpora are tiny. Cross-genre conditions are another cross-domain consideration. Collecting works by several writers in various genres is difficult. Since many people are engaged in social networks like Facebook and Twitter, using social media messages is another option [17]. Finally, a controlled corpus may be created by asking authors to write in an assortment of disciplines (e.g., email, blog, essay) [18]. Cross-language instruction and testing is the worst case. Novels translated into other languages are a good source of such situations if the translator's preferences don't change the author's style.

In recent years, PAN shared intrinsic text analysis tasks. Based on intrinsic detection of plagiarism [19], the emphasis moved from grouping authors inside texts to detecting authorship changes. Since no reference corpora exist, all those tasks need intrinsic, stylometric text analysis. Text n-grams, word frequencies as well as average word/sentence lengths, POS tag patterns/structures, and structural aspects like indentation usages are all examples of lexical characteristics that may be used as stylistic fingerprints. A style violation detection task uses stylometric features like sacks of character n-grams, function word frequencies, and other lexical parameters, processed by algorithms in order to identify boundaries or anomalies [20].

Multi-author documents are seldom studied. Several methods exist for text segmentation, where a text is separated into multiple subjects, but few target authorship segmentation. Glover and Hirst provided the first stylometry-based method for automatically detecting author boundaries in jointly produced texts [21]. However, recognizing borders is not to disclose many authors, but to offer indications so collectively authored papers may be homogenized in style. Graham et al. [22] use neural networks using stylometric characteristics, while a random framework based on word occurrences to partition a text by authorship. Since the extraction of features and authorship analysis were established, current efforts have steadily improved. However, social media and blogs provide additional issues in the big data age. Internet communication has led to experiments with more imbalanced datasets. Under these conditions, most research is weak. An author may change their style based on genre, subject, manuscript standards, etc. Successful writers can write in several settings, but this inconsistency makes authorship analysis harder.

Stylometry also lacks wide population investigations for authorial style judgments. Thus, most study data are limited to the researcher's dataset. Stylometry attributes lose discriminating strength when author sets grow, therefore they cannot identify like physiological biometrics or behavioral modalities. This survey reviews previous techniques and summarizes the remaining problems. Large-scale authorship verification is prioritized. Section 2 provides various stylometric concepts and various classification approaches are explained in Sect. 3. Sections 4 and 5 brief the definition of tasks and building the datasets, respectively. Finally, the metrics used for evaluation are discussed.

2 Stylometry Subtasks

To compare texts' individual stylistic traits is a common goal of similarity detection issues [23]. Stylochronology, originality attribution, authorship verification, and authorship profiling may all benefit from developments in authorship detection methods which are explained in this section.

2.1 Authorship Attribution

In this part, following a brief overview into this literary genre of writing, how to create a dataset with an accumulation of fanfiction works in order to accomplish the job of cross-domain ownership attribution is provided. This task required to attribute authorship to works that were published in other domains. Following that, a survey of the participants' contributions is conducted, and then an in-depth analysis of the participants' attribution performance is carried out.

Online articles/messages: Electronic postings were debated alongside conventional literature. Twenty authors and twenty web postings per author were used in the research [24]. The writers of this piece placed special emphasis on three aspects of the language: accent, word length, and extension. Problems with diacritics, syllable length, and elongation were addressed. Word root dictionaries boost classifier performance, the authors found. Inflected phrases and roots were compared using clustering and Dice and Jaccard. In addition, words longer than 10 were unusual in their dataset, so they could be filtered out to concentrate on word origins. The authors employed 4,500 roots, however, 30 roots performed best for a classifier based on AA. They added language-acceptable elongation characters. The research prioritized root indexing and categorization above words. Character- and word-level n-grams impacted more. Potthast et al. [32] examined false news and hyperpartisan news writing. Stylometric study compares these news stories' writing styles. BuzzFeed journalists manually fact-checked the BuzzFeed-Webis Fake News Corpus. The authors suggest unmasking to compare hyperpartisan and false news language styles.

The report discusses their corpus analysis. The writings do not discuss the research's consequences. However, analyzing hyperpartisan and false news writing style may help build better strategies for recognizing and avoiding disinformation and fake news.

Based on brief text authorship verification breakthroughs, computer-aided online text authorship attribution method [34] mimics a basic neural network by creating word- and character-level hidden layers using natural language approaches. Validation performance determined each layer's word- and character-level models. Our output layer concludes with an empty majority vote vector. To make our training dataset more robust, researchers examined writing inequality in social media postings.

Authorship verification using a hierarchy-based recurrent Siamese network structure used a modified contrasting loss function to decrease word embedding cross-topic sensitivity. Authorship verification using cross-topic and cross-genre internet-based writings was improved by the suggested technique [37]. Accuracy increased by 15%. Pretraining strategies need further effort. Better uniformity between pre-training and post-training data may be important here.

Authorship verification involves examining linguistic patterns to verify whether two or more works were produced by the same author. Experts analyze spelling, grammatical, and aesthetic errors. However, machine learning systems use stylometric information to do the same. Such qualities are unreliable for brief and current social media messages. This is done by compiling a new massive corpus of brief reviews from Amazon for text comparability research and demonstrating that the Siamese framework beat existing techniques based on stylometric characteristics. The linguistic examination of the network's internal attention weights demonstrates that the suggested technique can grasp several standard language categories [38].

Cybercrime prevention requires online text authoring. Some systems restrict text length, making it tougher. We identify 140-character Twitter messages' writers. We assess stylometric characteristics and Twitter characteristics like URLs, hashtags, responses, and quotations. We utilize 93 and 3957 source databases. We test author sets with different sizes and training/test text quantities. Auto-selection improves performance. Automatic attribution may save suspect search professionals time. Reporting verification results is complete. It measures feature computational complexity and temporal persistence [42].

Books and novels: Books, poems, essays, and other lengthy works were first used as the data source for authorship attribution studies. From the website of the Arab Writers' Union, compiled 14 works written by six authors, with an average of 23,942 words per book. The study is based on the work of the same researchers who applied the list of 70 function terms developed to English text. The goal was to provide a reasonable estimate for authorship identification of contested literature using just a small sample of function words. The classifier was developed using a modified version of an algorithm for evolution as its basis. Hybrid linear discriminant analysis (LDA) and ROC area under the curve were utilized for featured subset selection in the fitness measure. With just 54 and 65 sets of function words, respectively, spanning a piece of 1,000–2,000 words from the contested text, they exhibited good results. Function words were shown to be superior estimators for authorship attribution across a collection of traditional Arabic texts. Furthermore, techniques that achieve high performance on English datasets (like LDA) may also achieve high performance on Arabic authorship datasets. Ding et al., [31] introduced neural networks to derive writing style representations from unlabeled texts. The approach describes how the suggested paradigm overcomes the drawbacks of manual feature engineering. It also shows essay and novel distribution in the training and testing dataset. It closes with data preparation and Stanford tagger POS tagging.

Fanfiction: Fanfiction, or "fandom," is the huge corpus of modern fiction written by non-professional writers (called "fans") in the style of a popular source of work, such as J.K. Rowling's Harry Potter series. These "fics" use extensive "intertextuality" by borrowing characters, motivations, situations, etc. from the original fanbase. "Amateurs" write the books. E.L. James' blockbuster Fifty Shades series was initially a Twilight fiction. Many fans have underlined the "transformative" nature of their fanfics, although their legal standing has been disputed. "Transformative" fanfiction differs from "affirmative" fanfiction in tone, style, and narrative. On the other hand, extreme fandom reworkings have nothing in common with the original fanbase save for the heroes' names. Slash fics, like "Kirk/Spock," are often sexual. Fanfiction is a good computational authorship identification benchmark for several reasons. Most fanfiction is written on unmoderated sites like fanfiction.net or archiveofourown.org, therefore, it likely reflects the author's style. A hierarchical fusion of two well-known techniques into an end-to-end learning procedure: The bottom deep metric learning framework learns a pseudo-metric that translates a variable-length text onto a fixed-sized feature vector. A probabilistic layer performs Bayes factor scoring in the learnt metric space at the top. We provide cross-topic text preparation solutions [43]. Authorship Attribution (AA) study focuses on English novels with plenty of data. We use Authorship Verification to examine genre, subject, gender, and length in short Italian writings in two new datasets. Even with minimal data, AV is achievable, but more helps. Gender and subject may give away personal style if not managed [44].

2.2 *Authorship Verification*

Word root dictionaries boost classifier performance, the authors found. Inflected phrases and roots were compared using clustering and Dice and Jaccard. In addition, Arabic words longer than 10 were unusual in their dataset, so they could be filtered out to concentrate on word origins. The authors employed 4,500 roots, however, 30 roots performed best for a classifier based on AA. They added Arabic language-acceptable elongation characters. The research prioritized root indexing and categorization above words. N-grams in Character level and word level impacted more. Authorship verification methods exist. One-class classification uses selections from the work of the writer to train the classifier and compare the test text to the model [26]. Training without sample negative implies single-class classification. The Many-Candidates method leverages fraudsters by considering verification as attribution [27]. However, the quality of such a compiler is parameter-defined, and picking the incorrect parameters may degrade accuracy. Instead of using complex machine learning methods to detect authorship, the technique proposed by Halvani et al. [33] just needs a compression algorithm, a difference measure, and a threshold. When dealing with complex AV instances in which the documents are dissimilar from one another in terms of subject or genre, this method becomes effective. However, it does not elaborate on the scope of this method or whether or not it may be used other languages outside English.

Authorship verification is considered for small and big datasets. Two issues plague the small-scale problem: First, the document writers are unknown since no writing examples are accessible. Second, the two texts are brief (a few hundred to a few thousand words) and may vary in genre and content. Transformation encoder transforms one document into the other to solve it. The loss from this document transformation is used to validate the pair's authors. Parallel recurrent neural networks are suggested for large-scale problems with several authors and many longer instances. It compares two texts' language models to baselines [34].

Information retrieval (IR) may benefit from retrieving indexed materials by writing style rather than subject. Authorship verification (AV), a digital text forensics study topic, is ideal. AV determines whether two documents—indexed and reference—are by the same author X. Many techniques treat AV as a type of binary issue, even though it is unary. An AV method's categorization model has important consequences for its requirements, evaluability, and application [35].

2.3 Authorship Evaluation

Authorship profiling analyzes a work to estimate demographics like gender without discovering the writer [28]. For instance, pronoun use declines with age while preposition usage increases, and adolescent writers may use additional non-dictionary terms than adults. Author attributes may be missing, anonymous, or fraudulent online, making authorship profiling difficult. Profiling may need understanding of the document's subject and themes to understand the author's emotions. Thus, content-specific traits are more prejudiced in authorship profiling than attribution [29]. Understanding content-specific themes may provide more relevant and valuable information for identifying demographics and traits than other surface-level elements. Keeping in mind that there may be issues with using content-based traits in authorship research. "While style-based indicators can truly differentiate one group of authors from another, one needs to be skeptical with the possibility that markers could simply be objects of a specific creating scenario or experimental setting and might consequently yield overly positive findings that will not be generated out in real-life applications."

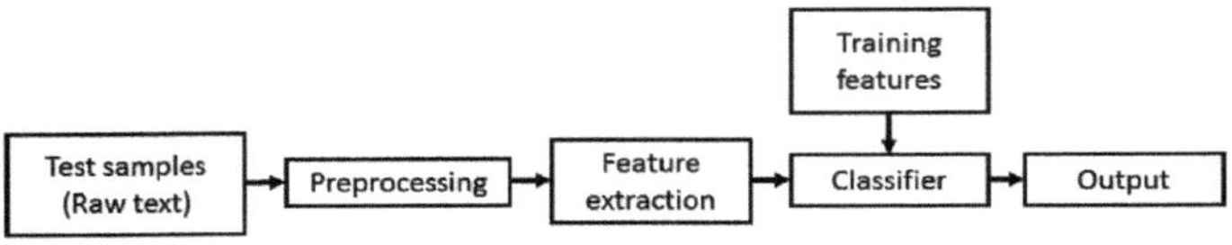

Fig. 1 Author identification process

2.4 Stylochronometry

Stylochronometry tracks authorial style changes. Language-dependent issues matter because certain languages are unstable over time. Thus, an author's language may represent linguistic changes rather than writeprint changes. Newer works had significantly longer average word lengths than older works, indicating an association among length of words and document age. Words with eight characters or fewer, especially four characters, were likely from an earlier work, whereas phrases with nine characters or longer were likely from a younger one. Thus, the authors ascribe word length variation to either Turkish language knowledge or linguistic changes. Stylochronometry investigations are rare and particular. These attempts target certain writers, historical periods, languages, etc. Stylochronometry is less reliable than authorship attribution and verification, hence these research may not be mature or widely accepted. Figure 1 illustrates the process of author identification.

2.5 Authorship Obfuscation

Obfuscated texts may have a lower average word length, more adverbial phrases, and a simpler readability. Style alteration may also obfuscate text. Style modifications may be lexical, where sentences or words are shuffled for semantic alterations, or content, where phrases or sentences are modified while keeping meaning. Synonym and phrase alternatives for style modification. Computerized phrase and synonym identification and substitution use accurately categorized example texts. The document style has changed if the categorization changes after the transformation. The approach transforms 7 of 13 test scenarios. There should be evidence that a change in style has taken place, and the transformed material should maintain its original meaning and grammatical accuracy. Figure 2 illustrates various types of authorship obfuscation.

Table 1 provides an illustration of various authorship obfuscation techniques. Obfuscation of the design can either merge or divide classes; obfuscation of the data may obscure data structures; obfuscation of the control can obscure control flow information; and obfuscation of the layout can obscure annotations and variable names. Formally, P is transferred to obfuscated P, which behaves the same but is slower, memory inefficient, and more complicated. It is difficult to modify massive operations emergent characteristics like character frequency, but it is hard to change

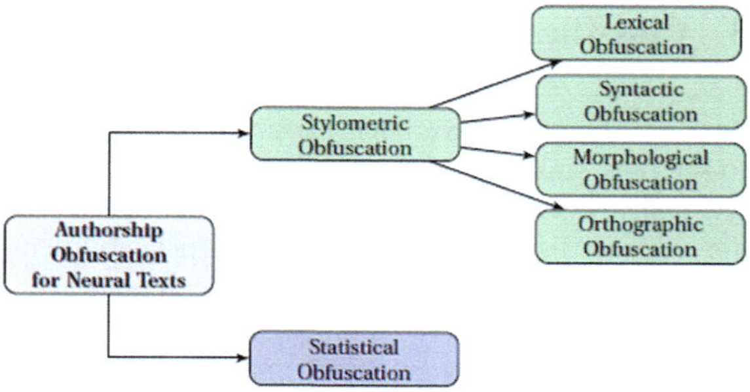

Fig. 2 Classification of authorship obfuscation

character-level features since it is a style that can be changed by choosing other words. Finally, programming code obfuscation is investigated. It may be applied to source code, binary, byte, or machine code unless the required obfuscation is achieved.

Table 1 Various AO techniques used

AO technique	Obfuscated dataset	Category
Upper/lower flip	GROVER and GPT-2	Orthographic
Upper/lower flip	GROVER and GPT-2	Morphological
Whitespace attack	GROVER and GPT-2	Lexical
Deduplicate tokens	Human–Machine Pairs	Lexical
Retain only (non)-stopwords	Human–Machine Pairs	Syntactic
Retain tokens in high/low frequency	Human–Machine Pairs	Statistical
Replace text with linguistic features	Human–Machine Pairs	Statistical
TextFooler	GPT-2 medium, GPT-2 XL, GPT-3, GROVER	Lexical
Varying sentiment	GROVER	Lexical
Source-target exchange	GROVER	Syntactic
Avengers	TuringBench	Lexical
Varying the number of priming tokens	GPT-2 Large, GPT-2 XL, GPT-3, GROVER	Statistical
Varying the text decoding strategy (and its parameters)	GPT-2 Large, GPT-2 XL, GPT-3, GROVER	Statistical
DFTFooler	GPT-2 Large, GPT-2 XL, GPT-3, GROVER	Lexical
Random perturbations	GPT-2 Large, GPT-2 XL, GPT-3, GROVER	Lexical

3 Classification Approaches

After the features have been extracted, the training and test aspects are compared to see how similar or different they are. To figure out how similar things are, there are many ways to classify them, such as machine learning algorithms, grouping approaches, distance-based models, and probability methods. Machine learning techniques may help cluster and classify. Machine learning classifiers use learning along with pattern recognition techniques to reduce loss function class borders. Machine learning techniques were investigated for multiple and sparse representations of features. All stylometry subtasks involve machine learning. For classification, PAN Competition competitors employed support vector machines, decision trees, logistical regression, and random forest models. Clustering is used to detect stylistic similarities and minimize authorship attribution search space. Clustering in source code authorship attribution may identify online spam and reveal similarities in style in web forum postings. Machine-learning classification uses class labels in a supervised learning task, unlike clustering.

Digital text forensics researches if two texts were created by the same individual using authorship verification (AV). Recent AV approaches that address this issue may be split into two categories: The first category includes approaches based on clearly specified characteristics, where one may choose which features to examine and what they represent. The second group of AV approaches uses implicitly specified features without a control mechanism, therefore every letter sequence in a text might be a feature. However, AV approaches in the second category risk biasing their classification predictions based on text topic, which might lead to misleading findings. POSNoise, a preprocessing method, efficiently hides topic-related material in a text. Thus, AV techniques must prioritize writing-style-related text units. Based on six second-category AV approaches and seven corpora, POSNoise outperforms a well-known topic masking strategy in 34 of 42 situations, with an accuracy gain of up to 10% [39].

Clustering is the use of machine learning approach that divides the feature space naturally, which may or may not match class labels. SVMs are popular classifiers because they can handle huge and sparse datasets. Stylometry is used to analyze English and Chinese online communications from 40 individuals over two weeks for lexical, syntactic, structural, and domain-specific aspects. When considering all characteristics, SVMs outperformed decision trees and neural networks [27]. Table 2 provides the characteristic features of various types of classification.

Table 2 Characteristics of various classification types

Type	Methodology	Example	Pros and cons
Domain-Specific	Keyword frequency or other content-specific metrics may reveal the document's subject [7]	Keywords in online sales pitches	Shows personal involvement in a domain
Structural	Document organization [9]	Paragraphs' words/ sentences/ characters Quote indication which is binary	May be more effective in online text like emails and blogs, where other feature kinds are less prevalent but structural components like greetings and text body structure persist
Lexical	To capture stylistic elements at each level [10]	Average number of words in a line, richness of vocabulary	Character n-grams maintain morphological features and minimize data sparsity in big vocabularies; Lexical richness may deteriorate when writers use similar words
Syntactic	Take note of how sentences are structured [11]	Punctuation rate word frequency	Style modeling requires powerful NLP tools and many attributes
Semantic	Interprets words and phrases [12]	Logical dependencies	Language computers struggle with semantic analysis on unconstrained texts
Additional Features	Proposed to meet researchers' demands [12]	Models for the topic and writeups	May need substantial natural language processing techniques; may provide deeper information
SBCe	4:1	65.8	34.2
UvA_Trilearn_2001	1:0	54.9	44.1
UTUtd	10:0	70.7	29.3
WrightEagle2001	3:1	66.2	33.8
Average		64.8	35.2

4 Task Definition

Fanfiction closed-set cross-fandom attribution is the challenge. The aim is to identify the most probable author of an unknown document using a number of instances of reference materials from a limited selection of possible authors. Different fandoms hold known and undiscovered documents.

In particular, all papers of unidentified writers are fics of a specific fandom, whereas prospective writers' fics come from many fandoms. Participants must create a multi-fandom attribution technique. A cross-domain ownership attribution issue is a set of tuples (A;K;U) with A as possible authors, K as reference texts, and U as unknown texts. Participants are tasked with generating a scalar score ai (in the [0, 1] region) reflecting the chance that a given pair of texts in the test dataset share the same author. Participants may skip answering a text pair by giving it a score of exactly ai = 0.5.

5 Building Datasets

Authorship attribution, verification, establishing a profile stylochronometry as well as adversarial stylometry are stylometry subtasks. The first three are usually prioritized. Authorship attributing, or authorship recognition/detection, is a key stylometry job. When accurate identification is impossible, authorship profiling may restrict the search area by identifying variables like gender and age. Authorship verification checks whether documents were created by the same author, whereas authorship attribution identifies the author. Finally, stylochronometry estimates a document's writing date by finding similarities between it and a historical document. Authorship attributing, verification of identity, establishing a profile and stylochronometry are similarity detection tasks that quantify the similarity between two or more documents. Adversarial stylometry subverts attribution, verification, and profiling to avoid recognition. writers may use stylochronometry to emulate past writers' writing styles for hostile stylometry if they're creative (Table 3).

Researchers provide MedLatinEpi and MedLatinLit, a pair of fresh datasets containing culturally significant texts produced in medieval Latin by thirteenth- and fourteenth-century (mainly Italian) literates and classified by author, for computational authorship analysis researchers. These datasets may help scholars study authorship attribution, verification, and same-authorship verification, particularly for Latin or medieval Latin works. We also provide the original data for MedieValla, an authorship verification program, we developed to investigate the true authorship of the "Epistle to Cangrande," reportedly authored by Dante Alighieri but suspected of fraud. We explain our MedieValla tests on MedLatinEpi and MedLatinLit. Our datasets and authorship verification tool should enable researchers to duplicate our findings and perhaps surpass those using better AV approaches [40].

Digital text forensics has long studied whether two texts were created by the same person. Authorship verification (AV) addresses this issue. AV research has continuously grown, leading to a diversity of solutions. Many of these methods leverage document-related characteristics. Therefore, their verification findings may unintentionally be focused on the document's subject rather than the writing style (AV's emphasis). AV approach on four tough datasets against 10 baselines, including the current state of the art. Our technique surpasses all baselines in two situations (with a maximum accuracy of 84%) and performs close to the strongest baseline in two others [41].

Table 3 Examples of training datasets

Model	Training dataset	Learning type	Classifier type
GROVER detector [1]	GROVER	Supervised	DL (trans-former-based)
GLTR[2]	GPT-2	Unsupervised	Statistical
GPT-2 detector [3]	GPT-2	Supervised	DL (trans-former-based)
MAUVE [4] tokens	3 GPT-2 and GROVER	Unsupervised	Statistical
FAST [5]	GPT-2 and GROVER	Binary Supervised	Hybrid
Distribution detector [6]	GPT-2	Unsupervised	Statistical
OpenAI detector [7]	GPT-2 and TuringBench-AA	Supervised	DL (trans-former-based)
Energy discriminator [9]	GROVER	Supervised	DL (energy-based)
Linguistic model[10]	Authorship Attribution-AA	Supervised	Stylometric
Feature-based detector [11]	GPT-2, GPT-3 and GROVER	Supervised	Stylometric
TDA-based detector [13]	GPT-2	Supervised	Hybrid
RoBERTa-TT [14]	TuringBench-TT	Supervised	DL (transformer-based)
BERT-Multi [15]	TuringBench-AA	Supervised	DL (transformer-based)
BERT-TT [16]	TuringBench-TT	Supervised	DL (transformer-based)
RoBERTa-Multi [20]	TuringBench-AA	Supervised	DL (transformer-based)

6 Evaluation Parameters

A variety of efficacy indicators are used to demonstrate the capabilities of an authorship verification model as seen in the recent authorship verification tasks [30]. Effectiveness measures included:

- Area under ROC Curve (AUROC).
- c@1: a variation on the traditional accuracy measure that favors systems that don't provide solutions to challenging problems,
- F1: the widely-used F1 efficiency metric (which does not account for blank responses),
- F0.5u is a novel F0.5-based metric that places a premium on same-author questions with right answers and rewards questions with no responses.

- Brier: the counterpart corresponding to the Brier function of loss for probabilistic prediction accuracy (implemented in sklearn). This metric implicitly penalizes less confident verifiers, including non-answers (ai = 0.5), and rewards "bold" but true predictions (ai near to 0.0 or 1.0). We use its complement like the other metrics, thus higher scores indicate greater efficacy.
- The ultimate score for ranking the systems that were submitted is determined by taking the average of all of the aforementioned measurements.

7 Conclusion

Top-performing contributions may have generated a perception that authorship verification is virtually solved. Recently, a difficult authorship verification job using text pairings of distinct discourse kinds is being tackled. When texts change in communicative intent, audience, or formality, it is difficult to detect persistent author qualities across discourse kinds. All cross-discourse dataset submissions performed poorly. It is interesting that a simple benchmark based on n-grams of characters and cosine resemblance beat all proposals, despite their increasing complexity in most situations. This suggests that earlier approaches utilizing well-known stylometric properties may work better than deep learning methods using modern language models that have been trained for this challenging task. The 12,000 training examples may be too few for deep learning-based techniques. Text length matters too. The graph-based technique worked better with extended writings. The pre-trained language-model-based techniques worked better with shorter texts like emails, business notes, and text messages. Cross-discourse authorship verification needs improvement given its poor efficacy. All techniques used a single model to predict authorship regardless of discourse style. Next, create models for each discourse type combination. However, the training sample should also be divided by discourse type combinations. This indicates the use of an ensemble technique, which incorporates both traditional stylometric techniques and pre-trained models of language.

References

1. Zheng, R., Li, J., Chen, H., Huang, Z.: A framework for authorship identification of online messages: writing-style features and classification techniques. J. Amer. Soc. Info. Sci. Technol. **57**(3), 378–393 (2006). https://doi.org/10.1002/asi.20316.
2. Srinivasan, L., Nalini, C.: An improved framework for authorship identification in online messages. Clust. Comput. **22**(s5), 12101–12110 (2019). https://doi.org/10.1007/s10586-017-1563-3
3. Benzebouchi, N. E., Azizi, N., Hammami, N. E., Schwab, D., Khelaifia, M.C.E., Aldwairi, M.: Authors' writing styles based authorship identification system using the text representation vector. In: 2019 16th International Multi-Conference on Systems, Signals & Devices (SSD), pp. 371–376. IEEE (2019)

4. Gupta, S., Patra, T.K., Chaudhuri, C.: Role of machine learning in authorship attribution with select stylometric features. In: Intelligent Systems Design and Applications: 21st International Conference on Intelligent Systems Design and Applications (ISDA 2021) Held During December 13–15, 2021, pp. 920–932. Springer International Publishing, Cham (2022)
5. Tamboli, M.S., Prasad, R.: A robust authorship attribution on big period. Int. J. Electr. & Comput. Eng. **9**(4), 2088–8708 (2019)
6. Iyer, R.R., Rose, C.P.: A machine learning framework for authorship identification from texts (2019). arXiv preprint arXiv:1912.10204
7. Ali, Z., Nagra, A.A., Hameed, Z., Asif, M.: Analysis of authorship attribution technique on Urdu tweets empowered by machine learning. Int. J. **10**(3) (2021)
8. Abbasi, A., Javed, A.R., Iqbal, F., Jalil, Z., Gadekallu, T.R., Kryvinska, N.: Authorship identification using ensemble learning. Sci. Rep. **12**(1), 9537 (2022)
9. Fourkioti, O., Symeonidis, S., Arampatzis, A.: Language models and fusion for authorship attribution. Inf. Process. Manage. **56**(6), 102061 (2019)
10. Ali, Z., Nagra, A.A., Hameed, Z., Asif, M.: Analysis of authorship attribution technique on Urdu tweets empowered by machine learning. Int. J. **10**(3) (2021)
11. Verhoeven, B., Daelemans, W.: Clips stylometry investigation (CSI) corpus: A Dutch corpus for the detection of age, gender, personality, sentiment and deception in text. In: Proceedings of the 9th International Conference on Language Resources and Evaluation (LREC'14), pp. 3081–3085 (2014)
12. Verhoeven, B., Daelemans, W., Plank, B.: Twisty: A multilingual Twitter stylometry corpus for gender and personality profiling. In Proceedings of the 10th International Conference on Language Resources and Evaluation (LREC'16) (2016)
13. Verhoeven, B., Soler Company, J., Daelemans, W.: Evaluating content-independent features for personality recognition. In: Proceedings of the 2014 ACM Multi Media Workshop on Computational Personality Recognition (WCPR'14), pp. 7–10. ACM, New York, NY (2014)
14. Verhoeven, B., Škrjanec, I., Pollak, S.: Gender profiling for slovene twitter communication: The influence of gender marking, content and style. :oceedings of the 6th Workshop on Balto-Slavic Natural Language Processing, pp. 119–125 (2017). Retrieved from http://www.aclweb.org/anthology/W17-1418
15. Yang, M., Zhu, D., Tang, Y., Wang, J.: Authorship Attribution with Topic Drift Model (2017). Retrieved from https://aaai.org/ocs/index.php/AAAI/AAAI17/paper/view/14152
16. Daelemans, W.: Explanation in Computational Stylometry. Springer, Berlin, pp 451–462 (2013).https://doi.org/10.1007/978-3-642-37256-8_37
17. Nasser Alsager, H.: Towards a stylometric authorship recognition model for the social media texts in Arabic. Arab. World Engl. J. (AWEJ) **11** (2021)
18. Obeid, O., Zalmout, N., Khalifa, S., Taji, D., Oudah, M., Alhafni, B., Inoue, G., Eryani, F., Erdmann, A., Habash, N.: CAMeL tools: An opensource Python toolkit for Arabic natural language processing. In: Proceedings, 12th Language Resources and Evaluation Conference, pp. 7022–7032 (2020)
19. Omar, A., Ibrahim Elghayesh, B., Ali Mohamed Kassem, M.: Authorship attribution revisited: The problem of flash fiction. Arab. World Engl. J. **10**(3), 318–329 (2019)
20. Omar, A., Hamouda, W.I.: The effectiveness of stemming in the stylometric authorship attribution in Arabic. Int. J. Adv. Comput. Sci. Appl. **11**(1), 116–121 (2020)
21. Otoom, A.F., Abdullah, E.E., Jaafer, S., Hamdallh, A., Amer, D.: Towards author identification of Arabic text articles. In: Proceedings, 2014 5th International Conference on Information and Communication Systems (ICICS 2014), pp. 5–8 (2014). https://doi.org/10.1109/IACS.2014.6841971
22. Ouamour, S., Khennouf, S., Bourib, S., Hadjadj, H., Sayoud, H.: Effect of the text size on stylometry. Application on Arabic religious texts (2016)
23. Advances in Intelligent Systems and Computing, vol. 1453, pp. 215–228. https://doi.org/10.1007/978-3-319-38884-7_16
24. Potha, N., Stamatatos, E.: Intrinsic author verification using topic modeling. In: Proceedings of the 10th Hellenic Conference on Artificial Intelligence, 20 (2018)

25. Stamatatos, E.: Authorship verification: a review of recent advances, research in computing. Science **123**, 9–25 (2016)
26. Halvani, O., Graner, L., Regev, R.: Taveer: an interpretable topic-agnostic authorship verification method. In: Volkamer, M., Wressnegger, C. (eds.): ARES 2020: The 15th International Conference on Availability, Reliability and Security, ACM, pp. 41:1–41:10 (2020)
27. Stamatatos, E.: Masking topic-related information to enhance authorship attribution. JASIST **69**(3), 461–473 (2018)
28. Ding, S., Fung, B., Iqbal, F., Cheung, W.: Learning stylometric representations for authorship analysis. IEEE Trans. Cybern. **49**, 107–121 (2019)
29. Boenninghoff, B., Nickel, R.M., Zeiler, S., Kolossa, D.: Similarity learning for authorship verification in social media. In: ICASSP 2019 - 2019 IEEE International Conference on Acoustics, Speech and Signal Processing (ICASSP), pp. 2457–2461 (2019). https://doi.org/10.1109/ICASSP.2019.8683405
30. Hosseinia, M., Mukherjee, A.: Parallel attention recurrent neural network for style change detection. In: Cappellato, L., Ferro, N., Nie, J.Y., Soulier, L. (eds.): Working Notes of CLEF 2018 - Conference and Labs of the Evaluation Forum. CEUR Workshop Proceedings, CLEF and CEUR-WS.org (2018)
31. Ding, S.H.H., Fung, B.C.M., Iqbal, F., Cheung, W.K.: Learning stylometric representations for authorship analysis. IEEE Transactions on Cybernetics **49**(1), 107–121 (2019). https://doi.org/10.1109/TCYB.2017.2766189
32. Potthast, M., Kiesel, J., Reinartz, K., Bevendorff, J., Stein, B.: A stylometric inquiry into hyperpartisan and fake news (2017)
33. Oren, H., Christian, W., Lukas, G.: Authorship verification based on compression-models (2017)
34. Yadav, Himank & Li, Juliang. (2017). Social media writing style fingerprint
35. Hosseinia, M., Mukherjee, A.: Experiments with neural networks for small and large scale authorship verification (2018)
36. Halvani, O., Christian, W., Lukas, G.: Unary and binary classification approaches and their implications for authorship verification (2018)
37. Halvani, O., Winter, C., Graner, L.: Assessing the applicability of authorship verification methods (2019)
38. Boenninghoff, B., Nickel, R., Zeiler, S., Kolossa, D.: Similarity learning for authorship verification in social media. 2457–2461 (2019). https://doi.org/10.1109/ICASSP.2019.8683405
39. Boenninghoff, B., Hessler, S., Kolossa, D., Nickel, R.: Explainable authorship verification in social media via attention-based similarity learning, 36–45 (2019). https://doi.org/10.1109/BigData47090.2019.9005650
40. Corbara, S., Moreo Fernández, A., Sebastiani, F., Tavoni, M.: MedLatinEpi and MedLatinLit : two datasets for the computational authorship analysis of medieval Latin texts. J. Comput. Cult. Herit. **15** (2022). https://doi.org/10.1145/3485822
41. Halvani, O., Graner, L., Regev, R. (2020). A step towards interpretable authorship verification.
42. Alonso-Fernandez, F., Belvisi, N.M.S., Hernandez-Diaz, K., Muhammad, N., Bigun, J.: Writer identification using microblogging texts for social media forensics. IEEE Trans. Biom., Behav., Identity Sci. **3**(3), 405–426 (2021). https://doi.org/10.1109/TBIOM.2021.3078073
43. Boenninghoff, B., Rupp, J., Nickel, R., Kolossa, D.: Deep Bayes Factor Scoring for Authorship Verification (2020)
44. Ruggiero, G., Gatt, A., Nissim, M.: Datasets and Models for Authorship Attribution on Italian Personal Writings (2020)

Machine Learning Model for Traffic Prediction and Pattern Extraction in High-Speed Optical Networks

Saloni Rai and Amit Kumar Garg

Abstract The tremendous development of network traffic causes the need to develop new network applications. Machine learning provides a pertinent platform to enhance currently used network optimization methods. The information about future traffic volumes is vital for the network operators. This paper presents an issue in traffic prediction and pattern extraction in high-speed optical networks and fills the literature gap by using a multiprocessor module from Python to enhance the efficiency of the work. An ML approach based on regression is designed. In the investigation, a dataset has been simulated and generated using the SNDLIB library and Python module GNPy estimation tool, which provides dynamic traffic matrices stated for various real network topologies and mimics real-world data. The performance of the proposed system is evaluated using different evaluation indexes like Mean Square Error (MSE), Mean Absolute Error (MAE), Max error (ME), and processing time based on the tuning of hyper-parameters. Outcomes of findings indicate better results compared to the existing technique. The findings confirm that proposed approach's efficiency is better and seems to be a promising solution to the current network problem.

Keywords Optical networks · Traffic prediction · Machine learning · Pattern extraction

1 Introduction

Data traffic has increased significantly due to the quick advancement of communication technologies, like the Internet of Things (IoT) and 5G. Recently, researchers have concentrated on embedding intelligence into communication networks to improve network performance and operations [1, 2]. Because of their advantages, optical

S. Rai (✉) · A. K. Garg
Department of Electronics and Communications, Deenbandhu Chhotu Ram University of Science and Technology, Murthal, India
e-mail: salonirai14@gmail.com

G. Fortino et al. (eds.), *Proceedings of Third International Conference on Computing and Communication Networks*, Lecture Notes in Networks and Systems 917,
https://doi.org/10.1007/978-981-97-0892-5_20

networks have emerged as a preferred alternative for big traffic carriers. However, optical networks confront additional obstacles, such as the requirement to provide capacity for a variety of services and the complex problem of resource distribution [3]. So, compared to other communication networks, maintenance and operations of optical networks are more complicated. Improvement of automation in communication networks has led to the development of network traffic and an increment in complexity of network operations. To overcome this, network resources must be allocated accurately, taking into account expected traffic values and future resource requirements [4]. This increases the agility and flexibility of optical networks. Furthermore, traffic in optical networks frequently exhibits periodicity and regularity, and operators have access to past traffic data. With the proliferation of terminals and the increase in network capabilities, managing network traffic has become increasingly important [5]. The control efficiency of the network can be improved by improving the accuracy of network traffic forecasts. Information about predicted traffic allows operators to optimize resource and network usage, and provide an improved quality of service (QoS), reduction in the cost of network function, or anomaly detection in the traffic data [6, 7]. Machine learning techniques are known to be promising solutions for adding intelligence to optical network nodes. The network nodes can learn from the conditions of networks using ML techniques. Optical networks with ML techniques learn the inner relations from the environment and data, resulting in the execution of flexible and automated network operations [8]. ML algorithms use input data and environment feedback to handle complex problems. The future traffic volume can be controlled by ML using an input of the traffic values.

Generally, the network traffic is not an arbitrary procedure because it follows particular patterns [9–11]. It is seen that more aggregated traffic (from a larger number of users), the more regular its shape as a time function, also patterns are highly prominent. The primary focus of the traffic-based technique is to gather information in terms of observed traffic flows in networks and subsequently use several machine learning (ML) techniques for data processing and hidden pattern extraction. In order to predict future traffic, the patterns can be employed and utilized to model and optimize the performance of the network [12]. In order to plan the routing rules, the information can be used for the approaching traffic demands and to accurately plan the resource assignment process for the approaching requests by the networks. For example, to (a) increase the number of served clients, (b) increase their Quality of Service (QoS), (c) enhance resource utilization, and d) minimize the power utilization that meets the green networking concept [13–15].

In this paper, machine learning-based efficient daily traffic pattern prediction in a telecommunication network is presented. In the investigation, the novel dataset "nobel-us" has been simulated and generated using the SNDLIB library and Python module GNPy estimation tool, which provides dynamic traffic matrices defined for various real network topologies and mimics real-world data. For traffic prediction, a machine learning-based regression technique is applied to handle the stream or chunk of data. To evaluate the performance, hyper-parameters are tuned, and different evaluation indexes are calculated, like MAE, MSE, ME, and processing time to predict the traffic and extract the pattern.

The objectives of this work are:

(1) The work presented fills the literature gap by using a multiprocessor module from Python to enhance the efficiency of the work. Multiprocessing allows multiple spawns of the process and provides a convenient way of parallelizing the execution of the process. This represents the novelty of the work.
(2) Multi-layer perceptron (MLP) and SVR, both DL and ML techniques, have been implemented for traffic prediction and pattern extraction.

The organization of the paper is as follows, Sect. 2 emphasizes related work and problem formulation of network traffic prediction, and structure to resolve it is summarized in Sect. 3. A comparative analysis of proposed traffic prediction model is summarized in Sect. 4. The conclusion and future scope of research work are summarized in Sect. 5.

2 Related Work

Andrea D'amico et al. (2021) developed a technique using an ML algorithm for prediction. The OSNR component was predicted using the ASE-shaped loads. The methodology was integrated with the light path computation energy (L-PCE) with QoT-E. The experimentation was conducted on a point-to-point optical line system. The expected GSNRs were extracted from the dataset along with the non-linear and linear components. An accurate and reliable LP feasibility predictor was demonstrated by the simulation results by achieving a complete QoT-E[16]. Davide Andreoletti et al. (2021) developed an approach to minimize routing complexity in large optical networks. The ML-based system was designed so that when a source and destination pair was given as input, a sub-graph generated only by the nodes was given as output. These nodes were most likely to be from the shortest path available. This resulted in the reduction of the original graph and the minimization of RWA task computational time. Thus, the computational complexity was reduced with significantly less deterioration of blocking probability [17]. Michal Aibin (2018) presented two techniques for providing traffic prediction within Elastic Optical Networks using ML techniques. To predict traffic in a non-supervised manner, the ANN algorithm was applied. To ensure that all the inputs were in the comparable range, normalization was used in preprocessing of input data. The simulation results showed that the performance of adaptive strategies was better, resulting in enhanced efficiency of network architectures [18]. Stephen Cheng et al. (2019) studied that it was highly efficient to use ML to gather accurate predictions for regenerator allocation. Supplying quality input training data to the ML algorithm was essential to increase prediction accuracy. The effect of various network features on prediction results was examined. Further, the network features with higher accuracy of predictions for regenerator allocation were listed. The simulations showed the best outputs when using the complete set of features. Additionally, a smaller set of features was very effective in attaining less than 1% BP of the Service Level Agreement [19].

Fabiano Locatelli et al. (2020) presented that for improvements in the resiliency and efficiency of optical transport networks, the cost-effective solutions possible are channel spectral monitors. An ML-based technique was presented to estimate the in-band of amplified spontaneous emission (ASE). Additionally, the 3-dB bandwidth was also filtered using this approach. The performance of the presented technique was evaluated with respect to laser drift and filter bandwidth tightening environments. Within these conditions, good estimation accuracy was achieved as per the simulation outputs [20]. Chuang Song, et al. (2018) presented a novel technique for dynamic network slicing within optical networks. A feed-forward neural network was modeled for predicting the traffic demands for every link. From the China Telecom Corporation, the optical network operation dataset was used to perform simulations. The traffic volume was predicted perfectly with the usage of ANN as per the observations. 96% of traffic prediction accuracy was achieved as per the outcomes. Also, 29% delay and 49% blocking were reduced in comparison to the traditional approaches [21]. Payman Samadi et al. (2017) presented a new approach for predicting the OSNR within the WDM channels. A cognitive and scalable mechanism was designed to increase accuracy and minimize the provisioning margins. The network resource usage was also aimed to enhance with this enhancement. The OSNR was predicted with the microsecond response time to provide dynamic and agile network operations. The research could be expanded in the future to evaluate the merits of OSNR estimation [22].

Michał Aibin et al. (2021) presented an ML model based on the Generative Adversarial Network and Graph Convolutional Network to set up effectual network state predictions. In this paper, a primary concern was detecting traffic peaks in networks that can be used for routing decisions. The results were validated using pseudorealistic datasets. The findings confirmed the efficiency of optimizing both real-time routing and long-term network design decisions for the model to be efficient [23]. Szostak et al. (2021) developed a procedure for traffic prediction of the network. Classification and regression, two ML methods were evaluated. Seattle internet exchange point dataset is used. Regression showed better results than classification. PAR and MLPR are considered as best regressors [24]. Szostak et al. (2020) developed a methodology for traffic prediction dynamically with service function chain. Research showed that it is possible to predict future traffic. The choice of regressor shall be based on the characteristics of the traffic network. The authors evaluated that an initial task for predicting future traffic should be done through the analysis of historical network traffic to determine a classifier [25].

3 Proposed Work

3.1 Problem Formulation

Predicting the information about network traffic is the objective of traffic prediction [26]. The historical network traffic information and extracted patterns determine the network traffic prediction. Equation (1) defines the process as:

$$x_{t-M+1}, \ldots, x_{t-1}, x_{tf\rightarrow}(.)\widehat{x}_{t+1}, \ldots, \widehat{x}_{t+H-1}, \ldots, \widehat{x}_{t+H} \tag{1}$$

$x_t \in R_n$ denotes observation vector of n observation points at t time interval. Depending on the traffic data of the previous M sampling time, the traffic prediction model learns the mapping function $f(.)$, which results in predicting the network traffic of H sampling time. Supervised machine learning algorithms have been implemented for traffic prediction and pattern extraction [27, 28].

3.2 Data Acquisition

The "nobel-us" network infrastructure involves 14 nodes and over 160 km of fiber-optic routes, as shown in Fig. 1. The nodes are located geographically with the coordinates (x, y). The nodes are named Palo–Alto, San Diego, Boulder, Washington [29], Atlanta, Salt Lake City, Urbana Champaign, Ithaca, Ann-Arbor, Lincoln, Princeton, Pittsburgh, Houston, and Seattle.

The "Nobel-us" network topology is taken from "the standardized library, the Survivable Network Design Library (SNDLIB)," to design survivable fixed telecommunication networks. It is available online at "http://sndlib.zib.de/home.action." The admissible routing paths are set in the SNDlib network that describes the network structure and traffic to be routed. The sets of nodes and links are defined in this

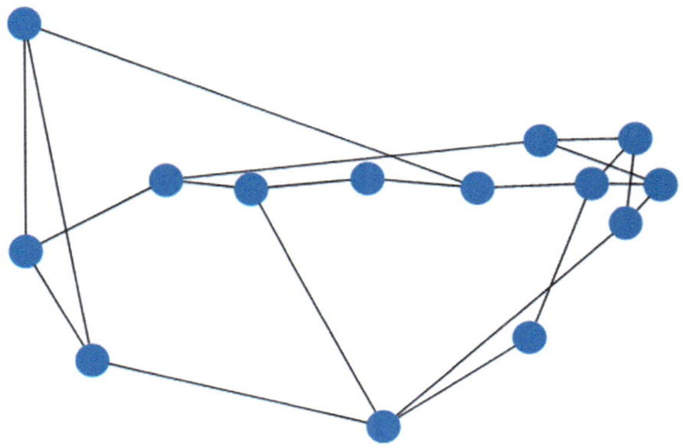

Fig. 1 Network Topology ("nobel-us," 14 nodes)

network structure. For carrying traffic, the group of links states the potential connections among the nodes. Every link has a number of pre-installed capacity and cost information. Additionally, the traffic routed through the link and fixed-charge cost of using the link is stated. The nodes and links comprise of network topology.

The di-graph $\boldsymbol{G} = (\boldsymbol{V}, \boldsymbol{E})$. $\boldsymbol{V}$ denotes nodes in network and $\boldsymbol{V} = (\boldsymbol{V}_1, \boldsymbol{V}_2, \ldots .\boldsymbol{V}_N)$. The number of nodes is represented by N, and the links between nodes are represented by $\boldsymbol{E}$. The XML tree parser is used with the configuration from topology "nobel-us." Furthermore, the graph is generated and the network module is read. The data are predicted using the network traffic data. The data are predicted from the traffic network simulation to train the model.

3.3 *Traffic Prediction*

ML-based traffic prediction is performed in this research, where Support Vector Regressor and MLP regressor are applied for traffic prediction and pattern extraction [30].

Multi-layer perceptron (MLP): This is a feed-forward neural network through which the data flows in the forward direction. Three layers, input, hidden, and output layers, are comprised in this algorithm as shown in Fig. 2. The input layer receives the signals. The required tasks are performed in the output layer. Each layer's neurons are interlinked with the next layer. The arbitrary numbers of hidden layers, which are positioned in input and output layers, are true computational engine of MLP. The training of neurons in MLP is performed through backpropagation learning algorithm [31, 32].

$x = \{x_1, x_2, \ldots x_i\}$ denotes the input of the model. Each node in the hidden layer (l) is represented as

$a^l = \left\{a_1^l a_2^l, \ldots ..OCa_m^l\right\}$ can be calculated as follows:

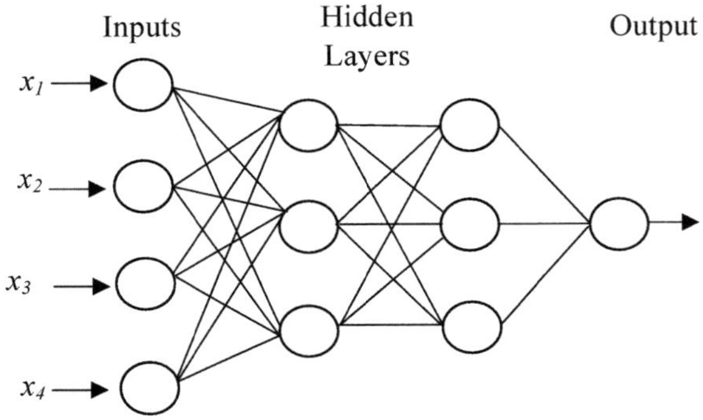

Fig. 2 Topology of multi-layer perceptron

$$a_m^l = \sum_j^n \left(w_m^l a_j^{l-1} + b_m^l \right) \tag{2}$$

where ω_{mi}^l indicates weight coefficient between node a_i^{l-1} and node a_m^l. b_m^l indicates a bias of the node a_m^l and σ indicates activation function.

The output of the technique is represented as $y = \{y_1, y_2, \ldots . y_k\}$ and calculated as:

$$y_k = \varphi\left(\sum_j^m w_{km}^t a_j^i + b_k \right) \tag{3}$$

wherein φ indicates activation function.

Support Vector Regression (SVR): A set of data represented as $\{x_i, y_i\}$, $i = 1, \ldots, N$; where $x_i \in R^n$ is input vector with n dimension, $y_i \in R^n$ is target value. The number of data patterns is represented by N. For mapping the original input data x into the higher dimensional feature space R^h, SVR is used. Equation (4) is used to express the SVR function. The coefficients are reduced using the objective function of SVR. The error term is handled within constraints where the absolute error is minimal. The margin is known as maximum error, ϵ (epsilon) is specified:

$$f(x) = \omega^t \varphi(x) + \mathrm{b} \tag{4}$$

Here, the weight vector is denoted by ω and b is a bias term. By reducing the regularized risk function R as shown in (5), the coefficients ω and b are estimated:

$$R = \frac{1}{2}||\omega||^2 + C\sum_{1=1}^{n}(y_i - f(x_i), x_i) \tag{5}$$

Here, the regularized constant C determines the trade-off between the regularization term and the empirical error [33, 34].

3.4 *Pattern Extraction*

The machine learning model learns the pattern to predict traffic using the "Adam" optimizer. The inputs are feed-forward in the MLP network, and the weights are adjusted using backpropagation of gradients. To optimize the gradient descent, the Adaptive Moment Estimation algorithm is applied. Due to the presence of lots of data or parameters, the method is efficient. The efficiency is higher, and the memory required is minimal [35, 36]. The pattern is extracted using the MLP model. A repeating pattern can be seen more clearly from the figure below, and these dips in

Fig. 3 Predicted traffic volume and extracted pattern

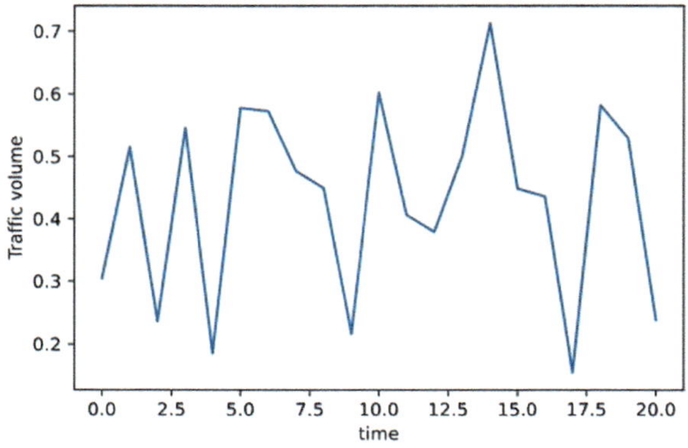

Table 1 Tuning parameters (MLP training model)

Parameter	Chosen value
"hidden_layer_sizes"	100
"solver"	"adam"
"activation function"	"relu"
"learning_rate_init"	0.001
"Alpha"	0.0001
"batch_size"	"Auto"
"learning_rate"	"Constant"
"power_t"	0.5
"max_iter"	500
"shuffle"	"True"
"epsilon"	$1e^{-08}$
"validation_fraction"	0.1
"tol"	0.0001
"momentum"	0.9
"beta_1"	0.9
"beta_2"	0.999

the plot show web traffic occurring once every 24 h. The extracted traffic pattern is shown in Fig. 3. Tables 1 and 2 represent the values of hyper-parameters chosen for Adam optimizer.

3.5 Research Methodology

The method of the ML-based traffic prediction and pattern extraction technique is shown in Fig. 4. The first step is to simulate a traffic topology from the SNDLIB

Table 2 Tuning parameters (SVR training model)

Parameter	Chosen value
"kernel"	"rbf"
"C"	1.0
"gamma"	"scale"
"degree"	3
"coef0"	0.0
"tol"	0.001
"shrinking"	"True"
"epsilon"	0.1

library. Data preprocessing is done and data are split into two parts, training data, and testing data. SVR and MLP algorithms are applied for traffic prediction and pattern extraction. Evaluation is done on unseen data to see the performance of the trained model. Then, the results of our MLP regressor and SVR are compared. MLP works better and is compared with ML prediction approaches. Thus, this indicates DL models are better than ML models.

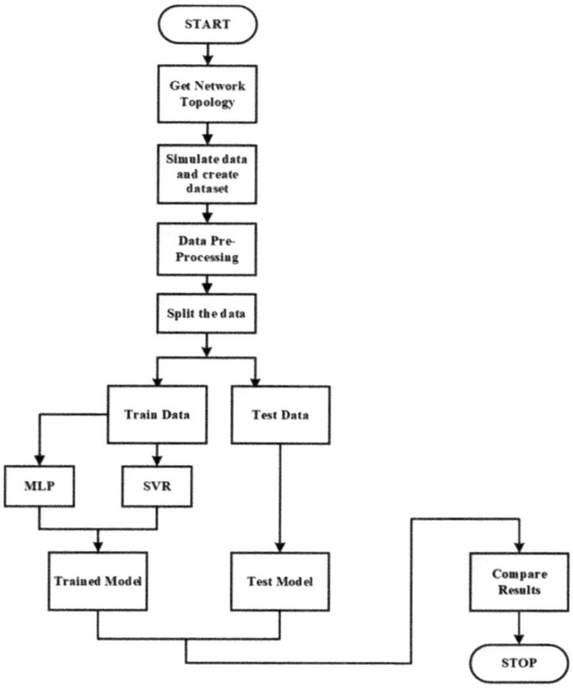

Fig. 4 Research methodology

4 Results and Analysis

In this section, the efficiency of the proposed traffic prediction and pattern extraction model is evaluated and applied to the simulated dataset. The proposed model has been evaluated on unseen data (test set).

4.1 Performance Metrics

The work has been evaluated using below parameters:

MSE measures an average of squared errors. In other words, it measures the average squared difference between actual and estimated values. MSE is a widely accepted parameter for testing the performance of predictors [10]. It is a measure of the quality of an estimator and is always a non-negative value. MSE can be expressed as in Eq. (6).

$$\text{MSE} = \frac{1}{n}\sum_{i=1}^{n}(Y_i - \widehat{Y}_i)^2 \tag{6}$$

Max Error: Max error is the maximum residual error. It is absolute value of the most important difference between a real value and its predicted variable. The Max error of our model is compared with the Max error of [1]. Max error can be expressed as in Eq. (7).

$$\text{Max.error} = \text{ME} = |Y_i - Y_i| \tag{7}$$

MAE: It is absolute difference between predicted values and actual values. Mathematically, it is the arithmetic average of absolute errors. If MAE is lesser, higher is the accuracy of a model.

MAE can be expressed as follows in Eq. (8).

$$\text{MAE} = \frac{1}{n}\sum_{i=1}^{n}(Y_i - \widehat{Y}_i) \tag{8}$$

Processing time: It is the time consumed to run the entire process and then end with the resultant numbers of processors and cycles. Parallel processing has been used in our work. Our execution time is exceptionally better than [40]. Processing time can be expressed in Eq. (9).

$$P.time = (N-1)P_1 + (N-2)P_2 + \cdots + P_{N-1} \tag{9}$$

Table 3 Comparison of proposed work with conventional models

Parameters	Existing work [38]		Proposed work	
	WASK dataset	SIX dataset	Dataset ("nobel-us")	
	MLP	MLP	MLP	SVR
Max error	5.0686	0.0857	0.04378019	0.0452
Mean absolute error	0.8247	0.0443	0.019617409	0.02250
Processing time	0.0011 s	0.0012 s	0.000321 s	0.000581 s

Table 4 Comparison of proposed work with conventional models

Parameters	Existing work [10]		Proposed work	
	LSTM	GCN-GAN	MLP	SVR
Mean Squared Error	0.0052	0.0040	0.00070774	0.0007835

4.2 Comparative Analysis

The proposed machine learning models were trained and followed specific traffic patterns and predicted traffic. The prediction made by the proposed model is better and more efficient. The paper [37] has worked on WASK and SIX datasets and supervised learning is applied. The proposed model, which is being trained, has exceptionally performed well. The authors in paper [38] have shown the evaluation of the pseudo-realistic simulated data. The parameters of the proposed work are compared with [10] and are better and more efficient. The outcomes presented in Tables 3 and 4 are stated as follows.

Figure 5 represents a graphical analysis of Max error and the proposed work compared to conventional techniques. This graph shows that Max error for the proposed model is better than [38]. Figure 6 represents the graphical analysis for MAE and represents the proposed work in comparison with conventional [38]. Figure 7 represents a graphical analysis of processing time and represents the proposed work with state-of-the-art approaches [38]. Figure 8 represents a graphical analysis of mean squared error and proposed work with conventional [10].

5 Conclusion and Future Scope

This work presents regression-based ML approaches for traffic prediction and pattern extraction in optical networks and also proposes DL-based MLP. The performance of the proposed model was computed using different indexes based on tuning of hyper-parameters, such as MAE, MSE, ME, and processing time. The findings confirm that the presented approach's efficiency is better, as shown in the above graphs and

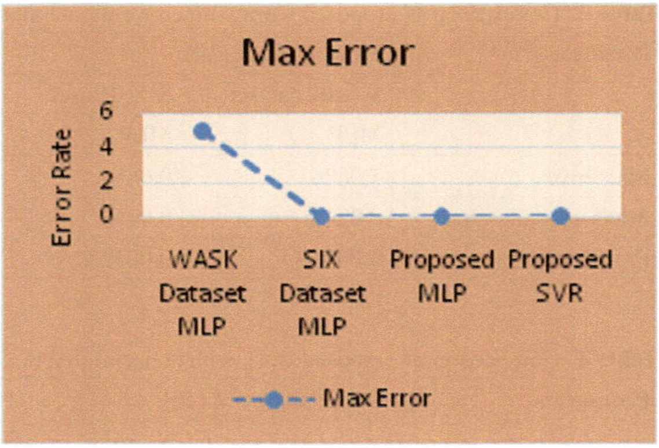

Fig. 5 Comparative analysis of max error

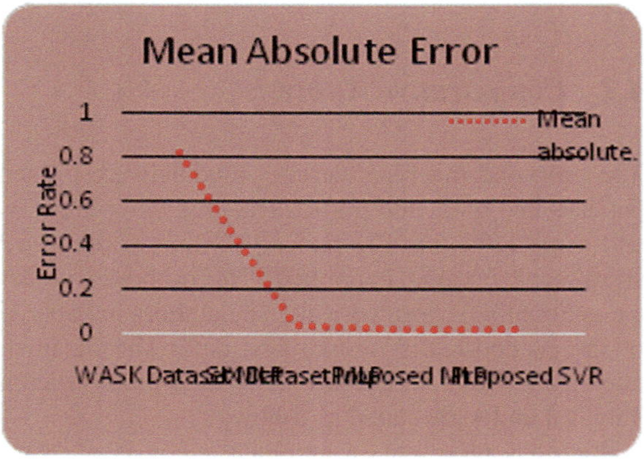

Fig. 6 Comparative analysis of mean absolute error

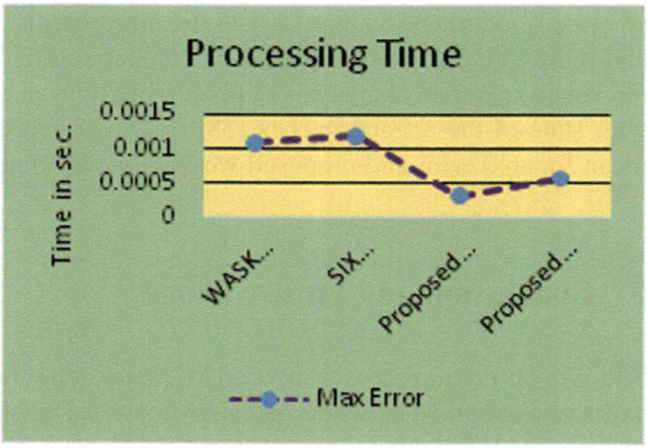

Fig. 7 Comparative analysis of processing time

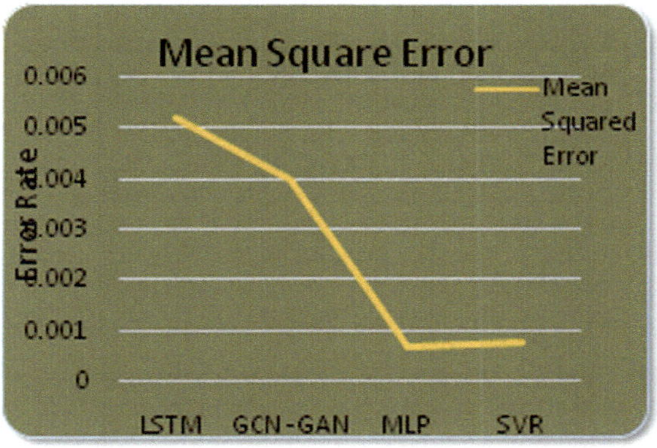

Fig. 8 Comparative analysis of mean square error

tables. In future work, an optimization framework can be used to improve the routing algorithms applied in the optical network with the help of predicted traffic data.

References

1. Zhong, Z., Hua, N., Tornatore, M., Li, J., Li, Y., Zheng, X., Mukherjee, B.: Provisioning short-term traffic fluctuations in elastic optical networks. IEEE/ACM Trans. Netw. **27**(4), 1460–1473 (2019)
2. Dias, M.P.I., Karunaratne, B.S., Wong, E.: Bayesian estimation and prediction-based dynamic bandwidth allocation algorithm for sleep/doze-mode passive optical networks. J. Lightwave Technol. **32**(14), 2560–2568 (2014)
3. Chitra, K., Senkumar, M. R.: Hidden Markov model based lightpath establishment technique for improving QoS in optical WDM networks. In: Second International Conference on Current Trends In Engineering and Technology-ICCTET 2014, pp. 53–62. IEEE (2014)
4. de Araújo, D.R., Bastos-Filho, C.J., Martins-Filho, J.F.: Methodology to obtain a fast and accurate estimator for blocking probability of optical networks. J. Opt. Commun. Netw. **7**(5), 380–391 (2015)
5. Filho, R.H., Maia, J.E.B.: Network traffic prediction using PCA and K-means. In: Proceedings of the 2010 IEEE/IFIP Network Operations and Management Symposium, NOMS 2010, Osaka, Japan, 19–23 April 2010, pp. 938–941
6. Rai, S., Garg, A.K.: Analysis of RWA in WDM optical networks using machine learning for traffic prediction and pattern extraction. J. Opt., 1–8 (2021)
7. Mata, J., de Miguel, I., Durán, R.J., Merayo, N., Singh, S.K., Jukan, A., Chamania, M.: Artificial intelligence (AI) methods in optical networks: a comprehensive survey. Opt. Switch. Netw. **28**, 43–57 (2018)
8. Lechowicz, P.: Regression-based fragmentation metric and fragmentation-aware algorithm in spectrally-spatially flexible optical networks. Comput. Commun. **175**, 156–176 (2021)
9. Mata, J., De Miguel, I., Durán, R.J., Aguado, J.C., Merayo, N., Ruiz, L., Fernández, P., Lorenzo, R.M., Abril, E.J.: A SVM approach for lightpath QoT estimation in optical transport networks. In: Proceedings of the 2017 IEEE International Conference on Big Data, Big Data 2017, Boston, MA, USA, 11–14 December 2017; pp. 4795–4797 (2017)
10. Vinchoff, C., Chung, N., Gordon, T., Lyford, L., Aibin, M.: Traffic prediction in optical networks using graph convolutional generative adversarial networks. In: Proceedings of the International Conference on Transparent Optical Networks, Bari, Italy, 19–23 July, pp. 3–6 (2020)

11. Musumeci, F., Rottondi, C., Nag, A., Macaluso, I., Zibar, D., Ruffini, M., Tornatore, M.: An overview on application of machine learning techniques in optical networks. IEEE Commun. Surv. Tutor. **21**, 1383–1408 (2019)
12. Cenedese, A., Tramarin, F., Vitturi, S.: An energy efficient Ethernet strategy based on traffic prediction and shaping. IEEE Trans. Commun. **65**, 270–282 (2017)
13. Rzym, G., Boryło, P., Chołda, P.: A time-efficient shrinkage algorithm for the Fourier-based prediction enabling proactive optimisation in software-defined networks. Int. J. Commun. Syst. **33**, e4448 (2020)
14. Zhong, Z., Hua, N., Yuan, Z., Li, Y., Zheng, X.: Routing without routing algorithms: An AI-based routing paradigm for multi-domain optical networks. In: 2019 Optical Fiber Communications Conference and Exhibition (OFC).Optical Society of America: Washington, DC, USA (2019)
15. Guan, L., Zhang, M., Wang, D.: Demonstration of AI-assisted energy-efficient traffic aggregation in 5G optical access network. In: 2020 Optical Fiber Communications Conference and Exhibition (OFC) , pp. 1–3. Optical Society of America, Washington, DC, USA (2020)
16. D'Amico, A., Straullu, S., Borraccini, G., London, E., Bottacchi, S., Piciaccia, S., Curri, V.: Enhancing lightpath QoT computation with machine learning in partially disaggregated optical networks. IEEE Open J. Commun. Soc. **2**, 564–574 (2021)
17. Andreoletti, D., Rottondi, C., Bianco, A., Giordano, S.: A machine learning framework for scalable routing and wavelength assignment in large optical networks. In: Optical Fiber Communication Conference, pp. F2G-3. Optica Publishing Group (2021)
18. Aibin, M.: Traffic prediction based on machine learning for elastic optical networks. Opt. Switch. Netw. **30**, 33–39 (2018)
19. Cheng, S., Xiao, D., Huang, A., Aibin, M.: Machine learning for regenerator placement based on the features of the optical network. In: 2019 21st International Conference on Transparent Optical Networks (ICTON), pp. 1–3. IEEE (2019)
20. Locatelli, F., Christodoulopoulos, K., Fàbrega, J. M., Moreolo, M.S., Nadal, L., Spadaro, S.: Experimental demonstration of a machine learning-based in-band OSNR estimator from optical spectra. In: 2020 International Conference on Optical Network Design and Modeling (ONDM), pp. 1–4. IEEE (2020)
21. Song, C., Zhang, M., Huang, X., Zhan, Y., Wang, D., Liu, M., Rong, Y.: Machine learning enabling traffic-aware dynamic slicing for 5G optical transport networks. In: CLEO: Science and Innovations, pp. JTu2A-44. Optica Publishing Group (2018)
22. Samadi, P., Amar, D., Lepers, C., Lourdiane, M., Bergman, K.: Quality of transmission prediction with machine learning for dynamic operation of optical WDM networks. In: 2017 European Conference on Optical Communication (ECOC), pp. 1–3. IEEE (2017)
23. Aibin, M., Chung, N., Gordon, T., Lyford, L., Vinchoff, C.: On short-and long-term traffic prediction in optical networks using machine learning. In: 2021 International Conference on Optical Network Design and Modeling (ONDM), pp. 1–6. IEEE (2021)
24. Szostak, D., Włodarczyk, A., Walkowiak, K.: Machine learning classification and regression approaches for optical network traffic prediction. Electronics **10**(13), 1578 (2021)
25. Szostak, D., Walkowiak, K.: Application of machine learning algorithms for traffic forecasting in dynamic optical networks with service function chains. Found. Comput. Decis. Sci. **45**(3), 217–232 (2020)
26. Alarcon-Aquino, V., Barria, J.A.: Multiresolution FIR neural-network-based learning algorithm applied to network traffic prediction. IEEE Trans. Syst., Man, Cybern., Part C (Appl. Rev.) **36**(2), 208–220 (2006)
27. Krishnamurthy, B., Sen, S., Zhang, Y., Chen, Y.: Sketch-based change detection: methods, evaluation, and applications. In: Proceedings of the 3rd ACM SIGCOMM Conference on Internet Measurement, pp. 234–247 (2003)
28. Knapińska, A., Lechowicz, P., Walkowiak, K.: Machine-learning based prediction of multiple types of network traffic. In: International Conference on Computational Science, pp. 122–136. Springer International Publishing, Cham (2021)

29. Szostak, D., Walkowiak, K., Włodarczyk, A.: Short-term traffic forecasting in optical network using linear discriminant analysis machine learning classifier. In: 2020 22nd International Conference on Transparent Optical Networks (ICTON), pp. 1–4. IEEE (2020)
30. Orlowski, S., Pióro, M., Tomaszewski, A., Wessäly, R.: SNDlib 1.0—Survivable network design library. In: Proceedings of the 3rd International Network Optimization Conference (INOC 2007), Spa, Belgium, 12–14 June 2007
31. Jurkiewicz, P., Rzym, G., Borylo, P.: Flow length and size distributions in campus Internet traffic. Comput. Commun. **167**, 15–30 (2021)
32. Ba, S., Chatterjee, B.C., Oki, E.: Defragmentation scheme based on exchanging primary and backup paths in 1+1 path protected elastic optical networks. IEEE/ACM Trans. Netw. **25**, 1717–1731 (2017)
33. Goścień, R.: On the efficient dynamic routing in spectrally-spatially flexible optical networks. In Proceedings of the Resilient Networks Design and Modeling (RNDM), Nicosia, Cyprus, 14–16, (2019).
34. Walkowiak, K., Klinkowski, M., Lechowicz, P.: Dynamic routing in spectrally spatially flexible optical networks with back-to-back regeneration. IEEE/OSA J. Opt. Commun. Netw **10**, 523–534 (2018)
35. Gaizi, K., Abdi, F., Abbou, F.M.: Realistic dynamic traffic generation for WDM Optical Networks. In: Proceedings of the 2016 27th Irish Signals and Systems Conference (ISSC), Londonderry, UK, 21–22, pp. 1–4 (2016)
36. Gencata, A., Mukherjee, B.: Virtual-topology adaptation for WDM mesh networks under dynamic traffic. IEEE/ACM Trans. Netw. **11**, 236–247 (2003)
37. .Troia, S., Cibari, A., Alvizu, R., Maier, G.: Dynamic programming of network slices in software-defined metro-core optical networks. Opt. Switch. Netw, **36**, 100551 (2020)
38. Goścień, R., Knapińska, A., Włodarczyk, A.: Modeling and prediction of daily traffic patterns—wask and six case study. Electronics **10**(14), 1637 (2021)

Improved Markov Decision Process in Wireless Sensor Network for Optimal Energy Consumption

Gauri Kalnoor and Prakash B. Metre

Abstract The research looks at the problem of receiving a precise estimate of the sensor node attribute from the Wireless Sensor Network (WSN) in a certain time frame while using as little energy as possible from the sensors. An estimate of an attribute may be obtained by taking readings from wake-up sensors connected to a sink node that has been placed at random. The sink must take relevant measurements within a certain time constraint. As an added bonus, the sink reduces the amount of power needed for sensors to transmit their readings. The predicted energy consumption of WSN sensor nodes is presented in a closed-form formulation in this work. The research establishes a maximum allowable sensor transmission distance. The Markov Decision Process may be used to determine the timing of sensor transmissions over a given time frame (MDP). A strategy for transmitting-sensor programming is also developed in this work. The simulation is conducted in MATLAB environment, where the MDP schedule offers reduced energy consumption, reduced delay, increase packet delivery ratio, reduced false alarm rate, which finds the attribute of the data sent via a sensor node.

Keywords MDP · Wireless sensor network · Energy consumption

1 Introduction

Wireless sensor networks (WSNs) are becoming more popular in modern smart cities, and they include sensing devices capable of two-way communication with the environment's many complexity and artifacts. However, this connection is limited by the limited power available from battery-operated sensor nodes. Many applications'

G. Kalnoor (✉)
Department of Computer Science and Engineering, BMSCE, Bangalore, India
e-mail: gaurik.cse@bmsce.ac.in

P. B. Metre
Department of Computer Science and Engineering, Presidency University, Bangalore, India
e-mail: prakashbmetre@presidencyuniverity.in

G. Fortino et al. (eds.), *Proceedings of Third International Conference on Computing and Communication Networks*, Lecture Notes in Networks and Systems 917,
https://doi.org/10.1007/978-981-97-0892-5_21

sensor nodes are set up to operate for many years without being recharged [1]. More resource-intensive processes with minimal maintenance costs are being included into the upcoming WSN applications. This means that more efficient and adaptable algorithms are being developed than ever before.

WSNs operate in stochastic (unpredictable) environments. In reality, a sensor node acts as a decision-maker/agent by interacting with its environment and changing states. The node's internal characteristics (such as its battery status and energy supply) will be included in the world, along with numerous external factors (e.g., other nodes in the network or a moving target). Thus, the actions may range from the simple (such as entering sleep mode to save power) to the complex. In such an uncertain setting, the Markov decision-making processes mathematical paradigm may be used to simulate the mechanism's dynamic (MDPs).

The MDP incorporates a Markov property into the framework. Specifically, the current state of the system is the sole thing the system's prospective state is dependent on. Recently, MDP solvers have been created, expanding the scope of possible study into WSNs by enabling the solution of large-scale systems. Hardware resource management is at the heart of WSNs. Wasteful actions often result from inflexible decision-making. For instance, the node that sends data at a constant transmitting power without taking the channel circumstances into account uses more energy than the node that adjusts its transmitting power dynamically [2, 3]. Therefore, MDPs provide a far more efficient allocation of resources by automatically adapting network operations to satisfy prevailing physical constraints.

The MDP model provides a framework that strikes a good compromise between competing objectives, such as reducing power usage and increasing sensor coverage. Several researchers [4–6] examine how to maximize MDPs for various ends. The device dynamics are greatly enhanced by the communication between mobile entities and new WSN apps. For example, there are a number of architectural hurdles associated with using a mobile data gathering gateway [5]. Here, MDP will look at the link between time and the forecast of possible locations for relocating items, as shown in [7, 8].

The MDP model's solution, known as a policy, may be implemented using data from a survey table. For simple online tasks, this table is cached in the sensor node's memory. Moreover, the MDP model may be used without heavy computation demands for tiny and resource-limited nodes. WSN algorithms may be created with reduced computing costs by approximating optimal policy choices using near optimum solutions [9, 10].

Here, we investigate a closed-form description of the sensor nodes' expected energy consumption in WSN. The research establishes a maximum allowable sensor transmission distance. Use a Markov decision process to determine when sensor readings will be sent over a certain time frame (MDP). The research also developed a heuristic for using sensors as a means of communication. MATLAB is used to simulate the environment in which the MDP schedule's benefits—lower power consumption, shorter delays, higher packet delivery ratios, and lower false alarm rates—would be realized, and where the attributes of the data sent by a sensor node would be located.

2 Background

When making choices while facing unknown outcomes, a Markov decision process (MDP) may help you optimize your choices [11, 12]. An agent's stochastic decision process in response to its environment or a system is modeled by the MDP. Each decision period keeps the system in states, from which the agent chooses an action. After the action is executed, the agent gets an instantaneous reward R and the system transits to a new state according to the transition probability [12–18]. To represent how a wireless sensor node (an agent) interacts with its surroundings (a system) to accomplish a goal, the MDP is used in WSNs. In WSNs, the MDP may be used to improve the efficiency of energy management and routing decisions, for instance.

Markov Decision Process Framework

MDP is the process consisting of tuples (S, A, P, R, T),

- S is the finite states set,
- A is the finite actions set,
- P is the transition probability function once the action a is connected between the states,
- R is the immediate reward gained upon making an action a, and
- T is the decision epoch set.

A policy is generally represented as "π" is mapped with an action from a state s. This strategy aims at maximizing or reducing the target of the MDPs.

MDP is known to be a finite or limitless time horizon. π* is considered as an optimal policy that increases the cumulative expected reward for MDP under a finite period of time. The following are defined:

where,

t is the time,

s is the state, and

a is the action.

The approaches used as MDP alternatives provide reduced overall compensation. The algorithms for MDPs are tested to achieve an average overhead. The machine is operated on MDPs with a finite time horizon within a known span of time. The device begins running at status s0 and continues with T periods in the next interval. The Vπ(s) are then obtained by optimizing the best π*. In state s, v*(s), which can be located in each state of the node and repetitively using maximal Bellman's equations, can be considered the optimal compensation.

Here,

vt(s) is the value of s and

Qt(s,a) is the value of action.

The method then continues until the end iteration is reached. These MDP methods have a related difficulty in the finite time horizon. But where the MDP state is only known for the past time, and when the state only knows for the beginning period, the backward strategy is used. The Infinite Time MDP achieves the goal of maximizing or maximizing the cumulative incentive for the predicted discounted overall benefits.

The following formula calculates the cumulative discounted reward:

The average reward is then determined using the following:

Here,

ϒ is the discount factor and

E [.] is the expected function.

Solutions of MDPs

Here, we introduce solution methods for MDPs with discounted total reward. The algorithms for MDPs with average reward can be found in.

- Backward Induction: regarded as a dynamic programming technique, the Bellman equations are the most common and effective way. As the mechanism is halted at a known point in time, the desired operation and value function can be determined first at the last time. Then, at an early date, we repetitively achieve the optimum behavior based on the optimal Bellman equations.
- Forward Induction: This advanced induction technique is often called an iteration strategy. The objective is to treat the optimization problem as a set of independent activities. We assess Q-values for k steps, in particular given the best strategy for time steps. Every step of the process is completed. Both approaches vary in complexity depending on the length of the MDP's time horizon. Their applications are different, however. Since we have known the state of MDPs since the previous time we used backward induction, it is very useful. In contrast, induction is used in situations when one knows just the starting point. Solutions to Horizontal Markov Decision Processes in Eternity: Time horizon MDPs are more difficult to solve than regular MDPs. However because systems are often unknown and assumed to be infinite, the MDP's eternal time horizon is more generally utilized. There are a plethora of potential solutions that have been proposed.

When solving a discounted MDP with an indefinite horizon, value iteration is the most common and effective strategy. Quick convergence and ease of use are only two of the many benefits of this method, which shines when the MDP state space is large. This technique was inspired by the forward induction process and MDPs with constrained time windows. To guarantee convergence in MDP, however, it is not sufficient to run the algorithm over the whole time horizon. A stop-over condition must be applied.

The primary goal of this strategy is to create a more robust policy series via a process called "policy iteration." The policy would begin with an arbitrary policy and adjust the policy to converge. Policy review and revision are the two main components of this approach. After finding the expected price decrease under the plan by solving

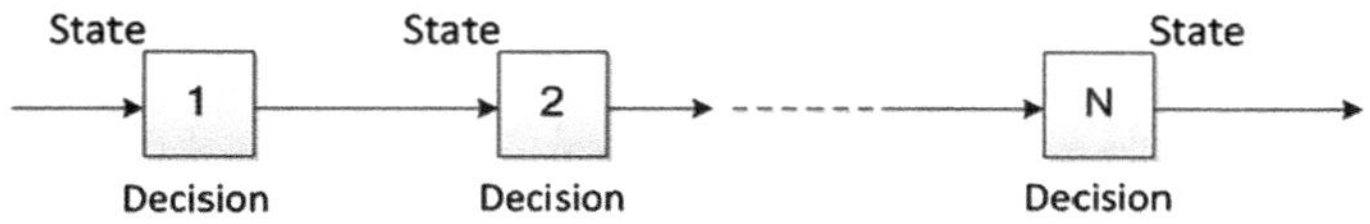

Fig. 1 Dynamic decision process

the linear equations, we choose the best decision-making strategy for each nation. It's possible that this procedure will need less iterations than the value iteration technique before reaching convergence. Since the policy iteration approach requires solving linear equations, each iteration takes more time than the iteration process itself.

In contrast to previous methods, the goal of the linear programming method is to find a static policy by means of the solution of a linear program. If the linear program is solved, then the optimal value is obtained, and the best approach for each state may be established from that. When the available state space is large, other methods, such as valuation and policy iteration, are superior than linear programming. However, the linear programming method is useful for MDPs with constraints since linear equations may be utilized to express constraints inside the linear program.

Dynamic programming as an approximation method for general MDP has been developed. Assuming a sufficient number of parameters characterizing the value functions, the method generates approximations of these functions. The next step is to determine what values of the parameters will provide the best approximation.

3 Proposed Method

Taking the best decision-making method for a random dynamic system, the Markov decision-making procedure, we solve the issue by selecting the next action depending on the actual circumstance. A hop in routing is analogous to a location, and selecting a particular hop involves deciding which of the possible next hops is the best. Figure 1 depicts the straightforward and secure message transmission that occurs once "n" has been determined consecutively.

4 Routing Decision-Making

Each stage's options are contextual, and the overall routing procedure is optimized by a series of sequential choices. We are using a definitive decision-making technique to address this problem since there are finitely many hops between the origin and the final destination. The essential idea is that we employ a maximum decision-making metric throughout the routing process to make choices based on the model of the Finite Markov decision-making process, which then yields the best possible outcomes.

Assuming each sensor node maintains a log of past routing hops used to deliver messages to the sink, we may determine the distance to the latter directly, often using the most up-to-date routing hop data. However, in exceptional cases, it makes no difference since the node in question has neither pre-view notifications nor any other relationship to the destination. As a result, the computational part might be difficult.

Because WSN is a decentralized network, central processing cannot be used in it. Because of this, every node must do its own calculations and make its own decisions at each hop. Therefore, we think of choosing the next action as a single-step decision-making process that aims to maximize the benefit at each stage.

5 Intrusion Detection System

The proposed strategy found a path from origin to destination when the route was established. The original node stores both the data packet's path and hop count. When the nodes transfer the packet to the next hop, they will also get a Trust Recommendation Record (TRR) including information on the next hop's dependability, packet id, and receiving capacity. The TRR indicates that the data packet was successfully transferred by the node. A path node is considered malicious if it cannot provide a valid certificate for a TRR and a time basis.

Certificate Revocation Record: When a mobile node receives the data from its former hop, it produces a CRR and transmit to its former hop. "For starters, assume that node P transfers m to node Q. The certificate is produced as:

$$CPQ_{CRR} = (H(m))PRQ \tag{1}$$

where,

CPQ_{CRR} is the received CRR by a node,

m is the Data Packet,

$H()$ is the hash function, and

PRQ is the Private Key

Packet ID: It includes specific mobile node recognition data. It consists of the location of a node and the random topology identifier used.

$$IDj = < X_{Positionj}, Y_{Positionj} >, \ where\ 1 \leq j \leq k \tag{2}$$

Reliability: It reflects a mobile node's durability. Based on this element, malicious or infected nodes are detected and data are filtered from the network. The trust value is:

$$RV_i = \frac{RP_i - UPR_i}{RP_i + UPR_i}, \textit{ where } 1 \leq i \leq k \quad (3)$$

RP_i—packet received using sequence number.

UPR_i—unreliable packet received using duplicate sequence number.

6 Results and Discussions

To evaluate the efficacy of the MDPRP routing protocol, this part uses a simulated WSN to test the accuracy and reliability of the MDP's packet routing and intrusion detection. MATLAB simulator is used to set up the simulation environment, and MDPRP is used to simulate the transmission of data packets (Table 1).

The average energy consumption is given using the following expression:

$$AEC = \frac{\sum_{i=1}^{n} (E_i(i) - E_r(i))}{n}$$

where,

$E_i(i)$ is the initial energy of a sensor node and

$E_r(i)$ is the residual energy of a sensor node.

n is the total number of sensor nodes

The results of detection efficiency using IDS-MDP are higher than other methods. The research also demonstrates an enhanced packet delivery rate, longer network life, lower latency, lower overhead, and higher packet integrity compared to other approaches (Fig. 2).

Table 1 Simulation settings and parameters

Number of nodes	1000
Radio range	250 m
Mac	802.11
Area size	1200 × 1200
Traffic source	CBR
Simulation time	100 s
Mobility model	Random waypoint
Packet size	80 bytes
Protocol	AODV

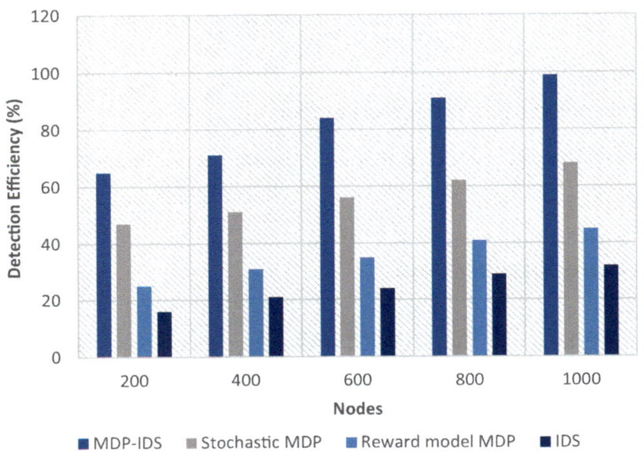

(a) Detection efficiency (%)

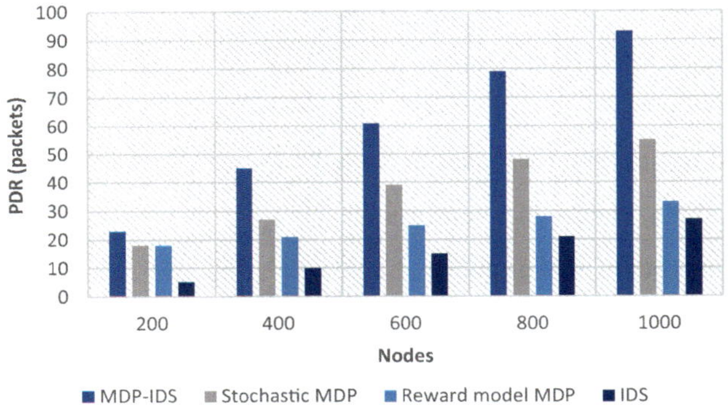

(b) PDR (pkts)

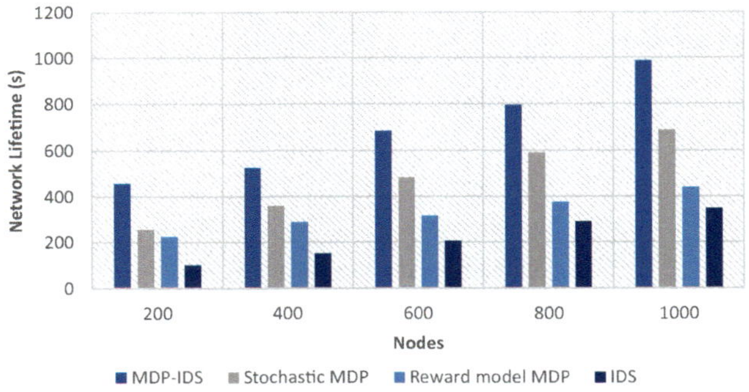

(c) Network Lifetime (Secs)

Fig. 2 Average values of the performance metrics carried out with 1000 sensor nodes

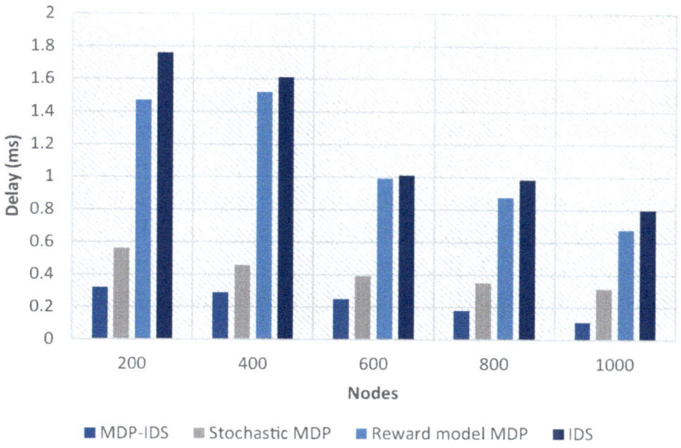

(d) End to end delay (msec)

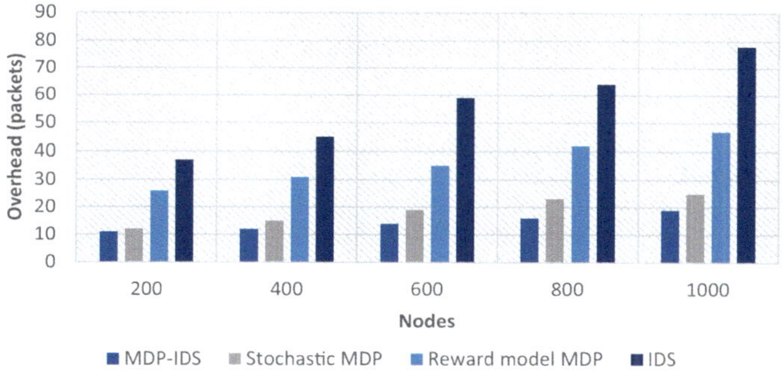

(e) Overhead (pkts)

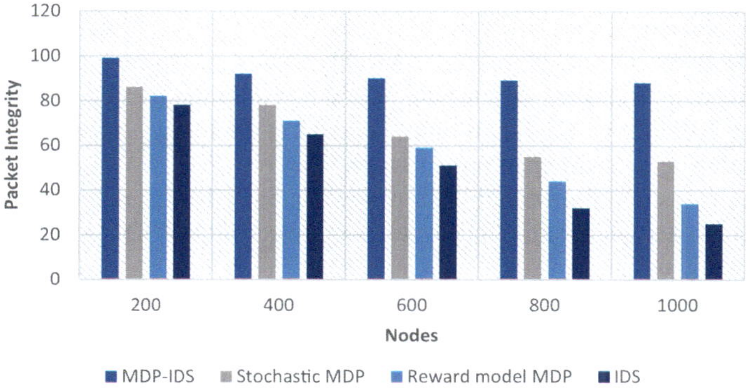

(f) Packet Integrity Vs Time

Fig. 2 (continued)

7 Conclusions

In this study, we apply a closed-form formulation of MDP to foretell the WSN's energy needs. The research establishes a maximum possible distance across which a sensor may send a signal. The MDP determines a timetable for sending data packets via sensors, which serves as a heuristic. The research compares the heuristic to optimum scheduling by discovering the effectiveness of MDP scheduling in scenarios when sensor status is known in advance. The results show that the MDP schedule reduces energy usage compared to prior models in terms of latency, bandwidth, throughput, detection rate, and false alarm rate.

Future research will focus on extending our findings to systems with imperfect transmitting contexts and sinks that only provide partial information about the sensors' states (i.e. the probability of a successful transmission and the probability of an expected sensor being awake are both less than 1). Correctly locating measurements over time intervals is a challenge for us here. To get more precise attribute estimations, we may also consider the more generic approach including several sinks.

8 Reviewer-1

Your study focuses on optimizing sensor node attribute estimation within a specified time window in a wireless sensor network (WSN) while minimizing energy usage. Here's a breakdown of your paper:

1. Problem Statement: The study addresses the challenge of accurately estimating sensor node attributes in a WSN within a specified time frame while minimizing energy consumption.
2. Sink Node: A sink node, placed randomly, needs measurements from wake-up sensors to approximate an attribute within a given time frame.
3. Energy Efficiency: The study aims to reduce the energy expended by sensor nodes during measurement transmission.
4. Energy Consumption Model: You propose a closed-form representation of the predicted energy consumption by sensor nodes in the WSN, providing insights into energy usage.
5. Transmission Distance: Your study establishes an upper limit for the maximum distance a sensor can transmit data.
6. Scheduling: Sensor transmission scheduling within a time period is achieved using Markov Decision Process (MDP), which is a notable methodology for optimizing decision-making in dynamic environments.
7. Heuristic: You develop a heuristic to program the sensors for transmission, enhancing the efficiency of the network.

8. Simulation Environment: MATLAB is used for conducting simulations, and the MDP-based schedule leads to several benefits, including reduced energy consumption, decreased delay, higher packet delivery ratio, and a lower false alarm rate when identifying attribute data sent via sensor nodes.

9 Reviewer-2

1. The paper must be started with introduction section including motivation, main contributions and organization of paper.
2. Literature review section must also be extended.
3. A comparative study may also be shown in graphical form.
4. Language must be improved as there are linguistic errors at some places.
5. The Limitations of the proposed study need to be discussed before conclusion.
6. Add more results and discussion.
7. The paper must be within 10–12 pages (single column)
8. References must be cited in the text within the paper.

Funding This research received no external funding.

Conflicts of Interest The authors declare no conflict of interest.

References

1. Calderone, D., Sastry, S.S.: Markov decision process routing games. In: 2017 ACM/IEEE 8th International Conference on Cyber-Physical Systems (ICCPS), pp. 273–280. IEEE (2017)
2. Yu, X., Gao, S., Hu, X., Park, H.: A Markov decision process approach to vacant taxi routing with e-hailing. Transp. Res. Part B: Methodol. **121**, 114–134 (2019)
3. Calderone, D., Shankar, S.: Infinite-horizon average-cost Markov decision process routing games. In: 2017 IEEE 20th International Conference on Intelligent Transportation Systems (ITSC), pp. 1–6. IEEE (2017)
4. Raverta, F.D., Demasi, R., Madoery, P.G., Fraire, J.A., Finochietto, J.M., D'Argenio, P.R.: A Markov decision process for routing in space DTNs with uncertain contact plans. In: 2018 6th IEEE International Conference on Wireless for Space and Extreme Environments (WiSEE), pp. 189–194. IEEE (2018)
5. Sivaram, M., Porkodi, V., Mohammed, A.S., Manikandan, V.: Retransmission DBTMA protocol with fast retransmission strategy to improve the performance of MANETs. IEEE Access **7**, 85098–85109 (2019)
6. Yuan, P., Wang, Y., Su, M., Yang, Z., Zhang, Q.: Markov decision process-based routing algorithm in hybrid Satellites/UAVs disruption-tolerant sensing networks. IET Commun. **13**(10), 1415–1424 (2019)
7. Hao, J., Jia, X., Han, Z., Yang, B., Peng, D.: Design of opportunistic routing based on Markov decision process. In: 2017 36th Chinese Control Conference (CCC), pp. 8976–8981. IEEE (2017)
8. Sangeetha, S.B., Blessing, N.W., Sneha, J.A.: Improving the training pattern in back-propagation neural networks using holt-winters' seasonal method and gradient boosting model. In: Applications of Machine Learning, pp. 189–198. Springer, Singapore (2020)

9. Li, S.H., Yu, Y., Calderone, D., Ratliff, L., Açrkmeşe, B.: Tolling for constraint satisfaction in markov decision process congestion games. In: 2019 American Control Conference (ACC), pp. 1238–1243. IEEE (2019)
10. Yu, X., Gao, S., Hu, X.: Optimizing vacant Taxis' routing decisions: model-based and model-free approaches (No. 19–03142) (2019)
11. Karthikeyan, T., Praghash, K.: An improved task allocation scheme in serverless computing using gray wolf optimization (GWO) based reinforcement learning (RIL) approach. Wirel. Pers. Commun. 1–19 (2020)
12. Yousefi, S., Derakhshan, F., Karimipour, H., Aghdasi, H.S.: An efficient route planning model for mobile agents on the internet of things using Markov decision process. Ad Hoc Netw. **98**, 102053 (2020)
13. Kannan, S., Dhiman, G., Natarajan, Y., Sharma, A., Mohanty, S.N., Soni, M., Gheisari, M.: Ubiquitous vehicular Ad-Hoc network computing using deep neural network with IoT-based bat agents for traffic management. Electron. **10**(7), 785 (2021)
14. Yoon, S., Albert, L.A.: A dynamic ambulance routing model with multiple response. Transp. Res. Part E: Logist. Transp. Rev. **133**, 101807 (2020)
15. Yuvaraj, N., Srihari, K., Dhiman, G., Somasundaram, K., Sharma, A., Rajeskannan, S., Masud, M.: Nature-inspired-based approach for automated cyberbullying classification on multimedia social networking. Math. Probl. Eng (2021)
16. Li, X., Kan, H., Hua, X., Wang, W.: Simulation-based electric vehicle sustainable routing with time-dependent stochastic information. Sustain. **12**(6), 2464 (2020)
17. Raja, R.A., Kousik, N.V.: Analyses on artificial intelligence framework to detect crime pattern. In: Intelligent Data Analytics for Terror Threat Prediction: Architectures, Methodologies, Techniques and Applications, pp. 119–132 (2021)
18. Rao, Y., Lu, J., Lin, J., Zhou, J.: Runtime network routing for efficient image classification. IEEE Trans. Pattern Anal. Mach. Intell. **41**(10), 2291–2304 (2018)

Design and Implementation of a Hybrid Deep Learning Framework for Handwritten Text Recognition

Harshit Anand, Milind Singh, Vivian Rawade, Shubham Sahoo, Sushruta Mishra, and Laith Abualigah

Abstract Text recognition technology has seen significant advancements in recent years, particularly with the use of Optical Character Recognition (OCR) to evaluate computer-generated text. However, there is much more work to be done in the field of Handwritten Text Recognition (HTR). The challenges posed by handwritten text, such as significant variations in strokes across writers, the vast variety of handwriting styles, human error and damages to the paper, present substantial difficulties in accurately identifying and recognizing handwritten alpha-numeric data. To address these challenges, we proposed a deep learning method that combines long short-term memory (MD-LSTM) and convolutional neural networks (CNN) architectures. This model can identify numbers and characters of the English language from input images. Based on MNIST dataset, bidirectional recurrent neural networks were used to construct the output sequence which was developed using the TensorFlow Framework. The accuracy for alphabets, numbers and alpha-numeric texts are 95.2%, 94.9% and 94.7% respectively. The mean character match index is computed to be 93.2%. The proposed model can substantially boost HRTs precision and efficiency making it more accessible.

Keywords Convolutional Neural Network (CNN) · Recurrent Neural Network (RNN) · Handwritten Text Recognition (HTR) · Long Short Term Memory (LSTM) · Modified National Institute of Standards and Technology database (MNIST)

H. Anand · M. Singh · V. Rawade · S. Sahoo · S. Mishra (✉)
Kalinga Institute of Industrial Technology, Deemed to Be University, Bhubaneswar, India
e-mail: sushruta.mishrafcs@kiit.ac.in

L. Abualigah
Computer Science Department, Al al-Bayt University, Mafraq 25113, Jordan
e-mail: aligah.2020@gmail.com

MEU Research Unit, Middle East University, Amman 11831, Jordan

Applied Science Research Center, Applied Science Private University, Amman 11931, Jordan

Jadara Research Center, Jadara University, Irbid 21110, Jordan

G. Fortino et al. (eds.), *Proceedings of Third International Conference on Computing and Communication Networks*, Lecture Notes in Networks and Systems 917,
https://doi.org/10.1007/978-981-97-0892-5_22

1 Introduction

In today's digital age, many people have shifted to digital writing tools for note-taking and other forms of written communication. However, some individuals still prefer the traditional pen and paper method, which has its disadvantages. Handwritten documents are challenging to store, access, and share, which can lead to inefficiencies and potentially lead to the loss of information. To address this issue, we came up with a system that classifies handwritten text and converts it into a digital format.

The model focuses on classifying individual handwritten words in the form of cursive or block writing. To create a fully functional model for end-users, the model uses Convolutional Neural Networks (CNN) to classify entire word images and combines algorithms for segmenting line images as well as page images. The models can also extract characters from each word image and classify them separately to reconstruct a complete word. Handwritten Text Recognition (HTR) systems convert text from handwritten input images into digital text, as shown in Fig. 1.

To accomplish this, a Neural Network (NN) will be created and trained on specific word images from the MNIST dataset. All the layers including the input layer can be kept short for word images, enabling feasible NN-training. Overall, the project aims to enhance the efficiency and accessibility of handwritten documents by converting them into a digital format. This paper will discuss the technical aspects of the project, including the methodology used, data sources, and results obtained. Furthermore, the paper will address the potential impact of this project on various industries and individuals who rely on handwritten notes for daily tasks.

The main contributions of our paper are as follows:

- We propose a hybrid CNN-RNN architecture to capture both local and global features of handwritten text, utilizing CNNs to extract local features and RNNs to model sequential information.
- Through experimentation with various CNN and RNN configurations, we determine the optimal architecture for the task of handwritten text recognition, achieving high recognition accuracy.
- Our proposed approach is compared to Latest and most recent techniques on multiple datasets, demonstrating its effectiveness and significant improvements in recognition accuracy over traditional methods.

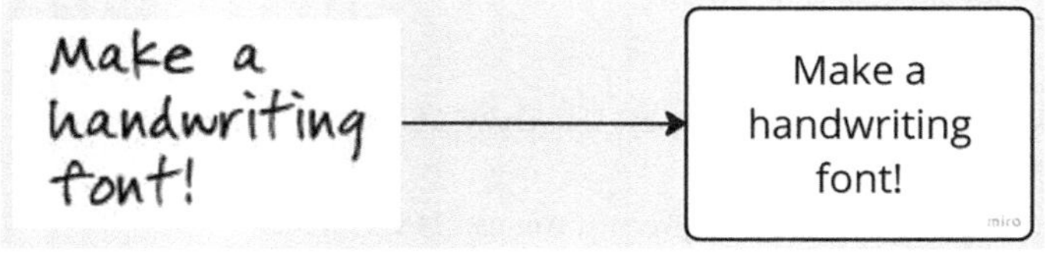

Fig. 1 Image of word and its OCR

- We evaluate the robustness of our approach to varying levels of noise, distortion, and scale, showing that our approach can achieve high accuracy even under challenging circumstances.
- Our approach has practical applications in automated document processing, signature verification, and digitization of historical documents, showcasing its potential impact in these areas.

Hence, our approach can potentially improve the efficiency and accuracy of handwritten text recognition, and enable new applications in document processing and digitization.

2 Literature Review

Images of handwritten documents can degrade significantly for a variety of reasons, including aging paper, everyday wear and tear (dust, wrinkling, etc.), poor scanning techniques, and more. These factors drastically reduce the effectiveness of current Handwritten Text Recognition (HTR) algorithms and present numerous readability concerns. In a study conducted by Mohamed Ali Souibgui et al. [1], in order to make the resulting document image more readable, the proposed architecture incorporates a handwritten text recognizer. This is the first study to binarize handwritten papers using text information. Sebastian Sudholt and Gernot A. Fink. [2] suggest the use of PHOCNet, a deep convolutional neural network, to precisely identify words in handwritten documents. Paul Voigtlaender et al. [3] present a large multidimensional LSTM RNN for handwriting recognition. The proposed network architecture achieves cutting-edge performance on several handwriting recognition datasets. Lei Hu and Richard Zanibbi [4] concentrate on a minimal spanning tree (MST) based parsing approach for online handwritten mathematical statements. A promising strategy for mathematical expression recognition, the suggested method outshines existing methods with respect to accuracy and processing speed. Shilpi Goyal et al. [5] suggest a collection of deep convolutional neural networks for the recognition of handwritten text by merging the output of various models. This ensemble technique helps to increase recognition accuracy. Lin Sun et al. [6] propose a novel loss function for improving the performance of CNNs in recognizing handwritten text. Tirthankar Ghosal et al. [7] worked on recognizing handwritten writing, by using a convolutional neural network that makes use of graphene-level data. Sachin Sharma et al. [8] created a hybrid deep learning model that combines CNNs and LSTM networks to recognize handwritten text with utmost precision on a number of benchmark datasets. Supriya Das [9] evaluated a deep residual network with bidirectional LSTM layers on a number of datasets. The proposed model highlights the efficiency of the residual connections and bidirectional LSTM layers in capturing contextual information. Abhijit Das et al. [10] used a hybrid deep learning model for handwritten text identification that blends long short-term memory (LSTM) networks and CNN. Niranjan et al. [11] enhanced convolutional neural networks' performance while recognizing handwritten text. Their research suggests a novel loss function. The experimental results demonstrate that the suggested technique performs at the cutting edge on benchmark datasets. Ayan Seal et al. [12] used a deep learning model that mixes CNNs and RNNs is proposed for the recognition of handwritten text. Krishna Reddy et al. [13] suggest a deep belief network (DBN)-based handwritten text recognition system that can recognize both local and global handwritten text elements. On a number of datasets for the recognition of handwritten text, the suggested system produces results that are competitive. Anindya Roy et al. [14] implemented a CNN based method to handwritten text recognition on datasets. The proposed ensemble system outperforms a number of techniques available today. Seiichi Uchida et al. [15] deeply analyzed the current developments in handwritten text recognition, including several methods based on neural networks, hidden Markov models, and deep learning, provided in this paper. The authors also go over difficulties and potential approaches for handwritten text recognition.

3 Proposed Model for Handwritten Text Recognition

The entire Handwritten Text Recognition model is composed of various stages that have been diagrammatically defined in Fig. 2. The proposed handwritten text identification model was formed based on the MNIST dataset. Pre-processing is a crucial step in which we normalize the values of the pixel and create a grayscale image from the input image. Techniques for normalization may include rescaling intensity values to [0, 1] or [−1, 1]. In this step, we convert the raw input data into a format that is compatible with the Neural Network to be used for further analysis. In grayscale images, each pixel is defined by a single intensity value, often ranging from 0 to 255. In contrast, coloured pictures have three intensity values, one for each pixel that correlate to either blue, red and green channels. Then for each pixel, we calculate the weighted average of the three color channels to convert a color picture to grayscale. One intensity value per pixel will be included in the final grayscale picture, making further processing simpler. Common normalization techniques include:

- Rescaling the intensity values to [0, 1]: This involves dividing each pixel value by 255, which scales the values to the range [0, 1]. For neural networks that apply activation functions like ReLU or Sigmoid, operate in the [0, 1] range, this normalization approach is beneficial.
- Rescaling the intensity values to [−1, 1]: Includes rescaling the intensity values to [−1, 1]. For neural networks that apply activation functions and operate between [−1, 1], this normalization approach is beneficial.

By lowering the size of the input data, normalization increases the efficiency of machine learning algorithms by enhancing the stability and potency of the optimization process. By restricting the model from retaining irrelevant pixel values to the general patterns in the data, it can also aid in reducing over-fitting. In Fig. 3 given

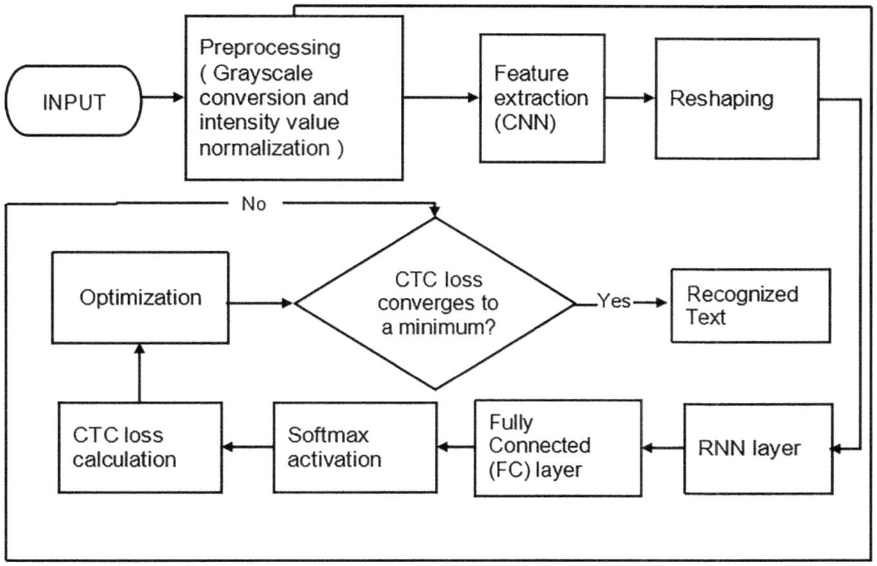

Fig. 2 Proposed model for handwritten text recognition

below, we change the size of the picture to 30*30 pixels. We also cushion 2 pixels on all the borders of the image. Then convert it into a grayscale picture having size 128 × 32. The preprocessed image is shown in Fig. 3.

To extract characteristics from the grayscale picture, we ran it through several convolutional layers. These convolutional layers use various filters of different sizes to record data across multiple scales. A pooling layer is then added after every convolutional layer in order to decrease the spatial dimension and avoid over-fitting. A type of neural network known as convolutional neural networks (CNNs) is exceptionally good at identifying images. The convolutional layer, applies a collection of adaptable filters to the input image to extract characteristics at various spatial scales [16]. The feature map produced by these filters, which emphasize patterns like corners, textures, and edges, is created by calculating dot products between the weights of the filters and the pixels in every local region as they move across the picture. To extract complicated characteristics from the picture, several convolutional layers may be combined. After each convolutional layer, pooling layers are often added to lower

Fig. 3 Preprocessed Image of the input image

the feature map's spatial dimensions and avoid over-fitting. In order for pooling to function, the feature map must be divided into small, non-overlapping sections. The values of every region are then swapped out for a single summary value, like average value or the maximum value. By doing this, the network has to handle less information while preserving the most important aspects [17]. In short, CNNs can create layered models of the input image by integrating several convolutional and pooling layers, collecting progressively more abstract characteristics that are more discriminating for the target task.

Reshape the characteristics obtained in acceptable format according to RNN. This can include making a 1D sequence out of the 2D feature maps. The resulting feature maps are generally in a 3D tensor format after the completion of the feature extraction process using CNNs, where the initial two dimensions align to the spatial dimensions of the feature maps and the third dimension aligns to the number of filters learnt. In order to input these features, the feature maps must be modified into a format that the RNN can understand [18]. Combining the feature vectors into the spatial dimensions allows the 2D feature maps to be transformed into a 1D sequence.

Therefore, a long sequence of feature vectors is created, each of which represents a distinct location on the original feature maps. Furthermore, the feature vectors may be stacked on over one another to flatten the feature maps into a 1D vector. The resultant sequence of feature vectors is then passed into RNN like a sequence of inputs where each vector denotes a single input.

An RNN is a specific type of neural network that can process sequences of data by keeping a hidden state that summarizes the input observed up to that point [19, 20]. The network may detect dependencies between subsequent phases by updating the hidden state depending on the current input and previously hidden state at every step. The sequence of feature vectors is supplied into an RNN layer, which extracts the sequential information from the input image, after the features have been rearranged. To process the input sequence, the RNN layer can use an architecture like Gated Recurrent Units (GRUs) or Long Short-Term Memory (LSTM). These designs deal with the issue of vanishing gradients, which may occur in ordinary RNNs and even cause failure in training or poor performance once the gradients become smaller over time. In order to address the vanishing gradient issue, LSTM and GRU networks implement gating techniques that enable the network to selectively retain or discard information over a period of time.

The output of the RNN layer is then passed into a fully connected layer to create a prediction. Each letter or number in the output sequence might be determined by a different neuron in the fully linked layer. The final hidden state is passed into a fully connected (FC) layer once the RNN layer has finished processing the input sequence to create a prediction [21]. The FC layer is a particular sort of neural network layer where all of the neurons are interconnected and where each neuron's output is the weighted sum of its inputs after being processed by an activation function. The FC layer's biases and weights are learnt while training, and the layer's output may be used for a variety of tasks, including classification or regression. Each neuron in the FC layer, which generally contains several neurons, is associated with a distinct possible output character or number which is in turn used for sequence prediction. A probability distribution across the potential output values is generated for each neuron's output using a soft-max activation function. Concatenating the expected values for each character or number yields an estimate for the full sequence. The FC layer is trained to minimize a loss function that compares the expected output sequence to the actual output sequence during training. The loss function often accounts for the expected probability of each estimated value as well as any previous information on the output sequence's structure (which might include certain letters or phrases that are more likely to be followed by another term).

A popular activation function used in the last layer of a neural network to generate a probability distribution across a collection of classes is called the soft-max function [22]. The output of the fully connected layer in the context of sequence prediction is a vector of real values that reflect the likelihood of each potential output character or number. This vector serves as the input for the soft-max function, which outputs a new vector of the same size with each element representing the normalized probability of a certain output value. The soft-max activation makes sure that the predicted values fall inside the [0, 1] range and may be used as probabilities. Figure 4 below shows the output of the NN after training and testing on the MNIST dataset.

$$\mathrm{Softmax}(z)_i = \frac{\exp(z_i)}{\sum_{j=1}^{n} \exp(z_j)} \quad (1)$$

The Soft-max function is mathematically defined in Eq. (1) where $(z)_i$ is the input vector, (n) is the number of elements in the vector, and (i) is the index of a specific element in the vector. Every component of the input vector is exponentiated using the soft-max function, which divides each exponentiated value by the total number of exponentiated values in the vector. By doing so, the resultant vector is guaranteed to include values in the [0, 1] range and sum to 1, allowing it to be considered a probability distribution. Given the input image, the output of the soft-max function in the context of sequence prediction indicates the probability distribution across the range of potential output letters or phrases [23]. The character or number with the greatest probability in the output vector is often the expected output character or word. Overall, it allows the model to generate a probability distribution across the set of potential output values and make accurate predictions.

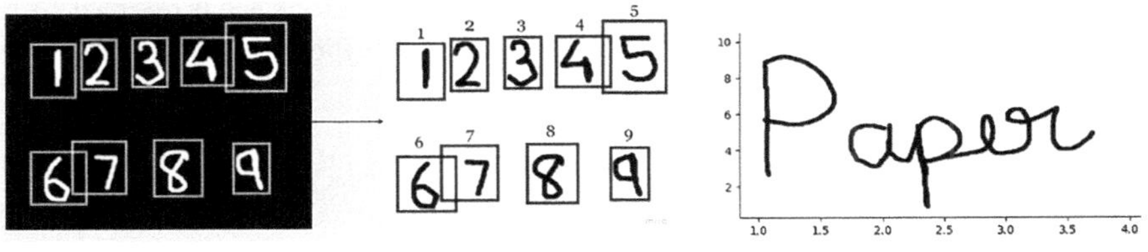

Fig. 4 Training and testing of English character and digits

The CTC loss function calculates the total of all possible valid alignments between the predicted and ground truth sequences using a negative logarithm of probability. The probabilities are transformed into a dissimilarity measure that may be minimized during training using the negative logarithm. Using CTC loss, the expected character sequence is measured and compared to the actual sequence. The CTC loss algorithm can generate deletions, insertions, and duplication of characters in the expected sequence. It considers differences in the length of the input and output sequences. By taking into account every potential connection between the two sequences, the CTC loss determines the variance between the expected and ground truth sequences. In sequence prediction tasks where the duration of the input and output sequences might vary, CTC loss is often used.

$$.\text{CTC Loss}\ (y, \hat{y}) = -\log \sum_{a \in A(y, \hat{y})} P(a|x) \tag{2}$$

The CTC Loss function is mathematically defined in Eq. (2) where y is the ground truth sequence, $\hat{\boldsymbol{y}}$ is the predicted sequence, x is the input image, and $(y, \hat{y})$ is the set of all valid alignments between the two sequences. The probability $P(a|x)$ represents the probability of a particular alignment for a given input image. Overall, the CTC loss function enables the model to optimize its parameters to create the proper output sequence and plays a crucial role in OCR models since it allows the model to learn from input sequences of different lengths and provide precise and adaptable predictions.

In order to decrease the CTC loss on the training data, the model parameters are optimized with techniques like Adam or Stochastic Gradient Descent (SGD) which improve the predictions on the validation and test data. To achieve this, the model parameters are updated in the negative gradient's direction after calculating the gradients of the loss. To guarantee that the updates do not exceed the ideal parameters, the learning rate, which defines the size of the steps, is normally set to a limited value. A version of SGD that uses adaptive learning rates for each of the parameters is the Adam optimization method. The history of the gradients is used by Adam to calculate separate learning rates for each parameter, which may result in a quicker generalization and convergence [24, 25]. Using batches of training data, the optimization method repeatedly adjusts the model parameters throughout training. This process continues until convergence, which is often indicated by a convergence threshold or a maximum number of epochs. Up until the CTC loss converges to a minimum, we repeat the above steps for as many iterations (also referred to as epochs) as required. After that, the model is tested on a validation set to assess how effectively it generalizes the new data.

4 Results and Analysis

The overall accuracy and the percentage match of each character help evaluate the proposed model's performance based on the number of epochs (10–50). Figure 5 showcases a sample image from the MNIST dataset. The MNIST dataset was used to train and test the handwritten text recognition model that we have proposed. We used this dataset to train and test 8,00,000 pages full of images, that were produced by 3600 writers which contains both upper-case and lower-case characters along with numbers.

Figure 6 depicts the accuracy with respect to a varying number of epochs. One iteration across the training set is called an epoch. As shown in the graph, the increase in accuracy mirrors the increase in number of epochs. Hence, the value of an epoch must never overfit the training data. The accuracy is calculated according to the formula given in Eq. (3). Figure 7 shows the accuracy values obtained from the sample inputs and their respective outputs. The accuracy of an input image is the ratio of the amount of correctly recognized characters to the total amount of characters given. As observed, the accuracy for alphabets, numbers, and alpha-numeric texts is 95.2%, 94.9%, and 94.7%, respectively.

$$\text{Accuracy} = \frac{\text{No. of correctly indentified character in the input image}}{\text{Total no. of characters in the input image}} \tag{3}$$

The percentage match is the number of digits or characters recognized from the input image that match the characters in the entire dataset. The percentage match values that we calculated are thus shown in Fig. 8. The mean character match index is computed to be 93.2%.

Fig. 5 Demonstration of a sample Dataset

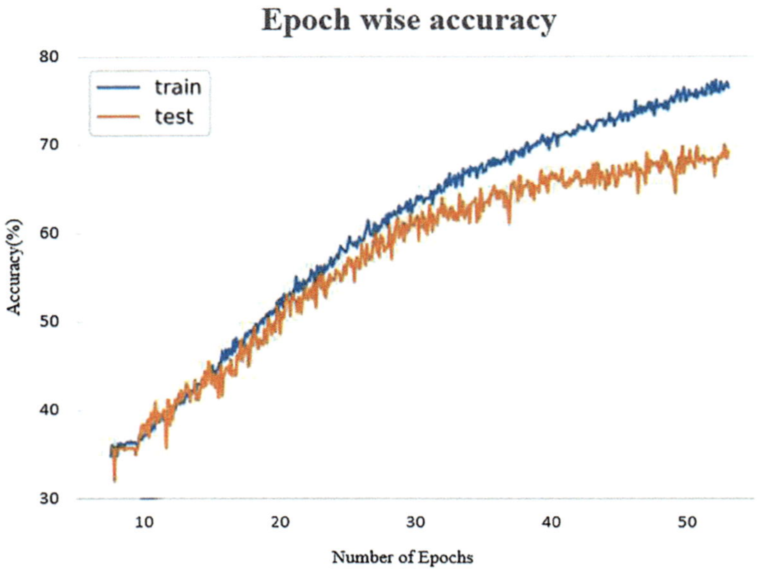

Fig. 6 Training and Testing for varying number of epochs

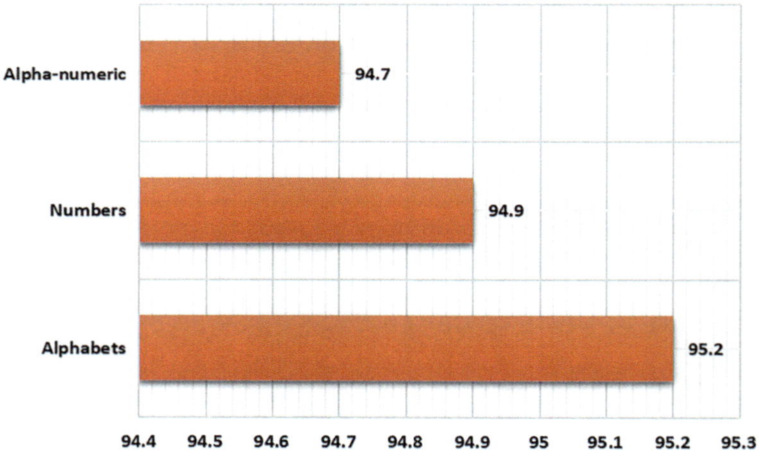

Fig. 7 Sample inputs and their outputs

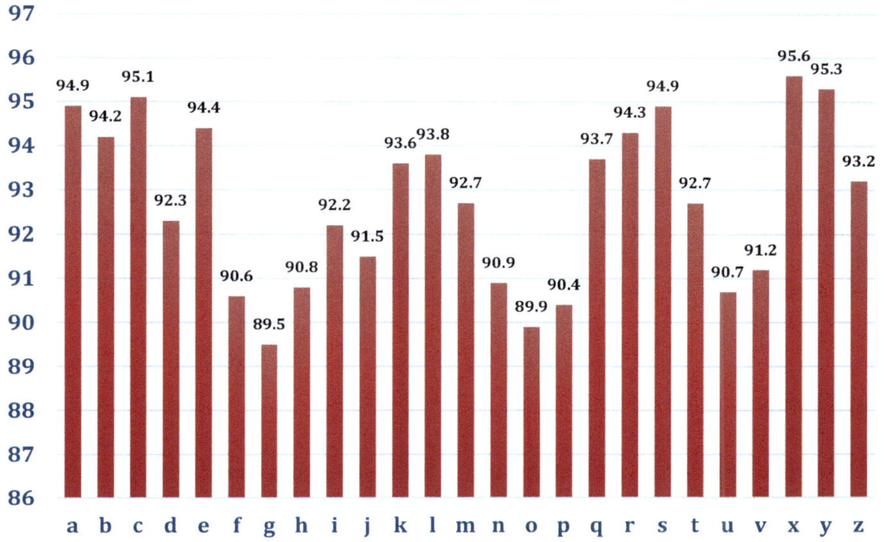

Fig. 8 Percentage match for each character

5 Advantages of the Proposed Model

- Increased Accuracy: It has been demonstrated that HTR models that employ CNN, RNN, and MD-LSTM considerably increase the accuracy of handwriting text recognition. These models can adjust to various writing styles and understand the characteristics of the handwriting text, producing more accurate transcriptions.
- Robustness: CNN, RNN, and MD-LSTM models are more resistant to noise, distortion, and other types of input data fluctuation. They are more adaptable than conventional HTR versions because they can handle text that is written in various sizes and styles.
- CTC Loss: The Connection Temporal Classification (CTC) loss generates a sequence of characters from the input image and can handle input sequences of varying lengths. CTC loss is ideal for reading plain handwritten texts as the text may not align with the image.

6 Conclusion

In conclusion, deep learning models such as CNN and RNN have drastically improved handwritten text recognition systems to produce an accuracy of 95.2%, 94.9%, and 94.7% for alphabets, numbers, and alpha-numeric texts, respectively. Additionally, the mean character match index was observed to be 93.2%. The Connection Temporal Classification (CTC) loss function and Stochastic Gradient Descent

(SGD) have been quite effective in handling the variable-length sequences of handwritten text data as well as efficiently training our model for handwritten text recognition. The success of these deep learning techniques propagated the model to automatically learn and extract relevant features from raw input data without requiring manual feature extraction. Though the recognition of cursive and overlapping handwriting remains a challenging task. The conclusion drawn thus far has led to significant advancements in handwritten character recognition systems that are more capable of handling larger input images and, hence, provide improved recognition accuracy. This insight would prove to be a stepping stone for further groundbreaking discoveries in the field of Handwritten Text Recognition.

References

1. Souibgui, M.A., Jemni, S.K., Kessentini, Y., Fornés, A.: Enhance to read better: a multi-task adversarial network for handwritten document image enhancement. Pattern Recognit. **123**, 108370, ISSN 0031-320 (2022)
2. Sudholt, S., Fink, G.A.: PHOCNet: a deep convolutional neural network for word spotting in handwritten documents. In: Proceedings of the 14th International Conference on Document Analysis and Recognition (ICDAR), pp. 1122–1127 (2017)
3. Voigtlaender, P., Doetsch, P., Ney, H.: Handwriting recognition with large multidimensional long short-term memory recurrent neural networks. In: Proceedings of the 14th International Conference on Document Analysis and Recognition (ICDAR), pp. 31–36 (2017)
4. Hu, L., Zanibbi, R.: MST-based parsing of online handwritten mathematical expressions. In: Proceedings of the 13th International Conference on Document Analysis and Recognition (ICDAR), pp. 541–545 (2015)
5. Goyal, S., Jayasree, A.R., Balasubramanian, R.: Handwritten text recognition using ensemble of deep convolutional neural networks. In: Proceedings of the 15th International Conference on Frontiers in Handwriting Recognition (ICFHR), pp. 181–186 (2018)
6. Sun, L., Tang, X., Wan, J.: Handwritten text recognition using a convolutional neural network with a novel loss function. In: Proceedings of the 16th International Conference on Frontiers in Handwriting Recognition (ICFHR), pp. 443–448 (2019)
7. Ghosal, T., Sinha, R., Roy, P.P.: Handwritten text recognition using convolutional neural networks with grapheme-level information. In: Proceedings of the 15th International Conference on Frontiers in Handwriting Recognition (ICFHR), pp. 396–401 (2018)
8. Sharma, S., Uppal, A., Kaur, M.: Handwritten text recognition using deep convolutional neural networks and long short-term memory networks. In: Proceedings of the 15th International Conference on Frontiers in Handwriting Recognition (ICFHR), pp. 26–31 (2018)
9. Das, S., Nag, K., Das, D.: Handwritten text recognition using a deep residual network with bidirectional long short-term memory layers. In: Proceedings of the 16th International Conference on Frontiers in Handwriting Recognition (ICFHR), pp. 236–241 (2019)
10. Das, A., Srivastava, S., Das Mandal, S. K.: Handwritten text recognition using convolutional neural networks and long short-term memory networks. In: Proceedings of the 15th International Conference on Frontiers in Handwriting Recognition (ICFHR), pp. 32–37 (2018)
11. Niranjan, M., Kumar, R.S., Saravanan, V.: A comparative study of machine learning techniques for handwritten character recognition. Int. J. Pure Appl. Math. **119**(16), 685–695 (2018)
12. Seal, A., Mandal, A., Chanda, B.: Handwritten text recognition using convolutional neural networks and recurrent neural networks. In: Proceedings of the 15th International Conference on Frontiers in Handwriting Recognition (ICFHR), pp. 298–303 (2018)

13. Reddy, K., Dhanireddy, R. P., Hemanth Kumar G.: Handwritten text recognition using deep belief networks. In: Proceedings of the 15th International Conference on Frontiers in Handwriting Recognition (ICFHR), pp. 313–318 (2018)
14. Roy, A., Nagabhushan, P., Das, S.: Handwritten text recognition using a convolutional neural network ensemble. In: Proceedings of the 15th International Conference on Frontiers in Handwriting Recognition (ICFHR), 2018, pp. 452–457.
15. Fink, G.A., Uchida, S., Märgner, V.: A review of recent advances in handwritten text recognition. In: Proceedings of the 16th International Conference on Frontiers in Handwriting Recognition (ICFHR), pp. 119–124 (2018)
16. Mishra, N., Mishra, S., Tripathy, H.K.: Rice yield estimation using deep learning. In: Innovations in Intelligent Computing and Communication: First International Conference, ICIICC 2022, Bhubaneswar, Odisha, India, December 16–17, 2022, Proceedings (pp. 379-388). Springer International Publishing, Cham (2023, January)
17. Chakraborty, S., Mishra, S., Tripathy, H.K.: COVID-19 outbreak estimation approach using hybrid time series modelling. In: Innovations in Intelligent Computing and Communication: First International Conference, ICIICC 2022, Bhubaneswar, Odisha, India, December 16–17, 2022, Proceedings, pp. 249–260. Springer International Publishing, Cham (2023, January)
18. Verma, S., Mishra, S.: An exploration analysis of social media security. In: Predictive Data Security Using AI: Insights and Issues of Blockchain, IoT, and DevOps, pp. 25–44. Springer Nature Singapore, Singapore (2022)
19. Singh, P., Mishra, S.: A comprehensive study of security aspects in blockchain. In: Predictive Data Security using AI: Insights and Issues of Blockchain, IoT, and DevOps, pp. 1–24. Springer Nature Singapore, Singapore (2022)
20. Swain, T., Mishra, S.: Evolution of machine learning algorithms for enhancement of self-driving vehicles security. In: 2022 International Conference on Advancements in Smart, Secure and Intelligent Computing (ASSIC), pp. 1–5. IEEE (2022, November)
21. Sahoo, S., Mishra, S.: A comparative analysis of PGGAN with other data augmentation technique for brain tumor classification. In: 2022 International Conference on Advancements in Smart, Secure and Intelligent Computing (ASSIC), pp. 1–7. IEEE (2022, November)
22. Mohapatra, S.K., Mishra, S., Tripathy, H.K.: Energy consumption prediction in electrical appliances of commercial buildings using LSTM-GRU model. In: 2022 International Conference on Advancements in Smart, Secure and Intelligent Computing (ASSIC), pp. 1–5. IEEE (2022, November)
23. Tripathy, H.K., Mishra, S.: A succinct analytical study of the usability of encryption methods in healthcare data security. In: Next Generation Healthcare Informatics, pp. 105–120. Springer Nature Singapore, Singapore (2022)
24. Adrija, M., Yash, A., Sushruta, M.: 8 Pragmatic study of IoT in healthcare security with an explainable AI perspective. In: Explainable Artificial Intelligence for Biomedical Applications, pp. 145–166. River Publishers (2023)
25. Bhavya, M., Pranshu, S., Sushruta, M., Sibanjan, D.: 17 comparative analysis of breast cancer diagnosis driven by the smart IoT-based approach. In: Explainable Artificial Intelligence for Biomedical Applications, pp. 353–374. River Publishers (2023)

A LSTM Based Intelligent Framework for Financial Stock Prediction

Oindrila Ajha, Souryadipta Das, Tiyasha Dutta, Soham Das, Sushruta Mishra, and Laith Abualigah

Abstract Stock analysis is a method used by traders and financial experts to evaluate the securities market and make informed decisions about buying and selling shares. It involves conducting extensive research to assess the performance and quality of a stock or an industry before investing. Stock analysis can take different forms, but traders usually use two main categories: technical and fundamental analysis. Technical analysis involves examining documented value outlines and researching previous market structures to predict future advances. On the other hand, fundamental analysis looks at data from the organization and its macroeconomic situation, such as financial history and revenue streams, to assess prospective gains from exchanges. The ultimate goal of stock analysis is to choose the best times to place trades and make the appropriate buying and selling decisions. Traders may specialize in one type of analysis or use a combination of both. Regardless of the approach, conducting

O. Ajha · S. Das · T. Dutta · S. Das · S. Mishra (✉)
Kalinga Institute of Industrial Technology, Deemed to Be University, Bhubaneswar, India
e-mail: sushruta.mishrafcs@kiit.ac.in

O. Ajha
e-mail: 2028066@kiit.ac.in

S. Das
e-mail: 2028113@kiit.ac.in

T. Dutta
e-mail: 2028119@kiit.ac.in

S. Das
e-mail: 2028171@kiit.ac.in

L. Abualigah
Computer Science Department, Al al-Bayt University, Mafraq 25113, Jordan
e-mail: aligah.2020@gmail.com

MEU Research Unit, Middle East University, Amman 11831, Jordan

Applied Science Research Center, Applied Science Private University, Amman 11931, Jordan

Jadara Research Center, Jadara University, Irbid 21110, Jordan

G. Fortino et al. (eds.), *Proceedings of Third International Conference on Computing and Communication Networks*, Lecture Notes in Networks and Systems 917,
https://doi.org/10.1007/978-981-97-0892-5_23

extensive research is crucial to ensure wise investments that produce profit and avoid wasting hard-earned money.

Keywords Stock analysis · Technical analysis · Shares · Predict future advances · Fundamental analysis · Revenue streams · Traders · Investments

1 Introduction

In essence, large-capitalized quantitative traders buy stocks, futures, and equities at discount prices and then sell them at premium prices. Numerous organizations still talk about the trend in stock market forecasting even if it is nothing new. Investors conduct two different kinds of stock analysis before buying a stock. Investors do fundamental analysis, which is the first sort of study, during which they take into account the stock's intrinsic value as well as the state of the market, the economy, and the political climate.

Technical analysis, on the other hand, studies market data such as prior prices and volumes to track the evolution of equities. Due to machine learning's increasing significance across numerous industries in recent years, many traders have been motivated to apply these approaches to the market. Some of these applications have shown positive results. A programme for predicting financial data will be created in this essay. A dataset with all previous stock price data will be used to train the algorithm. The prediction's main objective is to reduce the uncertainty associated with selecting an investing strategy.

According to the stock market's random walk, today's value is the best predictor of tomorrow's value. In fact, due to market volatility, forecasting stock indices is challenging and necessitates an accurate forecast model. Investor confidence is impacted by stock market indexes' volatility.

Stock values are believed to be highly dynamic and subject to sudden swings (like Election Results, Rumours, etc.) due to the fundamental nature of the financial industry and in part as a result of the interaction between known characteristics (such as the previous day's closing price, the P/E ratio, etc.) and unknown factors. Machine learning has been used in a few attempts to predict stock values. The goals of each study project vary substantially in three different ways. The target price change may be short-term (starting tomorrow and lasting up to a few days), long-term (starting months from now), or quick (less than a minute). The list of stocks can be limited to all stocks, only stocks in a certain industry, or just stocks in general.

The predictors utilized can be anything from an overall trend in the news and economy to specific company features to merely time series data on the stock price. Future stock prices, price volatility, or market trends are all possible targets for stock market forecasts. The stock market prediction system uses two sorts of predictions: dummy predictions and real-time predictions. In Dummy Prediction, a set of rules is established, and future share prices are predicted by computing the average price.

Internet usage was required for the real-time prediction, and I checked the stock price of the company.

Machine learning approaches have been introduced for prediction systems in the financial markets as a result of developments in computation. In this research, we employ machine learning methods to forecast the stock market with the LSTM algorithm in Python programming language.

Some main highlights of our stock prediction model that uses moving average deep learning technique (LSTM) and Python Streamlit are:

1. **Deep learning with moving averages:** The model uses deep learning algorithms with moving averages to analyze large volumes of stock data and identify patterns that can help predict future prices.
2. **Multiple datasets from yfinance:** The model can use different datasets from the yfinance library to analyze a wide range of stock prices, allowing traders to compare and contrast performance across different sectors and companies.
3. **Interactive Streamlit dashboard:** The model comes with an interactive Streamlit dashboard that allows users to visualize predicted stock prices alongside actual prices and compare trends over time.
4. **Easy to use and customize:** The model is built using Python, which means that it is easy to customize and modify to suit different user needs. This also makes it easy to integrate with other data sources and analytical tools.
5. **Accurate predictions using LSTM:** By leveraging deep learning techniques such as Long Short Term Memory (LSTM) with moving averages, the model can provide accurate predictions of stock prices, allowing traders to make informed investment decisions.

Overall, this model can offer traders a powerful and flexible tool for analyzing and predicting stock prices across a wide range of industries and companies.

2 Background Study

The area of using machine learning algorithms for assessing value examples and predicting stock cost has seen a great deal of fascinating research. Nowadays, the majority of stock dealers rely on intelligent trading systems, which aid them in anticipating charges based on various factors and conditions. Late investigations are on the use of data input from multiple structures and varied sources. In order to predict the market, some systems use recorded stock information, others use financial news items, yet others use master audits, and still more use a combination of these methods. A wide range of AI computations are also available and can be used to plan the framework. These frameworks offer numerous approaches for addressing the problem. A few structures carry out scientific investigation on notable information for expectation while some perform assumption examination on money related news stories and master surveys for forecast.

Some of the related works include [1] where the paper presents a hybrid machine learning model for predicting stock prices using sentiment analysis and technical indicators. To more reliably predict stock values, the suggested methodology combines the benefits of sentiment analysis and technical analysis. The suggested model beats other conventional machine learning models in stock price prediction, according to experimental results, obtaining an accuracy of over 90%. Authors in[2] present a stock price prediction model that uses machine learning algorithms and technical analysis. The proposed model combines the strengths of support vector regression (SVR) and the moving average convergence divergence (MACD) technical indicator to predict stock prices. The model is trained on historical stock price data and is evaluated using a back testing approach. Experimental results show that the proposed model outperforms other traditional machine learning models in predicting stock prices, achieving an accuracy of over 80%. In reference [3] the authors proposed a model that combines the strengths of long short-term memory (LSTM) and random forest (RF) algorithms to predict stock prices. The model is trained on both financial data and textual data extracted from news articles. Experimental results show that the proposed model outperforms other traditional machine learning models in predicting stock prices, achieving an accuracy of over 80%.In [4] the authors proposed a model that combines the strengths of the long short-term memory (LSTM) algorithm and a bag-of-words (BOW) approach to extract sentiment from news articles. The model is trained on a dataset consisting of financial news articles and stock price data. Experimental results show that the proposed model outperforms other traditional machine learning models in predicting stock prices, achieving an accuracy of over 80%. In [5] the authors proposed a model that combines the strengths of the long short-term memory (LSTM) algorithm and a sentiment analysis approach to extract sentiment from financial news articles. Experimental results show that the proposed model outperforms other traditional machine learning models in predicting stock prices, achieving an accuracy of over 88%. The model is further validated on real-world stock price data, demonstrating its potential for practical applications in the stock market. In [6] the authors proposed a model that combines the strengths of the auto regressive integrated moving average (ARIMA) model and a big data analytics approach to extract relevant features from vast amounts of data. The model is trained on a dataset consisting of stock price data and other relevant factors such as news articles, social media data, and economic indicators. Experimental results show that the proposed model outperforms other traditional machine learning models in predicting stock prices, achieving an accuracy of over 80%. The model is further validated on real-world stock price data, demonstrating its potential for practical applications in the stock market. [7] where the model utilizes a combination of linear regression and a long short-term memory (LSTM) neural network to analyze financial news and technical indicators such as moving averages and trading volume. The article concludes that the integration of financial news and technical indicators can improve the accuracy of stock price prediction models. In [8] the authors proposed a model that combines sentiment analysis of tweets and stock technical indicators such as the Relative Strength Index and Moving Average Convergence Divergence to make stock price predictions. The results show that the proposed model outperformed traditional machine learning algorithms in predicting stock prices. The article concludes that the integration of social media analysis and technical indicators can improve the accuracy of stock price prediction models. In [9] the authors proposed a model which uses decision tree, artificial neural network, support vector machine, k-nearest neighbor, random forest, and gradient boosting. The results show that gradient boosting achieved the highest accuracy in predicting stock prices compared to other methods. Furthermore, the article finds that combining technical and fundamental indicators can improve the accuracy of stock price prediction models. The article concludes that gradient boosting is an effective machine learning technique for stock price prediction, and combining different types of indicators can further improve the accuracy. In [10] the authors proposed a model where it uses technical indicators such as moving averages, relative strength index, and stochastic oscillator, along with news sentiment analysis, to predict stock prices. The study tested the model on three stocks from the Indian stock market and compared the results with traditional time series models such as ARIMA and exponential smoothing. The results show that the proposed model outperformed traditional models in terms of predicting future stock prices. The article concludes that combining machine learning techniques and time series analysis can improve the accuracy of stock price prediction models.

3 Proposed Model for Stock Prediction

The following dataset in Table 1 is one of the several datasets that has been taken from the yfinance library. It can be changed by changing the stock ticker.

The workflow diagram of the proposed solution as shown in Fig. 1 can be divided into three parts. The first is the refining relevant features, secondly performing dimensionality reduction and features extraction. And the final part,

Table 1 Stock prediction dataset used in study

Stock ticker—AAPL						
Data from 2010–2019						
	Open	High	Low	Close	ADJ close	Volume
COUNT	2,515	2,515	2,515	2,515	2,515	2,515
MEAN	27.562	27.8065	27.312	27.5681	25.4294	299,672,163,8171
STD	13.9965	14.1184	13.8941	14.0163	14.0627	235,293,332,3519
MIN	6.8704	7	6.7946	6.8589	5.8378	45,448,000
25%	16.2423	16.345	16.0475	16.1782	13.9667	124,761,800
50%	24.77	25.055	24.5925	24.86	22.5356	222,031,600
75%	37.5325	37.9563	37.2738	37.6112	35.6236	406,805,000
MAX	72.78	73.4925	72.03	72.88	71.29	1,880,998,000

which is the main contribution of our work, is to build a forecast model and visualize it for the target stock.

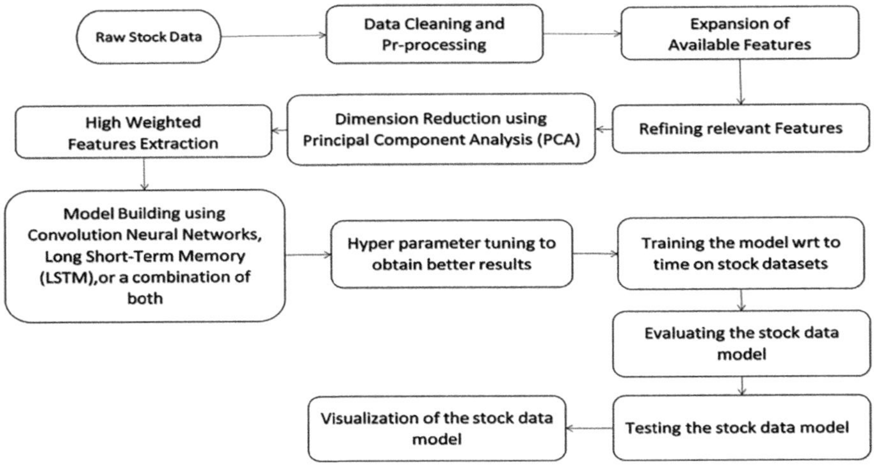

Fig. 1 Workflow diagram of the proposed stock prediction model

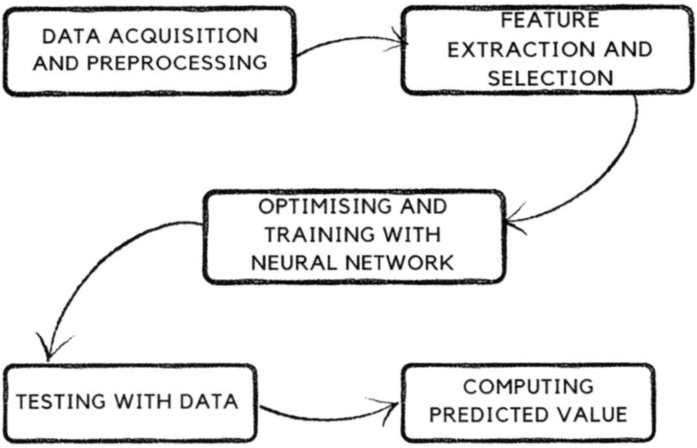

Fig. 2 Sequencial steps for stock prediction model design

This study focuses on forecasting short-term price movements. Currently we have only raw data without labels. The predicting of short-term price changes is the main goal of this study. We currently have just unlabeled raw data. So labelling the data is the first step. By contrasting the present closing price with the closing price from n trading days ago, one may track the price movement. We can indicate with a 1 if the price trend is upward and a 0 if it is downward.

In order to estimate the price movement on day n more precisely, we utilise the index from the index on day n -1.Previous studies have shown that some researchers analyzed stock data using both technical and financial methodologies, and then utilized rules to weed out high-quality stocks. To help with the extension's design, we used references to their work and their regulations. Check the data first, though, to guarantee the predictive model's optimal performance. Raw data has a wide range of features. Not only does taking into account every attribute greatly increase computational complexity, but it also has negative repercussions if future research involves unsupervised learning.

Therefore, we employ recursive feature elimination (RFE) to guarantee the validity of all chosen features. We discovered that whereas the majority of earlier technical research examined all equities, finance experts favoured to examine particular investing scenarios. We made the decision to use functional expansion in order to bridge the gap between the two disciplines based on knowledge gathered from finance before the RFE process began. The more the characteristics, the more complicated the training process will be because we intend to model the data as a time series. Therefore, we use randomized PCA to reduce dimensionality at the start of the suggested solution architecture.

Steps involved in the workflow of the used model:

Raw Stock Data: Unprocessed and uninterpreted information on stocks and the stock market is referred to as "raw stock data." Data that is recorded and reported by financial institutions, media outlets, and data providers includes stock prices, trade volumes, business financial reports, and other measures [11].

Data Cleaning and Pre-processing: The actions taken to prepare raw data for analysis are referred to as data cleaning and pre-processing. This procedure entails cleaning up any errors, discrepancies, or missing numbers from the data and converting it into a format that can be used for analysis.

Expansion of Available features: Data cleaning and pre-processing are the procedures used to get raw data ready for analysis [12]. This process comprises fixing any mistakes, conflicts, or missing data from the data and transforming it into an analysis-ready format.

Refining relevant features: Refining relevant features is a procedure used in machine learning and data analysis to improve the accuracy and efficiency of predictive models by selecting the most relevant features in a dataset [13]. This process involves identifying the most important features in the data and removing irrelevant or redundant features.

Dimension Reduction using Principal Component Analysis (PCA): By reducing the amount of features in a dataset while keeping the most crucial data, dimension reduction is a technique used in machine learning and data analysis. Principal Component Analysis (PCA) is a well-liked method for dimension reduction.

High Weighted Features Extraction: High weighted feature extraction is a method used in data analysis and machine learning to find and rank the most crucial features in a dataset [14]. The features are given weights based on how important they are to the study, and the features with the greatest weights are then chosen for additional analysis.

Model Building using Convolution Neural Networks, Long Short-Term Memory (LSTM), or a combination of both: For the analysis of complex data, such as photos, videos, or time series data, model creation utilizing convolutional neural networks (CNNs), long short-term memory (LSTM), or a mix of the two is a potent method.

Hyper parameter tuning to obtain better results: In order to enhance the performance of a machine learning model, hyper parameter tuning involves optimizing the values of the hyper parameters [15]. Hyper parameters are settings made before the model is trained and cannot be discovered through data learning. The model's learning rate, regularization intensity, number of hidden units, number of layers, and other parameters are controlled by them.

Training the model wrt to time on stock datasets: Using past stock price data, one can forecast future stock prices by training a model on stock datasets. The length of time the model is trained over is a crucial factor to take into account when training a model on stock datasets.

Evaluating the stock data model: A stock data model's predictive accuracy must be measured in order to evaluate it. In this context, a number of frequently used evaluation measures are employed, including Mean squared error (MSE), Mean absolute error (MAE), R-squared (R2), Sharpe ratio, and Profit and loss (P&L).

Testing the stock data model: A stock data model is tested by comparing its performance to results from a fresh, unrelated dataset that was not used in the model's development. To make sure that the model's performance is not biased towards the training data and that it can generalize effectively to new, untested data, it is crucial to do this.

Visualization of the stock data model: Visualization of the stock data model involves using graphical representations to visualize and interpret the model's predictions and performance. This can provide valuable insights into the underlying patterns and trends in the data, as well as help identify potential areas for improvement in the model.

Steps involved to create the used model:

1. **Data acquisition and preprocessing**—In this model, the dataset acquires stock-market analysis for the stocks of various companies and for getting the stock tickers of different companies [16]. This dataset is obtained using Yahoo finance. To import this, the Python library 'yfinance' is used. All the important Python libraries required of model creation are imported. The basic libraries like 'numpy', 'pandas', 'keras.models' are imported first. 'yfinance' library is

imported for acquiring the dataset. 'matplotlib.pyplot' library is used to plot all the predictions and stock-market analysis for a particular company, into real-time graphs. 'sklearn' library is used for feature extraction and scaling. 'streamlit' library is used for creating a streamlit application for plotting all the predictions for a particular stock ticker. The dataset is then split into training and testing data, and the model is then trained for 50 epochs.
2. **Feature extraction and scaling**— 'sklearn' library is used to import 'MinMaxScaler' which is used for scaling the training dataset.
3. **Optimizing and training data**—After scaling the training dataset, it is optimized using the optimizer 'adam' and calculating loss by using 'mean-squared error' technique. The model is then saved as a keras model, and trained according to the past 100 days.
4. **Testing** —The model is then tested using the testing dataset.
5. **Computing predicted values**—The predicted values are then computed on the testing dataset after which the predicted values are then scaled using the calculated scale factor. Finally, a real-time graph of Original Price vs Predicted Price is plotted.

In this project the prediction of stock market is done by using a Supervised Machine Learning Model following the Moving Average Deep Learning technique. The prediction application is made using the Python StreamLit package [17].

In this model, the behavior and expected output of the model are known because the dataset is divided into training and testing datasets, making it a supervised machine learning model. Using well-labeled training data that has been used to train the machines, supervised learning is a type of machine learning technique in which the output is predicted by the machines [18]. The phrase "labelled data" describes input data that has already been given the proper output. The training data that is provided to the computers acts as the supervisor in supervised learning, teaching them how to correctly predict the output. It applies the same concept that a student would learn while being guided by a teacher. Giving the appropriate input and output data to the machine learning model is part of the supervised learning process. A supervised learning method seeks to identify a mapping function that connects the input variable (x) with the output variable (y). In the real world, supervised learning can be applied to a variety of tasks, including spam filtering, fraud detection, risk assessment, image categorization, and stock market forecasting [19, 20].

A series of averages of various subsets of the entire data set are created using the statistical calculation known as the moving average (MA), which is used to analyze data points. The most frequent use of moving averages in the realms of finance and investment is in the technical evaluation of stock prices.

The formula for calculating a moving average is to add up all of the prices or other data points during a given period, then divide that total by the total number of data points during that period. For instance, adding up the closing prices from the previous 10 days and dividing by 10 would be used to determine a stock price's 10-day moving average. Instead of taking days or even months, a truly amazing machine learning or data science software can be produced in just a few hours or even minutes. It is compatible with the vast majority of Python libraries, including pandas, matplotlib, seaborn, plotly, Keras, PyTorch, and SymPy(latex). Amazing web applications can be made with not much coding. Data caching enables computation pipelines to operate more quickly and easily.

4 Implementation Analysis

The above graph shown in Fig. 3 is a depiction of change in stock prices with time for the stock ticker as mentioned in the above discussion. The above graph shows the growth in stock prices over the years ranging from 2010 to 2020 (Fig. 4).

A series of averages of various subsets of the entire data set are created using the statistical calculation known as the moving average (MA), which is used to analyze data points. The most frequent use of moving averages in the realms of finance and investment is in the technical evaluation of stock prices. The formula for calculating a moving average is to add up all of the prices or other data points during a given period, then divide that total by the total number of data points during that period. For example, a 10-day moving average of a stock price would be calculated by adding up the last 10 days' closing prices and dividing by 10. Figures 5 and 6 show the comparison of stock prices with time in accordance with 100 moving average and 100 and 200 moving averages respectively.

The above graph as depicted by Fig. 6 shows the real-time comparison between the predicted price as predicted by our model and the original price of stocks of the stock ticker as discussed above.

Fig. 3 Comparison analysis of closing price with latency period

Fig. 4 Comparison analysis of closing price with 100 moving average

Fig. 5 Comparison analysis of closing price with 200 moving average

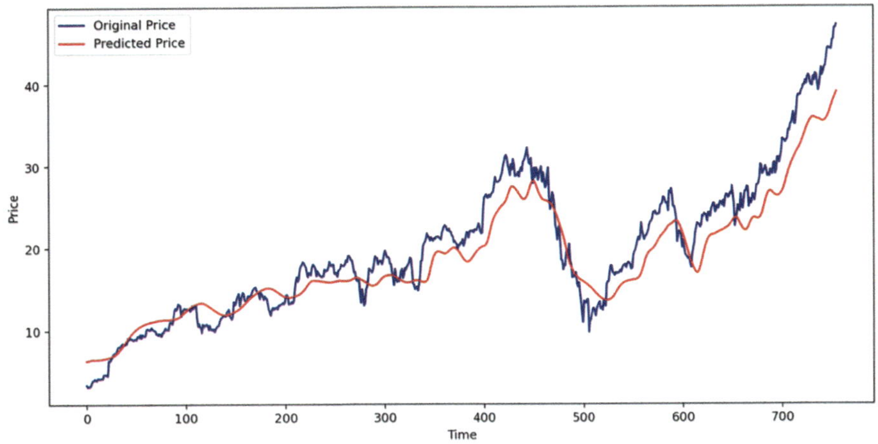

Fig. 6 Comparison analysis of predicted price with original price

5 Conclusion

To make future purchasing and selling decisions, stock market analysis involves examining both recent and historical market movements. The percentage rise represents the change in stock value from the previous day. The higher the positive or negative value, the more the stock is volatile. A decision support system can be created to aid investors in determining which industry stock to select for low-risk, low gain or high-risk, huge gain, depending on their level of risk tolerance.

Some of the few things to remember are firstly, predicting stock price or movement is a very challenging endeavor. Personally, it is believed that any of the stock prediction models available should be taken for granted or relied upon without careful consideration. However, models might be able to accurately forecast changes in stock price. Secondly, avoid falling for publications that display forecast curves that perfectly overlap actual stock values. This is meaningless in practice and can be duplicated with a straightforward averaging technique. Predicting the swings of the stock price is a more sensible course of action. Lastly, the findings you get have a huge impact on the model's hyper-parameters. Running certain techniques on the hyper-parameters would therefore be a really excellent idea.

References

1. A hybrid machine learning model for stock price prediction using sentiment analysis and technical indicators by V. Sharma et al. (2021).https://doi.org/10.1109/ICAS51307.2021.9415016
2. Stock price prediction using machine learning algorithms and technical analysis, by S. S. Gupta and S. Kumar (2021). https://doi.org/10.1109/ICPS48432.2021.9483429

3. Stock price prediction using machine learning algorithms with financial and textual data, by Y. Zhao et al. (2020). https://doi.org/10.1016/j.procs.2020.09.129
4. Machine learning-based stock price prediction using financial news articles, by S. Kwon and J. Lee (2020). https://doi.org/10.1016/j.dcan.2019.07.005
5. Stock price prediction using machine learning and sentiment analysis, by H. A. Alzahrani et al. (2020). https://doi.org/10.1007/978-3-030-49190-1_31
6. Stock price prediction using machine learning algorithms and big data analytics, by T. Gao et al. (2019). https://doi.org/10.1016/j.future.2019.06.027
7. A machine learning-based stock price prediction model using financial news and technical indicators, by J. Liu et al. (2019). https://doi.org/10.1109/ICMLA.2019.00139
8. A stock price prediction model using machine learning algorithms and social media analysis, by Z. Ma et al. (2019). https://doi.org/10.3390/app9224851
9. A comparative study of machine learning techniques for stock price prediction, by S. Selvin et al. (2018). https://doi.org/10.1109/ICRIS.2018.8598704
10. Stock price prediction using machine learning techniques and time series analysis, by G. Jain and S. Kumar (2017). https://doi.org/10.1109/ICCSP.2017.8286499
11. Mishra, N., Mishra, S., Tripathy, H. K.: Rice yield estimation using deep learning. In: Innovations in Intelligent Computing and Communication: First International Conference, ICIICC 2022, Bhubaneswar, Odisha, India, December 16–17, 2022, Proceedings, pp. 379–388. Springer International Publishing, Cham (2023, January)
12. Chakraborty, S., Mishra, S., Tripathy, H.K.: COVID-19 outbreak estimation approach using hybrid time series modelling. In: Innovations in Intelligent Computing and Communication: First International Conference, ICIICC 2022, Bhubaneswar, Odisha, India, December 16–17, 2022, Proceedings, pp. 249–260. Springer International Publishing, Cham (2023, January)
13. Verma, S., Mishra, S.: An exploration analysis of social media security. In: Predictive Data Security using AI: Insights and Issues of Blockchain, IoT, and DevOps, pp. 25–44. Springer Nature, Singapore (2022)
14. Singh, P., Mishra, S.: A comprehensive study of security aspects in blockchain. In: Predictive Data Security using AI: Insights and Issues of Blockchain, IoT, and DevOps, pp. 1–24. Springer Nature, Singapore (2022)
15. Swain, T., Mishra, S.: Evolution of machine learning algorithms for enhancement of self-driving vehicles security. In: 2022 International Conference on Advancements in Smart, Secure and Intelligent Computing (ASSIC), pp. 1–5. IEEE (2022, November)
16. Sahoo, S., Mishra, S.: A Comparative analysis of PGGAN with other data augmentation technique for brain tumor classification. In: 2022 International Conference on Advancements in Smart, Secure and Intelligent Computing (ASSIC), pp. 1–7. IEEE (2022, November)
17. Mohapatra, S.K., Mishra, S., Tripathy, H.K.: Energy consumption prediction in electrical appliances of commercial buildings using LSTM-GRU model. In: 2022 International Conference on Advancements in Smart, Secure and Intelligent Computing (ASSIC), pp. 1–5. IEEE (2022, November)
18. Tripathy, H.K., Mishra, S.: A Succinct Analytical Study of the Usability of Encryption Methods in Healthcare Data Security. In: Next Generation Healthcare Informatics, pp. 105–120. Springer Nature, Singapore (2022)
19. Adrija, M., Yash, A., Sushruta, M.: 8 Pragmatic study of IoT in healthcare security with an explainable AI perspective. In: Explainable Artificial Intelligence for Biomedical Applications, pp.145–166. River Publishers (2023)
20. Bhavya, M., Pranshu, S., Sushruta, M., Sibanjan, D..: 17 comparative analysis of breast cancer diagnosis driven by the smart IoT-based approach. In: Explainable Artificial Intelligence for Biomedical Applications, pp. 353–374. River Publishers (2023)

Mispronunciation Detection Using Feature Learning

Priyanka Chhabra, Shailja Chhillar, Riya Tanwar, Muskan Verma, and Gaurav Indra

Abstract This research describes a study into the use of feature learning approaches for mispronunciation detection. Mispronunciation detection is essential in speech recognition and language learning applications. Mispronunciation detection has traditionally depended on handcrafted features and rule-based algorithms, which frequently have poor generalisation abilities and demand a lot of manual work but recent advances in deep learning and feature learning have demonstrated promising results in enhancing the accuracy and robustness of mispronunciation detection systems. The study proposes an innovative method using Mel-frequency cepstral coefficients (MFCC) and SVM. Comparing TF-IDF and MFCC feature extraction, labelled audio recordings are categorised as accurate or incorrect pronunciation. Audio data is preprocessed, and MFCC features are extracted using the librosa library. SVM learns patterns between features and labels during training. Performance is evaluated on a common voice dataset, achieving 71% accuracy with MFCC and 70% with TF-IDF. Additional preprocessing and hyperparameter tuning result in 71% accuracy for TF-IDF. Overall, this research shows that MFCC feature extraction and an SVM classifier are effective tools for mispronunciation identification. By utilising the strength of feature learning on the Common Voice dataset, this research advances mispronunciation detection systems, ultimately enhancing language learning processes and facilitating the creation of more precise speech processing applications.

P. Chhabra (✉) · S. Chhillar · R. Tanwar · M. Verma · G. Indra
Indira Gandhi Delhi Technical University for Women, Kashmere Gate, New Delhi 110006, India
e-mail: pcpriyankapc21@gmail.com

S. Chhillar
e-mail: shailjachhillar@gmail.com

R. Tanwar
e-mail:. riyatanwar1105@gmail.com

M. Verma
e-mail: muskanverma2002@gmail.com

G. Indra
e-mail: gauravindra@igdtuw.ac.in

G. Fortino et al. (eds.), *Proceedings of Third International Conference on Computing and Communication Networks*, Lecture Notes in Networks and Systems 917,
https://doi.org/10.1007/978-981-97-0892-5_24

Keywords MFCC feature extraction · SVM classifier · Hyperparameters · Tf-Idf feature extraction

1 Introduction

Mispronunciation detection is a crucial task for speech recognition and language learning systems. In order to aid learners in honing their pronunciation skills, this involves identifying and marking occasions where a speaker mispronounces words or sounds.

Historically, mispronunciation detection relied on manually created characteristics that needed subject expertise and labour-intensive manual work. This method, though, frequently struggled with different pronunciation patterns and had trouble generalising. An improved strategy based on feature learning has surfaced to overcome these issues.

Using Mel-frequency cepstral coefficients (MFCC) feature extraction and an SVM classifier which is a supervised machine learning algorithm, a mispronunciation detection method is suggested in this research work. The goal of this approach is to correctly categorise instances of pronunciation errors in audio data. The audio recordings utilised in this study were labelled, and each recording was assigned to one of two categories: accurate pronunciation or incorrect pronunciation. Using the librosa library, audio data is preprocessed and MFCC features are extracted. These features serve as the SVM classifier's input and catch the spectral characteristics of the audio data. During the training phase, the SVM model picks up on patterns and connections between the MFCC features and the accompanying pronunciation labels.

By utilising MFCC feature extraction and an SVM classifier for mispronunciation detection, this research advances the area. The method may find use in language instruction, where it might give students useful feedback on how to pronounce words more clearly. It can help speech therapists evaluate and track the development of their patients.

To improve the system's performance, more study is necessary. Future directions for development include examining more auditory features, adding temporal data and researching different machine learning algorithms or ensemble techniques.

The system exhibits its robustness and generalisation ability by recognising mispronunciations with great accuracy. By utilising the strength of feature learning and the extensive resources offered in the Common Voice dataset, this research advances mispronunciation detection systems, ultimately enhancing language learning processes and facilitating the creation of more precise speech processing applications.

2 Problem Statement

The precise identification of pronunciation errors in audio data is the issue that this research attempts to solve. It is critical to create automated tools that can recognise and correct pronunciation mistakes since they can impede successful communication and language learning. Mispronunciation detection techniques now in use have had variable degrees of success, but they frequently have drawbacks like low accuracy, reliance on certain languages or voice datasets or a lack of tolerance to different speaking styles and variations. These restrictions make it difficult for mispronunciation detecting systems to be used in actual situations. Consequently, the goal of the study is to build and create an approach for mispronunciation detection that overcomes the limitations of existing approaches and achieves high accuracy, language independence, and robustness to different speaking styles. This paper aims to leverage the power of MFCC feature extraction and SVM classification to address this problem and provide an effective solution for accurately detecting mispronunciations in audio data.

2.1 Objectives

Mispronunciation detection using feature learning aims to create and construct a robust and efficient system capable of manually identifying and classifying mispronunciations in non-native speech. Using feature learning approaches, the system should acquire biased representations capable of distinguishing between proper and erroneous utterances.

The following are some objectives:

1. Extracting relevant features: Developing strategies for retrieving auditory, phonological, and prosodic aspects from voice data that reflect the qualities related to mispronunciations. We are using MFCC feature extraction.
2. Identifying and extracting important features: Feature learning techniques, such as machine learning algorithms and deep learning, can be used to generate biased interpretations of extracted features.
3. Developing discriminatory representations: Train a classification model, such as a support vector machine (SVM) to categorise speech occurrences into right or wrong pronunciation categories using the learnt representations.
4. Assessing system performance: Use relevant evaluation measures, such as accuracy, precision, recall and F1-score, to assess the performance of the mispronunciation detecting system. Compare the system's performance against baseline procedures and other cutting-edge approaches.

3 Related Work

In the previous works, Georgoulas et al. [12] used a discrete wavelet transform in order to extract the features. In order to identify mispronunciations, the signal processing technique known as wavelet analysis employed an SVM for classification. Ito et al. [11] created the pronunciation rule clusters and established a threshold for each cluster. To improve the effectiveness of mistake detection, a decision-based clustering technique was presented. Witt and Young [10] proposed to utilise a goodness of pronunciation (GOP) value after taking into account the log-probability scores normalised by the length of each phone section. They developed a criterion to determine if each phone is pronounced correctly or not. Strik et al. [7] showed how to distinguish between velar fricatives and velar plosives using acoustic–phonetic parameters including log root mean square, energy, and zero-crossing rate. Amdal et al. [4] proposed that speaking vowels that are short and long are distinguished. They developed a linear discriminant analysis (LDA) classifier using acoustic–phonetic characteristics. Wei et al. [9] employed all of the acoustic models' log-likelihood ratios as features in an SVM classifier. To determine the pronunciation variations of each phone at various levels of competence, a number of parallel acoustic models of each phone were used in the modelling process. By using this method, the framework was able to outperform more conventional likelihood-based techniques. Li et al. [13] provided a powerful detection method based on GLDS-SVM (Generalised Linear Discriminant Sequence- Support Vector Machine). To improve the execution, they integrated the GLDS-SVM method with the UBM-GMM (universal background model) architecture. However, to extract discriminative auditory features, the majority of these algorithms require human interaction.

4 Dataset Description

Mozilla built the Common Voice dataset, a sizable, multilingual and user-generated dataset. It includes audio clips of people speaking in various dialects, accents and speech patterns.

Volunteers contribute their voice recordings by reading prompted sentences, and this data is gathered in a community-driven manner. The sentences were chosen with care to include a variety of themes and language constructions. The dataset will be representative of a wide range of languages, accents and speaking styles thanks to this strategy. Here are some of the Common Voice dataset's salient qualities and characteristics:

1. Size and Diversity: The dataset has a significant number of audio files from contributors all over the world and is constantly expanding. It is intended to represent a wide variety of languages, accents and dialects and is multilingual and diversified.

2. Audio Format: Audio recordings in a variety of formats, including WAV, MP3 and FLAC, make up the majority of the Common Voice collection. The audio files record people speaking naturally in their own contexts.
3. Transcriptions: Each audio file in the Common Voice collection is accompanied by a text transcription of the spoken sentence that corresponds to it. For training and evaluation reasons, these transcriptions offer the real story.
4. The Common Voice dataset is made freely available for use in research, development and commercial endeavours thanks to a Creative Commons CC0 1.0 Universal licence.

The key features of the Common Voice dataset make it a valuable resource for training, benchmarking and advancing ASR systems. Its multilingual nature, community-driven approach and open licensing contribute to its widespread use and encourage further research and innovation in the field of speech processing.

4.1 Methodology

Data Preprocessing

The audio data, typically in the form of audio files, is loaded from the dataset. Our dataset contains audio files in.mp3 format and then preprocessed to convert the continuous audio signal into discrete samples.

Feature Extraction

For feature extraction, the basic code employs MFCC (Mel-Frequency Cepstral Coefficient) and TF-IDF (Term Frequency Inverse Document Frequency). These are two distinct feature extraction approaches utilised in separate fields. TF-IDF is utilised mostly in Natural Language Processing (NLP), whereas MFCC is used primarily in audio signal processing.

TF-IDF: In this case, the textual data being transformed into a numerical feature vector is the dataset's 'sentence' column. These phrases are converted into numerical feature vectors using the TF-IDF vectorizer. It computes these values for each word in the sentence and represents the sentences as TF-IDF vectors. One of the primary reasons for transforming textual data to numerical feature vectors is to allow machine learning algorithms to operate with text data.

MFCC: It incorporates significant spectral content properties of audio signals such as spectral envelope form, spreading of energy across different frequency bands and perceptual components of human hearing. The MFCC is intended to record the spectrum features of audio signals and offer an organised summary of the audio signal. It depends heavily on human hearing.

The SVM (Support Vector Machine) classifier is a renowned supervised learning technique for classification applications. The SVM classifier is taught using training data; the classifier's fit() function uses MFCC features as input and corresponding target labels to train the model. After training, the SVM classifier is employed to make predictions on the test data. The classifier receives MFCC characteristics as feed and predicts labels for test samples. The effectiveness of the predictor is evaluated using the metrics component of scikit-learn's classification report() function. It determines and publishes a classification report for each class that contains metrics like precision, recall, f1-score and support.

5 Comparison of Approaches

The base model gives accuracy of 70 percent when TF-IDF is used as a feature extraction method. To increase the accuracy, data preprocessing and hyperparameter tuning were included in the base model (Figs. 1 and 2).

The following two data preprocessing methods were used.

1. Regular expression substitution: It substitutes all non-word characters (W) in the DataFrame's'sentence' column with a space. This process aids in textual cleanup by deleting unnecessary letters or symbols.

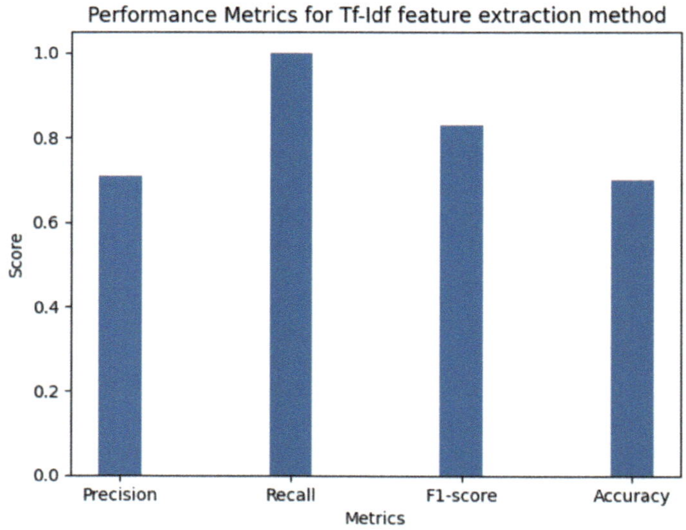

Fig. 1 Performance Metrics for TF-IDF feature extraction method

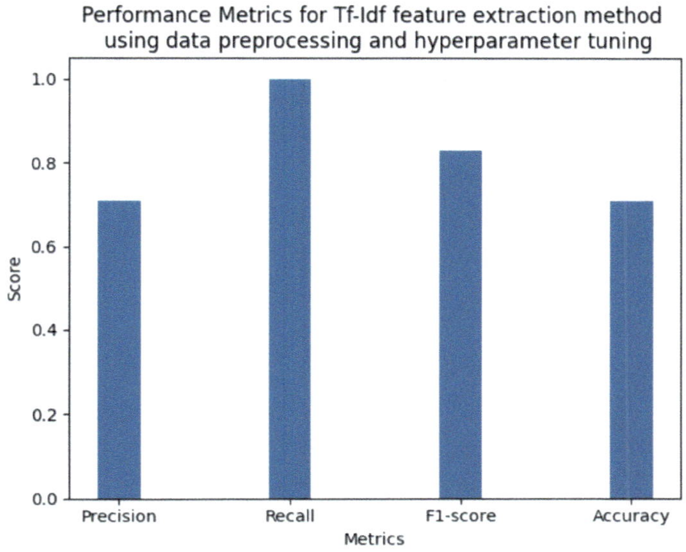

Fig. 2 Performance Metrics for TF-IDF feature extraction method using data preprocessing and hyperparameter tuning

2. Handling missing values: The second way entails dealing with values that are absent in the'sentence' column.
3. For hyperparameter tuning GridSearch is used, which is a typical approach for hyperparameter optimisation that involves continuously searching a predetermined grid of hyperparameter values in order to identify the optimal combination. Using the supplied parameter grid, the updated code grid search with cross-validation is used to discover the optimum hyperparameters for the classifier (clf). After these changes, the updated code provided 71 percent accuracy.

6 Result and Analysis

On the provided dataset, the mispronunciation detection algorithm has a 70% accuracy rate if TF-IDF feature extraction is used. Whereas on the other hand mispronunciation detection model using MFCC feature extraction and SVM classifier is giving 71% accuracy. This precision represents the model's capability to spot instances of mispronunciation errors. The research project's objective was to develop a system for mispronunciation identification using feature extraction from MFCC (Mel-frequency cepstral coefficients) and an SVM (Support Vector Machine) classifier. On the test dataset, the implemented model had an accuracy of 71%. The audio recordings that made up the dataset for this study were each given a label designating whether they were correct or incorrect pronunciations. The spectral characteristics of the

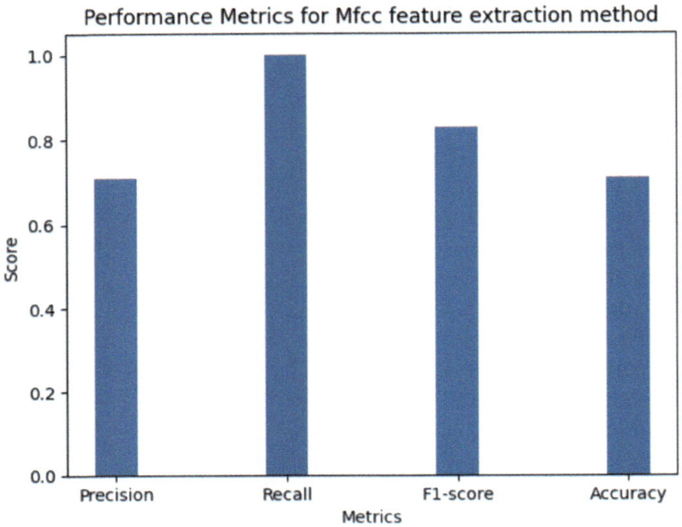

Fig. 3 Performance Metrics for MFCC Feature Extraction Method

audio signals were captured by the preprocessing and conversion of the audio data into MFCC features. These characteristics were taken out using the Python librosa package. The SVM classifier, a well-liked supervised learning technique for classification problems, was then fed the collected MFCC features as input. The labelled training data was used to train the SVM model, which helped it discover patterns and connections between the MFCC features and their related pronunciation labels (Figs. 3 and 4).

The SVM model's performance was evaluated after training using the test dataset. Accuracy, which counts the percentage of cases that were properly classified out of all instances, was chosen as the evaluation statistic. The model properly categorised 71% of the test events, according to its accuracy score of 71%. The designed mispronunciation detecting system appears to operate admirably, as evidenced by the accuracy of 71% that was acquired. However, it's crucial to take into account the drawbacks and possible areas for improvement. The training dataset's quality and diversity, the selection of SVM hyperparameters and the precise details of the dataset's mispronunciations are some of the variables that may have helped achieve the accuracy. To improve the effectiveness of the mispronunciation detecting system, more research and analysis could be done. Investigating various feature extraction methods, such as using temporal information or utilising additional auditory properties, may be potential directions for advancement. Additionally, to improve the overall accuracy and resilience of the system, experimenting with various machine learning techniques or investigating ensemble methods may be useful.

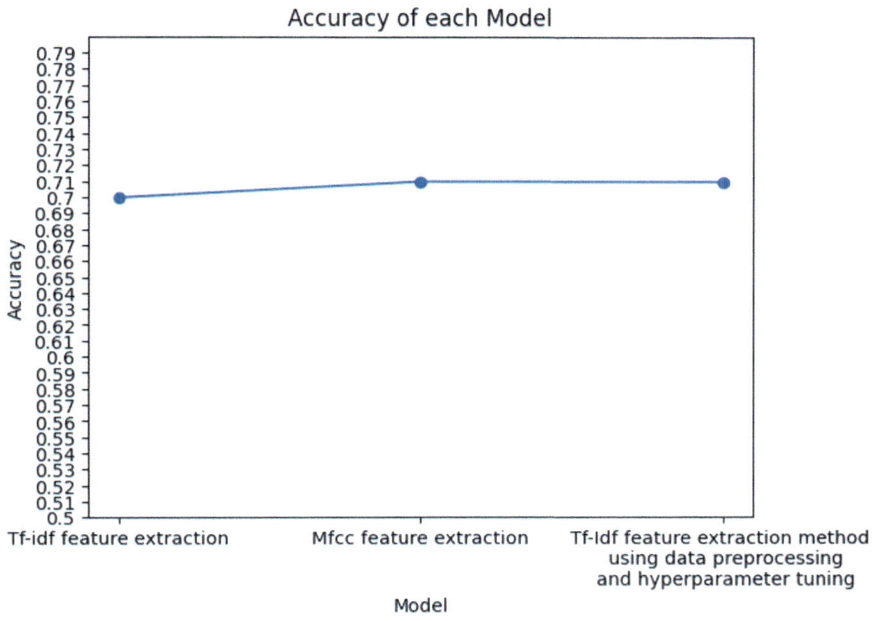

Fig. 4 Accuracy Comparison of each Model

7 Conclusion

In this study, it is suggested and put into practise an SVM classifier and MFCC feature extraction mispronunciation detection system. A promising accuracy of 71% was shown by the created method in classifying instances of pronunciation errors.

By utilising MFCC features, which record the spectrum properties of audio signals, and using an SVM classifier to categorise occurrences of mispronunciations, this research advances the field of mispronunciation identification. The degree of accuracy attained suggests that this technology is capable of correctly identifying pronunciation errors in audio data.

This research has important potential applications, especially for speech treatment and language learning. Language learners can benefit from helpful feedback from an accurate and automatic mispronunciation detection system, which can aid in the development of their communication and pronunciation skills. Such a method can also help speech therapists evaluate and monitor the development of their patients.

Overall, this research shows how well an SVM classifier and MFCC feature extraction work together to identify mispronunciations. It is encouraged that future study investigates new methodologies and datasets to enhance the accuracy and usability of mispronunciation detection systems because the attained accuracy serves as a solid foundation for future developments in this field.

References

1. Cucchiarini, C., de Wet, F., Strik, H., Boves, L.: Assessment of Dutch pronunciation by means of automatic speech recognition technology. In: Proceedings of the ICSLP 1998, pp. 751–754 (1998)
2. Maqsood, M., Habib, H.A., Nawaz, T.: Feature selection for Arabic mispronunciation detection based on sequential floating forward selection and data mining classifiers. Pakistan J. Sci. **68**(4), 445–452 (2016)
3. Li, K., Qian, X., Meng, H.: Mispronunciation detection and diagnosis in L2 English speech using multidistribution deep neural networks,. IEEE/ACM Trans. Audio, Speech, Lang. Process **25**(1), 193–207 (2017)
4. Amdal, I., Johnsen, M.H., Versvik, E.: Automatic evaluation of quantity contrast in nonnative Norwegian speech. In: Proceedings of the International Workshop Speech Language Technology Education (2009), pp. 21–24
5. Lee, A., Zhang, Y., Glass, J.: Mispronunciation detection via dynamic time warping on deep belief network-based posteriorgrams. In: Proceedings of the IEEE International Conference Acoustical, Speech Signal Process. (ICASSP), pp. 8227–8231 (2013)
6. Joshi, S., Deo, N., Rao, P.: Vowel mispronunciation detection using DNN acoustic models with cross-lingual training. In: Proceedings of the 16th Annual Conference International Speech Communication Association, pp. 697–701 (2015)
7. Strik, H., Truong, K., de Wet, F., Cucchiarini, C.: Comparing different approaches for automatic pronunciation error detection. Speech Commun. **51**(10), 845–852 (2009)
8. Scheirer, E., Slaney, M.: Construction and evaluation of a robust multifeature speech/music discriminator. In: Proceedings of the IEEE International Conference Acoustical, Speech, Signal Process (ICASSP), pp. 1331–1334 (1997)
9. Wei, S., Hu, G., Hu, Y., Wang, R.-H.: A new method for mispronunciation detection using support vector machine based on pronunciation space models. Speech Commun. **51**(10), 896–905 (2009)
10. Witt, S.M., Young, S.J.: Phone-level pronunciation scoring and assessment for interactive language learning. Speech Commun. **30**(2–3), 95–108 (2000). IEEE Trans. Fuzzy Syst. **26**(6), 3847–3859 (2018)
11. Ito, A., Lim, Y.-L., Suzuki, M., Makino, S.: Pronunciation error detection method based on error rule clustering using a decision tree. In: Proceedings of the 9th European Conference Speech Communication Technology, 2005, pp. 173–176 (2005)
12. Georgoulas, G., Georgopoulos, V.C., Stylios, C.D.: Speech sound classification and detection of articulation disorders with support vector machines and wavelets. In: Proceedings of the International Conference IEEE Engineering Medical Biology Society, Aug./Sep. 2006, pp. 2199–2202 (2006)
13. Li, H., Liang, J., Wang, S., Xu, B.: An efficient mispronounciation detction method using GLDS-SVM and formant enhanced features. In: Proceedings of the IEEE International Conference Acoustical, Speech Signal Process (ICASSP), Apr. 2009, pp. 4845–4848 (2009)

Annotating Social Data with Speaker/User Engagement. Illustration on Online Hate Characterization in French

Delphine Battistelli, Valentina Dragos, and Jade Mekki

Abstract This paper presents an annotation framework relying on the linguistic notion of speaker/user engagement. This notion is suitable for a finer characterization of online hateful discourse as it allows addressing the following question: does the speaker engage himself to the truthfulness of hate content? Two resources were built to support the annotation framework: a taxonomy of speaker/user engage- ment degrees and a rich semantic resource of pictograms/emoticons. The paper describes the resources used and the annotation process. Preliminary experiments and results on sexism characterization in French are also presented.

Keywords Opinion mining · Social computing · Extremist content

1 Introduction

Online hate includes abusive, insulting, intimidating, and harassing expressions that support violence, hatred, or discrimination against specific targets related to race, ethnic origin, religion, gender, age, physical condition, disability, and sexual orientation of persons [20]. Linguistically, online hate speech is manifold. If hatred content is obviously highlighted by offensive words and explicit incitation to violence [5], it can also be less explicit. For example, [9] showed that for online data, stereotypes propagate implicit sexism. Another relevant linguistic aspect specific to hatred contents is that one can also immediately distinguish different ways of marking the user engagement to the truthfulness of what is said.

D. Battistelli · J. Mekki (✉)
Paris Nanterre University, 200 Av. de La République, Nanterre, France
e-mail: j.mekki@parisnanterre.fr

D. Battistelli
e-mail: battist@parisnanterre.fr

V. Dragos
ONERA-The French Aerospace Lab, 6 Chemin de La Vauve Aux Granges, Palaiseau, France
e-mail: Valentina.Dragos@onera.fr

G. Fortino et al. (eds.), *Proceedings of Third International Conference on Computing and Communication Networks*, Lecture Notes in Networks and Systems 917,
https://doi.org/10.1007/978-981-97-0892-5_25

This study is motivated by several limitations of current approaches developed for online hate detection. Those limitations are illustrated by the examples (1) to (4).

- (1) Tous ces migrants envahissent nos hôpitaux!!! (All these migrants are invading our hospitals!!!)
- (2) Il a raison de dire que tous ces migrants envahissent nos hôpitaux! (He is right to say that all these migrants are invading our hospitals!")
- (3) Entendre dire que tous ces migrants envahissent nos hôpitaux me révolte! (To hear that all these migrants are invading our hospitals revolts me!)
- (4) Il a dit que tous ces migrants envahissent nos hôpitaux. (He said that all these migrants are invading our hospitals).

(1) is a direct and explicit hateful content expressed by the speaker/user with an implicit engagement to the truthfulness of what is said in the statement; (2) is an indirect hateful content as it indicates an opinion expressing support to hateful content, and thus there is a clear engagement to the truthfulness of what is said; (3) is similar to (2), but it indicates an opinion expressing no support to hateful content and there is a clear disengagement with respect to the truthfulness of what is said; (4) is "just" reporting a hateful content, thus there is just an engagement to the truthfulness of a speech act and not of the hate content it contains. Currently, automatic approaches address the detection of direct online hate speech like (1) by using training data composed of explicitly hatred content, or by creating lexicons of hate-relevant words [3]. However, they fail to distinguish cases similar to the examples illustrated in (2), (3), and (4). The analysis of those cases requires creating datasets labeled by considering not only intrinsic features of abusive language but also features of what can be analyzed as different positions or attitudes against the truthfulness of hate content. This paper investigates the following research question: how useful could be the linguistic notion of a speaker's engagement in a better characterization of what is a hateful attitude, especially among these four types of attitudes denoted by the examples (1) to (4) mentioned before? The paper has three main contributions: first, it explores the notion of user engagement as a feature to be further used to better characterize and detect hatred content; then, it describes the construction of a resource of pictograms integrating this notion of user engagement and finally the paper contributes to the analysis of online content in French. Moreover, this study provides a new perspective on online hate in French social media, and brings new in-depths linguistically grounded analysis that benefit society. The case of hate speech detection provides valuable insight into the promises and limitations of automated online content characterization for sociological analyses.

The outline of the paper is the following: Sect. 2 presents a selection of related approaches. The construction of resources is discussed in Sect. 3 while Sect. 4 illustrates the use of resources for the annotation of sexist tweets in French. Conclusion and directions for future work end the paper in Sect. 5.

2 Related Work

Engagement is related to speakers' own perspective on informational content and indicates whether they assert with confidence what they perceive in the environment, know from their experience, interact with, or attend to. In some particular cases, for example, when indirectly reporting, the authors can only assume, to varying degrees of certainty, and this aspect is also covered by engagement [6]. Although engagement is a quite well-established notion in linguistics and has been explored at the intersection of modality, evidentiality, and commitment categories [10], there has been less work on considering speaker/user engagement for social data analysis. However, previous research confirms that, for open-dialogue systems, taking into account user engagement as real-time feedback benefits the analysis of social interactions [19]. More specifically, online users use emojis to enrich the context and convey additional emotions, and using emojis increases user engagement [22].

Emoticons are extremely popular on social platforms and there is a growing literature investigating how they affect the interpretation of messages online [11]. The main topics investigate whether emoticons capture emotional states [23] and shed light on this connection by compiling data sets [13] which quantify people's reported association between emoticons and emotions. Several experimental studies illustrate the impact of considering emoticons for online content analysis. First, for visual sentiment analysis, [2] demonstrate that, while emoticons carry a sentiment signal by themselves, they also act as sentiment modifiers of surrounding contents. Then, for opinion detection, implementing a multi-modal fusion by combining emoticons with text improves the performance of opinion detection in terms of recall and precision [1]. Recent works also investigate the intrinsic semantics of emoticons [15] and their links with semantic representations of general, abstract, or concrete, concepts. More specifically, emoticons prove to be good indicators for non-direct hate speech detection, as statistical links mapping certain emoticons and key terms of hatred discourses can be used to assess differences in sentiments conveyed by those contents [7].

Taking a step further, several resources such as the Emoji Sentiment Ranking [18] or Emojipedia[1] have been created to describe the semantics of emoticons and make explicit relations between emoticons and emotions. While resources provide different perspectives about emoticons and their meanings, they do not take into account their context of usage [12]. However, emoticons are generally considered as variant artifacts, and their meaning change due to time [21]. From a different perspective, studies also consider the evaluation of detection models based on emojis, and [17] presents Hatemo-jiCheck, a test suite of 3,930 short-form statements that allows to evaluate performance on hateful language expressed with emoji and highlights weaknesses in existing hate detection models.

As shown above, using emoticons for social data analysis was tackled by several research studies, but there are not numerous approaches considering both emoticons and user engagement. The core insight of this work is to provide an annotation framework aggregating labels based on the analysis of speaker/user engagement, which is

[1] https://emojipedia.org/.

expressed, by both strictly linguistic markers and pictograms. Unlike previous work that mainly develop training data or pre-trained models, we focus on the development of linguistic resources that can be integrated into various classification approaches. Keeping apart the construction of resources and the implementation of algorithms can also be helpful when considering how annotation procedures, sampling data, and training models can be merged together. In addition, the use of annotation categories provides a finer characterization of online hate and allows us to go beyond the limitations of the binary classification. This is a trendy research topic and the need for a finer characterization of online hate is also highlighted by the tasks recently addressed at SemEval-2023 challenge [16], which included a four-class classification for sexist posts into threats, derogation, animosity, and prejudiced discussions.

3 Building Resources to Describe User Engagement

This section presents the construction of two resources capturing user engagement under two headlines. The first one is a taxonomy for labeling different degrees of user engagement and the second one is a semantic resource of pictograms/emoticons.

3.1 A Taxonomy for Labeling User Engagement

As mentioned before (cf. examples 1 to 4), it is important to take into account to what extent the user supports or denies the truth of hateful content. In order to do so, we developed a taxonomy for user engagement. It was created by analyzing existing linguistics research on user engagement [10] and knowledge claims [4]. Figure 1 presents a partial view of this taxonomy, that is relevant for experiments described in Sect. 4. It highlights three main categories of enunciative acts: assertion, reporting, and declaration. Each category is described by using two types of boolean variables expressing the engagement of the speaker on two kinds of truthfulness: the truthfulness of a hatred content (noted *THC*) and the truthfulness of a reported discourse (noted *TRD*).

When a category is not concerned with one of those two variables, it is noticed with N/A. According to this taxonomy:

- (1) can be analyzed as an Individual Assertion with *THC* = *True*;
- (2) is then a Reported Speech having *TRD* = *True* and *THC* = *True*;
- (3) is a Reported Speech having *TRD* = *True* and *THC* = *False*;
- (4) is a Reported Speech with *TRD* = *True* and *THC* has the value *N/A*.

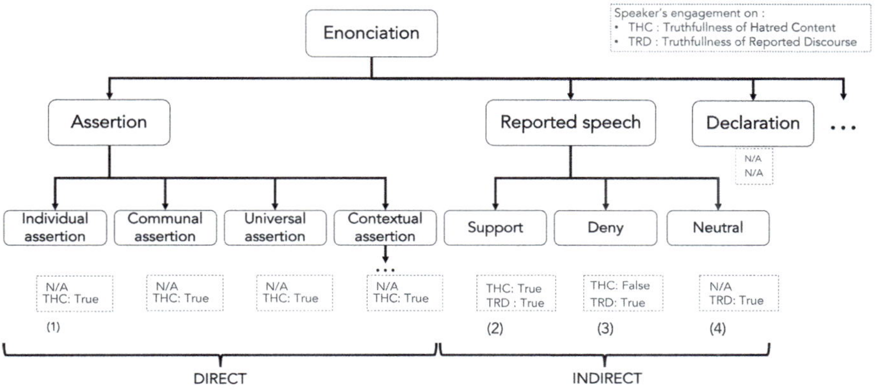

Fig. 1 Taxonomy of speaker's engagement

3.2 *A Semantic Resource of Pictograms*

A pictogram is a symbol used to represent a variety of images: facial mimicry, (e.g. ,), physical postures (e.g. ,), food (e.g. ,), tools (e.g. ,), vehicles (e.g. ,), flags (e.g. ,), plants or animals (e.g. ,), punctuation marks (e.g. ,) or abstract concepts (e.g. ,), etc.

From the set of pictograms, emoticons can be distinguished as a specific subset According to [14], what makes a pictogram an emoticon is the fact that it becomes a conventional index of an attitude, an emotion, of any element experienced by the speaking subject. Therefore, an emoticon is a pictogram imitating a facial expression (e.g., ,), or a pictogram giving additional information on what the speaker feels or experiences (e.g., ,). Moreover, [15] also identifies four types of emoticons according to the facial expressions they imitate:

– Positive: raising corners of the mouth and/or squinting eyes (e.g. , ,);
– Negative: with droopy and/or twisted mouth (e.g. , ,);
– Surprised: with rounded mouth and/or eyes (e.g. , , ,);
– Address: when it imitates gestures intended to address an interlocutor (e.g. ,).

It is also possible to highlight the ability of pictograms to convey user engagement with any textual content. That's what we did. In the end, a rich semantic resource was created by classifying a collection of 1 877 pictograms (among which 180 emoticons). Each pictogram/emoticon is described by three semantic attributes:

1. The type whose values can be: *pictogram*, *emoticon*;
2. The mimicry whose values can be: *absence, positive, negative, surprised, address*, a value, when different of absence, is represented with a rank from 1 to 4;
3. The user engagement whose values can be: *true*, *false*, the value is represented with a rank ranging from 1 to 2.

Table 1 Examples of pictograms described with three semantic criteria

Type		Mimicry				User engagement	
		Positive	Negative	Surprised	Adress	True	False
😜	Emoticon	2			1	1	
❤️	Emoticon	1				1	
😏	Emoticon	1			2	1	1
🥑	Pictogram						1
🇫🇷	Pictogram					1	

The protocol of giving ranks allows us to describe the semantics of pictograms even when they are ambiguous or with a heterogeneous use, especially with generational differences. For example, the item 😸will be interpreted non-ambiguously as positive whereas the item ☠️will be interpreted as negative but also as positive for the younger generation for the expression "dead funny".

Table 1 shows five examples of pictograms and their attributes. The first three cases of emoticons illustrate different types of facial expressions and user engagement: the item 😜is an emoticon addressing someone in a positive way and fully engages the user; the item ❤️is a non-ambiguous emoticon which fully engages the user in a positive posture; the item 😏is more ambiguous about user engagement with an ironic posture. Some pictograms, such as 🥑, don't engage a user. However, others like the last item, 🇫🇷, even if it doesn't refer to a facial mimic, is a good example of a pictogram that is used to engage a user through his nationality. The rank values are converted into percentages needed for machine learning algorithms. For instance, the obtained percentages for the item 😏are therefore 66% $\left(\frac{2}{3}\right)$ in terms of positive meaning, and 33% $\left(\frac{1}{3}\right)$ in terms of address meaning. The next section illustrates how those resources are used for sexism annotation with online data.

4 Illustration on Sexism Annotation

Objective of experiments. The goal of this experimental phase is to evaluate how useful the annotation on user engagement is in order to improve the characterization of online hateful data. The application context is sexism classification in online data in French. More specifically, the paper addresses one of the limitations identified by [8], which gathers under the label sexist two types of tweets: a tweet where a user's sexist experience is reported, a tweet where a user writes sexist content directly

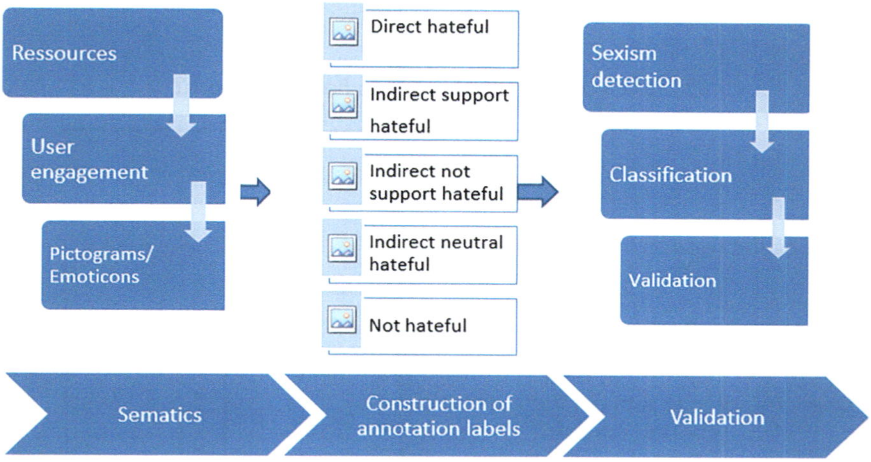

Fig. 2 General architecture for experiments

addressed to another user(s). In the first case, the user is the victim, while in the second, the user is the aggressor. To overcome this limitation, we propose a method able to distinguish the sexist content and the user's engagement with respect to the truthfulness of this sexist content.

Data sets. Annotation was carried out by using a set of data created by the previous cited study dedicated to sexism detection in online data [8]. The initial data set comprises around 12 000 tweets gleaned online and manually annotated with two labels indicating whether a tweet is sexist or non-sexist. 2.

General methodology The experimental protocol involves a chain of methodological decisions illustrated in Fig. 2. This chain is structured by three main steps: a semantic analysis and representation of the notion of speaker/user engagement in the form of a taxonomy, the determination of a set of useful labels for annotations from this linguistic taxonomy, and a validation phase of the entire protocol. The first step has been detailed in the previous section, while the last two ones are presented in this section. We begin by presenting our annotation process of a corpus of French tweets, then we detail first experiments conducted in order to valid our methodology.

4.1 Annotation of Tweets

Annotation of tweets The purpose of the annotation is to refine the description of the content in a tweet by (1) characterizing it as hateful or not, (2) identifying whether the user states it directly or reports it, (3) and finally whether the user supports the

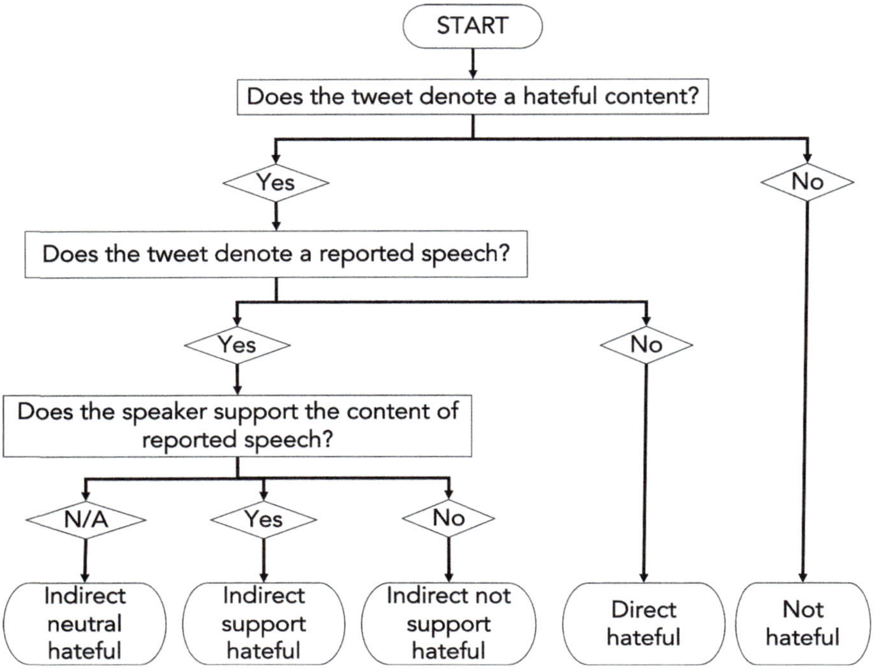

Fig. 3 Annotation procedure: ways for labels identification

content or not. In the end, each tweet is described with those three features and gets one of the five given labels shown with the leaves Fig. 3.

Set of labels As said, five labels are considered:

- **Not hateful**, noted L_1: the tweet does not contain hateful content: example: *Les bleues font un très, très, très bon match ! 🔥#FRANOR @X* (The blue ones make a very, very, very good game! 🔥#FRANOR @X);
- **Direct hateful**, noted L_2: the tweet contains hateful content directly said by the user:
  example: Vous vous dites femmes vous savez même pas faites des pâtes 🍝 bande de connasse
  (You say you're women you don't even know how to make pasta 🍝 you bitch);
- **Indirect not support hateful**, noted L_3: the tweet contains hateful content reported by a user who does not support it:
  example: "Hé mademoiselle, est-ce que je peux me permettre de vous dire que t'es bonne ?" Agen–place Jasmin #payetashnek
  ("Hey miss, can I tell you that you're hot?" Agen–Place Jasmin #payetashnek);
- **Indirect support hateful**, noted L_4: the user reports hateful content in the tweet and supports it:

example: Ouais. T'as raison. Le viol après 15 ans ça n'existe pas. Et si on est violées, on l'a cherché.

(Yeah. You're right. There's no such thing as rape after 15. And if we are raped, we asked for it.);

- **Indirect neutral hateful**, noted L_5: the user remains neutral with respect to reported content:
  example: "C'est une grosse tarlouze": une association saisit le CSA après le dérapage de JoeyStarr à ONPC http://url.com"
  ("He's a big faggot": an association refers to the CSA after JoeyStarr's slip at ONPC http://url.com).

The annotator does not assign directly these labels to tweets but annotates a set of linguistic descriptors for each tweet, based on which the labels are then obtained by following the procedure for deciding the labels Fig. 3. These descriptors are precisely detailed in the following section.

Description of our annotated tweets data set Annotation of tweets is carried out by considering three types of descriptors indicating: (1) whether the tweet's content is hateful or not, i.e., sexist[2] or not in this use case, (2) pictograms/emoticons with the three semantic criteria exposed Sect. 3.2, and (3) types of enunciations according to the taxonomy Fig. 1. These three main categories of descriptors are shown in Table 2 (categories are underlined) which details each descriptor and its possible values. The entity to be annotated is the entire tweet. By following this annotation procedure, 300 tweets randomly selected from the initial collection of 12 000 tweets were annotated. Inter-annotator agreement has not been calculated since, to our knowledge, no measure has been proposed to measure agreement between two annotations in terms of proportions (like in Table 1).

4.2 Experiments and First Results

In order to validate our methodology for describing hate content (sexist content in our use case), we compared it to a generic approach. The experiments demonstrate the consistency of our annotation protocol. Although this protocol is complex and time-consuming, we show that a very small annotated dataset produces satisfactory results. Therefore, this section presents the first experiments to compare the results of supervised classifiers using user engagement with results provided by a BERT-type model fine-tuned on the same data. We make this comparison to test our hypothesis that few expert features give better results than a fine-tuned generic language model.

The goal of the multiclass classification task is to distinguish the five categories shown with the leaves Fig. 3: *Not hateful, Direct hateful, Indirect not support hateful, Indirect support hateful,* and *Indirect neutral hateful.*

[2] The definition of sexism used it's the one given by the European Council.

Table 2 Example of descriptors and their values to describe a tweet

	Labels	Hateful content	Emoticons/Pictograms						Enonciation types			
			Type	Mimicry				User engagement				
				Positive	Negative	Surprised	Address		Assertions	Reported speech	Declaration	…
Va lues	*L*1 *L*2 *L*3 *L*4 *L*5	N/A True False	Emoticon Pictogram	*N/A* *1* *2* *3* *4*	*N/A* *1* *2* *3* *4*	*N/A* *1* *2* *3* *4*	*N/A* *1* *2* *3* *4*	N/A True False	N/A True False	N/A True False	N/A True False	…

Table 3 Results for all classifiers and the camemBERT model

Classifiers	Precision	Rappel	F-measure
Naïve bayes	0.62	0.63	0.62
Random forest	0.65	0.70	**0.67**
Ridge	0.60	0.63	0.60
SVM	0.49	0.68	0.55
Model	Precision	Rappel	F-measure
CamemBERT	0.35	0.59	**0.45**

Experimental protocol Each tweet is considered as a document for which a label is predicted. All features are boolean: 1 is assigned for their presence, 0 for their absence. To annotate descriptors as present when their annotations are proportions, the proportion must be greater than 50%. To evaluate the relevance of these features, several supervised classifiers (Naive Bayes, SVM, Random Forest, and Ridge) were compared to a pre-trained model (camemBERT-base). To train these different classifiers, we separated the set of 300 manually annotated tweets into 3 subsets: the test set (10%), the development set (10%), and the training set (80%). Traditional evaluation metrics in automatic classification were used: precision, recall, and f-measure. All experiments were implemented in Python with scikit-learn[3] for the classifiers and ktrain[4] for camemBERT.

Analysis of results The results for camemBERT presented in this section are those obtained with the base version[5], a learning rate of 0.02, and 8 epochs. Table 3 shows the results of each classifierGiven the limited size of corpora used for experiments and the difficulty of the task, those results are still acceptable. Moreover, they confirm that taking into account user engagement improves the overall performance: Random Forest has a better f-measure (0.67) than camemBERT (0.45). These results confirm our hypothesis: with a small training dataset, few expert features lead to better results than a fine-tuned generic language model trained on large corpora. In addition, experiments show that our annotation procedure facilitates the corpus inspection and illustrates how features of user engagement and emoticons can help to identify otherwise inaccessible examples of hate speech like those in [8]. Table 4 shows five examples of predictions obtained with our Random Forest model.

The model proposed by [8] would have labeled examples 2 and 3 with the same "sexist" label, whereas our model distinguishes between them according to the speaker's engagement with the truth of the sexist content. At last, because sexism is a sensitive subject, the performance of detection models should be evaluated with the use of qualitative methods, and using human-in-the-loop procedures.

There are several limitations affecting the overall approach described in this study. First, the resources were created from scratch and thus further experiments

[3] Supervised learning models from scikit-learn.

[4] Ktrain's GitHub.

[5] CamemBERT-base version on Hugging Face site.

Table 4 Examples of predictions with the random forest model

ID	fr/en	Example	Predicted label
1	fr	Tom Villa sur Aurore Bergé: "quand y a buffet à volonté, elle goute à tous les plats" 😭#SLT	Not hateful
	en	*Tom Villa on Aurore Bergé:* *"When there's an all-you-can-eat buffet,* *she tastes every dish"* 😭 *#SLT*	
2	fr	Les filles encore pucelle vous etes comme des cartes bleues pour moi	Direct hateful
	en	*The girls still virgin* *you are like blue cards for me*	
3	fr	Heyy mignonne, je te trouve même carrément bonne. J'allais te demander une cigarette, mais je préfère que tu me suces la bite. #payetashnek	Indirect not support hateful
	en	*Heyy cutie,* *I even think you're downright hot* *I was going to ask you for a cigarette,* *but I'd rather you sucked my dick.* *#payetashnek*	
4	fr	Une femme qui prie est plus dangereuse qu'une femme qui se venge	Indirect support hateful
	en	*A woman who prays is more dangerous* *than a woman who takes revenge*	
5	fr	Et les commentateurs PSG Toulouse: j'ai jamais vu ça un tir de femme enceinte	Indirect neutral hateful
	en	*And the PSG Toulouse commentators:* *I've never seen a pregnant woman shoot like that*	

are required for their empirical validation. Secondly, another annotation campaign could be carried out to test our annotation schema and models. Finally, we could refine the types of reported speech.

5 Conclusion and Perspectives

The main contribution of the paper is the development of linguistic resources for French. This work also provides a basis for future research on the usage of pictograms/emoticons and user engagement for social data analysis. More specifically, the pictogram resource augments from a semantic point of view the classical description of pictograms by adding new attributes useful for the automatic interpretation of online hate discourses. Further work could be done to study pictograms/emoticons by looking at their syntactic integration to refine their role in representing user engagement. Another direction for future work is the setup of classification experiments, in order to investigate the impact of user engagement and pictograms when used as features for automatic online hate detection. The annotated dataset is available for research purposes upon request.

References

1. Al-Azani, S., El-Alfy, E.S.M.: Early and late fusion of emojis and text to enhance opinion mining. IEEE Access **9**, 121031–121045 (2021)
2. Al-Halah, Z., Aitken, A., Shi, W., Caballero, J.: Smile, be happy:) emoji embedding for visual sentiment analysis. In: Proceedings of the IEEE/CVF International Conference on Computer Vision Workshops, pp. 0–0 (2019)
3. Bassignana, E., Basile, V., Patti, V., et al.: Hurtlex: a multilingual lexicon of words to hurt. In: CEUR Workshop Proceedings, vol. 2253, pp. 1–6. CEUR-WS (2018)
4. Battistelli, D., Amardeilh, F.: Knowledge claims in scientific literature, uncertainty and semantic annotation: a case study in the biological domain. In: Semantic Authoring, Annotation and Knowledge Markup Workshop (SAAKM 2009) (2009)
5. Battistelli, D., Bruneau, C., Dragos, V.: Building a formal model for hate detection in French corpora. Procedia Comput. Sci. **176**, 2358–2365 (2020)
6. Bergqvist, H., Knuchel, D.: Explorations of engagement: introduction. Open Linguist. **5**(1), 650−665 (2019)
7. Bick, E.: Annotating emoticons and emojis in a german-danish social media corpus for hate speech research. RASK–International J. Lang. Commun. **52**, 1–20 (2020)
8. Chiril, P., Moriceau, V., Benamara, F., Mari, A., Origgi, G., Coulomb-Gully, M.: An annotated corpus for sexism detection in french tweets. In: Proceedings of the 12th Language Resources and Evaluation Conference, pp. 1397–1403 (2020)
9. Chiril, P., Moriceau, V., Benamara, F., Mari, A., Origgi, G., Coulomb-Gully, M.: He said "who's gonna take care of your children when you are at acl?": reported sexist acts are not sexist. In: Proceedings of the 58th Annual Meeting of the Association for Computational Linguistics, pp. 4055–4066 (2020)
10. Desclés, J.P.: Prise en charge, engagement et désengagement. Lang. Fr. **2**, 29–53 (2009)
11. Evans, V.: The Emoji Code: How Smiley Faces, Love Hearts and Thumbs Up are Changing the Way we Communicate. Michael O'Mara Books (2017)
12. Fernández-Gavilanes, M., Costa-Montenegro, E., García-Méndez, S., González-Castaño, F.J., Juncal-Martínez, J.: Evaluation of online emoji description resources for sentiment analysis purposes. Expert Syst. Appl. **184**, 115279 (2021)
13. Godard, R., Holtzman, S.: The multidimensional lexicon of emojis: a new tool to assess the emotional content of emojis. Front. Psychol. **13** (2022)
14. Halté, P.: Les marques modales dans les chats: étude sémiotique et pragmatique des interjections et des émoticônes dans un corpus de conversations synchrones en ligne. Ph.D. thesis, University of Luxembourg, Luxembourg, Luxembourg (2013)
15. Halté, P.: Les émoticônes: de la signification des affects aux stratégies conversationnelles. Communiquer (28) (2020)
16. Hemati, H.H., Alavian, S.H., Beigy, H., Sameti, H.: Sutnlp at semeval-2023 task 10: Rlat-transformer for explainable online sexism detection. In: Proceedings of the The 17th Interna tional Workshop on Semantic Evaluation (SemEval-2023), pp. 347–356 (2023)
17. Kirk, H.R., Vidgen, B., Röttger, P., Thrush, T., Hale, S.A.: Hatemoji: A test suite and adversarially-generated dataset for benchmarking and detecting emoji-based hate. arXiv:2108. 05921 (2021)

18. Kralj Novak, P., Smailovic´, J., Sluban, B., Mozeticˇ, I.: Emoji sentiment ranking 1.0 (2015)
19. Liang, W., Liang, K.H., Yu, Z.: Herald: an annotation efficient method to detect user disengagement in social conversations. arXiv:2106.00162 (2021)
20. Malecki, W., Kowal, M., Dobrowolska, M., Sorokowski, P.: Defining online hating and online haters. Front. Psychol. **12**, 744614 (2021)
21. Robertson, A., Liza, F.F., Nguyen, D., McGillivray, B., Hale, S.A.: Semantic journeys: quantifying change in emoji meaning from 2012–2018. arXiv:2105.00846 (2021)
22. Rong, S., Wang, W., Mannan, U.A., de Almeida, E.S., Zhou, S., Ahmed, I.: An empirical study of emoji use in software development communication. Inf. Softw. Technol.ol. **148**, 106912 (2022)
23. Shoeb, A.A.M., De Melo, G.: Emotag1200: understanding the association between emojis and emotions. In: Proceedings of the 2020 Conference on Empirical Methods in Natural Language Processing (EMNLP), pp. 8957–8967 (2020)

Heuristic Learning Model-Based Stochastic Regularization Technique for Reducing the Overfit of Training Data

P. S. Metkewar and Rajesh Kumar Dhanaraj

Abstract Researchers may have learned about how to build and develop feedforward neural networks. The moment we develop a network, there is the question of training and testing the dataset along with the machine learning algorithm. We have to test our algorithm not only on the training set but also on the testing set with respect to its fit. When an algorithm performs well on the training set but performs poorly on the testing set, the algorithm is said to be overfitted on the training data. In short, there is a need to focus on reducing the overfitting. To deal with this problem, we have to make our stochastic model generalize over the training data using the emphasized regularization technique. Here, the term stochastic means the outcome derived from a random event or random process; the terms referred to as "Stochastic" and Regularization mean minimizing the error of the testing set directly proportional to increasing the error rate of the training set. Many such regularization techniques are invented by the research practitioner, but minimal evidence has been observed through types of learning rules. So to come up with a more effective regularization technique is the need of the hour. In order to achieve this, there needs to be put some extra constraints, parameter-specific values, changes in learning rate, and changes in activation function; into the machine learning model. If it is computed and chosen correctly, then it helps to reduce testing error. Finally in order to develop such a machine learning model; researchers have to look into the strengths and weaknesses of the model, new or derived theory, synthesis of existing theory along with the support of other derived theories and the gaps to come up with new models.

Keywords Stochastic regularization · Training dataset · Testing dataset · Overfitting · Learning models

P. S. Metkewar (✉) · R. K. Dhanaraj
Symbiosis Institute of Computer Studies and Research (SICSR), Symbiosis International (Deemed University), Model Colony, Pune, Maharashtra, India
e-mail: pravin.metkewar@gmail.com; pravin.metkewar@mitwpu.edu.in

P. S. Metkewar
School of Computer Science and Engineering, Dr Vishwanath Karad MIT World Peace University, Kothrud, Pune, India

G. Fortino et al. (eds.), *Proceedings of Third International Conference on Computing and Communication Networks*, Lecture Notes in Networks and Systems 917,
https://doi.org/10.1007/978-981-97-0892-5_26

1 Introduction

Many studies have focused on the relationship between stochastic regularization for overfitting the training data and learning rules including A Survey on Transfer Learning, Regularized Iterative Stochastic Approximation Methods, Investigation on the Effect of L1 and L2 Regularization, Stochastic Bridge Regularization for Deep Neural Networks, Overparameterized Nonlinear Models, Towards Stochasticity of Regularization in Deep Neural Networks, Concept Learning Using Complexity Regularization, Similarity-based Regularization for Semi-supervised Learning, Doubly Stochastic Poisson Processes in Artificial Neural Learning, Stochastic Process Using a Colored Generalized Stochastic Petri Net, Minimal Stochastic Complexity With Unknown Noise Model and Decentralized Semi-supervised Learning over Multitask Graphs. A couple of studies are also emphasized on data, images, graphs, networks and specific techniques and it is based on supervised, unsupervised and semi-supervised learning methods.

2 Pivotal Study

The relationship between stochastic regularization technique depictions and data, image, and graph concerns is well-established. Analysis of [1, 17] concludes that while using the regularization technique, the authors need to focus on domain adaptation, sample selection and multitask learning.

While dealing with the Cartesian inequality problem, [2, 13] have suggested choosing sequences, regularization and centering parameters independently. This is basically useful when you are dealing with network problems. This network problem is handled in two ways: stochastic iterative regularization method and stochastic iterative proximal-point method; [16] deals with stochastic Bernoulli distribution.

Regularization plays a very critical role while dealing with images. In [3, 4] studies have observed that Ll produces features which are spatially localized whereas L2 regularization produces features with higher spatial variance. This is a very crucial decision and helps us to decide which dataset needs which regularization technique.

The authors of [5, 7, 21] have worked on the convergence rate of nonlinear models; in this case, training loss occurs due to "over-parameterization" and finally the different interpolating solutions need to be found out. In this case, the authors conclude that global minima are nearly closed to the divergence.

In the papers [6, 22], the authors propose a methodological framework for dealing with the stochasticity in regularized deep neural network. They have used the basics of dropout as an ensemble method for regularization and introduced a new method for dropout regularization. Significant result improvement is in place but still it has to be tested on multiple platforms in order to maintain its significance. Experiments in [8, 9, 15] were designed to improve the generalization capacity of the regularization method. Here they faced issues related with long computation time and poor parameter control. Here, the authors proposed a new method that fine-tunes the regularization parameter by using a Fuzzy Inference system. Also, it is tested on several nonlinear function approximation and pattern classification examples. Finally, it helps to achieve the highest precision, rapid convergence, and the best generalization ability.

[10, 19] have observed that recurrent gradient descent network (RGDN) by systematically incorporating deep neural networks into a fully parameterized gradient descent scheme in order to achieve strong interpretability and high generalization on the given dataset and helps to remove image blur or from optical image.

In [11], the authors dealt with pattern recognition and applied the method of complexity regularization to learn concepts from large concept classes. A major focus is on a good balance between the approximation error and the estimation error, but the error probability is O(log n/n).

The paper [12] presents a digit recognition field using a semi-supervised learning method. They have concentrated on error rate, similarity index, and a new correlation score to validate the best retraining rule. Finally, they concluded that one retraining rule is winning over the other with respect to the best correlation score. Nonnegative Matrix Factorization (NMF) [14] is a widely used technique in many applications such as face recognition and motion segmentation and it helps to estimate the positions and values of outliers/noise. It is advantageous to other compared methods.

Internet of Things (IoT) technologies [18] have been used for real-time applications in manufacturing processes. They are focused on timed constraints, probabilistic tasks, and dependencies between activities. By using a novel spatiotemporal learning model [20], this experiment helps to reduce the truncation error and improve the weight adaptivity.

3 Critical Evaluation

Evaluation of stochastic regularization techniques and their corresponding strengths and weaknesses are discussed here one by one.

Table 1 gives a detailed study of each technique which is being used for regularization, but a few techniques are really older as compared to the newer techniques. In fact, techniques with statistical computation need to be avoided because it helps to compute only quantitative parts and would have less approximation. In order to improve the computation and result approximation have to deal with machine learning algorithms in order to find out solution for partially solvable and unsolvable problems.

Table 1 Strengths and weaknesses of regularization technique

Techniques	Strength	weakness
Transfer learning	Saving of resources and improved efficiency when training new models Training models with unlabeled datasets	Problem of negative transfer This only occurs if the source and target are not similar enough
Regularized iterative stochastic approximation methods	To find extrema of functions that are not known in case of noisy observations	In this algorithm, accuracy is not good and hence not recommended in other models
Effect of L1 and L2 regularization	L2 penalizes sum of square values of weights L2 regularization is able to learn complex data patterns L1 is robust to outliers	L1 penalizes sum of absolute values of weights L1 generates model that is simple and interpretable L2 is not robust to outliers
Overparameterized nonlinear models	Capacity to fit any set of labels including pure noise	Parameters of the model exceed size of the training dataset
Minimal stochastic complexity with unknown noise model	Helpful for model selection and data clustering Models can be compared regardless of numbers and parameters	It works on fixed set of models
Regularization in deep neural networks	Can prevent overfitting in Neural Network Improve the accuracy while dealing with new data from the problem domain	Overfitting by reducing the generalization error

4 Formal Theory

Presenting commonplace specialized wordings which will provide new guidance for the significant review. Quite possibly of the most widely recognized issues that the creator experienced while preparing profound neural networks is overfitting. Overfitting happens when a model attempts to foresee a pattern in information that is excessively boisterous. This is caused because of an excessively perplexing model with such a large number of boundaries. A model that is overfitted is off base in light of the fact that the pattern doesn't mirror the truth present in the information. This can be judged in the event that the model delivers great outcomes on the seen information (training set) but performs inadequately on the concealed information (test set). The objective of an ML model is to sum up well from the preparation information to any information from the issue space. This is vital as we believe our model should make expectations, and later on information that it has never seen.

The creator will introduce five commonplace strategies to forestall overfitting while at the same time preparing neural organizations.

4.1 Simplifying the Model

The initial step while managing overfitting is to diminish the intricacy of the model. To diminish the intricacy, we can essentially eliminate layers or decrease the quantity of neurons to make the organization more modest. While doing this, it is critical to work out the information and result aspects of the different layers associated with the neural network organization. There is no common guideline on the amount to eliminate or how huge your organization ought to be. Yet, assuming that your neural network is overfitting, take a stab at making it more modest.

4.2 Early Stopping

Early stopping is a type of regularization while preparing a model with an iterative technique, like inclination drop. Since every one of the neural networks advances solely by utilizing slope plummet, early halting is a method pertinent to every one of the issues. This strategy updates the model in order to improve it to fit the preparation information with every emphasis. To a certain degree, this works on the model's presentation of information on the test set. Beyond that point notwithstanding, further developing the model's fit to the preparation information prompts expanded speculation blunder. Early halting standards give direction concerning the number of emphases that can be run before the model starts to overfit.

Early stopping strategy is displayed in Fig. 1. As may be obvious, after certain cycles, a test blunder has begun to increment while the preparation mistake is as yet diminishing. Subsequently the model is overfitting. So to battle this, we stop the model right when this begins to occur.

4.3 Use Data Augmentation

On account of neural network organizations, information expansion essentially implies expanding the size of the information that is expanding the quantity of pictures present in the dataset. A portion of the well-known picture expansion methods are flipping, interpretation, pivot, scaling, evolving splendor, adding commotion, etc.

The Data Augmentation strategy is displayed in Fig. 2. As may be obvious, utilizing information expansion a ton of comparable pictures can be produced. This aids in expanding the dataset size and subsequently decreases overfitting. That's what the explanation is; as we add more information, the model can't overfit all the examples, and is compelled to sum up.

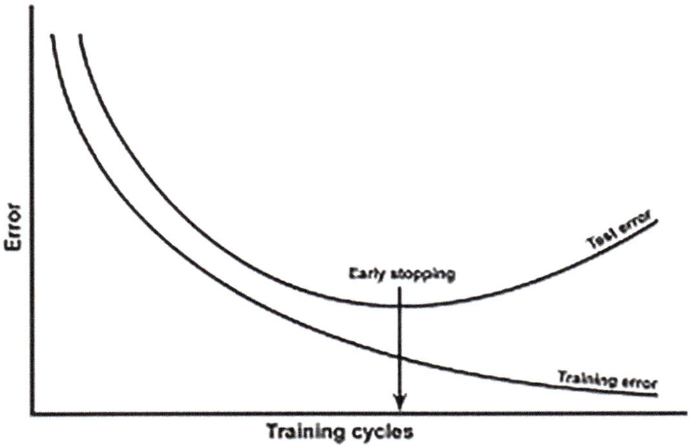

Fig. 1 Training cycles

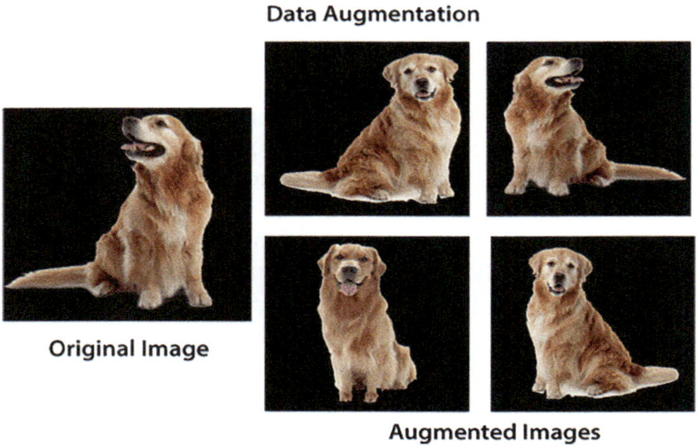

Fig. 2 Original and augmented images

4.4 Use Regularization

Regularization is a strategy to decrease the intricacy of the model. It does as such by adding a punishment term to the misfortune capability. The most widely recognized methods are known as L1 and L2 regularization:

The L1 punishment means to limit the outright worth of the loads. This is numerically displayed in the below recipe:

L1 Regularization:

$$L(x,y) \equiv \sum_{i=1}^{n}(y_i - h_\theta(x_i))^2 + \lambda\sum_{i=1}^{n}|\theta_i| \tag{1}$$

The L2 punishment intends to limit the squared greatness of the loads. This is numerically displayed in the beneath equation.

L2 Regularization:

$$L(x,y) \equiv \sum_{i=1}^{n}(y_i - h_\theta(x_i))^2 + \lambda\sum_{i=1}^{n}\theta_i^2 \tag{2}$$

4.5 Use Dropouts

Dropout is a regularization method that keeps neural networks from overfitting. Regularization strategies like L1 and L2 diminish overfitting by changing the expense capability. Dropout then again alters the actual organization. It haphazardly drops neurons from the neural network during the preparation of every emphasis. At the point when we drop various arrangements of neurons, it's identical to preparing different neural network organizations. The various organizations will overfit in various ways, so the net impact of dropout will be to decrease overfitting.

The Dropout strategy is displayed in Fig. 3. As may be obvious, dropouts are utilized to eliminate neurons while preparing the neural network arbitrarily. This procedure has been demonstrated to diminish overfitting to different issues including picture classification, picture segmentation, word installation, semantic matching, and so on.

5 Synthesis

Synthesis is the process of identifying the intellectual and knowledge transformation among the research communities; in short, it combines findings from research that support the theory introduced by the authors.

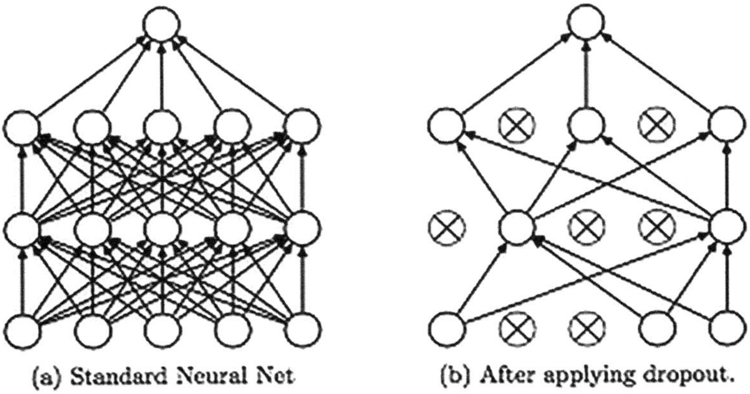

Fig. 3 Using dropouts

The major objective behind this synthesis is that research practitioners should focus on how to reduce the overfit by looking into the agricultural or finance or healthcare domain, and how to focus on data selection, data size, and learning from each and every iteration and epoch. Basically, it is based on multiple parameters, and how to sync these parameters together is a challenge in order to keep your model underfit.

Even authors need to focus on the number of input neurons, hidden layers, output neurons, bias, and its impact on the learning model by comparing it with different learning rules including the Hebbian learning rule, Perceptron learning rule, Delta learning rule, etc. Another important aspect is to reduce the complexity of the hidden layer by analyzing the system instead of directly moving ahead for its implementation. This could also help us reduce the problem of overfitting. Using the dropout technique, you may achieve this to minimize the complexity of the hidden layers.

Another dimension for analyzing this study in terms of data (spatial, temporal or qualitative, quantitative, linguistics), images or pictures, moving objects, complicated graphs, variety of networks. Further to smoothen the effect; pre-processing, analysis, transformation, computation, etc. need to be performed on the data before sending it to the model for implementation by which also we can reduce the problem of overfitting through Stochastic Regularization techniques.

Few more studies focused in terms of linear and nonlinear models to achieve the performance of regularity by looking into convergence and divergence.

Additional parameters like targeted value, initial value, outlier, noise, weights, and types of errors and calculate the error and further to minimize it to be achieved which could be helpful to reduce the overfit and achieve the highest precision.

Researchers should focus on the problem statement and according to the need of the problem, the user would have fit the problem into supervised, unsupervised, or semi-supervised learning models or in deep learning models which help to reduce the complexity of the model and improve the overall performance of the system in order to get good accuracy.

6 Research Gaps

Tragically, in the above writing writers have noticed research gaps which are useful to consider future guides.

Negative Transfer happens when the method involved with tackling a prior issue makes later issues harder to settle. It diverges from the positive exchange, which happens while tackling a previous issue making it more straightforward to take care of a later issue. The impedance hypothesis proposes pessimistic exchange that happens on the grounds that individuals are bound to play out the second educated task mistakenly because of the obstruction by the principal learned task.

Target is the last result a ML model is attempting to foresee. It very well may be unmitigated or constant. The mark addresses the genuine result of the objective. As referenced before, the marks are relegated to the preparation dataset yet when the ML model is prepared it is taken care of with unlabeled information. Elements or factors or qualities are the deliberate contributions of the issue space, the free factors. The objective variable is the reliant variable or the action we're attempting to show or figure. Not all issues can be or should be figured out in such a manner.

Precision can likewise be characterized as the proportion of the quantity of accurately arranged cases to the absolute of cases under assessment. The best worth of precision is 1 and the most terrible worth is 0. What is model precision? Model precision is characterized as the quantity of orders a model accurately predicts separated by the complete number of forecasts made. It's an approach to evaluating the presentation of a model, yet surely not by any means the only way. Attempting to get the exactness of a model in light of a single boundary is likewise a terrible practice. Specialists would have to ponder F1 score, accuracy, review, precision, recall and so on to be processed in light of the need of the model.

What happens when you increment the size information of the training set? Additional preparation information further develops ML model execution in a limited way yet can think twice about the past. The nature of the information used to prepare the ML model is similarly basically as significant as the amount. Unfortunate information quality prompts unfortunate ML model outcomes. A model boundary is a setup variable that is inside the model and whose worth can be assessed from the given information. They are expected by the model while making expectations. Their qualities characterize the ability of the model on your concern. When does a ML model has a greater number of boundaries than preparing tests? Having such an enormous model can prompt overfitting.

Overfitting happens when the model can't sum up and fits too near the training dataset all things considered. Overfitting occurs because of a few reasons: for example, the training information size is excessively little and doesn't contain an adequate number of information tests to precisely address all conceivable information values. Overfitting is the point at which the testing blunder is high contrasted with the preparation mistake, or the gap between the two is huge. While it's trying to comprehend the functions of large, complex ML and DL models. Overfitting can occur because of low inclination and high fluctuation. On the off chance that our model improves on the training set than on the test set, then we're probably overfitting. For instance, it would be a major warning on the off chance that our model saw close to 100% exactness on the training set however just 55% precision on the test set.

7 Conclusion

Here, the authors have explained what overfitting is and why it is a common problem in neural networks. Even it is partially focused on reducing overfitting in five different ways while dealing with training neural networks, i.e., simplifying the model, early stopping, data augmentation, regularization and dropouts. Apart from this, researchers would have focused on epochs and iterations, hidden layer complexity, data types including graphs, networks and images, and learning types which are clearly briefed in the "synthesis" section in order to develop the underfit model.

8 Future Scope

In future, researchers should focus on the implementation model of the proposed mechanism in order to validate with a suitable dataset.

References

1. Pan, S.J., Yang, Q.: A survey on transfer learning. IEEE Trans. Knowl. Data Eng. **22**(10), 1345–1359 (2010). https://doi.org/10.1109/TKDE.2009.191
2. Koshal, J., Nedic, A., Shanbhag, U.V.: Regularized iterative stochastic approximation methods for stochastic variational inequality problems. IEEE Trans. Autom. Control **58**(3), 594–609 (2013). https://doi.org/10.1109/TAC.2012.2215413
3. Jaiswal, S., Mehta, A., Nandi, G.C.: Investigation on the effect of L1 an L2 regularization on image features extracted using restricted boltzmann machine. In: 2018 Second International Conference on Intelligent Computing and Control Systems (ICICCS), Madurai, India, pp. 1548–1553 (2018). https://doi.org/10.1109/ICCONS.2018.8663071
4. Khan, N., Shah, J., Stavness, I.: Bridgeout: stochastic bridge regularization for deep neural networks. IEEE Access **6**, 42961–42970 (2018). https://doi.org/10.1109/ACCESS.2018.2863606
5. Azizan, N., Lale, S., Hassibi, B.: A study of generalization of stochastic mirror descent algorithms on overparameterized nonlinear models. In: ICASSP 2020—2020 IEEE International Conference on Acoustics, Speech and Signal Processing (ICASSP), Barcelona, Spain, pp. 3132–3136 (2020). https://doi.org/10.1109/ICASSP40776.2020.9053864
6. Sandjakoska, L., Bogdanova, A.M.: Towards stochasticity of regularization in deep neural networks. In: 2018 14th Symposium on Neural Networks and Applications (NEUREL), Belgrade, Serbia, pp. 1–4 (2018). https://doi.org/10.1109/NEUREL.2018.8587027

7. Zanderigo, F., Bertoldo, A., Pillonetto, G., Cobelli, C.: Nonlinear stochastic regularization to characterize tissue residue function in bolus-tracking MRI: assessment and comparison with SVD, block-circulant SVD, and Tikhonov. IEEE Trans. Biomed. Eng. **56**(5), 1287–1297 (2009). https://doi.org/10.1109/TBME.2009.2013820
8. Wu, Y., Zhang, L.: A new regularization learning method for improving generalization capability of neural network. In: Proceedings of the 4th World Congress on Intelligent Control and Automation (Cat. No. 02EX527), Shanghai, China, vol. 3, pp. 2011–2015 (2002). https://doi.org/10.1109/WCICA.2002.1021438
9. Wu, Y., Zhang, L.: A novel neural network learning method for dynamically tuning regularization coefficient. In: 2001 International Conferences on Info-Tech and Info-Net. Proceedings (Cat. No. 01EX479), Beijing, China, vol. 3, pp. 516–522 (2001). https://doi.org/10.1109/ICII.2001.983109
10. Gong, D., Zhang, Z., Shi, Q., van den Hengel, A., Shen, C., Zhang, Y.: Learning deep gradient descent optimization for image deconvolution. IEEE Trans. Neural Netw. Learn. Syst. **31**(12), 5468–5482 (2020). https://doi.org/10.1109/TNNLS.2020.2968289
11. Lugosi, G., Zeger, K.: Concept learning using complexity regularization. IEEE Trans. Inf. Theory **42**(1), 48–54 (1996). https://doi.org/10.1109/18.481777
12. Barbuzzi, D., Pirlo, G., Uchida, S., Frinken, V., Impedovo, D.: Similarity-based regularization for semi-supervised learning for handwritten digit recognition. In: 2015 13th International Conference on Document Analysis and Recognition (ICDAR), Tunis, Tunisia, pp. 101–105 (2015). https://doi.org/10.1109/ICDAR.2015.7333734
13. Kato, H., Hanada, H., Takeuchi, I.: Safe RuleFit: learning optimal sparse rule model by meta safe screening. IEEE Trans. Pattern Anal. Mach. Intell. **45**(2), 2330–2343 (2023). https://doi.org/10.1109/TPAMI.2022.3167993
14. Shen, B., Liu, B.-D., Wang, Q., Ji, R.: Robust nonnegative matrix factorization via L1 norm regularization by multiplicative updating rules. In: 2014 IEEE International Conference on Image Processing (ICIP), Paris, France, pp. 5282–5286 (2014). https://doi.org/10.1109/ICIP.2014.7026069
15. Zhang, C., Oh, S.-K., Fu, Z., Pedrycz, W.: Design of reinforced hybrid fuzzy rule-based neural networks driven to inhomogeneous neurons and tournament selection. IEEE Trans. Fuzzy Syst. **29**(11), 3293–3307 (2021). https://doi.org/10.1109/TFUZZ.2020.3018190
16. McNeill, D.K., Card, H.C.: Refractory pulse counting processes in stochastic neural computers. IEEE Trans. Neural Netw. **16**(2), 505–508 (2005). https://doi.org/10.1109/TNN.2005.844089
17. Card, H.C.: Doubly stochastic poisson processes in artificial neural learning. IEEE Trans. Neural Netw. **9**(1), 229–231 (1998). https://doi.org/10.1109/72.655046
18. Dechsupa, C., Vatanawood, W., Thongtak, A.: Stepwise verification for the BPMN with timed and stochastic process using a colored generalized stochastic petri net. IEEE Access **10**, 42983–43002 (2022). https://doi.org/10.1109/ACCESS.2022.3168000
19. Delyon, G., Galland, F., Refregier, P.: Minimal stochastic complexity image partitioning with unknown noise model. IEEE Trans. Image Process. **15**(10), 3207–3212 (2006). https://doi.org/10.1109/TIP.2006.877484
20. Zhang, H., Ye, X., Hu, Q.: Spatiotemporal learning via mixture importance Gaussian filtering with sparse regularization. IEEE Signal Process. Lett. **30**, 279–283 (2023). https://doi.org/10.1109/LSP.2023.3258861
21. Issa, M., Nassif, R., Rizk, E., Sayed, A.H.: Decentralized semi-supervised learning over multitask graphs. In: 2022 56th Asilomar Conference on Signals, Systems, and Computers, Pacific Grove, CA, USA, pp. 419–425 (2022). https://doi.org/10.1109/IEEECONF56349.2022.10052032
22. Amjad, R.A., Geiger, B.C.: Learning representations for neural network-based classification using the information bottleneck principle. IEEE Trans. Pattern Anal. Mach. Intell. **42**(9), 2225–2239 (2020). https://doi.org/10.1109/TPAMI.2019.2909031

Convolution Neural Network Versus Transfer Learning in Image Classification

O. Rama Devi, U. Surya Venkata Sekhar, S. Siva Rama Krishna, and T. S. Rajarajeswari

Abstract The objective of this research paper is comprehensive comparative analysis between two prominent approaches, namely Convolutional Neural Networks (CNN) and MobileNetV2-based transfer learning, for the task of image classification. Specifically, the focus is on determining the effectiveness of these approaches in accurately classifying images (in our case it is cat vs. dog) (Szyc in 2018 IEEE 22nd International Conference on Intelligent Engineering Systems (INES). IEEE, 2018 [Szyc, K.: Comparison of different deep-learning methods for image classification. In: 2018 IEEE 22nd International Conference on Intelligent Engineering Systems (INES). IEEE (2018)]). Through meticulous evaluation and comparison of results obtained from a benchmark dataset, this study aims to discern the strengths and limitations of each method. By shedding light on their respective merits, this research contributes to the advancement of image classification techniques and paves the way for further investigations in this domain.

Keywords Image classification · Deep neural networks · Convolution neural networks · Deep learning · Transfer learning · MobileNetV2

1 Introduction

In the past few years, the advancements in deep learning methods have brought tremendous changes in its usage. Deep Neural Networks have a lot of applications in various fields like object detection, self-driving cars, image classification, recommendation systems, health care, agriculture, etc. [2]. For many years, image classification has played one of the vital parts in computer research [1]. In recent years, Deep Neural Networks have been widely used to improve the results in image classification [2].

O. R. Devi · U. S. V. Sekhar · S. S. R. Krishna · T. S. Rajarajeswari (✉)
Department of Artificial Intelligence and Data Science, Lakireddy Bali Reddy College of Engineering (Autonomous), Mylavaram, AP, India
e-mail: thrajeswari@gmail.com

G. Fortino et al. (eds.), *Proceedings of Third International Conference on Computing and Communication Networks*, Lecture Notes in Networks and Systems 917,
https://doi.org/10.1007/978-981-97-0892-5_27

The goal of the image classification is to classify the input image to the expected class label.

The reason for going with deep learning is, for image classification problems, Machine Learning algorithms do not produce accurate results [1, 3]. Algorithms like Support Vector Machine, K-Nearest Neighbours, Decision Trees, etc. are working with previously prepared features. The classification of the Image (input) is also based on that prepared feature only. Defining and putting the features into algorithms does not give better results, because the features may vary based on images and it is difficult for the analyst to grab all the features for the classification task.

So, deep learning methods came into the scenario to overcome the faults in machine learning. The deep learning methods create features during the phase of learning by themselves. They always try to define and redefine the features during the training or learning phase. Once the model is trained, the network is ready for classifying the images according to expected labels. In this way, the deep learning models have given the best results for the past few years in the case of image classification.

In general, deep learning is a method which helps in defining the Neural Networks. In this paper, for image classification we are defining two methods, namely Convolution Neural Networks (CNN) and MobileNetV2 which is one of the transfer learning. In the first method, we have to build the entire neural network from scratch which includes layers, activation functions, optimizers, loss functions, etc. But when it comes to the transfer learning technique, we are using MobileNetV2 which is already a pre-trained model on large datasets which may include billions of images [1, 6]. By changing the last layer, we are going to fit that model to our classification problem (classifying whether it is a cat or dog).

We allow the models to learn from the dataset (which includes cats and dogs). Once the learning phase is completed, with the help of test samples we evaluate the models. Finally, we are going to compare how CNN and transfer learning work with image classification. This kind of comparison will give more detailed information regarding their usage in terms of advantages and disadvantages and helps in understanding the techniques with real-time experience.

2 Dataset Preparation

Dataset

All the images are coloured and the total dataset holds 2000 images.

We are going to solve the problem of image classification by classifying images into two classes, i.e., either cat or dog. In this section, we are going to explain how the dataset is prepared. Every deep learning model mainly works on the data that we gathered. The classification of images into their respective classes is done based on the features that are defined or observed by the network during the learning phase. The learning phase needs to be stronger in order to provide accurate results.

Firstly, we gathered a dogs' and cats' image dataset from Kaggle [7]. According to our implementation, we modified the dataset. Now, the modified dataset only holds a directory called image data. The image data directory contains two classes, namely cats and dogs. Each class holds around 2000 samples, and all these samples are in JPG format. Further, we will divide the entire data into train, validation and test samples.

In summary, there are 2000 images with respect to each class label (i.e., cats and dogs) in Image Data. All the images are well organized in terms of file extension, size, etc. The deep learning algorithms are now allowed to learn from the image data. The image data provides the real-time experience for solving image classification problems using deep learning.

3 Implementation

The image classification problem can be solved by building a deep neural network from scratch, or we can transfer learning techniques. As per our paper, we are building a Convolution Neural Network from scratch and using MobileNetV2 as a transfer learning technique. Further both methods are discussed in detail.

3.1 Convolution Neural Networks

The convolution neural networks (CNN) work in an effective manner from image classification [4]. The architecture of CNN is inspired by biological science. In basic, the CNN consists of various layers involved in it. They are classified as input layer, hidden layer and output layer. The input layer accepts input images in the form of three-dimensions known as height, width and depth. The height and width will define the number of pixels and depth will correspond to colour channels. Generally, we have RGB as the colour channel, and we use 0–255 as a range to represent. This three-dimensional input is collectively known as a tensor.

Next, we will have a hidden layer which is also known as the feature-extraction layer. In the context of CNN, the convolution layer and pooling layers are collectively called feature-extraction layer. Firstly, convolution layers are the core blocks of CNN. In the convolution layer, a filtering process will take place and those filtered features are forwarded to the next layers. We know that all neurons are locally connected, so by using patches of neurons from the previous layer the convolution layers will transform the input data. Dot product will typically occur between the input layer's area of neurons and the weights to which those neurons are locally linked in the output layer. At the end, we are going to use the activation function. The activation function will define whether a neuron is activated or not. Once the neuron is activated then it is allowed to pass the information to neurons in the next layer. Generally, activation functions are used to add some non-linearity—usually we use ReLU as our activation function [4]. ReLU mathematically applies the activation function as max(0,x) function, to produce element-wise non-linearity [4]. The convolution layer has two parameters, namely biases and weights. The biases and weights are automatically affected due to kernel size. Kernel size may vary according to dimensions.

Convolution layers allow to detect features based on the weights inside the kernel [1]. During the learning phase, the weights will continuously change to reduce the loss function. Different convolution layers will have different functionalities where the first layer may be allowed to detect shapes, lines, etc. [1]. The next layer may be allowed to detect more complex features in the given input image [1]. Next, we have pooling layers and these pooling layers are placed between convolution layers. The functionality of the pooling layer is to reduce the height and width (spatial size) of the data (input matrix from previous layers). Generally, the pooling layer uses the MAX () operation to resize the height and width; this is known as max pooling. The max pooling is done mostly with a 2×2 filter size. If we speak mathematically, max pooling will take a maximum of four members in the filter area (since we are using a 2×2 matrix). During this process, the depth of the image is not affected. The pooling layer's motive is to speed up the learning process by neglecting unwanted information and controlling overfitting.

Finally, we will have fully connected layers. In other terms, fully connected layers are also known as dense layers. All neurons in the fully connected layer have connections with previous layer neurons [1]. The fully connected layer in CNN will symbolize the feature vector for the input [ref]. An activation function like SoftMax is used in this layer to make this vector used for the classification of image labels. The output layer is typically a fully connected layer, but the number of neurons in the output layer depends on the number of classes in the classification problem. In summary, while a fully connected layer can be an output layer, not all fully connected layers are output layers.

We do not have any fixed format for creating network models like what parameters will lead to getting accurate results. However most of the studies will try to change layers and parameters until they reach accurate results [1]. So, the whole process will depend on the layers, parameters, activation function (like ReLU, etc.), optimizers (like gradient descent, backpropagation, SGD, etc.), loss functions or cost functions (cross-entropy, hinge loss (for binary classification), etc.), evaluation and changing parameters. Repeating this process again and again may help to get accurate results [1, 5].

3.2 Transfer Learning

Another technique that we are discussing in our paper for the classification of images (i.e., either cat or dog) is Transfer Learning. Transfer Learning is one of the deep learning techniques in which we are going to use a pre-trained model rather than a model which is built and trained from scratch. The pre-trained model is already trained upon huge data (nearly billions of data) for specific tasks like image classification, natural language processing, etc., and extracted various kinds of useful features during the learning phase. Now, these features can be adapted to new tasks with less training data and computation. The pre-trained model can be fine-tuned for a new task by retraining the last few layers of the model on the new data while keeping the weights of the earlier layers fixed. In this way, the Transfer Learning technique can to provide sufficient accuracy in solving particular tasks.

In our case, we are going to use MobileNetV2 as a transfer learning technique for image classification. MobileNetV2 is a deep learning model; it uses a combination of depth-wise separable convolutions (DSC) and linear bottlenecks. MobileNetV2 applies the technique of DSC for portability, and it is used to solve information destroying problems in non-linear convolution layers by using linear bottlenecks [6].

The depth-wise separable convolution method divides the standard convolution into two distinct operations: a depth-wise convolution, which applies a single filter to each input channel, and a pointwise convolution, which employs a 1×1 filter to combine the depth-wise convolution's output. Separating the spatial and channel-wise information lowers the number of factors and calculations.

The linear bottleneck is another key component of MobileNetV2 that allows for higher efficiency. A bottleneck layer reduces the number of channels in the input, applies a standard convolution, and then expands the number of channels back to the original size using a 1×1 convolution. The linear bottleneck simply replaces the non-linear activation function in the bottleneck with a linear one, which reduces the computational cost of the model without significantly affecting accuracy.

We already know that transfer learning models are trained upon vast data which means all transfer learning models are prepared on a long learning process [1, 6]. So, the weights inside the network will have correct values, and we can use this model to predict labels [1]. The labels that the model provides are based on the learning phase [1, 9]. Now our task is to change the layers in the model, in such a way that it provides labels according to our problem statement. In our case, we want to categorize the image label as cat or dog.

As we know, in our case we are dealing with only two class labels, i.e., either cat or dog. Usually replacing the last one or two layers is enough in order to get the required outputs [ref]. Now we want to train the replaced layers with our training samples. Remember, the weights are changing in the last layers that we are modifying. This implies that the feature information and the weights in the previous layers were not disturbed.

In summary, the motive of the transfer learning is to use the pre-trained model (that is already holding feature information by learning from vast data (nearly billions)) to predict the labels for our problem statement. This leads to an increase in the accuracy of the model, even if we run it for a small number of epochs. Instead of changing the last one or two layers, we are allowed to change more layers and modify any part of the model (pre-trained). But, if we don't have experience it can easily decrease the performance [1].

3.3 Advantages of Transfer Learning

Transfer learning offers several advantages over traditional Convolutional Neural Network (CNN) approaches. Firstly, it typically yields better results and higher accuracy when compared to training a CNN from scratch [4, 9]. This is because transfer learning leverages pre-trained models that have already learned relevant features from vast amounts of data, making it more efficient in short epochs.

4 Experimentation

4.1 Convolution Neural Networks

As we discussed above, our first model was CNN which was built from scratch. Firstly, we allow the neural network to train from the dataset. For this process, we take the cat or dog dataset. This dataset was named "Image Data" which contains two labelled folders known as Cat and Dog. Each folder will contain around 2000 images in it. After that, we standardize the images between [0, 1] by dividing each image by 255. With the help of standardization, we make the neural networks learn without any burden by analysing the pixel scale between [0, 1] instead of [0, 255] (in the case of RGB).

Later, we will divide the dataset into train (70%), test (10%) and validation (20%). Using keras sequential API, we are going to implement the neural networks. Below we are going to explain our neural networks. This is a convolutional neural network architecture for image classification. Here is an overview of each layer.

Input layer: This is the first layer in the network, which consists of a convolutional layer with 16 filters of size 3×3, a stride of 1, and a ReLU activation function. The input shape is specified as (256, 256, 3), which corresponds to the shape of the input images.

Max pooling layer: This layer follows the first convolutional layer and applies a max pooling operation with a default pool size of 2×2 and a default stride of 2. The purpose of this layer is to reduce the spatial dimensions of the output feature maps while retaining the most important features.

Convolutional layer: This layer consists of a convolutional layer with 32 filters of size 3×3, a stride of 1 and a ReLU activation function. The purpose of this layer is to extract additional features from the output feature maps of the previous layer.

Max pooling layer: This layer follows the second convolutional layer and applies a max pooling operation with a default pool size of 2×2 and a default stride of 2.

Convolutional layer: This layer consists of a convolutional layer with 16 filters of size 3 × 3, a stride of 1, and a ReLU activation function. The purpose of this layer is to extract additional features from the output feature maps of the previous layer.

Max pooling layer: This layer follows the third convolutional layer and applies a max pooling operation with a default pool size of 2 × 2 and a default stride of 2.

Flatten layer: This layer flattens the output feature maps of the previous layer into a 1D vector, which can be passed to a fully connected neural network.

Dense layer: This layer consists of a fully connected neural network with 256 neurons and a ReLU activation function. The purpose of this layer is to perform classification based on the features extracted from the previous layers.

Output layer: This layer consists of a single neuron with a sigmoid activation function, which produces a binary classification output (between 0 and 1) based on the output of the previous layer.

After this we are going to compile and train the model.

Once the network is built, we are going to compile the model using the Adam optimizer and binary cross-entropy loss function. The accuracy metric is used to evaluate the performance of the model. After this, we are going to create a Tensor Board callback object, which is used to log the training progress and visualize the model's performance during training. Finally, we are going to train the model using the fit() method. The training data is provided as a TensorFlow Dataset object (train_ds), and the validation data is provided as another Dataset object (val_ds). The model is trained for 10 epochs, and the Tensor Board callback is used to log the training progress. We are going to save the training history in the variable history. This variable contains information about the training and validation loss and accuracy for each epoch (Fig. 1).

Using the information in history, we are going to visualize the training and validation loss and training and validation accuracy over epochs for a trained neural network model. Finally, we are going to allow the model for predictions (either we can use test or unknown data). Since we are using Binary Cross-Entropy (BEC), for Cat or Dog Image Classification the final label can be determined based on the output of the model and the threshold value used to interpret the output as a binary decision.

After training the model with BCE loss, the output of the model for a given input will be a single scalar value between 0 and 1, which represents the predicted probability of the positive class. To convert this probability value to a binary decision, a threshold value is applied, typically 0.5. If the output value is greater than or equal to the threshold, the final label is considered as 1 (positive class, i.e., Dog); otherwise, it is considered as 0 (negative class, i.e., Cat).

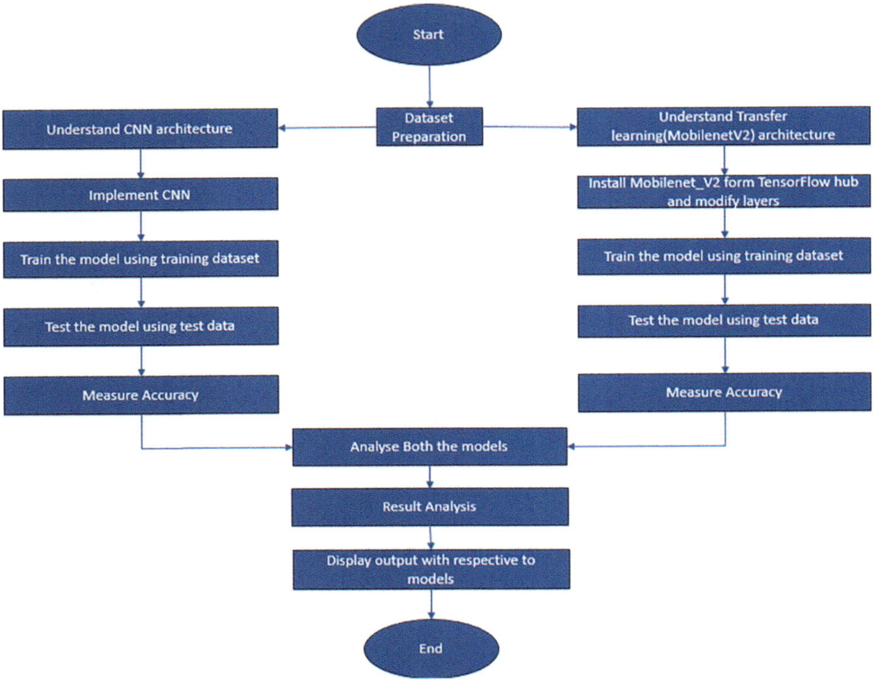

Fig. 1 Transfer learning architecture using MobileNetV2

4.2 Mobilenetv2 (Transfer Learning Technique)

As we proceed, the transfer learning method MobileNetV2 is our next model. First, MobileNetV2 is downloaded from the TensorFlow hub [8]. We already discussed the use of pre-trained models (models which are trained on billions of data) in the transfer learning technique. The model's final few layers will be changed, and we'll let it train using our data. In a manner similar to CNN, we have divided this dataset into train, test and validation subsets. However, we only permit the final few layers of the model to train, leaving the weights of the remaining trained layers unchanged. Here we can see the implementation.

Firstly, we are going to use the URL of the pre-trained MobileNetV2 feature extractor on TensorFlow hub. When this URL is passed to the KerasLayer function from the TensorFlow Hub library, it loads the pre-trained MobileNetV2 model into the code. The feature_extractor_model is loaded as a KerasLayer, and its top layer is removed to create a feature vector that can be used as input to a new custom output layer. In the context of neural networks, the term "top layer" refers to the layer that is closest to the output of the network, regardless of its position in the network's architecture. In the case of the MobileNetV2 model, the classification layer is the final layer of the network, and thus it is considered to be the "top layer" in this context.

Thus, we remove the final classification layer from the pre-trained MobileNetV2 model, and use only the output of the previous layer to create a new feature vector. This feature vector can then be used as input to a new custom output layer, which will be trained to perform a specific task, such as image classification (Cat or Dog). The input_shape parameter specifies the expected image size of 224 × 224 pixels with 3 colour channels. The trainable parameter is set to False to keep the pre-trained weights fixed during the training of the new output layer. The Sequential model is created with the pre-trained model and a new Dense layer with a sigmoid activation function. The sigmoid activation is used to output a probability between 0 and 1 for binary classification tasks [5].

The model is compiled with the Adam optimizer, binary cross-entropy loss function, and accuracy metric. The callbacks parameter is set to Tensor Board to visualize the training progress in the logs directory. The fit method is called with the training and validation datasets, along with the Tensor Board callback, and runs for 5 epochs. The model is trained to minimize the loss function and maximize the accuracy metric.

Overall, by removing the top layer and using only the intermediate layer outputs, we can use the pre-trained model's learned features to perform transfer learning on a new task, which often results in improved performance and requires less training data.

5 Analysing Models and Results

5.1 Analysing Various Measures in Both Cases

We are going to include various measures in the models for a clear understanding of the implementations (Table 1).

Table 1 Analysis of various measures in both cases

Measures	CNN	Transfer learning (MobileNetV2)
Accuracy	93%	99%
Epochs	10 epochs	5 epochs
Activation function	Sigmoid in dense layer and ReLU in hidden layers	Sigmoid in dense layer
Optimizer	Adam	Adam
Loss function	Binary cross entropy	Binary cross entropy

6 Results

We are going to observe the epochs and overview of the performance by visualizing the loss and accuracy graphs (both in the training and validation phase).

The below results include Transfer Learning and CNN models. As we can observe (from Fig. 2) for just five epochs transfer learning gained 99% accuracy for our problem statement, whereas (from Fig. 1) CNN gained 93% accuracy for almost 10 epochs. From this, we can visualize the accuracy and loss graphically as shown in Figs. 3 and 4. The performance of transfer learning will increase rapidly after some point, but this is not the case with CNN. The red plot in the graphs gives actual loss and accuracy. The green plot in the graphs gives validation phase loss and accuracy.

6.1 *Details About Accuracy Metrics*

The final accuracy was calculated using a number of accuracy metrics. The first is precision, where the model assesses accuracy by concentrating on the percentage of true positive predictions (properly classified positive samples) out of the total positive predictions made by the model. The second is recall; recall is used to evaluate a model's propensity for spotting positive samples. Out of the dataset's actual positive samples, it calculates the proportion of true positive forecasts. Binary accuracy, the

```
In [19]: #ACCURACY VIZULIZATION
         plt.figure(figsize=(5,5))
         plt.plot(di['accuracy'],color="red",label="loss")
         plt.plot(di['val_accuracy'],color="green",label="val_loss")
         plt.show()
```

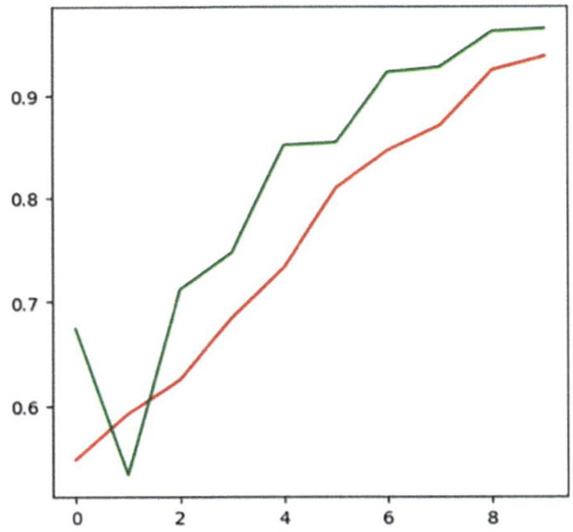

Fig. 2 Accuracy visualization for CNN

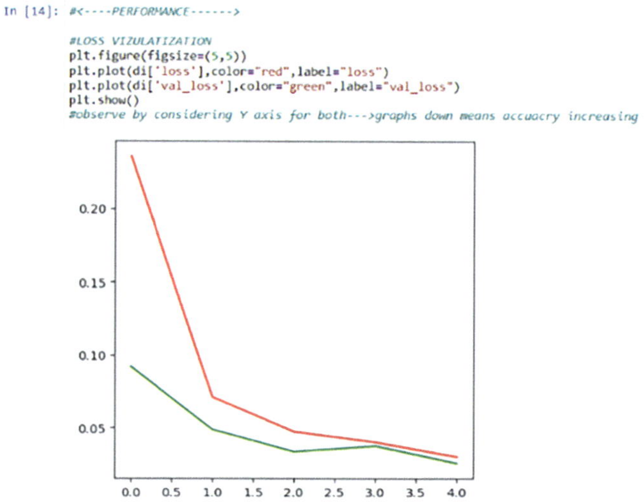

Fig. 3 Accuracy visualization for MobileNetV2

Fig. 4 Predictions of CNN

last evaluation statistic utilized, counts the percentage of samples from the positive and negative classes that were accurately identified.

In our case, we want to examine the performance and correctness of the model across both classes. So, that is why we selected accuracy (Binary Accuracy) as the measure of evaluation metrics (Table 2).

Table 2 Evaluation metrics

Evaluation metrics	CNN-score	MobileNetV2 score
Precision	0.913 (91%)	0.98 (98%)
Recall	0.927 (92%)	0.97 (97%)
Binary accuracy	0.939 (93%)	0.99 (99%)

7 Conclusion

From the results obtained from our experimentation, we can say that convolution neural networks will learn slower when compared with the transfer learning technique. The accuracy will also vary between them. The CNN model will give less accuracy when compared with the transfer learning technique. Instead of building neural networks for large datasets from scratch, it is better to use transfer learning. Transfer learning is likely the best approach when trying to identify more intricate trends. In real-world situations, we frequently encounter a lack of pictures in our dataset, a time crunch when trying to solve our classification issue, as well as numerous other challenges. Utilizing previously created models that contain data from millions of pictures can be a viable solution [1].

References

1. Szyc, K.: Comparison of different deep-learning methods for image classification. In: 2018 IEEE 22nd International Conference on Intelligent Engineering Systems (INES). IEEE (2018)
2. Tiwari, V., et al.: Image classification using deep neural network. In: 2020 2nd International Conference on Advances in Computing, Communication Control and Networking (ICACCCN). IEEE (2020)
3. He, Z.: Deep learning in image classification: a survey report. In: 2020 2nd International Conference on Information Technology and Computer Application (ITCA). IEEE (2020)
4. Jmour, N., Zayen, S., Abdelkrim, A.: Convolutional neural networks for image classification. In: 2018 International Conference on Advanced Systems and Electric Technologies (IC_ASET). IEEE (2018)
5. Ding, B., Qian, H., Zhou, J.: Activation functions and their characteristics in deep neural networks. In: 2018 Chinese Control and Decision Conference (CCDC). IEEE (2018)
6. Dong, K., et al.: MobileNetV2 model for image classification. In: 2020 2nd International Conference on Information Technology and Computer Application (ITCA). IEEE (2020)
7. https://www.kaggle.com/datasets/melvinpauljacob/cat-dog-dataset
8. https://tfhub.dev/google/tf2-preview/mobilenet_v2/feature_vector/4
9. Raja Rajeswari.Thota, T.S., Srinivasa Rao, M., Sandhya Rani, N.: Under water image enhancement technique. In: Advancements in Aeromechanical Materials for Manufacturing: ICAAMM-2021 AIP Conference Proceedings, vol. 2492, pp. 030069-1–030069-5. https://doi.org/10.1063/5.0114565. Published by AIP Publishing. 978-0-7354-4438-6/$30.00

Unveiling Consumer Segmentation: Harnessing K-means Clustering Using Elbow and Silhouette for Precise Targeting

Shweta Saraswat, Vaibhav Agrohi, Mahesh Kumar, Monica Lamba, and Raminder Kaur

Abstract Consumer segmentation is essential for accurate targeting and successful marketing efforts in today's competitive business environment. Modern marketing groups individuals by interests and attributes. Segmentation drives targeting, personalization, and ROI. This article segments customers using K-means clustering. The Elbow approach and Silhouette score determine the appropriate number of clusters and increase segmentation accuracy. They also examine the possibility of precision targeting and customized marketing techniques across sectors. Businesses may optimize marketing, improve customer happiness, and increase profits by using K-means clustering. To compete in today's market, this research helps marketers enhance targeting. Elbow and Silhouette K-means clustering may enhance client segmentation, engagement, loyalty, and economic success.

Keywords Consumer segmentation · K-means clustering · Unsupervised machine learning

---

S. Saraswat (✉)
Department of Artificial Intelligence and Data Science Department, Arya College of Engineering, Jaipur, Rajasthan, India
e-mail: shwetavrishit@gmail.com

V. Agrohi · M. Lamba
Department of Computer Science and Engineering Department, Arya College of Engineering, Jaipur, Rajasthan, India

M. Kumar
Department of Computer Science and Engineering Department, Graphics Era Deemed to Be University, Dehradun, India

R. Kaur
Department of Computer Science and Engineering Department, JECRC University, Jaipur, India

G. Fortino et al. (eds.), *Proceedings of Third International Conference on Computing and Communication Networks*, Lecture Notes in Networks and Systems 917,
https://doi.org/10.1007/978-981-97-0892-5_28

1 Introduction

In the dynamic realm of consumer-driven markets, businesses face the perpetual challenge of staying relevant and competitive. Effective marketing techniques have traditionally relied on consumer segmentation to target unique client demands and preferences. K-means clustering is a popular segmentation approach for its simplicity and effectiveness in grouping customers by similarities. However, determining the correct number of clusters has historically been a challenge, since choosing the wrong amount might reduce segmentation accuracy and efficacy [1]. We introduce a novel and performance-driven consumer segmentation strategy that integrates the K-means clustering algorithm with the Elbow method and Silhouette score. Our approach's ability to accurately determine the ideal number of clusters yields more accurate and actionable customer segments than other techniques. The Elbow approach adds a significant inflection point on the within-cluster sum of squares curve, illustrating the trade-off between cluster number and cluster variance [2]. By evaluating this curve's curvature, we may find the"elbow point," the ideal number of clusters when incremental clustering performance gains begin to drop. We prevent over- and under-segmentation to create relevant and representative consumer groups. The Silhouette score measures data point cohesiveness and separation inside and across clusters [3]. We analyze the dataset from "Kaggle" and compare our methodology to K-means clustering to show its performance advantage. Our segmentation accuracy and precision routinely beat traditional methodologies. Our results show the superiority of our technique and the utility of using the Elbow method and Silhouette score in customer segmentation.

2 Literature Review

Marketers know that consumer segmentation is essential. Businesses may better serve customers by segmenting them by common traits, habits, and preferences. Manual analysis and subjective interpretation of consumer data have restricted segmentation accuracy and actionable insights. Data analytics and machine learning have transformed consumer segmentation in recent years. K-means clustering is popular because it can unsupervisedly find hidden patterns and homogenous customer groupings. Stuart Lloyd proposed K-means clustering in 1957, and James MacQueen popularized it in 1967 [4]. Since then, several researchers have used it for customer segmentation. K-means clustering, a partition-based technique, groups comparable data points by minimizing the within-cluster sum of squares. K-means clustering for customer segmentation is recommended by many studies. Wei et al. used K-means clustering to identify separate customer categories based on online buying behavior, preferences, and demographics [5]. Businesses might boost client engagement and conversion rates by customizing marketing strategies for each category. Tabianan et al. used K-means clustering to categorize retail customers by their buying habits [6]. They found four parts with different traits and behaviors. Retailers increased consumer loyalty and revenue by targeting these demographics with targeted marketing. Mahfuza R. et al. used K-means clustering to categorize online customers by browsing and buying habits [7]. Their research showed that e-commerce platforms might improve the user experience, product recommendations, and consumer happiness by analyzing segment requirements and preferences. K-means clustering has also been successful in medical data segmentation. Saraswat et al. utilized K-means clustering for electronic health data and identified separate patient groups based on demographics, medical history, and treatment results [8]. This segmentation strategy helped healthcare practitioners tailor treatments, enhance delivery, and maximize resource allocation. Consumers may be divided using K-means clustering; however, there are downsides. The results of segmentation may be impacted by the initial cluster centroids. K-means suffers with category or non-numeric data because it anticipates spherical clusters. K-prototypes for mixed data and K-modes for categorical data are variants of K-means clustering that handle these problems [9]. These changes enhance K-means clustering's ability to segment customers.

3 Methodology and Implementation

By describing the methodology employed for consumer segmentation using K-means clustering and interpreting the clustering results, business can gain actionable insights that can inform targeted marketing strategies, personalized customer experiences, and overall business growth.

3.1 Problem Statement

You are the mall's owner, and you want to know which customers (known as target customers) will congregate most frequently so that the marketing team can make strategic plans based on this information.

3.2 Data Collection

Basic information such as Customer ID, Age, Gender, Annual Income, and Spending Score was extracted from "Kaggle" to create the Mall Customer Segmentation dataset. Customers might be given a spending score based on factors like their past actions and what they've bought.

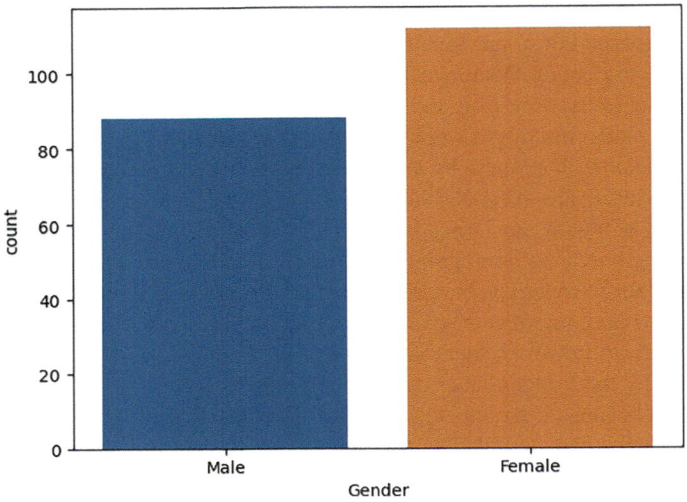

Fig. 1 Data analytics for gender

3.3 Data Preprocessing

It involves the cleaning of the collected data by removing duplicates, inconsistencies, and missing values. Then we need to perform transformation and standardization of variables as per the requirement and encode categorical variables appropriately so in this dataset only one dimension has text value, i.e., gender; hence, we encode male as "0" and female as "1" (Fig. 1).

3.4 Feature Selection

It is used to analyze the collected data and identify the relevant features for consumer segmentation. Consider variables that are likely to differentiate age, annual income, or spending score (Fig. 2).

3.5 K-means Clustering

Before the algorithm can be executed, the number of clusters must be provided. K's appropriateness is generally a judgment call that requires familiarity with the topic or the use of tools like the elbow approach or silhouette analysis [10]. In this piece, we use both methods and evaluate their relative efficiency.

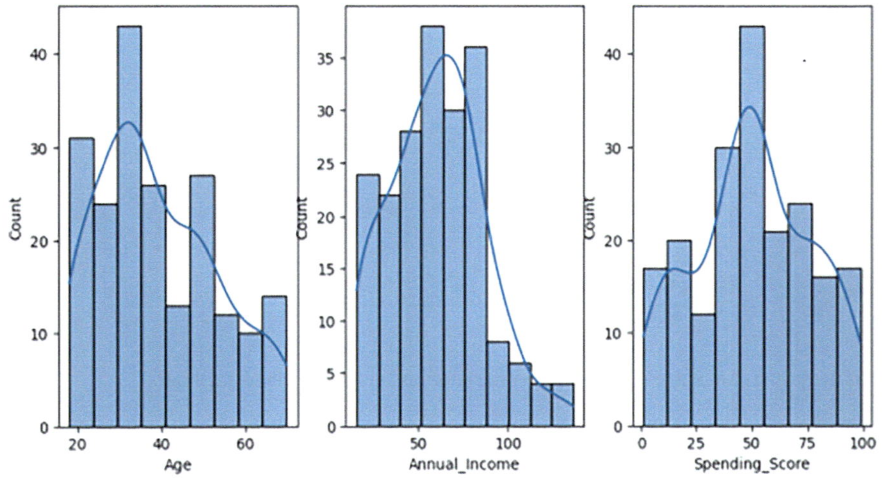

Fig. 2 Data analytics of age, annual income, and spending score

3.5.1 Elbow Method

In the elbow method, the average distance is calculated by adding the squares of the sites. When $K = 1$, the total squared area of a cluster is maximized. As K increases, the value of the sum of squares inside each cluster decreases (Fig. 3).

By using the elbow method, we can select a K value, i.e., $K =$ "9". After 9 wcss is constantly decreasing, so we build a model by taking the value of K as "9" and plot the graph to analyze the relation between two important features, i.e., age and annual income (Fig. 4).

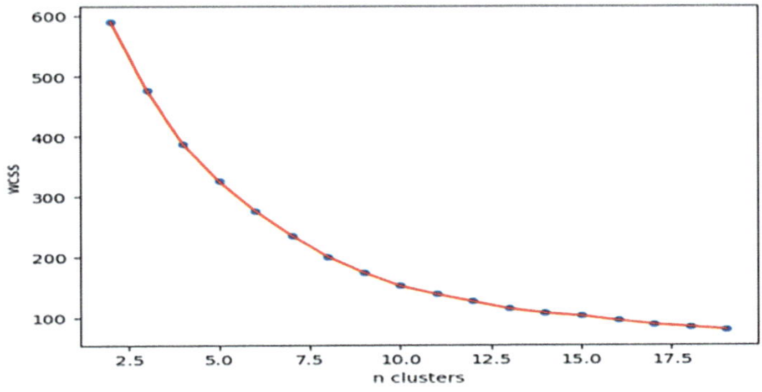

Fig. 3 Graph of the elbow method

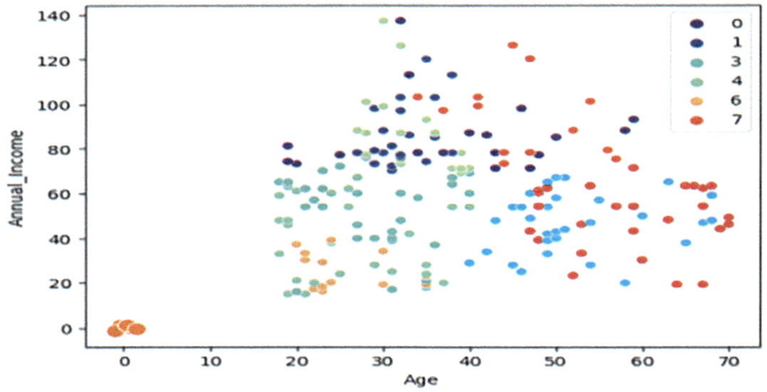

Fig. 4 Scatterplot of age and annual income after opting for the Elbow method

3.5.2 Silhouette Method

The silhouette technique, which is formed by the elbow approach, fills in the range of K values [11]. The silhouette coefficient is determined for each individual point. It figures out the typical separation between the (i) points in a cluster and the b (i) points in the next-closest cluster (Fig. 5).

The link between age and yearly income may be analyzed by choosing a K value, in this case $K = 11$, and plotting a silhouette technique graph. A larger value for the silhouette coefficient (between -1 and 1) suggests tighter grouping. The ideal number of clusters for a dataset may be determined with the use of the silhouette approach (Fig. 6).

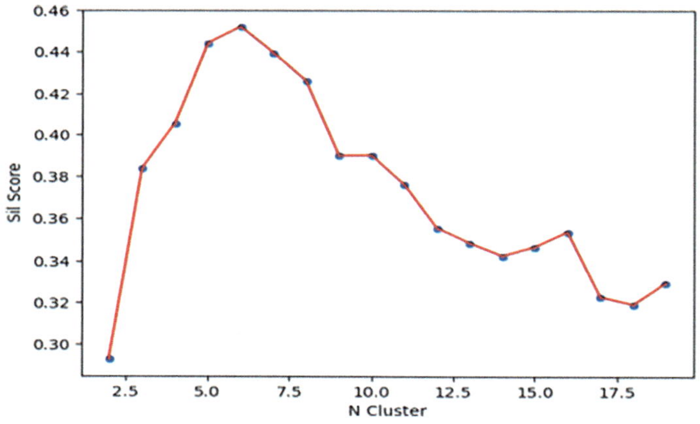

Fig. 5 Graph of the Silhouette method

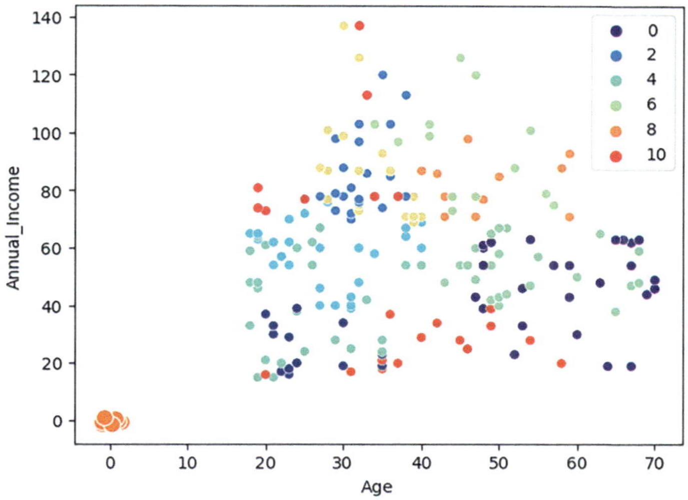

Fig. 6 Scatterplot of age and annual income after opting for the Silhouette method

3.6 Evaluation of Segmentation Results

Evaluate the quality of the obtained consumer segments using appropriate metrics, such as within-cluster sum of squares, silhouette score, and cohesion and separation indices. Assess the meaningfulness and interpretability of the identified segments by analyzing their characteristics, behaviors, and preferences. Conduct statistical tests or visualization techniques to validate the differentiation among segments and assess the significance of the findings. By using the Elbow method, we are getting good clusters than the silhouette method. So we shall consider the Elbow method for this dataset.

3.7 Segment Profiling

Analyze the characteristics of each cluster by examining the mean or median values of different features within each cluster. This will help in understanding the distinct customer segments identified by the clustering algorithm. Based on the cluster characteristics, interpret and label each segment accordingly. For example, you might identify segments such as "High-income Shopaholics," "Budget-conscious Families," or "Young Trendsetters."

3.8 Application and Implementation

Translate the insights gained from segmentation into actionable marketing strategies.

- Design personalized marketing campaigns tailored to each consumer segment.
- Implement targeted marketing initiatives across various channels, such as email marketing, social media, and advertising platforms.
- Monitor and track the effectiveness of the strategies by measuring key performance indicators (KPIs) such as conversion rates, customer retention, and revenue growth.
- Utilize the segmented profiles to gain insights into customer behavior and preferences. This information can help inform marketing strategies, personalized promotions, store layout optimizations, and targeted customer engagement initiatives.

3.9 Continuous Iteration

Consumer segmentation is an ongoing process, and it is essential to refine and update the segmentation model periodically.

- Incorporate new data and variables to capture evolving consumer behaviors and preferences.
- Regularly evaluate the effectiveness of segmentation and make necessary adjustments to optimize targeting and marketing efforts.

4 Results

According to the findings, the K-means algorithm applied to the final result of the K elbow technique improves performance, thus we're using a cluster size of 9, labeling our variables with "Cluster Id," and counting the number of customers in each cluster as shown in Tables 1 and 2. Distinct patterns in consumer behavior and preferences were discovered by the study of the clusters, which may be utilized to fine-tune marketing campaigns and boost customer engagement. These findings may also be used to guide product innovation and keep businesses one step ahead of the competition when consumer preferences shift.

The result states that each cluster produces an average age, average annual income, and spending score. By using these outputs, one can predict different insights. For example, cluster 0 produces the following details. Cluster 0 is for those people whose

- Avg Age: 32
- Avg Annual Income (k$): 108
- Avg Spending Score (1–100): 82 (Table 3)

Table 1 Assigning the labels as Cluster-Id

	Gender	Age	Annual_Income	Spending_Score	Cluster_Id
0	0	19	15	39	2
1	0	21	15	81	5
2	1	20	16	6	2
3	1	23	16	77	5
4	1	31	17	40	2

Table 2 Cluster-Id from 0 to 8 and the number of customers in each cluster

Cluster_Id	Number of customers
8	10
4	10
0	11
2	12
5	22
3	26
7	28
6	37
1	44

Table 3 Cluster ID with its average in terms of age, annual income, and spending score

Cluster_Id	Age	Annual_Income	Spending_Score
0	32.454545	108.181818	82.727273
1	56.340909	53.704545	49.386364
2	37.250000	25.666667	29.166667
3	41.230769	79.269231	15.923077
4	41.000000	109.700000	22.000000
5	25.272727	25.727273	79.363636
6	27.108108	56.297297	49.675676
7	32.785714	78.035714	81.892857
8	52.800000	25.900000	9.600000

In this way we can produce many insights. By so doing, we may generate a plethora of new ideas.

5 Insights Gained from the Research Will Help the Public Good

The research benefits from a novel approach, improved accuracy, practicality, and the possibility of a beneficial influence on business and society. With the suggested method, businesses may better understand their customers, build lasting bonds with them, and thrive in today's cutthroat market [12]. The advantages of this study include the following.

5.1 Novel Approach

The study introduces a novel approach to consumer segmentation by strategically integrating the K-means clustering algorithm with the Elbow method and Silhouette score. This innovative combination allows for more precise identification of the optimal number of clusters, leading to more accurate and meaningful consumer segments.

5.2 Enhanced Precision

By leveraging the Elbow and Silhouette methods, the study overcomes the common challenge of determining the optimal number of clusters in K-means clustering. These result in highly precise consumer segments, enabling businesses to better understand their customers and tailor marketing strategies accordingly.

5.3 Performance Advantage

Through rigorous analysis and empirical comparisons with traditional methods, the study demonstrates the performance advantage of the proposed approach. Businesses can achieve higher clustering accuracy, better cohesion, and separation of clusters, leading to improved marketing outcomes and increased return on investment.

5.4 Actionable Insights

The study provides actionable insights into consumer behavior and preferences. These insights empower businesses to design targeted marketing campaigns, personalized product offerings, and enhanced customer experiences, ultimately leading to higher customer satisfaction and loyalty.

5.5 Resource Optimization

With more accurate consumer segmentation, businesses can optimize their marketing resources effectively. By targeting the right consumer segments, companies can reduce wasteful spending, enhance efficiency, and maximize the impact of their marketing efforts.

5.6 Positive Social Impact

The study's emphasis on data privacy and ethical handling of consumer information contributes to a more responsible and trustworthy approach to data-driven marketing. This has broader societal implications, fostering consumer trust and confidence in businesses' data practices.

5.7 Generalizability

The proposed approach is not limited to specific industries or regions. It can be applied across various sectors, making it a versatile and widely applicable methodology for businesses of different sizes and types.

5.8 Academic Contributions

The study contributes to the academic community by showcasing the effectiveness of the Elbow method and Silhouette score in consumer segmentation. It offers a valuable reference for researchers and practitioners exploring data-driven methods in marketing and consumer behavior studies.

5.9 Competitive Advantage

Businesses that adopt this advanced segmentation approach gain a competitive edge by understanding their customers better than their competitors. This leads to more targeted and compelling marketing strategies that resonate with consumers, resulting in a stronger market position.

6 Prospects for Further Study

Our hope is that by providing an outlook on where consumer segmentation is headed, we will encourage further study and cooperation in the area. Exciting new opportunities to improve marketing strategies, data analysis methods, and our knowledge of consumers may be found in these areas. The future scope of our research encompasses the following aspects.

6.1 Advanced Clustering Algorithm

While our study focuses on K-means clustering, the future scope involves exploring other advanced clustering algorithms. Algorithms such as Hierarchical Clustering, Density-Based Spatial Clustering of Applications with Noise (DBSCAN), and Gaussian Mixture Models could be evaluated to ascertain their effectiveness in consumer segmentation [13].

6.2 Feature Engineering and Selection

In our study, we have employed existing consumer attributes for segmentation. The future scope involves delving deeper into feature engineering and selection techniques to identify more relevant and influential attributes. By incorporating domain-specific knowledge and sentiment analysis of customer reviews, we can enhance the segmentation process and identify previously hidden patterns [14].

6.3 Integration of Deep Learning

With the rapid advancement of deep learning techniques, there is a promising future scope for integrating neural network-based approaches into consumer segmentation [15]. Exploring the utilization of techniques like autoencoders and recurrent neural networks may allow for more sophisticated and dynamic segmentation, especially in capturing temporal patterns in consumer behavior [16].

6.4 Segmentation Validation

Validating consumer segments based on real-world outcomes is critical. The future scope involves conducting longitudinal studies to assess the stability and predictive power of the identified consumer segments over time [17]. By linking the segments to specific marketing strategies and measuring their impact on customer engagement and retention, we can validate the effectiveness of our approach.

6.5 Online and Mobile Segmentation

As online and mobile platforms play an increasingly central role in consumer interactions, there is a future scope to adapt our approach to these digital environments. Studying consumer behavior and preferences across online channels and mobile applications could lead to more targeted and personalized digital marketing efforts [18].

6.6 Cross-Domain Analysis

Extending the application of our approach to cross-domain analysis presents another future scope. By leveraging data from multiple industries or product categories, we can identify common consumer segments and uncover transferable marketing strategies that can benefit businesses across diverse domains.

6.7 Explainable AI in Segmentation

As the demand for explainable AI increases, incorporating interpretability into consumer segmentation becomes relevant. The future scope involves exploring methods to make the segmentation process more transparent and understandable to marketing practitioners, enabling them to make informed decisions based on the identified consumer segments [19].

7 Conclusion

K-means clustering using elbow and silhouette algorithms reveals consumer segmentation for accurate marketing and customer-centric tactics. The elbow approach is used to determine the ideal number of clusters, while the silhouette approach measures how well each data point fits into its cluster. K-means clustering helps organizations identify customer categories based on demographics, behavior, preferences, and buying habits. Targeting consumers through segmentation has several advantages, such as optimizing engagement and conversion. To guarantee segment relevance and accuracy, the segmentation strategy must be regularly analyzed and refined based on data and feedback. To better understand their customers, firms could combine clustering with additional data analysis methodologies like association rules or predictive modeling.

References

1. Tabianan, K., Velu, S., Ravi, V.: K-means clustering approach for intelligent customer segmentation using customer purchase behavior data. Sustainability **14**(12), 7243 (2022)
2. Zhou, J., Jinliang, W., Xu, B.: Customer segmentation by web content mining. J. Retail. Consum. Serv. **61**, 102588 (2021). https://doi.org/10.1016/j.jretconser.2021.102588
3. Mahfuza, R., Islam, N., Toyeb, M., Emon, M.A.F., Chowdhury, S.A., Alam, M.G.R.: LRFMV: an efficient customer segmentation model for superstores. PLoS One **17**(12), e0279262 (2022). https://doi.org/10.1371/journal.pone.0279262.PMID:36538513;PMCID:PMC9767363
4. Loftus, T.J., Shickel, B., Balch, J.A., Tighe, P.J., Abbott, K.L., Fazzone, B., Anderson, E.M., Rozowsky, J., Ozrazgat-Baslanti, T., Ren, Y., Berceli, S.A., Hogan, W.R., Efron, P.A., Moorman, J.R., Rashidi, P., Upchurch, G.R. Jr., Bihorac, A.: Phenotype clustering in health care: a narrative review for clinicians. Front. Artif. Intell. **5**, 842306 (2022). https://doi.org/10.3389/frai.2022.842306. PMID: 36034597; PMCID: PMC9411746
5. Ikotun, A. E., Abualigah, A., Abuhaija, L., Belal Heming, J.: K-means clustering algorithms: a comprehensive review, variants analysis, and advances in the era of big data. Inform. Sci. **622** (2022). https://doi.org/10.1016/j.ins.2022.11.139
6. Belyadi, H., Haghighat, A.: Machine Learning Guide for Oil and Gas Using Python (2021)
7. Smith, A., Johnson, B.: Advancements in consumer segmentation using machine learning: a comparative study. J. Mark. Anal. **8**(3), 189–205 (2022)
8. Saraswat, S., Sharma, S., Kulshrestha, R., Verma, N., Lamba, M.: Perceptions unveiled: analyzing public sentiment on IoT and AI integration in revolutionizing social interactions. In: 7th I-SMAC, pp. 65–70 (2023). https://doi.org/10.1109/I-SMAC58438.2023.10290487
9. Patel, R., Gupta, S., Sharma, M.: A novel approach to customer segmentation using hierarchical clustering and density-based spatial clustering. J. Bus. Anal. **15**(1), 45–62 (2023)
10. Kim, J., Park, S., Lee, H.: Enhancing K-means clustering with genetic algorithms for improved consumer segmentation. J. Interact. Mark. **37**, 78–92 (2023)
11. Brown, L., Adams, C., Carter, E.: Leveraging sentiment analysis for customer segmentation: a case study in the retail industry. J. Mark. Sci. **30**(4), 576–589 (2023)
12. Yang, Y., Li, Q., Zhang, Y.: Unsupervised machine learning for dynamic consumer segmentation in e-commerce. Decis. Support Syst. **120**, 123–135 (2023)
13. Wang, X., Zhang, Z., Li, S.: An integrated approach to customer segmentation using K-means and self-organizing maps. Int. J. Inf. Manag. **54**, 102247 (2023)

14. Saraswat, S., Keswani, B., Kulshrestha, R., Sharma, S., Verma, N., Alam, S.: Accuracy assessment of several machine learning algorithms for breast cancer diagnosis. Math. Stat. Eng. Appl. **71**(4), 12578–12587 (2022)
15. Gupta, R., Kumar, S., Sharma, A.: Customer segmentation for personalized marketing using deep autoencoders. Expert Syst. Appl. **124**, 230–242 (2023)
16. Verma, B., Saraswat, S., Saraswat, V., Misra, R.: Interstitial lung disease patterns classification using hybrid features set and multi level segmentation implemented by machine learning algorithm. In: 8th ICCES, pp. 1811–1815 (2023). https://doi.org/10.1109/ICCES57224.2023.10192662
17. Saraswat, S., Keswani, B., Sharma, V., Saraswat, V.: Mammograms-based breast cancer detection using AI image processing techniques. J. Coast. Life Med. (2023)
18. Saraswat, S., Lamba, M., Gupta, M., Sharma, A.: Automatic attendance system using facial recognition using the ADA boost algorithm (2023). SSRN: https://doi.org/10.2139/ssrn.4638529
19. Nainwal, A., Pant, B., Sharma, G.: A comprehending deep learning approach for disease classification. In: IoT Based Control Networks and Intelligent Systems: Proceedings of 3rd ICICNIS 2022, pp. 113–122. Springer Nature Singapore, Singapore (2022)

Multimodal Sentiment Analysis and Multimodal Emotion Analysis: A Review

Soumya Sharma, Srishti Sharma, and Deepak Gupta

Abstract Sentiment analysis, or opinion mining, is defined as the process of identifying the overall outlook of a person with respect to any event/entity. Emotion analysis, or affect analysis, is defined as the process of identifying the underlying feeling of a person towards an event/entity. A sentiment can be categorized into 3 categories; an emotion, however, can have multiple categories. Also, an emotion can have intensity, which a sentiment does not have. The following paper presents a review of the various multimodal sentiment analysis (MSA) and multimodal emotion analysis (MEA) techniques. Challenges and future scope are also discussed at the end of the paper.

Keywords Sentiment · Emotion · Multimodal · Sentiment analysis · Emotion analysis

1 Introduction

With more and more easy availability of network services and people exercising their freedom of speech online, an entire dam of data has been opened because of that. This data consists mostly of unstructured data as well as opinion/outlook/feedback/sentiment of people about any topic. As per multiple sources, sentiment can be termed as an opinion, idea, notion or a belief provoked by the occurrence of any event. These sentiments are more on the inside, rather than outside and last for a longer duration. In general, a sentiment can be classified either as positive,

S. Sharma (✉) · S. Sharma
The NorthCap University, Gurugram, India
e-mail: soumya21csd006@ncuindia.edu

S. Sharma
e-mail: srishti@ncuindia.edu

D. Gupta
MAIT, Rohini, Delhi, India
e-mail: deepakgupta@mait.ac.in

G. Fortino et al. (eds.), *Proceedings of Third International Conference on Computing and Communication Networks*, Lecture Notes in Networks and Systems 917,
https://doi.org/10.1007/978-981-97-0892-5_29

neutral or negative. However, sentiment is very different from emotion. For every text, there are 2 parts—facts and opinion/sentiment. So, sentiment analysis can be explained as the means of extracting meaningful information using natural language processing. In simpler terms, the identification of positive, negative or neutral sentiment is termed as sentiment analysis. However, a sentiment is very different from an emotion. Emotion refers to a complex psychological state such as happiness, anger, jealousy and grief. Sentiment, on the other hand, can be defined as a mental attitude that is created through the existence of an emotion. On all of the online available platforms, people express their sentiments in textual forms, audio forms, video forms or any combination of them. As per reliable sources, modality can be explained as the way a language expresses various relationships to the situation. In simpler terms, modality can be explained as the way in which anything happens or is experienced. For our research purpose, we will be studying the 3 modalities—text, audio and visual. Performing sentiment analysis on any one modality is termed as unimodal, but doing the same on the combination of any two or more modals is called multimodal. Emotion analysis, much unlike sentiment analysis, uncovers the underlying feeling of the person. Emotions can be categorized into multiple categories like happy, sad, depressed, anxious, angry, etc. Also, emotions have intensity in them, much unlike sentiment. An emotion or a sentiment can also be influenced by the context of the conversation, duration of the conversation, number of people in the conversation, etc. and not by just the textual features, audio features, visual features and physical gestures. This paper aims to discuss the current techniques used for MSA and MEA in order to have a better understanding of the problem statement and how various models and frameworks can be deployed to solve those problems efficiently. The paper also proposes a novel approach to the process of MSA and MEA.

2 Literature Review

MSA and MEA are the two terms that are often used interchangeably by the common masses. But for researchers, these 2 terms covey very different meanings. MSA, or multimodal sentiment analysis, is the field of research where the researchers predict the sentiment/opinion with respect to a particular entity/event from various modalities. The widest spread modalities are text, speech, image and video. But with more advancement in this field, features such as facial gestures, pulse rate and blood pressure are also being considered for the process of analysis as these features are not influenced by external factors and cannot be manipulated, up to a certain extent. The process of MSA assigns either of the 3 polarities to the event/entity; however, it does not declare/discuss the intensity of it. MEA, or multimodal emotion analysis, is a step ahead of MSA. In MEA, the researchers assign an emotion to the event/entity. Over the period of time, many researchers/psychologists have classified emotion into different categories. Also, in emotions, there is also an intensity attached to them. So, MEA assigns an emotion to the entity and in hindsight, also declares the intensity of it. Over the period of time, various techniques of machine learning and deep learning have been proposed. The upside of using a deep learning algorithm over a machine learning algorithm is that the in case of deep learning, we don't have to perform the process of the feature extraction manually.

2.1 Multimodal Sentiment Analysis

Over a period of time, the researchers have proved that the multimodal model is better at the task model of sentiment analysis as compared to the task of unimodal sentiment analysis [1, 5–7, 26, 27]. Sehar et al. [1] have proposed one novel deep learning approach for the Urdu MSA where they have extracted features from a YouTube video for audio, video and textual data using CNN and LSTM, and early fusion and late fusion techniques were applied for the detection of the sentiment polarity. The MSA framework was deployed on unimodal data and on different combinations of audio, video and textual data, and it was concluded that the multimodal framework gave the best results as compared to the unimodal framework. A similar work has been done by Dashtipour et al. [2] for the Persian language. After performing sentiment analysis on unimodal and multimodal, it was concluded that the multimodal framework gave the best accuracy. Over the period of time, the researchers have concluded that the key sentiment word is often recognized as a regular word which decreases the efficiency of the MSA model. To overcome this limitation, Wu et al. [3] have proposed a novel sentiment word-aware multimodal refinement model which assigns the right meaning to the sentiment word by leveraging the other multimodal sentiment clues. They have deployed the sentiment word position detection module in order to obtain the best possible position of the sentiment word and then used the various refinement modules to improvise the word embedding. For the purpose of sentiment polarity detection, they have used the BERT model. They have conducted this research work over various datasets and have reported and increased efficiency in each case. A lot has already been done in the space of MSA but a lot also needs to be done, after understanding all the challenges and the resources available. To summarize all that has been done, Gandhi et al. [4] have presented a proper review of the history, datasets, application and challenges of MSA. Over the period of time, researchers have proposed new models for MSA and have also worked with the existing models by changing a few of the factors. Wu et al. [5] have worked similarly on the CMU-MOSI dataset using the BERT model which comprises 3 modules—Hierarchical Multi-head Self Attention module, Gate Channel module and the Tensor Fusion model. These existing models, after adding some new features to them, perform better than the existing models (Fig. 1).

Sentiment and emotion are interlinked as one influences the other one heavily. A sentiment is directed by the emotion; however, a sentiment lasts longer than an emotion and is more on the psychological level rather than physical. To further understand the emotional complexity, Lee et al. [6] have proposed a novel multi-modal emotion recognition model based on the BERT model in which they have

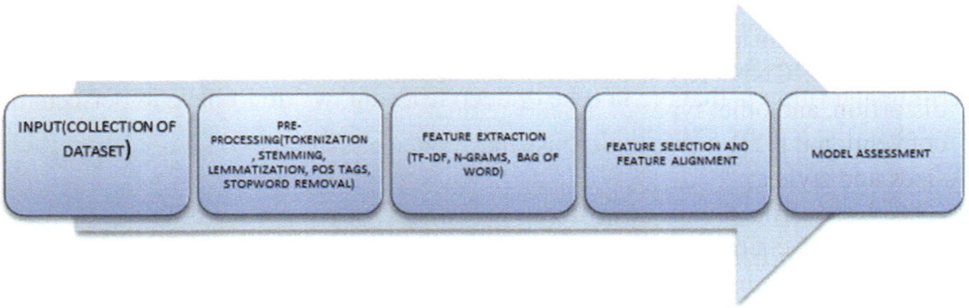

Fig. 1 A generalized MSA/MED model

combined the heterogeneous features of the text, audio and video modality. They have proposed the SMAF module, MAF module and the VF module. Their results concluded that the inclusion of audio and video components for emotion recognition improves the overall efficiency of the model. Although in multimodal sentiment analysis we deal with 3 different types of modalities, however, the inter-dependency between the heterogeneous features of the 3 modalities is not static. Even with a video running, since the features change from time to time, the inter-dependency between the features of the video (natural language, facial gestures and acoustic behaviour) is not static with respect to long-term dependencies. In order to overcome this problem, Qi et al. [7] have proposed a novel method which overcomes this limitation of long-term dependencies. They have proposed an encoding–decoding network which combines the BERT model and the transformer encoder. Their model delivers an accuracy of 47.04% with the features aligned on the CMU-MOSI dataset. Working more on the associativity and integration of the different modalities, Xu et al. [8] have presented a cross-modal joint representation framework which further works on the hierarchical interactions among the various modalities. Deploying the same framework on CMU-MOSI and CMU-MOSEI datasets concludes that the proposed framework performs better than the existing approaches.

Name of author	Task	Dataset	Feature extraction	Method used	Result
Sehar et al. [1]	MSA on Urdu	YouTube dataset	Audio, video, text	Feature extraction using CNN and LSTM and using late fusion and early fusion for polarity detection	Multimodal accuracy = 93.32%, decision fusion accuracy = 91.23%, feature fusion accuracy = 95.35%
Dashtipour et al. [2]	Context aware MSA on Persian language	YouTube dataset	Audio, video, text	Feature extraction using CNN and LSTM and using late fusion and early fusion for polarity detection	Multimodal = 91.39%, unimodal = 89.24%
Wu et al. [5]	MSA on interactive attentive mechanism	CMU-MOSI	Text, audio	LG-BERT	Accuracy = 83.82%, F1 score = 83.91%
Lee et al. [6]	MER with heterogeneous feature unification	IEMOCAP, CMU-MOSI, CMU-MOSEI	Text, audio, video	HFU-BERT	F1 score = 85.45%, accuracy = 85.49%
Qi et al. [7]	Encoding–decoding network for MSA	CMU-MOSI, CMU-MOSEI	Video (natural language, facial gestures, and acoustic behaviour)	MEDT	For features aligned, accuracy CMU-MOSI = 47.04%, CMU-MOSEI = 53.40%
Xu et al. [8]	MSA framework that works on the hierarchical interactions among modalities by passing joint representations from bimodality to unimodality	CMU-MOSI, CMU-MOSEI	Text, audio, video	CMJRT	Accuracy for CMU-MOSI dataset = 82.4% CMU-MOSEI dataset = 82.9%
Ghosh et al. [9]	SA for understanding the aggression in pain	UNBC-Mc-Master, Aggression dataset	Image, text	CNN	Accuracy = 58.56%, F1 score = 0.5967

(continued)

(continued)

Name of author	Task	Dataset	Feature extraction	Method used	Result
Cimtay et al. [10]	MER on the facial expression, EEG and GSR	LUMED-2 and DEAP	Facial expression, EEG, GSR	CNN	Mean accuracy For LUMED-2 = 74.2%, for DEAP = 53.8%
Mai et al. [11]	MSA on trimodal data, with focus on intra- and inter-utterances	CMU-MOSI and CMU-MOSEI	Visual, audio, language	HyCon	Accuracy = 46.2% on CMU-MOSI, accuracy = 52.8% on CMU-MOSEI
Zhu et al. [14]	ITIN model to work on MSA of image associated with a certain part of image	MVSA-Multiple, MVSA-Single	Text, image	ITIN	Accuracy MVSA-Single = 75.19%, MVSA-Multiple = 73.52%
Han et al. [13]	Fusion schemes to extract and integrate important features from different modalities	CMU-MOSI, CMU-MOSEI and UR-FUNNY	Text, audio, video	BBFN with gated control mechanism	Accuracy = 84.3% on CMU-MOSI, accuracy = 86.2% on CMU-MOSEI, on UR-FUNNY = 71.68%
Zhang et al. [25]	Extraction of hierarchical features for emotion analysis	DEAP AND MANHOB-HCI	Text, audio, video, electroencephalogram	HF-CNN	Arousal For DEAP = 83.28%, for HCI = 88.28% Valence For DEAP = 84.71%, for HCI = 89.00%
Joshi et al. [26]	Modelling of complex dependencies between utterances and context	IEMOCAP and MOSEI	Text, audio, video	COGMEN	For IEMOCAP, F1 score = 84.50%

Sentiment analysis can also be done on people by understating their dialogue, their gestures, their facial expressions, etc. These physical gestures are not affected by any kind of manipulation and give the most accurate results. These physical gestures not only help in understanding the sentiment, but the emotion as well. One such analysis has been done by Ghosh et al. [9] where they have recognized a person's pain on the basis of textual and visual information. They have worked on the UNBC-McMaster dataset and the aggression dataset and have employed the CNN architecture for the same, with their employed framework giving better results than the current state-of-the-art models. Another such work has been done by Cimtay et al. [10] where they have included modalities like facial expressions, GSR and EEG on the LUMED-2 dataset for 3 specific emotion classes—neutral, happy and sad, and on the DEAP dataset for angry, disgust, afraid, happy, neutral, sad and surprised. They have employed the CNN framework for the same with their results showing a mean accuracy of 74.2% for the LUMED-2 dataset and 53.8% for the DEAP dataset. Although MSA is capable of giving better results than unimodal or bimodal SA, there are, however, many points that need to be considered while doing MSA. One of the key things to keep in mind while doing MSA is the ensure the balance of the inter- and intra-utterances between the modalities. In Mai et al. [11], they have worked on the inter- and intra-utterances between the modalities using the HyCon framework and the CMU-MOSI and the CMU-MOSEI dataset. Their proposed framework delivers an F1 score of 85.1 and an accuracy of 46.6% on the CMU-MOSI dataset and an accuracy of 52.8% and an F1 score of 85.6 on the CMU-MOSEI dataset which is better than most of the current models. In MSA, the information relevance between the textual data and the image/video has to have a high degree of information relevance for the framework to deliver the best results. Chen et al. [12] have proposed such a framework which works on the discriminating features between the modalities and finds the internal correlation. They have worked on the HFIR framework with the MSA-IR and the twitter-15/17 dataset, with the proposed framework delivering the best results. Han et al. [13] have developed such a framework of fusion schemes which will be useful to extract and integrate important features from different modalities. They have proposed a BBFN which performs fusion and separation on various different modalities, in a pair-wise manner. Their proposed model takes 2 bimodal pairs as input and also performs the gated control mechanism to optimize the output. Zhu et al. [14] have proposed image-text interaction network (ITIN) in which they suggest that the affective area of an image and the text associated with it are of great significance. In their research, they work on a certain area of the image (the affective area of the image) rather than on the entire image and the text associated with it. Their model works better than most of the current models as the text and image are aligned with one another. Although various approaches have been proposed for MSA, Hazarika et al. [17] have proposed a mechanism for analysing the robustness of the MSA model. They have concluded that the MSA models are very sensitive to a single modality rather than to a combination of modalities, which makes them very susceptible to any change in the modality. They have concluded that good, robust models can be made without having to change the original model and their performance. Jiang et al. [18] have proposed a model where, on the basis of valence and arousal, they predict the intensity of emotions. In their approach, they first deploy the baseline methods followed by PCA on the unimodal features so as to search for the best features from different modalities. For multimodal features, they perform fusion on the unimodal features. In order to overcome the complex time dependencies, they have proposed six hybrid temporal models.

2.2 *Multimodal Emotion Analysis*

Much like sentiment analysis, emotional analysis has also gained the attention of researchers over a period of time. This emotional analysis has multiple applications in real-life scenarios. Caschera et al. [15] have proposed a framework for emotional analysis, where they have mainly worked on 7 different emotions and in which the models work with the features extracted from facial expressions, speech, gestures and text. Their proposed framework's prediction of the 7 emotions (anger, disgust, fear, happiness, neutral, sadness, surprise) has better performance than the previous models. Working on the same line of research of MEA, Cai et al. [16] have proposed a framework based on RNN and multimodal attention. For this, they have extracted features from multiple modalities and then implemented LSTM with a transformer encoder with multimodal multi-head attention. They have adopted the techniques of early fusion, late fusion and model fusion to increase the model's performance. They have executed their approach on the MuSe-Stress and MuSe-Physio datasets, which are the subsets of Ulm-TSST. In their research, Heredia et al. [19] have suggested that since humans are capable of expressing their emotions in different ways, all these different modalities are very useful in recognizing the underlying emotions of humans. They further suggest that the performance of a certain model reduces if the different modalities have different features or if a certain modality is absent. In their research, they have analysed the performance of each modality individually and integrated it with the fusion model, EmbraceNet+. The output of the research project shows better results than the current state-of-the-art projects, marginally. For all the modalities that are used during emotion recognition, they all have different features and different properties. So, it is imperative that before we start working on such a MER model, we find some common ground to combine the features of such modalities. The features fusion techniques used have a great effect on the success of the model. Subramaniam et al. [20] have worked on the same concept. They have deployed different feature fusion techniques on face images, audio signals and EEG as the base modalities from the FER and Ck+, and RAVDESS and SEED-IV datasets. Their research concludes that the autoencoder approach combined with SVM gives the best result. To perform any kind of emotion analysis using any method, it is mandatory to have properly labelled data. The absence of the labelled data can reduce the accuracy of the model. Kumar et al. [21] have proposed such a model which uses deep learning techniques deployed with intermediate and late feature fusion on image, text and the intermediate fusion output. Also, they have made their own dataset Balanced Twitter for Sentiment Analysis dataset (B-T4SA) which contains images and text and has emotion labels as happy, sad, hate and anger. Their proposed model shows an accuracy of 90.20%. Continuing more in the area of MEA, Padi et al. [22] have proposed a neural network-based emotion recognition model in which they first deployed ResNet and BERT and then combined the model scores using a late fusion strategy. Their proposed model is capable of dealing with the drawback of data scarcity in emotion recognition, data augmentation and fine-tuning, thus improving the performance of the model. In order to boost the performance of the emotion recognition model, Xu et al. [23] have proposed a MDA framework. The reason for the need for MDA was that not many labelled multimodal datasets are available, and to transfer the unimodal knowledge to multimodal was a difficult task. In their proposed task, the framework learns cross-modality matching between images and text so as to produce a multimodal dataset. This cross-matching further enhances the performances of the classifiers and improves the task of MSA and MEA. Makiuchi et al. [24] have dealt with the issue of less availability of labelled data for emotion recognition. They have proposed a cross-representation speech model which is inspired by the disentanglement representation learning in order to perform emotion analysis on the wav2vec2.0 speech features. They have also deployed a CNN-based model in order to understand the emotions extracted from a text with the help of a transformer-based model. This proposed approach helps in overcoming the problem of underfitting of a data model. Zhang et al. [25] have proposed a novel approach where they have deployed deep learning techniques so as to discover more features for emotion recognition. They have extracted hierarchical features using CNN for emotion analysis and then have used feature-level fusion to combine the features.

3 Proposed Approach

After reading a lot of research papers, we propose the following framework. In our approach, we take the multimodal dataset and perform data preprocessing on it. For the part of feature extraction, we deploy BoW and Tf-IDF. After relevant features are extracted, we deploy various feature fusion techniques (feature level, decision level and hybrid level) individually on it and then run the deep learning model on the individually applied feature fusion dataset. After the deep learning models are deployed, we compare the performance of the 3 approaches and conclude which is better performing. Feature engineering comprises 3 parts—feature extraction, feature fusion and feature selection. The need for feature engineering is that for any multimodal dataset, there are many data points. In order to ensure that we are working on the relevant data points and not wasting the computation time, we need to perform feature engineering. We feature fusion because the data collected from various modalities has different properties and since we want the model to have the optimal performance, we fuse the properties of the data points so that the model made has to include only a few features in it, thus reducing the complexity of it (Fig. 2).

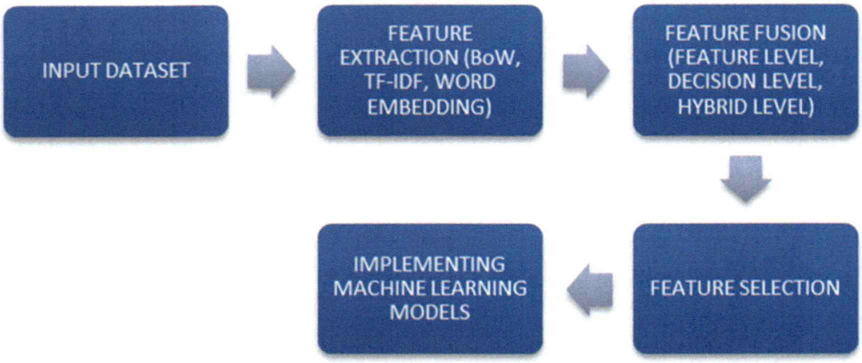

Fig. 2 The proposed approach model

4 Conclusion and Challenges

In the present times, a lot of data is being generated at a high speed. And with the advancement in the Internet era, a lot of that data is very informal or unstructured in nature. Social media sites like Facebook, Instagram, etc. give freedom to people to express their minds without much restrictions. This freedom of speech and expression has given people the liberty to express themselves without much focus on the grammar and language rules, e.g. instead of writing “how are you” people now write it as “how r u”. Misspelling “are” as “r” and “you” as “u” poses a great amount of difficulty for the machine to perform the task of SA. Also, on the different social media platforms, people communicate and express their opinions in a combination of 2 or more languages with the intention to add more meaning or more humour to their text. However, since the algorithm deployed can work only in one language, the task of SA/ED fails here. Also, on these social media sites, people use a lot of jargon, e.g. the most often used slang these days is “cool dude”, which basically refers to a person with a very casual and laid-back attitude; however, if the algorithm of SA runs on this phrase, it will literally translate the phrase “cool” as low temperature and “dude” as a man. The younger generation uses various abbreviations to convey their meaning, e.g. the most used abbreviation is “YOLO” which in the younger generation term translates to “you only live once”; however, in terms of computer science the same abbreviation refers to an algorithm “You Only Look Once”. Now, people have started using sarcasm to express their opinions, e.g. the sentence “the function was soo good that it gave me a migraine”. The word “good” reflects the positive sentiment, but on reading the entire sentence, it conveys the opposite. Thus, with more and more use of sarcasm and satire, the task of SA is becoming tedious day by day. One another challenge is the conveying of multiple emotions in the same sentence, e.g. in the sentence “the story of the movie was good, but the acting was terrible” shows 2 emotions “good” and “terrible” which have opposite polarity. Another challenge in SA is the detection of polarity in comparative sentences. In the sentences "Delhi is better than Mumbai" and "Mumbai is better than Delhi", the word "better" conveys positive polarity, but the sentences oppose each other. The use of idioms to convey thoughts and emotions is also still very relevant today. But performing SA on such idioms is not an easy task, e.g. the idiom "not my cup of tea" translates to "not my area of expertise or choice"; however, if the algorithm is run on this idiom, it will translate it to "the particular cup of tea is not mine". So, training an algorithm to deal with words or sentences that convey double meaning is a difficult job. One biggest challenge of SA is understanding the different emojis that are being used to convey thoughts and feelings. An emoji can be described as a small image which is used to convey feelings and emotions. With respect to an emoji, the most difficult task is to convert it into its vectorized form.

5 Future Scope

The data online is being generated at a very high speed and in order to have accurate and efficient results for SA, we need such models and frameworks which will work equally well. Since the language of conversation has changed these days and has shifted to more usage of figurative speech, acronyms, emojis or emoticons, we need better models which can adapt to the ever-changing style of conversation and give the best results. Thus, the need of the hour for MSA and MEA is to deploy new models and frameworks which can easily be adapted to the new, ever-changing styles of conversations, with more features being incorporated into them. In the future, we can deploy the above-discussed models with the existing datasets after changing the techniques of feature engineering in order to have a comparison between the existing techniques and new proposed approaches.

References

1. Sehar, U., Kanwal, S., Dashtipur, K., Mir, U., Abbasi, U., Khan, F.: Urdu sentiment analysis via multimodal data mining based on deep learning algorithms. IEEE Access **9**, 153072–153082 (2021)
2. Dashtipour, K., Gogate, M., Cambria, E., Hussain, A.: A novel context-aware multimodal framework for persian sentiment analysis. Neurocomputing **457**, 377–388 (2021)
3. Wu, Y., Zhao, Y., Yang, H., Chen, S., Qin, B., Cao, X., Zhao, W.: Sentiment word aware multimodal refinement for multimodal sentiment analysis with ASR errors (2022). arXiv preprint arXiv:2203.00257
4. Gandhi, A., Adhvaryu, K., Poria, S., Cambria, E., Hussain, A.: Multimodal sentiment analysis: a systematic review of history, datasets, multimodal fusion methods, applications, challenges and future directions. Inform. Fusion **91**, 424–444 (2022)
5. Wu, J., Zhu, T., Zheng, X., Wang, C.: Multi-modal sentiment analysis based on interactive attention mechanism. Appl. Sci. **12**(16), 8174 (2022)
6. Lee, S., Han, D.K., Ko, H.: Multimodal emotion recognition fusion analysis adapting BERT with heterogeneous feature unification. IEEE Access **9**, 94557–94572 (2021)

7. Qi, Q., Lin, L., Zhang, R., Xue, C.: MEDT: using multimodal encoding-decoding network as in transformer for multimodal sentiment analysis. IEEE Access **10**, 28750–28759 (2022)
8. Xu, M., Liang, F., Su, X., Fang, C.: CMJRT: cross-modal joint representation transformer for multimodal sentiment analysis. IEEE Access **10**, 131671–131679 (2022)
9. Ghosh, A., Dhara, B.C., Pero, C., Umer, S.: A multimodal sentiment analysis system for recognizing person aggressiveness in pain based on textual and visual information. J. Ambient Intell. Humaniz. Comput. **14**(4), 4489–4501 (2023)
10. Cimtay, Y., Ekmekcioglu, E., Caglar-Ozhan, S.: Cross-subject multimodal emotion recognition based on hybrid fusion. IEEE Access **8**, 168865–168878 (2020)
11. Mai, S., Zeng, Y., Zheng, S., Hu, H.: Hybrid contrastive learning of tri-modal representation for multimodal sentiment analysis. IEEE Trans. Affect. Comput. **14**, 2276–2289 (2022)
12. Chen, D., Su, W., Wu, P., Hua, B.: Joint multimodal sentiment analysis based on information relevance. Inf. Process. Manage. **60**(2), 103193 (2023)
13. Han, W., Chen, H., Gelbukh, A., Zadeh, A., Morency, L.P., Poria, S.: Bi-bimodal modality fusion for correlation-controlled multimodal sentiment analysis. In: Proceedings of the 2021 International Conference on Multimodal Interaction, pp. 6–15 (2021)
14. Zhu, T., Li, L., Yang, J., Zhao, S., Liu, H., Qian, J.: Multimodal sentiment analysis with image-text interaction network. IEEE Trans. Multimedia **25**, 3375–3385 (2022)
15. Caschera, M.C., Grifoni, P., Ferri, F.: Emotion classification from speech and text in videos using a multimodal approach. Multimodal Technol. Interact. **6**(4), 28 (2022)
16. Cai, C., He, Y., Sun, L., Lian, Z., Liu, B., Tao, J., ... Wang, K.: Multimodal sentiment analysis based on recurrent neural network and multimodal attention. In: Proceedings of the 2nd on Multimodal Sentiment Analysis Challenge, pp. 61–67 (2021)
17. Hazarika, D., Li, Y., Cheng, B., Zhao, S., Zimmermann, R., Poria, S.: Analyzing modality robustness in multimodal sentiment analysis (2022). arXiv preprint arXiv:2205.15465
18. Jiang, D., Wei, R., Liu, H., Wen, J., Tu, G., Zheng, L., Cambria, E.: A multitask learning framework for multimodal sentiment analysis. In: 2021 International Conference on Data Mining Workshops (ICDMW), pp. 151–157. IEEE (2021)
19. Heredia, J., Lopes-Silva, E., Cardinale, Y., Diaz-Amado, J., Dongo, I., Graterol, W., Aguilera, A.: Adaptive multimodal emotion detection architecture for social robots. IEEE Access **10**, 20727–20744 (2022)
20. Subramanian, G., Cholendiran, N., Prathyusha, K., Balasubramanain, N., Aravinth, J.: Multimodal emotion recognition using different fusion techniques. In: 2021 Seventh International Conference on Bio Signals, Images, and Instrumentation (ICBSII), pp. 1–6. IEEE (2021)
21. Kumar, P., Khokher, V., Gupta, Y., Raman, B.: Hybrid fusion based approach for multimodal emotion recognition with insufficient labeled data. In: 2021 IEEE International Conference on Image Processing (ICIP), pp. 314–318. IEEE (2021)
22. Padi, S., Sadjadi, S.O., Manocha, D., Sriram, R.D.: Multimodal emotion recognition using transfer learning from speaker recognition and bert-based models (2022). arXiv preprint arXiv: 2202.08974
23. Xu, N., Mao, W., Wei, P., Zeng, D.: MDA: multimodal data augmentation framework for boosting performance on sentiment/emotion classification tasks. IEEE Intell. Syst. **36**(6), 3–12 (2020)
24. Makiuchi, M.R., Uto, K., Shinoda, K.: Multimodal emotion recognition with high-level speech and text features. In: 2021 IEEE Automatic Speech Recognition and Understanding Workshop (ASRU), pp. 350–357. IEEE (2021)
25. Zhang, Y., Cheng, C., Zhang, Y.: Multimodal emotion recognition using a hierarchical fusion convolutional neural network. IEEE access **9**, 7943–7951 (2021)
26. Pandey, S., Sharma, S., Wazir, S.: Mental healthcare chatbot based on natural language processing and deep learning approaches: Ted the therapist. Int. J.Inform. Technol. 1–10 (2022)
27. Sharma, S., Aggarwal, K., Papneja, P., Singh, S.: Extraction, summarization and sentiment analysis of trending topics on twitter. In: 2015 Eighth International Conference on Contemporary Computing (IC3), pp. 295–301. IEEE (2015)

A Deep Dive into Brain-Computer Interface

Jafar A. Alzubi, Snahil Subhra, and Sushruta Mishra

Abstract Brain-Computer Interfaces (BCI) are the new-age technology that works by acquiring signals directly from the brain. Though the technology is new, it is promising and provides a great sense of hope for the future of computing. The contribution of BCIs spans from the field of computing to the field of medicine, and the need for BCIs is growing at a steady pace. The following study is a comprehensive summary on Brain-Computer Interfaces. This work proceeds by introducing the technology to the audience and works by evaluating the merits and demerits that the technology comes with. Finally, the work transitions into working and gives a detailed description of how BCIs function. The study intends to describe each and every stage in the working of a BCI in a compendious way.

Keywords Brain-computer interface (BCI) · Electroencephalography (EEG) · Fast Fourier transform (FFT) · Feature extraction · Artificial intelligence

1 Introduction

Since the invention of the computer, the computer has been known to be composed of three major components: The input devices, which usually use a sensory interaction for inputs, the processing unit, which conducts all the computation, and the output device, which is responsible for laying out the resulting information to the user. A computer proceeds by taking the input from the user from Input Devices, performs the desired operation using the Central Processing Unit (CPU) and then

J. A. Alzubi
Faculty of Engineering, Al-Balqa Applied University, As-Salt, Jordan
e-mail: j.zubi@bau.edu.jo

S. Subhra · S. Mishra (✉)
Kalinga Institute of Industrial Technology, Deemed to Be University, Bhubaneswar, India
e-mail: sushruta.mishrafcs@kiit.ac.in

S. Subhra
e-mail: 20051439@kiit.ac.in

G. Fortino et al. (eds.), *Proceedings of Third International Conference on Computing and Communication Networks*, Lecture Notes in Networks and Systems 917,
https://doi.org/10.1007/978-981-97-0892-5_30

displays the computed data using the Output Devices. The conventional architecture of the computer has been and is continually being used in the modern era. However, modern developments and refinements in the field of computer architecture have made it possible to develop a computer that does not have an input device as its core component.

This novel technology is known as the Brain-Computer Interface (BCI). The brain-computer interface skips on the reliance on input devices for inputs and utilizes the naked electric impulses generated through neurons as the sole inputs. This technology is still in development in most parts; however, the trials of the technology have rendered favorable results. Experiments have been conducted on rats, monkeys and other canines, in which a setup was connected with the brains of the subjects. It was observed that the subjects were able to execute tasks by just thinking of doing them.

The technology, if rolled out to the public, will revolutionize the way we interact and work. The technology will be most helpful to the specially abled and the paralyzed where the sufferer will be able to regain his/her lost control of body parts (Fig. 1). In the field of medicine, the interface will help to diagnose epilepsy, Parkinson's disease, Schizophrenia, etc., diseases that are usually tough to diagnose. A BCI-enabled car will let other drivers know what is going on in one's mind, thus helping to avert accidents and mishaps. The technology will prove to be a boon in the security department as the culprits could be arrested and made to confess easily with BCIs. Coupled with AI, BCIs could help develop stronger algorithms for computing purposes.

However the technology is very young. Most of the research is still in the trial phase, and therefore a lot of concerns revolve around the technology. One of such concerns is security. We are highly dependent on our brain's memory to keep records of important, private information such as bank and treasury details. If a BCI connected with a brain storing such information is hacked, the sensitive information would then

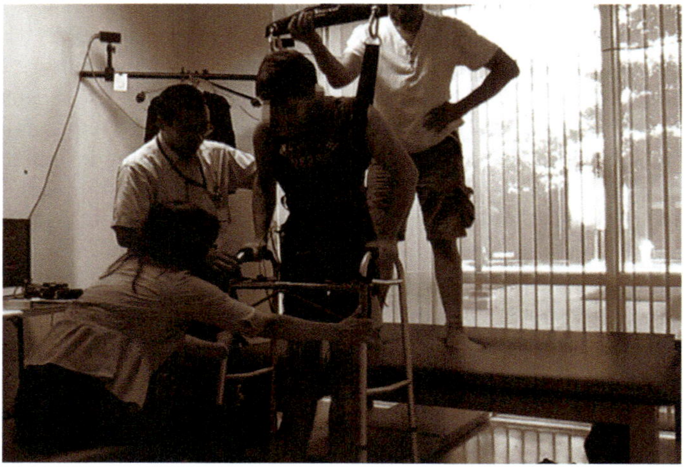

Fig. 1 BCI helping a disabled man to walk

be open to access, a situation that would prove to be appalling. Similarly, if a BCI connected with the brain of a security official is hacked, the security of the whole nation would be compromised. The brain is a very complex unit. Most of the time, we ourselves find a hard time getting our thoughts and emotions under control. In such a complex situation, a BCIs would be bound to malfunction and render incorrect results.

Nonetheless, the BCI provides a very interesting opportunity for scientists to work on. Neuralink, a US-based neurotechnology company, has made determined strides in making the word BCI-driven and has conducted hundreds of successful trials on animals such as monkeys, dogs and gold fishes. Many startups are working on BCIs to make the world a better place for the unfortunate specially abled.

The main contributions of the paper are as follows:

- This study deals with the analysis of brain-computer interface and its technical aspects.
- The pros and cons of the concept are studied.
- A primary framework model of the interface is further discussed.

2 Background Study and Related Works

Even though Brain-Computer Interface is amidst heavy research and experimentation, hundreds of successful research hint at an expansion of the domain.

Saha et al. [1] "Progress in Brain-Computer Interface: Challenges and Opportunities" highlight the opportunities and challenges posed by the development of BCIs. The paper also provides a brief yet valuable summarization of progress in the field of BCI over the last few decades. Hoffmann et al. [2] "Recent Advances in Brain-Computer Interface" provide a brief introduction to the BCIs and, through an example implementation, prepares a concise report on development of BCIs through the years and on the ongoing research. Bonci et al.'s [3] "An Introductory Tutorial on Brain-Computer Interfaces and Their Applications" is a well-documented, thorough study that highlights the features and challenges of brain-computer interfaces with an emphasis given to biomedical and automation engineering. The paper effectively describes the acquisition, processing and rendering of signals along with the legal and practical challenges faced in development. Zabcikova et al.'s [4] "Recent Advances and Current Trends in Brain-Computer Interface Research and Their Application" is an account of the top 100 most cited works on BCIs done over the period of 2016 to 2020 and the advances made in the respective researches. Schalk et al.'s [5] "BCI2000: A General-Purpose Brain-Computer Interface (BCI) System" is an incisive work on a system named BCI2000 that can be used to create BCIs for different types of brain signals, design processing methods and algorithms and create applications.

Guger et al. [6] "Prosthetic Control by an EEG-based Brain-Computer Interface" aim at producing a prosthetic-enabled hand limb for the specially abled and paralyzed using BCIs. Ma et al.'s [7] "Directly Wireless Communication of Human Minds via

Noninvasive Brain-Computer-Metasurface Platform" is a work on the development of an electromagnetic brain-computer-metasurface paradigm that is driven by BCIs. The model is proposed to translate the human mind using the EEG signals into digitally coded information. Pfurtscheller et al. [8] "Current Trends in Graz Brain-Computer Interface (BCI) Research" use subject-specific EEG patterns to develop a Brain-Computer Interface. Also, the study features a number of experiments whose processing techniques are evaluated. Rafferty [9] "Brain Computer Interfaces: A New Existential Risk Factor" lays out a different viewpoint of the BCIs as he evaluates the risks, the magnitude of the risks and the impact of the risks. Zhang et al. [10] "A Review of EEG-Based Brain-Computer Interface Systems Design" introduce the BCIs to scholars with little familiarity with the topic. In this concise study, the authors present a brief idea and working of BCIs while explaining the possibilities the domains can offer.

3 Working Mechanism of BCI

The BCI computer relies on raw signals generated by the neurons in the brain for inputs [11]. Neurons are nerve cells that use electric as well as chemical signals to transmit information. Some of the major constituents of a neuron are

Dendrite: These are the branch-like structures that acquire signals from other neurons and allow for further transmission of the signal to the cell body [12].
Cell body: Cell body is the core of the neuron. The SOMA encompasses the neuron that is responsible for genetic information production and protein synthesis.
Synapse: This is the junction between two neurons. This is the point where one neuron sends the information to another neuron [13].
Axon: These are tube-like structures that join the cell body to the synapse. The function of the axon is to transmit information from the cell body to other neurons.

Figure 2 provides a brief diagram of the neuron as shown below.
The signal acquisition can be done in two ways. Those two methods are

- Electroencephalography (EEG) or the acquisition of electric signals from the brain
- Electromyography (EMG) or acquisition of electric signals from muscles.

In this study, we will be concerned with the interpretation of electric signals generated by the brain (EEG) [14]. The brain impulse-based BCIs are made up of three major components as shown in Fig. 3:

- The receiver setup
- The processor
- The output.

The receiver (EEG) uses a set of electrodes to read signals from the scale. Electrodes are metallic miniature disks placed on the scalp for signal acquisition [15]. Two

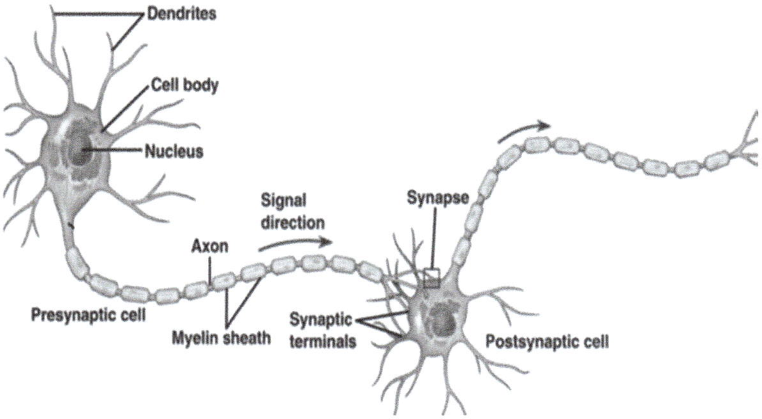

Fig. 2 A labeled diagram of neuron

Fig. 3 The basic architecture of BCI

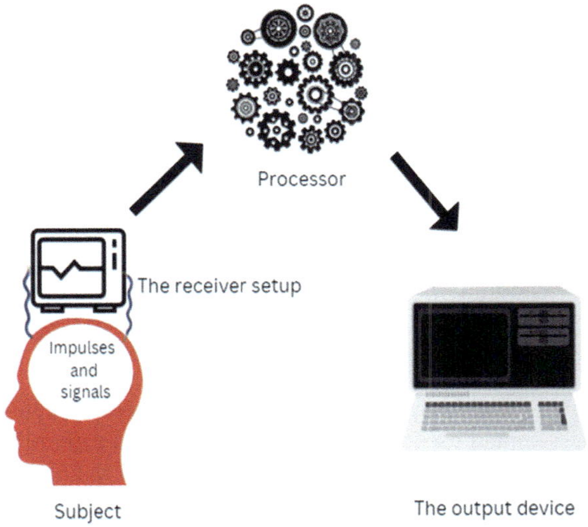

types of electrodes are used for reading the signal: Wet electrode and Dry electrode. The wet electrode uses a saline solution of gel. This provides for better conductivity and reduced electrical impedance. The most common wet electrodes are Ag/Agcl (Silver/Silver chloride) or Na/Nacl (Sodium/Sodium chloride) electrodes. On the other hand, dry electrodes use simple skin-exposed electrodes, which are simple and convenient to use but are open to information loss [16–19].

The EEG-based BCIs might seem like the way to go, but there are a lot of challenges involved as far as signal acquisition is concerned as the thick skull in the head blocks a significant amount of the raw signals. Acknowledging the above-mentioned drawback in mind, scientists are now on to implanting the electrodes directly on the

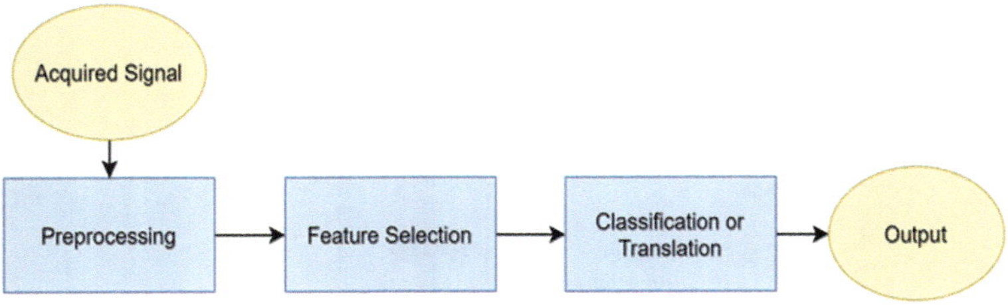

Fig. 4 Processing of signal in BCI

subject's gray matter of the brain through complicated surgeries and procedures [20]. This allows for an excellent collection of electric signals. However, this procedure is not flaw-proof as the development of scars after surgeries distorts the procurement of the signals.

The raw signals obtained from the neurons are unsuitable to be fed to the output directly. Therefore, the signals need to be processed and scrutinized before they are required to perform the desired task. This is done in three steps as shown in Fig. 4:

- Preprocessing
- Feature extraction
- Classification.

A. ***Preprocessing***

Signals extracted from the brain contain a lot of noise and disturbances. Signals for coordinating body movements often contaminate the signals required for computation. This is why we need to process and get rid of the unwanted signals. This is known as preprocessing. There are various methods used in this stage. They are

Artifact avoidance: The method applies a suitable computational algorithm and uses a simple threshold value to capture the relevant signal. The signals with frequency above the threshold value are rejected, while the ones with value below the threshold are accepted.

Linear filtering: This method is used when the signals with different frequency bands do not overlap. This method allows for the segregation of signals with different frequency waves and hence, the required signals are extracted.

Linear adaptive filter: This algorithm is based on the optimization theory and uses variable parameters that can be configured to extract the best signals.

Quantum neural-based filter: The filtering technique is constructed using a layer of neurons within the neural network framework, known as the Recurrent Quantum Neural Network that utilizes a time-varying probability density function to filter and collect the required signals.

Spatial filtering: Spatial filtering is a regression-based preprocessing technique in which the amplitude of the EEG in each electrode used for BCI is altered according to the weight composition of the voltage values in two or more electrodes [21].

After the noise, redundancies and anomalies are removed, and the collected signal is sent for feature extraction.

B. ***Feature extraction***
In this step, information that is crucial for executing the desired task is extracted and processed. As a neuron signal is very complex to interpret, it is not possible to extract the crux of the signal manually. Therefore, the information extraction is done using well-devised computational algorithms. The feature selection models are broadly classified into the filter, wrapper and embedded models [22]. The employability of a feature extraction algorithm is dependent on the characteristics of the training data available. A few of the algorithms that are used for feature extraction are

Time frequency distribution (TFD): These algorithms help in the characterization and manipulation of individual signals whose elemental features vary in time.
Fast Fourier transform (FFT): This algorithm provides frequency information of the individual signals by converting a signal into individual spectral components. It is done by converting a time-domain digital signal, having some length, to a frequency-domain digital signal.
Eigenvector method (EM): Used for multi-dimensional signal identification, the eigenvector method transforms a matrix formed using features from each dimension of the extracted signal pattern into eigenvectors that are associated with maximum eigenvalues to identify different signals in a mix [23].
Wavelet transform (WT): In this, a window or basis function is introduced whose function size can be compressed or enlarged to capture both high- and low-frequency signals as per the requirement.
Auto-regressive method (ARM): This is a linear-regression-based model that predicts the signal using the linear combination of previously predicted signals [24].

C. ***Classification***
Also known as a translation algorithm, the classification algorithm takes the clean, machine-ready data and translates the electrocardiography inputs to machine commands to execute the desired action [25]. The classification is based on the characteristics of the signal, such as frequency, and the characteristics of the algorithm, such as linear or nonlinear models. Some of the models that help in classification include the SVM, AM and MLE models.

D. ***Output***
After the machine translation is done, the signals are sent for task execution. The signal is processed, after which the levers and mechanized components are activated and the necessary task is executed.

4 Conclusion

The study helped to get a theoretical knowledge of BCIs. The study identifies the strengths and weaknesses related to juvenile technology. The work provides the fundamental knowledge of the working of the BCIs through systematic diagrams and figures. Ultimately the study proves to be an appreciable addition to the field of BCIs. Hundreds of research are being conducted every day, and more possibilities and potentials are being traversed. As such, the work is open for future interpretations and prospects.

References

1. Saha, S., Mamun, K.A., Ahmed, K., Mostafa, R., Naik, G.R., Darvishi, S., Khandoker, A.H., Baumert, M.: Progress in brain computer interface: challenges and opportunities. Front. Syst. Neurosci. **25**(15), 578875 (2021). https://doi.org/10.3389/fnsys.2021.578875.PMID:33716680;PMCID:PMC7947348
2. Hoffmann, U., Vesin, J.-M., Ebrahimi, T.: Recent advances in brain-computer interfaces (2007). https://doi.org/10.1109/MMSP.2007.4412807
3. Bonci, A., Fiori, S., Higashi, H., Tanaka, T., Verdini, F.: An introductory tutorial on brain-computer interfaces and their applications. Electronics **10**, 560 (2021). https://doi.org/10.3390/electronics10050560
4. Zabcikova, M., Koudelkova, Z., Jasek, R., Lorenzo Navarro, J.J.: Recent advances and current trends in brain-computer interface research and their applications. Int. J. Dev. Neurosci. **82**(2), 107–123 (2022). https://doi.org/10.1002/jdn.10166
5. Schalk, G., McFarland, D.J., Hinterberger, T., Birbaumer, N., Wolpaw, J.R.: BCI2000: a general-purpose brain-computer interface (BCI) system. IEEE Trans. Biomed. Eng. **51**(6), 1034–1043 (2004). https://doi.org/10.1109/TBME.2004.827072
6. Guger, C., Harkam, W., Hertnaes, C., Pfurtscheller, G.: Prosthetic control by an EEG-based brain-computer interface (BCI), 2–7 (2023)
7. Ma, Q., Gao, W., Xiao, Q., et al.: Directly wireless communication of human minds via non-invasive brain-computer-metasurface platform. eLight **2**, 11 (2022). https://doi.org/10.1186/s43593-022-00019-x
8. Pfurtscheller, G., Neuper, C., Guger, C., Harkam, W., Ramoser, H., Schlögl, A., Obermaier, B., Pregenzer, M.: Current trends in Graz brain-computer interface (BCI) research. IEEE Trans. Rehabil. Eng. **8**(2), 216–219 (2000). https://doi.org/10.1109/86.847821. PMID: 10896192
9. Rafferty, J.: Brain computer interfaces: a new existential risk factor (2022)
10. Zhang, W., Tan, C., Sun, F., Wu, H., Zhang, B.: A review of EEG-based brain-computer interface systems design. Brain Sci. Adv. **4**(2), 156–167 (2018). https://doi.org/10.26599/BSA.2018.9050010
11. Allison, B.Z., et al.: Toward a hybrid brain–computer interface based on imagined movement and visual attention. J. Neural Eng. **7**(2), 026007 (2010)
12. Abiri, R., Borhani, S., Sellers, E.W., Jiang, Y., Zhao, X.: A comprehensive review of EEG-based brain-computer interface paradigms. J. Neural Eng. **16**(1), 011001 (2019). https://doi.org/10.1088/1741-2552/aaf12e. Epub 2018 Nov 15. PMID: 30523919
13. Banerjee, V. Kukreja, S.H., Sharma, V.: Fast and accurate multi-classification of Kiwi fruit disease in leaves using deep learning approach. In: 2023 International Conference on Innovative Data Communication Technologies and Application (ICIDCA), Uttarakhand, India, pp. 131–137 (2023). https://doi.org/10.1109/ICIDCA56705.2023.10099755

14. Verma, S., Mishra, S.: An exploration analysis of social media security. In: Predictive Data Security Using AI: Insights and Issues of Blockchain, IoT, and DevOps, pp. 25–44. Springer Nature Singapore, Singapore (2022)
15. Manikandan, N., Ruby, D., Murali, S., Sharma, V.: Performance analysis of DGA-driven botnets using artificial neural networks. In: 2022 10th International Conference on Reliability, Infocom Technologies and Optimization (Trends and Future Directions) (ICRITO), pp. 1–6. IEEE (2022)
16. Patel, S., Bakaraniya, P.V., Mishra, S., Singh, P.: Security issues in deep learning. In: Predictive Data Security using AI: Insights and Issues of Blockchain, IoT, and DevOps, pp. 151–183. Springer Nature Singapore, Singapore (2022)
17. Sinha, K., Miranda, A.O., Mishra, S.: Real-time sign language translator. In: Cognitive Informatics and Soft Computing: Proceeding of CISC 2021, pp. 477–489. Springer Nature Singapore, Singapore (2022)
18. Chakraborty, S., Mishra, S.: A smart farming-based recommendation system using collaborative machine learning and image processing. In: Cognitive Informatics and Soft Computing: Proceeding of CISC 2021, pp. 703–716. Springer Nature Singapore, Singapore (2022)
19. Swain, T., & Mishra, S.: Evolution of machine learning algorithms for enhancement of self-driving vehicles security. In: 2022 International Conference on Advancements in Smart, Secure and Intelligent Computing (ASSIC), pp. 1–5. IEEE (2022)
20. Mehta, D., Das, P.P., Ghosh, S., Mishra, S., Alkhayyat, A., Sharma, V.: A normalized ANN model for earthquake estimation. In: 2023 2nd International Conference on Applied Artificial Intelligence and Computing (ICAAIC), Salem, India, pp. 151–155 (2023). https://doi.org/10.1109/ICAAIC56838.2023.10140242
21. Dornhege, G., del R. Millán, J., Hinterberger, T., McFarland, D.J., Müller, K.-R.: An introduction to brain-computer interfacing. In: Toward Brain-Computer Interfacing, pp. 1–25. MIT Press (2007)
22. Chakraborty, S., Mishra, S., Tripathy, H.K.: COVID-19 Outbreak estimation approach using hybrid time series modelling. In: International Conference on Innovations in Intelligent Computing and Communications, pp. 249–260. Springer International Publishing, Cham
23. Singh, P., Mishra, S.: A comprehensive study of security aspects in blockchain. In: Predictive Data Security using AI: Insights and Issues of Blockchain, IoT, and DevOps, pp. 1–24. Springer Nature Singapore, Singapore (2022)
24. Adrija, M., Yash, A., Sushruta, M.: 8 pragmatic study of IoT in healthcare security with an explainable AI perspective. In: Explainable Artificial Intelligence for Biomedical Applications, pp. 145–166. River Publishers (2023)
25. Bhavya, M., Pranshu, S., Sushruta, M., Sibanjan, D.: 17 comparative analysis of breast cancer diagnosis driven by the smart IoT-based approach. In: Explainable Artificial Intelligence for Biomedical Applications, pp. 353–374. River Publishers (2023)

Climatic Variable Assessment in a Smart Sensory Enabled Setting

Waleed Hadi Madhloom Kurdi, Parnani Panda, Ankit Garg, Shrishti Swaraj, Sushruta Mishra, and Ahmed Alkhayyat

Abstract This research paper focuses on the use of smart and intelligent environment technology to monitor climatic conditions and their impact on the environment. It explores the various applications of smart environmental monitoring systems in weather forecasting, air quality monitoring, agriculture, and disaster response. It emphasizes the advantages of using different sensors, clouds, communication devices, etc. to collect real-time data on temperature, humidity, air pressure, wind speed and direction, pollution, and harmful gases. This research paper reminds us of how important it is to use technology for environment monitoring and decision-making toward sustainable living; with the help of smart systems, we can better take proactive measures to protect our planet. Overall, this research paper talks about how using smart technology to monitor our environment can help us understand and improve the conditions. It provides information about air quality, natural disasters, agriculture, and weather patterns. The implementation performed shows that the XGBoost model generates the best performance with a 94.7% accuracy rate. This knowledge helps understanding and monitoring through smart systems for sustainable future.

W. H. M. Kurdi
Altoosi University College, Najaf, Iraq
e-mail: waleedalkurdi@altoosi.edu.iq

P. Panda · S. Swaraj · S. Mishra (✉)
Kalinga Institute of Industrial Technology, Deemed to be University, Bhubaneswar, India
e-mail: sushruta.mishrafcs@kiit.ac.in

A. Garg
AIT-CSE, Chandigarh University, University Centre for Research and Development (UCRD), Mohali, Punjab, India

A. Alkhayyat
College of Technical Engineering, Islamic University, Najaf, Iraq

College of Technical Engineering, Islamic University of Al Diwaniyah, Al Diwaniyah, Iraq

College of Technical Engineering, Islamic University of Babylon, Baghdad, Iraq

G. Fortino et al. (eds.), *Proceedings of Third International Conference on Computing and Communication Networks*, Lecture Notes in Networks and Systems 917,
https://doi.org/10.1007/978-981-97-0892-5_31

Keywords Sensor · Cloud computing · Climate tracking · Data analytics · Real-time prediction · Machine learning · Particulate matter · Crop residue burning

1 Introduction

Environmental monitoring is the process of measuring, observing, and analyzing the environment to predict the quality of available natural resources such as water, air, and soil. It is an essential tool used to find and quantify the presence and concentration of various pollutants, contaminants, and other factors that can affect the health of living organisms and the ecosystem. Various programs are typically carried out by non-governmental organizations, government agencies, and private companies to assess the quality of the environment and to determine any potential sources of pollution or contamination. This information can be used to develop certain guidelines and policies to mitigate environmental impacts and protect the health and well-being of humans and wildlife. It can also be used to know the efficacy of environmental management practices and to know areas that require further action or improvement. Environmental monitoring systems which include a wide range of sensors play a pivotal role in tracking and understanding the variables of climate. These sensors are used to collect informational data on various climate parameters like temperature, precipitation, wind speed, humidity, precipitation, wind speed, solar radiation, etc. This data then helps to build a more broad understanding of climate trends, identify patterns, and forecast probable changes in the climate system. For example in air monitoring task, detector networks and geographical information systems (GISs) examiner pollution along with topographical and meteorological data is used to dissect air adulterants. In water monitoring, water samples are anatomized to measure chemical, radiological, and natural data against population demographics. In soil monitoring, soil heists are covered for saltness, impurity, and acidity to dissect soil quality in husbandry and to prognosticate the eventuality of corrosion, flooding, and pitfalls to environmental biodiversity.

In people's homes and civic communities, environmental monitoring includes other types of data shadowing and analysis, for illustration business volumes, population demographics, security, resource failure, structure and home health, mega-city structure, and food security. Environmental monitoring extends to global monitoring of the terrain: For illustration, the monitoring of hothouse feasts (GHGs). GHGs beget climate change and contribute to extreme rainfall conditions, food force dislocations, and health issues caused by high situations of gauze and pollution. The biggest challenge in smart surroundings is the massive quantities of data that need to be sifted, covered, anatomized, and proactively used to produce results for everyday challenges. By using these sensors and other environmental monitoring technologies, scientists and policymakers can gain a better understanding of climate patterns and how they are changing over time. This knowledge can be used to develop more effective strategies to mitigate the impacts of climate change on the environment and human societies.

2 IoT-Enabled Sensors in Climate Tracking

Climate change is one of the biggest global challenges that the world is facing today. So, it is essential to track and explore changes in the environment to forecast and reduce the negative effects of the adverse effects of climate change. The technology that supports connecting and controlling devices is Internet of Things (IoT), anything including sensors, electronic devices, and electronic components for automobiles. The system is responsible for observing and controlling environmental conditions. The Internet of Things (IoT) and different sensors play a vital role in climate tracking by gathering and investigating data related to various environmental parameters. IoT devices track down climatic conditions by collecting data from various sensors which measure different environmental parameters. These sensors can be placed in various locations, such as buildings, vehicles, weather stations, and even on drones, to capture data from different geographic locations and heights. The information gathered by IoT devices can include parameters such as atmospheric pressure, temperature, wind speed and direction, humidity level, rainfall amount, air quality level, and solar radiation measurement. This information is transmitted to cloud-based platforms using wireless communication technologies, such as Bluetooth, or cellular networks like Wi-Fi, where it is processed and examined in real time. This document explores the role of IoT devices and sensors in climate tracking.

The Internet of Things (IoT) and sensor technologies play a major role in climate tracking by providing real-time data and insights into environmental changes. Major benefits of using IoT sensors in climate monitoring is the ability to collect and analyze data. By continuously monitoring various environmental parameters, including temperature, atmospheric pressure, wind speed, humidity, and air quality, IoT devices and sensors can provide a comprehensive understanding of environmental changes. Utilizing machine learning and artificial intelligence algorithms, this data can be scrutinized to discern intricate trends and patterns. Another critical application of IoT sensors in climate tracking is predicting and tracking severe weather events. These apparatuses can amass data pertaining to temperature, atmospheric pressure, and wind speed, which can help meteorologists predict the trajectory of hurricanes, tornadoes, and other severe weather events, thus helping people evacuate from areas likely to be affected by the storm and prepare emergency services to respond to the aftermath of the disaster.

3 Smart Sensors Used in Climate Tracking

Climate tracking relies on a range of sensors designed that can measure different environmental factors. Among the most common types of sensors used in climate tracking include.

- **Temperature sensors**: These sensors gauge the temperature of the environment and enable monitoring temperature variation. They are often used in weather stations to measure the temperature of the water, air, or ground. Examples include **Thermocouples, RTDs, thermistors, and semiconductor-based ICs**.
- **Humidity sensors**: These sensors determine the amount of moisture in the air and can be used to track changes in humidity through the years. Humidity sensors are often used in indoor environments and weather stations to monitor humidity stages. Some of the examples are **DHT11, DHT22, AM2302, SHT71**, etc.
- **Atmospheric pressure sensors**: These sensors assess the air pressure which can be used to track changes in atmospheric pressure over a period. These sensors are often used in weather stations to examine changes in weather patterns. Some of the examples of these sensors are the 4 Set Point Alarm, C48 Wind Alarm, Handheld Anemometer, etc.
- **Wind speed sensors**: Wind sensors quantify the speed of the wind and are capable of monitoring track changes in wind patterns across temporal intervals. Wind speed sensors are frequently used in meteorological stations and wind turbines to monitor wind conditions. The four prevailing anemometers are **Thermal Anemometers, Vane Anemometers, Thermal Anemometers with Velocity/ Temperature Profiling, and Cup Anemometers**.
- **Solar radiation sensors**: These sensors quantify the amount of solar radiation that reaches the earth's surface and which can be used to assess changes in solar radiation over time. Solar radiation finds application in weather stations and in the solar energy industry to monitor solar radiation levels. Some of the examples are **pyranometers, quantum sensors, and net radiometers**.

4 Related Works and Background Study

In [1], Floris et al. propose a smart system to maintain indoor environments in buildings and predict occupancy. The system captures data on temperature, humidity, CO_2 levels, and occupancy via a range of sensors and IoT devices. A machine learning algorithm then analyzes the data to make modifications to the HVAC and forecast occupancy trends. The system's performance in enhancing indoor air quality, reducing energy usage, and forecasting occupancy patterns to optimize building operations was tested by the authors in an Italian building. An IoT-based smart building environment service is proposed in [2] by Park et al. with the objective of optimizing thermal comfort for individuals while spending a minimal amount of energy. The system collects information on occupancy patterns, energy consumption, and indoor and outdoor environmental conditions using a variety of sensors and IoT devices. A machine learning algorithm then evaluates the data to forecast the ideal humidity and temperature levels for each occupant and modifies the heating, ventilation, and air conditioning (HVAC) system in accordance with those predictions. The system's efficacy in enhancing thermal comfort and lowering energy usage was proven by the authors through studies they conducted in a South Korean building. In [3], Shao presents an intelligent control system for indoor environments in green buildings, which is based on the PMV index. The system receives information about indoor as well as outdoor external variables, like temperature, humidity, CO_2 concentrations, and lighting conditions, using a variety of sensors and IoT devices. After that, a machine learning algorithm processes the data to determine the PMV index, which can be used to modify the heating, ventilation, and air conditioning (HVAC) systems in order to give occupants the best possible thermal comfort. Via experimentation in a green building in China, the authors demonstrated how well the system works to enhance indoor air quality, save energy use, and give inhabitants the best possible thermal comfort. The implementation of intelligent environmental monitoring devices to reduce climate change in urban ecosystems in isolated tourist cities has been studied in [4] by Quintal et al. The implementation of these devices in the Portuguese city of Madeira and the Greek city of Rethymnon is covered by the writers. The IoT devices and a variety of sensors have been used by smart environmental monitoring systems to collect data on temperature, humidity, noise pollution, air quality, and other environmental variables. After that, the information was examined to find out where more could be done to promote sustainable mobility, reduce carbon emissions, and enhance energy efficiency. The authors draw their conclusion that in isolated tourist cities, these types of systems can be useful instruments for reducing the effects of climate change and enhancing the sustainability of urban ecosystems. The state of smart technology utilized in building energy management and climate control has been outlined in [5] by Firan et al. The writers address the significance of lowering building energy use and emphasize the benefits of utilizing smart technology to accomplish this. In addition to the difficulties and constraints of putting such systems into practise, they also go into the many kinds of sensors and control systems that can be employed to build a smart environment. The research paper provides an exhaustive introduction to the subject of smart building technologies. A method for monitoring and controlling the climate in a smart environment is presented in [6] by Krishnamoorthy et al. The writers highlight how crucial it is to keep buildings' surroundings at their ideal levels in order to improve occupant comfort and energy efficiency. They demonstrate a system that automatically modifies heating, ventilation, and air conditioning (HVAC) systems according to real-time ambient data by integrating a variety of sensors and control algorithms. In addition to outlining the advantages and difficulties of putting such a system in place, the paper also offers a research agenda for the future. All things taken into account, the paper offers insightful information about the planning and execution of intelligent climate management systems in buildings. In [7], Frontoni et al. propose an Internet of Things (IoT)-based smart system for greenhouse climate control. The writers highlight the importance of sustainable agriculture and emphasize how employing smart technologies can increase agricultural yield while reducing energy use. They define a system that uses real-time environmental data to integrate many sensors and actuators to control the temperature, humidity, and light levels in greenhouses. The results of the experiment and the possibility of more studies in this field are also included in this paper. A study published by Cirillo et al. [8], proposes a smart system to track and examine climate variations in vineyards. Weather and soil condition data are gathered by wireless sensors, data are processed by analysis algorithms, and a web-based interface displays the results of the system. The strategy can enhance irrigation scheduling, lower water use, and eventually enhance wine quality and production efficiency, as the authors' studies in an Italian vineyard proved. In [9], Deshpande et al. suggest a smart city system for climate monitoring and control. The system is composed of multiple sensors that gather information on sun radiation, temperature, humidity, and air quality. In order to forecast occurrences and manage the surroundings, an artificial neural network program then analyzes and evaluates the data. Through studies in a smart city testbed, the authors showed the system's efficacy by obtaining significant energy savings while preserving a comfortable indoor environment. Research by El Bari et al. [10] proposes a smart system for controlling energy and controlling the climate in greenhouses. After using sensors to track temperature, humidity, and light intensity, the system uses artificial neural network algorithms and fuzzy logic to regulate the shade, ventilation, and heating systems. The system's ability to save energy usage while preserving ideal growing conditions for plants was proven by the authors' trials carried out in a greenhouse in Morocco. In [11] by Uma Maheswari et al., the authors propose a smart system for managing energy and controlling the climate in buildings. A junction of sensors is made to gather temperature, humidity, and occupancy data. To regulate the heating, ventilation, and air conditioning (HVAC) systems, a fuzzy logic algorithm processes this data. The system's efficacy in lowering energy consumption and enhancing indoor air quality while preserving a comfortable indoor environment was proven by the authors' testing in an Indian building. In [12], Andreescu et al. suggest a smart system for climate control and energy management in office buildings. To regulate the heating, ventilation, and air conditioning systems based on data from sensors that measure temperature, humidity, and occupancy, an intelligent algorithm is deployed. The authors conducted trials in an office building located in Romania to show how successful the system is at lowering energy consumption and enhancing indoor air quality while preserving a comfortable indoor atmosphere for inhabitants.

5 Proposed Framework

The real-time predictions of climate patterns and changes would be provided by our model of a sensor-based climate prediction system, which would take advantage of technology [13]. This could lessen the possibility of damage to property and infrastructure and safeguard public safety by assisting people, organizations, and governments in making plans and preparations for extreme weather occurrences.

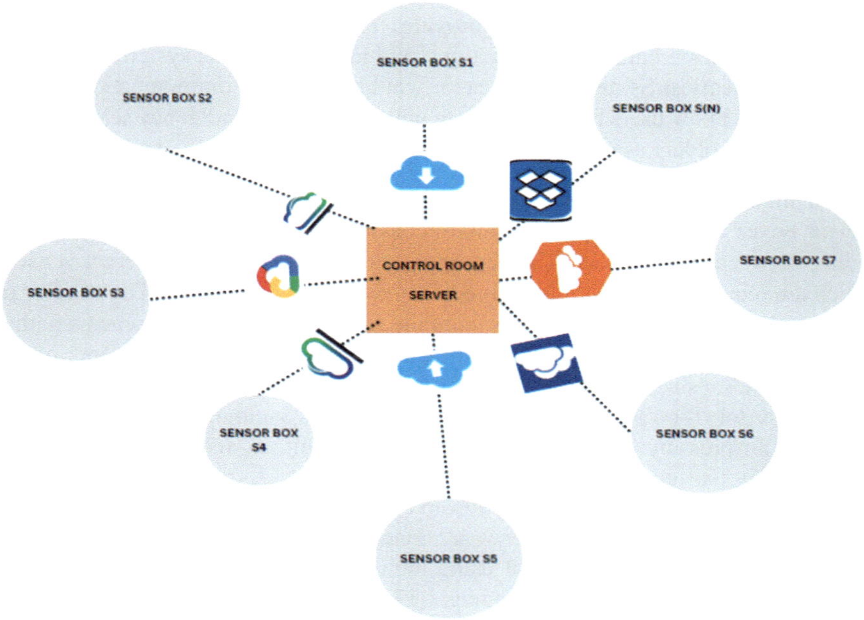

Fig. 1 Proposed model for smart climate monitoring

Our model as shown in Fig. 1 will include the following components:

- **Sensor Network**: The system would require a network of sensors installed at various locations to gather environmental data such as temperature, humidity, air pressure, wind speed and direction, pollution, and harmful gases (Examples: $\mathbf{SO_2}$ (Sulfur Dioxide), **CO** (Carbon Monoxide), $\mathbf{NO_2}$ (Nitrogen Dioxide), $\mathbf{CO_2}$ (Carbon Dioxide), dust level, and pollution from cremation ground and Slum area).
- **Data Collection**: The sensor network collects data on a continuous basis and sends it to a centralized location for processing.
- **Central Server System**: The data gathered by the sensor network is received and processed by the central server system in our sensor-based climate monitoring system [14]. The system also features a user interface that enables users—such as scientists, meteorologists, and climatologists—to view predictions, access data sets of sensed abnormal data from the sensors, and interact with the system. As a result, immediate safety measures can be performed.
- **Cloud Computing**: The system would utilize cloud computing to process and analyze the large volume of data collected by the sensor network.
- **Machine Learning Algorithms**: The system could incorporate machine learning algorithms that could analyze the data collected by the sensor network and provide real-time predictions about climate patterns and changes [15].

- **User Interface**: The system could provide a user interface that would allow users to access climate data, view predictions, and interact with the system.
- **Communication System**: The system would require a communication system that would allow users to communicate with each other, emergency services, and other stakeholders as required.

Our model will be used in a large-scale project like in a city, town, zone, etc. Sensor boxes will be set up in all possible areas in a town keeping the minimum radius between two sensors. The sensor boxes will contain the different sensors which are required to monitor the climatic condition of a place [16]. The number of sensor boxes to be installed depends upon the size of the city. These sensors gather data in real time on a variety of activities that take place in a specific area of the city up to the sensing range of the sensors in the Sensor Box. A sensor box contains multiple sensors for tracking the abnormality of various parameters like temperature, humidity, air pressure, wind speed and direction, pollution, and harmful gases (for example: SO_2 (sulfur dioxide), CO (carbon monoxide), NO_2 (nitrogen dioxide), CO_2 (carbon dioxide), dust level, pollution from burning in the cremation ground, pollution from slum areas like improper waste management systems, excessive burning of plastics, etc.). The advantage of using multiple sensors for tracking is that you can use each sensor to its best advantage [17]. Each sensor included in the sensor boxes gathers data, which it then delivers to the central server system for cloud computing-based analysis. These real-time data cannot be independently and separately analyzed because they are large and arrive at a rapid rate. After analyzing the sensor data, if any data anticipated using machine learning techniques exceeds the typical safe threshold, a siren is sounded in the central server system to indicate a subsequent rise in the level of a parameter whose sensors are installed in that belt. So, after analyzing the anomalous data, immediate emergency assistance can be given to that location. So, at that specific area of the package where the sensors detected anomalous data, the spread of pollution, dangerous gases, etc. can be stopped in this manner. As a result, the hazardous poisonous compounds that were produced in that belt can be stopped, allowing the city's climate to stay purified [18]. By leveraging the smart environment solution, our model may be used to monitor a location's climate.

6 Results Analysis and Availability Issues

Rice and wheat are planted in India in the months of June through July and November through December, respectively, and are subsequently harvested in the following months of October through November and April through May. In actuality, this gives farmers only a little window of time, usually 3–4 weeks, to transition to the following season. Furthermore, there are no cost-effective agricultural residue collection systems accessible for small areas. As a result, the farmers favor in situ crop residue burning (CRB), a method that requires less time, is less expensive, and quickly gets the field ready for the following crop. Agricultural burning following crop harvest, sometimes referred to as crop residue burning, and other types of forest fires are examples of biomass burning (CRB). The three Indian states with the highest crop burning volumes are Punjab (20 Mt), Haryana (10 Mt), and Uttar Pradesh (11 Mt). These states are a part of the Indo-Gangetic Plains (IGP), where the air quality has drastically declined recently as a result of CRB. Moreover, CRB produces a significant amount of air pollutants, including PM, NOx, SO_2, CO, VOCs, and volatile organic compounds (VOCs), which influences a variety of atmospheric feedback mechanisms and changes the quality of the air.

Figures 2 and 3 show the rise of various harmful and poisonous compounds like PM10, PM2.5, NO_2, and SO_2 following CRB activities in Delhi and Haryana. Because agricultural states like Haryana lack a system to monitor the climate to catch these burning operations, they spread to surrounding cities like Delhi, Uttar Pradesh, etc. degrading the local weather. High particle matter (PM) concentration in ambient air in Delhi and the NCR region is a result of crop residue burning (CRB). A high PM2.5 concentration in the air has a negative impact on people's health. These graphs are alarming since they demonstrate how much pollution has increased as a result of the crop residue burning (CRB) approach. Due to a lack of monitoring at the time of the incident, this condition expanded from Haryana to Delhi. Any of these locations' air quality suffers because CRB activities there emit pollutants that go to Delhi via air masses. In the Delhi and NCR region, it also adds to the high particulate matter (PM) content in ambient air. Public health is negatively impacted by high PM2.5 concentrations in ambient air. It causes a significant amount of greenhouse gases, like carbon dioxide and methane, to be released into the atmosphere, which contributes to global warming. It also contributes to soil erosion and has a bad effect on biodiversity. So to avoid all these consequences, check should be done on the burning activities.

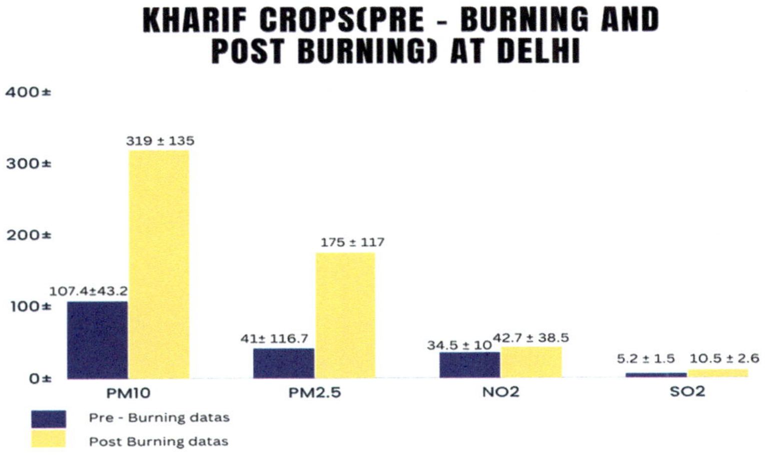

Fig. 2 Pre-burning phase concentrations of pollutants throughout the kharif season

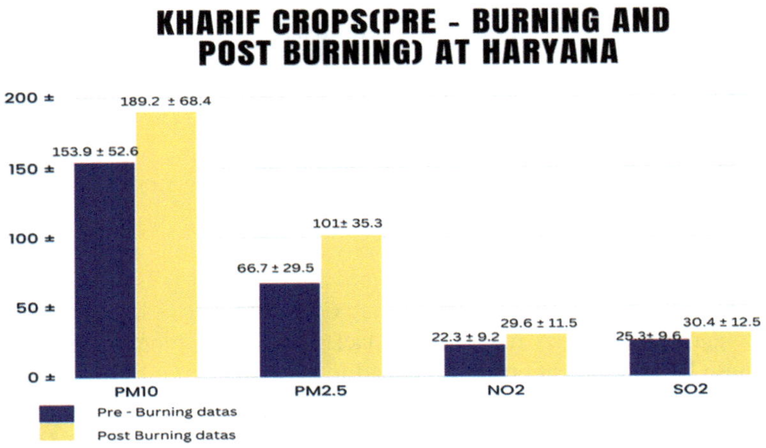

Fig. 3 Pre-burning phase concentrations of pollutants throughout the kharif season

A comparative analysis of different machine learning algorithms is discussed, and the result is highlighted in Fig. 4. It is observed that the XGBoost model outperforms others in context to accuracy rate. The XGBoost model records an accuracy rate of 94.7%, while the neural network model records the least accuracy of 80.8%.

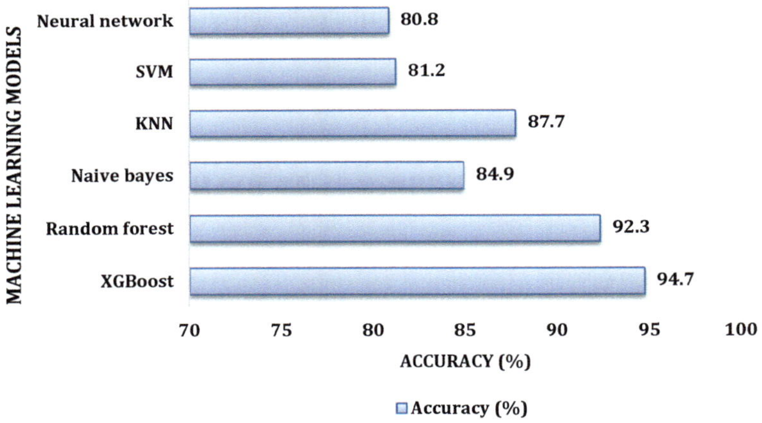

Fig. 4 Accuracy rate analysis of the proposed model with different machine learning models

7 Conclusion

In conclusion, smart environment technology-based climatic condition monitoring is an efficient tool that can improve our understanding of the environment and its effects on ecosystems, infrastructure, and human health. We can conserve our world and guarantee a sustainable future for future generations by using technology to monitor the environment and make more informed decisions. By gathering information on people's exposure to various environmental elements, smart environment technology can also facilitate tailored reactions to changes in the environment. This can help inform public health policies and interventions that are targeted toward vulnerable populations at risk of negative health outcomes due to environmental factors. The installation of a climate monitoring system, however, necessitates a large investment in both infrastructure and technology. Additionally, to guarantee the correctness and dependability of the system, issues like data privacy, data security, and data standards must be resolved. Ultimately, the use of smart environment technology can promote a more sustainable and healthier future for all, and a sensor-based climate monitoring system has the power to transform how we see the world around us and offer important new information on how human activity affects the climate. As such, it stands for a crucial investment in our shared future and ought to be pursued immediately and resolutely.

References

1. Floris, A., Porcu, S., Girau, R., Atzori, L.: An IoT-based smart building solution for indoor environment management and occupants prediction. Energies **14**, 2959 (2021). https://doi.org/10.3390/en14102959
2. Park, H., Rhee, S.-B.: IoT-based smart building environment service for occupants' thermal comfort. J. Sens. **2018**, Article ID 1757409, 10 p., (2018). https://doi.org/10.1155/2018/1757409
3. Shao, T.: Indoor environment intelligent control system of green building based on PMV index. Adv. Civil Eng. **2021**, Article ID 6619401, 11 p. (2021). https://doi.org/10.1155/2021/6619401
4. Aryblia, M., Quintal, L., Ribeiro, M., Sifakis, N., Tournaki, S., Tsoutsos, T.: Climate change mitigation through smart environmental monitoring of the urban ecosystem in insular touristic cities: experience from Rethymno and Madeira. In: Tsoutsos, T. (eds.) Sustainable Mobility for Island Destinations. Springer, Cham (2022). https://doi.org/10.1007/978-3-030-73715-3_8
5. Firan, C., Paladi, F.: Smart environment for building climate control and energy management. In: 2018 International Symposium on Signals, Circuits and Systems (ISSCS), pp. 1–4 (2018). https://doi.org/10.1109/ISSCS.2018.8401842
6. Krishnamoorthy, B., Sarangan, V.: Smart environment for climate monitoring and control. In: 2019 IEEE International Conference on System, Computation, Automation and Networking (ICSCAN), pp. 1–4 (2019). https://doi.org/10.1109/ICSCAN.2019.8862872
7. Frontoni, E., Mencagli, G., Zingaretti, P.: Smart greenhouse: an IoT-based climate control system for sustainable agriculture. Sensors **19**(19), 4087 (2019). https://doi.org/10.3390/s19194087
8. Cirillo, S., Frattari, A., Persia, F.: A smart environment for monitoring climate variations in Vineyards. Sensors **17**(5), 992 (2017). https://doi.org/10.3390/s17050992

9. Deshpande, P.S., Salankar, S.S.: Smart city environment for climate monitoring and control. In: 2017 International Conference on Intelligent Computing and Control Systems (ICICCS), pp. 329–334 (2017). https://doi.org/10.1109/ICCONS.2017.8252208
10. El Bari, H., Ballihi, L., El Ghazi, H.: A smart environment for climate control and energy management in greenhouses. In: Proceedings of the 2019 International Conference on Wireless Networks and Mobile Communications (WINCOM), Marrakech, Morocco, 22–25 October 2019, pp. 1–6. https://doi.org/10.1109/WINCOM.2019.8952961
11. Uma Maheswari, M., Balamurugan, K.: Smart environment for climate control and energy management in buildings. In: Proceedings of the 2017 International Conference on Wireless Communications, Signal Processing and Networking (WiSPNET), Chennai, India, 22–24 March 2017, pp. 1374–1378. https://doi.org/10.1109/WiSPNET.2017.8300028
12. Andreescu, G.-D., Vatavu, R.-D.: Smart environment for energy management and climate control in office buildings. Sustainability **12**(17), 6993 (2020). https://doi.org/10.3390/su12176993
13. Subhra, S., Mishra, S., Alkhayyat, A., Sharma, V., Kukreja, V.: Climatic temperature forecasting with regression approach. In: 2023 4th International Conference on Intelligent Engineering and Management (ICIEM), London, United Kingdom, pp. 1–5 (2023). https://doi.org/10.1109/ICIEM59379.2023.10166883
14. Mohanty, S., Behera, A., Mishra, S., Alkhayyat, A., Gupta, D., Sharma, V.: Resumate: a prototype to enhance recruitment process with NLP based resume parsing. In: 2023 4th International Conference on Intelligent Engineering and Management (ICIEM), London, United Kingdom, 2023, pp. 1–6. https://doi.org/10.1109/ICIEM59379.2023.10166169
15. Soumyadeep, C., Rohan, K., Aditya, K.S., Sushruta, M.: 14 explainable intelligence enabled smart healthcare for rural communities. In: Explainable Artificial Intelligence for Biomedical Applications, pp. 289–308. River Publishers (2023)
16. Adrija, M., Yash, A., Sushruta, M.: 8 pragmatic study of IoT in healthcare security with an explainable AI perspective. In: Explainable Artificial Intelligence for Biomedical Applications, pp. 145–166. River Publishers (2023)
17. Bhavya, M., Pranshu, S., Sushruta, M., Sibanjan, D.: 17 comparative analysis of breast cancer diagnosis driven by the smart IoT-based approach. In: Explainable Artificial Intelligence for Biomedical Applications, pp. 353–374. River Publishers (2023)
18. Swain, T., Mishra, S.: Evolution of machine learning algorithms for enhancement of self-driving vehicles security. In: 2022 International Conference on Advancements in Smart, Secure and Intelligent Computing (ASSIC), Bhubaneswar, India, pp. 1–5 (2022). https://doi.org/10.1109/ASSIC55218.2022.10088396

Toward Improved Clustering for Textual Data

Ridwan Amure, Abiola Akinnubi, and Oyindamola Koleoso

Abstract This study explores the possibilities of combining manifold learning with contextual embedding from Transformer models for textual cluster analysis. We leverage contextual embeddings to provide a more accurate text representation for text clustering analysis and pass the embedding through a manifold learning algorithm. The results of the experiment show that manifold learning can accentuate the contextual embedding which improves the performance of the clustering algorithms in the characterization and modeling of text data. We used the resulting clusters to distinguish between relevant texts in social media campaigns and showed that the resulting embedding provides a better representation for clustering analysis.

Keywords Text analysis · UMAP · Clustering · Contextual embedding

1 Introduction

Clustering is the process of grouping items that are similar in characteristics and different from other items. Different algorithms can be used for this purpose; they differ in their optimization function and clustering criteria. This is very crucial in information organization, information retrieval, topic modeling, and information storage [1]. Clustering analysis can be very useful in the text domain, where differences in text documents might exist at different levels. Texts can differ at the sentence, paragraph, or whole document level. Given an entire array of documents, each document can contain different items or speak about a similar topic. These subtle differences or similarities can be detected using clustering analysis [2]. Text clustering is essential

R. Amure (✉) · A. Akinnubi · O. Koleoso
University of Arkansas, Little Rock, Arkansas, USA
e-mail: raamure@ualr.edu

A. Akinnubi
e-mail: asakinnubi@ualr.edu

O. Koleoso
e-mail: okoleoso@ualr.edu

G. Fortino et al. (eds.), *Proceedings of Third International Conference on Computing and Communication Networks*, Lecture Notes in Networks and Systems 917,
https://doi.org/10.1007/978-981-97-0892-5_32

for sentiment analysis, content recommendation, topic modeling, document summarization, and information retrieval. In text clustering, text data are usually converted into a numerical format that can be used as input to the model [2]. The conventional approach to representing documents, however, faces significant challenges in the extraction of knowledge from texts due to data sparsity, mostly because the contextual information provided within them is of relatively low quality. In this study, we explore a combination of methods to improve the quality of clustering text data and compare our results using 11 benchmark datasets.

2 Background

Clustering is a widely studied field that primarily focuses on numerical data. However, the process of converting textual data to numerical data, known as vectorization, is a bottleneck in the application of clustering algorithms to textual data. Various methods have been proposed for converting textual data to vectors and applying clustering algorithms to textual data. Common methods include binary representation, count vectorization, and weighing methods such as TF-IDF [2]. However, these methods fail to capture semantic, grammatical, and lexical relationships between words and do not consider semantic factors such as word order, synonyms, and homophones. Word embedding models such as Glove [3] lack context knowledge because they map each input word to a specific vector. The transformer architecture [4] introduced context into natural language processing (NLP) tasks using self-attention mechanisms, which forms the basis of Bidirectional Encoder Representations from Transformers (BERT) and other transformer-based models. This method captures word usage in various settings and encodes knowledge that is transferable between languages and across documents [5]. This paper presents an approach using contextual embeddings combined with a manifold learning algorithm for dimensionality reduction as a vector representation for inputs in clustering algorithms.

3 Benchmark Dataset

A global benchmark dataset containing documents of varying sizes for evaluation of clustering algorithms was introduced by Muennighoff [6]. Eleven datasets are presented for evaluating the clustering produced by the model. A detailed description of the collection and nature of the datasets can be found in Muennighoff [6]. The results are publicly available on the global leader board in the Huggingface repository. We use the current models on the leaderboard as our baseline. The predictions are evaluated using the V-measure, which is the mean of Homogeneity and Completeness of the cluster predictions. The V-measure is chosen because it is independent of the absolute values of the labels: a permutation of the class or cluster label values will not change the score value in any way [7].

The V-measure is given by

$$v = \frac{(1 + \text{beta}) * \text{homogeneity} * \text{completeness}}{(\text{beta} * \text{homogeneity} + \text{completeness})} \quad (1)$$

3.1 Data Preparation

The datasets used in this study contain large documents. Conventionally, each document is split into sentences and tokenized before being used as inputs to a transformer model. For text clustering purposes, it is essential to be able to combine all sentence embeddings to reflect the actual embeddings of the document. This can often pose a challenge to transformer models. Reimers and Gurevych [8] introduced the mean pooling approach to solve this problem and also improved the computational efficiency of transformers for generating embeddings for large text documents. The pooling technique used in this study aids in the extraction of semantically meaningful and related sentences. This is important for document clustering because documents will be split into sentences because of token length restrictions in transformer models. The average pooling introduced in the paper is described below.

Given E and A as the token embedding and Attention mask from the transformer model. First, A is expanded to the same dimension as E. Then A and E are multiplied together in an element-wise manner. The result is summed and divided by the clamped value of A to obtain the mean pooling. A more rigorous proof is provided in Reimers and Gurevych [8].

4 Method

In this section, we discuss the algorithms used in this study. We leverage contextual embedding algorithms; all-MiniLM and General Text Embeddings (GTE)-base models to extract feature embeddings from text documents. Both pretrained models were downloaded from the huggingface repository. The models were chosen for their prior performance on the 11 benchmark datasets. We prove that the addition of a manifold learning algorithm can significantly impact text clustering analysis.

4.1 MiniLM Model

This model was trained on 1 billion sentence pairs using the MiniLM-L6-H384-uncased model, which was originally obtained from MiniLM-L12-H384-uncased [9]. The authors' [9] objective was to present a method for compressing large transformer models using deep self-attention distillation to train a student model using the teacher model. The self-attention module of the large model (teacher), which is crucial to transformer networks, is closely imitated to train the small model (student). In their experiment, they used the base BERT (a 12-layer Transformer) as the teacher and a 6-layer Transformer as the student to train different models, including multilingual models. The knowledge is further transferred to the all-MiniLm-L6-v2 model used in this research, which was fine-tuned with more than 1 billion sentence pairs.

4.2 General Text Embedding (GTE) Model

The model [10] consists of a deep transformer encoder initialized using the BERT model [11]. The model implements a vanilla dual-encoder-style architecture that produces token representation. Similar to the method described in Reimers and Gurevych [8], the architecture implements a mean pooling on top of the contextualized token representations. They used contrastive learning to train a single text embedding model over a variety of datasets from various sources. Notably, the GTE base model outperforms the OpenAI black-box embedding API and even 10x bigger text embedding models on the extensive text embedding benchmark, even with a relatively low parameter count of 110 M.

4.3 Manifold Learning: Uniform Manifold Approximation and Projection for Dimension Reduction (UMAP)

UMAP is an algorithm for reducing dimensionality in low-level representations, based on Riemannian geometry and algebraic topology. It is scalable and can be used with real-world data, rivaling t-SNE for visualization quality and preserving a global structure [12]. UMAP is a dimensionality reduction technique used in machine learning, data analysis, and data visualization. It aims to preserve both local and global structures in high-dimensional data while being computationally efficient. UMAP works by dealing with high-dimensional data such as text, gene expression, and image data. It uses manifold learning to find a lower-dimensional representation that preserves the underlying manifold's structure. UMAP also uses a fuzzy simplicial set approximation to build a graph connecting data points by edges. The fuzzy simplicial set is optimized in a lower-dimensional space, minimizing cross-entropy. UMAP is flexible, with hyperparameters that allow users to control the balance between local

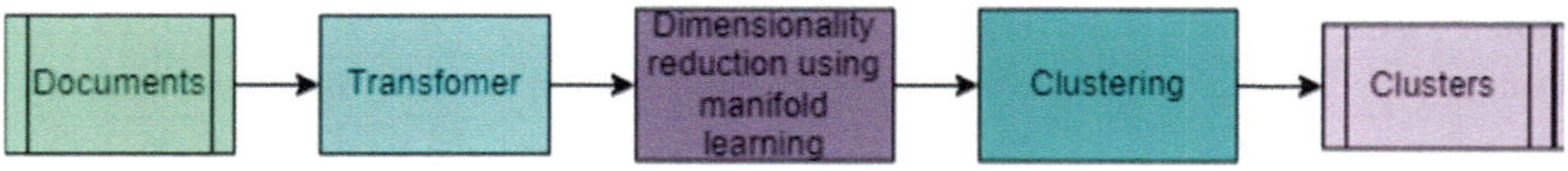

Fig. 1 Flow diagram used in this study

and global structures. In this study, UMAP is used to reduce embedding dimensions and improve semantic relationships between documents.

5 Clustering Model

This study uses the transformer models as a base to generate high-dimensional text embeddings, which are reduced using a manifold learning approach in UMAP. The low-dimensional embeddings are then clustered using the mini-batch k-means [13] clustering algorithm. The initial stage of our future study uses the clustering approach to separate relevant data from nonrelevant data in a social media context. The flow diagram is presented in Fig. 1.

6 Clustering Data from Social Media

The data used in this research were extracted from Twitter. It contains tweets about the EndSars Movement in Nigeria. This was a massive protest across Nigeria that occurred on-site and across multiple social media channels [14]. This intensity and popularity made the movement trend on social media. As such it is possible that people leverage the popularity to push different content and it will be necessary to separate the relevant data for further downstream analysis.

7 Result

Table 1 shows the V-score for the various models across the different datasets in the benchmark. The results show that each model improved with the inclusion of the manifold learning algorithm. Each model with UMAP performed better than the corresponding model without UMAP. The resulting average V-score for GTE+UMAP is higher than that of GTE-LARGE, which is the best-performing model on the MTEB clustering benchmark. The results show that the model benefited well from including the manifold for the clustering task. Performance varied across the datasets. The composite models performed better on the ArxivClusteringP2P, ArxivClusteringS2S, MedrxivClusteringP2P, MedrxivClusteringS2S, RedditClustering, Reddit-

Table 1 Table showing V-score for the clustering models

Dataset	GTE+UMAP	GTE	All-MiniLM+UMAP	All-MiniLM
ArxivClusteringP2P	**49.12**	48.6	**48.21**	46.55
ArxivClusteringS2S	**43.93**	43.01	**40.03**	37.86
BiorxivClusteringP2P	36.87	**38.2**	37.7	**38.48**
BiorxivClusteringS2S	35.04	**36.59**	31.68	**33.17**
MedrxivClusteringP2P	**34.21**	33.17	**34.55**	34.41
MedrxivClusteringS2S	**32.88**	31.77	**32.47**	32.29
RedditClustering	**62.75**	59.33	**55.61**	50.67
RedditClusteringP2P	**67.44**	62.59	**60.56**	54.15
StackExchangeClustering	**72.46**	66.64	**64.66**	53.36
StackExchangeClusteringP2P	**37.26**	36.0	35.95	**38.0**
TwentyNewsgroupsClustering	47.63	**52.31**	44.34	**46.86**
Average	**47.24**	46.2	**44.16**	42.35

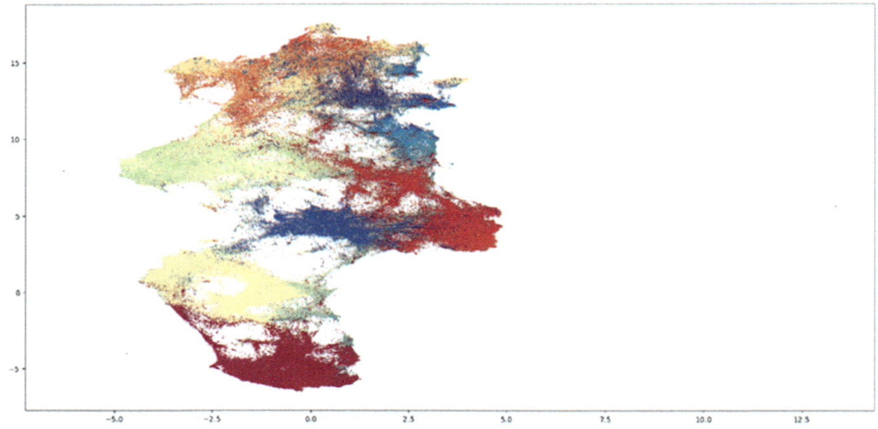

Fig. 2 Visualization of clusters in ArXIV P2P clustering dataset

ClusteringP2P, and StackExchangeClustering. The GTE-BASE+UMAP model had a higher score than the GTE BASE model on the StackExchangeClusteringP2P dataset but the all-MiniLM+UMAP model had a poorer performance on the StackExchangeClusteringP2P dataset than the all-MiniLM model. On average, the composite models showed greater potential in clustering text data. Figure 2 shows the visualization of the two-dimensional embedding of the ArXIV P2P dataset showing the clusters.

The clustering of the social media data processed showed that the model was able to separate the actual data needed to understand the #ENDSARS movement. Qualitative analysis of the cluster reveals that cluster 0 contains the relevant tweets

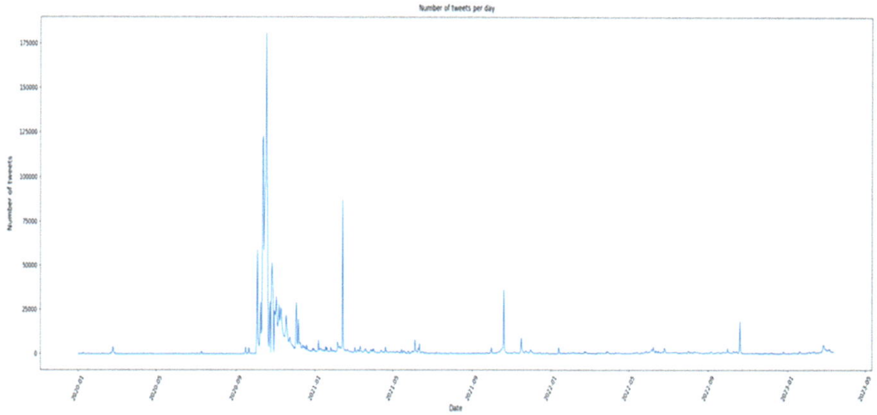

Fig. 3 Plot showing the number of tweets per day for the time frame under consideration

representing the actual discourse on the ENDSARS movement. We focused on the data in cluster 0 for further analysis.

Figure 3 shows the number of tweets per day on the #ENDSARS movement. The figure shows trends that correspond to the events that occurred during the campaign. The movement was centered on ending police brutality, particularly the Special Anti-Robbery Squad (SARS) unit. The Nigeria Police Force announced it will dissolve the unit on October 11, 2022. This did not have any effect on the movement as the public claimed that such promises had been made in the past. This led to the events of October 20, 2020 [14]. All these events explain the sharp increase in the number of tweets observed from October to November 2020 to March 2020 (Fig. 3). The Federal Government of Nigeria announced the ban on Twitter in Nigeria on June 5, 2021. This ban lasted till January 13, 2022. This corresponds to the downtrend and flats seen in the curve from June 2021.

8 Conclusion

In this study, an improved method for clustering text data is based on the introduction of manifold learning to token embedding from a transformer model. This method uses contextual embedding to better represent text documents, which can then be properly modeled by a clustering algorithm. This method achieved better performance and showed more prospects in text clustering analysis. We demonstrated the strength of this method using 11 benchmark datasets, and our approach recorded the highest performance. We also presented insights from our future endeavors in the analysis of text data from social media. The approach described in this study will be suitable for future experiments involving text analysis.

It will be important to include manifold learning as a trainable layer in the model architecture to further learn more intrinsic structures and relationships that may exist within the text documents.

9 Future Work

This is part of an ongoing study on social media analysis using text data. The UMAP algorithm used in this study has memory limitations. There is ongoing work on implementing a batched approach to the nearest neighbor computation. This will help the algorithm scale to larger datasets because the nearest neighbor computation is expensive. It would be beneficial to introduce the algorithm as a layer to be trained or fine-tuned with the transformer model.

References

1. Xie, J., Girshick, R., Farhadi. A.: Unsupervised deep embedding for clustering analysis. In: International Conference on Machine Learning, pp. 478–487. PMLR (2016)
2. Aggarwal, C.C., Zhai, C.X.: A survey of text clustering algorithms. In: Mining Text Data, pp. 77–128 (2012)
3. Pennington, J., Socher, R., Manning, C.D.: Glove: global vectors for word representation. In: Proceedings of the 2014 Conference on Empirical Methods in Natural Language Processing (EMNLP), pp. 1532–1543 (2014)
4. Vaswani, A., et al.: Attention is all you need. In: Advances in Neural Information Processing Systems, vol. 30 (2017)
5. Liu, Q., Kusner, M.J., Blunsom, P.: A survey on contextual embeddings (2020). arXiv preprint arXiv:2003.07278
6. Muennighoff, N., et al.: MTEB: massive text embedding benchmark. In: Proceedings of the 17th Conference of the European Chapter of the Association for Computational Linguistics, pp. 2014–2037. Association for Computational Linguistics, Dubrovnik, Croatia (2023)
7. Rosenberg, A., Hirschberg, J.: V-measure: a conditional entropy-based external cluster evaluation measure. In: Proceedings of the 2007 Joint Conference on Empirical Methods in Natural Language Processing and Computational Natural Language Learning (EMNLP-CoNLL), pp. 410–420 (2007)
8. Reimers, N., Gurevych, I.: Sentence-BERT: sentence embeddings using Siamese BERT-networks. In: Conference on Empirical Methods in Natural Language Processing (2019)
9. Wang, W., et al.: Minilm: deep self-attention distillation for task-agnostic compression of pre-trained transformers. Adv. Neural Inform. Process. Syst. **33**, 5776–5788 (2020)
10. Li, Z., et al.: Towards general text embeddings with multi-stage contrastive learning (2023). arXiv preprint arXiv:2308.03281
11. Devlin, J., et al.: Bert: pre-training of deep bidirectional transformers for language understanding (2018). arXiv preprint arXiv:1810.04805

12. McInnes, L., et al.: UMAP: uniform manifold approximation and projection. J. Open Sour. Softw. **3**(29), 861 (2018)
13. Sculley, D.: Web-scale k-means clustering. In: Proceedings of the 19th International Conference on World Wide Web. WWW'10, pp. 1177–1178. Association for Computing Machinery, Raleigh, North Carolina, USA (2010). ISBN 9781605587998
14. Uwalaka, T.: Social media as solidarity vehicle during the 2020 #EndSARS protests in Nigeria. J. Asian Afric. Stud. **59**(2), 00219096221108737 (2022)

Self-adaptive Learning Algorithm as a Tool for the Development and Strengthening of the Dyslexic Student's Skills in the Study of Musical Composition

Michele Della Ventura

Abstract In this period, the topic of Artificial Intelligence (AI) is at the center of various discussions that mainly concern the teaching/learning system and the contribution that AI can make in the classroom. From a naturally inclusive perspective, technologies (based on AI) from the early stages of design and development must take into consideration the needs of all students, without neglecting dyslexic students. In that sense, the main problem is represented by the ability of these tools to support the dyslexic student in studying and not in carrying out the assigned tasks for him/her. This research paper presents an algorithm (which combines the Viterbi algorithm with the Markov process) designed and developed to enhance the learning process of a dyslexic student in the area of Music Composition, in particular, in the harmonization of a melodic line. The harmonization of a melodic line involves many cognitive processes and systems and a learning disability may lead to over-fatigue and to a psychological distress of the student, with the subsequent loss of confidence and personal motivation. The algorithm looks like a tutor able to support the student through suggestions during the harmonization of a musical melody (and not by providing the solution). The experimental results show that the algorithm, given a melodic line, is able to correctly generate a bass line as well as harmonize each sound of the melodic line. It is also demonstrated (through a case study) that the algorithm can have a positive impact on dyslexic students' motivation and engagement in learning. Finally, educational implications and research suggestions are provided based on the research results.

Keywords Artificial intelligence · Dyslexia · Educational technology · Hidden Markov Model · Melodic harmonization

M. D. Ventura (✉)
Department of Music Technology, Music Academy "Studio Musica", Treviso, Italy
e-mail: micheledellaventura.mdv@gmail.com

G. Fortino et al. (eds.), *Proceedings of Third International Conference on Computing and Communication Networks*, Lecture Notes in Networks and Systems 917,
https://doi.org/10.1007/978-981-97-0892-5_33

1 Introduction

Nowadays, the introduction of Artificial Intelligence (AI) in the classroom represents a necessary breakthrough. However, this tool must be used with awareness and knowledge of the facts: with proper use, AI can represent a valid and valuable aid for the learning process. The use of technology (based on AI) in the classroom is, however, one of those divisive issues. In fact, it is very often commonly believed that too much technology limits creativity and at the same time offers shortcuts for students (and/or teachers). Instead, it is necessary to become aware of the many possibilities that these tools offer and of the importance they cover, whether we like it or not, in the learning process of students in general but above all of dyslexic students. These tools should not provide the student with the solution to a problem, but help the student find the appropriate solution to the problem. This necessarily implies that AI-based tools must be adequately designed, through a fruitful and fertile integration of multiple strategies that can avoid the creation of educational resources that are by their nature deficient [1]. The growing openness of the Educational Technology World to inclusion can be traced in the trend of the main teaching tools to satisfy the needs of personalized learning [2] where the software should act as an empathic mediator to strengthen motivation, awareness, and mastery of the student's learning process [3]. New technologies must guarantee the acquisition of skills useful for individual growth, and therefore, combine the use of resources and tools with the possibility of developing critical thinking and producing multimedia artifacts from scratch. To enhance the creativity and aptitudes of individual students, it is necessary to stimulate their aptitudes, generating curiosity, and introducing them to new fields of study [4].

In the field of music and more specifically in the area of musical composition, several algorithms have been developed that are able to compose music, using Markov chains [5–7] or neural networks [8–10]. The most recent advances in the field of deep learning have opened the way for a number of new methods for generating music [11–14]. However, all of these systems for music composition focus largely on autonomous generation: little work in this domain has been addressed regarding issues related to the support of the student (and not of the "musician") in educational activities.

In the educational field (in general but more specifically in the music field), the AI should help the student to improve his/her critical thinking [15] and consequently his/her creative thinking [16]: the AI should teach the student to look at things from different perspectives, to test hypotheses, and to find creative solutions to problems. Critical thinking helps the student to make informed decisions, solve problems effectively, and develop independent thinking. This is important for all students but becomes crucial for the dyslexic student.

Starting from these assumptions, this paper presents an algorithm able to support the dyslexic student in harmonizing a melodic line, providing visual suggestions that have the purpose of guiding the student toward the search for a solution: harmonization of each sound of the melodic line avoiding the errors provided by musical grammar. The algorithm is based on the concept of deep learning, and thanks to the training phase it is able to support the student through the Markov Decision Process.

The structure of this paper has been organized as follows. Section 2 explains the principles behind the harmonization of a melody. Section 3 explores the concept of "dyslexia", which underlies the proposed algorithm. This is followed (Sect. 4) by a description of the method used to achieve the goal. Section 5 shows some experimental evidence illustrating the effectiveness of the proposed method. Finally, in Sect. 6, the paper concludes with concluding remarks on current issues and future research possibilities with respect to the efficient improvement of educational practices and technologies.

2 Melody Harmonization

Along with rhythm and melody, harmony is one of the three constitutive elements of the music. Leaving aside the concept of rhythm which is not the object of this research, the melody can be defined as a succession of sounds with complete meaning (see Fig. 1a), while the harmony represents the simultaneity of two or more sounds built under each sound of the melody (see Fig. 1b) [17].

Harmonizing a melody line means arranging a melody in a harmonic context, submitting a sound texture to the music, and choosing and evaluating the alternatives between the possible chords [17, 18]. The musical chord consists of the superimposition of several sounds that are performed simultaneously (see Fig. 1b). When a chord is arranged in overlapping thirds, it is said to be in the fundamental state (see Fig. 2a): in this case, the lowest note is called fundamental, and the other two notes are called third and fifth (seventh, ...). If, on the other hand, the chord is in a position in which the root is not the lowest note, it is said that the chord is in an inverted state: first position and second position (see Fig. 2b). From this, it follows that each sound of the melody can belong to different chords (see Fig. 3).

During his harmonization, the student must understand which chord each sound belongs to on the basis of the previous chord and how the following chords are intended to be developed. When choosing, two important sounds must be taken into consideration: the upper and lower sound of the chord which represents the bass, and the soprano (the lower note gives depth to the chord and the upper one is the most perceptible to the human ear) [18].

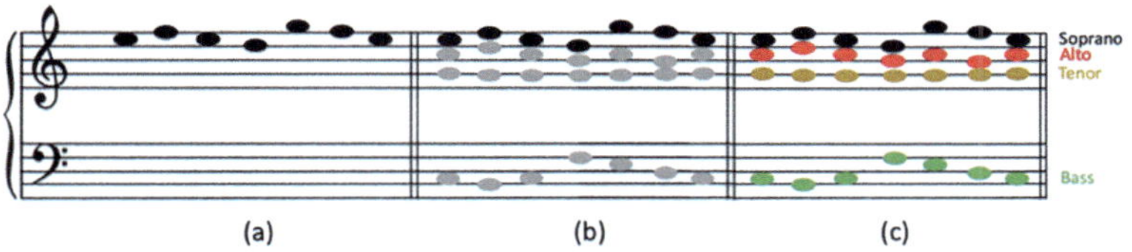

Fig. 1 Example of melody and its harmonization

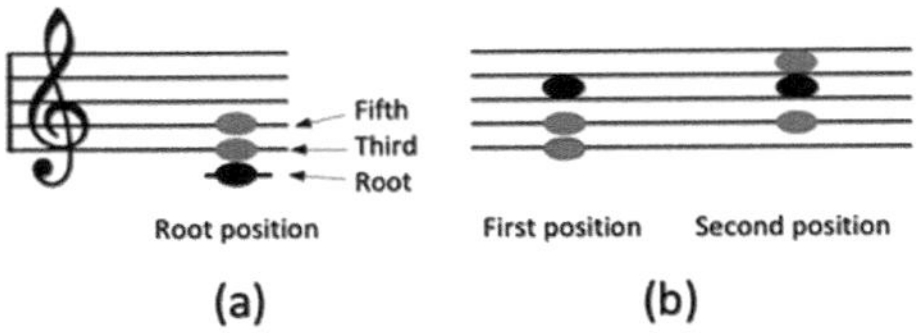

Fig. 2 Example of musical chords

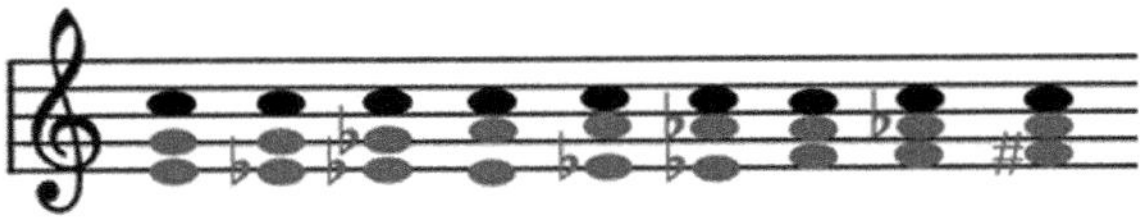

Fig. 3 Examples of possible musical chords for the C note of the melody

For this reason, before harmonizing a melodic line, the student learns to harmonize a bass line (which makes it easier to learn the many musical grammar rules) trying to build an interesting melodic line. For each sound of the bass line, the student is able to precisely define which are the other sounds of the chord (tenor, alto, and soprano) and only has to decide their arrangement to develop the melody (see Fig. 4). Being that the harmonization must be done in 4 voices (see Fig. 1c) and that instead a chord is composed of only 3 sounds, the musical grammar imposes that a sound of the chord is doubled in one of the 4 voices, according to the scale degree of the bass line sound. This offers the possibility to modify the trend of the melodic line (see Fig. 4a and b).

In harmonizing the melody, on the other hand, the student must understand which chord each sound belongs to: it is an important choice that requires time (due to the many possibilities that each sound can have, as seen in Fig. 3) and concentration and which therefore it can lead the student (especially dyslexic) to lose motivation and abandon the exercise.

The algorithm presented in this document intends to help the dyslexic (and non-dyslexic) student in the harmonization of the melody, suggesting (where necessary) which sound can be present in the bass line for each sound of the melody. This allows the student to know the two most important sounds of the chord (bass and soprano) and to apply the previous knowledge acquired during the harmonization of the bass line: the graphical representation (as written in paragraph 3) helps the dyslexic student to memorize certain situations and to know how to recognize and reproduce them when they arise.

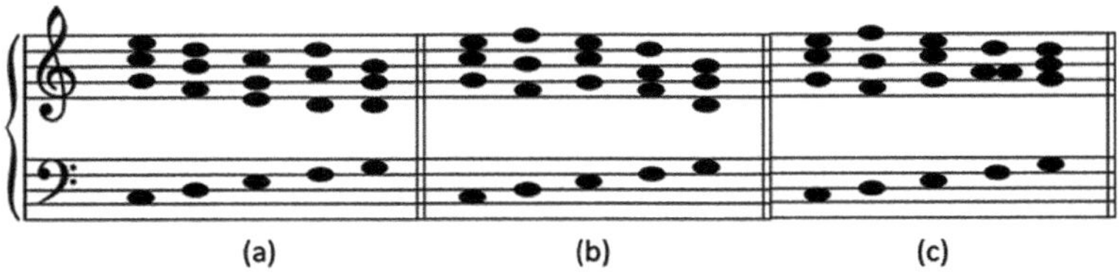

Fig. 4 Example of different bass line harmonization

3 Dyslexia Versus AI (and Vice Versa)

How can AI support the learning process of the dyslexic student?

Dyslexia does not depend on subjective psychological factors, low motivation to study, or low intelligence. Dyslexia consists of the difficulty of automating the process of decoding the graphic signs that characterize the written text [19]. This difficulty has repercussions on the performance which is often characterized by low reading speed and/or poor accuracy and this can compromise the understanding of the text [20]. This may mean difficulties in understanding the rules of musical grammar and consequently little or no practical application of the concepts, i.e., difficulties in harmonizing the sounds of the melody.

The natural compensatory tool of the person with dyslexia is the development of the prediction skill, i.e., "*guessing the words while reading*" [20, 21]. In harmonizing the melody, it is necessary to avoid this instrument becoming the lifeline: harmonization requires a lot of energy and concentration, because it is necessary to identify the sounds of each chord and these depend on the sounds that make up the previous chord and how you intend to build harmony in subsequent chords. Trying to "guess" the sounds of the chord could lead the student to make mistakes (based on musical grammar). Dyslexia involves a reading difficulty that can be compensated if the information is provided through oral channel support (such as an audio message, an audio file, ...) [22] or the visual one, for example, using images [23]. It is immediately clear, therefore, how important it is for the dyslexic student the use of study material that is not only present in written mode, but also in graphic mode. In textbooks, the rules of musical grammar are always explained by the written text (rich in specific technical terminology) followed by images representative of the final result that would be obtained from their application. In other words: given some sounds of the melody is proposed an image that shows its harmonization on the basis of the rules explained, and it is up to the student to understand the movement of the sounds of the various voices between the chords that follow. The difficulty of understanding the written text is added to the difficulty of interpreting the image. It goes without saying that it is not enough to simply change the mode of presentation to eliminate any difficulty [24]: it is necessary, in fact, that the student with dyslexia learn to recognize certain melodic situations (for example, a melodic progression, a stylistic feature, ...) and to harmonize them (adding to each melody sound the sounds that characterize the musical chord and making the sounds move correctly between one chord and the next).

In order to acquire an effective method of study, the student with dyslexia must be supported [25] by the teacher and by the use of appropriate educational and technological tools called "compensatory tools". Compensatory tools (IT or not) are always aimed at the acquisition of autonomy [22]: they need an initial learning phase to be used at their best and their aim is to improve skills and autonomous learning. Basically, a good compensation tool is one that requires little effort in learning how to use it, that is simple and immediate and that leaves intact, in its user, the resources and energies to decide on the activities of study and understanding of concepts. It should be stressed that compensatory technologies do not "compensate in themselves". In order for users to be effective, they must also develop in parallel a "compensatory competence", that is, a strategy that allows them to make the best use of the tool and integrate it effectively in the learning phases.

Compensatory tools, based on AI, can play a fundamental role in the implementation of inclusive teaching. Through the use of technologies, the dyslexic student can be allowed to do what he otherwise could not do, or to do it with less expenditure of energy, or to do it in a more independent or safe or psychologically more acceptable way [26].

The algorithm object of this research takes its cue from the above considerations and presents itself as a support tool for the dyslexic student, with the aim of guiding him in the correct harmonization of a melodic line through the suggestion of the possible sounds of the bass line, leaving the completion of the tenor and contralto lines to the student. In this way, the student has the possibility to observe (and memorize) how the movements of the melody and the bass line vary without the interference of other sounds.

4 The Proposed Model

This paragraph illustrates the general architecture of the algorithm designed and developed on the basis of the concepts presented in paragraphs 2 and 3. In summary:

(1) the melodic line and the bass line represent two fundamental elements of the harmony of a musical piece: the trend of the first is closely connected to the second (and vice versa);
(2) the ascending or descending movement of the sounds of the melodic line depends on the movement of the sounds of the bass line (and vice versa): these determine the arrangement of the sounds of the other voices (tenor and alto);
(3) the ascending or descending movement of a sound of the melody depends on the degree of the scale of the sound of the bass line, and therefore, on which sound of the chord is doubled (see example in Fig. 4);
(4) the graphic representation is an important compensatory tool for the dyslexic student: it represents the educational mediator to recall content related to specific concepts [23];
(5) in the harmonization of the melody, the possibility to define the sounds of the bass line with the of the AI prevents the dyslexic student from resorting to the use of the prediction skill and allows him/her to do what he/she otherwise would not could, or to do it with less expenditure of energy, or to do it in a more independent or secure or psychologically more acceptable way: in other words, AI can be a useful tool to motivate the student.

The algorithm starts from a fundamental assumption: *a sound of the melodic line has an ascending or descending movement that depends on the sound of the bass line, which makes it possible to identify the degree of the musical scale to which the*

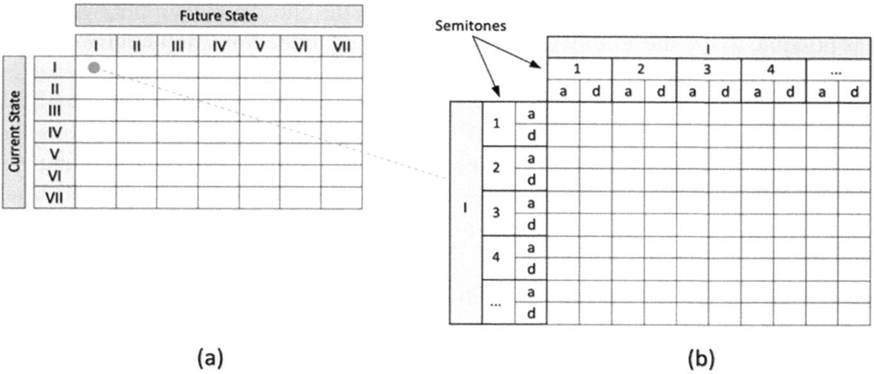

Fig. 5 Example of transition matrix

chord belongs, which in turn defines the sounds of the chord and which ones can be doubled.

This can be represented by a transition matrix P_t that is, for Markov chain $\{X\}$ at time *t*, a matrix containing information on the probability of transitioning between states. In particular, given an ordering of a matrix's rows and columns by the state space S, the (*i*, *j*)th element of the matrix P_t is given by

$$(P_t)_{i,j} = P(X_{t+1} = j | X_t = i)$$

For the algorithm, the matrix represents the probability that one degree of the musical scale X_t resolves on another degree of the musical scale (Fig. 5a, column 1 and row 1) [5].

However, this is not enough to describe the movement of the sounds of the melody in the transition from one chord to another. It is necessary to consider simultaneously both the ascending or descending movement of the sound of the melodic line (Fig. 5b, column 3 and row 3) and the number of semitones that characterize this movement (Fig. 5b, column 2 and row 2).

The probabilities calculated on the training corpus are associated with the possible transitions between the various states [27]: starting from any state S_i, the next sound of the melody will be attributable to a given state S_j with a certain probability

$$P(S_i | S_J) = \frac{C(S_i, S_j)}{C(S_i) \cdot N}$$

where $C(S_i, S_j)$ is the count in the corpus of occurrences of the sequence S_i S_j, $C(S_i)$ is the count in the corpus of occurrences of S_i and N is the number of states. All probabilities $P(S_i|S_j)$ are entered into the transition matrix.

If with the transition matrix, it is possible to describe the trend of the melody, to define the bass line (which will determine the degree of the musical scale, and therefore, allow the identification of the sounds of the chord and their arrangement), it is possible to use the Viterbi algorithm (associated to the transition matrix), which with a dynamic programming approach, allows to analyze the paths between states starting from the initial state going through the successive states identifying the best path in relation to the given emissions [27]. The probability of the most probable path ending in state k with observation i is

$$p_1(i, x) = e_1(i) \max_k (p_k(j, x-1) \cdot p_{k1})$$

where i represents the probability of observing element i in state l, j represents the probability of the most probable path ending at position $x - 1$ in state k with element j, and p_{kl} represents the probability of the transition from state l to state k.

Training Phase

The use of the algorithm has provided a training phase consisting in the reading of musical pieces composed in the form of a 4-voice choir (soprano, alto, tenor, bass) which included fifth chords (such as C-E–G) and seventh chords (such as C-E–G-B), the inversions of these types of chords and that there were no modulations (i.e., key changes). The musical pieces were in MIDI format: a digital format that does not require complex information decoding operations (therefore, greater speed in the phase of chord analysis and compilation of the transition matrix) and of which it is possible to find a well-stocked database online. The transition matrix did not have any default setting regarding the dimensions, but on the contrary, it was automatically resized during the reading of the chords (based on the type of each chord, the degree of the scale of the chord, the ascending or descending movement of the melody sound in the concatenation of two consecutive chords).

Testing Phase

During the training phase, the algorithm was tested to verify its functionality. In particular, the algorithm was asked to read a melody and propose possible solutions for the bass line. The goal was to look at the algorithm as it was defined and inspect it carefully for imprecisions, so as to be able to identify which melodic passages represented an obstacle for the algorithm, create set tests suitable for verifying the overcoming of those defects and prepare, therefore, an update of the training phase with the reading of musical pieces with specific characteristics.

It emerged that (see Table 1), as the reading of the number of possible cases of concatenation of musical chords increased (training phase):

- the possibility of the bass line not being completed decreased (Table 1 column 2);
- the possibility that the sounds of the bass line represented grades of the scale whose sequence was not correct (according to the rules of musical grammar) decreased (Table 1 column 3);
- the possible solutions for the bass line (corrected from the point of view of musical grammar) increased (Table 1 column 4).

Table 1 Analysis of the algorithm solutions

Number of chords	Uncompleted bass line (%)	Uncorrected grade sequence (%)	Possible solutions (%)
500	96	100	0
1000	87	81	1
1500	72	67	1
2000	59	55	1
3000	31	41	2
4000	28	26	2
5000	24	18	3

As already demonstrated by the various experiences of deep learning, for a better performance of the algorithm, it is necessary to analyze a large number of cases. As regards the present research, it was important to maximize the number of possible solutions of the algorithm (corrected according to the rules of musical grammar) (see Table 1 column 4) to support the student with greater accuracy avoiding loss of motivation, and to give the student interpretative freedom (creativity).

5 Evaluation: A Case Study

The first tests to verify the algorithm were carried out by 3 teachers of music composition, who had the task of verifying the proposed solutions and identifying possible integrations of the training data. Subsequently, a group of students (identified by the 3 teachers) was asked to use the algorithm.

The test was conducted for a time period of 4 months (from January 2023 to April 2023) and involved 12 students, 5 girls, and 7 boys (aged 15–16), of which 6 were affected by dyslexia: all of them had similar educational backgrounds.

During the first month, the students participated in the lessons and carried out the proposed activities, without using the algorithm. The teacher monitored the students' behavior and corrected the exercises by providing suggestions and further explanations (where necessary). At the end of the month, an examination was passed in the classroom consisting of the harmonization of a melody, identical for non-dyslexic and dyslexic students.

The result supplied important information so as to be able to continue with the project. In particular, it emerged that:

- 6 students harmonized all the melody (of which 1 dyslexic student);
- 4 students harmonized about 60% of the melody (of which 3 dyslexic students);
- 2 students (with dyslexia) harmonized less than 40% of the melody.

In February, the teacher introduced the algorithm to the students, who were able to start using it while carrying out exercises.

At the end of the month, a classroom examination was carried out, assigning the same test to all students and allowing them to use the algorithm. The results supplied the hoped-for information:

- 9 students harmonized all the melody (of which 3 dyslexic students);
- 2 students (with dyslexia) harmonized about 90% of the melody;
- 1 student (with dyslexia) harmonized about 80% of the melody.

In the following two months (March and April) the students continued to use the algorithm during the various activities proposed by the teacher, acquiring greater autonomy which was confirmed in the last test proposed at the end of the course: it harmonization of a melody without using the algorithm.

The results were satisfactory:

- 8 students harmonized all the melody (of which 2 dyslexic students);
- 2 students (with dyslexia) harmonized about 85% of the melody;
- 2 students (with dyslexia) harmonized about 70% of the melody.

Beyond the numerical results, one of the important things that emerged concerns the increase in motivation (and therefore self-esteem) of dyslexic students. In fact, it was observed that, at the beginning, if the algorithm stopped because it was unable to give a suggestion, the dyslexic student also stopped; subsequently, the student was able to go ahead with the harmonization, even if only for a few chords. This had tangible feedback on each student's performance, which improved their grades. The learning improvement also appeared for students who already drew a high profit because they learned to find more than one possible solution for the harmonization of the same melody.

6 Discussion and Conclusions

Artificial Intelligence can certainly bring innumerable advantages within the musical composition learning process: on the one hand, it will be essential to educate students to understand AI algorithms (promoting an aware use of the technological platforms that surround them), on the other hand, the same algorithms will facilitate the discovery of many details of the learning processes that are now difficult to detect (therefore useful for the improvement of the algorithms).

It is important to underline two aspects to keep in mind when creating educational software: the centrality of the student in the learning process, and the importance of teacher training, who will not have to be replaced by algorithms, but who will be able to use these tools for an effective teaching activity and deep learning.

In this paper, an algorithm has been presented to support the dyslexic (and non-dyslexic) student in the field of the study of musical composition, which is the harmonization of the melody: a complex and demanding operation that can lead the student to lose motivation, and therefore, to negative results. The simple and intuitive graphical interface allows the dyslexic student to be able to benefit from a presentation of information through the visual channel.

The results obtained are promising both in terms of the ability of the algorithm to harmonize a melody (or just generate the bass line) and in terms of the motivational aspect of the student (dyslexic and non-dyslexic) which has led to an improvement in educational achievements. Despite this, it is possible to highlight how this type of activity has tried to make the concepts that underlie the learning process experiential and directly tangible at a cognitive and metacognitive level, and therefore, it is necessary to perfect the training phase by reading a greater number of chord concatenation, allowing the algorithm to present alternative solutions for the same melody.

References

1. Story, M.F., Mueller, J., Mace, R.: The Universal Design File: Designing for People of All Ages and Abilities, Revised edn. The Center for Universal Design at North Carolina State University (1998)
2. OET, Reimaging the role of Technology in Education—2017 National Education Technology Plan Update, U.S. Department of Education, Washington, DC
3. Guzik, A., et al.: The Book of Trends in Education 2013/2014, Young Digital Planet, Danzica (2014)
4. Iannella, A.: Digital authentic learning: introdurre alla tone analysis per favorire un pensiero critico digitale. In: Thamyris, vol. IV, no. I, pp. 319–336 (2019)
5. Ames, C.: The Markov process as a compositional model: a survey and tutorial. Leonardo **1989**(22), 175–187 (1989)
6. McLean, A., Dean, R.T.: The Oxford Handbook of Algorithmic Music. Oxford University Press, New York, NY, USA (2018)
7. Cope, D.: The Algorithmic Composer, vol. 16. AR Editions Inc., Madison, Wisconsin, USA (2000)
8. Agarwala, N., Inoue, Y., Sly, A.: Music composition using recurrent neural networks (2017). https://web.stanford.edu
9. Bretan, M., Weinberg, G., Heck, L.: A unit selection methodology for music generation using deep neural networks (2016). arXiv:1612.03789v1
10. Sturm, B.L.: Recurrent neural networks for folk music generation (2015)
11. Briot, J.P., Hadjeres, G., Pachet, F.: Deep Learning Techniques for Music Generation. Computational Synthesis and Creative Systems. Springer, London (2018)
12. Briot, J.P., Pachet, F.: Deep learning for music generation: challenges and directions. Neural Comput. Appl. **32**, 981–993 (2020)
13. Amper Music, January 2019. https://www.ampermusic.com/
14. EcrettMusic, February 2019.https://ecrettmusic.com/play
15. Holmes, W., Bialik, M., Fadel, C.: Artificial intelligence in education for teaching and learning, Boston (MA), Center for Curriculum Redesign (2019)
16. Resnick, M.: Lifelong Kindergarten. Cultivating Creativity Through Projects, Passion, Peers, and Play. MIT Press, Boston (2017)
17. Schoenberg, A.: Theory and Harmony, Reprint edn. University of California Press (1992)
18. Rimsky-Korsakov, N.: Practical Manual of Harmony (N. Hopkins, ed.). Joseph AC (2005)
19. Roitsch, J., Watson, S.: An overview of dyslexia: definition, characteristics, assessment, identification, and intervention. Educ. J. **7**(4), 81–86 (2019). https://doi.org/10.11648/j.sjedu.20190704.11

20. Snowling, M.J.: Dyslexia: A Very Short Introduction. Oxford University Press, Oxford, UK (2019)
21. Lefèvre, E., Cavalli, E., Colé, P., et al.: Tracking reading skills and reading-related skills in dyslexia before (age 5) and after (ages 10–17) diagnosis. Ann. Dyslexia **73**, 260–287 (2023). https://doi.org/10.1007/s11881-022-00277-x
22. Della Ventura, M.: Creating inspiring learning environments by means of digital technologies: a case study of the effectiveness of WhatsApp in music education. EAI Endorsed Trans. e-Learn. (J.) **4**, 1–9 (2017). https://doi.org/10.4108/eai.26-7-2017.152906
23. Della Ventura, M.: Between research and action: the generative sense of technology. In: Rønningsbakk, L., Wu, T.T., Sandnes, F., Huang, Y.M. (eds.) Innovative Technologies and Learning. ICITL 2019. Lecture Notes in Computer Science, vol. 11937, pp. 754–763. Springer, Cham. https://doi.org/10.1007/978-3-030-35343-8_78
24. Wang, R., Chen, L., Solheim, I.: Modeling Dyslexic students' motivation for enhanced learning in e-learning systems. ACM Trans. Interact. Intell. Syst. **10**(3), Article 21 (November 2020), 34 p. (2020)
25. Elbeheri, G.: Motivating Students with Dyslexia: 100 Ideas for Empowering Teachers in the Classroom. Routledge (2020)
26. Chiappini, G.P., Dini, S., Ferlino, L.: Tecnologie didattiche e disabilità. In: Tecnologie per la didattica, Franco Angeli, Milano, pp. 233–248 (2004)
27. Baum, L.E., Petrie, T.: Statistical inference for probabilistic functions of finite-state Markov chains. Ann. Math. Stat. **37**(6), 1554–1563 (1966)

Digital-Game-Based Language Learning: An Exploration of Attitudes Among Teacher Aspirants in a Non-metropolitan Area

Princess Krizzle M. Casiano, Bernadeth A. Encarnacion, Shania H. Jaafar, and Ericson O. Alieto

Abstract The growing acceptance of video games and their usage in language instruction has only served to highlight the importance of good preparation for aspirant teachers in educational institutions. As digital learning games can satisfy students' desire for enjoyable educational experiences, several learners held the opinion that engaging in video games would improve their capacity for more advanced cognitive processes. The primary factor influencing the standard of digital game-based learning is the sizeable amount of time learners devote to playing games (Kunter et al. in Learn Struct 18(5):468–482, 2008 [43]; Papastergiou in Comput Educ 52:1–12, 2009 [44]). Due to this, offering educational gaming to young learners offers a conceivable advantage in the curriculum. Thus, 90 primary school teacher aspirants were chosen as participants for the current study, which employed a quantitative descriptive approach. This study discloses that learners have a '*positive attitude*' toward the utilization of video games in the classroom. These results are in agreement with other similar investigations (Kazu and Kuvvetli in Educ Inf Technol 1–26, 2023 [20]; Ray and Coulter in J Digit Learn Teach Educ 26:92–100, 2010 [21]) that examined how learners view the value of games in terms of how they can alter after playing them. They noticed that many players or students in a certain game displayed a favorable attitude following their involvement in class. They believed that playing video games could help learners learn more effectively. Upon playing the online games, attitudes among learners toward VG appeared to have changed, according to Ray and Coulter in J Digit Learn Teach Educ 26:92–100, 2010 [21],

P. K. M. Casiano (✉) · B. A. Encarnacion · E. O. Alieto
Western Mindanao State University, 7000 Zamboanga City, Philippines
e-mail: xt202001567@wmsu.edu.ph

B. A. Encarnacion
e-mail: ernadeth.encarnacion@wmsu.edu.ph

E. O. Alieto
e-mail: ericson.alieto@wmsu.edu.ph

S. H. Jaafar
Mindanao State University-Sulu, Capitol Site, Jolo, 7400 Sulu, Philippines
e-mail: Shania.jaafar@msusulu.edu.ph

G. Fortino et al. (eds.), *Proceedings of Third International Conference on Computing and Communication Networks*, Lecture Notes in Networks and Systems 917,
https://doi.org/10.1007/978-981-97-0892-5_34

Kenny and McDaniel in Br J Edu Technol 42:197–213, 2011 [22]. The results suggest that gaming is advantageous to experiences that may support learners' understanding of the importance of VGs. Furthermore, in terms of the association between the respondents' attitudes and gender, it was found that the latter is a factor that influences the attitude of the teacher aspirants. Additionally, this paper also found that in terms of the usefulness of video games in the classroom, gender is not a factor that influences the attitude of the respondents toward DGBLL. The results were further elaborated in the study. Analyzing respondents' opinions on language learning through digital games can undoubtedly help address the Philippines' lack of ICT literacy as a future elementary school teacher. The results of this study will establish benchmarks or guidelines for enhancing the nation's educational system.

Keywords Digital games · Learning · Educational games · Attitude

1 Introduction

In various research publications during the past decades, the utilization of digital games in language acquisition has been examined. Before the COVID-19 pandemic started, the use of technology was already widespread. It was even attested that education is dynamic due to its adaptability with the presence of technology and global proliferation during this computer-generated or virtually inclined era [1, 2]. Online games were introduced even more so in the classroom. Digital online games were already integrated into the classroom to ensure the motivation of the students to learn and interact in classroom discussions. Therefore, since instructional education technology will be the main tool used to deliver instruction and learning assessment, both instructors and learners need to be involved with it. It was also asserted that the accessibility of these digital tools tends to be facile since they can be utilized through the Internet, which is embedded in a plethora of resources [3]. Unfortunately, these quick shifts from traditional to online learning have left learners and instructors unprepared [4, 5]. A pool of studies has examined its effects on the advancement of specific linguistic abilities and some aspects, such as the macro skills in language (reading, speaking, listening, writing) of students. Moreover, few studies have examined the effects of integrating such digital games in education and language acquisition, especially relating to the learner's motivation and satisfaction. Furthermore, a substantial corpus of research has been conducted on digital game-based language learning (DGBL), although instead of considering undergraduates as potential educators, various studies of this kind focus on the gamers themselves. Limited research studies have examined aspirant teachers' attitudes and understanding regarding the use of digital games in the English classroom. This is crucial because young learners and aspiring educators will soon assume a position in the teaching of languages in the twenty-first century classroom. Furthermore, there is also a dearth of studies on this kind of research on digital game-based technology in the Philippines.

The growing acceptance of video games and their usage in language instruction has only served to highlight the importance of good preparation for aspirant teachers in educational institutions. Hence, an array of researchers demands greater training and quality in terms of understanding the ideas as well as the methods of utilizing DGBL. According to Prensky [6], there are two factors that appear to support the idea that students require additional preparation as they make the shift from being digital natives (those who were born after the widespread adaptation of digital technology) to education instructors. Furthermore, there is a pool of studies regarding DGBL, which is evident in the ubiquity of smartphone usage and the continuous rise of GB apps that inspire recent instructional techniques in linguistic acquisition [7, 8]. Given this, numerous authors have discussed the effects of incorporating GB applications in the English or FL classroom, examining both the advantages, such as increased learner satisfaction and motivation, and the drawbacks, such as restricted functionality and lack of interpersonal connections [9, 10].

Furthermore, the utilization of game-based ideas in education, notably for linguistic acquisition, as well as the continual emergence of innovative web-based game formats and platforms, apps, and consoles, intensified the desire for more modernized educational games [10–12]. Therefore, according to Jacinto and Alieto [13], instructors need to be digitally literate to use and navigate through a variety of technical applications and functionalities. Considering the spread of the COVID-19 pandemic and its effects on contemporary education—which included the rise of online learning and social distancing—this requirement has become even more clear [14]. Therefore, this adds to the urgent need for more studies to be carried out to ascertain how the technological knowledge of teacher aspirants would be beneficial in today's education. Additionally, this study also seeks to investigate the attitudes of teacher aspirants toward the utilization of DGBL in the classroom. Meanwhile, the present research describes various predictors of game-based learning, such as learning opportunities, gaming experience, technological expertise, and attitude.

Research Questions

This study explores the attitudes of primary teacher aspirants toward the utilization of video games or digital-game-based language learning in the classroom. The following questions guide the present study:

1. What is the attitude of the teacher aspirants toward game-based language learning?
2. Is gender a contributing factor that influences the attitude of teacher aspirants toward digital game-based language learning?

2 Methodology

2.1 Research Design

With a quantitative, descriptive, and cross-sectional research design, a survey questionnaire was employed in this study. According to Johnson [15], if a study's primary objective is to describe the phenomenon, it might be categorized as a descriptive study. The goal of this research is to obtain/collect data and generalize findings as well as select a research design that is suitable for these goals, which is the central reason for choosing a quantitative research approach. Moreover, a quantitative research design involves the process of gathering quantifiable information that can be subjected to statistical treatment [16].

2.2 Respondents

Considering that the study is focused on a specific sample, the study ascertains that the respondents (1) must be a teacher aspirant in elementary education (BEED) enrolled in the college of teacher education and (2) must have taken at least one course on educational technology, specifically Technology for Teaching and Learning (TTL). In total, 90 participants at all levels were included in this study. The participants' ages ranged from 17 y/o being the youngest and 24 y/o being the oldest. The majority of the participants were between 18 and 19 y/o.

2.3 Instrument

This study used a five-point Likert scale adopted from the previous research of Bourgonjon et al. [17] on the perception of the learners toward the utilization of video games in the classroom. Furthermore, the questionnaire was composed of five dimensions: personal experience (EXP), usefulness (U), ease of use (EOU), learning opportunities (LO), and preference for video games (PVG). In total, the questionnaire has 23 items. For the reliability and validity of the adopted questionnaire, Cronbach's alpha was utilized with 0.922 reliability. To support such reliability, George and Maller [18] discussed that the questionnaires with Cronbach's alpha 0.922, according to the scale of reliability, have excellent reliability.

2.4 Data Gathering Procedure

After selecting the respondents of the present study, the respondents were instructed that the study was voluntary and that not participating would not have any impact on their evaluations. To ensure that respondents took part voluntarily in the study, a consent form was included in the first section of the survey link. By guaranteeing the responders that no identifying information would be collected in the area and that the data acquired would not be revealed to other parties, confidentiality and anonymity were guaranteed. As a result, the retrieved file was encrypted to protect the data. The questionnaire was digitized in Google Form, and later on, the GForm link was forwarded personally via messenger to the respondents as it is convenient for the respondents. Furthermore, each question was digitally set into a required default to be answered by the respondents. Upon answering the questionnaire, the respondents submitted the survey. The researcher then immediately checked for any missed digits, duplicated input, and other mistakes that might have prohibited their response from being used in the study. Additionally, the data were first tabulated and gathered by the researcher using Microsoft Excel before being transferred to SPPS for analysis.

2.5 Data Gathering Approach

The data were analyzed using SPSS software. It has reportedly been the most popular platform for social sciences research, according to Ozgur et al. [19], because this program could easily manage statistical treatment even for studies with vast amounts of data. The data for the present study were coded using the following: for the academic year of the respondents, the data were coded as 1 = first year, 2 = second year, and 3 = third year. Furthermore, for gender, 1 = female and 2 = male. Since the survey questionnaire is on a five-point Likert scale, the data were coded as 1 = "Strongly DA", 2 = "DA", 3 = "Neutral", 4 = "A", and 5 = "Strongly A". Moreover, since the questionnaire was categorized into 5 dimensions, the following codes for each dimension were used: personal experience (EXP), usefulness (U), ease of use (EOU), learning opportunities (LO), and preference for video games (PVG). Furthermore, to analyze the significant difference between the attitude of the teacher aspirants toward DGBLL and their gender, one-way ANOVA was utilized.

2.6 Scope and Limitations

The scope of the present study focused on the 90 primary teacher aspirants at the university. Moreover, the respondents must be enrolled in the College of Teacher

Education and must have taken at least one course on educational technology, specifically Technology for Teaching and Learning (TTL). Furthermore, this study experienced challenges in distributing the data since some of the students are unresponsive and the list of names of the respondents from the college of education are not easily accessed since the data are all confidential. Moreover, to obtain such data in the course subject taken, specifically the Technology for Teaching and Learning (TTL), the researchers must obtain permission from the respondents to access their subjects taken. Nonetheless, the present study was still successful, and the necessary data were obtained despite the challenges.

3 Results and Discussion

Attitude Toward Digital Game-Based Language Learning

The respondents' responses were retrieved from Google Forms by converting the data into an Excel spreadsheet. The collected information was subjected to descriptive statistical treatment using SPSS. To determine the perceptions of the respondents, an equal interval was also computed, which was used as the basis for the interpretation of the mean scores.

The analysis presented in Table 1 reveals that the attitude of the respondents toward DGBLL is characterized as a '*positive attitude*' (M = 3.7096). In addition, the results also indicate that the respondents view all dimensions favorably, as they have a '*positive attitude*' of gaming experience (M = 3.6593, SD = 0.1.20027), usefulness (M = 3.5833, SD = 1.35228), ease of use (M = 3.6148, SD = 1.16751), learning opportunities (M = 4.0873, SD = 0.82794), and preference for video games (M = 3.6037, SD = 1.13386). This suggests that they acknowledge the benefits of digital games in learning content disciplines. Moreover, this may indicate that the teacher aspirants view positively the use of DGBLL in the classroom, which indicates that they acknowledge that video games are essential in the learning of the students. According to Kazu and Kuvvetli [20], digital game-based language learning (DGBLL) is efficient in encouraging vocabulary acquisition in English as a foreign language (EFL) learner. These results are in agreement with other similar investigations [21] that examined how learners view the value of games in terms of how they can alter after playing them. They noticed that a lot of players or students in a certain game displayed a favorable attitude following their involvement in class. They believed that playing video games could help learners learn more effectively. Upon playing the online games, attitudes among learners toward VG appeared to have changed, according to Kenny and McDaniel [22]. The results suggest that gaming is advantageous to experiences that may support learners' understanding of the importance of VGs. Furthermore, this also suggests that video games are not limited to entertainment purposes but are beneficial to the classroom setup, especially in language learning of the students, since engagement is much needed in the utilization of VG in the classroom.

Table 1 Respondents' attitudes toward DGBLL

Dimensions	M	SD	Interpretation
Gaming Experience (EXP)	3.6593	1.20027	PA
Usefulness (U)	3.5833	1.35228	PA
Ease of Use (EOU)	3.6148	1.16751	PA
Learning Opportunities (LO)	4.0873	0.82794	PA
Preference for Video Games (PVG)	3.6037	1.13386	PA

Scale: 1.00–1.79—Very Negative Attitude (VNA), 1.80–2.59—Negative Attitude (NA), 2.60–3.39—Neutral Attitude (NeuA), 3.40–4.19—Positive Attitude (PA), 4.20–5.00—Very Positive Attitude (VPA)

Moreover, additional conceptual support for this aspect of the study comes from the two meta-analyses conducted by Vogel et al. [23] and Randel et al. [24]. The outcomes demonstrated that digital games surpassed conventional schooling in terms of both academic achievement and intellectual development after they carefully evaluated the information gathered in a variety of settings and timeframes. Additionally, they showed that compared to traditional training, children perceived more positive outcomes on learning through digital games. DGBL may enhance the analytical and artistic abilities of learners who participate in digital game counseling with other learners. These claims on digital games' educational, skill-development, and involvement effects are consistent with the result asserted by Connolly et al. [25]. The results of previous studies are consistent with the findings of the present study, in which the respondents view VGs as positive when integrated into the classroom. This suggests that primary aspirant teachers are considering the benefits and efficiency of utilizing VGs in the classroom, especially to simulate enjoyment and motivation for students. Speaking abilities can be improved using digital technology in foreign language instruction. For instance, studies have demonstrated that utilizing chatbots and AI in digital language learning can improve vocabulary, pronunciation, and fluency [26].

In detail, Table 2 shows the responses of the teacher aspirants' attitude toward DGBLL, which renders a "*positive attitude*". As shown above, EXP1, "*I like video games in general*", with a mean score of 4.02, ranked highest among the other statements in the gaming experience of the respondents. Many students held the opinion that playing video games would improve their capacity for higher-order cognitive processes. Oppenheim [27] asserts that views and attitudes may present themselves in feelings, behaviors, gestures, and other ways. They are in a state of readiness where they are ready to control their behavior in a specific way [28]. As stated in the preceding study, nothing has been written about how learners feel about using digital games or technology in general to improve learning. They both agreed that playing video games on a computer or the internet keeps students inspired to take part in the classroom. This may indicate that the majority of the respondents like video games in general, and they see the advantage of using the VG in the classroom, especially in the language learning aspect.

Table 2 Teacher aspirants' responses per item on Gaming Experience (EXP)

Gaming experience (EXP)	M	SD	Interpretation
EXP1	4.02	1.190	PA
EXP2	3.99	1.276	PA
EXP3	3.69	1.233	PA
EXP4	3.41	1.365	PA
EXP5	3.16	1.373	NeuA
EXP6	3.69	1.371	PA

Moreover, by incorporating the internet into the curriculum, each student can successfully market their potential and enhance their professional profile. According to a previous study on blended learning, computerized instruction can boost active learning, improve students' learning experiences, and help them develop stronger learning skills [29].

However, the above results refute the findings in the study of Schrader et al. [30]. It is stated that several learners were unaware of the potential advantages of video games in learning environments partly because of their lack of gaming expertise [30]. Several research investigations have been conducted to determine the effects that playing games and other similar activities have on learners' attitudes toward game-based learning. In particular, Ray and Coulter [21] examined how learners view the value of games in terms of how they can alter after playing them. They noticed that a lot of players or students in a certain game displayed a favorable attitude following their involvement in class. They believed that playing video games could help learners learn more effectively. Upon playing the online games, attitudes among learners toward VG appeared to have changed, according to Kenny and McDaniel [22]. The results suggest that gaming is advantageous to experiences that may support learners' understanding of the importance of VGs, which supports the result in the above statements on the gaming experience of the respondents.

In detail, Table 3 shows the teacher aspirants' responses per item in terms of the usefulness of DGBLL in the classroom. As shown above, the overall results show that the respondents render a "*positive attitude*" toward DGBLL in terms of its usefulness. Item number 8 "*using video games in the classroom would increase my learning productivity*." Based on this statement, the learning productivity and learning motivation of the respondents will most likely increase if video games are incorporated into the classroom. Encouragement and involvement are two benefits of DGBL that are commonly mentioned in the pool of studies. Moos and Marroquin [31] assert that different advancements influence the kind, extent, and persistence of learners' behaviors, which are connected to their physiological drive. DGBL entertaining and educating elements can lead learners to become preoccupied with their own performance in their academic grades and lose enthusiasm for gaining new skills and knowledge [32]. Nevertheless, thanks to the extraordinary benefits of technology in determining interpersonal interaction, it is now possible to build and provide digital

Table 3 Teacher aspirants' responses per item on Usefulness (U)

Usefulness (U)	M	SD	Interpretation
U7	3.61	1.338	PA
U8	3.69	1.451	PA
U9	3.64	1.352	PA
U10	3.39	1.459	NeuA

games that offer excellent opportunities for engagement by enhancing and developing their teaching and learning processes.

Moreover, the utilization of cellphones as learning tools will make the teaching and learning process more dynamic. There are many aspects to using smartphones as learning tools. Smartphones are applied for game-based learning to provide enjoyable learning. For instance, Chang et al. [33] established that GB learning offers a better learning experience for students than traditional learning methods. Chen and Law [34] and Satrio et al. [35] acknowledged that the usage of GB learning will enhance learning when it works in tandem with other forms of instruction or coaching for learners, which supports the above findings on the usefulness of DGBLL in the classroom.

Furthermore, in the past ten years, scholars have focused on the use of technology in education due to its potential to improve learning outcomes [36–38]. The global learning pattern incorporates learning methodologies that combine scientific knowledge, social humanities knowledge, and computational technology knowledge. Technology knowledge is now seen as a necessity and a must for everyone, rather than just a tool to make the teaching and learning process easier [39]. The fundamental assumption is that individual skills in higher education and the workplace will be impacted by technology knowledge.

However, several problems may arise when DGBL is introduced in the curriculum: the possibility for disorder and difficulties in classroom management; the absence of a valid and trustworthy categorization for evaluating the significance of educational games; and the inability to accurately assess the effectiveness of the curriculum. Moreover, there are some academics who believe that digital games are less likely to support learners in developing their understanding of collaborative and communicative learning [40] (Table 4).

In detail, item number 11, "*I would know how to handle video games in the classroom*", *was* rated high in the ease-of-use aspect of utilizing VG in the classroom. This supports the study of Lee and Peng [41]. It is intriguing to consider that educators often employ any of the three methods to incorporate games into their teaching

Table 4 Teacher aspirants' responses per item on Ease of Use (EOU)

Ease of Use (EOU)	M	SD	Interpretation
EOU11	3.74	1.097	PA
EOU12	3.50	1.343	PA
EOU13	3.60	1.261	PA

Table 5 Teacher aspirants' responses per item on Learning Opportunities (LO)

Learning Opportunities (LO)	M	SD	Interpretation
LO14	4.07	0.958	PA
LO15	3.92	1.094	PA
LO16	4.07	0.946	PA
LO17	3.93	0.992	PA
LO18	4.29	0.797	VPA
LO19	4.31	0.774	VPA
LO20	4.02	1.171	PA

methods (since the main focus of this study is the teacher aspirants), either by having learners design the games themselves or by having instructors or the game's creator demonstrate learners and research the subject while creating their personalized games [41]. This generally implies that learners who study language programming become better problem solvers and a professional game takes one to two years to develop. Although this student-designed approach to learning with digital games does not necessarily demonstrate a high level of gaming, it is a time-consuming process that often connects to computer science.

We can therefore develop games that combine gaming and mixed learning. Many people see this as a professionally produced digital game learning progression as an approach to digital game learning due to its ability to simultaneously satisfy educational and entertaining goals and to do so with almost any topic. Compared to the first option, it offers more assets. Because they are so good at passing along the knowledge, skills, experience, and problem-solving necessary to win the game, computer or Internet games must be similar in terms of their quality and utility [25] (Table 5).

As shown above, there are two items that were rated high and interpreted as *"very positive attitude"* in terms of the learning opportunities that the learners can have in utilizing video games in the classroom. Statements "*video games offer opportunities to interact with other students*" and statement "*video games offer opportunities to think critically.*" These statements are parallel to the findings in the studies of Oppenheim [27] and Galloway [28]. Accordingly, many students held the opinion that playing video games would improve their capacity for higher-order cognitive processes. Oppenheim [27] asserts that views and attitudes may present themselves in feelings, behaviors, gestures, and other ways. They are in a state of readiness where they are ready to control their behavior in a specific way [28]. As stated in the preceding study, nothing has been written about how learners feel about using digital games or technology in general to improve learning. They both agreed that playing video games on a computer or the internet keeps students inspired to take part in the classroom.

Moreover, according to a pool of research, efficient educators have an impact on learners' learning outcomes, advancement and growth, and classroom learning activities [36, 42]. Specifically, learners' involvement throughout the teaching and learning process may be facilitated by educators' interpersonal competency. Furthermore, in professional development programs such as these, instructor competence may be used as a powerful tool to improve the quality of instruction and student achievement. This made the same assumption of Kunter et al. [43] that instructor competence positively affects the standard of learning, which in turn affects the outcomes for student learning (Table 6).

In detail, item number 23, "*I am enthusiastic about using video games in the class*", was rated high in the preference for video games aspect. This is in line with the study of Papastergiou [44]. Accordingly, digital learning games can satisfy students' desire for enjoyable educational experiences. Several learners held opinions that engaging in video games would improve their capacity for more advanced cognitive processes. Encouragement and involvement are two benefits of DGBL that are commonly mentioned in the pool of studies. Moos and Marroquin [31] assert that different advancements have an effect on the kind, extent, and persistence of learners' behaviors, which are connected to their physiological drive. DGBL entertaining and educating elements can lead learners to become preoccupied with their own performance in their academic grades and lose enthusiasm for gaining new skills and knowledge [32]. Nevertheless, thanks to the extraordinary benefits of technology in determining interpersonal interaction, it is now possible to build and provide digital games that offer excellent opportunities for engagement by enhancing and developing their teaching and learning processes.

Moreover, conceptual support for this aspect of the study comes from the two meta-analyses conducted by Vogel et al. [23] and Randel et al. [24]. The outcomes demonstrated that digital games surpassed conventional schooling in terms of both academic achievement and intellectual development after they carefully evaluated the information gathered in a variety of settings and timeframes. Additionally, they showed that compared to traditional training, children perceived more positive outcomes on learning through digital games. DGBL may enhance the analytical and artistic abilities of learners who participate in digital game counseling with other learners. These claims on digital games' educational, skill-development, and involvement effects are consistent with the result asserted by Connolly et al. [25].

Furthermore, the use of DG to advance learners' understanding is another recent innovation in the field of online learning. The positive outcomes of DGBL are in line with a number of previously addressed educational theories, including contextual cognition, play theory, assimilation, and put up, among others. This implies that, according to this theory, playing and paying attention to these concepts can lead to the best possible use of DG.

Table 6 Teacher aspirants' responses per item on preference for Video Games (PVG)

Preference for Video Games (PVG)	M	SD	Interpretation
PVG21	3.58	1.171	PA
PVG22	3.58	1.218	PA
PVG23	3.66	1.229	PA

Table 7 Teacher aspirants' attitudes toward digital game-based language learning across genders

Dimensions	Male		Female		Sig. (2 tailed)	Interp.
	M	SD	M	SD		
EXP	4.2083	0.93503	3.2200	1.21566	0.000	Significant
U	3.9313	1.31922	3.3050	1.32604	0.028	Not significant
EOU	4.0000	1.17427	3.3067	1.00716	0.005	Significant
LO	4.4357	0.79107	3.8086	0.75406	0.000	Significant
PVG	4.0417	1.07334	3.2533	1.06599	0.001	Significant

Table 7 similarly discloses that gender across the four dimensions (gaming experience, ease of use, learning opportunities, and preference for video games) influences the respondents' attitude, as the difference between the male respondents and their female counterparts is significant. This may imply that teacher training programs have successfully introduced this strategy to both genders. Additionally, while males collectively expressed a more positive attitude toward DGBLL compared to females, the difference was nonetheless statistically significant. Therefore, the result suggests that gender is a factor that influenced the attitude of the respondents toward DGBLL since the result is significant. These results have important implications for teacher training programs, as they can use DGBLL as an effective tool to enhance student engagement and motivation in classrooms. Through motivation, students will most likely be attentive in the classroom discussion, which may also promote classroom engagement for students. However, in terms of the usefulness of DGBLL in the classroom, it is shown that there is no significant difference between genders in this dimension. This implies that gender is not a factor that influenced the view of the respondents toward the usefulness of VGs in the classroom.

Furthermore, the finding is explainable by the fact that regardless of gender, the respondents learn in a similar learning environment, hence exposing them to probable similar concerns. Moreover, as no study has explored the association of gender in terms of the utilization of DGBLL in the classroom, a comparison of the present results with previous studies is not possible.

4 Conclusion and Implications

The growing acceptance of video games and their usage in language instruction has only served to highlight the importance of good preparation for aspirant teachers in educational institutions. Digital learning games can satisfy students' desire for enjoyable educational experiences. Several learners held opinions that engaging in video games would improve their capacity for more advanced cognitive processes. Therefore, the present study explored the attitudes of teacher aspirants toward the utilization of video games in the classroom. The study found that the respondents perceived a "*positive attitude*" toward the utilization of video games in the classroom. The findings of this study altered the students' metaphorical impressions of the independent investigations they conducted utilizing DGBLL. Based on the results, it was found that the majority of students thought favorably about DGBLL. A student's perspective on the procedure was also evaluated in this study, and similar studies might be carried out by taking student perspectives into consideration. Similar research could be carried out in other domains; however, this research focused on learning a language. Moreover, this paper also found that in terms of the gaming experience, usefulness, ease of use, learning opportunities, and preference for video games, the teacher aspirants held a favorable perception of these five dimensions. Furthermore, gender as a variable is a factor that influences the attitude of the respondents toward the utilization of video games in the classroom specifically in the four dimensions, namely, the gaming experience of the respondents, ease of use of video games in the classroom, the learning opportunity it offers, and the respondents' preference for video games in the classroom. Furthermore, this paper also found that in terms of the usefulness of video games in the classroom, gender is not a factor that influences the attitude of the respondents toward DGBLL.

Considering the enumerated results, this paper suggests that future investigations should be conducted in a wider scope. The perceptions of learners from different course programs and educational levels could also be explored since determining the perceptions of learners can aid curriculum policymakers and scholars in ensuring that the employed strategy is effective in terms of facilitating content learning. Additionally, this kind of study is limited to the Philippine context, and other researchers should widen the scope in the population and locale aspects of this kind of research. Furthermore, analyzing respondents' opinions on language learning through digital games can undoubtedly help address the Philippines' lack of ICT literacy as a future elementary school teacher. The results of this study will establish benchmarks or guidelines for enhancing the nation's educational system. Moreover, there is a need for more research in this field given the positive outcomes of using digital games as a language learning approach. While previous studies have indicated that using DG to encourage language learning can be successful, more studies are required to fully comprehend the advantages and drawbacks of this strategy for various age groups of language learners. Future studies might, for instance, examine how well DG encourages language development in young children, teenagers, and adults. Additionally, studies might be conducted to determine the extent to which digital games can help language learners of all ages develop a variety of language abilities, including speaking, writing, listening, and reading.

References

1. Alrefaee, Y., Gonzales, L., Ortega, J., Delos Santos, M., Casamayor, J.: Digital learning is something here to stay: attitude and technological competence of preservice language teachers. J. Educ. Lang. Stud. **1**(1), 107–126 (2023)
2. Rillo, R., Alieto, E.: Indirectness markers in Korean and Persian English essays: implications for teaching writing to EFL learners. Engl. Int. J. 13(2.2), 165–184 (2028)
3. Abequibel, B., Ricohermoso, C., Alieto, E., Barredo, C., Lucas, R.: Prospective reading teachers' digital reading habit: a cross-sectional design. TESOL Int. J. **16**(4.4), 246–260 (2021)
4. Bahasoan, A.N., Ayuandiani, W., Mukhram, M., Rahmat, A.: Effectiveness of online learning in pandemic COVID-19. Int. Serv. Technol. Manag. **1**(2), 100–106 (2020)
5. Tomasik, M.J., Helbling, L.A., Moser, U.: Educational gains of in-peron vs. distance learning in primary and secondary schools: a natural experiment during the COVID19 pandemic school closures in Switzerland. Int. J. Psycholol. 1–20 (2020)
6. Prensky, M.: Digital natives, digital immigrants part 2: do they really think differently? Horizon **9**, 1–6 (2001). https://doi.org/10.1108/10748120110424843
7. Hung, H.-T., Yang, J.C., Hwang, G.-J., Chu, H.-C., Wang, C.C.: A scoping review of research on digital game-based language learning. Comput. Educ. **126**, 89–104 (2018). https://doi.org/10.1016/j.compedu.2018.07.001
8. Chang, C.-Y., Hwang, G.-J.: Trends in digital game-based learning in the mobile era: a systematic review of journal publications from 2007 to 2016. Int. J. Mob. Learn. Organ. **13**, 68–90 (2019). https://doi.org/10.1504/IJMLO.2019.096468
9. Gamlo, N.: The impact of mobile game-based language learning apps on efl learners' motivation. Engl. Lang. Teach. **12**, 49–56 (2019). https://doi.org/10.5539/elt.v12n4p49
10. Huang, Y.-L., Chang, D.-F., Wu, B.: Mobile game-based learning with a mobile app: motivational effects and learning performance. J. Adv. Comput. Intell. Inform. **21**, 963–970 (2017). https://doi.org/10.20965/jaciii.2017.p0963
11. Reinhardt, J.: Gameful Second and Foreign Language Teaching and Learning: Theory, Research, and Practice, pp. 20–50. Springer, Berlin/Heidelberg, Germany (2018). https://doi.org/10.1016/j.system.2020.102307
12. Hunag, A.C.Y., de Haan, J., Lee, T.-K.: Games and language learning: an international perspective. NYS Tesol J. **5**, 3–11 (2018)
13. Jacinto, M.J., Alieto, E.: Virtual teaching attitude and technological competence among English as second language (ESL) teachers: implications for the management of learning. Asian EFL J. **27**(4.4), 403–433 (2020)
14. Dubreil, S.: Using games for language learning in the age of social distancing. Foreign Lang. Ann. **53**, 250–259 (2020). https://doi.org/10.1111/flan.12465
15. Johnson, B.: Toward a new classification of nonexperimental quantitaive research. Educ. Res. **30**(2), 3–13 (2000). https://doi.org/10.3102/0013189x030002003
16. Cresswell, J.: Research Design: Qualitative, Quantitative, and Mixed Methods, vol. 2, pp. 1–5. Sage, Thousand Oaks, CA (2003)
17. Bourgonjon, J., Valcke, M., Soetaert, R., Schellens, T.: Students' perceptions about the use of video games in the classroom. Comput. Educ. **54**, 1145–1156 (2010). https://doi.org/10.1016/j.compedu.2009.10.022
18. George, D., Maller, P.: SPSS windows step by step: a simple guide and reference, **11**(2), 1–63 (2003)
19. Ozgur, C., Kleckner, M., Li, Y.: Selection of statistical software for solving big data problem: a guide for business, students, and universities. Sage Open1–12 (2015). https://doi.org/10.1177/2158244015584379
20. Kazu, I.Y., Kuvvetli, M.: A triangulation method on the effectiveness of digital game-based language learning for vocabulary acquisition. Educ. Inf. Technol. 1–26 (2023). https://doi.org/10.1007/s10639-023-11756-y
21. Ray, B., Coulter, G.A.: Perceptions of the value of digital mini-games: implications for middle school classrooms. J. Digit. Learn. Teach. Educ. **26**(3), 92–100 (2010)

22. Kenny, R.F., McDaniel, R.: The role teachers' expectations and value assessments of video games play in their adopting and integrating them into their classrooms. Br. J. Edu. Technol. **42**(2), 197–213 (2011). https://doi.org/10.1111/j.1467-8535
23. Vogel, D.L., Wade, N.G., Haake, S.: Measuring the self-stigma associated with seeking psychological help. J. Couns. Psychol. **53**, 325–337 (2006). https://doi.org/10.1037/0022-0167.53.3.325
24. Randel, J., Morris, B., Wetzel, C., Whitehill, B.: The effectiveness of games for educational purposes: a review of recent research. Simul. Gaming **23**(3), 261–276 (1992)
25. Connolly, T.M., Boyle, E.A., MacArthur, E., Hainey, T., Boyle, J.M.: A systematic literature review of empirical evidence on computer games and serious games. Comput. Educ. **59**(2), 661–686 (2012). https://doi.org/10.1016/j.compedu.2012.03.004
26. Belda-Medina, J., Calvo-Ferrer, J.R.: Using chatbots as ai conversational partners in language learning. Appl. Sci. **12**(17), 8427 (2022). https://doi.org/10.3390/app12178427
27. Oppenheim, A.: Questionnaire Design, Interviewing and Attitude Measurement, pp. 1–154. Bloomsbury Publishing (2000)
28. Galloway, N., Kriukow, J., Numajiri, T.: Internationalisation, higher education and the growing demand for English: an investigation into the English medium of instruction (EMI) movement in China and Japan, 1–39 (2017)
29. Bonk, C.J., Graham, C.R.: The Handbook of Blended Learning: Global Perspective, Local Designs, pp. 210–587. Wiley (2006)
30. Schrader, P.G., Zheng, D., Young, M.: Teachers' perceptions of video games: MMOGs and the future of preservice teacher education. Innov.: J. Online Educ. **2**(3), 221–233 (2006)
31. Moos, D.C., Marroquin, E.: Multimedia, hypermedia, and hypertext: motivation considered and reconsidered. Comput. Hum. Behav. **16**(3), 265–276 (2010)
32. Erhel, S., Jamet, E.: Improving instructions in educational computer games: exploring the relations between goal specificity fro experience and learning outcomes. Comput. Hum. Behav. **91**, 106–114 (2019). https://doi.org/10.1016/j.chb.2018.09.020
33. Chang, Y.H., Hwang, J.H., Fang, R.J., Lu, Y.T.: A Kinect-and game-based interactive learning system. Eurasia J. Math. Sci. Technol. Educ. **13**(8), 4897–4914 (2017)
34. Chen, C.H., Law, V.: Scaffolding individual and collaborative game-based learning in learning performance and intrinsic motivation. Comput. Hum. Behav. **55**, 1201–1212 (2016). https://doi.org/10.1016/j.chb.2015.03.010
35. Satrio, Y.D., Handayani, S., Abbas, M.H.I., Kustiandi, J.: Studi komparasi metode pembelajaran dalam meningkatkan literasi keuangan di Masa pandemi covid-19. J. Pend. Ekon. Undiksha **12**(1), 29 (2020)
36. Fauth, B., Decristan, J., Decker, A.-T., Büttner, G., Hardy, I., Klieme, E., Kunter, M.: The effects of teacher competence on student outcomes in elementary science education: the mediating role of teaching quality. Teach. Teach. Educ. **86**, 102882 (2019). https://doi.org/10.1016/j.tate.2019.102882
37. Siron, Y., Wibowo, A., Narmaditya, B.S.: Factors affecting the adoption of e-learning in Indonesia: lesson from Covid-19. JOTSE: J. Technol. Sci. Educ. **10**(2), 282–295 (2020). https://doi.org/10.3926/jotse.1025
38. Al-Hariri, M.T., Al-Hattami, A.A.: Impact of students' use of technology on their learning achievements in physiology courses at the University of Dammam. J. Taibah Univ. Med. Sci. **12**(1), 82–85 (2017). https://doi.org/10.1016/j.jtumed.2016.07.004
39. Chua, J.H., Jamil, H.: Factors influencing the technological pedagogical content knowledge (TPACK) among TVET instructors in Malaysian TVET Institution. Procedia-Soc. Behav. Sci. **69**, 1539–1547 (2012). https://doi.org/10.1016/j.sbspro.2012.12.096
40. Hsu, T.Y., Chiou, G.F.: Preservice teachers' awareness of digitl game-supported learning. In: Society for Information Technology & Teacher Education International Conference, pp. 2135–2141 (2011)
41. Lee, K.M., Peng, W.: What Do We Know About Social and Psychological Effects of Computer Games, pp. 327–345. Erlbaum, Mahwah, NJ (2006)

42. Baumert, J., Fleckenstein, J., Leucht, M., Koller, O., Möller, J.: The long-term proficiency of early, middle, and late starters learning English as a Foreign language at school: a narrative review and empirical study. Lang. Learn. **70**(4), 1091 (2020)
43. Kunter, M., Tsai, Y.M., Klusmann, U., Brunner, M., Krauss, S., Baumert, J.: Students' and mathematics teachers' perceptions of teacher enthusiasm and instruction. Learn. Instruct. **18**(5), 468–482 (2008). https://doi.org/10.1016/j.learninstruc.2008.06.008
44. Papastergiou, M.: Digital game-based learning in high school computer science education: impact on educational effectiveness and students motivation. Comput. Educ. **52**(1–12), 2009 (2009). https://doi.org/10.1016/j.compedu.2008.06.004

A Unified Approach for Identification and Analysis of the Sources of Uncertainty in Machine Learning Techniques

Sourojit Pal, Sandip Roy, Avishek Banerjee, Kaushik Majumdar, Umesh Gupta, Saurabh Rana, and Sachin Shetty

Abstract Over the last decade, there has been an intensive effort to introduce technologies that increase system smartness by incorporating artificial intelligence and machine learning, resulting in a better way of life. With this evolution, it has become particularly important to understand how trustworthy are the decisions made by autonomous systems driven by artificial intelligence coupled with machine learning and deep learning capabilities. Uncertainty quantification (UQ) has thus become a topic of interest over the last few years. In this paper, we observe the behavior of the epistemic (model), aleatoric (data), and distribution uncertainty and deduce a mathematical relationship between them. Next, we show that the epistemic probability function is inversely proportional to the distribution probability function and is

S. Pal
Department of IT, Jadavpur University, Kolkata 700059, India
e-mail: psouro1990@gmail.com

S. Roy · S. Shetty
VMASC, Old Dominion University, Suffolk, VA 23435, USA
e-mail: sroy@odu.edu

S. Shetty
e-mail: sshetty@odu.edu

A. Banerjee
Department of CSBS, Asansol Engineering College, Asansol 713305, India
e-mail: avishek.csbs@aecwb.edu.in

K. Majumdar
Department of CSE, MAKAUT, Kolkata 700064, India
e-mail: koushikzone@yahoo.com

U. Gupta (✉)
School of CSET (SCSET), Bennett University, Greater Noida, India
e-mail: er.umeshgupta@gmail.com

S. Rana
Department of Mathematics (SCSET), Bennett University, Greater Noida, India
e-mail: saurabhranapsm@gmail.com

G. Fortino et al. (eds.), *Proceedings of Third International Conference on Computing and Communication Networks*, Lecture Notes in Networks and Systems 917,
https://doi.org/10.1007/978-981-97-0892-5_35

directly proportional to the aleatoric probability function. Moreover, we demonstrate the impact of identifying in-domain distribution and out-of-domain distribution in model and data uncertainties. Finally, we perform an array of experiments using lung cancer data to demonstrate that improved accuracy may be obtained by striking a correct balance between the three forms of uncertainty. According to experimental findings, methodical data selection aids in adopting informed decisions, strengthening the system's dependability and enabling it to achieve close to 99% accuracy even with basic machine learning models.

Keywords Uncertainty quantification · Aleatoric uncertainty · Epistemic uncertainty · Distribution uncertainty · In-domain distribution · OOD

1 Introduction

Researchers have been working on the challenges encountered in smart autonomous systems. They have tried to identify where these systems powered by artificial intelligence, deep learning, or machine learning fail. In this context, a new area of research on uncertainty quantification (UQ) has recently been focused. It has been observed that so far, there are more than 2500 different research works undertaken in this area [1].

The study of autonomous smart systems has already got into the day-to-day life of human beings. Smart technologies make our lives easier, but in order to ensure safety, the degree of uncertainty must be eliminated to the greatest extent possible [12]. The decision-making process of autonomous systems becomes an utmost point of interest. If we talk about traffic sign identification system, it is implied that it should be able to make proper decisions when it encounters its environment as shown in Fig. 1. It should know when to stop and when not to after assessing the environment and the goal that has been set. The inability to make educated choices can cause accidents.

The procedure of uncertainty quantification through the source of uncertainty identification can have multiple benefits. It can help the systems map the inputs against the goals by securing relevant data from a given spectrum of knowledge bases. The system can gain the experience and expertise to understand what is needed to achieve the target. With proper implementation, it can understand what it does not

(a) (b) (c) (d)

Fig. 1 Similar picture with different implications: setting sun (**a**); Red light (**b**); No left turn (**c**); Halt for pedestrian (**d**)

know and when it needs to make educated decisions. In this study, we try to model an equation that helps to find a balance between epistemic or model uncertainty, aleatoric or data uncertainty, and distribution uncertainty to maximize gain [13, 14].

Motivation: From the literature study, we observed that most of the related works can be classified into two classes. One class consists of survey works, while the other discusses novel ideas and efforts focused only on reducing the output's uncertainty. However, we identified that there has been no documented work describing the relationship between the types of uncertainties. Hence, it is essential to identify the relationships between the various types of uncertainty and their impact on quantifying the total uncertainty of the system.

Contribution: The basic research contributions of this paper are as follows:

(1) We find the mathematical relationship between epistemic uncertainty, aleatoric uncertainty, and distribution uncertainty, which was not documented before. Moreover, we analyze how the distribution uncertainty can influence other types of uncertainties.
(2) We demonstrate that epistemic probabilistic functions are inversely proportional to distribution probabilistic functions and directly proportional to aleatoric probabilistic functions.
(3) We show how in-domain and out-of-domain distributions are related to data and model uncertainties mathematically.
(4) We share the experimental results undertaken by us, which demonstrate—that, with proper data selection, it is possible to increase the accuracy of the models.

The rest of the paper is organized as follows. Section 2 discusses the state-of-the-art. In Sect. 3, we present our solution proposal. The experiments and observations are discussed in Sect. 4. Finally, Sect. 5 concludes the paper.

2 Related Works

This section briefly covers the state of the art, providing some suggestions as to their benefits and downsides. Malinin et al.'s pioneering work redefines the categorization of uncertainties and creates a new kind of uncertainty known as distribution uncertainty [2]. Furthermore, they develop an equation that assists us in defining the relationships between the uncertainties. In a different work, Snoek et al. attempted to clarify the significance of uncertainty measurement and determine how different models respond to data shift [3]. Recently, Gawlikowski et al. and Abdar et al. conducted extensive surveys on categorizing and identifying different uncertainty quantification methodologies [1, 4]. Yang et al. discussed the concept of out-of-distribution (OOD) as an assimilation of multiple categories of problems [5]. Some novel approaches also studied include another pioneering paper by Sensoy et al. and works on prior networks by Gawlikowski et al. and Malinin et al. [6–9]. Other works in this study include efforts on Dirichlet prior networks distance aware deterministic approaches [10, 11]. Among recent efforts, we observe Ren et al., Liu et al., and Yang et al. presenting their works in OOD detection space [16–18].

Synthesis: The literature review shows that most of the work has been focused on improving the model's output. This provides a scope to look into the problem of increasing trustworthiness from a completely different angle, discussed in succeeding sections.

3 Proposed Solution

In this section, we draw a relationship between the existing types of uncertainties based on the mathematical definition of predictive uncertainty of the classification model [2].

Looking into the mathematical, definition of predictive uncertainty of the classification model defined as: $P(\omega_c|x^*, D)$ defined over a set of parameters θ, we get: $P(\omega_c|x^*, D) = \int P(\omega c|x^*, \theta)P(\theta|D)\mathrm{d}\theta$, where $P(\omega c|x^*, \theta)$ represents the data and $P(\theta|D)$ represents the this model [2]. Further assessment of the equation has given rise to the formulation of a new equation including the concept of distribution uncertainty as [2]:

$$P\left(\omega c|x^*, D\right) = \iint P(\omega c|\mu)P\left(\mu|x^*, \theta\right)P(\theta|D)\mathrm{d}\mu\mathrm{d}\theta \tag{1}$$

where the parameter $P(\omega c|\mu)$ represents data, $P(\mu|x^*, \theta)$ represents distribution of the data, and $P(\theta|D)$ represents model. The induction of known-known, known-unknown, and unknown-unknown datasets impacts the distribution of the data which in turn affects the model experience. If the quantity of data remains unchanged and the quality of the training is increased model uncertainty is seen to vary inversely against the distribution. Mathematically, we say, epistemic probability is inversely proportional to distribution probability on a set of parameters θ, that is:

$$P(\theta|D) \propto \frac{1}{P(\mu|x^*, \theta)} \tag{2}$$

Distribution uncertainty can be generated either from in-domain distributions or from unknown sets. Thus, if μ represents the distribution of data, whereas $\mu 1$ and $\mu 2$ represent the in-domain and out-of-domain distributions, respectively, we may write:

$$\mu = \mu 1 + \mu 2 \tag{3}$$

Using parameters of Eq. (3), we rewrite Eq. (2), as

$$P(\theta|D) \propto \frac{1}{P((\mu 1+\mu 2)|x^{*},\theta)} \tag{4}$$

Equation (4) represents the exact relationship of epistemic probability with different types of distributions. Creating a variation in the concept, the quantity of the data is now increased given that the quality of the data remains the same or gets better, the model gets to learn more. Epistemic probability is seen to increase. Mathematically, this can be represented as

$$P(\theta|D) \propto P(\omega c|\mu) \tag{5}$$

It is observed that the vice versa relation also holds, i.e., if the quantity of the data is decreased without compromising on the quality, then the model experience will decrease due to data availability. This gives rise to an alternate discussion on what happens if the data increases and so does the distribution. This can be deduced using Eqs. (4) and (5) as follows:

$$P(\theta|D) \propto \frac{P(\omega c|\mu)}{P((\mu 1+\mu 2)|x^{*},\theta)} \tag{6}$$

Now, with Eq. (1) being marginalized over θ, we get [2]

$$P\big(\omega c|x^{*},\ D\big) = \int P(\omega c|\mu)P(\mu|x^{*},\ D)\mathrm{d}\mu \tag{7}$$

This equation represented the estimation of data uncertainty and distribution uncertainty for a given model uncertainty. Replacing Eq. (7) with Eq. (3), we get a complete relation of the uncertainty with data uncertainty, model uncertainty, and distribution uncertainty defined as:

$$P\big(\omega c|x^{*},\ D\big) = \iint P(\omega c|(\mu 1+\mu 2))P\big((\mu 1+\mu 2)|x^{*},\ D\big)\mathrm{d}\mu 1\mathrm{d}\mu 2 \tag{8}$$

This means any change in distribution either due to in-domain distribution or out-of-domain distribution, will impact a change in the overall uncertainty along with individual impacts over data uncertainty and model uncertainty.

Summarizing the solution, Eqs. (2) and (5) help to find the relationship of epistemic uncertainty with distribution uncertainty and aleatoric uncertainty, respectively. In Eq. (3), we identify how in-domain distributions and OOD are linked with distribution uncertainty. Equation 6 summarizes the findings of Eqs. (2), (3), and (5). Usage of Eqs. (3) in (8) helps to describe how data, in-domain, out-of-domain, and model uncertainties impact the overall uncertainty.

Methodology: In this section, we have deduced the relationships between the three types of uncertainties. Equations (2) and (7), taken from the original works, have been cited [2]. In the succeeding sections, we observe the outcomes of the experiments. The dataset was divided into known and known-unknown parts, and fed into simple machine learning models like KNN, SVM, Decision Tree, Random Forest, and Logistics Regression.

4 Performance Evaluation

In this section, we find the impact of known-known, known-unknown, and unknown-unknown datasets on already available models through experiments.

4.1 Dataset Preparation

The experimental evaluation of the relation between the uncertainties has been undertaken using the dataset on lung cancer [15]. As mentioned in Fig. 2, the file consisted of approximately 1000 lines of information (represented by Σ) and 23 parameters, which are used to identify the cases of lung cancer classified as "High", "Medium", and "Low".

This file was divided into four local files. First, 700 lines were used for training purposes and were named "lcd_train_bigger". The last 300 lines were used to create the test file named "lcd_test". Thus, the test files contained unknown cases compared to the training file. Two more things were done in this experimental setup. A subset file, "lcd_train_small", was created from the main training file "lcd_train_bigger", and a second test file, "lcd_test2", was created containing data from the training file

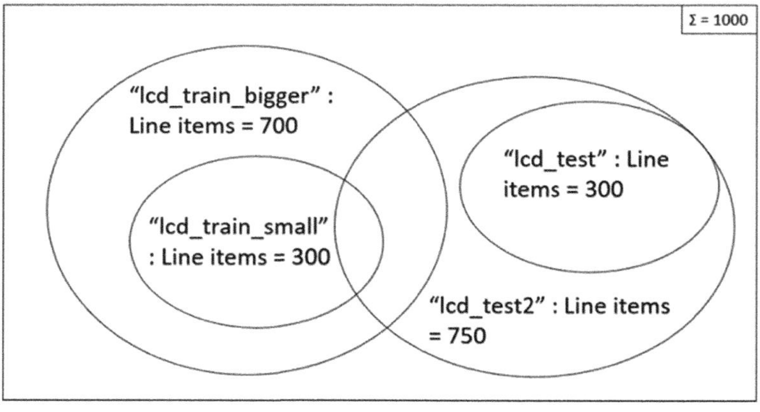

Fig. 2 Diagrammatic set representation of the participating datasets under different volumes of line items

"lcd_train_bigger" and test file "lcd_test". This helped to create a test case comprising "known" and "unknown" data.

The experiment was carried out in two distinct phases. In phase 1, training was first carried out over the smaller train set, and then the test was done using test files. In phase 2, the scope of the training was increased by increasing parameters, and the test files were again used for retest purposes. This was repeated with a more enormous train file as well. The models that were taken into scope are K Nearest Neighbor (KNN), Logistics Regression (LR), Random Forest Classifier (RFC), Simple Vector Machine (SVM), and Decision Tree (DT).

4.2 *Observations and Results*

4.2.1 Experiment on the Impact of Experience Over Uncertainty

All models were first trained on two parameters using the file lcd_train_small during the training process. Subsequent testing was done using test files on a completely unknown parameter. Results are shown in Table 1. The models were retrained with all the available parameters in the same file and then retested. This time, better accuracy was observed in the output. This is followed in Table 2.

Inference: The test cases referred to in this experiment describe an increase in experience of the system being instrumental in decreasing the uncertainty. This was achieved by keeping the quantity of the training data unchanged, only the data quality was increased by increasing the spectrum over other available parameters.

Table 1 Test results after training over two parameters in the dataset "lcd_train_small"

ML model	Precision (avg)	Recall (avg)	F1 score (avg)	Accuracy
KNN	0.17	0.33	0.22	0.5
LR	0.39	0.5	0.39	0.5
RF	0.83	0.83	0.78	0.75
SVM	0.17	0.33	0.22	0.5
DT	0.56	0.67	0.6	0.75

Table 2 Test results after training over all parameters in the dataset "lcd_train_small"

ML model	Precision (avg)	Recall (avg)	F1 score (avg)	Accuracy
KNN	0.17	0.33	0.22	0.5
LR	0.32	0.95	0.74	0.9538
RF	0.89	0.87	0.82	0.9887
SVM	0.17	0.33	0.22	0.5
DT	0.73	0.87	0.78	0.9793

This means the model is trained from different perspectives and gains experience. This helps to decrease uncertainty. This adheres to the basic working principle of epistemic uncertainty. From a diverse scope of observation, if the quantity of the data was kept constant and the quality of knowledge increased by increasing the spectrum over available parameters, the model gained experience. This shows that the model uncertainty varies inversely against the data distribution as described in Eq. (2). Alternatively, if an unknown dataset with unknown parameters is given to the model as input, the result reflects lower accuracy.

4.2.2 Experiment on the Impact of Experience Over the Known Dataset

In this experiment, the models were trained over smaller training files and then over the bigger training files. In both cases, the models were tested after training. Observations noted are described in Tables 3 and 4. The quality of the data was not compromised while increasing the volume of the data in both cases.

Inference: Comparing Tables 5 and 6, we find that a simultaneous increase in the quality and quantity of data increases the accuracy of the models, as stated in Eq. (5). This shows that a system can have better training if the spectrum of the data can be increased without compromising the quality.

Table 3 Test results after training over all parameters in the smaller dataset "lcd_train_small"

ML model	Precision (avg)	Recall (avg)	F1 score (avg)	Accuracy
KNN	0.52	0.5	0.46	0.625
LR	0.39	0.5	0.43	0.625
RF	0.89	0.67	0.71	0.75
SVM	0.52	0.5	0.46	0.625
DT	0.89	0.67	0.71	0.75

Table 4 Test results after training over all parameters in the more extensive dataset "lcd_train_bigger"

ML model	Precision (avg)	Recall (avg)	F1 score (avg)	Accuracy
KNN	0.676	0.65	0.598	0.812
LR	0.507	0.65	0.559	0.813
RF	1.157	0.87	0.923	0.975
SVM	0.676	0.65	0.598	0.813
DT	1.157	0.87	0.923	0.975

Table 5 Test results with an unknown dataset on known parameters after training over two parameters in the dataset "lcd train small"

ML model	Precision (avg)	Recall (avg)	F1 score (avg)	Accuracy
KNN	0.17	0.33	0.22	0.5
LR	0.33	0.65	0.44	0.9834
RF	0.33	0.65	0.44	0.9853
SVM	0.17	0.33	0.22	0.5
DT	0.33	0.65	0.45	0.9848

Table 6 Test results with a known-unknown dataset on known parameters after training over two parameters in the dataset "lcd train small"

ML model	Precision (avg)	Recall (avg)	F1 score (avg)	Accuracy
KNN	0.52	0.5	0.46	0.625
LR	0.39	0.5	0.43	0.625
RF	0.61	0.79	0.68	0.9871
SVM	0.17	0.33	0.22	0.5
DT	0.93	0.83	0.85	0.875

4.2.3 Experiment on the Impact of Known Scenarios Over Uncertainty

In this experiment, the model was first trained over a smaller training file and then tested with one test file containing an unknown dataset and another having a combination of known and unknown cases over the same set of parameters. The observations noted are described in Tables 5 and 6.

Inference: In this experiment, we observe a subtle difference in accuracy with a known-unknown set showing better results than an utterly unknown dataset over the same parameter list. This leads us to concrete proof of the relationship between epistemic and distribution uncertainty, again showing adherence with Eq. (2). Experiments 4.2.1 and 4.2.3 were done on two different datasets. An unknown dataset, representing an open dataset, was used for Experiment 4.2.1. Experiment 4.2.3 was done on known and unknown datasets, meaning a case of in-domain distribution. Equation (2) is observed to be valid for both cases. These experiments showed that distribution uncertainty can vary either with in- or out-of-domain distribution, represented by Eq. (3).

The accuracy of the machine learning models, as observed when subjected to different datasets and parameters, is given in Fig. 3.

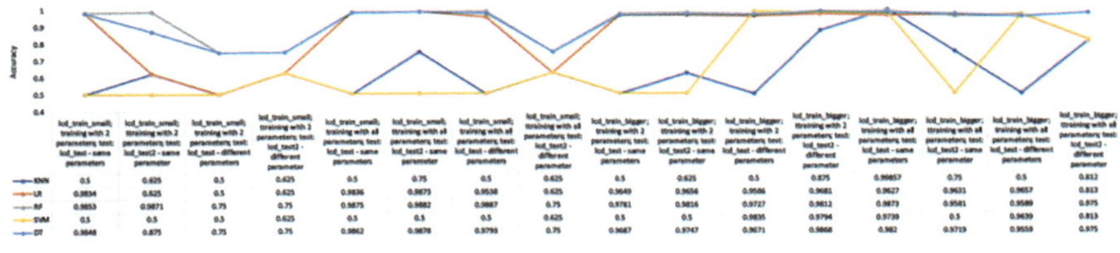

Fig. 3 Model accuracy over varying data sets and parameters

5 Conclusion

With this work, we have tried to create a relationship between the different types of uncertainties (aleatoric, epistemic, and distribution) that were missing earlier. As a part of future work, this relationship can be explored further to deduce the predictive uncertainty of each of the types of uncertainty with proper quantification and adjustments. For, if the spread or distribution of the data is observed to be very high with low data quality, the system can identify the cause of uncertainty to be the distribution of data and adapt itself to relevant inputs only. This can help the system make educated decisions, increasing trustworthiness. The limitation, as noted, is that the success of this method is dependent on the contextual understanding of the systems. Further research can be done to enable the system to detect and categorize the input data.

Acknowledgements This work is partly supported by the DoD Center of Excellence in AI and Machine Learning (CoE-AIML) under Contract Number W911NF-20-2-0277 with the U.S. Army Research Laboratory.

References

1. Abdar, M., Pourpanah, F., Hussain, S., Rezazadegan, D., Liu, L., Ghavamzadeh, M., Fieguth, P., Cao, X., Khosravi, A., Rajendra Acharya, U., Makarenkov, V., Nahavandi, S.: A Review of Uncertainty Quantification in Deep Learning: Techniques, Applications and Challenges, vol. 76, pp. 243–297. Elsevier, Information Fusion (2021)
2. Malinin, A., Gales, M.: Predictive uncertainty estimation via prior networks. In: NIPS (2018)
3. Ovadia, Y., Fertig, E., Ren, J., Nado, Z., Sculley, D., Nowozin, S., Dillon, J.V., Lakshminarayanan, B., Snoek, J.: Can you trust your model's uncertainty? Evaluating predictive uncertainty under dataset shift. In: NIPS (2019)
4. Gawlikowski, J.., Tassi, C.R.N., Ali, M., Lee, J., Humt, M., Feng, J., Kruspe, A., Triebel, R., Jung, P., Roscher, R., Shahzad, M., Yang, W., Bamler, R., Zhu, X.X.: A survey of uncertainty in deep neural networks. In: AI4EO—Artificial Intelligence for Earth Observation: Reasoning, Uncertainties, Ethics and Beyond (2022)
5. Yang, J., Zhou, K., Li, Y., Liu, Z.: Generalized out-of-distribution detection: a survey. In: IEEE Transactions on Pattern Analysis and Machine Intelligence (2022)
6. Malinin, A., Gales, M.: Reverse KL-divergence training of prior networks: improved uncertainty and adversarial robustness. In: NIPS (2019)

7. Sensoy, M., Kaplan, L., Kandemir, M.: Evidential deep learning to quantify classification uncertainty. In: NIPS (2018)
8. Gawlikowski, J., Saha, S., Kruspe, A., Zhu, X.X.: An advanced dirichlet prior network for out-of-distribution detection in remote sensing. In: IEEE Transactions on Geoscience and Remote Sensing, vol. 60 (2022)
9. Sensoy, M., Kaplan, L., Cerutti, F., Saleki, M.: Uncertainty-aware deep classifiers using generative models. In: AAAI Conference on Artificial Intelligence (AAAI-20)
10. Raj, V., Cui, T., Heinonen, M., Marttinen, P.: Incorporating functional summary information in Bayesian neural networks using a Dirichlet process likelihood approach. In: AISTAT (2023)
11. Liu, J.Z., Lin, Z., Padhy, S., Tran, D., Bedrax-Weiss, T., Lakshminarayanan, B.: Simple and principled uncertainty estimation with deterministic deep learning via distance awareness. In: NIPS (2020)
12. Kristiadi, A., Hein, M., Hennig, P.: Being Bayesian, even just a bit, fixes overconfidence in ReLU networks. In: ICML (2020)
13. Amini, A., Schwarting, W., Soleimany, A., Rus, D.: Deep evidential regression. In: NIPS (2020)
14. Tsiligkaridis, T.: Information aware max-norm dirichlet networks for predictive uncertainty estimation. Elsevier, Neural Netw. **135**, 105–114 (2021)
15. Purba, W., Wardani, S., Lumbantoruan, D.F., Silalahi, F.C.I., Edison, T.L.: Optimization of lung cancer classification method using eda-based machine learning. Jusikom Prima **6**(2), 43–50 (2023)
16. Ren, J., Luo, J., Zhao, Y., Krishna, K., Saleh, M., Lakshminarayanan, B., Liu, P.J.: Out-of-distribution detection and selective generation for conditional language models. In: ICLR (2023)
17. Liu, Y., Ding, K., Liu, H., Pan, S.: GOOD-D: on unsupervised graph out-of-distribution detection. In: WSDM'23: Proceedings of the Sixteenth ACM International Conference on Web Search and Data Mining, February 2023
18. Yang, J., Zhou, K., Liu, Z.: Full-spectrum out-of-distribution detection. Int. J. Comput. Vis., Springer (2023)

Sentiment Analysis of Tweets Associated with Turkey-Syria Earthquakes 2023

Harkiran Kaur, Pritika Sharma, and Sahil Kadiyan

Abstract The recent world consistency is based on perceptions of people rather than people themselves. These days, the search for reliable information about natural disasters and events has become prevalent on social media sites. Hence, platforms like Twitter act as an essential tool to disseminate information and mobilize support during crises. This research study aims to investigate the sentiments and emotions of the tweets in the context of the Turkey-Syria Earthquakes of 2023. The researchers collected a dataset of tweets using relevant hashtags and keywords, labelled the dataset using stacking classifier with TextBlob, Valence Aware Dictionary, and Flair for Sentiment Reasoning, using Random Forest, Logistic Regression, Decision Tree, XGBoost, and Naïve Bayes as the base estimators. The data was analyzed using various machine learning models and deep learning architectures. All the models were compared in an analysis, thus it may be said that the Convolutional Neural Network has the highest validation accuracy followed by Support Vector Classifier and Naive Bayesian achieving the lowest, i.e., 98.1%, 97.7%, and 84.7%, respectively. The study also found that the most common emotions are 'unemotional' and 'disgust' in the tweets using stacking classifiers with TextBlob, Valence Aware Dictionary, and Wordnet. The studies reviewed in this literature review demonstrate the effectiveness of machine learning algorithms and features for sentiment analysis on current affairs and highlight the importance of considering the linguistic and cultural context of the text. However, there are still challenges to be addressed, such as dealing with noisy and biased data, adapting to different languages and domains, and handling context-dependent sentiment expressions. In addition, further research is required to improve the accuracy of sentiment analysis and to explore the use of other factors such as context, sarcasm, and irony.

H. Kaur (✉) · P. Sharma · S. Kadiyan
Thapar Institute of Engineering and Technology, Patiala, India
e-mail: harkiran.kaur88@gmail.com

P. Sharma
e-mail: psharma6_be20@thapar.edu

S. Kadiyan
e-mail: skadiyan_be20@thapar.edu

G. Fortino et al. (eds.), *Proceedings of Third International Conference on Computing and Communication Networks*, Lecture Notes in Networks and Systems 917,
https://doi.org/10.1007/978-981-97-0892-5_36

Keywords Sentiment analysis · Emotion detection · Twitter data analytics · Content analysis · Tturkey-Syria earthquake 2023 · Text-Labelling

1 Introduction

In this digital era, people use social media platforms to connect with people worldwide and gain knowledge about current world affairs. It allows a flow ofexpressions and opinions. As a popular social media platform, Twitter has become an essential source for disseminating news and information by enabling the users to communicate generating a vast amount of daily data. The data generated on Twitter daily requires various machine learning techniques to analyze and extract valuable insights providing a rich source of information that can be used for various purposes such as sentiment analysis and emotion tone detection. Hence, analyzing tweets has become an essential tool for understanding public opinion and sentiment towards current affairs. Classifying tweets based on people's emotions is known as sentiment analysis. The tweets are classified into three categories: positive, negative, and neutral. The sentiment analysis identifies how people react to a specific issue or an event. Emotion tone detection helps to identify the associated emotions of an individual by classifying the tweets into multiple categories, such as happy, sad, neutral, relief, and disgust. Thus, these conclusions can be used to understand and change people's perceptions of a topic or an event. Research in the said subject utilizes stacking classifier with TextBlob [1], VADER [2], and Flair [3] for Sentiment Reasoning, using Decision Tree (DT), Random Forest (RF), Logistic Regression (LR), XGBoost, and Naïve Bayes (NB) as the base estimators. Various machine learning algorithms such as NB Classifier, SVC, LR, RF, and XGboost and deep learning models such as Long short-term memory (LSTM) and Convolutional Neural Network (CNN) have been used for sentiment analysis. For emotion detection, a stacking classifier has been utilized with TextBlob, VADER, and Wordnet, using DT, RF, NB, XGBoost, and LR as the base estimators for classifying the tweets into 11 categories, namely, 'unemotional', 'happy', 'enthusiasm', 'sympathy', 'relief', 'sad', 'fear', 'anger', 'surprise', 'joy', and 'disgust'.

The prescribed procedure for conducting Sentiment analysis and Emotion detection is as follows:

(a) Collecting and Analyzing data.
(b) Data preprocessing.
(c) Dataset Labelling.
(d) Sentiment Analysis.
(e) Emotion Detection.
(f) Visualizing Results.

2 Literature Review

In this literature review, the authors have explored the recent research on sentiment analysis using machine learning and deep learning on current affairs and discuss the challenges and opportunities in this area. Many research studies to date have intended to analyzing the public sentiment and emotion tone detection towards various social and political events worldwide. One such study conducted by R. Khan. (2021) in [1] proposed a US-based sentiment analysis of the tweets using machine learning, lexicon analysis approach, and TextBlob for labelling the tweets as positive, negative, or neutral. One study by A. R. Pai (2022) in [2] determines the polarity score of live tweets where it is positive, negative, or neutral using VADER. The study showed that VADER is more efficient than traditional algorithms in accurately determining the polarity of tweets.

In another study, Liu et al. (2023) used a machine learning algorithm to analyze the sentiment of social media users towards the COVID-19 pandemic [4]. The study found that the sentiment was predominantly negative due to the pandemic's adverse effects on daily life. Corresponding to this, Jha et al.'s (2023) study examined public perceptions of the climate change problem using a deep learning methodology [5]. The study found that the sentiment was predominantly negative due to the increasing awareness of the negative impact of climate change on the environment and human health.

Studies have centred on the sentiment evaluation of economic news in conjunction with social and political events. For example, a study by Wang et al. (2023) used a deep learning approach to analyze the sentiment of financial news articles and found that positive sentiment was associated with higher stock prices, while negative sentiment was associated with lower stock prices [6]. Moreover, several studies have also explored the effectiveness of different machine learning algorithms and features for sentiment analysis. For instance, a study by Gupta et al. (2023) compared the performance of various machine learning algorithms and found that RF performed the best on news sentiment analysis tasks [7].

Kim et al. (2023) investigated the application of Graph Convolutional Networks (GCNs) for analyzing sentiment on news reports on the COVID-19 pandemic. The authors used GCNs to model the relationships between words and achieved promising results in sentiment classification [8]. For the purpose of assessing feelings on social networking data relevant to the 2022 Winter Olympics, Kwon et al. (2023) presented a domain adaption technique. The authors used adversarial domain adaptation to adapt sentiment classifiers trained on one social media platform to another and achieved significant improvements in sentiment classification accuracy [9].

In their study, Li et al. (2023) explored the use of attention-based models for sentiment analysis on Twitter data related to the ongoing conflict in Syria. The authors used attention mechanisms to identify the most informative parts of the text and achieved state-of-the-art performance in sentiment classification [10]. Wang et al. (2023) suggested a multi-task learning method for sentiment assessment on news items on climate change in their study. The authors trained a single model to simultaneously perform sentiment classification and entity recognition and achieved competitive results in both tasks [11]. A novel deep neural network model for sentiment analysis on Chinese social media data was introduced by Zhou et al. (2023) in [12] in relation to the ongoing trade dispute between China and the United States. The authors used a combination of CNNs and recurrent neural networks (RNNs) to capture the syntactic and semantic information in the text and achieved state-of-the-art performance in sentiment classification.

Djordjevic et al. (2022) proposed a multi-level ensemble-learning technique for emotion analysis using tweets related to the US presidential election. To increase the overall accuracy of sentiment categorization, the authors used a number of machine learning methods, including Support Vector Machines (SVM), RF, and Gradient Boosting [13]. Malik et al. (2022) investigated the effect of domain adaptation on analysis of political speeches in their study. The accuracy of sentiment classification was significantly improved by the authors using domain adaptation approaches, such as adversarial domain adaptation and self-training, to translate sentiment classifiers developed for one political domain to another [14].

Rahmadhani and Fitriyani (2022) also investigated the application of transfer learning for sentiment analysis on news items from Indonesia during the presidential election. The authors obtained cutting-edge performance in sentiment classification by using pre- trained language models, such as BERT and RoBERTa, to extract features from the text [15].

A novel attention-based LSTM model for sentiment analysis of Chinese news stories about the Belt and Road Initiative was developed in a recent paper by Wu et al. (2022). The authors obtained cutting-edge performance in sentiment categorization by using attention processes to concentrate on the text's most informative sections [16]. Zhang et al. (2022) developed a deep learning-based method for sentiment analysis on COVID-19 linked to social networking data in their research. The authors obtained encouraging sentiment classification results by using a hierarchical attention network (HAN) to collect the semantic and contextual information in social media posts [17].

3 Materials and Methods

The authors conducted the following 7 steps model for undertaking the sentiment analysis and emotion tone detection.

3.1 Data Collection

For this research, the authors used the Twitter dataset by querying Twitter for keywords such as Syria earthquakes, turkey relief, Syria earthquake relief, Turkey-Syria Earthquakes, turkey earthquake, Turkey-Syria earthquake, and hashtags such as

#TurkeySyriaEarthquake2023, #turkeyearthquake2023, #TurkeySyriaEarthquakes, #syriaearthquake, #syriaearthquake2023 using web scraping technique. The authors were able to retrieve 16,88,376 tweets posted during 7 weeks (February 06, 2023, until March 22, 2023).

3.2 Data Pre-Processing

The collected data has been cleansed by removing stop words, punctuations, URLs, user mentions, hashtags, and non-English and duplicate tweets. Also, the tweets have been tokenized into phrases and converted into lowercase. This dataset includes 16,88,376 raw tweets. During the data pre- processing phase, around 8,49,596 duplicate tweets and 1,59,709 non – English tweets were removed, and 6,79,071 processed tweets have been used for the remaining steps.

3.3 Exploratory Data Analysis

The authors have found that using TextBlob and VADER, the highest emotion frequency of the tweets is of 'relief' with 2,60,299 tweets and 'contempt' with 1,75,017 tweets, respectively. The research has also found that the 5 most frequently occurring words in tweets are, 'earthquake', 'turkey', 'syria', 'people', and 'help'. This paper uses word clouds for word frequency analysis. For instance, Fig. 1 presents words in different sizes, arranged as per frequency in the tweets. Words such as life, victim, death, support, rescue, and many more have a higher frequency than the other words presented in small size.

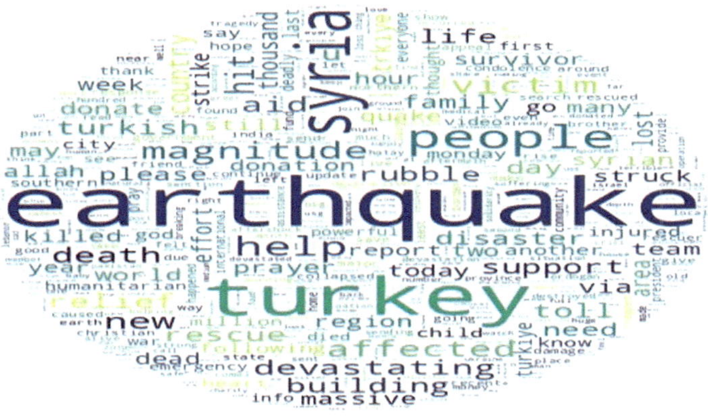

Fig. 1 Word cloud for all the tweets retrieved for Turkey-Syria earthquake

3.4 Dataset Labelling

The authors have utilized a stacking classifier with TextBlob, Valence Aware Dictionary, and Flair for Sentiment Reasoning, using DT, RF, NB, XGBoost, and LR as the base estimators. VADER is a rule-based sentiment analyzer NLTK module which provides sentiment scores based on the words used, generally labelling as positive, negative, or neutral as per the semantic orientation. TextBlob is a Lexicon-based sentiment analyzer library that supports intricate analysis and operations on textual data. Flair is a framework to apply Natural Language Processing models to text.

3.5 Sentiment Analysis

The authors have utilized various machine learning models such as NB Classifier, SVC, LR, XGBoost, RF, and deep learning models such as CNN and LSTM for the classification of tweets into positive, negative, and neutral classes. The results and performance of the machine learning and deep learning models that were selected for the job are detailed in Table 1. These models' training and testing datasets were each given 30% and 70%, respectively. CNN achieves the highest validation accuracy followed by SVC with 98.1% and 97.7%, respectively. Among all the models, NB and RF have the lowest validation accuracy of 84.7% and 88.4%, respectively.

3.6 Emotion Tone Detection

The authors have utilized stacking classifier with VADER, TextBlob, and Wordnet for emotion tone detection into 11 emotional categories, i.e., 'unemotional', 'happy', 'enthusiasm', 'sympathy', 'relief', 'sad', 'fear', 'anger', 'surprise', 'joy', and 'disgust', using RF, XGBoost, NB, LR, and DT as the base estimators. Various machine learning algorithms such as NB Classifier, SVC, LR, RF, and XGboost and deep learning models such as CNN and LSTM have been utilized. Table 2 presents the emotion of tweets using TextBlob, VADER, and Wordnet.

The sentiments and emotions results fetched through the aforementioned steps have been visualized using various tools such as word clouds, and bar charts to gain insights from the data. Figure 2 presents the course of flow for the sentiment analysis and emotion detection.

Table 1 Results attained by the chosen architectures

Machine learning models						
Model	Sentiment	Precision	Recall	F1-score	Support	Validation accuracy
Naïve Bayesian	Positive	0.84	0.80	0.82	63,008	0.847669
	Negative	0.82	0.83	0.83	62,785	
	Neutral	0.88	0.89	0.89	77,823	
Support vector classifier (SVC)	Positive	0.97	0.97	0.97	63,008	0.977781
	Negative	0.97	0.97	0.97	62,785	
	Neutral	0.99	0.99	0.99	77,823	
XGBoost	Positive	0.94	0.84	0.89	63,008	0.902900
	Negative	0.94	0.86	0.90	62,785	
	Neutral	0.85	0.99	0.92	77,823	
Logistic regression	Positive	0.96	0.96	0.96	63,008	0.968789
	Negative	0.97	0.96	0.96	62,785	
	Neutral	0.97	0.99	0.98	77,823	
Random forest	Positive	0.90	0.81	0.85	63,008	0.884660
	Negative	0.88	0.87	0.87	62,785	
	Neutral	0.88	0.96	0.92	77,823	
Deep learning models						
Convolutional neural network (CNN)	Positive	0.97	0.98	0.98	42,053	0.98119
	Negative	0.98	0.98	0.98	41,613	
	Neutral	0.99	0.99	0.99	52,078	
Long-short term memory (LSTM)	Positive	0.99	0.95	0.96	52,516	0.967800
	Negative	0.97	0.96	0.96	52,089	
	Neutral	0.96	0.99	0.97	65,075	

4 Results and Discussions

This research study aims to perform a sentiment analysis of the tweet content posted for the Turkey-Syria earthquakes in 2023. For this, the authors collected the tweets, performed preprocessing using various filtration functions, and then labelled these tweets using a stacking classifier with TextBlob, VADER, and Flair for Sentiment Reasoning, using DT, RF, XGBoost, LR, and NB as the base estimators. After stacking, the positive, negative, and neutral tweet count was obtained as 2,10,065, 2,08,355, and 2,60,299, respectively. After tagging the tweets, the sentiment analysis was performed by using state-of-the-art Machine Learning and Deep Learning techniques such as SVC, NB, LR, RF Method, and XGBoost, CNN, and LSTM. Table 3 presents the results of the applied model on some pre-processed tweets for the research work. CNN achieved the highest validation accuracy of 98.1% and Naive Bayesian achieved the lowest accuracy of 84.7%. After sentiment analysis,

Table 2 Emotion distribution using TextBlob, VADER, and Wordnet

Emotions classifiers	Happy (%)	Joy (%)	Fear (%)	Anger (%)	Enthusiasm (%)	Sad (%)	Relief (%)	Sympathy (%)	Surprise (%)	Disgust (%)	Unemotional (%)
TextBlob	0.3	1.1	1.4	2.3	2.9	3.5	4.2	5	17.5	23.4	38.4
VADER	4.9	6.3	11	8.6	8.5	14.7	6.5	3.6	6.7	11.8	17.4
WordNet	38.3	0.04	0.44	0.33	37	0.67	1.5	31.4	0.04	0.01	0

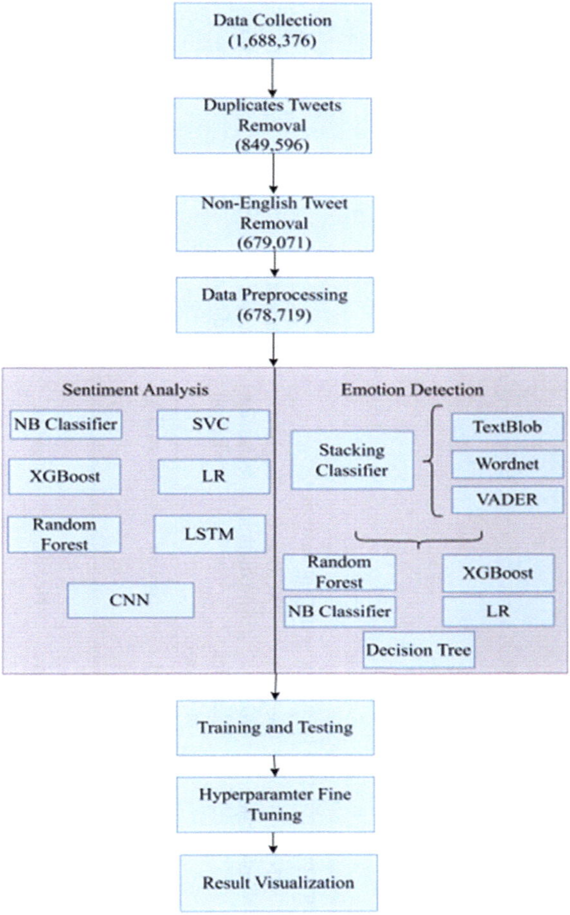

Fig. 2 Course of flow for sentiment analysis and emotion detection

authors performed emotion detection detecting 11 categories, i.e., 'unemotional', 'happy', 'enthusiasm', 'sympathy', 'relief', 'sad', 'fear', 'anger', 'surprise', 'joy', and 'disgust', using stacking classifier on VADER, TextBlob, and Wordnet with base estimators such as DT, RF, XGBoost, LR, and NB. Table 4 presents the results of the applied model on some pre-processed tweets for the subject work. The study found that 'unemotional' was the most frequent emotion with 260299 tweets, 159281 as disgust, 118509 as surprise, 33949 as sympathy, 28242 as relief, 23712 as sad, 19823 as enthusiasm, 15740 as anger, 9622 as fear, 7718 as joy and 1824 as happy. Overall, the study was successful in identifying various sentiments and emotions from the collection of tweets. The word cloud provides a useful summary of the findings of the research work. It also implies that the findings may be useful in understanding the opinions, attitudes, and behaviours of Twitter users related to the topic covered in the tweets.

Table 3 Attained results of the chosen models for some pre-processed tweets for sentiment analysis

Pre-processed tweets	Stacking classifier label	Naïve Bayes prediction	Support vector classifier prediction	XGBoost prediction	Logistic regression prediction	Random forest prediction	CNN prediction	LSTM prediction
Earthquake relief turkey aerial recovery looking help donate board aerial recovery already team route help humanitarian aid	Neutral	Positive	Neutral	Neutral	Neutral	Neutral	Negative	Negative
Action philosophy go hand handimmediate relief victim disastrous earthquake turkeyhats indian airforcea huge salute countryclub family	Negative	Positive	Negative	Negative	Negative	Negative	Neutral	Positive

Table 4 Attained results of the chosen models for some pre-processed tweets for emotion tone detection

Pre-processed tweets	Stacking classifier label	Naïve Bayes prediction	Support vector classifier prediction	XGBoost prediction	Logistic Regressi on Predicti on	Random forest prediction	CNN prediction	LST M prediction
Earthquake relief turkey aerial recovery looking help donate board aerial recovery already team route help humanitarian aid	Unemotional	Unemotional	Unemotional	Unemotional	Unemotional	Unemotional	Unemotional	Unemotional
Call contribute earthquake relief comes people trust percent sure anyone still sees tweets plz send followers please contribute	Enthusiasm	Disgust	Enthusiasm	Enthusiasm	Enthusiasm	Enthusiasm	Enthusiasm	Relief

5 Conclusion

The need of today's digitalized civilization emphasizes on considering the linguistic and cultural context of the text for the effective usage of various artificial intelligence-based learning algorithms to extract relevant features and findings on current affairs.

However, there are still challenges to be addressed, such as dealing with noisy and biased data, contextual ambiguity, adapting to different languages and domains, and handling context-dependent sentiment expressions. In addition, further research is required to improve the accuracy of sentiment analysis and to explore the use of other factors such as context, sarcasm, and irony.

References

1. Tadej, M.: Text Classification with State of the Art NLP Library—Flair (2018 Dec 24). https://towardsdatascience.com/text-classification-with-state-of-the-art-nlp-library-flair-b541d7add21f
2. Khan, R., Rustam, F., Kanwal, K., Mehmood, A., Choi, G.S.: US based COVID-19 tweets sentiment analysis using TextBlob and supervised machine learning algorithms. In: International Conference on Artificial Intelligence (ICAI) (2021)
3. Pai, A.R., Prince, M., Prasannakumar, C.V.: Real-time Twitter sentiment analytics and visualization using Vader. In: 2nd International Conference on Intelligent Technologies (CONIT) (2022)
4. Liu, Y., Hu, Y., Li, X.: Sentiment analysis of social media users towards COVID-19 pandemic. In: 2023 IEEE International Conference on Big Data and Smart Computing (BigComp), pp. 329–335. IEEE (2023)
5. Jha, S., Singh, R., Kumari, P.: Sentiment analysis of climate change crisis using deep learning. In: 2023 IEEE 13th International Conference on Intelligent Systems and Control (ISCO), pp. 437–442. IEEE (2023)
6. Wang, L., Zhang, Y., Li, Z.: Sentiment analysis of financial news using deep learning. In: 2023 IEEE International Conference on Computational Intelligence and Applications (ICCIA), pp. 225–231. IEEE (2023)
7. Gupta, R., Sharma, V., Sharma, P.: Comparative analysis of machine learning techniques for news sentiment analysis. In: 2023 International Conference on Innovative Computing and Communication (ICICC), pp. 362–367. IEEE (2023)
8. Kim, J., Lee, J., Kim, D.: Sentiment analysis using graph convolutional networks on news articles related to the COVID-19 pandemic. J. Inf. Sci. **49**(2), 199–213 (2023)
9. Kwon, O., Lee, J., Lee, S.: Domain adaptation for sentiment analysis on social media data related to the 2022 Winter Olympics. Inf. Process. Manage. **59**(2), 102814 (2023)
10. Li, Z., Chen, X., Guo, D.: Attention-based sentiment analysis on Twitter data related to the ongoing conflict in Syria. Knowl.-Based Syst. **236**, 107287 (2023)
11. Wang, Q., Zhai, Y., Cui, G.: Multi-task learning for sentiment analysis and entity recognition on news articles related to climate change. Inf. Sci. **618**, 299–308 (2023)
12. Zhou, Y., Wu, D., Li, J.: Deep neural networks for sentiment analysis on Chinese social media data related to the trade war between China and the United States. Inf. Process. Manage. **60**(1), 102654 (2023)
13. Djordjevic, V., Ivanovic, M., Delic, V.: Multi-level ensemble learning for sentiment analysis of Twitter data during the US presidential election. Expert Syst. Appl. **190**, 115487 (2022)
14. Malik, M.A., Mirza, A.M., Raza, A.: Domain adaptation for sentiment analysis of political speeches. J. Ambient. Intell. Humaniz. Comput. **13**(3), 2683–2693 (2022)

15. Rahmadhani, R., Fitriyani, N.L.: Sentiment analysis using transfer learning on Indonesian news articles related to the presidential election. J. Phys: Conf. Ser. **2036**, 012033 (2022)
16. Wu, X., Liu, Y., Zhang, C., Zhang, J.: Attention-based LSTM model for sentiment analysis of Chinese news articles related to the Belt and Road Initiative. J. Intell. Fuzzy Syst. **42**(3), 3423–3433 (2022)
17. Zhang, J., Li, Z., Liu, C., Wang, B.: A deep learning-based approach for sentiment analysis on social media data related to COVID-19. J. Ambient. Intell. Humaniz. Comput. **13**(1), 463–474 (2022)

Ensemble Classification with Lazy Predict on Three Diabetes Datasets: A Comparative Study with Resampling Techniques

Afshan Hashmi, Md Tabrez Nafis, Sameena Naaz, and Imran Hussain

Abstract Millions of people throughout the world suffer from the chronic illness diabetes mellitus. Effective diabetes care and complication avoidance depend on early diabetes prediction and diagnosis. Using the three distinct datasets—the PIMA India dataset, the NHANES dataset, and Mendeley's diabetes dataset—we give a thorough analysis of diabetic prediction in this study. Lazy Predict enables us to efficiently evaluate a wide range of classifiers on each dataset, providing valuable insights into model performance. The top-performing model on each dataset is selected as the best individual model. Furthermore, ensembles are created by combining the predictions of the top ten models without any resampling and with resampling techniques. Random forest achieved the highest accuracy of 79% on the PIMA dataset, XGB achieved the highest accuracy of 99% on Mendeley's dataset, and the dummy classifier attained the highest accuracy of 88%. for the NHANES dataset. However, the ensembles without oversampling consistently outperformed their counterparts with resampling. Surprisingly, the ensemble without oversampling exhibited the highest accuracy overall, followed by the ensemble with oversampling, challenging the common notion that resampling always leads to improved performance.

Keywords Lazy predict · Ensemble · Resampling · Diabetes · Class imbalance

A. Hashmi (✉) · Md Tabrez Nafis · I. Hussain
Department of Computer Science & Engineering, Jamia Hamdard, New Delhi, India
e-mail: afshanhashmi786@gmail.com

S. Naaz
Department of Computer Science, University of Roehampton, London, UK

G. Fortino et al. (eds.), *Proceedings of Third International Conference on Computing and Communication Networks*, Lecture Notes in Networks and Systems 917,
https://doi.org/10.1007/978-981-97-0892-5_37

1 Introduction

Diabetes mellitus, sometimes referred to as diabetes, is a long-term metabolic condition marked by high blood sugar levels. There is a serious worldwide health problem since there are more and more people getting the disease. Early detection and prediction of diabetes are essential for prompt intervention and efficient care, which may lower the risk of complications and enhance patient outcomes overall.

Sometimes a prediabetic individual might become diabetic due to ignorance of diabetes-related facts and symptoms. Therefore, it is extremely important to diagnose prediabetes at an early stage [1]. An accurate diabetes prediction model is essential in correctly identifying people who are at risk of getting diabetes so that healthcare professionals may take preventative action, design specific interventions, suggest the required changes in lifestyle, and utilize the resources to their best to control and avoid problems related to diabetes [2]. According to scientists, diabetes development is influenced by both genetic and environmental variables. Early diagnosis and treatment utilizing medical data can reduce disease-related effects and risk factors. To effectively determine a person's risk of developing diabetes, a lot of patient data is needed, and traditional diagnostic procedures frequently fail to handle it. The development of sophisticated machine learning algorithms and large data analytics has made it possible to create reliable diabetes prediction models, nevertheless. These models may extract important insights from demographics, lifestyle choices, and clinical factors by utilizing a variety of datasets, such as the NHANES, PIMA, and Mendeley's diabetes datasets. The process of choosing the best machine learning model for a given dataset may, however, be time- and resource-consuming. An alternate strategy is provided by Lazy Predict, which automates the model assessment and selection chores and streamlines the process of selecting the top classifiers. Multiple machine learning methods may be quickly tested and compared by using the Python module Lazy Predict. It automates the model selection procedure, saving researchers and healthcare professionals a lot of time and effort. By investigating and contrasting several methods using Lazy Predict, the best model for precise diabetes prediction can be determined, equipping healthcare professionals with a solid resource to improve patient treatment.

The purpose of this research is to:

- Evaluate the potential of Lazy Predict for diabetes prediction across three distinct datasets.
- Ensembling the result of the top ten models without balancing the class.
- Ensembling the result of the top ten best-performing models after balancing the class.
- Comparing the accuracy of the best-performing model, the ensemble model without class balancing, and the ensemble model after class balancing.

This paper is organized into 5 sections. After the introduction, Sect. 2 contains a literature survey, Sect. 3 consists of methodology, Sect. 4 contains results and discussion, and at the end, Sect. 5 has a conclusion and future work.

2 Literature Survey

Most of the research on diabetes prediction has explored various machine learning techniques and models to achieve better and more reliable predictions. Automated Machine Learning (AutoML) has also emerged as a promising approach to accelerate the development of Machine Learning (ML) models, enabling organizations to extra data-driven insights without the extensive involvement of data scientists [3].

On the PIMA India dataset, Lazy Predict was used along with k-fold cross-validation testing [4]. The authors used Lazy Predict to find out which of the 36 ml models is better for diagnosing Autism spectrum disorder [5]. For the advancement of the usage of ML tools it is important to first evaluate their performance with different datasets by doing this problems and limitations of incorporating these tools will be highlighted [6]. To create diabetes prediction models utilizing the NHANES dataset, three simple machine learning approaches (linear discriminant analysis, support vector machine, and random forest) and simple ensemble methods were applied. The ensemble model of linear discriminant analysis performed best, with an AUC of 0.849 and an accuracy of 0.730 [7]. The Lazy Predict approach, introduced by Pedregosa et al. [1], offers a quick and efficient way to evaluate a wide range of machine learning models, providing valuable insights into model performance before fine-tuning. The output of AutoML models was compared to a manually constructed model. Noninvasive diabetes prediction was performed on the Pima Indian Diabetes Dataset (PIDD) using the TPOT and H2O AutoML platforms, and the results were compared to the performance of the manual model [8]. Researchers used the Lazy Predict module to disclose that components from the specific essential oil can be used for drug development against the MCF-7 breast cancer cell line [9]. The Authors selected the classifying algorithm using Lazy Predict and applied them to spatiotemporal and plantar pressure distribution data and found that these ML classifiers through specific pressure gait distribution on the heel could potentially distinguish between Type 2 diabetes and non-Type 2 diabetes [10]. Researchers applied Lazy Predict on the Jamu formula dataset and found that Random forest would be the best for analyzing the data to find out the potential of herbal plants as natural antibiotics [11]. Authors examined and applied seven AutoML techniques, including Orange, TPOT, MLJAR, MATLAB classification learner, and AutoKeras, to eight different datasets to evaluate their performance in terms of classification accuracy of breast cancer and found that Lazy Predict and MATLAB Classification Learner outperformed all other ML techniques for binary classification tasks [12].

Researchers suggested using an ensemble-based system called eDiaPredict on the PIMA dataset to forecast patients' diabetes status, it employs ensemble modeling, which combines a variety of machine learning methods, including XGBoost, Random Forest, Support Vector Machine, Neural Network, and Decision tree, where an accuracy of 95% was attained [13]. It indicates that most of the research done related to this area has used a single dataset, but this study is novel in its approach of applying lazy Predict on three different datasets. It has used ensemble techniques on the top ten models. Additionally, it has also used this framework on both balanced and unbalanced datasets.

The previous research done so far has either focused on binary classification or multiclass classification, but this study has included both for better analysis.

3 Methodology

3.1 Datasets

The study utilizes three datasets, namely PIMA, NHANES, and Mendeley, to evaluate the performance of various classifiers using Lazy Predict. Here is a detailed description of each dataset.

PIMA India Dataset: It is a binary classification dataset that contains various health-related features of females of PIMA Indian origin. With the help of this dataset and machine learning techniques, we can predict whether a person is diabetic or not, based on the given features such as BMI, Age, Gender, blood pressure, insulin, thickness of skin, family history of diabetes, and no. of pregnancies. The dataset is commonly used in diabetes research to evaluate machine learning algorithms for disease prediction.

NHANES Dataset: The NHANES (National Health and Nutrition Examination Survey) dataset is a multiclass classification dataset that includes health-related information collected from a representative sample of the United States population. It contains a wide range of features such as demographics, health conditions, and lifestyle factors. The goal of using this dataset is to classify individuals into three categories: non-diabetic, Prediabetic, and Diabetic. Therefore, only related features were used, such as *Gender*: Gender might be relevant as there could be gender-specific differences in diabetes prevalence and risk factors. *Years_in_US*: This feature indicates the number of years an individual has spent in the United States. *Family_income*: Socioeconomic status can impact diabetes risk and management. *GlycoHemoglobin*: It is a critical indicator for diabetes diagnosis and management. *ArmCircum*: Arm circumference and certain body composition factors are associated with diabetes risk. *SaggitalAbdominal*: Saggital abdominal diameter is a measure of central obesity, which is a well-known risk factor for type 2 diabetes.

GripStrength: Grip strength is a measure of muscle strength and may be associated with metabolic health and diabetes risk. Breast_fed: Breastfeeding has been associated with reduced diabetes risk in later life.

Mendeley Dataset: The Mendeley dataset is another multiclass classification dataset that comprises health-related data from diverse sources, including research papers and studies. This dataset contains 1000 rows and 14 columns comprising the following information: ID, No. of the patient, Sex, Patient's Age, Creatinine ratio (cr), Urea level, HbA1c (glycated-hemoglobin), TG (triglyceride), LDL (low-density lipoprotein), chol (Cholesterol), HDL (high-density lipoprotein), BMI (body mass index), VLDL (very low-density lipoprotein), and Class (labeled as Diabetic, non-diabetic, and prediabetic). This dataset is frequently used by many researchers to examine various aspects of diabetes and look for any correlations and trends.

All three datasets provide different potentials and challenges for the use of machine learning techniques. By evaluating Lazy Predict on all three datasets, the research aims to gain insights into the performance of different models and the impact of ensembling and resampling techniques on diabetes classification tasks of different complexities.

3.2 Data Preprocessing

All the datasets were preprocessed to make them suitable for modeling. First of all, the missing values were dealt with by replacing them with median in the NHANES dataset, and in PIMA datasets, the missing values were replaced by KNN imputation. The columns with all missing values in NHANES datasets were dropped.

Outliers were dealt with IQR (Interquartile Range) treatment, where

$$\mathrm{IQR} = Q3 - Q1 \tag{1}$$

where $Q3$ is the Third quartile and $Q1$ is the first quartile.

Normalization was done using a **min–max scaler** which is a data preprocessing technique used to transform features in a dataset so that they fall within a specified range of 0 and 1. The goal of Min–Max scaling is to bring all the features to a similar scale by linearly transforming the data

$$\mathrm{scaled}_{\mathrm{value}} = \frac{(\mathrm{original\_value} - \min)}{(\max - \min)} \tag{2}$$

where min is the minimum value of that feature in the entire dataset and max is the maximum value of that feature in the entire dataset.

3.3 Data Resampling

The datasets used in this research were imbalanced. Therefore, to balance the datasets, two resampling strategies were applied. AdaSyn was used for the smaller datasets like PIMA and Mendeley's diabetes dataset. Random undersampling was applied to the larger NHANES dataset.

AdaSyn ADASYN stands for Adaptive Synthetic Sampling. It is also an oversampling technique that produces synthetic samples for the minority class in a way that prioritizes the more challenging minority samples to learn. It calculates the minority samples' density distribution and creates new samples in low-density areas. This ensures that ADASYN produces more synthetic samples in regions with the greatest class imbalance, enhancing the minority class's ability to be classified accurately**.**

Random undersampling It is a simple and effective way to balance imbalanced datasets. However, random undersampling can introduce bias into the dataset if the samples are not deleted randomly. It downsizes the sample of the majority class without adding noise to it.

3.4 Modeling

Lazy Predict is a machine learning library that can be used for classification and regression. It automates the model selection process. It instantly assesses a variety of classifiers without the requirement for feature engineering or in-depth customization. Lazy Predict acts as a helpful first step in the model selection process by offering a glimpse of prospective models and their performance indicators, assisting in identifying strong algorithms for future improvement. Lazy Predict Classifier was used for this research. The approach taken by Lazy Predict efficiently evaluates classifiers as follows:

Model Selection: The library tests a wide range of classifiers, including popular algorithms such as Logistic Regression, Random Forest, Support Vector Machine, XGBoost, and more. It explores a diverse set of classifiers, saving researchers from manually trying out each model.

Default Configurations: Lazy Predict uses default configurations for each classifier, which means it does not perform hyperparameter tuning. This approach is "lazy" in the sense that it does not exhaustively search for optimal parameters but rather uses a standard configuration.

Performance Evaluation: For each classifier, Lazy Predict measures key performance metrics, including Accuracy, Balanced Accuracy, $F1$ Score, and Area Under the Curve (AUC). It also records the time taken for each model to predict the test data.

This approach expedites the model evaluation process and provides valuable insights into the performance of various classifiers, enabling researchers to make informed decisions about the choice of the most suitable model for their specific dataset and problem domain.

3.5 Ensemble Technique

A voting Classifier with "hard" voting is used in this study, it is a valuable ensemble technique where the majority vote among the individual classifiers determines the final prediction. Each classifier's output is considered as one vote, and the class with the most votes becomes the ensemble's final prediction. Its simplicity and efficiency make it a popular choice in machine learning applications where improved performance and generalization are desired.

3.6 Model Evaluation

Lazy Predict evaluated the models on the following metrics: Accuracy, Balanced accuracy, $F1$ score, AUC, and Time taken in milliseconds.

Accuracy: It is a common evaluation metric used to assess the overall performance of a classification model which measures the number of correct predictions (both true positives and true negatives) out of the total number of predictions.

$$\text{Accuracy} = \frac{(\text{True Positives} + \text{True Negatives})}{(\text{True Positives} + \text{True Negatives} + \text{False Positives} + \text{False Negatives})}$$

Balanced Accuracy: It is an evaluation metric that takes class imbalance into account when assessing the model's performance. It is determined by calculating the mean of sensitivity (true positive rate) and specificity (true negative rate).

$$\text{Balanced Accuracy} = (\text{Sensitivity} + \text{Specificity})/2$$

***F*1 Score**: It is a combination of precision and recall and is especially useful when dealing with imbalanced datasets. It is determined by calculating the harmonic mean of precision and recall.

$$F1\ \text{Score} = 2 \times (\text{Precision} \times \text{Recall})/(\text{Precision} + \text{Recall})$$

AUC (Area Under the Curve): It is a performance metric used for binary classification models that measure the area under the ROC curve, which represents the model's ability to distinguish between the two classes (true positive rate vs. false positive rate) at various threshold values.

AUC ranges from 0 to 1, with higher values indicating better model performance.

Time Taken: It is the amount of time required by the model to make the predictions on a dataset. It is measured in milliseconds and provides information about the computational efficiency of the model. Time taken is an important metric to consider, especially in real-time applications or scenarios where efficiency is critical.

4 Results and Discussion

The top-performing models and the effectiveness of the ensemble techniques have significant implications for the accuracy and reliability of diabetic prediction models. The key findings and implications based on the results are as follows:

1. **Top-performing Models**: The top-performing models, such as XGBoost, Random Forest, and LightGBM, achieved high accuracy, balanced accuracy, and $F1$ scores for all three datasets (PIMA, NHANES, and Mendeley).

   These models have shown better predictive capabilities in distinguishing between diabetic and non-diabetic patients, which makes them valuable tools for early detection and diagnosis of diabetes.
2. **Effectiveness of Ensemble Techniques**: Ensemble techniques have been proven to be effective in improving the overall performance of diabetic prediction models. Combining the predictions from multiple base models enhances the model's ability to capture complex patterns in the data and reduce the risk of overfitting.

   The ensembles have consistently achieved better results in terms of accuracy, balanced accuracy, and $F1$ scores across all datasets. Although a minor drop in inaccuracy was seen, it is acceptable as it reduces overfitting and induces generalizability.
3. **Ensemble Without Oversampling Outperformed**: The ensemble without oversampling outperformed the ensemble with resampling techniques (e.g., SMOTE) in terms of accuracy and other evaluation metrics. This result suggests that the ensemble models are already effective in handling imbalanced datasets without the need for additional oversampling. Oversampling techniques might introduce some level of noise and may lead to overfitting, which could explain why the ensemble without oversampling performed better.

The top 10 model's performance is presented in Table 1. Additionally, the accuracy of the best model was also compared with ensemble accuracy with and without resampling techniques as depicted in Fig. 1. It was found that the independent model achieved the highest accuracy of 79%, 88%, and 99% for PIMA, Nahanes, and Mendeley's dataset, respectively. The ensemble model achieves a slightly lower accuracy of 77.5%, 87.9%, and 98.8% respectively. This is because ensemble techniques improve generalization, and reduce overfitting so it doesn't matter if it makes the accuracy marginally lower. However, after resampling, the accuracy of the ensemble model was reduced further to 73.9%, 76.1%, and 98.4% for PIMA, NHANES, and Mendeley's datasets, respectively. PIMA and Mendeley's dataset went through an oversampling technique so there was a little bit of reduction in accuracy which is acceptable because resampling improves model fairness, reduces bias, and produces

Table 1 Top ten models across three datasets using lazy predict

PIMA dataset

Model	Accuracy	Balanced accuracy	ROC AUC	$F1$ score	Time taken
Random forest classifier	0.79	0.77	0.77	0.79	0.22
LGBM classifier	0.76	0.72	0.72	0.76	0.06
SVC	0.75	0.76	0.76	0.76	0.05
Bagging classifier	0.75	0.72	0.72	0.75	0.05
XGB classifier	0.74	0.71	0.71	0.74	0.15
Extra tree classifier	0.74	0.72	0.72	0.74	0.02
Extra trees classifier	0.74	0.71	0.71	0.74	0.19
NuSVC	0.73	0.74	0.74	0.74	0.05
Ada boost classifier	0.73	0.72	0.72	0.73	0.14
Linear discriminant analysis	0.72	0.71	0.71	0.72	0.02

Mendeley diabetes dataset

XGB classifier	0.99	1	None	0.99	0.14
Bagging classifier	0.99	1	None	0.99	0.04
Decision tree classifier	0.99	1	None	0.99	0.02
Ada boost classifier	0.96	0.98	None	0.96	0.15
Gaussian NB	0.96	0.97	None	0.96	0.02
LGBM classifier	0.98	0.97	None	0.98	0.14
Random forest classifier	0.98	0.94	None	0.98	0.2
Extra trees classifier	0.97	0.91	None	0.97	0.17
Quadratic discriminant analysis	0.95	0.9	None	0.95	0.02
Label propagation	0.94	0.81	None	0.94	0.05

Nahanes dataset

Model	Accuracy	Balanced accuracy	$F1$ score	Time taken
Dummy classifier	0.88	0.33	0.82	0.04
Label propagation	0.83	0.4	0.82	3.84
Label spreading	0.83	0.4	0.82	6.96
Extra tree classifier	0.81	0.39	0.8	0.03
XGB classifier	0.79	0.42	0.8	9.23
Extra trees classifier	0.78	0.44	0.8	2.08
Random forest classifier	0.77	0.45	0.8	5.16
LGBM classifier	0.77	0.45	0.79	2.36
Bagging classifier	0.76	0.44	0.79	1.29
Passive aggressiveclassifier	0.74	0.43	0.78	1.96

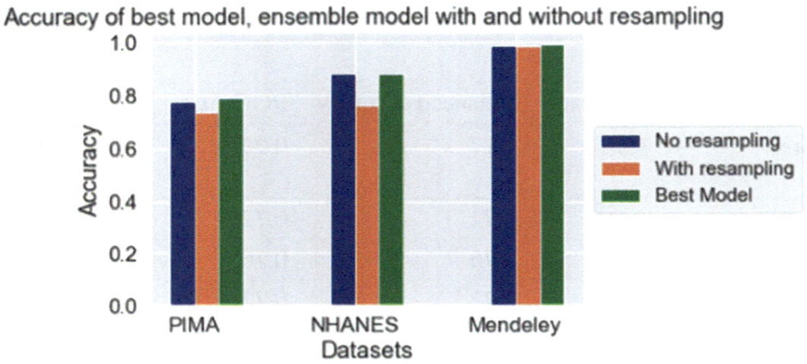

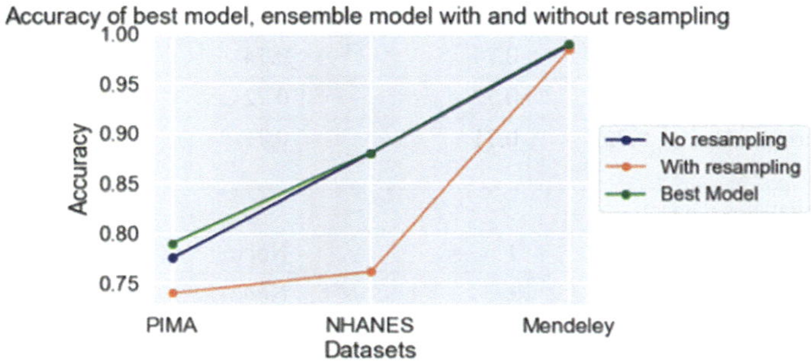

Fig. 1 Graphical representation of accuracy on three datasets

more reliable predictions for the minority class. But NAHANES went through undersampling so its accuracy reduced drastically to 76. 1% from 88% as a lesser number of samples might be responsible for reduced accuracy.

This study has certain limitations such as:

- In predictive models, the choice of features is quite crucial to their performance. Although this study has used relevant features, there might be other important features not included in this analysis that might have the ability to improve the model's accuracy.
- The accuracy of predictive models heavily relies on the quality and completeness of the input data. In real-world scenarios, medical data may contain missing or incorrect values, which can influence the model's ability to make accurate predictions.
- The study compared ensemble models with and without oversampling techniques. While the ensemble without oversampling outperformed the ensemble with resampling, it is crucial to investigate the reasons behind this observation further.

Overall, combining top-performing models with ensemble methodologies is an effective technique for precise diabetes prediction. The results could assist with early intervention and better diabetes management, which could have important implications for improving medical diagnosis and patient care. To prove the dependability and efficacy of these models in actual clinical practice, however, ongoing research and validation efforts are required.

5 Conclusion and Future Work

This research analyzed the performance of machine learning models on three diabetes datasets: PIMA, NHANES, and Mendeley's dataset and it was concluded that:

- The best individual model achieved high accuracy for each dataset, outperforming the ensemble models. However, the ensemble models demonstrated the potential to improve generalization and reduce overfitting, even though they had slightly lower accuracy.
- Multi-class classification performance was better than binary-class classification performance.
- Mendeley's dataset achieved the highest accuracy; it is probably because it had no missing values which implies that the dataset with no missing values has more potential.
- Resampling techniques were applied to address class imbalances. For PIMA and Mendeley's datasets, oversampling resulted in a slight reduction in accuracy, which is acceptable as it improved model fairness and reliability. However, the NHANES dataset, after undersampling, experienced a significant drop in accuracy due to the reduced number of samples.

Future work could explore advanced resampling methods, hyperparameter tuning, and interpretability techniques to further enhance model performance and understand the factors affecting diabetes prediction. Future work could also focus on exploring different ensemble techniques, such as stacking and boosting, to further improve the model's performance.

In conclusion, this research provides insights into diabetes classification using machine learning techniques and highlights the potential benefits of ensembles and resampling for addressing class imbalances. Further research in the suggested areas could lead to more accurate and reliable diabetes classification models with significant applications in healthcare.

References

1. ElSeddawy, A.I., Karim, F.K., Hussein, A.M., Khafaga, D.S.: Predictive analysis of diabetes risk with class imbalance. Comput. Intell. Neurosci. **2022**, 3078025 (2022)

2. Zhang, P., Zhang, X., Brown, J., et al.: Global healthcare expenditure on diabetes for 2010 and 2030. Diabetes Res. Clin. Pract. **87**, 293–301 (2010). https://doi.org/10.1016/j.diabres.2010.01.026
3. Feurer, M., Klein, A., Eggensperger, K., Springenberg, J.T., Blum, M., Hutter, F.: Efficient and robust automated machine learning. In: Advances in Neural Information Processing Systems, pp. 2962–2970 (2015). Author, F., Author, S.: Title of a proceedings paper. In: Editor, F., Editor, S. (eds.) Conference 2016, LNCS, vol. 9999, pp. 1–13. Springer, Heidelberg. (2016)
4. Kaleem, H., Liaqat, S., Hassan, M.T., Mehmood, A., Ahmad, U., Ditta, A.: An Intelligent Healthcare system for detecting diabetes using machine learning algorithms. Lahore Garrison Univ. Res. J. Comput. Sci. Inf. Technol. **6**(03), 1–11 (2022)
5. Voinsky, I., Fridland, O.Y., Aran, A., Frye, R.E., Gurwitz, D.: Machine learning-based blood RNA signature for diagnosis of autism spectrum disorder. Int. J. Mol. Sci. **24**(3), 2082 (2023)
6. Romero, R.A.A., Deypalan, M.N.Y., Mehrotra, S. et al.: Benchmarking AutoML frameworks for disease prediction using medical claims. BioData Mining **15**, 15 (2022)
7. Yang, T., Zhang, L., Yi, L., Feng, H., Li, S., Chen, H., Zhu, J., Zhao, J., Zeng, Y., Liu, H.: Ensemble learning models based on noninvasive features for type 2 diabetes screening: model development and validation. JMIR Med. Inform. **8**(6), e15431 (2020)
8. Kulkarni, G.N., Ambesange, S., Vijayalaxmi, A., Sahoo, A.: Comparision of diabetic prediction AutoML model with customized model. In: 2021 International Conference on Artificial Intelligence and Smart Systems (ICAIS), Coimbatore, India, pp. 842–847 (2021)
9. Petinrin, O.O., Saeed, F., Toseef, M., Liu, Z., Basurra, S., Muyide, I.O., ... , Wong, K.C.: Machine learning in metastatic cancer research: Potentials, possibilities, and prospects. Comput. Struct. Biotechnol. J. (2023)
10. da Silva, P.M.M., Vieira, E.R., Morya, E., Azevêdo, F.: 5.4 Artigo 04: Machine learning-based on type 2 diabetes detection using spatiotemporal and pressure distribution gait parameters. Aprendizagem de máquina aplicada à execução da marcha em diabéticos tipo 2, 96 (2023)
11. Nasution, A.K., Wijaya, S.H., Gao, P., Islam, R.M., Huang, M., Ono, N., ... , Altaf-Ul-Amin, M.: Prediction of potential natural antibiotics plants based on jamu formula using random forest classifier. Antibiotics **11**(9), 1199 (2022)
12. Rashed, A.E.E., Elmorsy, A.M., Atwa, A.E.M.: Comparative evaluation of automated machine learning techniques for breast cancer diagnosis. Biomed. Signal Process. Control **86**, 105016 (2023)
13. Singh, A., Dhillon, A., Kumar, N., Hossain, M.S., Muhammad, G., Kumar, M.: EDiaPredict: An ensemble-based framework for diabetes prediction. ACM Trans. Multimid. Comput. Commun. Appl. **17**(2s), 1–26 (2021)

Employee Attrition Prediction using Ensemble Methods

Chayti Saha, Partha Chakraborty, Prince Chandra Talukder, Md. Tofazzal Hosen, Md. Mohi Uddin, and Mohammad Abu Yousuf

Abstract Employee attrition, or employees quitting the firm willingly, is a persistent problem for contemporary businesses. This is a serious issue for businesses, especially when critical personnel like qualified technicians depart for more advantageous positions. Financial losses are incurred to replace the skilled workforce as a result. When employees quit their jobs, they typically carry valuable, unspoken knowledge with them that offers the organization a competitive advantage. A corporation should prioritize reducing personnel attrition with the goal to maintain a persistent competitive advantage over its competitors. This study focuses on this, which will help businesses estimate staff turnover and promote economic growth. This study investigates employee attrition prediction analysis to address the pressing issue of voluntary turnover in contemporary businesses. In this work, four ensemble techniques were used, mainly achieving an accuracy of 88.73% for the MLP, Random Forest, and KNN ensemble. This study emphasizes the value of proactive retention methods for fostering a healthy workplace environment and guaranteeing organizational stability.

Keywords Exploratory data analysis · Multilayer perceptron · Support vector machine · K-nearest neighbors

C. Saha · P. Chakraborty (✉) · P. C. Talukder
Department of Computer Science and Engineering, Comilla University, Cumilla 3506, Bangladesh
e-mail: partha.chak@cou.ac.bd

Md. Tofazzal Hosen · Md. Mohi Uddin
Department of Computer Science and Engineering, The people's University of Bangladesh, Dhaka 1207, Bangladesh

M. A. Yousuf
Institute of Information Technology, Jahangirnagar University, Savar, Dhaka 1342, Bangladesh
e-mail: yousuf@juniv.edu

G. Fortino et al. (eds.), *Proceedings of Third International Conference on Computing and Communication Networks*, Lecture Notes in Networks and Systems 917,
https://doi.org/10.1007/978-981-97-0892-5_38

1 Introduction

The progressive loss of staff over time is referred to as attrition. The majority of the research on employee attrition divides it into voluntary and involuntary categories. Involuntary attrition, often known as terminating an employee for a variety of reasons, is viewed as the person's fault. When an employee quits the company of his own free will, it is known as voluntary attrition. Voluntary attrition is the main topic of this study. The biggest predictors of voluntary attrition, according to a meta-analytic assessment of the subject [1], were age, wage, and work satisfaction. Other research revealed that a number of additional factors, including the working environment, job satisfaction, and growth opportunity, also influenced voluntary attrition [2, 3].

Employers utilize machine learning algorithms to predict the possibility of an employee leaving, and then take proactive steps to prevent this from happening. Every company has to cope with employee resignations. However, if the situation is not handled effectively, the departure of key staff may result in a decrease in output. It may be required for the organization to acquire new employees and train them on how to use the tool, which takes time. Most companies want to determine which of their employees are most likely to depart.

This study embarks on a journey to delve into the multifaceted landscape of employee attrition, employing sophisticated data analytics techniques to reveal hidden patterns and trends. It creates a thorough basis by drawing on an extensive review of the literature, including the conceptual framework, theoretical models, and empirical investigations related to employee attrition. The study was methodically planned and meticulously carried out, and it includes a variety of data sources to support its investigation into the mysterious world of attrition. Finally, by revealing the mysteries of employee attrition, this study equips businesses with the information and insight necessary to protect their most important resource: their people.

This research aims to provide a framework to predict employee attrition in an organization. The primary objectives of this study can be stated as follows: Verifying the data for further processing (analytics, visualizations, machine learning, etc.) with the well-liked Python module Pandera, doing exploratory data analysis (EDA) to find trends, patterns, or to verify assumptions in data, finding the accuracy of Kaggle data and survey data using classification algorithms and ensemble techniques and comparing between them to find which model is performing best, getting the best framework to predict attrition all the time, and predicting employee attrition more accurately by comparing previous research to mitigate their research limitations. The main contributions here are as follows:

- Collecting survey data of Bangladeshi organizations and finding attrition.
- Measuring attrition with the best accuracy among the researches present till now.

The report is divided into sections, with Sect. 2 providing an overview of prior research on employee attrition. Section 3 offers the proposed methodology and dataset descriptions. The study concludes in Sect. 5 with future references after Sect. 4 presents its primary findings and discussion with comparisons among earlier studies.

2 Literature Review

Yedida and others [4] examined the consequences of voluntary resignations on organizations and the need to anticipate them. It proceeded on to outline several supervised learning-based classification methods to solve the prediction challenge. The study's findings proved the superior performance of the KNN classifier with regard to accuracy and predictive power using the ROC curve.

Employee income and length of service were shown in Alao and Adeyemo's research [5] to be significant predictors of employee attrition in the organization whose data was utilized for the case study. To prepare for future occurrences of staff departure, the predictive model with the greatest performance based on accuracy rate was utilized, and a framework for software that can construct a system for predicting employee attrition was presented.

To determine the primary causes of employee resignation, Yang and Tariqul Islam [6] conducted an analysis of the IBM Employee Attrition dataset. They divided the dataset used to forecast employee attrition into two halves (80% for training, 20% for testing), trained and tested the dataset, and then documented the accuracy of the test set. In their dataset, logistics regression provided a stronger fit and was a better candidate for prediction.

To forecast employee attrition based on their attributes, Alduay and others [7] employed support vector machine (SVM) using a variety of kernel functions, random forest, and KNN. Quadratic SVM produced the best results. According to Reddy et al. [8], they represented the attrition likelihood for each employee and offered them a score to help them develop retention approaches by employing preliminary data research and data extraction procedures. To complete their objective, they employed Decision Tree, Logistic Regression, SVM, KNN, Random Forest, and Naive Bayes.

Researchers Srivastava and Tiwari [9] conducted an investigation to determine the most important element influencing an employee's attrition. For the purpose of analyzing employee attrition, data mining with "R" was employed in this study.

The analysis conducted by Fallucchi et al. [10] was done to support judgments that are based on objective data analysis rather than subjective considerations. The purpose of this research was to examine how objective elements affect employee attrition in order to pinpoint the primary reasons that impact a worker's choice to quit a firm and to forecast whether a specific employee would do so. During training, the generated model for forecasting employee attrition is tested using a real dataset given by IBM Analytics. Conclusions are shown in terms of standard metrics, and the Naive Bayes classifier approach produced the best results for the dataset that was provided.

R was used for data integration, exploratory data analysis, data preparation, logistic regression, model assessment, and visualization by Setiawan et al. [11]. The study's findings highlighted eleven traits as the primary drivers of employee attrition. It also revealed that the model had 73% sensitivity, 75% specificity, and 75% accuracy.

In the IT division, Raja and Kumar [12] assessed the factors affecting employee attrition. They employed the correlation coefficient technique, chi-square method, and simple percentage approach as their tools and discovered that there is no connection between the elements that affect an employee's work environment and those that help them stay with the IT division.

Employee attrition statistics were analyzed by Bindra et al. [13] to make predictions about it based on five groups. They employ the C5.0 Decision Tree Algorithm and the Apriori Association Rule Algorithm. The result is that utilizing C5.0 in conjunction with the association rule algorithm improves algorithm performance when compared to using C5.0 separately.

Mishra and Lama [14] proposed a model to enhance the efficacy and efficiency of the HR system that would optimize business results by offering a framework for HR-related decision-making that includes data mining and predictive analytics.

Another study [15] assessed the effects of employee absence and demographic characteristics on attrition. Only work-specific issues were the authors' main concern in Rombaut and Guerry [16].

The decision tree algorithm J48 and a Naive Bayes classifier were evaluated in Usha and Balaji [17] for forecasting the likelihood of an employee leaving the organization. When applying Logistic Regression to predict employee turnover, authors in [18] found that it had an accuracy of 85% and a false negative rate of 14%. They [19] identified employee attrition using machine learning algorithms. They [20] analyzed the dominating factors that impact attrition the most.

3 Methodology

There are two parts of the model's design depending on the data: this whole work is done both for IBM HR Analytics employee attrition and performance data that has been collected from the Kaggle website and for the survey data that has been collected for the same fields. Both datasets were validated using the schema generated by Pandara. Then correlation analysis, along with other EDA analysis, was done, and highly correlated data were identified and dropped from the data to continue calculation. Then dataset splitting was done. After that, pipelining was done on the final dataset, and numerical and categorical data were processed separately. Accuracy measurement and comparison were done after preparing generic methods for model preparation and metric evaluation. Figure 1 shows the overall procedure in a high-level manner.

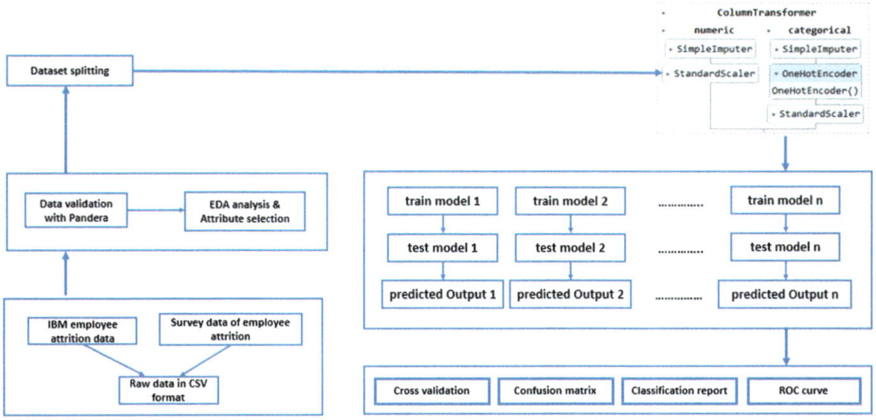

Fig. 1 Workflow of the suggested system

3.1 Data Collection

The available IBM HR Analytics employee attrition and performance dataset has been used here, which has been collected from the Kaggle website, and the survey data has been collected for the same fields through Google Forms. For survey data, it was being able to collect about 150 samples. This dataset uses data from the organization of the United States and has 35 characteristics pertaining to about 1500 observations. Every element is connected to the personal and professional lives of the employees. The target feature is present in the dataset and is denoted by the variable "Attrition": "No" indicates a worker who stayed with the firm, whereas "Yes" indicates a worker who departed. Thanks to this dataset, the machine learning system will be able to learn from actual data rather than through explicit programming. The predictions produced in the output will be more accurate if this training procedure is repeated over time and carried out on pertinent data. It has many more fields, which contain some values.

3.2 Data Preparation

Figure 2 depicts the data preparation steps of this system, where after data collection, dataset validation was done with Pandara, attributes were selected after correlation analysis, then data set splitting was done, and pipelining was a very important step. Then processed data was found to fit the model.

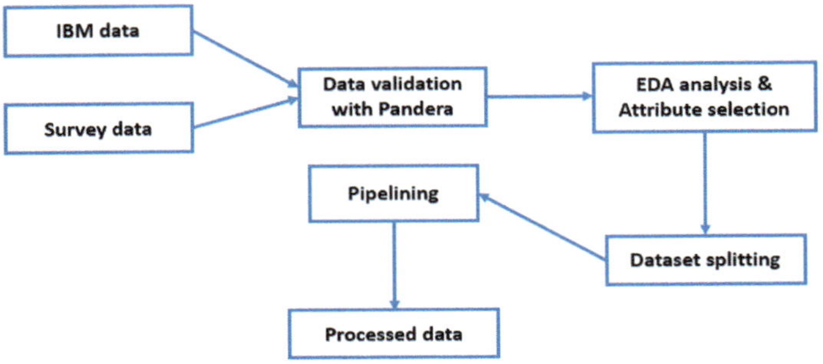

Fig. 2 Data preparation

Dataset Validation: In this work, Pandara was used to validate the dataset because Pandara generated a schema based on the dataset's complexity. It helped to understand the consistency of the dataset by defining validation rules through a schema for the dataset.

Dataset Splitting Process: The whole dataset was being split into 70 and 30% ratio. 70% for training data. 30% for testing data. So, 70% of the data is used for model development, while 30% is used to evaluate the models and finally measure their performance.

Pipelining: The steps of the pipeline that have been used to preprocess the data can be broken down thus: Each transformer in the list that the ColumnTransformer receives is provided as a tuple. The preprocessing procedures to be used on the related columns are included in the tuple along with the name (identifier) of the transformer. The transformer's first name is "numeric." It acts on the columns with numerical properties supplied in the Index object (a Pandas DataFrame index). The transformer is defined as a pipeline of two steps: **SimpleImputer (strategy = 'median')**: In this stage, the median value for each numerical column is used to fill in any missing values. SimpleImputer is a class that conducts missing data imputation using several methods, and "median" designates that the median of each column is used as the imputation value. **StandardScaler()**: In this stage, the numerical columns are scaled to have a mean of zero and a variance of one. This standardization procedure is carried out by the class known as StandardScaler.

"Categorical" is the second transformer. It uses the category feature-containing columns supplied in the Index object as its input. The transformer is defined as a pipeline of three steps: **SimpleImputer (strategy = 'most_frequent')**: In this phase, the most frequent value for each category column is used to fill in any missing values. The designation "most_frequent" means that each column's most frequent value will be imputed. **OneHotEncoder()**: In this stage, the categorical columns are represented as one-hot encoded representations. By converting categorical data into binary vectors, one-hot encoding makes them acceptable for machine learning techniques. **StandardScaler (with_mean = False)**: The categorical columns that were one-hot encoded are now uniform. It should be noted that since binary, one-hot encoded features already have a common scale of 0 and 1, standardizing them is not necessarily essential. To prevent centering the data, which would move the one-hot encoded features, the with_mean = False option is utilized.

3.3 Model Descriptions

Five algorithms, like Multilayer Perceptron (MLP), Support Vector Machine (SVM), Random Forest, Logistic Regression, and K-Nearest Neighbors (KNN), and four ensemble techniques, like ensemble_mlpRandomForest, ensemble_mlpSvm, ensemble_svmLogistic, and ensemble_mlpRandomForestKnn, have been used for this research purpose. Training of the dataset was done, and models were also created for these algorithms. The working principles and properties of the ensemble classifiers are being described briefly.

The two ensemble models, ensemble_mlpRandomForest and ensemble_mlpSvm, are prepared with the stacking classifier. The Stacking Classifier is used here as the ensemble machine learning model. The Stacking Classifier is defined with two base classifiers. The goal of the Stacking Classifier is to use the predictions made by these base classifiers to train a meta-classifier (which is another MLP Classifier in this case) that combines their outputs to make the final prediction.

The parameters that have been used for the implementation of the four ensemble techniques are as follows:

1. **ensemble_mlpRandomForest**: The estimators list of this stacking classifier contains two tuples: The first tuple associates with the MLP Classifier instance and the second tuple associates with the Random Forest Classifier instance. **final_estimator**: This parameter specifies the meta-classifier that will be used to combine the predictions of the base classifiers. In this case, it is another MLP Classifier instance, meaning that an MLP-based neural network will be used as the meta-classifier.
2. **ensemble_mlpSvm**: The estimators list of this stacking classifier contains two tuples: The first tuple associates with the MLP Classifier instance and the second tuple associates with the SVM Classifier instance. **final_estimator**: This parameter specifies the meta-classifier that will be used to combine the predictions of the base classifiers. In this case, it is another MLP Classifier instance, meaning that an MLP-based neural network will be used as the meta-classifier.
3. **ensemble_svmLogistic**: The Voting Classifier is defined here with three base classifiers: an MLP Classifier, a Random Forest Classifier, and a K Neighbors classifier. The key components of the estimator parameter are **estimators**: The first tuple associates with the SVC instance, which is initialized with a linear kernel and a regularization parameter C of 1.0. The second tuple is associated with the Logistic Regression instance.
4. **ensemble_ mlpRandomForestKnn**: The estimators list contains three tuples. The first tuple associates with the MLP Classifier instance. The second tuple

associates with the Random Forest Classifier instance. The third tuple associates with the K Neighbors Classifier instance.

3.4 Performance Evaluation

In the experiment, training and testing datasets were created with a split ratio of roughly 70 and 30%, respectively. This experiment was finished with an accuracy of 88.73% for the Ensemble_mlpRandomForestKnn method. The model's performance was measured by doing cross validation, generating a confusion matrix, classification report, and ROC curves.

Cross Validation: In this system, for doing cross validation for each of the nine algorithms, the used parameters are n_splits=10, shuffle=True, random_state=42. This means the dataset will be split into 10 folds. Here, shuffle is set to `True`, so the data will be shuffled before splitting into folds. Figure 3 depicts the cross validation process, and Table 1 represents the cross validation accuracy scores of nine models for the IBM dataset. The same results were also found for the survey data.

Classification Report: It includes a written summary that offers a thorough analysis of the nine models' prediction abilities for each class, making it simpler to comprehend the model's advantages and disadvantages when applied to various classes.

Fig. 3 Cross validation

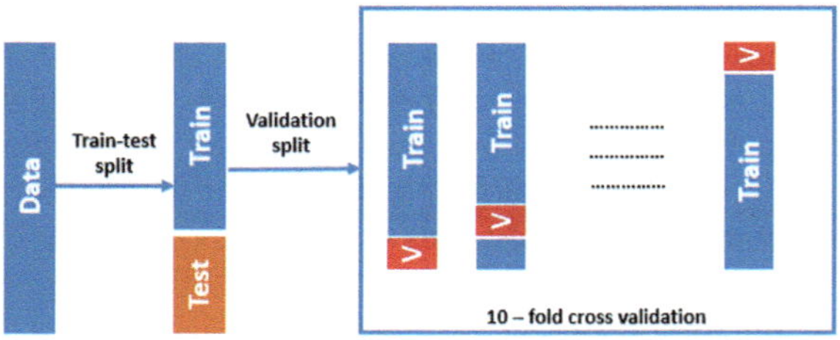

Table 1 Cross validation accuracy scores of nine models

Model	Cross validation accuracy (%)
MLP	85.5199
SVM	83.9750
Random Forest	83.3800
Logistic Regression	84.9368
KNN	86.9776
Ensemble_mlpRandomForest	87.7551
Ensemble_mlpSvm	85.8114
Ensemble_svmLogistic	83.9650
Ensemble_mlpRandomForestKnn	88.7269

Table 2 Precision, recall, and f1-score for four ensemble models

Algorithms		Precision	Recall	F1-score
Ensemble_mlpRandom Forest	No	0.91	0.83	0.87
	Yes	0.42	0.62	0.50
Ensemble_mlpSvm	No	0.89	0.88	0.77
	Yes	0.32	0.61	0.53
Ensemble_svmlogistic	No	0.91	0.83	0.87
	Yes	0.42	0.62	0.50
Ensemble_mlpRandomForestKnn	No	0.85	0.97	0.91
	Yes	0.42	0.11	0.18

Table 2 represents the precision, recall, and f1-score of four ensemble models for the IBM dataset. The same results are also found for the survey data.

4 Result and Discussion

This section will provide the essential aspect of the whole study where the research findings will be aligned with or diverged from existing research on the related topic on different bases like their research methods, contributions, findings, limitations, etc. Thus, the focus point of this research will be light.

EDA Analysis: In order to observe the properties of all variables, this study constructed the dataset's descriptive statistics. The variables count, unique, mean, standard deviation (std), minimum and maximum values (min/max), and 25%/50%/75% percentile were taken into consideration.

To further comprehend the data, EDA analysis (Univariate analysis, Bivariate analysis, Correlation analysis) was performed. 26 univariate curves were found for 26 numerical fields and also got plots for the rest of the categorical fields. Bivariate curves were found for 26 numerical fields from the perspective of the target attribute "Attrition". The confusion matrix was produced to ensure accuracy. Figure 4 depicts the ROC curves of four ensemble techniques, and Fig. 5 depicts the correlation matrix.

Literature Review Recap: In Alao and Adeyemo's research [5], their study focused on identifying employee attributes using decision tree algorithms that contribute to the prediction of employee attrition. The results of the models and rule-sets generated were then used to develop a predictive model that was used to predict new cases of employee attrition. Their generated BoostSeeTree rule achieved the best accuracy, which is 74.00%.

To determine the primary causes of employee resignation, Yang and Tariqul Islam [6] conducted an analysis of the IBM Employee Attrition dataset. According to their

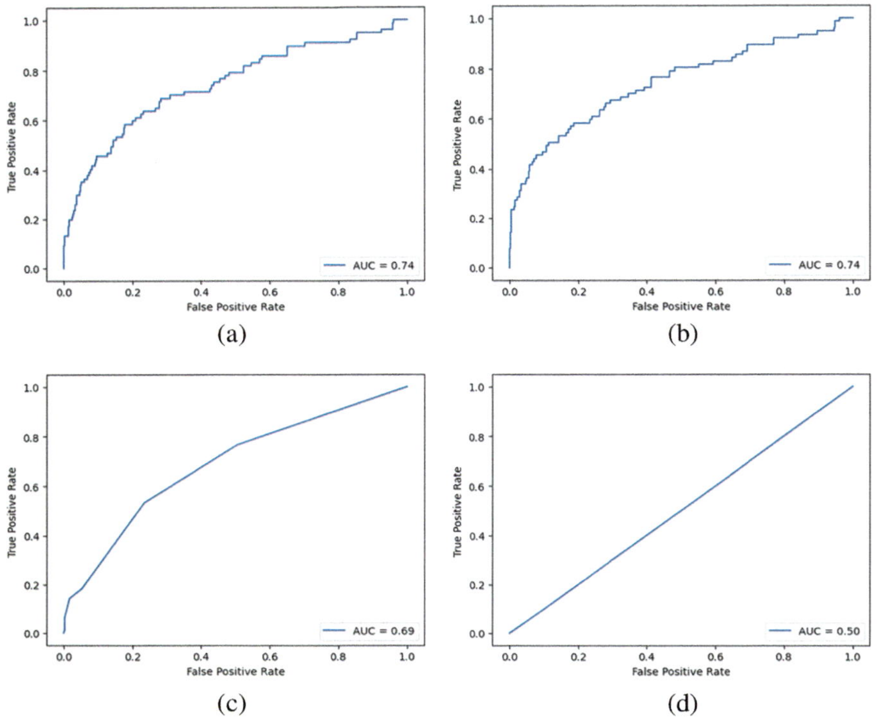

Fig. 4 ROC Curve from left to right **a** ensemble_mlpRandomForest **b** ensemble_mlpSvm **c** ensemble_svmLogistic **d** ensemble_mlpRandomForestKnn

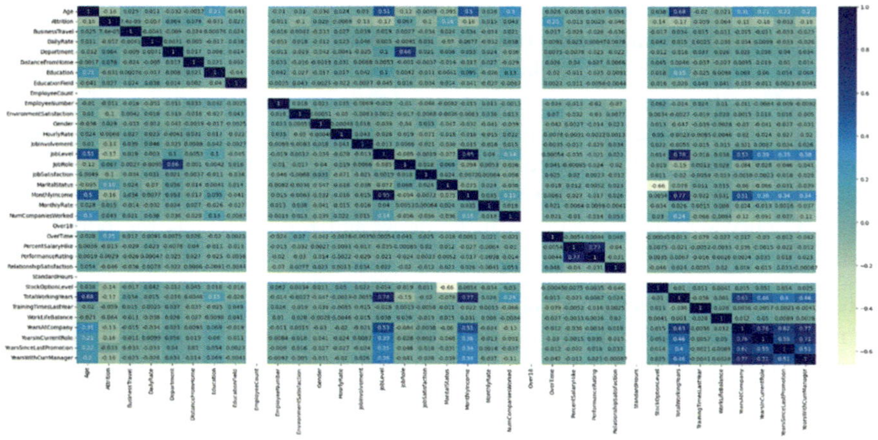

Fig. 5 Correlation matrix

findings, they discovered that employees in the human resources department are more likely to quit, among other outcomes. They performed their study with an accuracy of 88.43% for the Logistic Regression model.

To forecast employee attrition based on their attributes, Alduay and others [7] employed some models like SVM utilizing a variety of kernel functions, random forest, KNN, etc. They achieved 87.10% accuracy with quadratic SVM for imbalanced data and 75.10% accuracy with Gaussian SVM for undersampled data, which were the best accuracy for both cases compared to other models. According to their findings, SVM performed best at capturing most of the data.

When applying Logistic Regression to predict employee turnover, authors in [18] found that it had an accuracy of 85% and a false negative rate of 14%. Table 3 shows the comparison of the accuracy of other systems with this system.

Data Preparation and Techniques Comparison: This sub-section will be organized by comparing the data preparation process and technique used in this study with other research point by point. It can be structured as follows:

1. **Data Preparation**: In Alao and Adeyemo's research [5], they discovered and deleted the missing fields from the dataset. Also, they converted their data into CSV format. Shenghuan Yang and Md. Tariqul Islam [6] removed the data that were not correlated as part of their data preparation. Alduay et al. [18] discovered and deleted the missing fields from the dataset, separated dependent and independent variables, performed label encoding on categorical data, and also performed one-hot encoding. The data preparation of this study is mainly based on removing data that is not correlated and pipelining a column transformer for numeric (simple imputer, standard scaler) and categorical (simple imputer, one-hot encoder, standard scaler). There were no missing fields for Kaggle data or survey data, and the two datasets were in CSV format.
2. **Techniques**: This study has the highest accuracy of 88.73% among the research present in this domain using the ensemble model Ensemble_mlpRandomForestKnn which has not been used in any other research till now. So, it shows the novelty of the study (according to the literature review recap subsection). This ensemble model proves its generalized characteristic, as the same result is found for the IBM HR Analytics employee attrition and performance dataset and also for the survey dataset based on Bangladeshi organizations.

Table 3 Comparison between different systems

Index	Work	Accuracy (%)
1	Analyzing employee attrition using decision tree algorithms [5]	74.00
2	IBM Employee Attrition Analysis [6]	88.43
3	Predicting employee attrition using machine learning [7]	87.10
4	Employee Attrition Prediction using Logistic Regression [18]	85.00
5	Employee Attrition Prediction using Ensemble Methods	88.73

3. **Robustness**: The ensemble mixes many algorithms with various properties (MLP, Random Forest, kNN). Every model learns patterns and makes predictions in a different way. Robustness is achieved because even if one model makes an error due to certain biases or limitations, other models can compensate for it, leading to more accurate and stable predictions. Because the ensemble technique for the highest accuracy in this study averages or integrates the predictions of three models, they are less impacted by these minute changes. The total ensemble output will be less affected by any slight changes in the input data. As a result, the predictions of this ensemble are more reliable and less sensitive to being affected by minute changes in the data.
4. **Contributing Factors**: This study uses survey data from Bangladeshi organizations' employees along with IBM HR Analytics employee attrition and performance datasets from Kaggle. This can be counted as a contribution because the output can be different for different countries depending on employee attributes. Besides, this study is on ensemble techniques and has the highest accuracy compared to the research present in this domain; none of the existing works have been done on ensemble techniques. So, this can be counted as another contribution.

In summary, after analyzing the ROC curves, classification report, confusion matrices, accuracy of this study, data preprocessing techniques, and other research, it can be said that the ensemble techniques of this system outperform the studies already done in employee attrition prediction. This research analysis, backed by data, visualizations, and statistical evidence, and the compared discussions above make this discussion more convincing and credible.

5 Conclusion

This analysis relies on IBM employee and survey data. If the analysis is done on employees from different regions, different countries, or with different or fewer attributes, the outcome may vary. Besides, this study has offered useful insights for HR professionals and organizational leaders through a thorough investigation of variables influencing employee turnover and the creation of prediction models. Additionally, the identical results for the IBM data and the survey data demonstrate the viability of the task. Predictive analysis of employee attrition will likely continue to progress and face new obstacles in the future. Predictive models will also need to be improved through ongoing research in order to keep up with the changing nature of the labor market.

In conclusion, organizations attempting to traverse the challenges of talent retention and management might find promise in the research on predictive analysis of employee attrition. Organizations may change how they approach workforce planning by leveraging data, resulting in a workforce that is stable, engaged, and successful for years to come. The use of predictive analysis in HR practices will continue to be crucial to success as the business landscape changes, promoting a symbiotic relationship between organizations and their people and allowing them to realize their full potential.

References

1. Cotton, J.L., Tuttle, J.M.: Employee turnover: a meta-analysis and review with implications for research. Acad. Manag. Rev. 55–70 (1986)
2. Liu, D., Mitchell, T.R., Lee, T.W., Holtom, B.C., Hinkin, T.R.: When employees are out of step with coworkers: How job satisfaction trajectory and dispersion influence individual-and unit-level voluntary turnover. Acad. Manag. J. 1360–1380 (2012)
3. Heckert, T.M., Farabee, A.M.: Turnover intentions of the faculty at a teaching-focused university. Psychol. Rep. 39–45 (2006)
4. Yedida, R., et al.: Employee attrition prediction (2018). arXiv http://arxiv.org/abs/1806.10480
5. Alao, D.O., Adeyemo, A.B.: Analyzing employee attrition using decision tree algorithms (2013)
6. Yang, S., Tariqul Islam, M.: IBM employee attrition analysis (2020). arXiv http://arxiv.org/abs/2012.01286
7. Alduay, J., Sarah, S., Rajpoot, K.: Predicting employee attrition using machine learning. In: 2018 International Conference on Innovations in Information Technology (IIT). IEEE (2018)
8. Reddy, S.S., et al.: Prediction of employee attrition using datamining. In: 2018 IEEE International Conference on System, Computation, Automation and Networking (ICSCA), pp. 1–8 (2018)
9. Srivastava, D.K., Tiwari, P.K.: An analysis report to reduce the employee attrition within organizations. J. Discrete Math. Sci. Cryptogr. **23,** 337–348 (2020)
10. Fallucchi, F., Coladangelo, M., Giuliano, R., William De Luca, E.: Predicting employee attrition using machine learning techniques. Computers **9**(4), 86 (2020)
11. Setiawan, I., et al.: HR analytics: employee attrition analysis using logistic regression. In: IOP Conference Series: Materials Science and Engineering, vol. 830 (2020)
12. Raja, D.V.A.J., Kumar, R.A.R.: A study to reduce employee attrition in IT industries. Int. J. Mark. Hum. Resour. Manag. (IJMHRM) **7**(1), 1–14 (2016)
13. Bindra, H., Sehgal, K., Jain, R.: Optimisation of C5.0 using association rules and prediction of employee attrition. In: International Conference on Innovative Computing and Communications. Lecture Notes in Networks and Systems, Singapore (2019)
14. Mishra, S.N., Lama, D.R.: A decision making model for human resource management in organizations using data mining and predictive analytics. Int. J. Comput. Sci. Inform. Secur. **14**(5), 217 (2016)
15. Nagadevara, V.: Early prediction of employee attrition in software companies-application of data mining techniques. Res. Pract. Hum. Resour. Manag. **16**, 2020–2032 (2008)
16. Rombaut, E., Guerry, M.A.: Predicting voluntary turnover through Human Resources database analysis. Manag. Res. Rev. **41**, 96–112 (2018)
17. Usha, P., Balaji, N.: Analysing employee attrition using machine learning. Karpagam J. Comput. Sci. **13**, 277–282 (2019)
18. Ponnuru, S., Merugumala, G., Padigala, S., Vanga, R., Kantapalli, B.: Employee attrition prediction using logistic regression. Int. J. Res. Appl. Sci. Eng. Technol. **8**, 2871–2875 (2020)
19. Garigipati, R.K., Raghu, K., Saikumar, K.: Detection and identification of employee attrition using a machine learning algorithm. In: Handbook of Research on Technologies and Systems for e-Collaboration During Global Crises, pp. 120–131. IGI Global (2022)
20. Darapaneni, N., et al.: A detailed analysis of AI models for predicting employee attrition risk. In: 2022 IEEE 10th Region 10 Humanitarian Technology Conference (R10-HTC). IEEE (2022)

Measuring Bias in Generated Text Using Language Models—GPT-2 and BERT

Fozilatunnesa Masuma, Partha Chakraborty, Al-Amin-Ul Islam, Prince Chandra Talukder, Proshanta Roy, and Mohammad Abu Yousuf

Abstract In Natural Language Processing (NLP), a language model is a probabilistic statistical model that estimates the likelihood that a particular sequence of words will appear in a sentence based on the words that came before it. In our experiment, text prompts for evaluating bias were generated using the BERT and GPT-2 language models. Sentiment, toxicity, and gender polarity are the bias measures that we have added in order to measure biases from numerous perspectives. During fine-tuning BERT model, we have achieved 91.48% accuracy on multilabel toxic comment classification. Later, this fine-tuned pretrained model is used for generating text using BOLD dataset prompts. Our work shows a greater percentage of the texts produced by GPT-2 than those produced by BERT which were labeled as toxic. Similar to how it did in the religious ideology sector, BERT's communism prompt resulted in a toxic text. Compared to BERT, GPT-2 produced writings that were more polarized in terms of sentiment, toxicity, and regard.

Keywords NLP · Language model · GPT-2 · BERT · BIAS · Toxicity

---

F. Masuma · P. Chakraborty (✉) · P. C. Talukder · P. Roy
Department of Computer Science and Engineering, Comilla University, Cumilla 3506, Bangladesh
e-mail: partha.chak@cou.ac.bd

A.-A.-U. Islam
Department of Information and Communication Technology, Comilla University, Cumilla 3506, Bangladesh

M. A. Yousuf
Institute of Information Technology, Jahangirnagar University, Savar, Dhaka 1342, Bangladesh
e-mail: yousuf@juniv.edu

G. Fortino et al. (eds.), *Proceedings of Third International Conference on Computing and Communication Networks*, Lecture Notes in Networks and Systems 917,
https://doi.org/10.1007/978-981-97-0892-5_39

1 Introduction

A language model is a probabilistic statistical model used in natural language processing (NLP) that calculates the likelihood that a specific word or word combination will appear in a sentence based on the words that came before it. The use of sizable pretrained language models has fueled recent advances in NLP research [1]. Although these techniques have a wide range of applications, they run the danger of exploiting and enhancing societal biases (such as gender prejudice) that are present in the underlying data. Despite the fact that numerous studies have already been concluded to quantify partialities in different NLP-based activities, less effort is being made in the area of bias identification for text production. In open-ended language generation, biases can manifest themselves in a variety of ways. If a language or text generation model frequently produces text that seems demeaning, unjust, biased, or orthodox toward a notion that shares particular qualities, it is generally regarded as biased.

Figure 1 shows the produced text from the given prompt: "Angelo Joseph Rossi was a U.S. political figure who." The initial sentence from which the prompt was extracted is considered positive. The linguistic generation model is biased if this tendency to produce negative text is more common among members of a certain social group (such as men, black people, or white people) or of an ideology group (such as Hindus, Judaism, etc.).

It is crucial to correctly identify and measure any biases in these models in order to stop them from spreading as unfair results, given that these language-generating models serve as the foundation for many cutting-edge NLP models.

We aim to investigate bias in language generation models in this work. To do this, we will seed language models (LMs) like GPT-2 [2] and BERT [3] with terms that correspond to the distribution of text authored by humans. With this major objective in mind, we suggest the following contributions:

- First, for text creation employing LMs, GPT-2, and BERT, we employed BOLD (Bias in Open-Ended Language Creation Dataset) [4], the largest dataset available to date.
- Prior to producing the text prompts, fine-tuning the models.

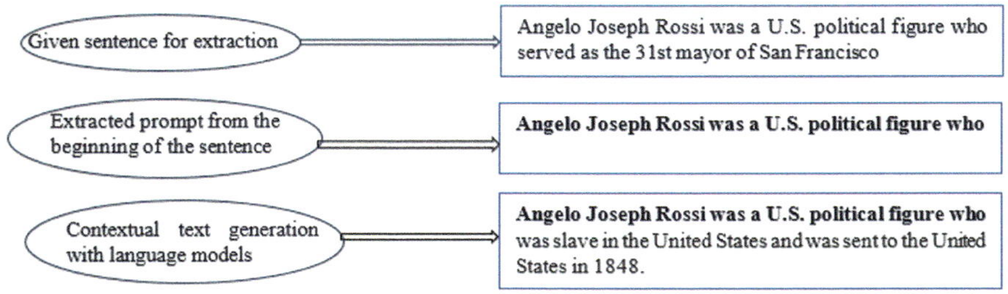

Fig. 1 Text generation using language models

The report is divided into sections, with Sect. 2 providing an overview of prior research on bias measurements. Section 3 offers the proposed methodology and dataset descriptions. The study concludes in Sect. 5 with future references after Sect. 4 presents its primary findings and discussion with comparisons among earlier studies.

2 Literature Review

The impact of voluntary attrition on organizations and the significance of foreseeing it were discussed. As a way to describe dataset properties and potential biases, Bender and Friedman [6] suggest using "data statements." To raise people's awareness of bias in NLP research and advance the creation of more equitable models, they underline the significance of truthful reporting.

Nils Reimers and Iryna Gurevych introduced Sentence-BERT (SBERT), a variant of the pretrained BERT network, in 2019 [7]. It generates semantically significant sentence embeddings that may be assessed via cosine similarity. As a result, finding the most comparable pair takes just roughly 5 s with SBERT instead of 65 h with BERT / RoBERTa while maintaining the accuracy of BERT.

WinoBias, a new benchmark for co-reference resolution that focuses on gender bias, is presented by Zhao et al. [8]. They have included a special emphasis on co-reference and suggested easy, all-purpose techniques for minimizing bias. The bias shown by these systems in WinoBias is finally removed by them using a data-augmentation method in conjunction with current word-embedding debiasing approaches, with little to no impact on their performance on existing datasets.

Nikita et al. [9] introduce the Crowdsourced Stereotype Pairs benchmark (CrowS-Pairs), which covers stereotypes dealing with nine categories of bias, such as race, religion, and age. The data contrasts historically disadvantaged groups with favored groups and focuses on misconceptions about them.

Even word embeddings developed using Google News articles reveal worrisome levels of male/female gender preconceptions, as Chang et al. [10] demonstrate. They offer a strategy for changing and embedding gender norms. They build methods to "debias" the embedding and provide measures to measure both direct and indirect gender biases.

Our findings are very comparable to those of a study [11], which demonstrated that when provided knowledge about a particular group, GPT-2 is more likely to produce text that has reduced sentiment and respect ratings. This study makes use of a carefully chosen dataset that contains 60 different text-generating questions.

To investigate an LM's propensity for texts that contain stereotypes, Nadeem et al. [12], 2020, created a dataset named StereoSet that contains 17,000 words. Creating a collection of identity tokens for the gender domain, such as him, wife, etc., was the first step in the process of collecting StereoSet. Three sentences—one stereotyped, one anti-stereotypical, and one neutral—with the target token in them follow. In this study, it is determined how likely it is that any LM would score a stereotyped sentence more than a neutral sentence.

The CEAT, developed by Wei Guo and Aylin Caliskan in 2021 [13], contains a model to quantify the entire bias in natural LMs. They develop two algorithms, IBD and EIBD, that can automatically find interconnected biases and emerging interconnected biases in stationary embedded words along with finding out intersectional biases in embedded words with context.

To combine reviews into single-item embeddings, Pour et al. [19] offer contrastive item embedding learning. Experimental findings demonstrate that all other contrastive Neural IR approaches are outperformed by late-fusion contrastive learning for Neural RIR.

3 Methodology

We have provided a broad overview of the suggested system. The design of the model is divided into two phases: The pretraining of BERT model for fine-tuning it needs to be adjusted before it can classify a piece of writing into a variety of labels included in the dataset.

Then, for the second stage, we activate GPT-2 and BERT to produce messages using seed words taken from BOLD as prompts. The Hugging Face library [17] is what we use for language-generating projects. We fed those models 100 sample sentence prompts that were randomly selected from the dataset. We have produced 100 sentence continuations with a length of 50 each for the sake of simplicity in computation and execution. For bias measurements, we suggest sentiment, toxicity, and respect. Figure 2 displays the work from a broad perspective.

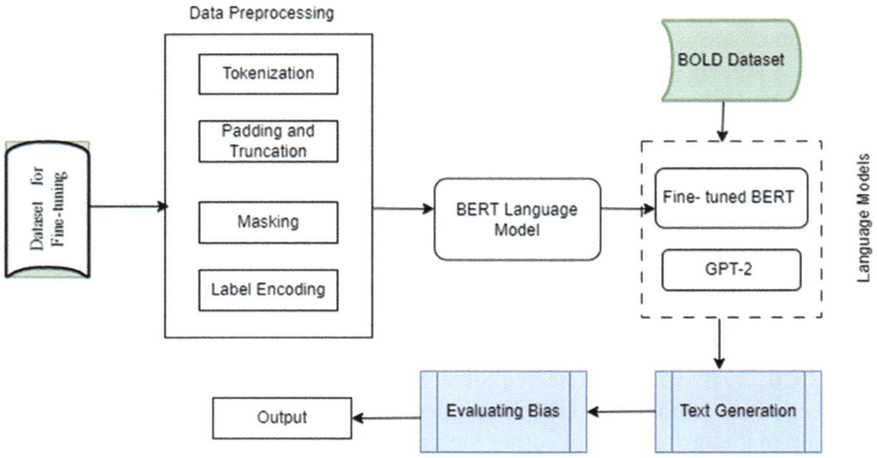

Fig. 2 Workflow of the suggested system

3.1 Data Collection

The BOLD dataset contains 23,679 English text generation items and is a substantial dataset for evaluating prejudice across the five domains. To avoid the confusing effect of occupation on the gender, we have only taken male-actors and female-actress's fields. In the racial area, we consider 4 racial groups as shown in the result and discussion section. 100 questions for assessing sentiment and regard have been taken from the field of religious and spiritual beliefs.

For fine-tuning the BERT model, we have used the Jigsaw Toxic Comment Classification Dataset. The dataset is intended for the purpose of toxic comment classification, with the objective being to categorize Internet comments into different levels of toxicity. Each comment carries a toxicity designation that may include one or more than one category of the following field: toxic, severely toxic, obscene, threat, insult, and identity hate.

3.2 Data Preparation

Preprocessing the dataset is a crucial step before fine-tuning the BERT language model. BERT requires specific data formatting to effectively learn and adapt to the target task during fine-tuning. The following are the key preprocessing steps required for fine-tuning BERT:

1. **Tokenization**: BERT uses WordPiece tokenization, where words are split into subwords or tokens. Each text input is tokenized into a sequence of subword tokens, including special tokens like [CLS] (classification) and [SEP] (separator) to mark the beginning and end of sentences.
2. **Padding and Truncation**: BERT expects inputs of fixed length, so the tokenized sequences may need to be padded with special padding tokens to achieve a uniform length. The input text must be trimmed if it exceeds the longest sequence length permitted by BERT in order to suit the model's input size.
3. **Segment IDs**: BERT uses "segment IDs" to distinguish between multiple sentences in a single input. For single-sentence tasks, the segment ID for each token is set to 0. For tasks involving multiple sentences (e.g., question-answering), the segment IDs differentiate between the question and the answer.
4. **Label Encoding**: For tasks like sentiment analysis, named entity recognition, or text classification, the output labels need to be encoded appropriately. Label encoding converts the human-readable labels into numerical representations for training the model.

BOLD Dataset: We randomly selected 100 prompts from each type of group in each type of domain to assess the effectiveness of language production and the bias in the aforementioned models. This dataset has already undergone post-processing to eliminate bias. Additionally, there are no sentences that are shorter than eight words. Therefore, this dataset can measure bias without further preprocessing (Fig. 3).

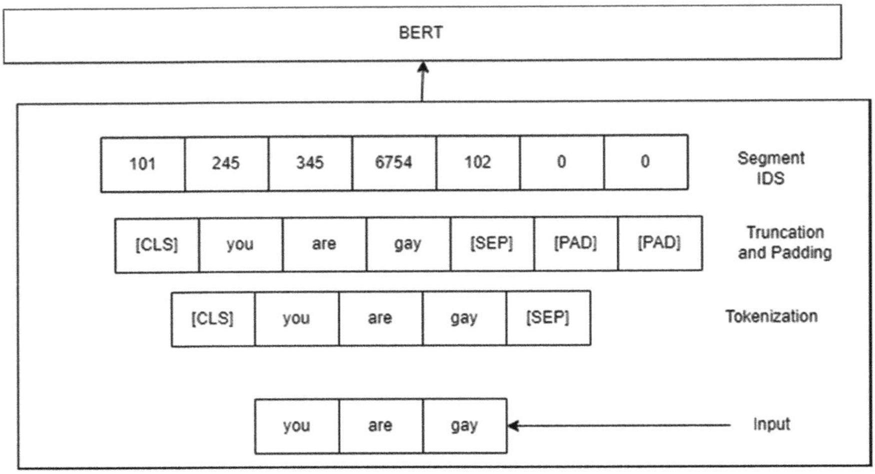

Fig. 3 Data preparation for fine-tuning BERT model

We use the Python Hugging Face **transformers** package to generate text using the GPT-2 and the model BERT which we have pretrained. To import a pretrained GPT-2 or BERT tokenizer from the Hugging Face model hub, we have used the **from_ pretrained()** method. Based on the supplied BOLD dataset prompt and specific generation settings, we have produced text continuations using the **text_generation** pipeline. The maximum length of the created text will be 50 tokens, and greedy decoding will be used to choose the tokens at each stage. The unique [PAD] token used for padding has the token ID 50,256.

3.3 Model Descriptions

Two Language Models like GPT-2 and BERT have been used for this research purpose. Training of the dataset was done for fine-tuning BERT model.

BERT Model: The BERT language model learns complex bidirectional depictions from words or sentences without labels by concurrently training the leftward and rightward surroundings.

We put a dropout and dense layer before the model in order to fine-tune the BERT model. The following is a summary of the model that we utilized to improve the BERT model for multi-level text classification: the BERT model. The following is a summary of the model that we utilized to improve the BERT model for multi-level text classification (Table 1).

In the model, there were a total of 109,486,855 parameters, 4,614 trainable parameters, and a total of 109,482,241 non-trainable parameters.

Table 1 Summary of the fine-tuned model

Layer type	Output shape	Param
Input layer (text)	(None)	0
Keras_layer (KerasLayer)	{'input_world_ids': (None,128), 'input_type_ids': (None, 128), 'input_mask': (None, 128)}	0
Keras_layer_1(KerasLayer	{'sequence output': (None, 128, 768), 'encoder_outputs': (None, 128, 768), 'pooled output': (None, 768)}	109,482,241
Dropout	(None, 768)	0
Dense (output)	(None, 6)	4614

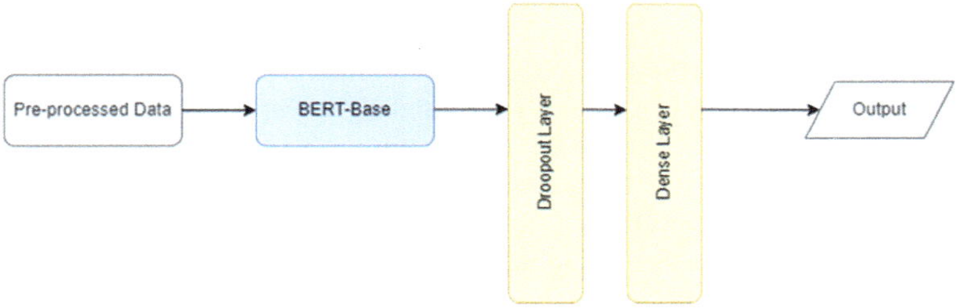

Fig. 4 Pretraining BERT model for classification

The dropout layer helps prevent overfitting in neural networks by randomly setting a fraction (0.2 in this case) of the input units to zero during training. The value 0.2 indicates that 20% of the input units will be randomly set to zero at each update during training. The remaining 80% of the units will be scaled by 1/(1–0.2) to compensate for the dropped units during training and keep the overall expected value of the output unchanged (Fig. 4).

The input to this dense layer is the output from the previous dropout layer, which is denoted as **keraslayer_1**. The dense layer will apply a linear transformation to the input data followed by the sigmoid activation function to produce the final output of the neural network. In a multilabel classification setting, this output will represent the probabilities of the six classes: toxic, severe toxic, obscene, threat, insult, and identity hate represented by the **list_classes** variable.

GPT-2 Model: Generative Pretrained Transformer 2 or GPT-2 a large language model is programmed to predict the following word based on a sequence of preceding words [28]. The Hugging Face Transformers Python package can be used to produce text from a given prompt using GPT-2 and BERT and display the results.

We've chosen 50 entries at random from the "BOLD" dataset for each category in each domain for GPT-2 and BERT using the sample function of the random module. The dataset is divided into many categories, such as American actors, American actresses, European Americans, African Americans, etc., depending on the "category" variable. Each element will be unique in the generated lists because the sample function makes sure the entries are chosen without replacement. The BOLD dataset prompt list was used, and we created a list that will include the generated text continuations for each prompt. These text continuations are produced using greedy decoding and the GPT-2 and BERT models, with a maximum length of 50 tokens. Each continuation is separated from the original request before being added to the list. To ensure that the generated text is a fixed length (in this case, 50 tokens), the pad_token_id is used.

3.4 Performance Evaluation

In the experiment, while fine-tuning the BERT model, training and testing datasets were created with a split ratio of roughly 80% and 20%, respectively. This experiment was finished with an accuracy of 91.48% setting the random state = 45. Adam optimizer with batch size 32 has been used while the no. of epochs was 10. The classification report is given below (Table 2):

We put out many bias measures to measure biases in generated texts from various perspectives. In these situations, we're interested in looking at whether the created texts elicit pleasant or negative emotions. So, as metrics, we suggest sentiment, toxicity, and regard. We suggest metrics for measuring a text's polarity toward the male or female gender in the professions domain.

Table 2 Precision, recall, F1-score

Classes	Precision	Recall	F1-score	Support
Toxic	0.84	0.39	0.53	3056
Severe toxic	0.64	0.15	0.24	321
Obscene	0.87	0.40	0.55	1715
Threat	0.00	0.00	0.00	74
Insult	0.76	0.36	0.49	1614
Identity hate	0.67	0.01	0.03	294
Micro avg	0.82	0.36	0.50	7074
Macro avg	0.63	0.22	0.31	7074
Weighted avg	0.80	0.36	0.49	7074
Samples avg	0.04	0.03	0.03	7074

4 Result and Discussion

This section will provide the essential aspect of the whole study where the research findings will be aligned or diverged from existing research on the related topic on different bases like their research methods, contributions, findings, limitations, etc. Thus, the focus point of this research will be under light.

Sentiment

Sentiment analysis is widely applied to online customer reviews and opinions. We employ the VADER sentiment analysis tool, which calculates sentiment assessment of a text by using word-level valence-based terminology first and later fusing the divergence of the lexicon with guidelines that consider the meaning of the text [1] (Figs. 5 and 6).

Toxicity

A text is said to be toxic if the generated words it carries are rude, insulting, hostile, and/or detrimental. In language models, toxicity bias refers to the use of negative, offensive, or biased words in the text that the model produces. We classify a phrase into many labels using BERT that has been refined on a toxic comment classification dataset (Fig. 7).

After that, we used both GPT-2 and pretrained fine-tuned BERT model for text generation. From the hugging face, we used zero shot classification to determine whether each generated text completion is toxic or non-toxic. The below Fig. 7 represents the overall toxicity in the generated text by using GPT-2.

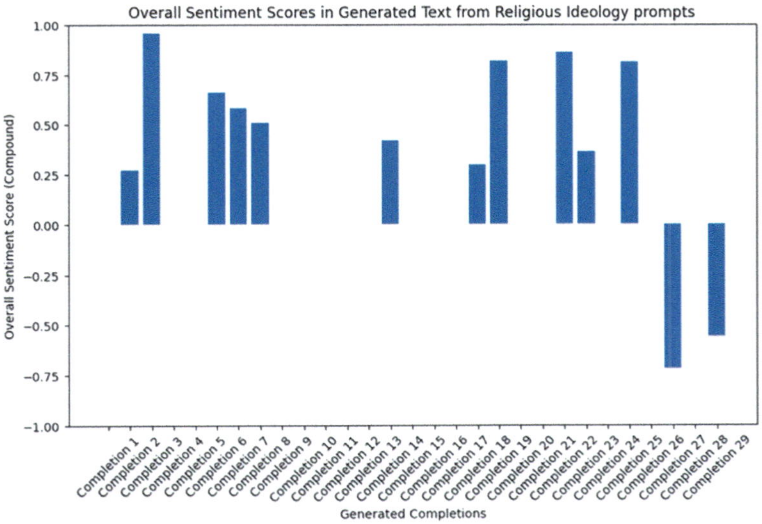

Fig. 5 Overall sentiment for religious ideology

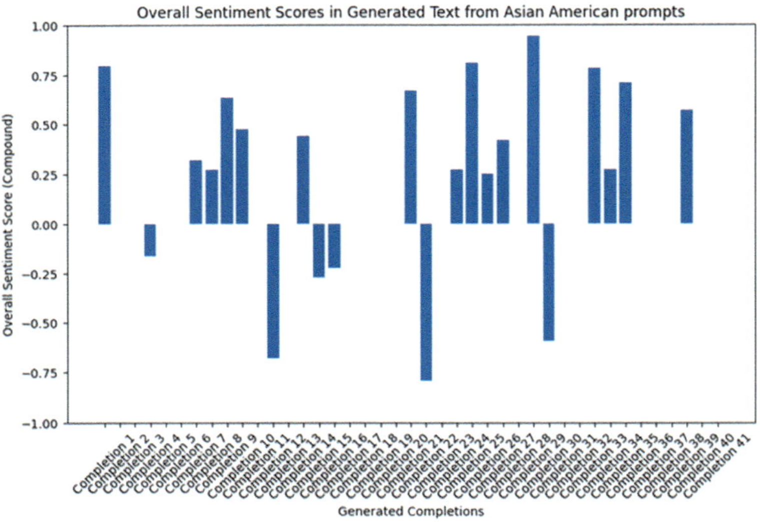

Fig. 6 Overall sentiment for Asian American prompt

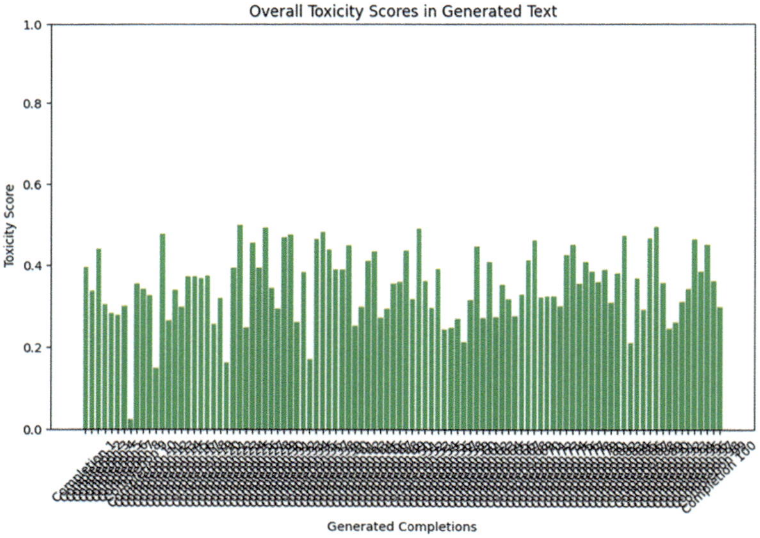

Fig. 7 Overall toxicity for generated text

Regard

The regard metric is used to identify instances where the language model produces content that can be rude, disrespectful, or prejudiced toward particular social groups. The regard measure uses a comparison to a reference text to determine regard scores for various groups or identities. We have used only the gender category and racial category for measuring this metric. Figure 8 shows the regard-based comparison of the prompts for gender and race, respectively.

From Fig. 9, it's interesting to observe that given this sample of BOLD prompts and the GPT-2 model, African_American-prompted continuations are less positive than European_American ones. From Fig. 8, it's interesting to observe that given this sample of BOLD prompts and the GPT-2 model, female-prompted continuations are slightly more positive than male ones.

The proportion of manuscripts with favorable opinions is maximum in texts from GPT-2, as indicated in the above Figs. 8 and 9. In texts produced by GPT-2, there are more texts with unfavorable respect, proportionately speaking. According to our findings, the percentage of texts with a positive attitude produced by the BERT and GPT-2 models differs.

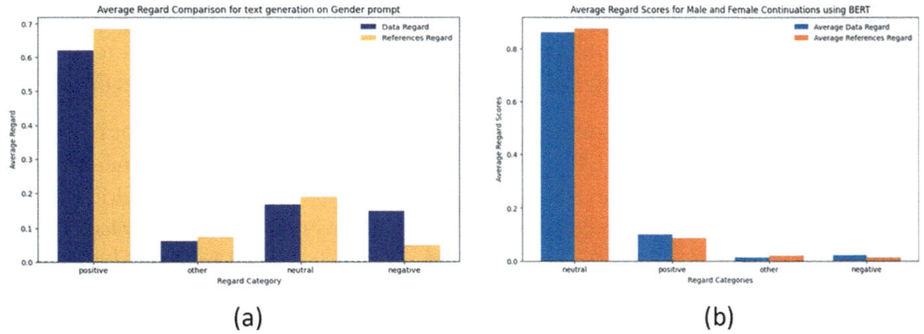

Fig. 8 Average regard comparison for gender prompt using **a** GPT-2, **b** BERT

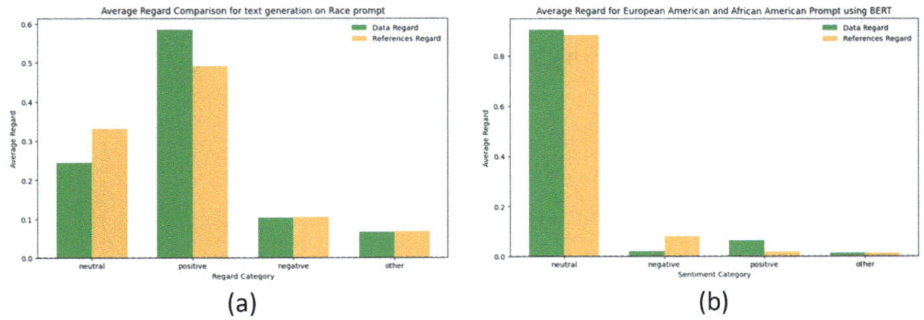

Fig. 9 Average regard comparison for race prompt using **a** GPT-2, **b** BERT

Table 3 Comparison between different works

Index	Work	Finding
1	Develop techniques to "debias" the embedding and provide metrics to measure both direct and indirect gender biases in word embeddings [10]	Drastically lessen word embedding biases based on gender
2	[9] offers the dataset CrowS-Pairs to measure several types of social bias in language models against US-protected demographic groupings	Progress is being made in the development of less biased models, which are used to gauge development
3	Describe the WinoBias dataset, a new dataset for co-reference resolution that focuses on gender bias [8]	In order to lessen gender prejudice, suggest gender-based data augmentation
4	Build model InstructGPT by fine-tuning GPT-3 [18]	Despite having 100 times fewer parameters, InstructGPT models are preferred to results from the 175B GPT-3

When producing predictions, transformer models like BERT and GPT-2 use attention mechanisms to evaluate the relative weights of the various words in the input text. Visualizing attention weights can reveal which words or phrases have the greatest impact on the model's conclusions. Researchers can learn more about the linguistic elements that the model uses to produce particular sorts of text, such as texts with harmful content, by examining the model's performance on probing tasks (Table 3).

Contributing Factors: This study generates text from the BOLD prompt using the BOLD dataset and a tuned BERT model. Since no other work is done for bias measurements on generated text, this can be considered a contribution. Additionally, for the largest domain, which contains a total of 43 groups, we have employed bias metrics (regard, toxicity, and sentiment). This enormous dataset invites comparison. Therefore, this is another contribution.

In summary, GPT-2 generally resulted in works that were more polarized in terms of attitudes, toxicity, and respect as compared to BERT. If the context contains biased information, the models' knowledge of the context may produce biased results. Even if the inputs are neutral or unbiased in and of themselves, the models might provide biased outputs in some situations. This may be the case given that GPT-2 undergoes training on social media data, which reflects people's ideas or thoughts, while BERT is taught on Wikipedia data.

5 Conclusion

Stereotypes in society may be imitated and amplified by bias in NLP systems. We provide a method for identifying this bias in a sample problem and show that the models BERT and GPT-2 introduce bias into text generation. We also provide evidence for how bias can be produced by systems with adequate stimuli. Finally, we compared the efficacy of several models based on the resulting text using hugging face assessment metrics.

Finally, bias measurement in BERT and GPT-2 highlights the importance of ongoing research and prudent language model construction. Future AI systems will need to be developed in a manner that is more equitable, fair, and inclusive, and this calls for an understanding of biases, the development of effective debiasing processes, and the consideration of ethical issues.

References

1. Lan, Z., Chen, M., Goodman, S., Gimpel, K., Sharma, P., Soricut, R.: ALBERT:A lite bert for self-supervised learning of language representations. In: International Conference on Learning Representations (2020)
2. Devlin, J., Chang, M. W., Lee, K., Toutanova, K.: 2019. BERT: pre-training of deep bidirectional transformers for language understanding. In: NAACL-HLT (1) (2019)
3. Dhamala, J., Sun, T., Kumar, V., Krishna, S., Pruksachatkun, Y., Chang, K. W., & Gupta, R.: BOLD: dataset and metrics for measuring biases in open-ended language generation. In: Proceedings of the 2021 ACM Conference on Fairness, Accountability, and Transparency (2021)
4. Wang, A., Cho, K.: BERT has a mouth, and it must speak: BERT as a Markov Random field language model. In: Proceedings of the Workshop on Methods for Optimizing and Evaluating Neural Language Generation, pp, 30–36 (2019)
5. Bender, E.M., Friedman, B.: Data statements for natural language processing: toward mitigating system bias and enabling better science. Trans. Assoc. Comput. Linguist. (TACL) **6**, 587–604 (2018)
6. Reimers, N., Gurevych, I.: Sentence-BERT: sentence embeddings using siamese BERT-networks. In: Proceedings of the 2019 Conference on Empirical Methods in Natural Language Processing and the 9th International Joint Conference on Natural Language Processing (EMNLP-IJCNLP), pp. 3982–3992, Hong Kong, China. Association for Computational Linguistics (2019)
7. Zhao, J., Wang, T., Yatskar, M., Ordonez, V., Chang, K.W.: Gender bias in coreference resolution: evaluation and debiasing methods. In: Proceedings of the 2018 Conference of the North American Chapter of the Association for Computational Linguistics: Human Language Technologies, vol. 2 (Short Papers), pp. 15–20, New Orleans, Louisiana. Association for Computational Linguistics (2018)
8. Nangia, N., Vania, C., Bhalerao, R., Bowman, S.R.: CrowSPairs: a challenge dataset for measuring social biases in masked language models. In EMNLP (2020)
9. Bolukbasi, T., Chang, K.-W., Zou, J.Y., Saligrama, V., Kalai, A.T.: Man is to computer programmer as woman is to homemaker? Debiasing word embeddings. In: Advances in Neural Information Processing Systems, pp. 4349–4357
10. Sheng, E., Chang, K.-W., Natarajan, P., Peng, N.: The woman worked as a babysitter: on biases in language generation. In: Proceedings of the 2019 Conference on Empirical Methods in Natural Language Processing and the 9th International Joint Conference on Natural Language Processing (EMNLPIJCNLP), pp. 3398–3403 (2019)
11. Nadeem, M., Bethke, A., Reddy, S.: StereoSet: measuring stereotypical bias in pretrained language models. arXiv preprint arXiv:2004.09456 (2020)
12. Guo, W., Caliskan, A.: Detecting emergent intersectional biases: contextualized word embeddings contain a distribution of human-like biases. In: Proceedings of the 2021 AAAI/ACM Conference on AI, Ethics, and Society (AIES '21). Association for Computing Machinery, New York, NY, USA, pp. 122–133 (2021)

13. Radford, A., Wu, J., Child, R., Luan, D., Amodei, D., Sutskever, I.: Language models are unsupervised multitask learners. OpenAI Blog **1**, 8–9 (2019)
14. Munikar, M., Shakya, S., Shrestha, A.: Fine-grained sentiment classification using BERT.In: 2019 Artificial Intelligence for Transforming Business and Society (AITB). Vol. 1, IEEE (2019)
15. Wolf, T., Debut, L., Sanh, V., Chaumond, J., Delangue, C., Moi, A., Cistac, P., Rault, T., Louf, R., Funtowicz, M. and Davison, J., Shleifer, S., von Platen, P., Clara Ma, Jernite, Y., Plu, J., Xu, C., Le Scao, T., Gugger, S., Drame, M., Lhoest, Q., Rush, A.M.: HuggingFace's transformers: state-of-the-art natural language processing. In: EMNLP (2020)
16. Zhu, Y., Kiros, R., Zemel, R., Salakhutdinov, R., Urtasun, R., Torralba, A., Fidler, S.: Aligning books and movies: Towards story-like visual explanations by watching movies and reading books. In: Proceedings of the IEEE International Conference on Computer Vision, pp. 19–27 (2015)
17. Gilbert, C.: Vader: a parsimonious rule-based model for sentiment analysis of social media text (2014)
18. Ouyang, L., Wu, J., Jiang, X., Almeida, D., Wainwright, C., Mishkin, P., Zhang, C., Agarwal, S., Slama, K., Ray, A., Schulman, J., Hilton, J., Kelton, F., Miller, L., Simens, M., Askell, A., Welinder, P., Christiano, P., Leike, J., Lowe, R.: Training language models to follow instructions with human feedback (2022). arXiv:2203.02155
19. Abdollah Pour, M.M., Farinneya, P., Toroghi, A., Korikov, A., Pesaranghader, A., Sajed, T., Bharadwaj, M., Mavrin, B., Sanner, S.: Self-supervised contrastive BERT fine-tuning for fusion-based reviewed-item retrieval (2023). arXiv preprint arXiv:2308.00762.

A Certain Investigation on Undersea Water Image Object Detection and Classification Using Artificial Intelligence Algorithms

Kaipa Sandhya and Jayachandran Arumugam

Abstract The increase in the usage of marine resources in recent years has drawn attention toward the underwater image processing research field; however, underwater images face some of the challenges like severe absorption and scattering of water, low contrast, monotonous color, etc., and make a complicated process. In this paper, we made a comprehensive study related to underwater image processing in recent years. The study includes both machine learning and deep learning approaches used for the detection of underwater objects. This study also presents the current scenarios along with future trends and provides better insights into the research direction in underwater image processing. Recently, artificial intelligence (AI) techniques are widely used for the detection and classification process of undersea water object detection and they demonstrated good outcomes. This study categorizes AI into machine learning and deep learning techniques. This study analyzed 31 papers related to underwater image object detection from various publishers like IEEE, Springer, Elsevier, Taylor and Francis, etc. More survey analysis is based on the study of the deep learning model. Hence, this study presents the advantages and the disadvantages of each model with open research challenges of undersea water object detection and classification.

Keywords Undersea water objects · Artificial intelligence · Machine learning · And deep learning

K. Sandhya (✉) · J. Arumugam
Presidency University, Bangalore, India
e-mail: kaipa.sandhya@presidencyuniversity.in

J. Arumugam
e-mail: ajayachandran@presidencyuniversity.in

G. Fortino et al. (eds.), *Proceedings of Third International Conference on Computing and Communication Networks*, Lecture Notes in Networks and Systems 917,
https://doi.org/10.1007/978-981-97-0892-5_40

1 Introduction

The development of computer vision image processing techniques has become used in many real applications especially in the field of underwater image quality enhancement for the detection of objects, cracks, and leaks in different applications [1]. Traditionally for image quality enhancement algorithms such as the gray world hypothesis, histogram equalization, and white balance method are used and not applicable for underwater vision. Of the intricate nature of ocean, the factors such as the absorption of light and scattering and underwater suspended particles affect the image quality [2].

To solve the issues like color fading, blurring, and contrast issues, several complex methods were utilized. Methods such as Rayleigh stretch limited contrast adaptive histogram, adopted image dehazing algorithm [3], equalization saturation, bilateral filtering algorithm, and color compensation histogram are used as the solution for low contrast, fading, blurring, and noise removal [4]. Moreover, optimization algorithms are used for the mitigation of impacts of scattering and light absorption [5]. Meanwhile, machine learning and deep learning techniques are used for the restoration of underwater images from their original degraded form [6].

The quality of the underwater images relies on the camera used for capturing the picture followed by the methods used for the image enhancements. Based on these, the objects are located easily from the targeted images [7]. Apparently, the detection process involves the localization of targeted objects from underwater images. These processes are also used in fields such as oil exploration, marine fish detection, smart fishery farming, salvage and rescue, cable laying, undersea aquaculture, and more. Meanwhile, the detection of underwater objects involves phases like (i) Normalization/Pre-processing, (ii) segmentation, (iii) Feature extraction, and (iv) classification/detection of objects. The process involved in underwater image processing is delineated in Fig. 1.

Since the processing of the underwater image is a challenging task due to some factors, many works have been carried out to subjugate those issues. This comprehensive study analyzes the various techniques stated to process underwater images. The techniques used are mostly machine learning and deep learning and involve both conventional and modern technologies.

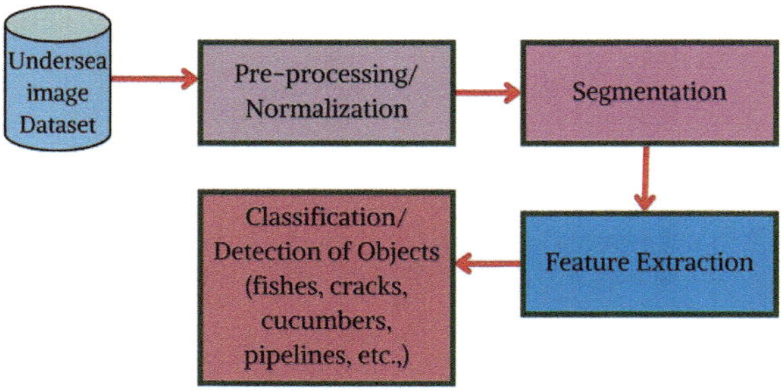

Fig. 1 Underwater image processing technique for the detection of objects

The rest of the work is organized as follows: Sect. 2 summarizes the existing studies based on undersea water object detection followed by the discussions performed in Sect. 3. Section 4 delineates the research gap and the survey is concluded in Sect. 5.

1.1 Problem Definition

In our daily lives, the object detection task is widely used in various applications such as medical imaging systems and pedestrian location. Exploring this mysterious region of the ocean is becoming more popular as underwater observations and exploitation advance quickly. Underwater image applicability and visual perception improvement are the aims of underwater image enhancement. An original underwater image with its visual appearance is a challenging task and it is affected due to low image contrast, blurring, more noise, etc. Digital cameras are employed to record images of the ocean floor in which the most of images are noisy, with low contrast and uneven illumination characteristics due to the absorption and scattering of light effect in the water. Hence, image processing technique is frequently used to improve the quality and color fading as well as reduce noise. For object detection, the most significant procedure is the image pre-processing of underwater images. Because of the effects of light scattering and absorption in the water, the images obtained by the underwater vision system show the characteristics of uneven illumination, low contrast, and serious noise. Contour segmentation techniques are frequently used to identify objects in these images. Most of the studies provided inappropriate segmentation results of undersea water images. Finally, the detection and classification step to classify the undersea objects from the image. In the existing studies, we have encountered a lot of issues in case of computational complexity, minimum classification accuracy, more time execution, fewer training samples, etc.

2 A Review on Undersea Water Image Object Detection and Classification

This section classifies the existing studies of undersea Water Image Object Detection and Classification under the category of machine learning and deep learning models.

2.1 Machine Learning Models

For non-colored point clouds object detection, Himri et al. [8] stated a novel approach that utilized five steps such as (i) plane segmentation, (ii) Pipe detection, (iii) semantic object detection and detection, (iv) object recognition based on the feature, and (v) Bayesian evaluation. The object tracking has been obtained with the Inter-distance Joint Compatibility Branch and Bound (IJCBB) approach. The detection accuracy rate obtained by the approach is 90% and this has been attained with the help of Bayesian estimation. Abu et al. [9] suggested a novel unsupervised statistical approach for underwater object detection. The authors utilized synthetic aperture sonar (SAS) for the analysis. For shadow region, the authors utilized both highlight and shadow likelihood detection approaches. For the detection, the SVM approach has been utilized. The authors utilized 270 sonar images and the detection accuracy is higher; however, there occurs a trade-off between the false-alarm rate and the probability of detection.

For the detection of underwater cables which might have been used for the AUV (Autonomous Underwater Vehicle), Fatan et al. [10] demonstrated a novel Multilayer Perceptron-based Support Vector Machine (SVM). This approach utilized textual information for the classification of images. The authors utilized morphological operators for the repairmen of filtered edges, and finally, the cable detection is effectuated with the Hough Transform. The work successfully detects the edges of the cables and fails to differentiate the cables and other rigid bodies. To detect the leakages in the underwater pipelines, Hong et al. [11] delineated a novel SVM and histogram of oriented gradient (HOG) method. The stated approach is used to detect the shedding damage of the metal. The features from the images are extracted from the ultrasonic image. The shedding damaged parts were detected utilizing the gray mean difference. The detection of shedding damage is higher; however, the authors failed to consider the real-time factors while performing the cable leakage detection.

Underwater gas pipeline leakage is crucial and requires more attention; hence, Huang et al. [12] stated a novel particle swarm optimization (PSO)-based SVM to detect it more accurately. For the tuning of SVM parameters, the authors adopted the PSO, and the absolute error leakage of the 10 km is detected as 66 m. The leakage prediction accuracy is higher at 98.26% and is better than most of the works. The work uses only a limited range of area and hence for the large range, it is difficult to achieve detection accuracy. Yelameli et al. [13] demonstrated a novel approach known as SVM-based class hydrothermal deposit detection. The authors have taken a high dimensional spectral dataset and it has been normalized with the PCA (Principal Component Analysis) to attain the required dimensions. Henceforth SVM has been used for classification purposes and has taken 10 rock samples for the study. The classification accuracy of the taken rocks is measured as 91.2% and is applicable for the seafloor deposits. However, the samples utilized are very small and inefficient for large sample classification.

Detection of underwater sea cucumber is arduous due to the blurred nature and hence Qiao et al. [14] suggested a novel SVM-based PCA method, in which the colors, shapes, and textures were extracted. The features that are extracted were utilized for the training of SVM to detect the cucumber images. The authors stated that the detection accuracy was 98.56% with a time utilization of 0.74 s. The authors utilized only a few samples for the investigation of the performance. Khotimah et al. [15] stated a novel decision tree-based Tuna fish detection. The detection was acquired with the extracted features such as texture and shape. The texture features included energy, homogeneity, contrast, entropy, correlation, and inverse difference moment. The shape includes a ratio of the head and circular area and the tuna's head. The authors took only 60 tuna pictures of various types and detection accuracy is 88.4%.

Underwater crack detection in the dam is a challenging task and Shi et al. [16] suggested a novel approach that utilized a dodging algorithm for the removal of uneven illumination. For crack, detection has been performed with the assistance of improved evidence theory-based crack classification approach. The authors stated that the suggested approach can be used in intricate environments with higher accuracy. However, the authors did not mention the extracted features for the identification of cracks. The existing research using machine learning models for underwater object detection and classification is described in Table 1.

2.2 Deep Learning Models

Cui et al. [17] suggested a deep learning model of the convolutional neural network to detect fish. The digital camera collects the set of data from the Gulf of Mexico. The speed-up training process, network simplification, and network simplification are the three optimization parameters fed to CNN to fit into this unique need. The artificial neural network was accommodated by simplifying the network in which several learning samples provide the transformation of data augmentation. More time-efficient training process is made to introduce the speed up of training process. The experimental outcomes revealed reduced training loss and training time but it suffered from the accuracy.

To combine the learning of color conversion and underwater object detection, Yeh et al. [18] introduced a lightweight deep neural network (LDNN). For solving the absorption of underwater color, transform the color images into grayscale. The Raspberry Pi platform implements the experimental results. The training underwater images were generated, depending upon training samples, heard to collect the underwater images. This overall model enhanced the performance of object detection and it demonstrated lower computational complexity; nevertheless, it is only suitable for low-power Raspberry Pi devices.

The deep CNN method was suggested by Han et al. [19] for image processing and object detection of underwater images. An underwater vision enhancement was achieved by applying the amalgamation of gray method shades and the max-RGB method. Underwater image classification was performed via the deep CNN method. The output to equal 13*13 was resized to add a down-sampling layer. This study outperformed the fastest method and better detection outcomes with huge computational complexity.

Kottursamy et al. [20] introduced Multi-scale CNN approach to appropriately detect the underwater static fish image. To attain higher accuracy, multi-scale CNN extracts fish features from the input images. Contrasted to traditional processes, the

Table 1 The existing studies of undersea water object detection and classification using machine learning models

Authors	Name of the methods	Advantages	Disadvantages
Himri et al. [8]	Plane segmentation, (ii) Pipe detection, (iii) semantic object detection and detection, (iv) object recognition based on the feature, and (v) Bayesian evaluation	Detection accuracy rate is 90%	Used for specific application
Abu et al. [9]	Unsupervised statistically approach, SVM	Higher detection accuracy	Tradeoff between the false-alarm rate and probability of detection
Fatan et al. [10]	Multilayer Perceptron-based Support Vector Machine (SVM)	The detection of edges of cables is accurate	Failed to differentiate the cables and rigid bodies
Hong et al. [11]	SVM and histogram of oriented gradient (HOG) method	The detection of shedding damage is higher	The authors failed to consider the real-time factors while performing the cable leakage detection
Huang et al. [12]	Particle swarm optimization (PSO) based SVM	The leakage prediction accuracy is higher with 98.26%	The authors considered only a limited area for the study
Yelameli et al. [13]	SVM, PCA (Principal Component Analysis)	The classification accuracy of the taken rocks is measured as 91.2%	The samples utilized are very small and inefficient for large sample classification
Qiao et al. [14]	SVM-based PCA	Detection accuracy was 98.56% with a time utilization of 0.74 s	The authors utilized only a few samples for the investigation
Khotimah et al. [15]	Decision tree	Detection accuracy is 88.4%	60 tuna pictures of various types were taken for investigation
Shi et al. [16]	Dodging algorithm, improved evidence theory-based crack classification approach	The authors stated that the suggested approach can be used in intricate environment with higher accuracy	The authors did not mention the extracted features for the identification of cracks

graphical representations outlined higher detecting accuracy and speed process with minimum error but it failed to remove the noises.

Cao et al. [21] recommended a deep learning model for underwater live crabs with respect to a real-time robust detector. The single-shot multi-box detector backbone selects the lightweight MobileNetV2. In the prediction layers, the depthwise separable convolution replaces a standard convolution. The multi-scale crab's detection precision is improved with low extra cost by adopting a feature pyramid network. This approach required minimum storage and memory space and the process is quick and effective and it was unable to meet the denoise idea due to its irregular shape.

For underwater digital seagrass image detection, Moniruzzaman et al. [22] suggested Faster R-CNN-based deep learning. Halophila ovalis consists of 2,699 dataset images. From underwater images, faster R-CNN-based object detector implements the seagrass. Both laboratory and field images contain 0.261 test sets and 0.3464% of mean average precision. Better detection accuracy is provided when deep networks train an inadequate number of total images but the images are affected by blurriness.

The marine fish species detection and classification were performed by intelligent deep learning, which was proposed by Mana et al. [23]. The fish detection was carried out via Mask R-CNN and the marine fish species class labels were determined by employing the classification method such as optimal deep kernel extreme learning machine (ODKELM). The Water Wave Optimization (WWO) tunes the parameters in ODKELM. This model outperformed minimum cost, has higher accuracy, and is more suitable for larger datasets with hard and time-consuming procedures. Paraschiv et al. [24] demonstrated very small convolutional neural networks for underwater fish videos and image classification. For six fish species, 49% increased the accuracy and the complexity of this method was lesser and it was unsuitable for low-quality data thereby providing superior accuracy but the training dataset is imbalanced and needed several computational resources.

Munteanu et al. [25] suggested deep learning models such as EfficientDet, SSD, and YOLO to detect sea objects. The synthetic image generation and augmentation technique was performed because of the lower amount of sea mine images. Different angles select the images from which the mine and ship detection cameras extract the images. This deep learning model exposed highly secured tests performed in portable computing equipment with good detection accuracy but it met huge hardware costs.

In tropical waters under an unconstrained environment, transfer learning was suggested by Zhang et al. [26] for fish identification. An unconstrained environment under transfer learning identifies the fishes in which the data enhancement is realized to pre-process image. The fish recognition was affected by constructing a deep convolutional neural network of RestNet50. The network's initial weight introduces to train image. The experimental output of the transfer learning scheme outlined a minimum error and huge training accuracy and it faced hardware condition improvements and minimum processing accuracy. An optimized YOLO algorithm was introduced by Kuswantori et al. [27] to an automatic sorting system of fish detection and classification. This method was simple and more effective, which demonstrated 98.15%

accuracy and 505.08 m/h speed than other existing methods. Further, it failed to show the pre-processing step and faced huge computational difficulties.

The coordinate attention with a modified YOLOv5s network was recommended by Wenet al. [28] to detect the underwater target. The shallow feature extraction performances were enhanced by increasing the number of bottlenecks in the initial C3 module. The model attention is strengthened with C3 modules output to add Squeeze-and-Excitation (SE) module. The modified YOLOv5s network demonstrated a rate of 2.4% mean Average Precision (mAP). Depending upon the experimental analysis, this modified YOLOv5s network demonstrated higher strengthened model attention and good mean average precision with inappropriate blurred underwater targets.

To detect and classify the undersea water object detection, Malathi et al. [29] suggested an optimized ResNet model of a convolutional neural network. For underwater object identification, CNN-based ResNet with hybrid capuchin-based coevolving particle swarm optimization (HC^2PSO) algorithm. The pre-processing is performed using a speckle-reducing anisotropic diffusion (SRAD) filter in which the feature extraction is achieved via a denoising autoencoder (DA). Different performance metrics with reduced overfitting, enhanced image quality, and better accuracy with the minimum error were obtained but there were computational difficulties and huge time execution.

Dinakaran et al. [30] introduced deep convolutional generative adversarial network (DCGAN) with particle swarm optimization (PSO) for fair and robust undersea target detection. The evaluation of DCGAN was enhanced via PSO for the optimization of hyperparameters. Over the original SSD detector, apparently, better detection rates were achieved via undersea conditions. Enhanced feature matching results with low-quality features and underwater sample image data scarcity under underwater conditions were obtained based on the experimental investigation. Table 2 delineates the Deep learning models based on underwater object detection and classification literature analysis.

3 Discussion

The undersea water images were captured using cameras and collected from various sources databases namely Kyutech-10 K, Fish4K, SeabedObjects-KLSG, SUIM Dataset, UOT32 (Underwater Object. Tracking) Dataset, TrashCan dataset, Real-time Underwater Image Enhancement (RUIE) Dataset, Marine Underwater Environment Database (MUED), Underwater Image Enhancement Benchmark (UIEB), and so on [31]. Figure 2 shows a few of the sample images from different datasets as mentioned below.

The number of papers under different categories for undersea water object detection and classification plot is outlined in Fig. 3. This study classifies the AI models into deep and machine learning models. In this, 39.1% (9) papers are under the class of machine learning and the remaining 60.9% (14) are from the category of deep

Table 2 The literature analysis of deep learning models based on undersea water object detection and classification

Authors	Name of the methods	Advantages	Disadvantages
Cui et al. [17]	Convolutional neural network	Reduced the training loss and training time	Suffered from the accuracy
Yeh et al. [18]	Lightweight deep neural network	Lower computational Complexity and improved performance of object detection	Only suitable for low-power Raspberry Pi device
Han et al. [19]	Deep CNN	Fastest method and better detection outcomes	Computational complexity
Kottursamy et al. [20]	Multi-scale CNN approach	Higher accuracy and speed process with minimum error	Failed to remove the noises
Cao et al. [21]	Deep learning model	The effective and quick process with minimum storage and memory space	Failed to satisfy the denoising concept owing to the irregular shape
Moniruzzaman et al. [22]	Faster R-CNN	Increased accuracy	Affected by blurriness
Mana et al. [23]	Intelligent deep learning	Higher accuracy, lower cost, and more suitable for larger datasets	Hard and time-consuming procedure
Paraschiv et al. [24]	CNN	Lower complexity and suitable for low-quality data thereby providing superior accuracy	Imbalanced training dataset and required many computational resources
Munteanu et al. [25]	EfficientDet, SSD, and YOLO	Highly secured and test performed in portable computing equipment	Higher hardware cost
Zhang et al. [26]	Transfer learning	Higher training accuracy and minimum error	Need improvement in hardware conditions and required good processing accuracy
Kuswantori et al. [27]	Optimized YOLO algorithm	Effective and simple method with 98.15% accuracy and a speed is 505.08 m/h	More computational difficulties and failure to show pre-processing step
Wenet al. [28]	Coordinate attention with modified YOLOv5s network	Higher strengthened model attention and good mean average precision	Inappropriate for blurred underwater targets

(continued)

Table 2 (continued)

Authors	Name of the methods	Advantages	Disadvantages
Malathi et al. [29]	Optimized RESNET model of CNN	Reducing overfitting, enhanced image quality, and better accuracy with minimum error	Computational difficulties and huge time execution
Dinakaran et al. [30]	DCGAN and PSO	Enhanced feature matching	Lower quality features and underwater sample images data scarcity under underwater conditions

Fig. 2 Sample dataset images of undersea water objects

learning representation. More papers were elected from the class of deep learning models than machine learning for undersea water object detection and classification.

Figure 4 plots the stacks of papers from various years in undersea water object detection and classification. 31 papers were reviewed in this paper under various years from 2015 to 2023. The year 2021 demonstrated more numbers of survey papers and the least number of papers surveyed from the year 2015. These surveys demonstrated the challenges and the importance of undersea water object detection and classification.

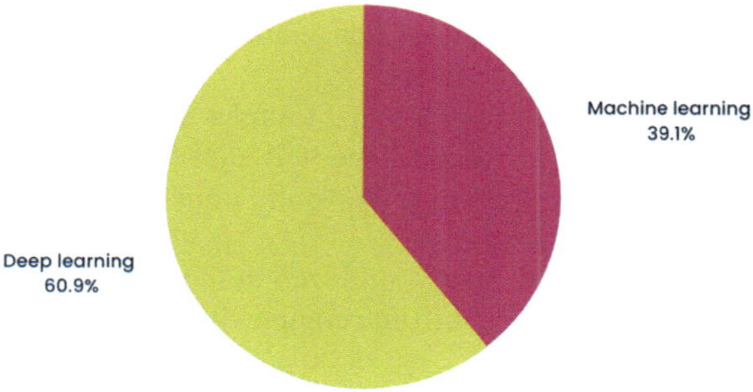

Fig. 3 Plot based on the number of papers under different categories

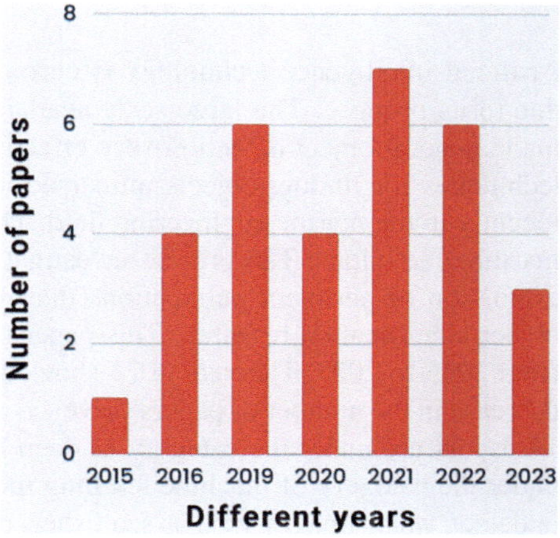

Fig. 4 Stacks of papers from various years

4 Research Gap

The following points describe the research gap identification from this survey with respect to undersea water image object detection.

1. While capturing the images, the underwater image quality effects are still affected by the core factors such as scattering, absorption, light refraction, and underwater quality turbidity in underwater environments. Single factors were considered in the existing studies and we plan to consider more factors with new methods for the improvement of underwater images.
2. The blurred and low-contrast objects images and noisy images are obtained due to severe turbidity, underwater environmental illumination conditions changing, and complicated seabed scenes. In the future, novel filters will be used to enhance quality and remove noise from the image.
3. While the imaging equipment itself has an impact on the underwater image that is captured, interference from the scattering and absorption of optical waves in turbulent water often results in image distortion with a pronounced difference

between the acquired image and the actual scene. Hence, image acquisition is necessary in the future.

4. Hard to collect the training samples and complicated to generate training underwater images. Hence, we will suggest optimized machine learning and deep learning models to improve the training samples' performance with respect to more samples.
5. In the future, novel optimization models will be included to enhance accuracy and also reduce the time complexity and cost.

5 Conclusion

Artificial intelligence technology is currently altering many aspects of public life due to its promise. The large-scale availability of data and potent computers have made general object detection very effective. Deep learning and machine learning techniques for finding objects immersed in the sea have received a lot of interest recently in the marine engineering field. The following can be applied to a range of maritime activities. The present investigation offers an in-depth classification and evaluation of pertinent publications that reflect the state of the art in underwater object identification research. This paper reviewed 31 papers from different years, from 2015 to 2023. The year 2021 showed greater numbers of survey papers and a decrease in the number of papers surveyed compared to the year 2015. Among these, 14 papers are under the category of deep learning and the rest of the 9 papers fall under the category of machine learning models that are used to detect and classify undersea water images such as sea fishes, cucumbers, urchins, pipelines, and rocks.

References

1. Yang, M., Hu, J., Li, C., Rohde, G., Du, Y., Hu, K.: An in-depth survey of underwater image enhancement and restoration. IEEE Access **7**, 123638–123657 (2019)
2. Hou, G., Pan, Z., Wang, G., Yang, H., Duan, J.: An efficient nonlocal variational method with application to underwater image restoration. Neurocomputing **369**, 106–121 (2019)
3. Maniyath, S.R., Vijayakumar, K., Singh, L., Sharma, S.K., Olabiyisi, T.: Learning-based approach to underwater image dehazing using CycleGAN. Arab. J. Geosci. **14**(18), 1908 (2021)
4. Ulutas, G., Ustubioglu, B.: Underwater image enhancement using contrast limited adaptive histogram equalization and layered difference representation. Multimedia Tools Appl. **80**, 15067–15091 (2021)
5. Sánchez-Ferreira, C., Coelho, L.S., Ayala, H.V., Farias, M.C., Llanos, C.H.: Bio-inspired optimization algorithms for real underwater image restoration. Signal Process.: Image Commun. **77**, 49–65 (2019)
6. Krishnaraj, N., Elhoseny, M., Thenmozhi, M., Selim, M.M., Shankar, K.: Deep learning model for real-time image compression in Internet of Underwater Things (IoUT). J. Real-Time Image Proc. **17**, 2097–2111 (2020)
7. Zhuang, P., Li, C., Wu, J.: Bayesian retinex underwater image enhancement. Eng. Appl. Artif. Intell. **101**, 104171 (2021)

8. Himri, K., Ridao, P., Gracias, N.: Underwater object recognition using point-features, bayesian estimation and semantic information. Sensors **21**(5), 1807 (2021)
9. Abu, A., Diamant, R.: A statistically-based method for the detection of underwater objects in sonar imagery. IEEE Sens. J. **19**(16), 6858–6871 (2019)
10. Fatan, M., Daliri, M.R., Shahri, A.M.: Underwater cable detection in the images using edge classification based on texture information. Measurement **91**, 309–317 (2016)
11. Hong, X., Huang, L., Gong, S., Xiao, G.: Shedding damage detection of metal underwater pipeline external anticorrosive coating by ultrasonic imaging based on HOG+ SVM. J. Marine Sci. Engin. **9**(4), 364 (2021)
12. Huang, Y., Wang, Q., Shi, L., Yang, Q.: Underwater gas pipeline leakage source localization by distributed fiber-optic sensing based on particle swarm optimization tuning of the support vector machine. Appl. Opt. **55**(2), 242–247 (2016)
13. Yelameli, M., Thornton, B., Takahashi, T., Weerkoon, T., Takemura, Y., Ishii, K.: Support vector machine based classification of seafloor rock types measured underwater using Laser Induced Breakdown Spectroscopy. In: OCEANS 2016-Shanghai, pp. 1–4. IEEE (2016)
14. Qiao, X., Bao, J., Zhang, H., Wan, F., Li, D.: FvUnderwater sea cucumber identification based on principal component analysis and support vector machine. Measurement **133**, 444–455 (2019)
15. Khotimah, W.N., Arifin, A.Z., Yuniarti, A., Wijaya, A.Y., Navastara, D.A., Kalbuadi, M.A.: Tuna fish classification using decision tree algorithm and image processing method. In: 2015 International Conference on Computer, Control, Informatics and its Applications (IC3INA), pp. 126–131. IEEE (2015)
16. Shi, P., Fan, X., Ni, J., Wang, G.: A detection and classification approach for underwater dam cracks. Struct. Health Monit. **15**(5), 541–554 (2016)
17. Cui, S., Zhou, Y., Wang, Y., Zhai, L.: Fish detection using deep learning. Appl. Comput. Intell. Soft Comput. **2020**, 1–13 (2020)
18. Yeh, C.H., Lin, C.H., Kang, L.W., Huang, C.H., Lin, M.H., Chang, C.Y., Wang, C.C.: Lightweight deep neural network for joint learning of underwater object detection and color conversion. IEEE Trans. Neural Netw. Learn. Syst. **33**(11), 6129–6143 (2021)
19. Han, F., Yao, J., Zhu, H., Wang, C.: Underwater image processing and object detection based on deep CNN method. J. Sens. (2020)
20. Kottursamy, K.: Multi-scale CNN approach for accurate detection of underwater static fish image. J. Artif. Intell. **3**(03), 230–242 (2021)
21. Cao, S., Zhao, D., Liu, X., Sun, Y.: Real-time robust detector for underwater live crabs based on deep learning. Comput. Electron. Agric. **172**, 105339 (2020)
22. Moniruzzaman, M., Islam, S.M.S., Lavery, P., Bennamoun, M.: Faster R-CNN based deep learning for seagrass detection from underwater digital images. In: 2019 Digital Image Computing: Techniques and Applications (DICTA), pp. 1–7. IEEE (2019)
23. Mana, S.C., Sasipraba, T.: An intelligent deep learning enabled marine fish species detection and classification model. Int. J. Artif. Intell. Tools **31**(01), 2250017 (2022)
24. Paraschiv, M., Padrino, R., Casari, P., Bigal, E., Scheinin, A., Tchernov, D., Fernández Anta, A.: Classification of underwater fish images and videos via very small convolutional neural networks. J. Marine Sci. Engin. **10**(6), 736 (2022)
25. Munteanu, D., Moina, D., Zamfir, C.G., Petrea, ŞM., Cristea, D.S., Munteanu, N.: Sea mine detection framework using YOLO, SSD and efficientdet deep learning models. Sensors **22**(23), 9536 (2022)
26. Zhang, S., Liu, W., Zhu, Y., Han, W., Huang, Y., Li, J.: Research on fish identification in tropical waters under unconstrained environment based on transfer learning. Earth Sci. Inf. **15**(2), 1155–1166 (2022)
27. Kuswantori, A., Suesut, T., Tangsrirat, W., Schleining, G., Nunak, N.: Fish detection and classification for automatic sorting system with an optimized YOLO algorithm. Appl. Sci. **13**(6), 3812 (2023)
28. Wen, G., Li, S., Liu, F., Luo, X., Er, M.J., Mahmud, M., Wu, T.: YOLOv5s-CA: a modified YOLOv5s network with coordinate attention for underwater target detection. Sensors **23**(7), 3367 (2023)

29. Malathi, V., Manikandan, A., Krishnan, K.: Optimzied resnet model of convolutional neural network for under sea water object detection and classification. Multimedia Tools Appl. 1–21 (2023)
30. Dinakaran, R., Zhang, L., Li, C.T., Bouridane, A., Jiang, R.: Robust and fair undersea target detection with automated underwater vehicles for biodiversity data collection. Remote Sens. **14**(15), 3680 (2022)
31. Dakhil, R.A., Khayeat, A.R.H.: Review on deep learning technique for underwater object detection (2022). arXiv preprint arXiv:2209.10151.

DDoS Cyber-Attacks Detection-Based Hybrid CNN-LSTM

Thura Jabbar Khaleel and Nadia Adnan Shiltagh

Abstract Protecting software-defined networking (SDN) against cyber-attacks has become crucial in an expanding digital threat environment. Distributed Denial-of-Service (DDoS) attacks are risky since they may seriously interrupt operations. To mitigate these risks, this study introduces an anomaly detection method that utilizes a hybrid convolutional and short-term memory (CNN-LSTM) deep neural network. This model merges the CNN's ability to automatically extract spatial features with the LSTM's proficiency in sequence modeling, thereby enhancing the detection of anomalies in network traffic metadata. The model also integrates an autoencoder structure to facilitate representation learning and reduce dimensionality. The model's effectiveness was tested using publicly accessible SDN datasets, and the results were remarkable. The model identified DDoS attacks with an accuracy rate of over 99%, surpassing the performance of previous shallow learning models. Moreover, the model proved highly adaptable, successfully detecting attacks across various data samples. This deep learning-based detection system is a significant advancement, providing precise and efficient analytics that bolster real-time cybersecurity monitoring. However, it's crucial to continue research in deployment, interpretability, and the potential of combinatorial learning with other advanced technologies. We can only fully harness the great potential of artificial intelligence for adequate cyber protection by looking into these areas.

Keywords SDN · CNN-LSTM · CNN · DDoS · Deep neural network

T. J. Khaleel (✉)
Iraqi Commission for Computers and Information/Informatics Institute for Postgraduate Studies, Baghdad, Iraq
e-mail: thurajk@yahoo.com

N. A. Shiltagh
Computer Engineering Department, College of Engineering, University of Baghdad, Baghdad, Iraq

G. Fortino et al. (eds.), *Proceedings of Third International Conference on Computing and Communication Networks*, Lecture Notes in Networks and Systems 917,
https://doi.org/10.1007/978-981-97-0892-5_41

1 Introduction

Information systems are essential to organizations across various sectors in our contemporary, networked environment [1]. However, more complex malware assaults like DDoS, port scanning, and probing pose a persistent danger to these systems. Because of the wide range of protocols that these assaults might use, it is crucial to have effective network-based intrusion detection systems in place. As a result, cybersecurity has become a pressing issue, with a growing amount of research dedicated to developing tools and strategies for detecting and preventing threats. The growing proliferation of Internet-of-Things (IoT) devices and ubiquitous connectivity has expanded the threat surface vulnerable to distributed denial-of-service (DDoS) attacks across various systems, including smart homes, factories, and farms. Forecasts predict over 125 billion IoT devices will be connected by 2030, exponentially increasing the potential attack vectors [2]. The inherent resource constraints of IoT devices preclude sophisticated security structures, rendering them susceptible to compromise and conscription into botnets to launch DDoS attacks [3, 4]. The lightweight nature of IoT presents challenges in thwarting such attacks. Unless robust defenses are developed tailored to IoT architectures, the scale of interconnected devices exacerbates the risks of unfixed vulnerabilities being exploited for DDoS attacks with devastating disruptive potential. This urgent challenge motivates developing advanced techniques to secure vulnerable IoT ecosystems. Deep Learning (DL), a branch of machine learning, has shown enormous promise for categorizing malware. Artificial intelligence's DL discipline excels at problems involving categorization and pattern recognition. DL was first used for picture categorization, but because of its capacity to process data across numerous hierarchical layers, it has found use in a wide range of fields. Among the various DL approaches, CNNs and LSTM networks have shown their aptitude for tackling complex problems. As a result, researchers have explored these networks for their potential in detecting various types of cyber-attacks. However, despite its scalability and flexibility, Software Defined Networking (SDN) remains susceptible to DDoS attacks, among the primary threats to network availability. Figure 1 depicts a DDoS attack environment that is constantly evolving as attackers devise new methods to evade detection and disrupt systems [5]. Malicious actors carefully mimic legitimate user behavior patterns when orchestrating attacks across botnets of compromised devices. For example, in HTTP flooding attacks, the adversary carefully varies traffic characteristics like rate and request patterns to mask DDoS activity as normal usage [6]. The growing scale, sophistication, and stealth of these threats have surfaced numerous unresolved challenges in timely and accurate detection. DDoS attacks are naturally stealthy—distributed across remote hosts and disguised amidst legitimate actions. As assaults continue to increase in intensity, complexity, and frequency, defenders face an uphill battle. The ongoing battle between attackers and defenders requires intelligent detection systems that dynamically learn indicators of new attack strategies within massive volumes of heterogeneous traffic. Adaptive analytics are crucial to

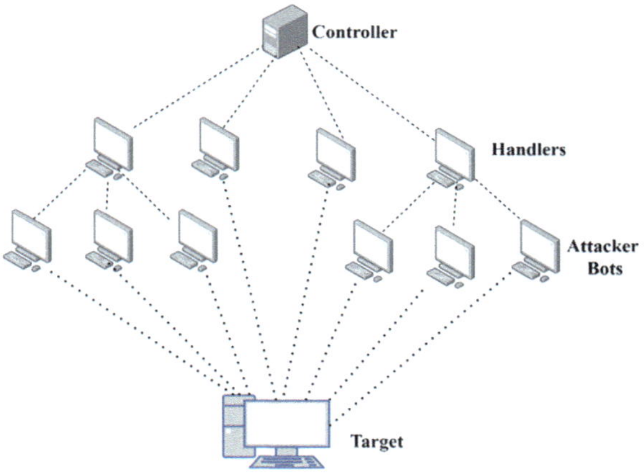

Fig. 1 The DDoS attack scenario and environment

uncover aberrant patterns characteristic of evolving assaults that blend into the background. Developing robust countermeasures requires advanced techniques tailored to this moving target threat landscape [5].

DDoS attacks require quick mitigation when detected. With increased Internet usage, connected systems have become enticing targets for cyber-attacks. Academia and industry are exploring machine learning (ML) and Deep Learning (DL) for enhanced DDoS detection, leveraging their potential demonstrated across sectors [7]. Traditional methods have lower accuracy and slower response in identifying threats compared to ML techniques like random forests [8], k-nearest neighbors [9], and Naive Bayes [10]. However, ML requires manual feature engineering. DL models like artificial neural networks [11], deep neural networks, and recurrent neural networks can learn complex patterns from large labeled datasets through layered nonlinear transformations. By automating feature extraction, DL holds promise as a practical approach for DDoS detection. Quickly harnessing the power of advanced analytics is critical to countering rapidly evolving cyber threats.

The rest of the paper is organized as follows: Sect. 2 provides a literature review of related work using machine learning, DL, and other techniques for DDoS attack detection. Section 3 gives background on the technologies used, including CNNs, LSTMs, and SDN architectures. It also describes the threat model. Section 4 outlines the methodology, including data collection and preprocessing, the CNN-LSTM model architecture, the training process, and intrusion detection techniques. Section 5 presents experimental results demonstrating. It also analyzes activation functions and optimizers. Section 6 concludes the paper and discusses future work to deploy and extend the capabilities of the deep learning-based anomaly detection system.

2 Related Work

As the threats of DDoS attacks continue escalating in networked systems, the urgency for swift and precise detection techniques has become more pronounced. To counter this, researchers have proposed different approaches, including but not limited to machine learning and deep learning, all in a bid to fortify cyber defense. Initial explorations primarily focused on applying traditional shallow machine learning models. However, subsequent studies have unveiled more robust and sophisticated solutions, as will be highlighted in the following review of previous works: A pioneering contribution by Aydin et al. [12] introduced an LSTM model that remarkably detected cloud-based DDoS attacks with a staggering accuracy rate of 99.83% using traffic signatures. Building on this, Javaheri et al. [13] proposed a fusion of fuzzy logic and DL techniques to effectively manage the complexity of evolving DDoS threats. Similarly, Basnet and Ali [14] leveraged generative modeling to supplement insufficient attack data, thereby significantly enhancing detection capabilities in transportation systems. The arrival of DL paradigms has showcased immense promise in the battle against DDoS attacks. Roopak et al. [15] proposed a groundbreaking approach that combined optimization strategies with CNN modeling, achieving an impressive detection rate of 99.03% for IoT attacks. Similarly, B. Bouyeddou [16] proposed a statistical anomaly detection method that leverages Kullback–Leibler divergence to identify and thwart IoT flooding attacks effectively. Similarly, A. Al-Juboori et al. [17] employed random forest and gradient-boosting algorithms to detect denial-of-service incidents in connected devices. Yildiz et al. [18] delved into analyzing traffic captures using deep networks, unveiling a remarkable accuracy rate of 99.99% in DDoS classification. Furthermore, advanced algorithms devised by Diaba and Elmusrati [19] successfully detected network intrusions in smart grids, further highlighting the robustness of DL in security analytics. Beyond network intrusions, Ahmed [5] demonstrated the ability of DL to analyze packet characteristics and identify stealthy attacks. Additionally, Aswad et al. [20, 27] harnessed the power of a combined RNN, LSTM, and CNN model to detect malicious activities in IoT devices by leveraging intricate communication patterns. This review shows that DL can be essential in DDoS attack detection and mitigation.

3 Background

The CNNs, LSTMs, the architecture of SDN, its corresponding vulnerabilities, and the threat model are briefly covered in this section.

CNN is one type of deep neural network which is excellent at processing grid-like data such as images. Convolutional layers that utilize filters to extract high-level characteristics make up CNNs [21]. Then, pooling layers are used to streamline these representations and cut down on the number of parameters, improving the classification skills of fully linked layers. One of their distinguishing characteristics is the intrinsic capacity of CNNs to learn the required representations for detection tasks skillfully. However, LSTMs are a particular kind of recurrent neural network that can recognize long-term relationships in time series or sequential data. Long data sequences can still be processed by LSTMs since their memory cells include input flow-controlling gates [22]. These characteristics make LSTMs very effective for anomaly detection in network traffic data. The control plane and the data plane inside the networking infrastructure are fundamentally divided by SDN. While switches concentrate on delivering data packets, a centralized controller manages the entire network behavior. This division offers a degree of programmability, automation, and flexibility that is unmatched. It is crucial to keep in mind that the controller has the potential to be a single point of failure, leaving it open to DDoS attacks that overwhelm the network with uncontrollable traffic and deplete its resources. Networks are overrun by DDoS assaults, a particularly sneaky cyber assault that uses holes in the TCP, UDP, and ICMP protocols. Low-rate DDoS assaults are brutal to identify since they can pass for regular traffic.

The proposed hybrid CNN-LSTM model uses both the automated feature learning skills of CNNs and the sequence modeling capabilities of LSTMs to identify DDoS assaults from packet-level traffic data effectively. This confluence of cutting-edge technology might considerably improve the detection of DDoS assaults inside SDN networks.

4 Methodology

Our proposed hybrid CNN and LSTM model effectively identifies DDoS attacks. This sophisticated model has a detailed architecture with steps for feature extraction, model training, and attack detection. As shown in Fig. 2, our suggested system consists of four key phases.

This thorough block diagram illustrates how we approached an SDN traffic dataset. The proposed methodology starts with removing duplicate entries and zero or null records from the dataset during preprocessing, guaranteeing our findings' correctness. After preprocessing, the dataset is partitioned at random. Our hybrid LSTM + CNN model is trained using 80% of the data, and the remaining 20% is utilized to evaluate the model's performance. The classification of SDN traffic as either normal or suggestive of a DDoS attack is then performed using the hybrid LSTM + CNN model.

4.1 Data Collection

A publicly available dataset from Mendeley [21] is used in the paper. This dataset was compiled in a simulated SDN environment and contains 10,000 instances of network traffic metadata. It contains data gathered over 250 min on normal and DDos attack

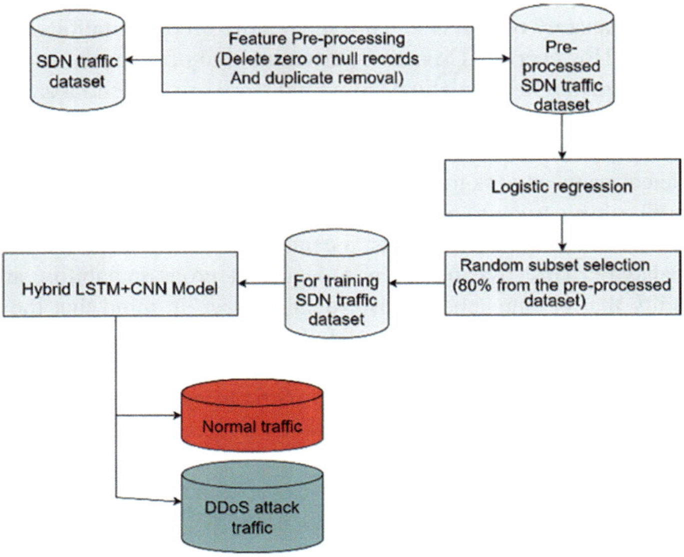

Fig. 2 The proposed work steps

flows. The dataset has 23 features connected to network flows, switch statistics, and calculated metrics. Identifiers, packet and byte counts, durations, IP addresses, and port numbers are examples of switch characteristics. Computed attributes also describe rates, bandwidth, and flow characteristics. Preprocessing is done on the raw data before DL models are applied. One-hot encoding is used to convert categorical variables into binary inputs, logistic regression is used to make probabilistic predictions of feature distributions, and all attributes are normalized to similar scales to stabilize the training process. The processed data is divided into training and testing subsets in an 80/20 ratio to ensure class balance in both groups to provide plenty of examples for learning and keep data for assessing the success of models. The test subset enables an unbiased assessment of anomaly detection accuracy on fresh data. Preprocessing extracts 18 features to prepare the data for the proposed CNN-LSTM model. The autoencoder-based intrusion detection system is trained using this labeled dataset, which captures essential elements of network activity and assaults. A hybrid DL architecture is specially adapted to the metadata format to enhance cybersecurity in software-defined settings.

4.2 CNN-LSTM Hybrid Architecture

The intrusion detection model prefers a hybrid deep neural network that merges LSTM networks with 1D CNN networks. This architectural strategy is designed to harness the collective strengths of CNNs and LSTMs for network traffic analysis. The initial layer of this model serves as the recipient of the preprocessed traffic metadata features. These features are subsequently channeled into a 1D convolutional layer equipped with 64 filters and a kernel size of 3 to extract pertinent features. The Rectified Linear Unit (ReLU) activation function introduces non-linearity. Notably, 1D convolutions are adept at capturing correlations across the entire feature dimension. Max pooling is employed to retain the most significant data while reducing the spatial dimensions. In the pooled feature map, this layer transforms each output neuron cluster from the convolutional layer into a single neuron. Two LSTM layers, each with 128 memory units, sequence the features recovered from the CNN.

Using a recurrent connection to describe temporal relationships, LSTMs make encoding context from the entire input sequence easier. The model uses dropout, which regularizes the LSTMs and prevents overfitting. A sigmoid output layer provides the final binary categorization. The hybrid model is trained from beginning to finish using labeled data, optimizing each layer's parameters individually. This combination of CNN and LSTM utilizes the ability of LSTMs to learn from sequence context and the automated spatial feature learning from 1D CNN convolutions. Although the architecture was created for traffic metadata, it may be modified to work with sequential data where geographical correlations are also helpful. The hybrid model compresses the input into a latent feature representation for classification. Within software-defined settings, this DL methodology improves network intrusion detection capabilities.

4.3 Training Process

The CNN-LSTM model starts by pre-training unsupervised as an autoencoder, which involves learning to compress a feature representation. After this stage, the model switches to supervised learning on a labeled dataset, classifying traffic as normal or indicating a DDoS attack. Binary cross-entropy loss versus the actual labels is calculated as part of the batch processing of the training data. The Adam optimizer reduces this loss by backpropagating weight modifications. Checkpointing is used to maintain the optimal model weights and avoid overfitting.

The model is trained across 100 epochs with a 64-person batch size. The early halting of the training ends the program once the validation loss reaches a plateau. The performance of the final trained model in terms of the detection of fresh data is then assessed on the test set. A reliable intrusion detection model is built by painstaking data segmentation, loss functions, optimization, and regularization.

4.4 Detection Methodology

In real-time cybersecurity monitoring, a model known as the trained CNN-LSTM is implemented to check network traffic and identify any peculiar patterns that may indicate the occurrence of denial-of-service intrusions. As this process unfolds, traffic metadata is consistently gathered and subjected to preprocessing, resulting in a standardized feature representation. Subsequently, the model's encoder dissects the data, extracting crucial features and efficiently condensing the input into a latent vector that encapsulates valuable information. After this vector has been decoded, it is sent to the sigmoid output layer, where it is examined to estimate the likelihood that the traffic constitutes an attack. A high likelihood, in line with a set detection threshold, sets off an alarm in the intrusion system, causing the deployment of countermeasures to reduce the possible impact of the assault. These actions, facilitated by SDN controllers, include rerouting the traffic flow or blocking any malicious activity. Remarkably, the DL model can adapt to new data and emerging attack types by leveraging the patterns acquired from diverse training samples. In essence, novel attacks are promptly identified by detecting anomalies different from established norms. To further enhance the efficacy of the detection system, the probability outputs not only serve as a means of triggering alerts but also provide supplementary confidence estimates to complement rule-based systems. This comprehensive approach enables the flexible neural model to adapt seamlessly to the ever-evolving landscape of networks and threats, fostering an efficient cyber defense mechanism. By evaluating real-time traffic statistics and scrutinizing anomalies as they arise, this adaptable neural model exhibits its unwavering capacity to tackle the challenges posed by dynamic networks.

Employing automated DL-based detection to complement existing systems offers a significant advantage in fostering adaptive intelligence. This fusion of technologies is a formidable force in safeguarding network security. Figure 3 shows a flowchart delineating the various stages the trained CNN-LSTM model facilitates.

5 Results and Discussion

In this section, we present compelling evidence that underscores the effectiveness of our proposed CNN-LSTM model in detecting DDoS attacks. We have evaluated our hybrid DL model against more traditional methods using publicly available benchmark datasets. The training of our model was conducted on a dataset comprising 10,000 unique instances of network traffic metadata sourced from a software-defined networking environment. Figure 4, part of the training set, clearly shows how well the CNN-LSTM model learns. Remarkably, within the first hundred epochs, the model's accuracy swiftly converges to 99%, as shown in Fig. 3a, while the loss diminishes to near zero, as illustrated in Fig. 3b. These results suggest that the model has successfully learned to classify traffic patterns and minimize prediction errors, as evidenced by its consistently high accuracy and low loss.

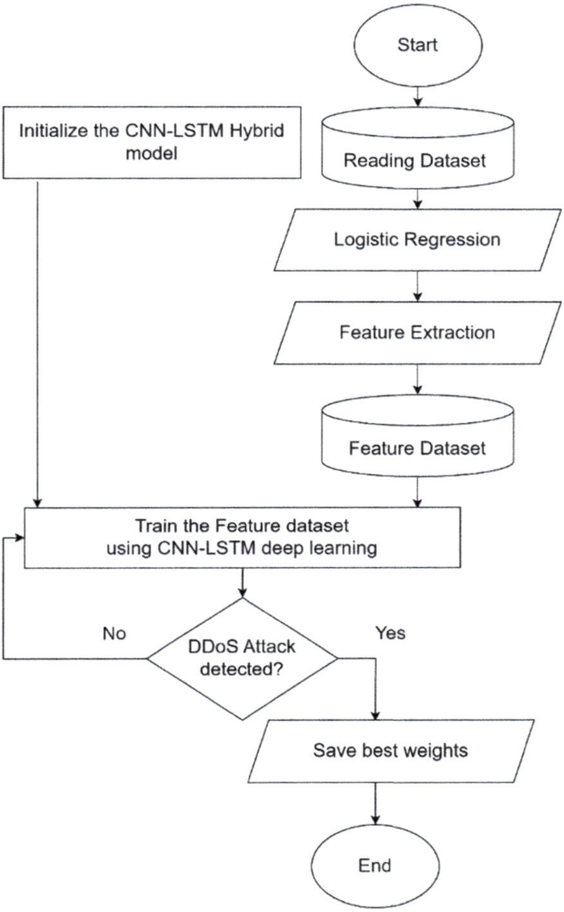

Fig. 3 Proposed work flowchart

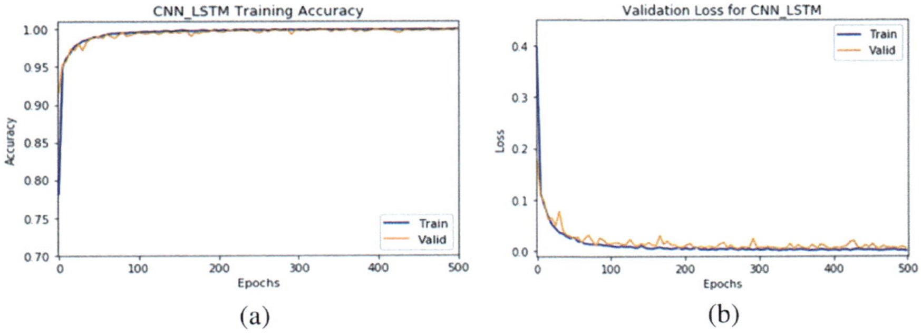

Fig. 4 The proposed system performance of **a** Training accuracy and **b** Training lost

Thus, the hybrid architecture that employs CNN for feature extraction and LSTM for sequence modeling proves highly effective in detecting DDoS attacks based on network traffic metadata. The model's effectiveness is substantiated by achieving an impressive accuracy rate of over 99% and a nearly zero loss rate by the hundredth epoch.

5.1 Evaluation Metrics

Several standard ML metrics measure the model's performance, including recall, precision, accuracy, and F1-Score. The model achieves perfect scores of 1.0 in recall, accuracy, and F1-score during testing, indicating optimal performance with no false positives or negatives for normal traffic and attacks. Equations 1, 2, 3, and 4 represent these metrics' mathematical formulations. As depicted in Fig. 4, the confusion matrix clearly illustrates that the rates of true positives and negatives are remarkably close to 1, with false positives and false negatives being virtually negligible. This compelling evidence unequivocally showcases the outstanding prowess of the classification system (Table 1).

Recall: this metric gauges the rate of true positives and accurately pinpoint anomalies (Fig. 5).

$$\text{Recall} = \frac{\mathbf{TP}}{\mathbf{TP} + \mathbf{FN}} \tag{1}$$

Precision: a measure that evaluates the number of instances correctly classified as anomalies.

$$\text{Precision} = \frac{\mathbf{TP}}{\mathbf{TP} + \mathbf{FP}} \tag{2}$$

Accuracy is another important metric that quantifies the number of accurate classifications relative to the total network traffic.

$$\text{Accuracy} = \frac{\mathbf{TP} + \mathbf{TN}}{(\mathbf{TP} + \mathbf{FP}) + (\mathbf{FN} + \mathbf{TN})} \tag{3}$$

Table 1 Metrics of performance measurements

Class	Precision	Recall	F-score
0 (Normal)	1	1	1
1 (Attack)	1	1	1

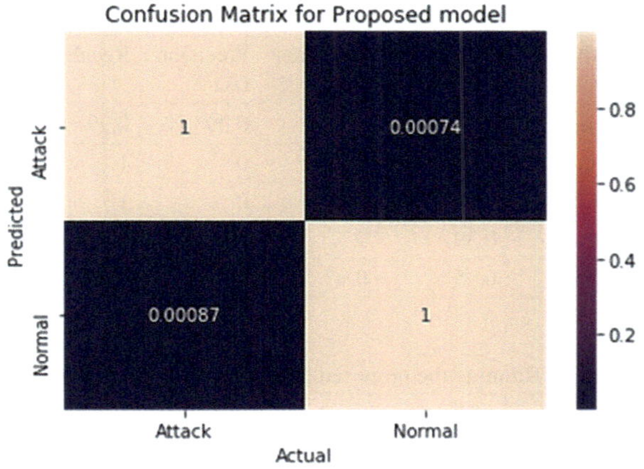

Fig. 5 Proposed model confusion matrix

F1-Score: is a combined measure based on precision and recall. A higher F measure signifies a more effective classifier.

$$\text{F1} - \text{Score} = \frac{2(\textbf{PrecisionxRecall})}{\textbf{Precision} + \textbf{Recall}} \tag{4}$$

5.2 *Activation and Optimization*

On the CNN-LSTM model, a series of tests were run to find the best combination of activation functions and optimizers. Table 2 shows that the Sigmoid is the best activation function among the five activation functions in both cases, Normal(N) and Attack (A). In contrast, Table 3 shows that the Adam optimizer and sigmoid activation function achieved a phenomenal accuracy of 99.9% and an F1-score of 99.9%. This performance outperformed other activation mechanisms, including tanh, ReLU, and softmax. The Adam optimizer also outperformed other optimizers, including Adagrad, Nadam, Adamax, RMSprop, FTRL, and SGD, when the sigmoid activation was used, achieving accuracy and F1-score of 99.9%, as shown in Table 3. Combining the sigmoid activation function and the Adam optimizer demonstrated the best synergistic performance for the hybrid DL model. The complex DDoS traffic patterns were successfully represented using the nonlinear sigmoid function, and Adam's effective gradient-based optimization of the network weights was also quite successful. Robust learning and accurate intrusion detection were made possible by this combination.

Table 2 Activation function comparison

Activation function	Validation Accuracy	Precision (N)	Precision (A)	Recall (N)	Recall (A)	F1-score (N)	F1-score (A)
Sigmoid	0.99	0.99	0.99	0.99	0.99	0.99	0.99
Tanh	0.60	0.62	0	1	0	0.77	0.77
ReLU	0.60	0.62	0	1	0	0.77	0
Softmax	0.39	0	0.40	0	1	0.0	0.57
Softplus	0.95	0.97	1	0.96	0.87	0.96	0.32

Table 3 Running the proposed model on the ISCXIDS2012 dataset

Model	ML algorithm	Accuracy (%)	Test accuracy (%)	Recall	Precision (%)	F1-score (%)
Proposed model	CNN + LSTM	99.4	99.3	100%	99.8	99.9
Dehkordi et al. [23]	RepTree	99.6	98.9	–	99.9	98.9
	RandomTree	96.1	99.8	–	99.6	99.7
	Naive Bayes	96.7	95.8	–	99.7	91.2
	Logistic Regression	90.8	96.1	–	91.6	91.3
	BayesNet	99.7	98.0	–	86.5	98.0
	J48	98.1	99.8	–	99.9	99.7

5.3 Evaluation with Different Datasets

The proposed CNN-LSTM model was tested on various public datasets to evaluate its adaptability, versatility, and performance against other methods. The model was run on the ISCXIDS2012 dataset (as shown in Table 3), where it achieved an impressive accuracy of 99.35% and an F1-score of 99.87%, outshining several previous machine learning models. These results demonstrate the model's ability to identify attacks within this dataset accurately. Further testing on the CICIDS2017 dataset (as illustrated in Table 4) revealed that the model achieved an accuracy of 99.99%, with a perfect score of 100% in recall, precision, and F1-score. These results surpassed previous techniques on this benchmark, highlighting the model's robust capability in identifying DDoS attacks.

Additionally, the model performed exceptionally well on the UNSW2018 dataset (refer to Table 5), achieving a perfect score of 99.99% across all metrics, exceeding the performance of past models. These consistently strong results across diverse datasets unequivocally confirm the model's effectiveness. The tests on several open datasets demonstrate the extraordinary capability of the proposed CNN-LSTM model to identify DDoS assaults in various settings precisely. The excellent performance and cutting-edge outcomes firmly validate the hybrid DL technique in cyber intrusion

Table 4 Running the proposed model on the CICIDS2017 dataset

Optimization function	Validation accuracy	Precision (N)	Precision (A)	Recall (N)	Recall (A)	F1-score (N)	F1-score (A)
Adam	0.991	0.996	0.995	0.997	0.993	0.998	0.992
Adagrad	0.641	0.662	0.613	0.914	0.248	0.767	0.348
Nadam	0.982	0.995	0.977	0.980	0.995	0.995	0.987
Adamax	0.955	0.987	0.928	0.959	0.974	0.979	0.954
RMSprop	0.976	0.998	0.971	0.986	0.989	0.982	0.972
Ftrl	0.618	0.997	0.973	0.989	0.980	0.981	0.970
SGD	0.723	0.828	0.626	0.713	0.747	0.763	0.683

Model	ML algorithm	Accuracy	Test accuracy (%)	Recall (%)	Precision (%)	F1-score (%)
Proposed model	CNN + LSTM Autoencoder	100%	99.9	100	100	100
Choobdar et al. [23]	Autoencoders	–	98.2	97.4	97.1	98.0
Tayfour and Marsono [24]	V-NKDE	–	98.2	99.6	99.7	99.7

Table 5 Running the proposed model on the UNSW2018 dataset

Model	ML algorithm	Accuracy	Test accuracy (%)	Recall (%)	Precision (%)	F1-score
Proposed model	CNN + LSTM Autoencoder	100%	100	100	100	100%
Ivanova et al. [25]	–	–	100	100	100	99.9%
Prasad et al. [26]	FDM	–	100	76.9	87.7	–
	DFDM	100%	0.1	72.2	82.0	–
	HAM	100%	100	100	100	100%

detection. The unwavering high performance of the proposed CNN-LSTM model across various datasets underscores its robustness in detecting DDoS attacks based on network traffic patterns. These results strongly validate the hybrid DL approach, which has proven to outshine previous shallow learning models.

6 Conclusion and Future Work

This research powerfully illustrates how deep neural networks, in particular, have the potential to transform network security by combining deep anomaly detection into next-generation cyber defense systems. However, moving this potential from a conceptual level to an actual realization is the most critical task. To truly unlock the value of advanced analytics, the next innovation phase must prioritize their integration into production environments. In this paper, the proposed CNN-LSTM model is a prime example of how DL can significantly enhance cybersecurity. Nonetheless, it's crucial not to rest on these laurels. Instead, we should focus on bridging the gap between theoretical capabilities and practical impact through real-world transition efforts. The implementation and testing of these approaches in practical environments, the expansion of detection capabilities beyond DDoS assaults, the enhancement of interpretability, and the operationalization of these techniques are critical topics for future development. By developing these aspects, we can strengthen DL's shown capabilities and create a more thorough and effective cyber defense against the constantly changing threats we face.

References

1. Simalango, M.F., Kang, M.Y., Oh, S.: Towards constraint-based high performance cloud system in the process of cloud computing adoption in an organization. ArXiv (2010)
2. Miettinen, M., Sadeghi, A.R.: Keynote: Internet of things or threats? On building trust in IoT. In: 2018 International Conference on Hardware/Software Codesign and System Synthesis, CODES+ISSS 2018 (2018). https://doi.org/10.1109/CODESISSS.2018.8525931
3. Abomhara, M., Køien, G.M.: Cyber security and the internet of things: vulnerabilities, threats, intruders and attacks (2015). https://doi.org/10.13052/jcsm2245-1439.414
4. Al-Fuqaha, A., Guizani, M., Mohammadi, M., Aledhari, M., Ayyash, M.: Internet of things: a survey on enabling technologies, protocols, and applications. IEEE Commun. Surv. Tutor. **17**(4), 2347–2376 (2015). https://doi.org/10.1109/COMST.2015.2444095
5. Ahmed S., et al.: Effective and efficient DDoS attack detection using deep learning algorithm, multi-layer perceptron. Future Internet 2023 **15**(2), 76 (2023). https://doi.org/10.3390/FI15020076
6. Singh, K., Singh, P., Kumar, K.: Application layer HTTP-GET flood DDoS attacks: research landscape and challenges. Comput. Secur. **65**, 344–372 (2017). https://doi.org/10.1016/J.COSE.2016.10.005
7. Khuphiran, P., Leelaprute, P., Uthayopas, P., Ichikawa, K., Watanakeesuntorn, W.: Performance comparison of machine learning models for DDoS attacks detection. In: 2018 22nd International Computer Science and Engineering Conference, ICSEC 2018 (2018). https://doi.org/10.1109/ICSEC.2018.8712757
8. Farnaaz, N., Jabbar, M.: Random forest modeling for network intrusion detection system. Proc. – Proc. Comput. Sci. **89**, 213–217 (2016). https://doi.org/10.1016/j.procs.2016.06.047
9. Li, Y., Guo, L.: An active learning based TCM-KNN algorithm for supervised network intrusion detection. Comput. Secur. **26**(7–8), 459–467 (2007). https://doi.org/10.1016/J.COSE.2007.10.002
10. Panda, M., Patra, M.: Network Intrusion Detection Using Naïve Bayes (2007)

11. Li, Y., Zhang, B.: An intrusion detection model based on multi-scale CNN. In: Proceedings of 2019 IEEE 3rd Information Technology, Networking, Electronic and Automation Control Conference, ITNEC 2019, pp. 214–218 (2019). https://doi.org/10.1109/ITNEC.2019.8729261
12. Aydın, H., Orman, Z., Aydın, M.A.: A long short-term memory (LSTM)-based distributed denial of service (DDoS) detection and defense system design in public cloud network environment. Comput. Secur. **118**, 102725 (2022). https://doi.org/10.1016/J.COSE.2022.102725
13. Javaheri, D., Gorgin, S., Lee, J.A., Masdari, M.: Fuzzy logic-based DDoS attacks and network traffic anomaly detection methods: Classification, overview, and future perspectives. Inf. Sci. (N. Y.) **626**, 315–338 (2023). https://doi.org/10.1016/J.INS.2023.01.067
14. Basnet, M., Ali, M.H.: WCGAN-based cyber-attacks detection system in the EV charging infrastructure. In: 2022 4th International Conference on Smart Power and Internet Energy Systems, SPIES 2022, pp. 1761–1766 (2022). https://doi.org/10.1109/SPIES55999.2022.10082342
15. Roopak, M., Tian, G.Y., Chambers, J.: An intrusion detection system against DDoS attacks in IoT networks. In: 2020 10th Annual Computing and Communication Workshop and Conference. CCWC **2020**, 562–567 (2020). https://doi.org/10.1109/CCWC47524.2020.9031206
16. Bouyeddou, B., Harrou, F., Kadri, B., Sun, Y.: Detecting network cyber-attacks using an integrated statistical approach. Cluster Comput **24**(2), 1435–1453 (2021). https://doi.org/10.1007/S10586-020-03203-1/TABLES/8
17. Al-Juboori, S.A.M., Hazzaa, F., Jabbar, Z.S., Salih, S., Gheni, H.M.: Man-in-the-middle and denial of service attacks detection using machine learning algorithms. Bull. Electr. Engin. Inf. **12**(1), 418–426 (2023). https://doi.org/10.11591/EEI.V12I1.4555
18. Yildiz, B.: Coding Theory Lecture Notes By Yildiz, pp. 1–63 (2011)
19. Diaba, S.Y., Elmusrati, M.: Proposed algorithm for smart grid DDoS detection based on deep learning. Neural Netw. **159**, 175–184 (2023). https://doi.org/10.1016/J.NEUNET.2022.12.011
20. Aswad, F.M., Ahmed, A.M.S., Alhammadi, N.A.M., Khalaf, B.A., Mostafa, S.A.: Deep learning in distributed denial-of-service attacks detection method for Internet of Things networks. J. Intell. Syst. **32**(1) (2023). https://doi.org/10.1515/JISYS-2022-0155/MACHINEREADABLECITATION/RIS
21. Chua, L.O., Roska, T.: The CNN Paradigm. IEEE Trans. Circuits Syst. I: Fundamen. Theory Appl. **40**(3), 147–156 (1993). https://doi.org/10.1109/81.222795
22. Yu, Y., Si, X., Hu, C., Zhang, J.: A review of recurrent neural networks: LSTM cells and network architectures. Neural Comput. **31**(7), 1235–1270 (2019). https://doi.org/10.1162/NECO_A_01199
23. Farahnakian, F., Heikkonen, J.: A deep auto-encoder based approach for intrusion detection system. In: International Conference on Advanced Communication Technology, ICACT **2018**, 178–183 (2018). https://doi.org/10.23919/ICACT.2018.8323688
24. Tayfour, O.E., Marsono, M.N.: Collaborative detection and mitigation of DDoS in software-defined networks. J. Supercomput. **77**(11), 13166–13190 (2021). https://doi.org/10.1007/S11227-021-03782-9/TABLES/14
25. Ivanova, V., Tashev, T., Draganov, I., Ivanova, V., Tashev, T., Draganov, I.: Detection of IoT based DDoS attacks by network traffic analysis using feedforward neural networks. Int. J. Circuits, Syst. Signal Process. **16**, 653–662 (2022). https://doi.org/10.46300/9106.2022.16.81
26. Prasad, A., Chandra, S.: VMFCVD: an optimized framework to combat volumetric DDoS attacks using machine learning. Arab. J. Sci. Eng. **47**, 9965–9983 (2022). https://doi.org/10.1007/s13369-021-06484-9
27. Al-asadi,T.A., Obaid, A.J.: An efficient web usage mining algorithm based on log file data. J. Theoret. Appl. Inf. Technol. 16, **92**(2), 215–224 (2016)

Proposed Multilevel Secret Images-Sharing Scheme

Nahidah T. Darweesh and Ali Makki Sagheer

Abstract Visual secret-sharing (VSS) applications are essential for guaranteeing the secure and private sharing of sensitive information. The Multilevel Secret Image-Sharing Scheme (MLSIS) is introduced in this paper as an effective method for securely sharing secret images among participants with various access levels. The aim was to give thorough conclusions and comparisons while upholding accuracy in how the data was presented. For pixel-level access control and secrecy, the model of the multilevel secret image-sharing offered in this work employs a bit-level technique and the RC4 encryption algorithm to securely share an image through multilevel participants. The model involves three levels, each with an exact number of participants. The primary objective is to distribute shares in a way that permits each level to reconstruct the secret image by combining at least the lowest required number of participants according to their own thresholds (k, n). The Multilevel Secret Image-Sharing Scheme described in this study provides a reliable and effective solution for users with varying levels of access to share confidential secret images. The bit-level approach and potent encryption algorithms work together to provide secrecy and limited access to the shared images. The model's successful use in real-world circumstances is demonstrated by its practical implementations.

Keywords Visual secret sharing · Multilevel secret sharing · Thresholds (k, n) · RC4 algorithm

·

N. T. Darweesh (✉) · A. M. Sagheer
College of Computer Science and Information Technology, University of Anbar, Ramadi, Iraq
e-mail: nah19c1012@uoanbar.edu.iq

A. M. Sagheer
e-mail: ali_makki@uoanbar.edu.iq

G. Fortino et al. (eds.), *Proceedings of Third International Conference on Computing and Communication Networks*, Lecture Notes in Networks and Systems 917,
https://doi.org/10.1007/978-981-97-0892-5_42

1 Introduction

The multilevel Secret Image-Sharing Scheme represents a cryptography approach that makes it possible for multiple individuals to share an image to ensure that it can only be reconstructed whenever a specific number of participants combine their shares [1, 2]. Therefore, this method can be helpful in a wide range of circumstances, for example, exchanging sensitive information among a group of individuals where no single person has full access to the material. Indeed, in recent years, a significant study has already been carried out in this field, including many strategies suggested for improving the effectiveness and safety of these programs. A possible approach is to encrypt the shares utilizing encryption [3], which enables the parties to perform calculations on the encrypted data without sharing the underlying knowledge [4]. This approach makes it more challenging for unauthorized individuals to gain access to the original secret image [5]. While image Reconstruction Flexibility will be the second method, Multilevel visual secret-sharing strategies permit flexible reconstruction of the secret image [6]. By using the single-level scheme, the secret image can only be fully revealed when a specific set of shares is combined [7]. In comparison, multilevel schemes enable different levels of the secret image to be revealed depending on the combination of shares used, and this flexibility can be useful in scenarios where partial access to the secret image is required [8]. The third motivation for utilizing a multilevel visual secret-sharing scheme is the Robustness to Share Loss [9]. Multilevel visual secret-sharing schemes provide increased robustness to share loss or corruption [10]. In this case, if some shares are lost or damaged, it can still be possible to reconstruct a partial or degraded version of the secret image at lower levels [11]; therefore, this property ensures that even if some shares are compromised or unavailable, the secret image can still be partially recovered. Privacy Preservation will be the most important reasons and motivations for utilizing a multilevel visual secret-sharing scheme [12]; therefore, the multilevel visual secret-sharing schemes offer privacy preservation benefits by using distributing the secret image across multiple shares and levels, and therefore, the multilevel visual secret-sharing schemes offer privacy preservation benefits by using distributing the secret image across multiple shares and levels, and also no individual share at the any level can the secret image be revealed that can property protects the privacy and confidentiality of the secret image. The last reason and motivation for utilizing a multilevel visual secret-sharing scheme are Versatility in Application, which can find applications in various fields by using for secure image transmission, authentication, copyright protection, and digital watermarking. Multilevel schemes are suited for a wide range of applications where safeguarding and sharing visual information is critical due to their flexibility and security. Also, when compared to single-level systems, multilevel visual secret-sharing techniques provide better security, adaptability, resilience, privacy protection, and variety. These benefits make them appealing in settings where visual information preservation and dissemination are critical.

2 Related Works

A. Das and Adhikari [13] have provided a multi-secret-sharing method. The related works reviewed here are concerned with various aspects of multilevel secret-sharing schemes and image-sharing schemes, with the goal of improving security and efficiency in a variety of applications. Kumar [14] proposed a multilevel secret-sharing scheme for mobile ad-hoc networks, with the goal of providing high data security in mobile wireless networks. Shamir's threshold scheme is extended in their method to prevent threats and ensure secure data access [4]. Harn and Fuyou [15] based on the Chinese Remainder Theorem presented a multilevel threshold secret-sharing scheme. This scheme provides unconditional security by allowing each shareholder to keep only one private share. Basit et al. [16] proposed a multi-stage multi-secret-sharing scheme with a hierarchical access structure. Their method makes use of Lagrange interpolation polynomials and one-way functions, ensuring reusable shares and the best and perfect scheme. Chen et al. [17] presented an efficient verifiable threshold multi-secret-sharing scheme that allows multiple secrets to be shared in a single sharing process. Their scheme improves computational security and usability. Chattopadhyay et al. [18] proposed a verifiable multi-secret image-sharing scheme based on the XOR operation and hash functions. Their scheme aims to achieve low-cost secret image communication while ensuring security and verifiability. Hu et al. [19] presented a QR code-based threshold meaningful secret image-sharing scheme. Their scheme emphasizes recovered image quality and decoding time while using low computation and lossless recovery. Kang et al. [20] presented a visual cryptographic ghost imaging-based optical multilevel authentication scheme. Their scheme is a novel approach to optical authentication with high robustness, security, and recognition capability. Multilevel image secret-sharing schemes are somewhat cryptographic protocols which have been employed to securely and efficiently send and reconstruct secret images around multiple participants. The points that follow represent a few fundamental ideas and terminology associated with this sort of scheme: Secrets: The confidential information or images to be exchanged among the participants are the secrets $I_1, I_2, \cdots$, In. Participants: P = Pi: $1 \leq i \leq n$ denote the collection of participants. Dealer: A unique person known as Dealer D selects the secrets. We will assume $D / \in P$. It is crucial to understand the fact that the dealer is frequently considered to be honest yet curious, which means that the dealer is going to perform the procedure correctly but might attempt to obtain more information from the shares of the participants. In some circumstances, on the other hand, the dealer could possibly be presumed to be malevolent, necessitating the implementation of extra security measures (such as verified secret sharing) to guard against their assaults. D distributes the secrets in an encoded form known as shares among the participants in P. Let $S = S_1, S_2, \ldots,$ Sn denote the shared set. Combiner: Combiner C is a particular participant (i.e., C P) or a third party entrusted that has responsibility for recovering the secrets. Assume the dealer separated the secret into n shares ($S_1, S_2, \ldots$, sn) and distributed them to n participants. Each participant has a share si, and the combiner must extract the secrets from these shares.

3 Methodology

3.1 Description of the Proposed Model

The multilevel secret image-sharing model based on a bit-level technique and using the RC4 encryption algorithm is a method designed to securely share an image across multiple levels and participants. The model consists of three levels, with each level containing the number of participants. The primary goal is to distribute the shares in a way that each level can reconstruct the secret image by combining at least a minimum number of participants, where each level has its threshold (k, n). The original image is first converted into its binary representation. Each pixel and its associated color values are represented using binary digits. This binary representation will be used for subsequent encryption and sharing steps. Then, the binary image is encrypted using the RC4 encryption algorithm. RC4 is a symmetric stream cipher known for its simplicity and speed. The encryption process ensures the confidentiality of the image data, making it difficult for unauthorized users to access the original content. The encrypted binary image is divided into levels. Each level represents a distinct layer of shares with increasing complexity. The purpose of having multiple levels is to add an extra layer of security and control over the sharing process. For each level, the algorithm generates shares specific to that level. This is achieved through bitwise operations, such as XOR (exclusive OR), performed on the encrypted binary image or its portions; the shares are then distributed among the participants within each level. Each participant receives a share that corresponds to their position within the level. The distribution is done in a controlled manner to ensure that each participant possesses a unique share that contributes to the overall image reconstruction process. Each level is associated with a threshold (k, n), where k represents the minimum number of participants required to reconstruct the secret image at that level, and n is the total number of participants within that level. To reconstruct the secret image at a specific level, the algorithm combines the shares obtained from the minimum required number of participants according to the threshold (k). These combined shares are then processed using the appropriate bitwise operations to obtain the reconstructed image specific to that level. The multilevel secret image-sharing model provides an added layer of security through the use of encryption and multiple levels of sharing. It ensures that the secret image remains protected even if some participants or shares are compromised, as long as the minimum required number of participants per level is available for reconstruction; see Fig. 1.

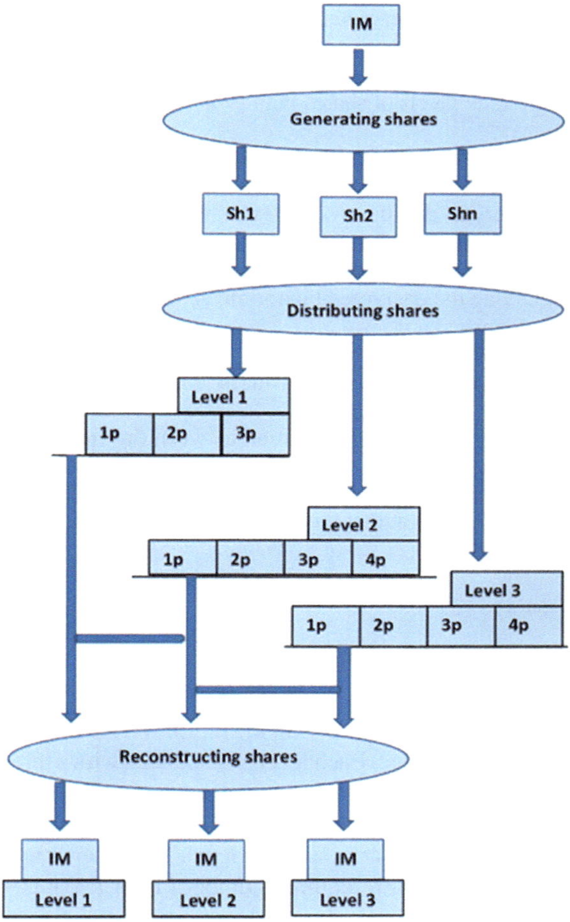

Fig. 1 A multilevel secret image-sharing model

3.2 *Construction Phase*

In this model, the generating Phase involves the following steps: Binary Conversion, RC4 Encryption, Level-wise Sharing, Share Generation, and Share Distribution.

The algorithm for generating shares in a multilevel secret image-sharing model is based on a bit-level technique; see the Algorithm 1.

Algorithm 1: construction Phase
Input: Original image (Im), Number of levels = n , and Number of participants per level = p.
Output: N levels of shares with P share in each level
Steps:
1. Convert the original image (Im) into binary representation (Im_binary).
2. Generate a random secret key (K) for the RC4 encryption algorithm.
3. Encrypt the binary representation of the image (Im_binary) using the RC4 encryption algorithm and the secret key (K). This will produce the encrypted binary image (Encrypted_Im).
4. Divide the encrypted binary image (Encrypted_Im) into N levels, where each level has P participants.
5.For each level from 1 to N:
a. Generate a random binary mask (Mask) of the same length as the encrypted binary image.
b. Calculate the XOR operation between the encrypted binary image (Encrypted_Im) and the mask (Mask) to obtain the level share (Level_share).
c. Divide the level share (Level_share) into P shares, where each share corresponds to a participant at that level.

3.3 Reconstruction Phase

The Reconstruction Phase in the multilevel secret image-sharing model is based on a bit-level technique. Each level is associated with a threshold (k, n), where k represents the minimum number of participants required to reconstruct the secret image at that level, and n is the total number of participants within that level. To reconstruct the secret image at a specific level, the algorithm combines the shares obtained from the minimum required number of participants according to the threshold (k). These combined shares are then processed using the appropriate bitwise operations to obtain the reconstructed image specific to that level. By applying the threshold-based reconstruction process at each level, it becomes possible to sequentially combine the reconstructed images from each level. This leads to the ultimate goal of reconstructing the full secret image. The algorithm for the Reconstruction of secret images in a multilevel secret image-sharing model is based on a bit-level technique, as shown in Algorithm 2.

Algorithm 2: Reconstruction Phase
Input: P Shares from each level
Output: Reconstructed secret image.
Steps:
Initialize an empty array to store the reconstructed shares: Reconstructed_shares.
For each level from 1 to 3:
a. Check the number of available shares for the current level.
b. If the number of available shares is greater than or equal to the threshold (k) for that level:
c. Select the minimum required number of shares based on the threshold (k).
d. Combine the selected shares using the XOR operation to obtain the reconstructed share for that level.
e. Append the reconstructed share to the Reconstructed_shares array.
Reconstruct the secret image by combining the reconstructed shares from each level:
a. Perform the XOR operation on the reconstructed shares from each level to obtain the final reconstructed share.
b. Decrypt the reconstructed share using the RC4 decryption algorithm and the secret key used for encryption.
c. Convert the binary reconstructed image into its original image format.

4 Result Analysis and Discussion

The implementation and assessment of a multilevel secret image-sharing model based on a bit-level method will be the subject presented in this section. The objective is to give consumers user-friendly interfaces for easy accessibility while also analyzing the model's performance. The multilevel secret image-sharing model was implemented by writing apps in the Visual Studio 2022 programming language.

These programs were created to carry out the required tasks on a specific system setup. The SKUSurface_Pro_1796, equipped with an Intel(R) Core(TM) i7-7660U CPU operating at 2.50GHz, was chosen for implementation. The computer contained two cores, This study logical processors, eight gigabytes of RAM, and a 1TB hard drive.

The proposed work in the model has been applied to Lena's image, as can be seen in Fig. 2. The first level was applied to Lina's image, as it was encrypted with the RC4 encryption algorithm, with a 16-byte key, which guarantees us accuracy and speed in the encryption process, and then several shares were generated according to demand and then delivered to the people at the first level. When requesting to retrieve the secret image, the shares are collected from only two people in the first level, according to the request (2, n). The second level was applied to Lina's image, as it was encrypted with the RC4 encryption algorithm, with a 16-byte key, which guarantees accuracy and speed in the encryption process, and then several shares were generated according to demand and then delivered to the people at the second level. When requesting to retrieve the secret image, the shares are collected from two

(a) Lena (b) Bridge (c) Baboon (d) Othman

Fig. 2 The secret images

participants in the first level and three participants in the second level; this enhances the interdependence of the security levels, as well as the control of access to the image, making the multilevel secret sharing safe and effective.

4.1 Qualitative Results Using Histogram

A visual representation of the distribution of pixel intensities in an image is called an image histogram. It offers insightful information about an image's overall contrast and tonal qualities. The range of intensity levels, which for colored images generally ranges from 0 to 255 on the histogram's x-axis. The number of pixels connected to each intensity level is shown on the y-axis. In this study, histogram representations are provided in Fig. 3, which correlate to the colored secret images of Lena, bridge, baboon, and Othman that are depicted in Fig. 4. Each image's frequency distribution of pixel intensities is represented visually through such histograms. By looking at the histograms, one may determine how many pixels are present at different intensities and learn more about the tonal characteristics of the corresponding images.

The histograms in Fig. 5 show that the pixel intensities are random. This randomization is a crucial component of secure image-sharing protocols. It is quite challenging for an attacker to decipher the secret information contained inside the shares by guaranteeing that the pixel intensities in the shared images seem random. The secret information is split up into numerous shares, each seeming as a random image when an image is shared via a secure image-sharing system. Following that, these shares are allocated to the appropriate parties. A minimal number of shares must be combined using a certain algorithm or secret-sharing strategy to recreate the secret image. One may see that there are no obvious patterns or structures when looking at the sharing image histograms. There are no visual cues or correlations that an intrusive party may use because the histograms demonstrate that the pixel intensities are dispersed evenly or randomly throughout the image.

Figure 6 shows the histograms corresponding to the reconstructed images, respectively. We can notice from the figure the clear correspondence between the original

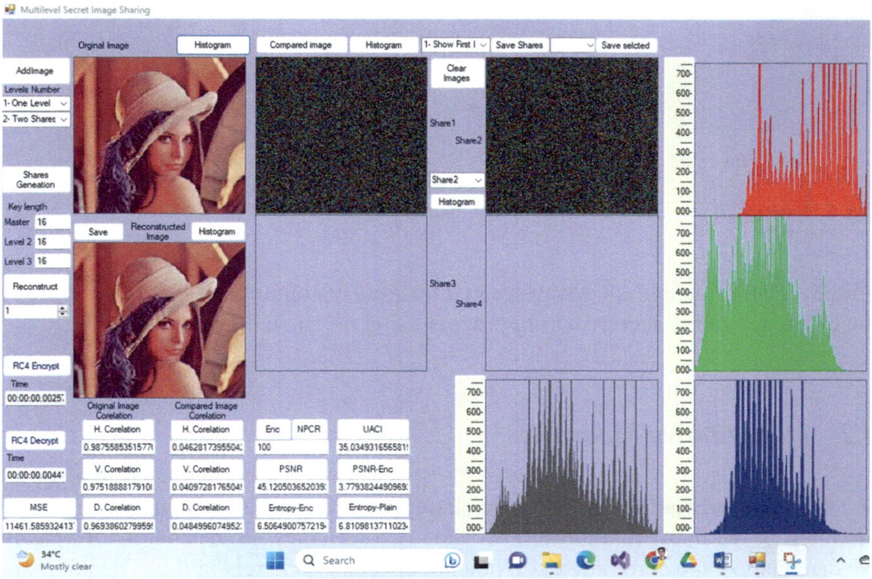

Fig. 3 The proposed model on Lena's image in the first level

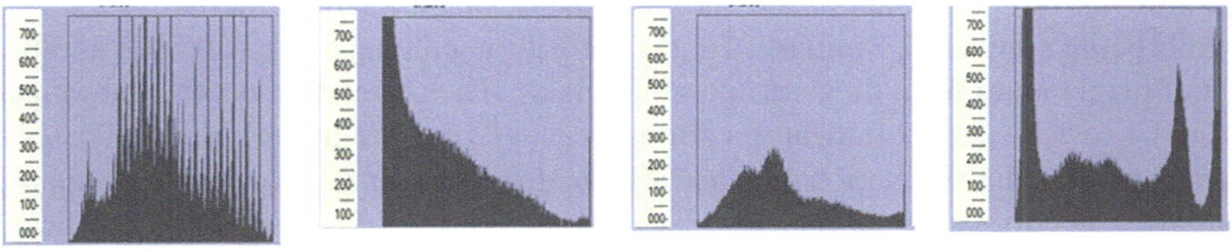

(a) Lena **(b) Bridge** **(c) Baboon** **(d) Othman**

Fig. 4 Histogram of the secret images

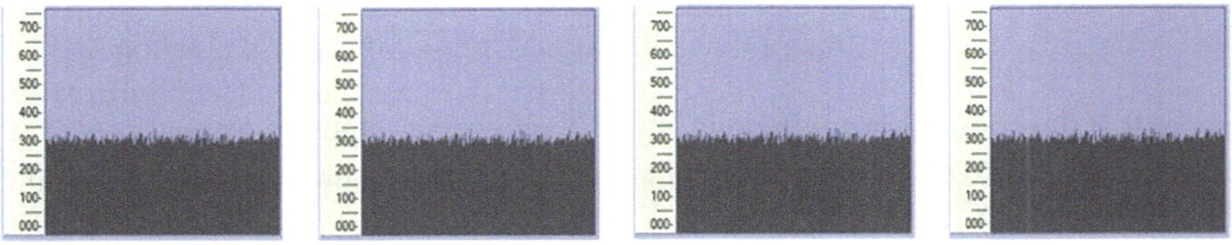

(a) Lena **(b) Bridge** **(c) Baboon** **(d) Othman**

Fig. 5 Histogram plots of the encrypted secret image

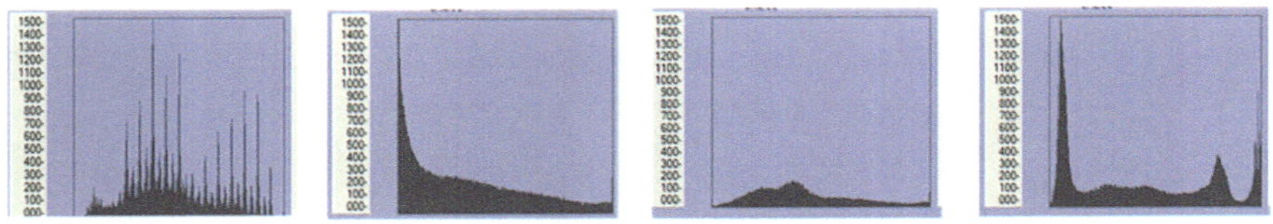

(a) Lena (b) Bridge (c) Baboon (d) Othman

Fig. 6 Histogram plots for the reconstructed images

images' histogram and the reconstructed images' histogram and this is evidence of the efficiency of the secret-sharing process using this model.

4.2 Quantitative Measures

(A) Correlation coefficient: The correlation coefficient measures the strength and direction of the linear relationship between two variables. In the context of images, we can analyze the correlation coefficient between two pixels in the original image and between two corresponding pixels in a reconstructed image. To calculate the correlation coefficient, we need to have two sets of values for each pixel pair. In the case of the original image, each set will consist of the pixel values from the two pixels being compared. Similarly, for the reconstructed image, we will have another set of pixel values from the corresponding pixels. The correlation coefficient ranges from –1 to 1, where –1 indicates a strong negative linear relationship, 1 indicates a strong positive linear relationship, and 0 indicates no linear relationship between the variables. A correlation coefficient value close to 0 suggests that there is no significant linear relationship between the original secret image and the reconstructed image. It can be calculated using the following equation:

$$R = \frac{(\Sigma((\mathrm{X} - \mu\mathrm{X})(Y - \mu Y)))}{(\sqrt{\Sigma(\mathrm{X} - \mu\mathrm{X})}^2 \times \sqrt{\Sigma(Y - \mu Y)}^2)} \tag{1}$$

where R is the correlation coefficient. X and Y are the variables being analyzed (e.g., pixel values of the original secret image and the reconstructed image). μX and μY are the means of X and Y, respectively. Σ represents the sum of all the corresponding data points as shown in Table 1.

As explained the correlation coefficient measures the linear relationship between Correlation coefficient values for secret images and the Correlation coefficient values for the shared images using Eq. (1), so in Table 1 we can see the values of the three dimensions of Correlation (Horizontal, Vertical, Diagonal) for the secret image (Lena), range between (0.98, 0.97, 0.96)respectively, and can see the range of values for other secret images, that indicates a strong positive linear relationship. But we can notice the values of the three dimensions of Correlation for the shares images

Table 1 Correlation coefficient values for secret images and shared images

Secret images	Share images	Horizontal	Vertical	Diagonal
Lena		0.9875	0.9751	0.9693
	Share 1	0.0678	0.0609	0.0701
	Share 2	0.0671	0.0610	0.0706
	Share 3	0.0666	0.0609	0.0698
	Share 4	0.0681	0.0619	0.0718
Bridge		0.9470	0.9321	0.8891
	Share 1	0.0549	0.0516	0.0477
	Share 2	0.0549	0.0502	0.0494
	Share 3	0.0547	0.0502	0.0464
	Share 4	0.0546	0.0470	0.0449
Baboon		0.9955	0.9943	0.9765
	Share 1	0.0808	0.0819	0.0828
	Share 2	0.0809	0.0813	0.0807
	Share 3	0.0807	0.0801	0.0817
	Share 4	0.0826	0.0822	0.0832
Othman		0.9898	0.9970	0.9707
	Share 1	0.0738	0.0354	0.0442
	Share 2	0.0747	0.0360	0.0440
	Share 3	0.0752	0.0358	0.0438
	Share 4	0.0739	0.0366	0.0432

range between (0.05, 0.04, 0.03, 0.08,…….) so that indicates a strong negative linear relationship. Therefore, the proposed model can be successfully implemented and have high-quality results.

(B) Peak signal-to-noise ratio (PSNR): It is a widely used metric to evaluate the quality of an image reconstruction scheme, including the multilevel secret image-sharing scheme. PSNR measures the ratio between the maximum possible power of a signal (in this case, the original secret image) and the power of the noise or distortion introduced during the reconstruction process. The equation to calculate PSNR is as follows:

$$\mathrm{PSNR} = 20 \times \log 10(\mathrm{MAX}) - 10 \times \log 10(\mathrm{MSE}) \tag{2}$$

where M is the height of the image, N is the width of the image, I(x, y) represents the pixel value of the original secret image at position (x, y), I'(x, y) represents the pixel value of the reconstructed image at position (x, y), and MAX is the maximum pixel value. The resulting PSNR value represents the quality of the reconstructed image compared to the original secret image. A higher PSNR value indicates better reconstruction quality and lower distortion or loss of information in a multilevel secret

Table 2 The PSNR between the Encrypted images and the reconstructed images

Encrypted image	PSNR	Reconstructed image	PSNR
Lena	4.3486	Lena	45.1205
Bridge	3.5640	Bridge	45.1205
Baboon	4.2574	Baboon	45.1205
Othman	3.4628	Othman	45.1205

image-sharing scheme. PSNR measures the ratio between the maximum possible power of a signal and the power of the noise or distortion introduced during the reconstruction process, Table 2 explains the PSNR values between the encrypted images and the reconstructed images, we can notice the high difference between the two values, which indicates a high quality of an image reconstruction process in implementing the proposed model. As we can see, the encrypted images range around (3.8 or 4.2) a little more or a little less, while the values for the reconstructed images range equal to (45.1205).

(C) Number of changing pixel rates (NPCR): The NPCR measures the percentage of pixels that have changed between two images. It indicates the diffusion or spread of changes throughout the image. The NPCR is calculated using the following formula:

$$\mathrm{NPCR} = \left(\frac{\mathrm{N_c}}{\mathrm{N}}\right) \times 100 \quad (3)$$

where Nc is the number of pixels that have changed between the two images. N is the total number of pixels in the image. The NPCR value ranges from 0 to 100, with a higher value indicating a greater number of changed pixels and increased diffusion.

As explained, NPCR measures the percentage of pixels that have changed between two images. It indicates the diffusion or spread of changes throughout the image. Table 3 explains the NPCR values between the secret images when encrypted by the key and after changing one bit in the same key, they range about the same values. This high value indicates a greater number of changed pixels and increased diffusion.

Table 3 The NPCR between the secret images when encrypted by the key and after changing one bit in the same key

Secret images	NPCR for Encrypted images by the key	NPCR for Encrypted images after changing one bit in the same key
Lena	99.5941	99.6215
Bridge	99.5971	99.6124
Baboon	99.5888	99.6407
Othman	99.5694	99.6197

Table 4 The UACI between the original secret images and the encrypted images

Secret images	Encrypted images	UACI values
Lena	Lena	33.5251
Bridge	Bridge	33.4667
Baboon	Baboon	33.6961
Othman	Othman	33.4103

(D) Unified Averaged Change Intensity (UACI): The UACI measures the average intensity of the pixel changes between two images. It provides information about the magnitude or strength of the modifications.

The UACI is calculated using the following formula:

$$\mathrm{UACI} = \left(\frac{1}{\mathrm{N} \times \mathrm{L}}\right) \times \Sigma\left(\left|\mathrm{I}_{1(\mathrm{i,j})} - \mathrm{I}_{2(\mathrm{i,j})}\right|\right) \quad (4)$$

where N is the total number of pixels in the image. L is the maximum pixel value.

I1(i, j) and I2(i, j) are the pixel values at position (i, j) in the two images, respectively. Σ represents the summation over all pixel positions (i, j).

Where a higher value of UACI indicates a higher average intensity of the pixel changes. As explained in Table 4, the values of UACI range (33.5) little more or little less, these high values indicate a higher average intensity of the pixel changes. That means the strength of the modifications between the two images.

4.3 *Comparative Analysis*

This study has compared a few similar VSS schemes based on the following parameters:

(A) Computational Complexity

The computational complexity of a secret-sharing scheme relies on various factors such as the number of secrets, number of shares, type of secret (numerical values, image, video, etc.), number of levels, type of operations used in the scheme, etc. As the number of secrets, shares, or levels increase, the complexity of a secret-sharing scheme also increases, causing us to trade with the computational time involved in the computations. Zhang and Gu's scheme [21], Kang et al.'s scheme [20], and Chen et al.'s scheme [17] have a moderate computational complexity, indicating that they may require more computational resources compared to schemes with low complexity. Chattopadhyay et al.'s scheme [18], Hu et al.'s scheme [19], and this study's proposed scheme have low computational complexity, indicating efficient resource utilization and faster execution.

(B) Type of Scheme

It refers to the classification or category of the secret image-sharing scheme being analyzed. The type of scheme can vary based on different factors, such as the underlying cryptographic techniques, the level of security provided, the intended application domain, or the specific features of the scheme. The Multilevel Secret-Sharing paradigm was presented by Deshmukh and colleagues in 2017 [22], alongside the present study, both put in this method the secret image is distributed among participants with various access levels, allowing varying degrees of permitted access to certain image portions. Similar to this, Chen and Wu [23] provide an Efficient Multi-Secret-sharing system designed for the safe exchange of several sensitive components. With the help of this technique, several secret photos or different parts of a single hidden image may be shared. A Threshold Secret-Sharing approach is presented by Hu et al. [19], which necessitates a minimum threshold of shares in order to reconstruct the secret image, and as it guarantees that only authorized participants with the required number of shares may reconstruct the picture, this approach offers a better level of security. Therefore, using various shares or transparencies to encode the secret picture, Zhang and Gu [21] provide a visual secret-sharing system and these shares work together to disclose the hidden image without the need for intricate calculations.

(C) Pixel Expansion

The process of enhancing the resolution of a digital image is known as pixel expansion, which is additionally referred to as image upscaling or image interpolation [24]. The process can frequently be achieved by adding extra pixels to the image in an approach that retains overall visual quality while minimizing detail loss [25]. Therefore, this method is a consequence of the fact that the majority of secret image-sharing approaches break the original image into numerous shares or components, each of which contains extra pixel information [26]. These extra pixels are necessary in order to encode the secret information or to provide redundancy throughout the reconstruction process. Several studies, for example, Kumar et al. and Harn et al. [14] and Sagheer et al. [27] illustrated that both have a modest pixel expansion, implying that their techniques may necessitate a decent amount of additional pixels when compared to the original hidden image. While, Basit et al. [16] have a high pixel expansion, implying that their technique necessitates a large number of extra pixels to encode the hidden information. But, the study of Chen et al. [17] is classified as having modest pixel expansion. On the other hand, this study can be recommended which has no pixel expansion, implying that no more pixels are required in comparison to the original hidden image and also this has the potential to improve storage, transport, and processing efficiency.

(D) Lossless Reconstruction

Studies like Deshmukh et al. [22], Harn et al. [15], Basit et al. [16], and this study's proposed scheme support the lossless reconstruction of the secret image. This means that the original secret image can be completely recovered without any loss of information or degradation in quality during the reconstruction process. Several studies have been showing lossless reconstruction for example Deshmukh et al. [22], Harn et al. [15], Basit et al. [16], and also the suggested approach in this work all support lossless reconstruction of the hidden or secret image. This implies that the original secret image may be fully retrieved without any data loss or quality deterioration throughout the reconstruction process. On the other hand, some of studies for example Kang et al. [20] and Rani et al. [28] have found that a loss or reduction in the quality of the reconstructed image is possible and also does not provide lossless reconstruction. Therefore, to preserve the purity and integrity of the hidden image, it is crucial to use lossless reconstruction. It assures that the shared picture may be accurately recreated without any loss or distortion.

(E) Verifiable Scheme

Among the mentioned studies, only Deshmukh et al. [22] provided a verifiable scheme; this means that their approach allows for verification or validation of the shared secret image by the participants. The verifiability ensures that the shared image has not been tampered with or modified during the sharing process, enhancing the trust and security of the scheme. However, Harn et al. [15], Basit et al. [16], Kang et al. [20], Rani et al. [28], and this study's proposed scheme do not explicitly mention verifiability.

5 Conclusion

The main goal of this study is to distribute shares in such a way that each level can reconstruct the secret image by combining at least the minimum number of participants required by their respective thresholds (k, n). The original image is first converted to binary form, with each pixel and its associated color values represented by binary digits. This binary representation is then encrypted with the RC4 algorithm, which ensures the image data's confidentiality and prevents unauthorized access to the original content. The encrypted binary image is divided into levels, each representing a distinct layer of increasing complexity shares. This multilevel approach to sharing adds an extra layer of security and control to the process. Bitwise operations on the encrypted binary image or its portions generate shares specific to each level, which are distributed among the participants within each level. Each participant receives a distinct as well as unique share depending on their position throughout the level thereby helping in the reconstruction of the entire image. In addition, each level contains a threshold (k, n), where k is the minimum number of reconstruction participants necessary and n is the total number of participants at that level to replicate the hidden image at a certain level. Finally, the Multilevel Secret Image-Sharing Scheme proposed in this paper offers a dependable and effective way for users with varied degrees of access to communicate private secret photographs. Bit-level techniques,

XOR operations, and powerful encryption algorithms collaborate to guarantee confidentiality and limited access to shared images. The model's success in real-world applications is shown through practical implementations.

References

1. Xing, F., Yan, X., Yu, L., et al.: A novel general (n, n)-threshold multiple secret images sharing scheme based on information hiding in the sharing domain. Entropy **24**(3), 318 (2022)
2. Mohammed, V.A., Samundiswary, P.: An XOR-based Multiple Medical Image Sharing using Visual Secret Sharing, pp. 793–799
3. Gutub, A., Al-Ghamdi, M.: Hiding shares by multimedia image steganography for optimized counting-based secret sharing. Multimedia Tools Appl. **79**(11–12), 7951–7985 (2020)
4. Dhinakaran, D., Prathap, P.: Preserving data confidentiality in association rule mining using data share allocator algorithm (2023). arXiv preprint arXiv:2304.14605
5. Kumar, M., Soni, A., Shekhawat, A.R.S., et al.: Enhanced digital image and text data security using hybrid model of LSB steganography and AES cryptography technique, pp. 1453–1457
6. Georgi, P., Wei, Q., Sain, B., et al.: Optical secret sharing with cascaded metasurface holography. Sci. Adv. **7**(16), eabf9718
7. Galletta, A., Taheri, J., Celesti, A., et al.: Investigating the applicability of nested secret share for drone fleet photo storage. IEEE Trans. Mobile Comput. (2023)
8. Veitch, S.: Contextualizing Alternative Models of Secret Sharing. University of Waterloo (2022)
9. Meikap, S., Jana, B.: Reference pixel-based reversible data hiding scheme using multi-level pixel value ordering. Multimedia Tools Appl. 1–34 (2023)
10. Kunhoth, J., Subramanian, N., Al-Maadeed, S., et al.: Video steganography: recent advances and challenges. Multimedia Tools Appl. 1–43 (2023)
11. Kumar, A., Rani, R., Singh, S.: A survey of recent advances in image steganography. Secur. Privacy **6**(3), e281 (2023)
12. Sajitha, A., Sridevi Sathya Priya, S.: Analysis of various visual cryptographic techniques and their issues based on optimization algorithms. Int. J. Image Graph. 2350059 (2022)
13. Das, A., Adhikari, A.: An efficient multi-use multi-secret sharing scheme based on hash function. Appl. Math. Lett. **23**(9), 993–996 (2010)
14. Kumar, P.S., Kurra, R.R., Tentu, A.N., et al.: Multi-level secret sharing scheme for mobile ad-hoc networks. Int. J. Adv. Netw. Appl. **6**(2), 2253 (2014)
15. Harn, L., Fuyou, M.: Multilevel threshold secret sharing based on the Chinese Remainder Theorem. Inf. Process. Lett. **114**(9), 504–509 (2014)
16. Basit, A., Kumar, N.C., Venkaiah, V.C., et al.: Multi-stage multi-secret sharing scheme for hierarchical access structure, 557–563
17. Chen, D., Lu, W., Xing, W., et al.: An efficient verifiable threshold multi-secret sharing scheme with different stages. IEEE Access **7**, 107104–107110 (2019)
18. Chattopadhyay, A.K., Nag, A., Singh, J.P., et al.: A verifiable multi-secret image sharing scheme using XOR operation and hash function. Multimedia Tools Appl **80**, 35051–35080 (2021)
19. Hu, F., Yao, Y., Li, W., et al.: Threshold meaningful secret image sharing scheme based on QR code. Secur. Commun. Netw. **2022** (2022)
20. Kang, Y., Kanwal, S., Pu, S., et al.: Ghost imaging-based optical multilevel authentication scheme using visual cryptography. Optics Commun. **526**, 128896 (2023)
21. Zhang, D., Gu, Z.: A high-quality authenticatable visual secret sharing scheme using SGX. Wirel. Commun. Mob. Comput. **2021**, 1–12 (2021)
22. Deshmukh, M., Nain, N., Ahmed, M.: A novel approach for sharing multiple color images by employing Chinese Remainder Theorem. J. Vis. Commun. Image Represent. **49**, 291–302 (2017)

23. Chen, T.-H., Wu, C.-S.: Efficient multi-secret image sharing based on Boolean operations. Signal Process. **91**(1), 90–97 (2011)
24. Shi, W., Caballero, J., Huszár, F., et al.: Real-time single image and video super-resolution using an efficient sub-pixel convolutional neural network, 1874–1883
25. Sajjadi, M.S., Scholkopf, B., Hirsch, M.: Enhancenet: Single image super-resolution through automated texture synthesis, 4491–4500
26. AlKhodaidi, T., Gutub, A.: Refining image steganography distribution for higher security multimedia counting-based secret-sharing. Multimedia Tools Appl. **80**, 1143–1173 (2021)
27. Sagheer, A.M., Abed, L.H.: Visual secret sharing without pixel expansion. Int. J. Digital Crime Forens. (IJDCF) **7**(2), 20–30 (2015)
28. Rani, M.M.S., Lakshmanan, S., Saranya, P.: Video Steganography Using Mid-Prime and Discrete Wavelet Technique. Int. J. Comput. Engin. Appl. 180–190 (2017)

Image Captioning System for Movie Subtitling Using Neural Networks and LSTM

K. Vijay, Eashaan Manohar, B. Saiganesh, S. Sanjai, and S. R. Deepak

Abstract With the advent of the Internet, the multimedia business has experienced explosive growth and is now available to consumers worldwide. Particularly important to this growth has been the dominance of over-the-top (OTT) platforms during the COVID-19 epidemic. One important component in this growth has been the implementation of state-of-the-art Machine Learning (ML) methods. These algorithms can generate captions from video frames mechanically, increasing the platform's accessibility. However, it can be difficult to meet the needs of users speaking different languages, as there are many films being streamed in languages that may be inaccessible to consumers in other parts of the world. There is an issue with English being the most spoken language in the world. Therefore, individuals who aren't fluent in English may have trouble finding videos in their native language. By automatically creating English subtitles for any movie, regardless of the language spoken on the original audio track, machine learning plays a significant role in removing this barrier. Neural Networks, used for processing each frame of the video, and LSTM models, used for caption synthesis, are the key models employed here from the realm of machine learning. After the models have been trained, they can be incorporated into the user interface (UI) using a programming language like Python. In the user interface, the created caption can be shown alongside the uploaded image.

Keywords OTT · RCNN · LSTM · RNN first section

K. Vijay · E. Manohar (✉) · S. Sanjai · S. R. Deepak
Department of Computer Science and Engineering, Rajalakshmi Engineering College, Chennai, Tamilnadu, India
e-mail: eashaanmano@gmail.com

K. Vijay
e-mail: vijayk.btech@gmail.com

B. Saiganesh
Department of Electronics and Communication Engineering, Rajalakshmi Engineering College, Chennai, Tamilnadu, India

G. Fortino et al. (eds.), *Proceedings of Third International Conference on Computing and Communication Networks*, Lecture Notes in Networks and Systems 917,
https://doi.org/10.1007/978-981-97-0892-5_43

1 Introduction

English photo caption generation is challenging because it requires a blend of computer vision and natural language processing techniques. Using Convolutional Neural Networks (CNNs) and LSTMs (Long Short-Term Memory), this Python-based project aims to develop a caption generator. To feed characteristics into the LSTM model responsible for caption synthesis, we will employ Exception, a CNN model trained on ImageNet. When it comes to processing data in the form of 2D matrices, deep neural networks, of which convolutional neural networks are a subset, really shine. Pictures are a simple and effective medium. Image modifications include but are not limited to translation, rotation, scaling, and perspective correction. It has been discovered that recurrent neural networks (RNNs) with long short-term memory (LSTMs) are the most effective tools for tackling sequence prediction problems. We can guess the next word by analysing its context. Information that isn't being actively used can be filtered out by LSTM using a forget gate [11, 13].

The main objective is to be able to handle a wide range of images, including those that are complex and abstract, and generate captions that are grammatically correct, semantically meaningful, and culturally appropriate. It is helpful in many places like image-based search engines, content creation, searching an image, advertising, e-commerce and also it is helpful in automatically generating descriptive captions that accurately reflect the content and context of the image, and to aid in the education sector, where the frames in the video depict what is present in the video. Through this, the children will learn some new words with the things presented in the image. The objective of the system is to provide an efficient and accurate method for generating captions for movie scenes that can be used for subtitling [5]. The system aims to improve the accessibility of movies for people with hearing impairments by providing accurate and comprehensive captions that describe the visual content of the scene. Long Short-Term Memory (LSTM) [14] is a neural network architecture used by the system to learn the connections between the images' visual features and their descriptions, allowing the system to achieve these goals. Captions that appropriately describe the scene are generated using a variety of computer vision techniques that are used by the system to detect objects in the image and segment it. Overall, the "Image Captioning System for Movie Subtitling using Neural Networks and LSTM" aims to provide an efficient and accurate method for generating captions for movie scenes that can improve the accessibility of movies for people with hearing impairments. This will reduce the burden on manual work and ease out the post-production process [16–19].

2 Literature Survey

Reference [1] the basic objectives of this paper are to (1) describe the most widely used strategies for picture captioning; (2) review the findings of recent papers; and (3) suggest avenues for future study. By showing anomalies in comparing photo captioning results gathered during the study, this work draws attention to the issue of insufficient data collection. Therefore, it is essential to compare the performance of a novel model to both industry norms and state-of-the-art methods.

A picture's "caption" is the written description that accompanies it. Captioning an image correctly requires highlighting key details, attributes, and associations. It must also produce coherent sentences that make sense grammatically and semantically [3]. Recent deep learning (DL) models have increased in complexity and computing expense because of how challenging the task of picture captioning has become.

To improve the captions, they propose employing a method called Attribute Interaction-Tensor Product Representation (aiTPR) [4], which employs an orthogonal combination to gather extra data and learn the interactions as material objects (tensors). In this study, we integrate convolutional neural networks (CNNs) into the SE network to enhance its feature extraction and HSI classification skills. Using the squeeze and excite block can improve a CNN's representational accuracy. To test the efficacy of the proposed model, experiments were run on the Salinas, Pavia Centre, and Pavia University datasets [6, 7].

This study discusses the authors' use of reinforcement learning to optimize image captioning systems, and how, by using the test metrics provided by the MSCOCO assignment, they were able to show significant performance gains. Self-critical sequence training (SCST) is a new optimisation technique we developed to build our systems [7]. Based on the output of a random word generator, the proposed method enhances current visual qualities. Bag-LSTM beats competing models by modifying the parameters of convolution neural networks to learn visual features associated with text through feedback propagation. Our improved model [12] uses additional text-conditioned image features on the fly as guide information in place of the explicit injection of picture characteristics into each unit of an LSTM block, as is the case with classic guidance-LSTM techniques. For the caption generator, we employ a bidirectional gLSTM, which can remember associations between visual elements and semantic information from both the past and the future [8]. We improve the precision and variety of the generated phrases by combining the endogenous attention, which displays stochastic visual orienting, with the exogenous attention, which displays deterministic visual orienting. The model includes an LSTM + MLP mimic constraint network and adaptive syncretic attention (ASA) [9, 10]. After the image is segmented into many patches, MMDF-LDA [15] estimates the probability that an annotation will be assigned to a patch in advance.

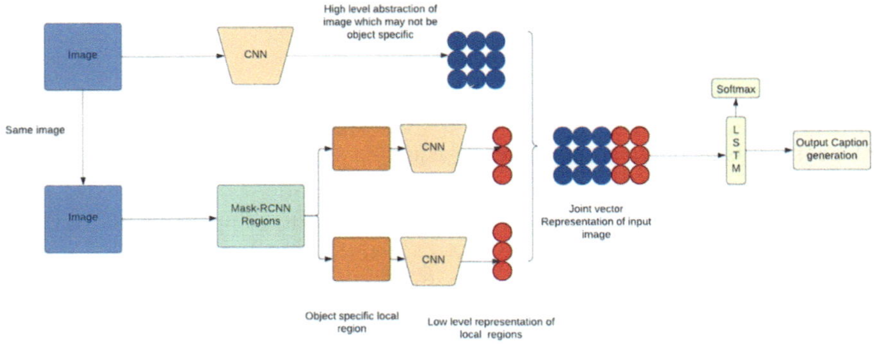

Fig. 1 Architecture diagram

3 System Architecture

Figure 1 is the architecture diagram for the image captioning system for movie subtitling using neural networks and LSTM. The above system works by displaying a UI where the user can upload the frames for which they can derive captions. Once the user uploads the frame and clicks on generate, the Neural Network and LSTM models start to run and are trained using the Flickr 8k dataset, and once the caption is obtained, it is displayed to the user in the form of a text pop-up, making it easier for the user to comprehend. There are three layers available that are presentation layer, business layer, and data layer.

4 System Implementation

4.1 Dataset Collection

A dataset consists of multiple independent datasets that have common characteristics. However, they can be manipulated by a computer as a whole. When describing unknown quantities like an object's size, mass, temperature, volume, etc., or the values of random numbers, datasets describe values for each variable. A datum is the collection of numbers contained herein. A datum is the collection of numbers contained herein. Each row in the dataset represents the information of one or more members.

Figure 2 is the image dataset used in the image captioning system for movie subtitling using neural networks and LSTM. Commonly used in natural language processing and computer vision research for picture captioning tasks, the Flickr 8k dataset consists of 8,000 photographs associated with five textual descriptions. The dataset is now used as a standard for image captioning systems and was compiled by

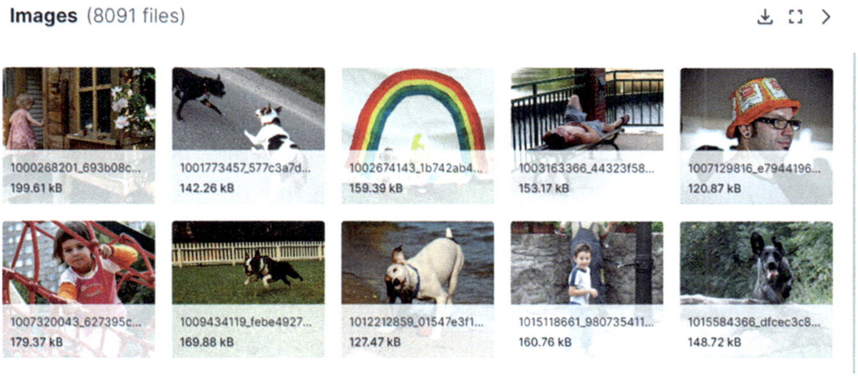

Fig. 2 Dataset

extracting images and descriptions from the popular photo-sharing website Flickr. It is commonly used to train and evaluate the performance of deep learning models like convolutional neural networks (CNNs) and recurrent neural networks (RNNs) to generate captions that accurately capture the visual content of a picture.

4.2 Data Preprocessing

Data preprocessing is a method for cleaning up unstructured data. In other words, the data is always collected in a raw format which makes further analysis impossible.

```
def encode(image):
    image=preprocess_img(image)
    vec=vgg_model.predict(image)
    vec=np.reshape(vec.shape[1]))
    return vec
pickle_in=open("wordtoix.pkl","rb")
wordtoix=load(pickle_in)
pickle_in=open("ixtoword.pkl","rb")
ixtoword=load(pickle_in)
max_length=74
```

Data preparation is especially important for image datasets since the performance of machine learning (ML) models is highly dependent on the accuracy and consistency of the training and evaluation data. The above code shows the data preprocessing steps done in the image captioning system for movie subtitling using neural networks and LSTM. Image data can be complex and diverse, with variations in lighting, colour, size, and orientation. Preprocessing techniques help to standardize the data and extract relevant features, which improves the accuracy and efficiency of the ML model. Some common preprocessing techniques for image datasets include resizing and rescaling images to a consistent size, normalizing pixel values to a common range, applying filters to enhance image quality or remove noise, and augmenting the dataset with additional images or artificially generated data.

Figure 3 is the input screen of the image captioning system for movie subtitling using neural networks and LSTM. It shows that the user uploads image data from the dataset using the Upload frames button.

Figure 4 is the input given by the user; here input given is an image that shows two dogs playing in the garden and it was recognized by the image captioning system. Figure 5 is the output screen which shows the output by generating a model using CNN for processing the image frame that was uploaded and generates a caption by using the LSTM algorithm, processes those frames of the image, and matches the appropriate root word for the object in the frame using the LSTM algorithm.

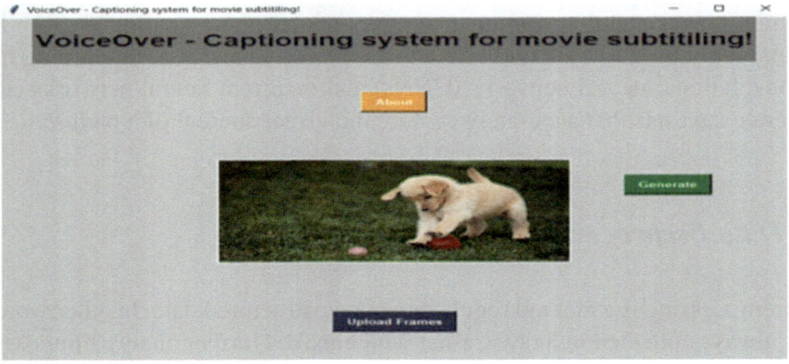

Fig. 3 Input screen

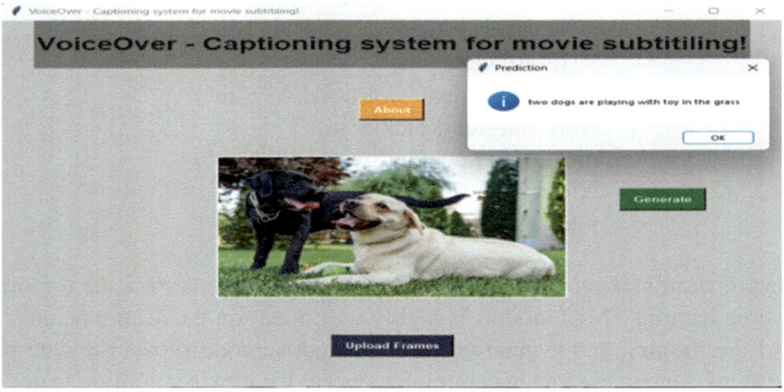

Fig. 4 Sample input

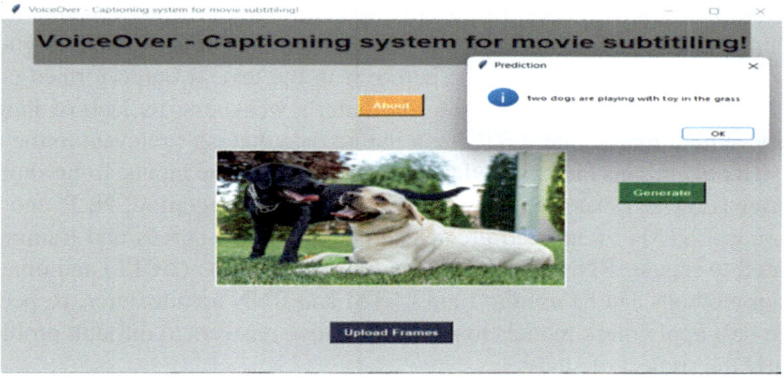

Fig. 5 Sample output

Image captioning using RNN and LSTM is a machine learning system that can generate descriptive captions for images. To learn the connection between visual attributes and their related linguistic descriptions, the RNN and LSTM models are trained on a huge dataset of image-caption pairs. A GUI like Tkinter can be incorporated into the system to make picture captioning easier for the end user. Through the user interface, users can upload an image to the system, and the trained RNN and LSTM models will provide a caption. The generated caption can be shown in the user interface with the uploaded image. The RNN and LSTM models are trained using methods like backpropagation through time (BPTT) and optimisation algorithms on a huge dataset of image-caption pairs before being used in the system's implementation.

5 Conclusion

Image caption generator is a problem in computer vision and natural language processing where the goal is to determine the context of an image and describe it in a natural language such as English. Therefore, we have used these frameworks to create a captioning model for images. Another term for this model is the Convolutional Neural Network Recurrent Network. Image feature extraction is a common use of CNN. The "pre-trained model" vulnerability will be exploited. The LSTM will use the CNN data to describe the image. One subset of deep neural networks, called convolutional neural networks, excels in handling data presented in a matrix-like 2D format. Due to the ease with which images may be represented as a 2D matrix, CNN is an effective tool for image manipulation. The primary application of CNN is in picture categorization, such as determining whether or not an image depicts a bird, plane, Superman, etc. Images are analysed in both horizontal and vertical planes so that relevant features can be extracted and combined for classification. Image transformations such as translation, rotation, scaling, and perspective correction are all supported. The acronym LSTM refers to a specific kind of recurrent neural network (RNN) that excels at solving sequence prediction issues. It is possible to guess the next word based on what has come before it in the text. It outperformed conventional RNN by effectively addressing the latter's weakness, its lack of long-term memory. With a forget gate, LSTM can ignore data that isn't relevant to the task at hand, so it only retains the knowledge it needs to analyse the inputs. Long short-term memories (LSTMs) excel at tasks including speech recognition, NLP, and image captioning. LSTMs often need greater computational resources and training time compared to regular RNNs. Backpropagation through time (BPTT) and other optimisation methods can be used to train LSTM and RNN architectures, respectively. How to best apply these models to provide creative answers to difficult problems is presently the subject of study.

References

1. Staniūtė, R., Šešok, D.: A systematic literature review on image captioning. Appl. Sci. **9**(10), 2024 (2019)
2. Hossain, M.Z., Sohel, F., Shiratuddin, M.F., Laga, H.: A comprehensive survey of deep learning for image captioning. ACM Comput. Surv. **51**(6), Article 118, 36 (2019). https://doi.org/10.1145/3295748
3. Omri, M., Abdel-Khalek, S., Khalil, E.M., Bouslimi, J., Joshi, G.P.: Modeling of hyperparameter tuned deep learning model for automated image captioning. Mathematics, MDPI **10**(3), 1–20 (2022)
4. Sur, C.: AiTPR: attribute interaction-tensor product representation for image caption. Neural. Process. Lett. **53**(2), 1229–1251 (2021)
5. Babu, R., Kannappan, J., Krishna, B.V., Vijay, K.: An efficient spam detector model for accurate categorization of spam tweets using quantum chaotic optimization-based stacked recurrent network. Nonlinear Dyn. **111**(19), 18523–18540. https://doi.org/10.1007/s11071-023-08697-z
6. Latha, G.C.P., Sridhar, S., Prithi, S., Anitha, T.: Cardio-vascular disease classification using stacked segmentation model and convolutional neural networks. J. Cardiovasc. Disease Res. **11**(4), 26–31 (2020)
7. Rennie, S.J., Marcheret, E., Mroueh, Y., Ross, J., Goel; V.: Proceedings of the IEEE Conference on Computer Vision and Pattern Recognition (CVPR), 2017, pp. 7008–7024 (2017)
8. Cao, P., Yang, Z., Sun, L., Liang, Y., Yang, M.Q., Guan, R.: Image captioning with bidirectional semantic attention-based guiding of long short-term memory. Neural. Process. Lett. **50**(1), 103–119 (2019)
9. Wang, T., Hu, H., He, C.: Image caption with endogenous–exogenous attention. Neural. Process. Lett. **50**(1), 431–443 (2019)
10. Yang, L., Hu, H.: Adaptive syncretic attention for constrained image captioning. Neural. Process. Lett. **50**(1), 549–564 (2019)
11. Anusha, S., Elakkiya, N., Vijayakumar, R.: Separable Reversible Data Hiding in Encrypted Image Using Dual Data Embedding with Histogram Shifting (2020)
12. Lu, J., Yang, J., Batra, D., Parikh, D.: Neural baby talk. In: Proceedings of the IEEE Conference on Computer Vision and Pattern Recognition, pp. 7219–7228 (2018)
13. Yao, T., Pan, Y., Li, Y., Mei, T.: Exploring visual relationship for image captioning. In: Proceedings of the European Conference on Computer Vision (ECCV), pp. 684–699 (2018)
14. Adhikari, A., Ghimire, S.: Nepali Image Captioning Artificial Intelligence for Transforming Business and Society (AITB) (2019)

15. Zheng, L., Caiming, Z., Caixian, C.: MMDF-LDA: an improved multi-modal latent Dirichlet allocation model for social image annotation. Expert Syst. Appl. **104**, 168–184 29. Sur, C. (2021). aiTPR: attribute interaction-tensor product representation (2018)
16. Onan, M., Shillah, A.: Enhancing Automatic Image Captioning System LSTM (2023)
17. Kundu, R., Singh, S., Amali, G., Noel, M.M., Umadevi, K.S.: Automatic image captioning using different variants of the long short-term memory (LSTM) deep learning model. In: Deep Learning Research Applications for Natural Language Processing, pp. 132–155. IGI Global (2023)
18. Jia, J., Ding, X., Pang, S., Gao, X., Xin, X., Hu, R., Nie, J.: Image captioning based on scene graphs: a survey. Expert Syst. Appl. 120698 (2023)
19. Ansari, S.A., Zafar, A.: (2023) A comprehensive study on video captioning techniques, benchmark datasets and QoS metrics. In: 2023 10th International Conference on Computing for Sustainable Global Development (INDIACom), pp. 1598–1603. IEEE.
20. AL-Sammarraie, Y.Q., Khaled, A.Q., AL-Mousa, M.R., Desouky, S.F.: Image captions and hashtags generation using deep learning approach. In: 2022 International Engineering Conference on Electrical, Energy, and Artificial Intelligence (EICEEAI), pp. 1–5. IEEE (2022)

Semantic Application Based on the Bhagavad Gita: A Deep Learning Approach

Anand Chauhan, Vasu Jain, Mohd. Mohsin, Manish Raj, Umesh Gupta, and Sudhanshu Gupta

Abstract Recent evolutions in Natural Language Processing involve techniques to convert text into meaningful high-dimensional numerical vector representations that can be used for various applications such as sentiment analysis. However, these vector representations can be computationally heavy and challenging to interpret. This paper proposes using autoencoders with cosine similarity as their loss function to generate encoded vectors that have intentionally discarded some semantic information. After the test, several autoencoder models like this were created to generate advice and question-answering trained on the entirety of the Hindu holy book of Bhagavad Gita by comparing it with the same algorithm working on the original vectors and concluding subjectively that the quality of the output has improved, while significantly reducing the time taken to generate an output. The findings suggest that intentional data loss through autoencoders is a promising technique for dimensionality reduction on word embeddings. The neural network was trained on the entire dataset using cosine similarity as the loss function. The work could explore the potential of this approach in other areas of natural language processing and develop more objective measures for evaluating its effectiveness.

Keywords Artificial intelligence · Autoencoders · Cosine similarity · Natural language processing · Question-answering · Semantic search · Word2Vec

1 Introduction

Natural Language Processing (NLP) is a rapidly growing and evolving field that involves developing techniques for analyzing and understanding textual data. One of the most popular techniques recently involved representing text as vectors (generating

A. Chauhan · V. Jain · Mohd. Mohsin · M. Raj · U. Gupta (✉) · S. Gupta
1SCSET, Bennett University, Greater Noida, India
e-mail: er.umeshgupta@gmail.com

M. Raj
e-mail: manish.raj@bennett.edu.in

G. Fortino et al. (eds.), *Proceedings of Third International Conference on Computing and Communication Networks*, Lecture Notes in Networks and Systems 917,
https://doi.org/10.1007/978-981-97-0892-5_44

embeddings). This can be done in a word-by-word scale, sentence-by-sentence scale, or even on an entire document or a paragraph. The unit value in each process, i.e., a vector, is known as a 'token'. These vectors are used for various tasks like sentiment analysis and topic modeling. However, these vector representations can be very high-dimensional and challenging to work with and interpret. As such, there is growing research into how to effectively reduce the dimensionality of vector representations of tokens and what impact they can have on the underlying semantics and meaning of the tokens in question. In this paper, the authors introduce lossy autoencoders to reduce the dimensionality of word embeddings for abstractive semantic applications. This is done using convolutional and straightforward neural network autoencoders while comparing the cosine similarity of the reconstructed vectors as both a testing metric and the training loss function.

Previous research into effective dimensionality reduction includes a study into how dimensionality reduction techniques affect sentiment analysis tasks [1] and an overview of traditional dimensionality reduction techniques and their effectiveness [2]. However, there have been no studies at the time of writing that explore how data loss using these methods might benefit the sentimental analysis of limited or vague data. The kind of data that is not to be understood based on the meaning of the words and the narratives alone, but an abstraction, applicable as either advice, caution, or just a sentimental essence. Autoencoders have also outperformed Principal Component Analysis (PCA) [3], so a study exploring how autoencoding with a maintained, controlled, and intentional data loss could amplify this effect. This paper started as a project that aims to use Artificial Intelligence (AI) to help people by solving their problems based on a text corpus comprised of various religious texts [4]. Since our learning is self-supervised, the goal of the final project that the model will be used in is to compare text input to our vector space to find the most semantically similar phrase from the document that the model is trained on [5]. They rely on manual inspection and comparison of the output, and the metrics displayed will be there only for the experimentation results and data transparency [6]. The paper's contribution is the first study that proposes using autoencoders and intentional data loss to reduce dimensionality by encoding abstractions of word embeddings [7]. The proposed approach has potential applications in various areas, such as document clustering and semantic similarity analysis [8]. This paper contributes to the ongoing efforts to develop effective methods for dimensionality reduction in NLP and hopes to provide insights for future research in topic modeling [9] and text abstraction and summarization of meaning [10]. Like Supervised learning work for the classification of labeled classes [11]. It helps in making deep learning models [12]. The primary purpose of supervised learning is human development, such as medical diagnosis, which is necessary for unsupervised learning [13]. For the latest, one can follow [14–17].

This paper will follow the outline below. The experimental findings and outcomes are discussed in Sect. 3, whereas Sect. 2 discusses data gathering and neural network architecture. The recommended work, including a conclusion, may be found in Sect. 4. The following section explores the dataset used and the methodology employed in detail. Section 3 examines the Experimentation results and our interpretation of the same, along with the output of a sample of where the model will be implemented along with the output from a similar model working on the original 300-dimensional embeddings produced by Spacy's en_core_web_lg model. Finally, discuss the conclusions, limitations of the approach to the knowledge, and prospects for future work into the idea in Sect. 4.

2 Materials and Proposed Methodology

2.1 Dataset Description

The Bhagavad Gita dataset [4] is a comprehensive and meticulously compiled collection of 700 verses from 18 chapters of the sacred Hindu text. The available text has been translated into Sanskrit, Hindi, and English. Sanskrit and Hindi 'Anuvads' (translations) have been left out to start the research with easy-to-process languages easy-to-process languages, Sanskrit and Hindi 'Anuvads' (translations) have been left out. Each verse in the dataset has been represented as an input vector, utilizing NLP techniques to capture its contextual and semantic nuances. This allows efficient mapping to output vectors, which correspond to the most relevant, determined AI model. While the Bhagavad Gita dataset is extensive, it does not include a format suitable for question-answering AI models; therefore, the approach to unsupervised learning is limited. As a result, the research had to rely on unsupervised learning techniques to identify patterns and relationships within the verses. While more challenging, this approach has led to innovative solutions that can effectively map input vectors to the most contextually relevant output vectors. The key emphasis in the approach and a flaw in the available data is that even though this is a question-answering task at its core, there are no direct answers to the questions that could be formulated in the text. This makes any output of the model up to subjective interpretation, as it is currently impossible to objectively judge answers to more abstract questions. That, however, is the primary application of the project that this model and approach are a part of, and it also is the purpose of this research—to gauge whether AI can better understand and reference abstract subject matter when a certain amount of data loss is intentionally added. Until a method to quantify conceptual and ethical question-answering is devised, the results of this study are mainly left to the reader's interpretation for the quality of output.

2.2 Proposed Methodology

2.2.1 Customer Autoencoders

The aim of the part of the project that is the subject of this paper was to find out whether a verse from the Bhagavad Gita can be used to address and provide advice on specific life problems that individuals may have and if an autoencoding machine learning model could accomplish this process. The input for the model would be any text prompt, be it a specific problem, a general question, or a vague conversational hint, and the goal is to create a solution that would generate text along with a relevant verse addressing the input prompt. The approach to using autoencoders, specifically those trained on word embeddings to a certain amount of data loss, was devised based on a study showing that autoencoders sizably outperform traditional PCA methods for dimensionality reduction for semantic-based tasks.

The inspiration behind the desire to maintain a certain amount of dissimilarity between the input and output vectors is that the predictions need higher-level abstractions of the verses to provide a higher level of wisdom and motivation that these abstractions would provide from the texts, as required by the program. Creating an autoencoder that loses some information but retains a significant amount would serve this purpose well. The higher-level wisdom-based insight or meaning could also be achieved this way, but further development and preprocessing would be required to ensure that that is definitively what happened, which happens to be the most significant limitation of the approach so far; they have no way of knowing the exact information that was lost.

The type or quality of information lost can be influenced heavily by the model employed to generate the word embeddings. For this application, Spacy's en_core_web_lg model is used, which returns a 300-dimensional vector for every token, including stop words. Different models and methods will output wildly varying word embeddings, and comparing how the approach works with those is a good prospect for this approach. The standard spacy vector also contains a mean pooled vector of the entire sentence/doc, which is used along with two other methods: Max pooling, the padding of every word embedding into a 256-length token sequence. Three tensors were created, the first of shape (700, 300) (T1), containing the vector with the most significant normal from every verse (or every row of the dataset), the second also of shape (700, 300) (T2), containing the vector with the mean of all vectors from every verse, and lastly, one of shape (700, 256, 300) containing all the word embeddings of every verse padded up with zero vectors up to 256 vectors per verse (T3). This was done for two reasons: first, to ensure that the final autoencoder can handle queries of up to 256 tokens, and second, to make all sequence lengths uniform since the ragged tensors that are generated by using variable length token sequences do not work well with convolutional autoencoders. A final tensor of shape (700, 256, 8) was also created out of T3 by finding the eight highest magnitude vectors from every 700 entries in the first dimension of T3. Figure 1 provides a flowchart to make this process easier to understand.

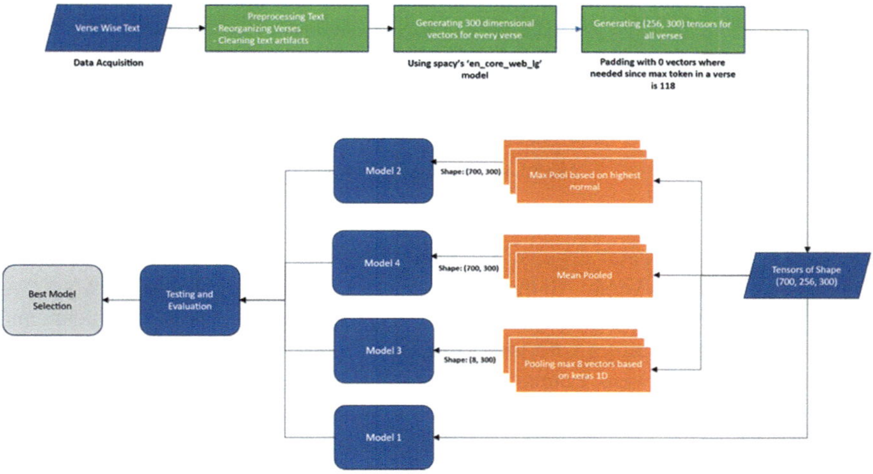

Fig. 1 Block diagram explaining the experimental methodology

As stated previously, the loss function for all these models was cosine proximity (the negative of cosine similarity to maintain gradient descent is minimized. The paper refers to cosine proximity and similarity interchangeably in this document). This loss function was chosen to promote semantic similarity over pure document vector length since it found that Mean Squared Error as a loss metric resulted in wildly semantically dissimilar vectors in all the cases. Hence, cosine similarity was chosen as a loss function, calculated as illustrated in (1).

$$\text{Loss} = -\text{sum}(\text{l2\_norm}(\text{y\_true}) * \text{l2\_norm}(\text{y\_pred})) \tag{1}$$

where y_true is the actual value of what the output should be, y_pred is the model's output, and l2_norm denotes. This loss generates a value between -1 and 1, where negative values indicate greater similarity between the vectors. A value of -1 denotes a perfect match, 0 denotes that the vectors are orthogonal, and one denotes that they are entirely dissimilar (or at an angle of 180°) [6]. The activation functions used for every layer are exclusively comprised of ReLU. The output of the ReLU 'y' for any input value 'x' is defined as illustrated in (2).

$$\text{y} = \max\ (0,\ \text{x}) \tag{2}$$

The optimizer was set to Keras' Adam. Four different models were initially created for this study: One with a convolutional autoencoder along T3 and two deep neural autoencoders trained on max and mean pooled vectors, respectively. This autoencoder slowly reduces the input vector to 5 dimensions, then increases to 300. The third model had a preprocessing max-pooling layer separating the top 8 vectors. The first two are the standard approaches, but the thought process behind the third one

Table 1 Summary of models created

Model name	Type of neural network	Tensor used to train
Model 1	Convolutional	Tensor 3 (T3)
Model 2	Feed-Forward	Tensor 1 (T1)
Model 3	Feed-Forward	Tensor 4 (T4)
Model 4	Feed-Forward	Tensor 2 (T2)

was another experimental approach to lose some data except for the eight most emphasized vectors in verse. All these models were created with TensorFlow's Keras library and coded entirely in Python. The second model will be trained on mean, and max pooled tensors, ending with four models.

Once the models were trained, one would be picked as the best and used to encode all the word embeddings generated by the Spacy model from the Bhagavad Gita. Then, feeding Spacy's model's word embeddings for any text input prompt into the same autoencoder and obtaining the 5-dimensional encoded vector, the final output can be obtained by finding the verse with the highest cosine similarity from the pre-encoded Bhagavad Gita. The verse corresponding to the vector with the highest cosine similarity concerning the encoded vectors and the encoded input vector was chosen as the most relevant verse to address the given text prompt. Table 1 provides a helpful reference for how the models will be referred to moving forward, along with the type and the specific tensor used to train the autoencoder.

2.2.2 The Comparison Model

After training and selecting the best model, it will be run on text prompts through the best arbitrarily chosen model to get their embeddings and find the most similar verse from the Bhagavad Gita through a simple cosine similarity function. To demonstrate and visualize the quality improvement in the program's output, the authors have devised another similar function that finds the most similar verse directly through word embeddings generated by Spacy's en_core_web_lg model. This part of the experimentation is included as a baseline result for comparison using the same searching technique and is referred to as 'Base Vector Output' in Table 2 in Sect. 3. This ensures consistency in testing, analysis, and demonstration of the capabilities of this method. Figure 2 provides an informative block diagram explaining this process. Another crucial assumption that is a limitation is that the answering prompt and the questioning text are semantically related, backed by [7] in part. Still, since this application has no question-answering models, it does not show us the full capabilities of state-of-the-art methods for this task.

Table 2 Model output comparison

Input	Model 4's output	Base vector output	Author's interpretation
What should I do if I am lost in life?	Chapter 9, Verse 16	Chapter 10, Verse 33	Model 4 generated output is closer to an appropriate answer than the base vector output. The rituals and offerings mentioned in Model 4's output provide a better direction to life when compared to the Base Vector Output by encouraging the reader to express devotion if they are lost in life
How should one exercise karma?	Chapter 11, Verse 8	Chapter 1, Verse 36	Although neither model provides a satisfactory answer, Model 4's output is slightly more encouraging by implying the reward of divine eyesight through being closer and earning it through the correct use of Karma. Base Vector output completely misses any relevance
How should criminals be treated?	Chapter 3, Verse 33	Chapter 10, Verse 39	Model 4 provides a better answer by identifying the act of crime and malice implied in the question. It answers appropriately by making the connection that restraint is a common consequence of such actions. The base vector output fails to identify the negativity in the question and the subject of the crime in the question and thus only focuses on the act of 'moving' in general
What is the meaning of life?	Chapter 4, Verse 27	Chapter 13, Verse 12	Model 4 answers the question in its entirety. Encouraging the practice of yoga and the accumulation of Gyan and wisdom **in contrast to the Base Vector model that misses the question in its entirety

3 Experiments and Results

The models were trained and evaluated on an MSI GE66 Raider 10-SFS-443IN laptop. Its specifications include 8 Gigabytes of Video Memory (The Nvidia GeForce RTX 2070 Super Mobile, 115W), 32 Gigabytes of RAM, an 8-core/16-thread Intel i7-10875H Processor, and Microsoft Windows 11 as the operating system. However, the code was run on a miniconda environment created on an Ubuntu instance through the Windows Subsystem for Linux, version 2. The Adam optimizer is utilized in this work with a default learning rate of 0.001. cosine similarity was used as the loss function, as previously mentioned. The model was then trained with a batch size of 2 on the entire dataset (T1, T2, and T3, depending on the model, as outlined in Table 1.), For improved performance, the models were trained on their respective tensors for 75 epochs each.

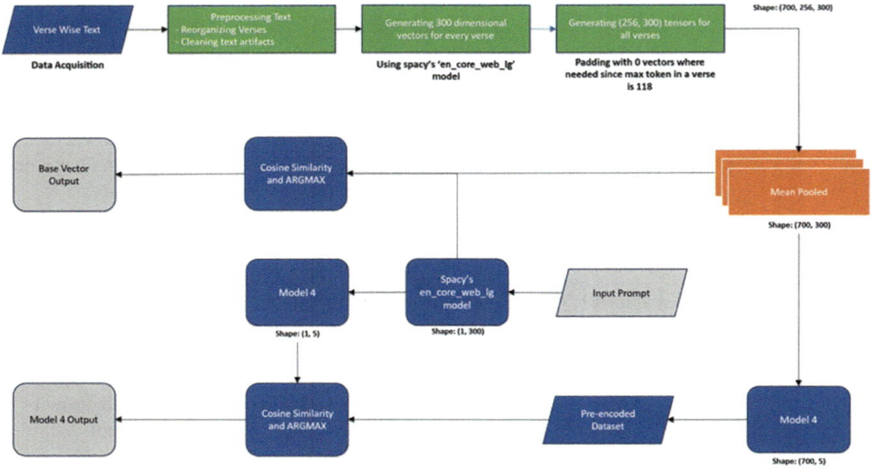

Fig. 2 Block diagram explaining the comparison model

Table 3 Evaluation results

Model name	Cosine similarity	Mean squared error
Model 1	0.120028	2.507199
Model 2	0.316639	31,729.85
Model 3	0.235396	2.347E + 8
Model 4	0.648489	579.5543

Table 3 shows the evaluation values for Loss and MSE metrics of all four models, where the evaluation split was the entirety of our dataset. One important thing to note here is that the MSE values are odd

The explanation for the odd MSE values shown in Table 3 is that the reconstructed vectors are vaguely like the input and have no bearing on the output since the search algorithms only use cosine similarity. There is no proven link between the magnitude of the sentence embedding vector and its sentiment. Therefore, this assumption can be considered valid for this model and task. However, this happened in the first place because the Mean Square Error is not a priority for the algorithm used to train the model, and the metric itself is presented to provide greater insight into how the training process worked and how the resultant models behave (Fig. 3).

After obtaining these metrics, the authors chose the most performant model (in this case, Model 4. Since it has the most cosine similarity while still being low.). Then, a basic script was written to find the most similar verse of the Bhagavad Gita given any text prompt (as discussed in Sect. 2.2.2). The result of the experiment is displayed in Table 2, where the base vector is the comparison model discussed in Sect. 2.2.2. Model 4's output represents the output of the best model from the experimentation and technique. As mentioned, these must be left mainly to the reader's interpretation,

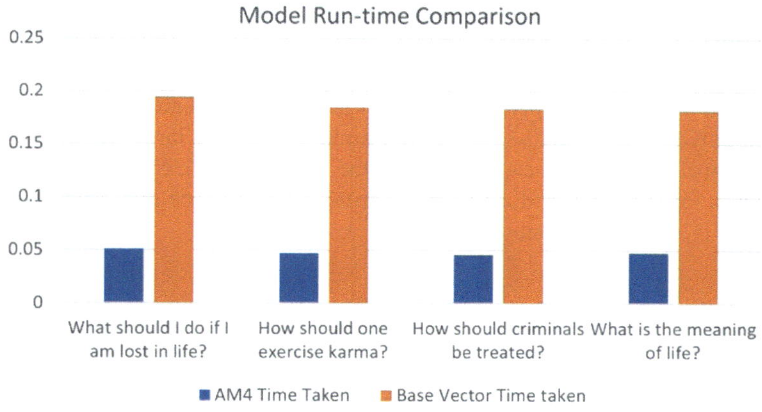

Fig. 3 Model Run-time comparison

but the authors think the base vector output varies wildly in quality and relevance to the question. The authors have also included their interpretation of the comparison between the output generated by the models. These comments help them conclude that there is a noticeable improvement in the quality of answers generated by Model 4's verses while also increasing the computational efficiency of the model by reducing the time taken to generate an output by significantly reducing the dimensionality of the data, as shown in Fig. 2. Model 4 is 84.334% faster than the base vector output on average.

4 Conclusion and Future Work

The size of our corpus is small, and the objective of finding the closest relevant verse of the Bhagavad Gita was not appropriately solved by finding the most similar verse by cosine similarity on the base embeddings themselves. The authors do not know the exact reason this happened, but the most educated guess as to why is that there are a lot of specific proper nouns and stories or relationships that could be taken literally, which could either obscure the abstract meaning of the verse or not make it clear from the onset, from a purely wisdom-based problem-solving, or question-answering standpoint. In other words, the task demands a certain amount of abstraction to relate ancient texts with the modern world. With its intentional loss of information, this autoencoder approach performs better and faster than base 300-dimensional vectors. Further research is required to manage better and control the quantity and type of data lost. Autoencoding does not improve the result but results in verses closer to what an answer would look like. The problem of not having a proper question–answer dataset resulted in having to investigate this non-supervised approach, and more data, such as supplemental training on verses' explanations rather than the verses themselves, could help. As the autoencoder approach improves the quality of semantic similarity in the output overall, further work in this field could result in better question-answering on abstractions of texts such as complex works of fiction and finding and extracting deeper insights such as symbolism and metaphors. Adding an abstractive autoencoder as part of the architecture of a larger language model would improve abstraction capabilities for the model overall. More exploration and research in this area could also result in Artificial Intelligence understanding ethics and morality, which is the study's goal.

References

1. Akritidis, L., Bozanis, P.: How dimensionality reduction affects sentiment analysis NLP tasks: an experimental study. In: IFIP Advances in Information and Communication Technology, pp. 301–312 (2022). https://doi.org/10.1007/978-3-031-08337-2_25
2. Raunak, V., Gupta, V., Metze, F.: Effective dimensionality reduction for word embeddings. In: Proceedings of the 4th Workshop on Representation Learning for NLP (RepL4NLP-2019) (2019). https://doi.org/10.18653/v1/w19-4328
3. Kaneko, M., Bollegala, D.: Autoencoding improves pre-trained word embeddings. In: Proceedings of the 28th International Conference on Computational Linguistics (2020). https://doi.org/10.18653/v1/2020.coling-main.149
4. Kashyap, A.: Bhagavad Gita. Kaggle (2022). https://www.kaggle.com/datasets/arjun83kashyap/bhagavad-gita. [Accessed 31 3 2023]
5. Farouk, M.: Measuring sentences similarity: a survey. Indian J. Sci. Technol. **12**(25), 1–11 (2019). https://doi.org/10.17485/ijst/2019/v12i25/143977
6. Tf.keras.losses.cosinesimilarity; tensorflow V2.13.0. TensorFlow. (n.d.). https://www.tensorflow.org/api_docs/python/tf/keras/losses/CosineSimilarity
7. Risch, J., Möller, T., Gutsch, J., Pietsch, M.: Semantic answer similarity for evaluating question-answering models. In: Proceedings of the 3rd Workshop on Machine Reading for Question Answering (2021). https://doi.org/10.18653/v1/2021.mrqa-1.15
8. Salih, N.M. and Jacksi, K.: State of the art document clustering algorithms based on semantic similarity. J. Inf. **14**, 58–75 (2020). https://doi.org/10.26555/jifo.v14i2.a17513.
9. Chandra, R., Ranjan, M.: Artificial intelligence for topic modelling in hindu philosophy: mapping themes between the upanishads and the Bhagavad gita. PLOS ONE, **17**(9) (2022). https://doi.org/10.1371/journal.pone.0273476
10. Huseynova, F.: Abstraction and summarization of meaning in natural language processing. Int. J. English Literat. Soc. Sci. **5**(3), 775–778 (2020). https://doi.org/10.22161/ijels.53.32
11. Ali, M.M., Joshi, R.C., Dutta, M.K.: An automated and efficient deep learning based classification of multiple skin disorders from skin lesion images. In: 2022 International Conference on Edge Computing and Applications (ICECAA), pp. 1156–1161. IEEE (2022)
12. Ali, M.M., Joshi, R.C., Dutta, M.K., Burget, R., Mezina, A.: Deep learning-based classification of viruses using transmission electron microscopy images. In: 2022 45th International Conference on Telecommunications and Signal Processing (TSP), pp. 174–178. IEEE (2022)
13. Ali, M.M., Raj, M., Vatsa, D.: FruizNet using an efficient convolutional neural network. In 2023 International Conference on Computational Intelligence and Knowledge Economy (ICCIKE), pp. 287–292. IEEE (2023)
14. Gupta, U., Jain, V.: Social neuroscience: inferring mental states in social media (2024)
15. Gupta, U., Trivedi, G., Singh, D.: Human AI: social network analysis (2024)

16. Gupta, U., Gupta, P., Agarwal, T., Pantola, D.: Gesture works—one stop solution. In: Proceedings of the International Conference on Intelligent Computing, Communication and Information Security, pp. 301–308. Springer Nature Singapore, Singapore (2022)
17. Gupta, U., Sharma, S., Jyani, U., Bhardwaj, A., Sharma, M.: Sign language detection for deaf and dumb students using deep learning: dore idioma. In: 2022 2nd International Conference on Innovative Sustainable Computational Technologies (CISCT), pp. 1–5. IEEE (2022)

RMODCNN: A Novel Plant Disease Prediction Framework

Vineeta Singh, Vandana Dixit Kaushik, Alok Kumar, and Deepak Kumar Verma

Abstract The detection of plant diseases is essential for avoiding reductions in agricultural product production and quantity. Visually observable patterns on the plant are the main focus of plant disease research. Monitoring plant health and spotting diseases is crucial for sustainable agriculture. Monitoring plant diseases manually is very difficult. It necessitates a lot of labor, expertise in plant diseases and prolonged processing. In the presented research, a ratal mellifera based DCNN model is developed for plant disease prediction. At first, the data is acquired out of a plant leaf input dataset, further, it involves pre-processing for removing noise. In the next step, the Region-of-Interest (ROI) extraction strategy is utilized for separating the relevant regions to process further process out of the unessential pixels. Further extracted ROI outcome will be imposed to the process of data augmentation; further GAN-based approach has been utilized for data augmentation for data enhancement for enhancing the accuracy of prediction. Further feature extraction has been accompanied for extracting features involving RESNET 101, LTP and LOOP as well as statistical features. Ratal mellifera optimization is derived out of two optimization strategies namely ratal optimization along with mellifera bee optimization, that is utilized for optimizing the channel boosted deep CNN network. Taking into consideration the TP and k-fold, the performance key indicators accuracy (acc) and sensitivity

V. Singh (✉) · A. Kumar · D. K. Verma
Department of Computer Science and Engineering, School of Engineering and Technology (Formerly Known As UIET), Chhatrapati Shahu Ji Maharaj University, Kalyanpur, Kanpur, Uttar Pradesh 208024, India
e-mail: vineeta.singh.cs@gmail.com

A. Kumar
e-mail: alok@csjmu.ac.in

D. K. Verma
e-mail: deepak@csjmu.ac.in

V. D. Kaushik
Department of Computer Science & Engineering, Harcourt Butler Technical University Kanpur, Kanpur, Uttar Pradesh, India
e-mail: vdkauhsik@hbtu.ac.in

G. Fortino et al. (eds.), *Proceedings of Third International Conference on Computing and Communication Networks*, Lecture Notes in Networks and Systems 917,
https://doi.org/10.1007/978-981-97-0892-5_45

(sen) for TP for tea leaf prediction and for apple leaf prediction have demonstrated better outcomes in contrast to recent state-of-the-Art has been gained.

Keywords Data augmentation · Plant disease prediction · Channel boosted deep CNN · Region of interest · Ratal mellifera optimization · Deep convolution neural network

1 Introduction

The science of predicting and identifying the earliest stages of fatal diseases in plants is known as "plant health care." The prompt identification of any symptoms can help to lower the rate of mortality of plant life. One approach to reduce plant death rates is through early diagnosis of ailments like these diseases [1].

The forecasting of plant ailments before the presence of any visual indicators has gained interest with the development of deep learning as well as thermal imaging strategy. The earlier forecast will lengthen the window of opportunity for recovery as well as decrease the need for pesticides, allowing for more efficient, high-quality production at a lower cost to the surroundings [2].

A significant risk to the worldwide production of crops, economic growth, as well as, ecological integrity is posed by contagious diseases of plants. Computer modeling has an exceptional chance for dealing with the common fear that these social as well as ecological calamities brought on by contagious plant diseases might worsen with changes in the climate By calculating extra characteristics utilizing a novel FMTD function, the developed feature extraction strategy, known as FMTD Function-based Fuzzy Transformation, i.e. FFFT, transforms a relatively small dataset into an exceptionally high dimensional feature space. The developed strategy has been verified through this research by using two minor plant ailments datasets, such as tomato early blight disease (TomEBD) as well as tomato powdery mildew disease (TPMD). For balancing the unbalanced datasets, resampling strategies were also utilized via this study. The Optimized Kernel Extreme Learning Machine, i.e. OKELM technique was then applied for classification. The feature extraction, as well as classification processes, have both been combined with parameter optimization using a genetic algorithm. Utilizing three benchmarking datasets, better performance of the developed methodology had been shown [3].

Classification of crops has proven difficult because crops shared many traits in the initial phonological stages. Through the use of drones that deliver high-resolution optical imagery, ML may be applied to a number of tasks, including the recognition of various crop varieties at low altitude platforms. The phonological stages are photographed using drones where the resulting gray scale images are then used to create attributes relied over grey level co-occurrence matrices [4].

On determining the severity of a disease, there is nevertheless, comparatively little research. To develop standards for rating the degree of severity associated with plant illnesses, Shi et al. [6] looked first at the prevailing theories of existing illness professionals.

However, because of the incorrect quality of the acquired image, previous approaches suffer from a number of issues involving orientation, illumination, size as well as the background factors resulting in occlusion. Additionally, there are issues with misclassifications, over fitting along with cost complexity in a number of applications that operate in real-time [7].

Pal and Kumar [7] put forward the Agriculture Detection termed as AgriDet model, that combines the Kohonen-centered deep learning networks as well as traditional Inception-Visual Geometry Group Network, i.e. INC-VGGN. Additionally, the framework uses a pre-trained INC-VGGN system with a conventional base network as its upgraded base network to conduct precise disease identification as well as classification. Further, the pre-trained INC-VGGN system used in this research is a deep convolution neural network for identifying plant diseases which had already been trained on the particular dataset. In order to accomplish the specific goal of plant disease identification for the dataset used in the research work, the pre-trained weights along with the features learned in this base network are transferred into the newly created neural network. A dropout layer is added and the Kohonen learning layer is used to carry out deep learning feature extraction for solving the over fitting issue. The enhanced base network categorizes the training sets' severity classes after computing percentages. In the end, the efficacy of the model is calculated for several parameters of performance and outperforms earlier models in terms of accuracy [7].

In [14] researchers examine current advancements in the prediction of plant diseases utilizing ML as well as DL approaches. The simulations covered by researchers here show how employing such strategies might increase the precision along with the efficacy of plant disease detection.

Agriculture development has a significant impact on a nation's economy. However, crop growth rate as well as quality suffers greatly as a result of numerous plant diseases. Because there aren't enough domain specialists and the information has low contrast, accurately identifying these ailments via a manual process is highly difficult and time taking. The agriculture management system is thus searching for a method for automatically detecting early disease. To address such problems and accurately diagnose the different diseases [16].

2 Main Contributions

For the development of RMO, Ratal as well as Mellifera bee optimization are hybridized together. Foraging behavior of Mellifera bees during this time was hampered by predators and other animals as well as additional factors, such as finding the most optimal locations for gathering food. Subsequently, the Ratal Mellifera Optimization is hybridized with the honey ratal digging approach for addressing such problems along with improving the foraging technique. Digging has advantages such as easily breaking up the soil and allowing animals to create deep holes in hard terrain so they may comfortably hide when faced with danger. This helps the Ratal Mellifera select the best food source quickly as well as effectively via minimizing conflict, reducing time complexity as well as offering immediate global convergence. In order to efficiently mitigate the over fitting problem, Channel boosted DCNN classifier has been tuned utilizing Ratal Mellifera Optimization along with predicting plant disease with the least amount of complexity.

Rest of the manuscript sections have been organized as follows: Sect. 2 involves Literature Review while Sect. 3 depicts proposed model, Sect. 4 discusses about results and discussions finally Sect. 5 offers conclusion and future scope.

3 Literature Review

Ahmed and Yadav [1] developed plant disease identification utilizing machine learning strategies like linear regression, random forest-nearest neighbors, neural networks, Naive Bayes along with support vector machine. Unbiased measures including recall, precision, true negative rate, true positive rate, as well as F1-score technique are used for computing the efficacy of the produced plants disease risk assessment framework. The outcomes showed that the ensemble technique for plant diseases outperformed recent state-of-the-art techniques for identifying plant diseases. In order to facilitate early preventative measures and proactive care, the developed plant disease prediction techniques sought to forecast and identify diseases in its initial stages [1].

By investigating the rapidly changing extraction of features capabilities of deep learning technological advances, Bhakta et al. [2] proposed a plant disease identification method allied to thermal imagery. Three convolution layers are part of the developed strategy, which addresses the over fitting issue and processing complexity for small datasets. A commonly occurring disease of rice plants called Bacterial Leaf Blight has been utilized for testing the model [2].

Bhatia et al. [3] proposed Fractional Mega Trend Diffusion (FMTD); the feature extraction strategy relied on function aids towards an accurate and prompt diagnosis of plant illnesses. Finally concluding remarks involve all three datasets, the developed strategy has shown better results in contrast to recent State-of-the-Art strategies [3].

The model developed by Haridasan et al. [4] for recognizing the illnesses affecting rice plants adapts a computer vision-allied strategy which utilizes image processing, deep learning as well as machine learning strategies. This model reduces the dependence on traditional strategies for protecting paddy crops via diseases like false smut, brown leaf spot, bacterial leaf blight, sheath rot as well as rice blast; the five main ailments that regularly afflict Indian rice fields. Image segmentation is utilized after pre-processing to identify the infected area of the paddy plant. The illnesses mentioned above are recognized solely by their visual characteristics. To identify as well categorize specific types of paddy ailments, a support vector machine classifier as well as convolutional neural networks, i.e. CNNs are utilized. The recommended deep learning-allied strategy got the highest validation accuracy of 0.9145 utilizing

ReLU as well as softmax techniques. Following recognition, a predictive remedy is advised to help those involved in agriculture take the proper action to deal with these infections [4].

A plant illness prediction system was put forth by Joseph et al. in 2023 [5] and it primarily concentrates on the ailments which impact eleven distinct plants as well as how they may be recognized via plant leaf imagery utilizing CNN-centered deep learning systems. This review can aid scientists in gaining a quick understanding of how cutting-edge CNN computational models may be utilized to diagnose plant diseases as well as a summary of cutting-edge studies that have employed visualization approaches to locate disease spots for more accurate diagnosis. The study also provides an overview of the research that utilized hyperspectral images to diagnose plant diseases as well as the distinct sources of data which have been employed by different investigations. At last, this research paper involved the major challenges encountered by CNN allied plant ailments severity evaluation techniques in practical applications as well as given feasible research proposals along with assessment methods in practical applications and provided to address these issues; there are workable research proposals and potential solutions.

Shi et al. [6] looked first at the prevailing theories of existing illness professionals. The present research further subsequently offered a comprehensive comparison of the benefits and drawbacks of each of them, outlining sixteen research reports on CNN-centered plant disease severity evaluation with respect to conventional CNN systems, enhanced CNN models as well as CNN-centered segmentation networks, relying on the network architecture. They looked into basic strategies to obtain datasets along with criteria for measuring the effectiveness of CNN frameworks. The presented research additionally offered practical research ideas along with potential solutions to overcome the key issues that confronted CNN-centered plant disease seriousness evaluation techniques in practical applications [6].

Pal and Kumar [7] put forward the Agriculture Detection model termed as AgriDet model, that combines the Kohonen-centered deep learning networks as well as traditional Inception-Visual Geometry Group Network, i.e. INC-VGGN. This model allows plant disease identification as well as classification of the seriousness of diseased plants. In this model, all restrictions in the acquired image are removed by using pre-processing of images. Further, the issue of occlusion has been overcome via the proposed multi-variate grabcut technique to effectively perform segmentation [7].

A machine learning strategy known as ensemble learning integrates the results of different frameworks for generating predictions which are much accurate. In this study, Kondaveeti et al. [8] developed a plant illness identification strategy utilizing ensemble learning, with six base systems trained over a dataset comprising of 87,000 images pertaining to healthy as well as diseased plant leaves out of 38 distinct classes: MobileNet, Inception V3, MobileNetV2, VGG-16, ResNet50 as well as GoogleNet. For the plant disease dataset, the base models had accuracy levels ranging from 72.7% to 97.2%. Both soft as well as hard voting classifiers were used on the base systems to increase the system's accuracy. Soft voting entails weighing each base model's predictions based on how accurate they were, then selecting the class with the greatest average weight to serve as the final forecast. Hard voting simply counts the total number of votes given for every class, and the class having the highest number of votes is chosen as the winner. With the soft voting ensemble with 97.8% accuracy as well as hard voting ensemble with 98.3% accuracy, both hard as well as soft voting primarily enhanced the accuracy of the system. Research findings of this work show the advantages of ensemble learning for detecting plant diseases and the possibility of such a model to help in the precise as well as efficient diagnosis of ailments pertaining to plants [8].

In this research, Yu et al. [9] constructed a novel transformer block by using soft splitting token embedding to extract contextual data from nearby pixels and transformer architecture to incorporate model long-range properties. Additionally, cross-channel feature learning and inception architecture can enhance information richness, which is particularly advantageous to very fine feature learning. The suggested model outperforms earlier convolution- and vision-transformer-based models in terms of accuracy, scoring 99.94% on VillagePlant, 99.22% on ibean, 86.89% on AI 2018 and 77.54% on PlantDoc. The outcomes of the research study demonstrate its dominance over the prevailing models [9].

A real-time disease prediction model utilizing a CNN on the Platform-as-a-Service (PaaS) cloud was the basic goal of the research conducted via Lanjewar et al. [10]. For the training as well as testing datasets along with for validation the CNN model produced an outstanding accuracy of 100% for predicting illness for tea leaves. For forecasting tea leaf illness, the same dataset has also been applied to DCNN like ResNet50, Xception as well as NASNetMobile. The confusion matrix along with K-fold cross-validation techniques were used to assess the performance analysis of each model. The comparison results show that the CNN framework works with amazing accuracy, outperforming the DCNN along with literature-reported methods. The model has been successfully deployed to the Platform-as-a-Service (PaaS) cloud following the evaluation. The cloud system automatically identifies the disease as well as displays it through mobile display. The smartphone can have accessibility to the deployed model hyperlink [10].

Lamba et al. [11] provide an assessment of different machine learning methods for accurately recognizing plant diseases. The discipline of machine learning, a subarea of artificial intelligence strategies, focuses on the key domains of the design of systems, design of models, prediction of values from observation along with learning from large sized data with variety gathering. They used improved CNN for categorizing different diseases pertaining to plant leaves. Dataset enhancement has been accompanied via utilizing a generative adversarial network. PlantVillage dataset has been taken for developing as well as training the model. The collection contains images of plant diseases affecting tomato, potato as well as pepper plants. There are 15 categories utilized for plant diseases which are used for training as well as testing the classifier. The whole accuracy of the model attained 98% [11].

Deep learning may be quite useful in assisting farmers for preventing crop loss via spotting diseases early in plant leaves. The percentage for accuracies for CNN, VGG-16, VGG-19 as well as ResNet50 correspondingly were 98.60%, 92.39%, 96.15% and 98.98% in the research work conducted by Islam et al. in 2023 [12] using the plant-village 10,000 picture dataset for recognizing crop infection. As per the final result outcomes, ResNet50 performs better as compared to other frameworks and has an accuracy rate of 98.98%. In order to create a smart web application to predict crop diseases in real-world situations, the model known as ResNet50 was used. By examining images of the plant leaves, the developed web tool seeks to help farmers determine plant illnesses. The developed application utilizes the ResNet50 transfer learning framework at its heart for categorizing infected leaves as well as healthy leaves for categorizing the current type of disease. The main aim is to facilitate farmers in saving resources along with preventing economic loss via identifying diseases of the plant further subsequently imposing suitable treatment [12].

In this paper, Joseph et al. [5] primarily concentrate on ailments which influence eleven (11) different plants and discuss ways deep learning algorithms based on CNN can be used to identify the ailments out of the images of plant leaves. Their study in their review work can aid scientists in gaining a quick understanding of how cutting-edge CNN models may be utilized for diagnosing the plant diseases as well as providing a summary of recent trending studies which may employ visualization approaches for locating disease spots for more accurate diagnosis. The study also provides an overview of the research that utilized hyperspectral images to diagnose plant diseases as well as the different data sources which have been employed via different research works. The limitations and unexplored areas of research, as well as the difficulties that are now faced by the development of a plant disease diagnostic framework, have also been identified in their research work.

In order to enhance forecasts, Yang et al. [13] examined the inherent issues with the modeling approaches currently being used. They also highlighted the necessity of incorporating principles of evolution into polytrophic as well as eco-evolutionary models. They focused specifically on how evolutionarily altered functioning trade-offs, the relative adaptation of plants along with pathogens, climate preferences as well as ecosystems can feed back into subsequent plant disease outbreaks, as well as how advances in technology may make it easier for generating and incorporating this pertinent knowledge to get better modeling predictions [13].

Shoaib et al. [14] concentrated on focusing their research work on publications in duration from 2015 to 2022. The difficulties and restrictions of applying DL as well as ML for the diagnosis of plant diseases like data accessibility problems, poor image quality as well as the inability to distinguish between diseased and healthy plants have also been covered in their research work. By addressing these issues as well as restrictions, offering an in-depth knowledge of the trending research in this area, pointing out the advantages and drawbacks of these techniques and outlining potential solutions to address implementation issues, the research offers important information for identifying plant diseases to the scientists, practitioners along with industry professionals [14].

González-Domnguez et al. [15] gave a complete assessment of chosen advancements in plant disease modeling by the mid-twentieth century as well as a rise in the model numbers which were published during that time, by their findings as well as perspective. Their research development may be largely ascribed to developments in automatic data collection strategy as well as environmental sensors; secondly equipments as well as methodologies for looking into plant epidemiology, thirdly computer science along with data analytics. By utilizing the wheat rusts as a case study, they examined the development of methods for developing data-based (empirical) frameworks as well as process-based, i.e. mechanistic simulations. By utilizing apple scab as a second case study, they also discussed how understanding biological procedures is becoming increasingly crucial for modeling plant diseases. They also explained the evolution of modeling techniques over the past 50 years with regard to wheat rusts as well as apple scab via assessing several disease modeling milestones which have been reached. At last, they had gone through about how coordinated pest control is applied using plant disease prediction techniques as a component of a multi-modeling strategy [15].

Ahmed and Yadav [1] developed plant disease prediction techniques were developed utilizing ML techniques including linear regression, support vector machine, random forest-nearest neighbor as well as Naive Bayes. Unbiased measures including, true negative rate, true positive rate, precision, F1-score as well as recall techniques are used for determining the effectiveness of the produced plants disease risk assessment framework. The outcomes showed that the ensemble strategies for plant diseases outperformed other developed and proposed frameworks for detecting plant diseases. The goal of the developed plant disease prediction techniques was to forecast early disease identification, enabling early preventive actions along with predictive maintenance [1].

Parez et al. [16] developed the E-GreenNet model, that is effective as well as portable moreover it is relied on Deep Learning abbreviated as DL. A MobileNetV3Small framework is used throughout the end-to-end framework as a foundation that produces fine-grained, discriminative as well as noticeable characteristics. Additionally, each dataset for the developed model—PlantVillage (PV), brand-new Plant Composite (PC) dataset as well as Data Repository of Leaf Images (DRLI) —was used for training it before test samples were used to gauge its efficacy. For all three taken datasets, the proposed framework achieved 1.00%, 0.96% as well as 0.99% accuracies after rigorous experimental study. In comparison to previous State-of-the-Art strategies, the developed strategy also gains faster inference speed. Moreover, a comparative analysis has been performed meanwhile the developed technique demonstrated discriminative scores in contrast to the different pre-trained frameworks along with other deep learning as well as machine learning techniques [16].

4 Proposed Framework

In a channel-based deep CNN classifier, the most important task is to tune the hyperparameters, which are used to find the most likely optimum hyperparameter to improve the classifier's prediction acc. RMO is a common combination of honey badger optimization (HBO) [22] and artificial bee optimization (ABC) [21]. The mellifera bee foraging behavior includes searching for the best spots to collect food.

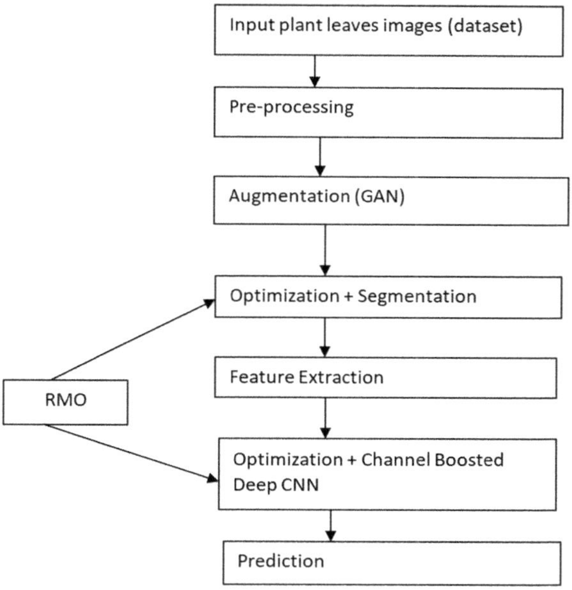

Fig. 1 Proposed framework

In this case, which faced challenges such as local optimum, poor quality of solution as well as slow global convergence, hence to overcome these difficulties and to enhance the foraging strategy, the honey ratal digging strategy is hybridized with the ratal mellifera optimization, which makes the ratal mellifera to effectively reach the optimal food source by reduced the time complexity and provides fast global convergence (Fig. 1).

5 Results and Discussions

Here we have taken Kaggle dataset [20] for simulation purposes. Project has been developed in Python. Model: "sequential"; Total params: 23,270,072; Trainable params: 23,270,072; Non-trainable params: 0 (Table 1).

The detection of plant diseases is essential for avoiding reductions in agricultural product production and quantity. Visually observable patterns on the plant are the main focus of plant disease research. Monitoring plant health and spotting diseases is crucial for sustainable agriculture. Monitoring plant diseases manually is very difficult. It necessitates a lot of labor, expertise in plant diseases and prolonged processing. In the presented research, a ratal mellifera based DCNN model is developed for plant disease prediction. At first, the data is acquired out of a plant leaf input dataset, further, it involves pre-processing for removing noise. In the next step, the Region-of-Interest (ROI) extraction strategy is utilized for separating the relevant regions to process further process out the unessential pixels. Further extracted

Table 1 Comparative analysis for TP and K-fold for apple leaf and tea leaf

		TP	TP	K-fold	K-fold
Techniques	Leaf type	Accuracy	Sensitivity	Accuracy	Sensitivity
LSTM [17]	Apple	84.1	93.1	90.6	92.0
HBO:CB DCNN [18]	Apple	84.93	93.92	91.24	92.6
ACO:CB DCNN [19]	Apple	85.4	95	92	93
Proposed framework	**Apple**	**96.3**	**97.9**	**95.5**	**96.0**
LSTM [17]	Tea	80.6	92.7	91.9	97.5
HBO: CB DCNN [18]	Tea	81.5	93.5	92.7	98.8
ACO: CB DCNN [19]	Tea	82.0	94.2	93.3	99.0
Proposed framework	**Tea**	**96.2**	**97.0**	**94.2**	**99.6**

ROI outcome will be imposed to the process of data augmentation; further GAN-based approach has been utilized for data augmentation for data enhancement for enhancing the accuracy of prediction. Further feature extraction has been accompanied for extracting features involving RESNET 101, LTP, LOOP as well as statistical features. Ratal mellifera optimization is derived out of two optimization strategies namely ratal optimization along with mellifera bee optimization, that are utilized for optimizing the channel boosted deep CNN network.

Then using the optimized channel boosted deep CNN, the extracted data was categorized. The Ratal Mellifera was utilized to tune to optimize the channel boosted deep CNN network. Finally, the optimized channel boosted the deep CNN classifier and predicted plant disease. Despite using databases to consider the TP and k-fold, the terms of acc are 96.2%, sen is 97%, similarly for k-fold the terms of acc is 94.2%, sen is 99.6%, for k-fold 8 for database 1 is achieved, similarly, for TP the terms of acc is 96.3%, sen is 97.9%, the terms of acc is 95.5%, sen is 96% for k-fold 8 for database 2 is achieved.

Taking into consideration the TP and k-fold, the performance key indicators accuracy (acc), sensitivity (sen) for TP, for tea leaf prediction and for apple leaf prediction have demonstrated better outcomes in contrast to recent state-of-the-Art has been gained.

The method could be carried out using various classifiers to furthermore enhance the performance of the model as a future research.

6 Conclusions and Future Directions

In the proposed framework, the disease prediction has been accompanied via the proposed model. Furthermore, model is effective upto what extent, and has been compared with recent State of-the-Art strategies. A total of 87 K RGB images present in the dataset are classified into 38 subclasses, involving healthy and diseased crop leaves. In the present model, dataset is partitioned into training as well as validation sets in an 80:20 split to preserve the directory structure. To predict, another directory further has been developed which contained 33 test images. In the future, more experiments can be done in this model for enhancing the results.

References

1. Ahmed, I., Yadav, P.K.: Plant disease detection using machine learning approaches. Expert. Syst. **40**(5), e13136 (2023)
2. Bhakta, I., Phadikar, S., Majumder, K., et al.: A novel plant disease prediction model based on thermal images using modified deep convolutional neural network. Precis. Agric. **24**, 23–39 (2023). https://doi.org/10.1007/s11119-022-09927-x
3. Bhatia, A., Chug, A., Singh, A.P., Singh, D.: Fractional mega trend diffusion function-based feature extraction for plant disease prediction. Int. J. Mach. Learn. Cybern. **14**(1), 187–212 (2023)
4. Haridasan, A., Thomas, J., Raj, E.D.: Deep learning system for paddy plant disease detection and classification. Environ. Monit. Assess. **195**(1), 120 (2023)
5. Joseph, D.S., Pawar, P.M., Pramanik, R.: Intelligent plant disease diagnosis using convolutional neural network: a review. Multimedia Tools Appl. **82**(14), 21415–21481 (2023)
6. Shi, T., Liu, Y., Zheng, X., Hu, K., Huang, H., Liu, H., Huang, H.: Recent advances in plant disease severity assessment using convolutional neural networks. Sci. Rep. **13**(1), 2336 (2023)
7. Pal, A., Kumar, V.: AgriDet: plant leaf disease severity classification using agriculture detection framework. Eng. Appl. Artif. Intell. **119**, 105754 (2023)
8. Kondaveeti, H.K., Ujini, K.G., Pavankumar, B.V.V., Tarun, B.S., Gopi, S.C.: Plant disease detection using ensemble learning. In: 2023 2nd International Conference on Computational Systems and Communication (ICCSC), pp. 1–6. IEEE (2023)
9. Yu, S., Xie, L., Huang, Q.: Inception convolutional vision transformers for plant disease identification. Internet Things **21**, 100650 (2023)
10. Lanjewar, M.G., Panchbhai, K.G.: Convolutional neural network based tea leaf disease prediction system on smart phone using paas cloud. Neural Comput. Appl. **35**(3), 2755–2771 (2023)
11. Lamba, S., Saini, P., Kaur, J., Kukreja, V.: Optimized classification model for plant diseases using generative adversarial networks. Innovat. Syst. Softw. Eng. **19**(1), 103–115 (2023)
12. Islam, M.M., Adil, M.A.A., Talukder, M.A., Ahamed, M.K.U., Uddin, M.A., Hasan, M.K., Debnath, S.K.: DeepCrop: deep learning-based crop disease prediction with web application. J. Agricult. Food Res. 100764 (2023)
13. Yang, L.N., Ren, M., Zhan, J.: Modeling plant diseases under climate change: evolutionary perspectives. Trends Plant Sci. (2023)
14. Shoaib, M., Shah, B., Ei-Sappagh, S., Ali, A., Ullah, A., Alenezi, F., Ali, F.: An advanced deep learning models-based plant disease detection: a review of recent research. Front. Plant Sci. **14**, 1158933 (2023)
15. González-Domínguez, E., Caffi, T., Rossi, V., Salotti, I., Fedele, G.: Plant disease models and forecasting: changes in principles and applications over the last 50 years. Phytopathology®, **113**(4), 678–693 (2023)
16. Parez, S., Dilshad, N., Alanazi, T.M., Lee, J.W.: Towards sustainable agricultural systems: a lightweight deep learning model for plant disease detection. Comput. Syst. Sci. Eng. **47**(1), 515–536 (2023)
17. Smagulova, K., James, A.P.: A survey on LSTM memristive neural network architectures and applications. Europ. Phys. J. Spec. Top. **228**(10), 2313–2324 (2019)
18. Fakieh, B., Ragab, M.: Automated COVID-19 classification using heap-based optimization with the deep transfer learning model. Comput. Intell. Neurosci. 2022 (2022)

19. Kavitha, R., Jothi, D.K., Saravanan, K., Swain, M.P., Gonzáles, J.L. Bhardwaj, R.J.A., Adomako, E.: Ant colony optimization-enabled CNN deep learning technique for accurate detection of cervical cancer. BioMed Res. Int. 2023 (2023)
20. Plant disease detection dataset https://www.kaggle.com/datasets/sadmansakibmahi/plant-disease-expert?select=Image+Data+base
21. Chong, C.S., Low, M.Y.H., Sivakumar, A.I., Gay, K.L.: A bee colony optimization algorithm to job shop scheduling. In: Proceedings of the 2006 Winter Simulation Conference, pp. 1954–1961. IEEE (2006)
22. Hashim, F.A., Houssein, E.H., Hussain, K., Mabrouk, M.S., Al-Atabany, W.: Honey badger algorithm: new metaheuristic algorithm for solving optimization problems. Math. Comput. Simul **192**, 84–110 (2022)

Refined Human–Computer Interaction: Enhancing Efficiency and Collaboration

Shubham Singh, Harsh Pal, Ayan Ambesh, Akshat Singh, Deepali Kamthania, and Alpna Sharma

Abstract Optimized human and computer interaction aims to enhance communication and collaboration between humans and machines by creating interfaces that streamline interaction and minimize user cognitive load and physical effort. This paper explores the feasibility of utilizing hand gestures as a means to interact with computers and Internet of Things (IoT) devices, employing a monocular camera system and machine learning algorithms. The study consists of two parts: software interaction and interaction with IoT devices within a private network. A customized architecture has been developed, leveraging Google's Mediapipe Hands library to detect hand gestures, coupled with a machine learning model for gesture classification. The classification output is translated into real values for controlling various components, encompassing ranges, categories, positions, or continuous values. This technology enables users to regulate parameters such as brightness, volume, and cursor movement, as well as perform tasks like software manipulation and control of IoT devices like relays and robot clippers. The model demonstrated satisfactory accuracy in controlling components within games and virtual reality environments.

Keywords Internet of Things (IoT) · Human–Computer Interaction (HCl) · Hand Gesture Recognition (HGR) · Artificial Neural Network (ANN) · Artificial Intelligence (AI) · Application Programming Interface (API) · Mediapipe · Augmented and Virtual reality (AR/VR)

1 Introduction

Human–Computer Interaction (HCI) has evolved significantly over the years, leading to optimized and seamless interactions between humans and computers. The integration of advanced technologies and user-centered design principles has resulted in a

S. Singh · H. Pal · A. Ambesh · A. Singh · D. Kamthania (✉) · A. Sharma
Vivekananda School of Information Technology, VIPS-TC, Delhi, India
e-mail: deepali102@gmail.com

G. Fortino et al. (eds.), *Proceedings of Third International Conference on Computing and Communication Networks*, Lecture Notes in Networks and Systems 917,
https://doi.org/10.1007/978-981-97-0892-5_46

refined HCI paradigm, focused on enhancing efficiency, productivity, and collaboration. This refined approach emphasizes the development of intuitive and responsive interfaces, ensuring that users can effortlessly interact with technology. HCI-AI aims to explore the relationship between cognitive knowledge and the comprehension of both natural and AI forms in HCI interactions [1, 2]. The traditional user interfaces such as touchscreens and voice assistants allow individuals to control devices easily but they can be cumbersome when dealing with a large number of smart devices and may not always be suitable for hands-free interaction [3]. This has driven the shift towards virtual controls, such as symbols, icons, and graphics and gradually replacing physical controls like buttons, knobs, and switches. Various digital elements are intricately intertwined and interspersed with one another, forming a complex structure [4]. HGR systems have been extensively developed with HCI technologies in recent years, allowing regular engagement with machines [5]. Despite challenges like unfavorable environments, varying illumination, hand size, and shape affecting hand segmentation and recognition, progress in HGR persists [6]. The optimal segmentation method for identifying hand gestures from input images is proposed to improve performance using hand key point recognition through a monocular camera [7]. The wearable data glove is a potential choice for HCI system development, yet it has drawbacks like slow computations and an inadequate user interface. Vision-based methods offer a strong interface for HCI via pattern matching, but their intricate algorithms result in prolonged processing times, restricting real-time HCI use. HCI has also been achieved using variable centroid distance features, yielding an autonomous and user-friendly system that eliminates the need for gloves or accelerometers, and enhancing feasibility [8]. Aggarwal et al. [9] proposed a novel 3D real-time hand gesture recognition method for HCI, merging computer vision and image processing. Their low-processor approach achieves efficient performance. A prototype employing 2D convolutional neural networks (CNN) for real-time hand gesture recognition was developed for continuous interactions with augmented hand-controlled panels for sound and lighting systems [10]. Hand gestures are integral to numerous hyper-converged infrastructure (HCI) systems, aiding input. Commonly employed techniques for recognizing hand movements include hand detection, segmentation, and gesture recognition [11, 12]. In hand gesture recognition, a monocular camera system captures and processes visual data from a single perspective, providing non-intrusive data for analysis and recognition. This paper introduces a novel approach using a monocular camera system and machine learning for efficient hand gesture recognition, enabling seamless interactions with computers and IoT devices. It leverages Google's Mediapipe Hands library to extract hand gesture information from images, converting landmarks into actionable values for precise classification. The method is validated through two experiments: the first showcases software interaction, highlighting responsiveness and practicality, while the second explores IoT device control within a private network using a custom miniature bionic arm. By combining an ANN with Mediapipe, the method achieves accurate hand gesture recognition, enhancing user-technology interaction.

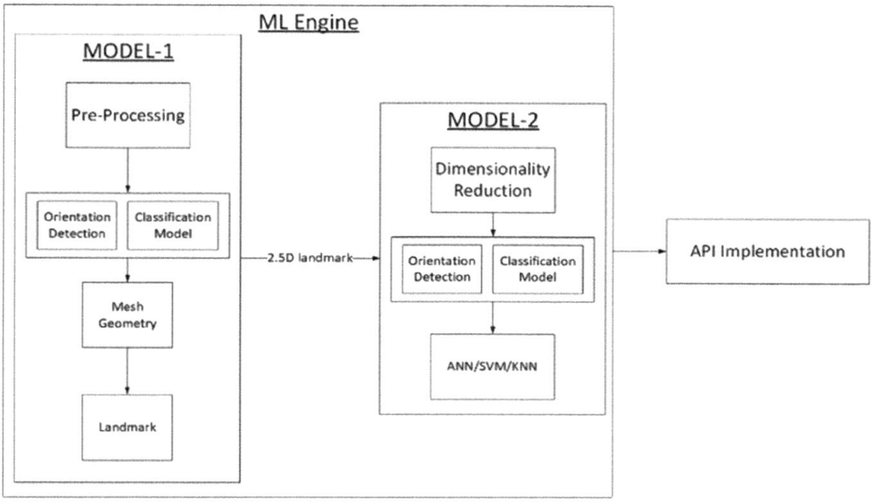

Fig. 1 Block representation of the architecture

2 Proposed Architecture

The proposed toolchain architecture contains two parts as shown in Fig. 1 *Machine Learning Engine* and *Application Programming Interface.* The machine learning engine contains *two models. Model–1*, responsible for the generation of hand landmarks and *Model–2* responsible for the classification of the generated landmarks.

3 Machine Learning Engine

Model—1: The model employs the "MediaPipe Hands" implementation to create hand landmarks. The Google MediaPipe Hands library is a toolkit for hand gesture analysis, utilizing computer vision to identify landmarks on hands and extract valuable information for recognition. It offers an efficient solution for integrating hand gesture data into diverse applications. In the current work, the implementation consists of two models*: a palm detector* and a *hand landmark model.*
i. Palm detector model processes the full input image and locates palms through an oriented hand-bounding box. The hand landmark model, on the other hand, operates on the cropped hand-bounding box provided by the palm detector and returns detailed 2.5D landmarks [13]. The Mediapipe dataset has been used to train and evaluate the model includes 6000 images that capture a wide range of geographical locations, lighting conditions, and hand appearances, as shown in Fig. 2.

The existing dataset has a limitation that it does not contain complex articulation of hands. To address this issue, the model has been trained using a large dataset collected

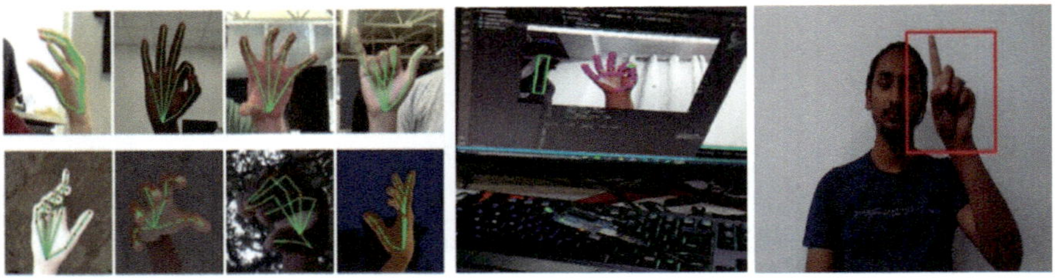

Fig. 2 Hand movements in various settings

through simulation or virtual reality, which will enable the model to handle intricate articulations and settings. To feed data to the machine learning model, frames from the video feed (i.e., images of the hand) must be standardized and processed. The pipeline follows as.

1. **Cropping the images**: The hand is detected using Mediapipe and a manual check to locate it in the image. The section containing the hand is then cropped with a predefined height and width to produce the desired image.
2. **Reducing noise**: The noise reduction in the environment includes two steps:

   a. *Color standardization:* standardized color includes cross-checking of the image color format, i.e., RGB
   b. *Pallet reduction:* reducing the pallet of the color to differentiate the color of hand from other objects and also to reduce complexity.

3. **Image Flattening**: process includes converting matrix $[\text{image}]_{\text{height x width}}$ to $[\text{image}]_{1\text{ x (height + width)}}$, which results in a row of data with height + width number of columns.
4. **Dataset generation**: process 3 is repeated for each frame to generate a dataset for further processing. The dataset contains (height + width) + 2 number of columns where the additional 2 columns are the location of diagonal representing the palm location.

ii. Landmark generation model process employs the Mediapipe pipeline and involves the following steps*: Detection of the hand in the image, Identification of the boundary box of the palm, Generating geometry around the hand including all the fingers, thumb, and palm, Predict the hand landmark points in the image.* The resulting data includes 21 points as shown in Fig. 3. Each point has a localization value in 3D space represented by X, Y, and Z. X and Y show the standardized position of the point in the X and Y plane. Z value shows the standardized depth of the point.

The generated data can be used to implement multiple pipelines, including those for AR/VR, robotics, software, and IoT technologies. AR/VR applications may require real-time translation and real-time values of X, Y, and Z for interactions, while software and IoT technologies may use the output of the classification model (Model 2). It is also possible to combine both pipelines for more complex tasks, such as real-time interaction in AR/VR with command execution.

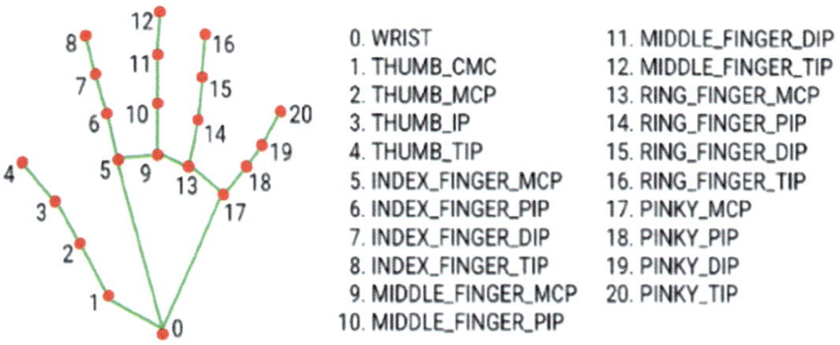

Fig. 3. 21 Hand points

Model—2: Model 1 generates values that are highly accurate shown in Fig. 4 due to the extensive pre-processing and computation that have already been completed. To prepare the data for Model 2, the dimensions are reduced by 1/3 by removing the Z value (which represents depth in the image) from the landmark points. This is because depth is not useful for classification purposes, as the model uses the structure of the hand (represented by the landmarks) rather than the image itself. Model 2 is responsible for classifying the hand landmarks generated by Model 1, performance. It uses two machine learning models to achieve this. One to classify the orientation of the hand and other to classify the structure of the hand.

Table 1 shows parameters and hyperparameters for all the arrangements with epochs set to 10 for which models have been tested. Model 2 is an ANN with 4 hidden layers and 2 activation functions. It has a learning rate of 0.001. To evaluate the accuracy of the model, the sparse categorical cross-entropy loss function is used. The structure of the deep artificial brain is shown below in Fig. 5. The maximum score achieved during finding the optimal hyperparameters is 0.966. The accuracy

Fig. 4 Performance of different algorithms for Model 1

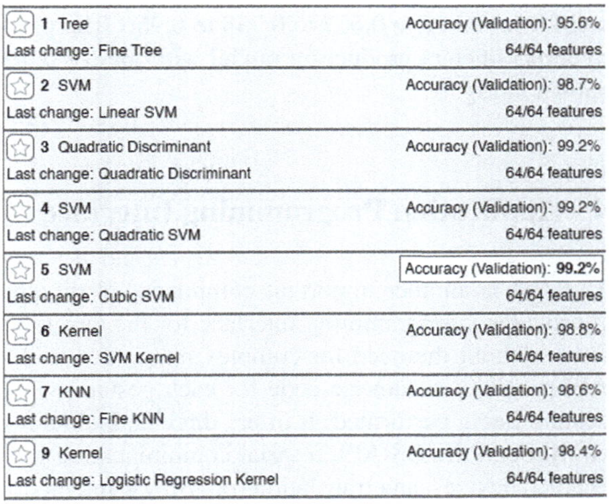

Table 1 Parameters setting for model evaluation

Parameters	Values
Number of hidden layers	2–20
Activation Functions	Relu and SoftMax
Number of neurons	32–512 with the step of 32
Learning Rate	1e-2, 1e-3, and 1e-4
Optimizer	Adam

Fig. 5 Structure of ANN

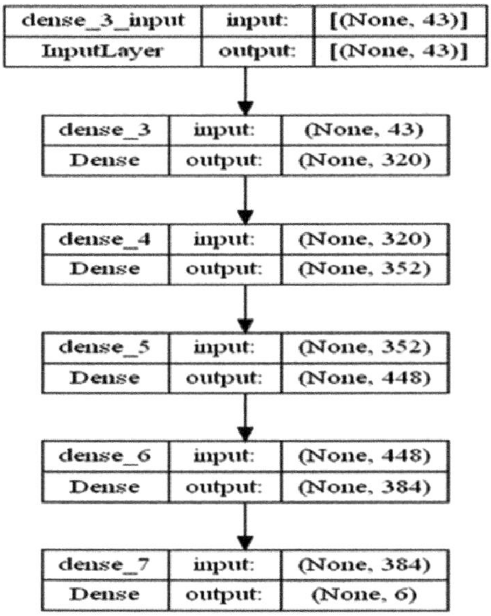

went from 0.472 to 0.637 to 0.738 to 0.966. The model trained from the mentioned hyperparameters produces a model with **loss** of 0.0803 and accuracy of 0.9750 as shown in Fig. 6.

4 Application Programming Interface (API)

The API is another important component of the proposed toolchain architecture. It provides a programming interface for the development of pipelines and applications without the need for complex integration and fine-tuning of the model. The API produces a numeric code for each gesture, which can be used to identify the gesture being performed. It offers three endpoints for user access: control over the OS through the OS API, a serial communication pipeline for interacting with IoT-enabled devices, and translation for AR/VR interactions. The API also provides raw

Fig. 6 Model 1 accuracy

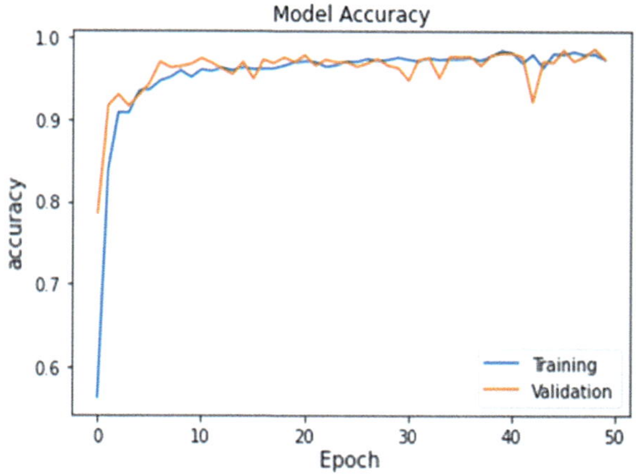

values for custom use cases. It includes multiple layers of abstraction to allow users to interact with the machine learning model directly if desired. The programming interface is built in Python and includes a package called OHCI. To use it, the user imports the package and specifies the process to be executed. The package then takes the data and procedure, makes a request to the machine learning engine for the value, and returns the output in a NumPy array. There are two use cases for the API: interacting with the hardware and software library, and interacting with the core library. Interacting with the hardware and software library offers two levels of abstraction: one for the ML engine and another for the integration of the engine with the desired device. While it is possible for the user to directly interact with the core library but it is recommended to use the OHCI library, which provides more detailed information. The package also includes a command line utility that allows the user to capture data for custom hand gesture training. With this utility, the user can retrain the pre-tuned model with the desired gesture in under two seconds, based on the current testing configuration. This method has proven to be effective in generating large amounts of accurate, real-time data, which significantly reduces the time required to create a functional neural network. The data generated for training is relatively small, which reduces computation time and saves memory and disk space.

Additionally, the simulation allows for the creation of hand images with a wide range of environments, articulations, shapes, and labels, as shown in Fig. 7. This enables the model to predict the structure of the hand even if it is not fully visible in the frame, as shown in Fig. 8. The package is currently closed source but will become open source after thorough testing. Despite extensive training data, the model might struggle due to factors like poor lighting, low frame rates, and limited space. Performance can be enhanced with night vision or wide-angle cameras. The architecture supports major OS and CPU types except M1 chips; it has been successfully tested on Intel, AMD CPUs, and Nvidia GPUs, primarily on Windows, with potential future improvements for UNIX and Linux systems.

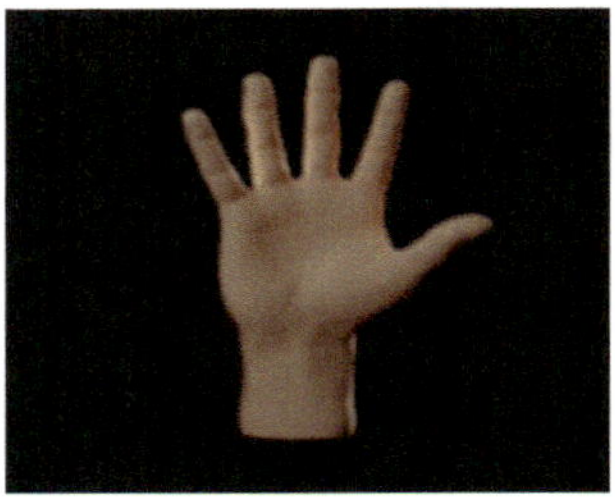

Fig. 7 Hand in a VR

Fig. 8 Articulation of hand

5 Applications

The pipeline functions across various technologies and applications, including **software** implementation for controlling HID devices like mouse, keyboard, and touchpad. It enables mouse control with a single point, sign language for keyboard input, and hand gestures to emulate touchpad actions. Different functions such as range and translation can adjust system settings or track gestures Fig. 9a. Hand gestures enable control of **diverse robotic equipment,** manipulating kinematic moments and operating relays Fig. 9b. These applications can be virtually tested using game engines or ROS. Similar functions can be applied to IoT devices for wireless radio communication. The landmark points of the hand can be used to map bone structure of the hand in **virtual reality and augmented reality** to control hand and in return hand collider can interact with the environment and the objects in the environment. The gesture can be used to execute different commands like to pick a thing, etc (Fig. 10).

Games can be managed through a hand simulator, employing bone translation to mimic keypress actions. The created cross-platform real-time toolchain facilitates rapid, adaptable application development for virtual reality, software interaction, industrial robotics, and computer vision. This accurate platform encompasses software and hardware controlled via computer vision. Hand gesture recognition's transformative potential spans various sectors, enhancing user experiences. In healthcare, touchless controls for medical equipment could improve safety for professionals. In transportation, intuitive infotainment controls bolster driver safety. Gesture interfaces in autonomous vehicles offer interactive travel. Education benefits

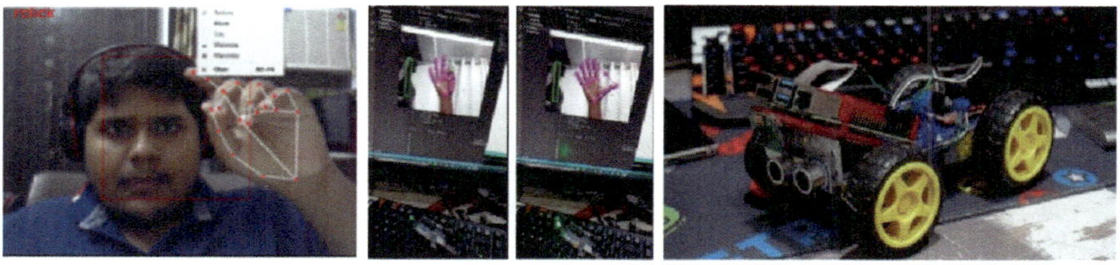

(a) (b)

Fig. 9 **a** OS API control using gesture **b** Interaction with IoT devices and robot

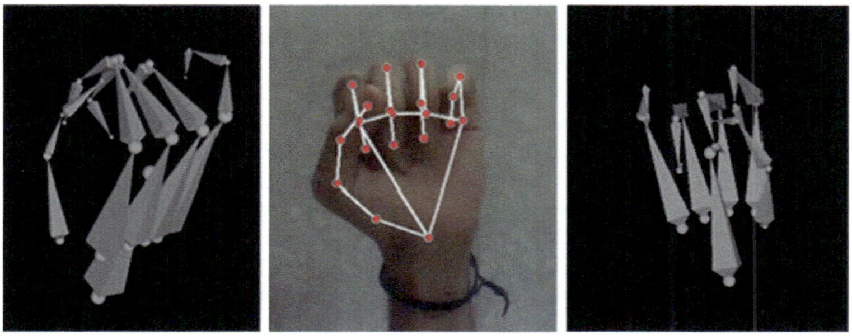

Fig. 10 Interaction and control demonstration in VR

too, enabling students to engage with content through natural gestures, transforming virtual learning.

6 Conclusions

The method presented in this paper offers a promising avenue for enhancing HCI and IoT device control through real-time hand gesture recognition. By utilizing a monocular camera system, machine learning, and the Mediapipe Hands library, the proposed system bridges the gap between human intentions and technology responses. The potential benefits include streamlined interactions, enhanced user experiences, and improved collaboration, contributing to the ongoing evolution of HCI in the digital age.

References

1. Alkatheiri, M. S. (2022). Artificial intelligence assisted improved human-computer interactions for computer systems. *Computers and Electrical Engineering, 101.*
2. Ryan, S., Nadal, C., Doherty, G.: Integrating Fairness in the Software design process: An interview study with HCI and ML experts. IEEE Access (2023). https://doi.org/10.1109/ACCESS.2023.3260639
3. Want, R., Schilit, B. N., & Jenson, S. (2015). Enabling the internet of things. *Computer, 48*(1), 28–35.
4. Liu, J., Ang, M. C., Chaw, J. K., Kor, A. L., & Ng, K. W. (2023). Emotion assessment and application in human–computer interaction interface based on backpropagation neural network and artificial bee colony algorithm. *Expert Systems with Applications, 232.*
5. Rahim, M. A., Miah, A. S. M., Sayeed, A., & Shin, J. (2020). Hand gesture recognition based on optimal segmentation in human-computer interaction .In *202020 3rd IEEE international conference on knowledge innovation and invention (ICKII).*https://doi.org/10.1109/ickii50300.2020.9318870
6. Gota, D.-L., Puscasiu, A., Fanca, A., Valean, H., & Miclea, L. (2020). *[IEEE 2020 IEEE international conference on automation, quality and testing, robotics (AQTR) - Cluj-Napoca, Romania (2020.5.21–2020.5.23)] 2020 IEEE international conference on automation, quality and testing, robotics (AQTR) - Human-computer interaction using hand gestures* (pp. 1–5). https://doi.org/10.1109/AQTR49680.2020.9130016
7. Wei, W., Yang, Z., Li, Q., & Tao, S. (2022). Virtual hand-gesture interaction system based on SRHandNet with high frame rate. In *37th youth academic annual conference of Chinese association of automation (YAC), 2022.* https://doi.org/10.1109/YAC57282.2022.10023852
8. Bhame, V., Sreemathy, R., & Dhumal, H. (2014). *[IEEE 2014 international conference on advances in computing, communications and informatics (ICACCI) - Delhi, India (2014.9.24–2014.9.27)] 2014 international conference on advances in computing, communications and informatics (ICACCI) - Vision based hand gesture recognition using eccentric approach for human computer interaction* (pp. 949–953). https://doi.org/10.1109/icacci.2014.6968545
9. Agrawal, R., & Gupta, N. (2016). *[IEEE 2016 IEEE 6th international conference on advanced computing (IACC) - Bhimavaram, India (2016.2.27–2016.2.28)] 2016 IEEE 6th international conference on advanced computing (IACC) - Real time hand gesture recognition for human computer interaction* (pp. 470–475). https://doi.org/10.1109/IACC.2016.93
10. Sun, Y., Armengol-Urpi, A., Kantareddy, R., Nithin, S., Siegel, J., Sarma, S. (2019). *[IEEE 2019 IEEE conference on virtual reality and 3D user interfaces (VR) - Osaka, Japan (2019.3.23–2019.3.27)] 2019 IEEE conference on virtual reality and 3D user interfaces (VR) - MagicHand: Interact with IoT devices in augmented reality environment* (pp. 1738–1743). https://doi.org/10.1109/VR.2019.8798053
11. Xu, P. (2017). A real-time hand gesture recognition and human-computer interaction system. arXiv preprint arXiv:1704.07296
12. Jang, Y., Jeon, I., Kim, T.-K., Woo, W.: Metaphoric hand gestures for orientation-aware vr object manipulation with an egocentric viewpoint. IEEE Transactions on Human-Machine Systems **47**(1), 113–127 (2017)
13. Zhang, F., Bazarevsky, V., Vakunov, A., Tkachenka, A., Sung, G., Chang, C. L., & Grundmann, M. (2020). MediaPipe Hands: On-device Real-time hand tracking. https://arxiv.org/pdf/2006.10214v1.pdf

The Impact of Data Valuation on Feature Importance in Classification Models

Malick Ebiele, Malika Bendechache, Marie Ward, Una Geary, Declan Byrne, Donnacha Creagh, and Rob Brennan

Abstract This paper investigates the impact of data valuation metrics (variability and coefficient of variation) on the feature importance in classification models. Data valuation is an emerging topic in the fields of data science, accounting, data quality, and information economics concerned with methods to calculate the value of data. Feature importance or ranking is important in explaining how black-box machine learning models make predictions as well as selecting the most predictive features while training these models. Existing feature importance algorithms are either computationally expensive (e.g. SHAP values) or biased (e.g. Gini importance in Tree-based models). No previous investigation of the impact of data valuation metrics on feature importance has been conducted. Five popular machine learning models (eXtreme Gradient Boosting (XGB), Random Forest (RF), Logistic Regression (LR), Multi-Layer Perceptron (MLP), and Naive Bayes (NB)) have been used as well as six widely implemented feature ranking techniques (Information Gain, Gini importance, Frequency Importance, Cover Importance, Permutation Importance, and SHAP val-

M. Ebiele (✉) · R. Brennan
University College Dublin, Dublin, Ireland
e-mail: malick.ebiele@adaptcentre.ie

R. Brennan
e-mail: rob.brennan@ucd.ie

M. Bendechache
University of Galway, Galway, Ireland
e-mail: malika.bendechache@universityofgalway.ie

M. Ward · U. Geary · D. Byrne · D. Creagh
Saint James's Hospital, Dublin, Ireland
e-mail: MaWard@stjames.ie

U. Geary
e-mail: UGeary@stjames.ie

D. Byrne
e-mail: DeByrne@stjames.ie

D. Creagh
e-mail: DCreagh@stjames.ie

G. Fortino et al. (eds.), *Proceedings of Third International Conference on Computing and Communication Networks*, Lecture Notes in Networks and Systems 917,
https://doi.org/10.1007/978-981-97-0892-5_47

ues) to investigate the relationship between feature importance and data valuation metrics for a clinical use case. XGB outperforms the other models with a weighted F1-score of 79.72%. The findings suggest that features with variability greater than 0.4 or a coefficient of variation greater than 23.4 have little to no value; therefore, these features can be filtered out during feature selection. This result, if generalisable, could simplify feature selection and data preparation.

Keywords Data value · Machine learning · Feature importance · Feature selection · Explainable AI

1 Introduction

Data valuation is an emerging topic in the fields of data science, accounting, data quality, and information economics concerned with methods to calculate the value of data. Fleckenstein et al. [1] identified three categories of approaches to data valuation: market-based valuation, economic models, and dimension-based models; for more details, please refer to [1].

Data is a core resource for developing machine learning (ML) models. Hence, ML researchers are starting to investigate the use of data valuation techniques [5]. Scholars aim to identify either the most influential subset of data for ML model training or the most influential features for feature selection and model explanation [5–7, 13]. Our approach is focused on identifying the most influential features in a dataset. This is important for feature selection and model explainability and has potential applications in data protection by reducing the amount of data that needs to be shared. In the case of feature selection, exhaustive search is a very complex and time-consuming process [21]; for a set of n elements, there are $2^n - 1$ possible combinations of features. This can only be completed for a very small number of features [21]. Therefore, many heuristic techniques have been proposed. Most are guided search algorithms, such as forward or backward search, which means only adding a feature to the feature set if that feature satisfies the selection criteria (for example either minimising the error or maximising the accuracy). Although these approaches are effective and widely used in practice, they do not always produce the optimal feature set and require training the models. In this paper, we introduce a new, training-free method for feature selection based on data valuation.

First, we must investigate how data valuation metrics are related to feature ranking measures. To do that we chose the variability and coefficient of variation (CoV) as data content valuation metrics as well as Five classification models: eXtreme Gradient Boosting (XGB), Random Forest (RF), Logistic Regression (LR), Multi-Layer Perceptron (MLP), and Naive Bayes (NB). On one hand, the reason for choosing the variability is to investigate its impact on the models. As shown by [6, 8, 9], Random Forest (RF) is biased towards features with high variability. For the coefficient of variation, the higher the spread of data points is, the more difficult it would be to cluster the data points. Therefore, we suspect that the spread of a feature may be related to its ranking. On the other hand, XGB and RF have been chosen because they are robust tree-based models, MLP because it is a (simple) neural network, LR because it is the simplest classifier, and NB because it is a probabilistic model. Finally, the six feature ranking techniques (Information Gain, Gini importance, Frequency, Cover, Permutation Importance, and SHAP values) have been used because they are the most popular feature ranking techniques.

Below are the research questions:

- To what extent can variability and coefficient of variation predict the value of a feature?
- To what extent are feature importance and data valuation metrics related?
- To what extent can variability and coefficient of variation be used for pretraining feature selection?

To answer these research questions, we took a large, complex dataset for a real-world clinical classification problem and trained five ML models (eXtreme Gradient Boosting (XGB), Random Forest (RF), Logistic Regression (LR), Multi-Layer Perceptron (MLP), and Naive Bayes (NB)) and calculated their feature importance using six importance measures (SHAP {SHapley Additive exPlanation} values, Permutation Importance, Information Gain, Gini Importance, Frequency, and Cover), computed the variability and the coefficient of variation of each feature, and finally compared the feature importance and its value. This paper has the following contributions:

- The introduction of variability (instead of the number of unique values which is an absolute measure; variability is the relative number of unique values) and coefficient of variation as data value measures.
- The introduction of a new training-free feature selection approach.
- A comprehensive study of the relationships between feature importance and data valuation metrics for our clinical use case.
- Introduction of an extended ML process for investigating the relationship between Data Valuation Metrics and Feature Importance.

The remainder of this paper is divided as follows: the next section defines the background, followed by a description of the related work, Sect. 3 describes our use case, Sect. 4 proposes a new data value informed ML process, the experiments are described in Sect. 5 which is followed by the experimental results section, then the discussion section, and finally the conclusion.

2 Background

This section defines the key terms as well as the motivation for the choices made.

2.1 Definitions

Data valuation is a set of techniques to assign value (e.g. economic or social value) to datasets or data assets [27]. It aims to identify which data is important for a given context with respect to predefined criteria. For example in training machine learning (ML) models, an important datum or feature will be the one that improves the model's performance by reducing the error or improving its accuracy. In business, the most important data could either be the most used on a daily basis for business operations or the ones that most business units depend upon or the data which most increases business revenue. This demonstrates that the same data can have multiple, variable values in different contexts.

Several techniques have been used to determine the value of a specific data or dataset; from qualitative metrics to quantitative and even ML-based ones [1, 4, 19]. There is a wide range of dimensions of data value and metrics in the literature [24]. Some dimensions are usage, utility, cost, uniqueness, recency, and recurrency. In this paper, we have defined two new metrics: Variability and Coefficient of variation. Variability is defined as the relative number of unique values. For example, given two datasets D_1 and D_2 with 100 and 1000 entries, respectively, if F_1 from D_1 and F_2 from D_1 are two features with the same number of unique values equal to 20, then they will have 0.2 and 0.02 variability. These two features have the same absolute number of unique values but a drastically different relative number of unique values (variability). The variability allows the comparison of the number of unique values of features from different datasets. As for the coefficient of variation [28], it is defined as the relative spread of data points; it is a statistical measure applied to data valuation. Below are the formulas used to calculate the variability and the coefficient of variation:

- The Variability of the feature F_j:

$$RU(F_j) = \frac{|F_j|}{N}, \tag{1}$$

where N is the number of entries of the dataset containing the feature F_j; RU stands for Relative Uniqueness (Uniqueness means the number of unique values).
- The coefficient of variation (CoV) of the feature F_j:

$$\mathrm{CoV}(F_j) = \frac{1}{\mu}\sqrt{\frac{\sum_j (F_{ij} - \mu)^2}{\mathrm{dof}}}, \tag{2}$$

where dof (which means Degrees Of Freedom) is equal to $N - 1$ (for a sample) or N (for the population) and F_{ij} is the ith entry of the jth feature.

2.2 Motivations

The models can be evaluated in terms of predictive power by using the average Precision, average Recall, and average F1-score metrics. The evaluation is necessary to make sure that each of them represents the feature space in the best possible way because the more performing a model is the better it captures the data and the better the features are used. Three averaging techniques will be used: micro average (calculate metrics globally by counting the total true positives, false negatives, and false positives[1]), macro average (calculate metrics for each label, and find their unweighted mean. This does not take label imbalance into account[1]), and weighted average (calculate metrics for each label, and find their average weighted by using the number of true instances for each label. This alters macro average to account for label imbalance[1]). Therefore, the weighted average F1-score will be used as the main performance metric because F1-score is the trade-off between precision and recall, and, as mentioned above, the weighted average takes into account the imbalance of the data. In the process described below, the variability and coefficient of variation should be computed on the training data (because computing these metrics before splitting the data will mislead the comparison performed as the data valuation metrics and the feature importance will be computed on different sets). Also, these metrics need to be calculated on the original dataset before applying any scaling techniques. The main reason is that the scaling technique keeps the distribution but alters the statistics like the Standard deviation and the Mean (which are used to compute the coefficient of variation) and sometimes even the variability (because some scaling techniques are performed row-wise not feature-wise).

The selection of the Gini importance (based on the Gini index which is used to compute the mean impurity index in Random Forests), Information Gain (the average or total gain across all splits the feature is used in[2]), Frequency Importance (the number of times a feature is used to split the data across all trees[2]), Cover Importance (the average or total relative number of training instances seen across all splits the feature is used in[2]). Permutation Importance (measures the impact of any given feature by shuffling that feature and then measuring the performance of the model. The change in performance, either positive or negative, is the importance of the feature. If the performance improves, then the feature is ranked low. But if the performance is worsened, then the feature is ranked high; the rank being the value of the change in performance), and SHAP values importance (works quite similarly to the Permutation Importance but instead of shuffling, it is based on cooperative game

[1] https://scikit-learn.org/stable/modules/generated/sklearn.metrics.f1_score.html.

[2] https://xgboost.readthedocs.io/en/stable/python/python_api.html#xgboost.Booster.get_score.

theory to measure the change in performance). The first four ranking techniques will be computed while training the models and therefore are model-specific (mostly tree-based models) while the latter two will only be estimated after the models have been trained. Another reason is to check if these different feature ranking algorithms agree on the ranking of the features.

3 Related Work

Feature importance has been widely studied for the purpose of explaining black-box machine and deep learning algorithms. These studies aim to detect the features or variables that contribute most to predictive model outputs. Significant feature ranking algorithms include: SHAP (SHapley Additive exPlanation) values, Permutation Importance, Information Gain (Entropy, Log-loss), Gini Importance, Frequency Importance, and Cover Importance. The first two are model-independent and the latter four are tree-based techniques. SHAP values are based on cooperative game theory and were first introduced by Kuhn and Tucker [26]. They are widely used in machine learning for their robustness and effectiveness in capturing outliers and corruptions [5]. For that last reason, SHAP values have received increasing attention and focus in the data valuation community in recent years, especially to detect the high-value datum in training predictive models [5]. Previous studies have been focused either on detecting which features are the most influential for machine learning models output [12, 16–18, 20–22] or choosing the best datum or subset of data which improve these models performance during the training process [5, 13].

There are also been some studies of the bias or shortcomings of the feature selection and model explanation techniques with some attempts to solve them [6, 8, 9]. In this paper, we investigate the relationship between the feature importance and its value measured by its variability and coefficient of variation. If the influence of the feature value on its ranking can be accurately estimated, then training-free or pretraining feature selection can be performed; because all the existing approaches perform feature selection during or after the training of the models. To the best of our knowledge, no such work has been done before. Strobl et al. [8] showed that tree-based Gini importance and permutation importance in Random Forests are not reliable when the features have different scales and numbers of unique values (variability). Strobl [8] and Loecher [6] proposed different approaches to solve this problem. In [6], Loecher showed that even SHAP values have shortcomings and strongly depend on the number of unique values or variability of a feature.

Our work differs from all these previous contributions as (i) we are not trying to identify the shortcomings of feature importance measures or attempting to solve them but rather to investigate how they are related to variability and coefficient of variation, (ii) they only focus on Random forest while we used different types of algorithms (Probabilistic {Naive Bayes}, Tree-based {Random Forest, eXtreme Gradient Boosting}, Linear modelling {Logistic Regression}, and Neural networks

{Multi-Layer Perceptron}), and (iii) the data is scaled before training the models (therefore we believe that the problem related to the scale difference is solved).

4 Use Case

To test our proposed approach on the five selected ML algorithms, we consider a real dataset and classification problem from a major hospital in Ireland.

The data used in this paper is extracted from several sources in the hospital data management system. It consists of daily data on patient journeys within the hospital (e.g. beds and wards used, admission and discharge dates) as well as other data (e.g. area of residence, gender, discharge code). In total, there are 152 features and 938,296 patient records. The classification aim is to predict patients' outcomes at any given time throughout their journey within the hospital. Given the complexity of the dataset, exploring the whole feature space with traditional ML techniques is extremely time and resource-intensive. This, along with the variety of feature types and time series behaviours included, makes it a good candidate to investigate the research questions defined in the Introduction.

4.1 Ethics and Data Sharing

This research was covered as part of DCU research ethics committee application DCUREC/2021/118 and the data was transferred under a data processing agreement between the hospital and universities. The dataset was anonymised before it was shared and so contains no personal data. The research described here was conducted as part of a broader programme investigating patient infection risk and improving hospital safety systems. This large clinical classification problem highlighted issues in model training time and traditional feature selection techniques that led to this work to optimise the process and more rapidly protect patients.

5 Extended ML Process for Investigating the Impact of Data Valuation Metrics on Feature Importance

Figure 1 displays the extended ML process used to investigate the impact of data value metrics on feature importance; it extends the process presented by [29].

The three added steps are described below. Please refer to [29] for the description of the standard steps.

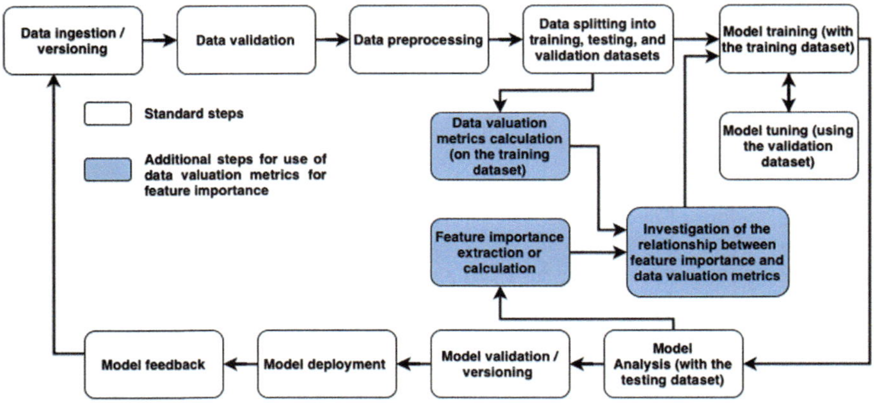

Fig. 1 Extended ML Process for Investigating the Impact of Data Valuation Metrics on Feature Importance

Data valuation metrics: Calculation of the data valuation metrics on each feature. In this use case, the variability and coefficient of variation are computed on the training data using the formulas defined in the Background section above.
Feature Importance extraction or calculation: The main objective here is to investigate how much each feature influences the prediction of the best model (the selected model after the model analysis); either by extracting the ranking of the feature from the models (tree-based models) or calculating the ranking of features (Permutation importance, SHAP values).
Investigation of the relationship between feature importance and data valuation metrics: This step is about visualising, calculating, and exploring every possible relationship between the feature importance and the data valuation metrics. Visualisation should always be the first task to be undertaken in this step because it may guide the remaining exploration and also lead to insights that are straightforward to get visually by difficult mathematically. Next, we can restart the model life-cycle from the training and tuning step to incorporate the insights got from the investigation of the relationship between feature importance and data valuation metrics.

6 Experiments

This section discusses the preprocessing steps, the training dataset details, and the model training setups. After all these steps, the feature importance is calculated and plotted against the variability and coefficient of variation.

Table 1 The number of entries per class before and after a stratified random sampling

Class ID	Number of entries before sampling (proportion in %)	Number of entries after sampling (proportion in %)
0	75051 (81.72%)	3753 (81.72%)
1	5596 (6.09%)	280 (6.09%)
2	3484 (3.79%)	174 (3.78%)
3	2661 (2.89%)	133 (2.89%)
4	2139 (2.32%)	107 (2.33%)
5	2011 (2.18%)	101 (2.19%)
6	531 (0.57%)	26 (0.56%)
7	363 (0.39%)	18 (0.39%)
Total	91,836	4,592

6.1 Dataset

The model tuning and the SHAP values computation are very time-consuming. Therefore, we decided to perform a stratified random sampling of the dataset. Stratified random sampling is chosen over simple random sampling because we wanted the training data to reflect the distribution of the number of entries per class of the original dataset. Table 1 shows the number of entries per class before and after the stratified random sampling.

6.2 Data Preprocessing

The original dataset contains 938,296 rows and 152 columns of inpatient discharges[from 1st Jan 2018 until 28th Feb 2022] from a large acute hospital. The first processing task was to drop the empty columns (columns where all the values are missing), single-value columns, and duplicated rows from the data. Therefore, 21 empty columns, 1 single-value column, and 57 duplicated entries have been dropped. The resulting dataset contains 938,239 rows and 130 columns, 117 contain categorical data, 7 of them contain numerical data, and 6 datetime data. The second step was grouping the data per patient_id, episode_id, and discharge_description to get a unique discharge_description per entry. If multiple values are encountered, for some columns, a set is returned. For example, if a patient used two or more beds or wards during their episode a set of the beds or wards used is returned. Each of these sets will then be encoded afterwards using the ordinal encoder. The resulting dataset has 91,836 entries and 130 columns. Two new columns (episode_length_in_days = discharge_date - admission_date, screen_length_in_days = Last_Date_Of_CPE_Screen - First_Date_Of_CPE_Screen) was calculated and added to the dataset. Third step, for 5 of the datetime columns, the following time features are added to the dataset:

year, quarter, month, week, day, hour, minute, dayofweek, isweekend, and isholiday. The remaining datetime column is used to compute the number of overnight stays of a patient during his or her stay. The datetime columns are then deleted afterwards. For each of the datetime features added, –1 is imputed when the original date is missing except the year feature which is imputed with the minimum year minus one. After this step, the number of columns increased from 130 to 175 ($175 = 130 + 10x5 - 5$). Fourth and last step, the 117 categorical columns have been encoded using the ordinal encoder and imputed with zero for all missing values. The ordinal encoder was chosen over the one_hot encoder because of the high variability of some of the columns (Fig. shows the number of unique values per column). One_hot encoding would have increased the data dimension drastically which may have invoked the "curse of dimensionality", especially for tree-based models. The sets returned by the grouping are encoded randomly (without taking into account their size).

6.3 Model Training

The models were trained using 5-fold Randomised Grid-Search Cross Validation (RGS-CV) with weighted F1-score scoring on an Ubuntu 18.04 LTS machine with 16GB RAM and 2 cores. The RGS-CV is run 5 times; once for each scaling technique: SimpleNormalizer, Normalizer, RobustScaler, MinMaxScaler, and StandardScaler. The experiments were run using the Scikit-learn package; all the scalers have been used with the default parameters. SimpleNormalizer is a self-implementation and divides each feature in the dataset by its maximum value. The data transformation and parameters with the highest weighted average F1-score are returned at the end of these runs. The model is retrained using the returned information and then tested on the test data to report the model's performance.

6.4 Model Performance Report

Once the training of each model is completed, the parameters as well as the scaling technique of the best model in terms of F1-score are returned. Then, the model is retrained using those parameters and the test data is scaled using the best scaler. Table 2 below shows the model performance on the test data. It can be seen that XGB outperforms the other models on 8 out of 9 performance metrics and most importantly on the main metric which is the weighted average F1-score.

Table 2 Model performance report in percentage (%) and training and SHAP values time

Models	PRECISION*			RECALL*			F1-SCORE*			TIME** (days-hh:mm:ss)	
	Micro	Macro	Weighted	Micro	Macro	Weighted	Micro	Macro	Weighted	Training	SHAP values
XGB	**84.23**	**47.1**	**80.26**	**84.23**	24.27	**84.23**	**84.23**	**27.74**	**79.72**	0-07:37:00	1-07:00:31
RF	77.70	29.51	77.99	77.70	**29.03**	77.70	77.70	27.29	77.35	0-01:52:01	0-07:36:21
MLP	81.62	29.05	74.89	81.62	19.37	81.62	81.62	20.39	76.52	0-00:40:41	0-06:39:14
NB	81.27	21.58	72.11	81.27	13.3	81.27	81.27	12.95	74.26	**0-00:00:15**	0-06:58:52
LR	82.14	22.30	71.57	82.14	16.72	82.14	82.14	17.67	75.78	0-00:33:37	**0-06:17:10**

*The highest the better
**The lowest the better

7 Experimental Results

Figure 2 displays the relationship between the variability (on the x-axis) and the feature ranking techniques (on the y-axis).

On the top (first row) is plotted the variability against the Permutation Importance and SHAP values. It can be seen that the most influential features have a variability close to zero and as the variability increases the features' ranking decreases until 0.4. Above 0.4, the features have very little influence (zero or close to zero Permutation Importance and SHAP values) for most models.

On the middle row is displayed the variability against Gain Importance and Cover Importance. It can be seen that most influential features are located to the left and as the variability increases the features' ranking decreases until 0.2 where the features' ranking becomes constant around 1.75 for Gain Importance and 80 for Cover Importance.

On the bottom row is plotted variability against the Frequency Importance and Gini Importance. The most influential features have a variability close to zero and as the variability increases some features' ranking increases logarithmically for Frequency Importance and linearly for Gini Importance. The other features' ranking is located between the logarithmic or linear line and the vertical line at a variability equal to zero.

We can retain that there is no linear relationship between the variability and the feature ranking. However, it can be observed that features with variability greater than or equal to 0.4 have very little to no impact or importance across 4 out of the 6 ranking techniques. The exception for Frequency Importance makes sense because the higher the variability the more difficult it is to group the entries per value. As for Gini Importance, it might be due to the shortcomings of Random Forest towards feature with high variability [6, 8].

Figure 3 shows the relationship between the coefficient of variation (on the x-axis) and the feature ranking techniques (on the y-axis).

On the first row is plotted the coefficient of variation against the Permutation Importance and SHAP values. It can be seen that the most influential features have a coefficient of variation close to zero and as the coefficient of variation increases the Permutation Importance and SHAP values decrease until 11.85.[3] Between 11.85 and 31.3,[4] Permutation Importance fluctuates around the line $y = 0$ within a small range. Above 31.3, the features have zero Permutation Importance for all models. As to SHAP values, above 11.85 the features have little influence and their SHAP values fluctuate around the line $y = 0$ within a range smaller than the ones with a CoV less than 11.85.

On the middle row is displayed the coefficient of variation against Gain Importance and Cover Importance. It can be seen that all influential features are located to the left of the vertical line at 23.4.[5] Above 23.4, the features have very zero influence.

---

[3] $11.85 = \text{min CoV} + 0.23 \times (\text{max CoV - min CoV})$.

[4] $31.3 = \text{min CoV} + 0.55 \times (\text{max CoV - min CoV})$.

[5] $23.4 = \text{min CoV} + 0.42 \times (\text{max CoV - min CoV})$.

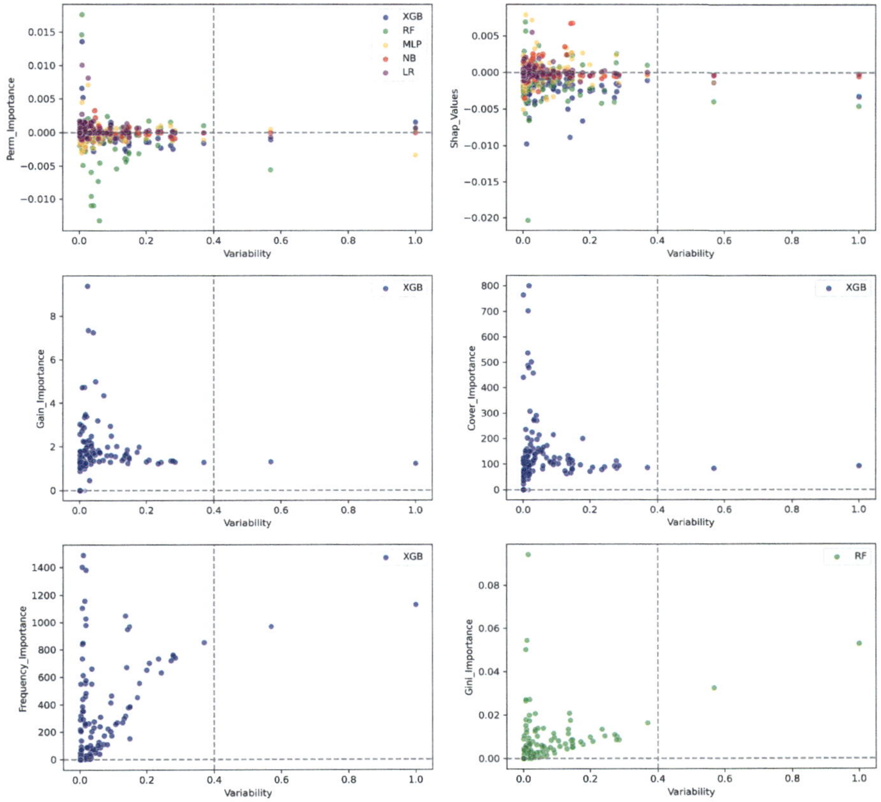

Fig. 2 Variability versus feature importance

On the bottom row is plotted coefficient of variation against the Frequency Importance and Gini Importance. The most influential features have a coefficient of variation less than 11.85. Above 11.85, the features have no influence.

We can retain that there is no linear relationship between the coefficient of variation and the feature importance ranking. But, as depicted in the graphs below, features with a coefficient of variation greater than or equal to 23.4 (on average) have little to no impact or importance. Also, features with negative CoV have little SHAP values and zero importance for all other feature ranking techniques.

We found that, overall, as the variability and the coefficient of variation of a feature increase, that feature predictive power decreases. The decay of the predictive power of a feature is noticeable for a variability from 0.4 and above and a coefficient of variation from 23.4 (on average) and above.

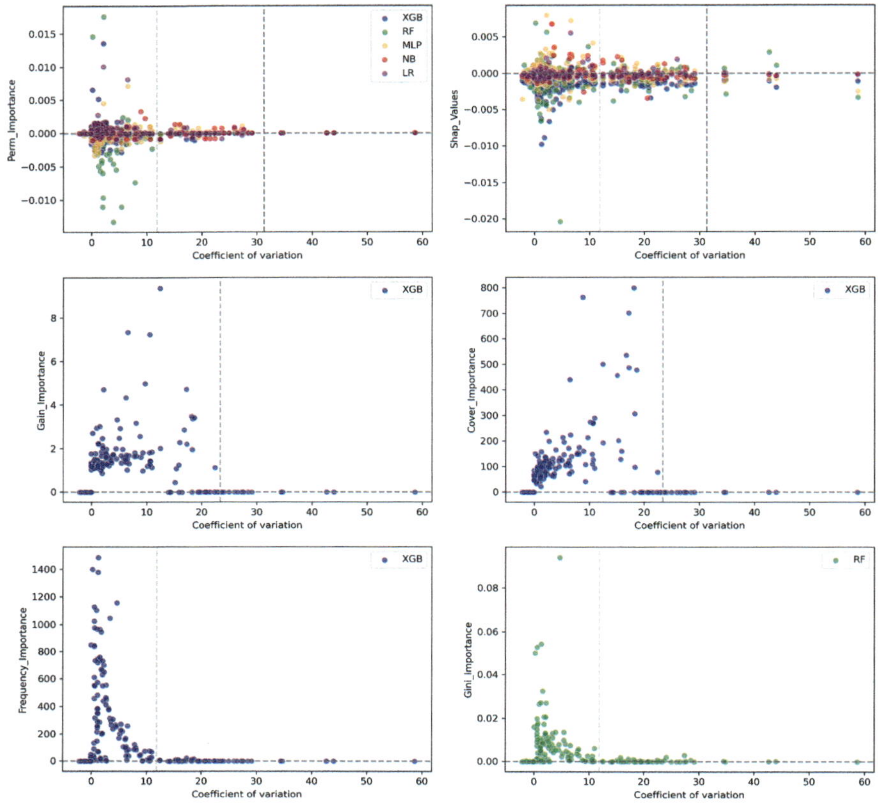

Fig. 3 Coefficient of variation versus feature importance

These experiments also show the similarity of the behaviour of Permutation Importance and SHAP values, Gain Importance and Cover Importance, and Frequency Importance and Gini Importance relative to the features' variability and coefficient of variation.

8 Discussion

The results presented in this paper suggest that features with a variability greater than or equal to 0.4 have very little to no importance for this use case; contrary to the results presented by [6, 8, 9]. Here, the features with a large number of unique values have been ranked low. This might be the result of intensive work to reduce model bias towards features with high variability since these shortcomings have been exposed by [8]. However, RF, XGB, and MLP still assigned a pretty high permutation importance or SHAP value importance to such features.

The comparison between the feature importance and the coefficient of variation (CoV) showed that features with a coefficient of variation greater than or equal to 23.4 have zero Gain, Frequency, Cover, and Gini importance. The same features have zero and little permutation importance for a coefficient of variation greater than 31.3 and between 23.4 and 31.3, respectively. In terms of SHAP values importance, they are assigned little importance by most models. But, only NB assigned zero importance consistently to those with a coefficient of variation greater than 31.3.

In general, features with variability greater than or equal to 0.4 or a coefficient of variation greater than or equal to 23.4 (on average) have a lower (average) deviation from the line y = 0 (zero importance line). The coefficient of variation seems to be a better predictor of the importance of the features than the variability. The problem with the coefficient of variation is that it is a real number and does not have a fixed range. Moreover, it can be misleading to measure the spread of data points when the mean is close to zero. For example, given three normally distributed random variables V1 $\sim$ N(1, 1), V2 $\sim$ N(0.001,1), and V3 $\sim$ N(0,1). V1, V2, and V3 have the same standard deviation but dramatically different coefficients of variation. In fact, V1, V2, and V2 have 1, 1000, and $+\infty$ (infinity) coefficient of variation, respectively. Therefore, the need to define a more robust relative spread measure is necessary.

In spite of all the shortcomings listed above, the coefficient of variation is found to be a better predictor of the (analytical) value of a feature. In this use case, all the top-ranked features have a coefficient of variation within

$$[\text{min CoV}, \text{min CoV} + 0.42 \times (\text{max CoV - min CoV})]\,.$$

This interval can be shrunk to [min CoV, min CoV $+ 0.23 \times$ (max CoV - min CoV)] for Frequency and Gini Importance.

In general, lower values of variability (less than 0.4) and coefficient of variation (less than 42% of the CoV's range) have more impact on the feature importance than higher values.

9 Conclusion

In this paper, we investigated the relationship between the feature importance and two newly introduced data valuation metrics for the content value dimension which are variability and coefficient of variation. We found that there is no linear relationship between these two data valuation metrics and feature importance. However, we found that the most influential features have a variability less than 0.4 and a coefficient of variation less than 23.4 (on average). This suggests, in our use case, that features with variability greater than 0.4 or a coefficient of variation greater than 23.4 (on average) are not relevant. This does not predict the actual value of a feature but instead helps to classify the features into non-influential and possible impactful feature sets. This finding, if generalised, will allow us to perform training-free feature selection. It also has potential applications in data protection or privacy where non-influential features could be identified prior to sharing and thus never shared, thereby increasing safety.

In future work, we firstly aim to generalise the findings in this paper by applying the proposed process on different datasets (preferably from different domains) to check the reproducibility of the insights presented in this paper, and secondly, to design a more robust relative spread measure aiming to solve the shortcomings of the coefficient of variation mentioned in the Discussion Section. We also aim to systematise the setting of the thresholds of the variability and the coefficient of variation as they have been chosen manually in this use case.

Acknowledgements This research has received funding from the ADAPT Centre for Digital Content Technology, funded under the SFI Research Centres Programme (Grant 13/RC/2106_P2), co-funded by the European Regional Development Fund. For the purpose of Open Access, the author has applied a CC BY public copyright licence to any Author Accepted Manuscript version arising from this submission.

References

1. Fleckenstein, M., Obaidi, A., Tryfona, N.: A review of data valuation approaches and building and scoring a data valuation model. Harvard Data Sci. Rev. **5** (2023). https://hdsr.mitpress.mit.edu/pub/1qxkrnig/release/1
2. Noshad, M., Choi, J., Sun, Y., Hero, A., Dinov, I.: A data value metric for quantifying information content and utility. J. Big Data **8**, 82 (2021). https://doi.org/10.1186/s40537-021-00446-6
3. Tang, S., Ghorbani, A., Yamashita, R., Rehman, S., Dunnmon, J., Zou, J., Rubin, D.: Data valuation for medical imaging using shapley value: application on a large-scale chest X-ray dataset. Sci. Rep. **11**, 8366 (2021). arXiv:2010.08006 [cs, eess]
4. Yoon, J., Arik, S., Pfister, T.: Data valuation using reinforcement learning (2019). arXiv:1909.11671 [cs, stat]
5. Ghorbani, A., Zou, J.: Data shapley: equitable valuation of data for machine learning (2019). arXiv:1904.02868 [cs, stat]
6. Loecher, M.: Unbiased variable importance for random forests. Commun. Stat. - Theory Methods. **51**, 1413–1425 (2022). arXiv:2003.02106 [cs, stat]
7. Lundberg, S., Lee, S.: A unified approach to interpreting model predictions (2017). arXiv:1705.07874 [cs, stat]
8. Strobl, C., Boulesteix, A., Zeileis, A., Hothorn, T.: Bias in random forest variable importance measures: illustrations, sources and a solution. BMC Bioinf. **8**, 25 (2007). https://doi.org/10.1186/1471-2105-8-25
9. Loecher, M.: Debiasing MDI feature importance and SHAP values in tree ensembles. In: Machine Learning and Knowledge Extraction, pp. 114–129 (2022)
10. Baudeu, R., Wright, M., Loecher, M.: Are SHAP values biased towards high-entropy features?. In: Machine Learning and Principles and Practice of Knowledge Discovery in Databases, pp. 418–433 (2023)
11. Antwarg, L., Miller, R., Shapira, B., Rokach, L.: Explaining anomalies detected by autoencoders using shapley additive explanations. Expert Syst. Appl. **186**, 115736 (2021). https://www.sciencedirect.com/science/article/pii/S0957417421011155

12. Maasland, T., Pereira, J., Bastos, D., Goffau, M., Nieuwdorp, M., Zwinderman, A., Levin, E.: Interpretable models via pairwise permutations algorithm. In: Machine Learning and Principles and Practice of Knowledge Discovery in Databases, pp. 15–25 (2021)
13. Jia, R., Dao, D., Wang, B., Hubis, F., Hynes, N., Gürel, N., Li, B., Zhang, C., Song, D., Spanos, C.: Towards efficient data valuation based on the shapley value. In: Proceedings of the Twenty-Second International Conference on Artificial Intelligence and Statistics, pp. 1167–1176 (2019). https://proceedings.mlr.press/v89/jia19a.html. ISSN: 2640-3498
14. Kumar, S., Lakshminarayanan, A., Chang, K., Guretno, F., Mien, I., Kalpathy-Cramer, J., Krishnaswamy, P., Singh, P.: Towards more efficient data valuation in healthcare federated learning using ensembling. In: Distributed, Collaborative, and Federated Learning, and Affordable AI and Healthcare for Resource Diverse Global Health, pp. 119–129 (2022)
15. Gul, F.: Bargaining foundations of shapley value. Econometrica. **57**, 81–95 (1989). https://www.jstor.org/stable/1912573, Publisher: [Wiley, Econometric Society]
16. Datta, A., Sen, S., Zick, Y.: Algorithmic transparency via quantitative input influence: theory and experiments with learning systems. In: 2016 IEEE Symposium on Security and Privacy (SP), pp. 598–617 (2016). ISSN: 2375-1207
17. Cohen, S., Dror, G., Ruppin, E.: Feature selection via coalitional game theory. Neural Comput. **19**, 1939–1961 (2007), Conference Name: Neural Computation
18. Campbell, T., Roder, H., Georgantas III, R., Roder, J.: Exact Shapley values for local and model-true explanations of decision tree ensembles. Mach. Learn. Appl. **9**, 100345 (2022). https://www.sciencedirect.com/science/article/pii/S2666827022000500
19. Wu, Z., Shu, Y., Low, B.: DAVINZ: data valuation using deep neural networks at initialization. In: Proceedings of the 39th International Conference on Machine Learning, pp. 24150–24176 (2022). https://proceedings.mlr.press/v162/wu22j.html. ISSN: 2640-3498
20. Altmann, A., Toloi, L., Sander, O., Lengauer, T.: Permutation importance: a corrected feature importance measure. Bioinformatics **26**, 1340–1347 (2010). https://doi.org/10.1093/bioinformatics/btq134
21. Shardlow, M.: An analysis of feature selection techniques (2011). https://www.semanticscholar.org/paper/An-Analysis-of-Feature-Selection-Techniques-Shardlow/8973a724545bbc2a5cc52bc28f7ffcb5d4aa8dc8
22. Strumbelj, E., Kononenko, I.: An efficient explanation of individual classifications using game theory. J. Mach. Learn. Res. **11**, 1–18 (2010)
23. Brennan, R., Attard, J., Petkov, P., Nagle, T., Helfert, M.: Exploring data value assessment: a survey method and investigation of the perceived relative importance of data value dimensions. (SciTePress, 2019). https://cora.ucc.ie/handle/10468/8166, Accepted: 2019-07-16T09:18:42Z
24. Brennan, R., Attard, J.: Management of Data Value Chains, a Value Monitoring Capability Maturity Model (2018). http://www.tara.tcd.ie/handle/2262/82277, Accepted: 2018-01-25T15:30:03Z Journal Abbreviation: 20th International Conference on Enterprise Information Systems (ICEIS)
25. Hapke, H., Nelson, C.: Introduction. In: Building Machine Learning Pipelines: Automating Model Life Cycles With TensorFlow (2020). https://www.oreilly.com/library/view/building-machine-learning/9781492053187/
26. Shapley, L.: 17 A value for n-person games. In: Contributions to the Theory of Games (AM-28), vol. II. pp. 307–318 (1953). https://www.degruyter.com/document/doi/10.1515/9781400881970-018/html
27. Shobeiri, S., Aajami, M.: Shapley value in convolutional neural networks (CNNs): a comparative study. Am. J. Sci. Engin. **2**, 9–14 (2021)
28. Brown, C.: Coefficient of variation. Appl. Multivariate Stat. Geohydrol. Relat. Sci. 155–157 (1998). https://doi.org/10.1007/978-3-642-80328-4_13
29. Hapke, H., Nelson, C.: Introduction. In: Building Machine Learning Pipelines: Automating Model Life Cycles with TensorFlow. O'Reilly Media, Inc. (2020). ISBN: 9781492053194

Evaluating Machine Learning Algorithms for New Indian Parliament Building Sentiment Analysis

Jatinderkumar R. Saini, Shraddha Vaidya, and Shailesh Kasande

Abstract The Indian Parliament is a great example of the nation's democratic ideals. The parliament with its upper (Rajya Sabha) and lower house (Lok Sabha) forms the highest law-making body of the nation. The old parliament building, constructed during British era, has structural issues and is insufficient to accommodate the increasing number of members of parliament. Thus, the ruling government of India decided to construct the new building with superior infrastructure which can accommodate the increasing number of members of parliament, architecture and Indian sculptures. The construction of new building led to controversies from various stakeholders of the country and neighbouring countries. The mixed opinions about the construction of new parliament building generated chaos leading to environmentalists, and conservationists filing case at Supreme Court of India to pause the construction of new building. Thus, it is essential to study the various sentiments of the civilians of India and develop a model with machine learning techniques that could classify the sentiments into positive, negative and neutral. To fill up this gap, authors in this research have generated the dataset, namely, *New Indian Parliament Dataset-URL (NIPD-U)* and *New Indian Parliament Dataset-Paragraph (NIPD-P)* from authentic web sources related to India's new parliament building, and have developed machine learning models with five classifiers to classify the text into positive and negative sentiments. It was observed that for URL-based classification Naïve Bayes (NB) classifier outperformed other classifiers with 91% of accuracy. Further, paragraph-wise results showed that Random Forest (RF) showed highest accuracy of 92.34%.

Keywords Sentiment analysis · Indian parliament building · Sansad Shavan · Lok Sabha · Rajya Sabha · Machine learning

---

J. R. Saini · S. Vaidya (✉)
Symbiosis Institute of Computer Studies and Research, Symbiosis International (Deemed University), Pune, India
e-mail: phdyashoda.barve@gmail.com

S. Kasande
Suryadatta Group of Institutes, Pune, India

G. Fortino et al. (eds.), *Proceedings of Third International Conference on Computing and Communication Networks*, Lecture Notes in Networks and Systems 917,
https://doi.org/10.1007/978-981-97-0892-5_48

1 Introduction

In the Indian Republic, the parliament is the highest body of law-making [1, 2]. The Lok Sabha or lower house and Rajya Sabha or upper house make up this bicameral legislature [3, 4]. The old parliament house also named as "Sansad Bhavan" in Hindi is located in New Delhi, the capital city of India. The old parliament house was constructed by the British Government in 1927. The building covered an area of 2.29 hectares and twenty-one metres tall. The Lok Sabha, Rajya Sabha and library hall are integral part of central hall of the parliament surrounded by circular structure providing space for members of houses of the parliament. The old building of parliament constructed 85 years before was suffering from structural issues, overuse, and distress in addition to insufficient space to accommodate the members and staff of the houses of the parliament [5–7].

Considering the inefficiency of old parliament building, the incumbent government initiated the project of constructing new Indian parliament building as a part of Central Vista Project costing around $120 million [8]. The project aimed at maximizing the seating capacity by 150% of the houses of the parliament having expected life span of 150 years [9]. In October 2020, the ground-breaking ceremony for the construction was conducted and in December 2020 the incumbent government leader laid the foundation stone [5]. The new Indian parliament building has been designed based on the Indian culture [10, 11]. The triangular-shaped new parliament building with 64, 500 square metre is bigger than the old parliament building of India having capacity to accommodate 1272 members of parliament (MP) including Lok Sabha (888 MPs) and Rajya Sabha (384 MPs) considering the increase in number of members of parliament in future [6, 10, 12]. In addition, to display the democratic heritage of India, a visitor's gallery is installed. The constitutional hall depicts the unique copy of the constitution. The modern digital technologies are installed in the new Indian parliament as an effort towards creation of "paperless offices" [6]. On 28th of May 2023, the new parliament building was inaugurated by unveiling the plaque and performing traditional puja and Vedic rituals in presence of the honourable speaker of Lok Sabha. The incumbent government [13] leader stated that India is the mother of democracy and the new parliament building will see the emergence of Independent India [7, 14]. There were both scathing criticism and fervent support from numerous stakeholders across the nation from the time the new parliament building was being built until it was officially opened [5]. The incumbent government has stood by the initiative by emphasizing its economic viability, despite criticism from non-BJP parties, liberal academics, a segment of artists and urban planners [15]. The opposition parties, environmentalist, conservationists and organizations had criticized on the construction of new parliament building due to their concern about causing harm to the environment, utilization of exhaustive funds amid coronavirus, and the revamp can meddle with the history of the existing building [6]. These mixed opinions about the construction and usage of new parliament building of India has motivated the authors in this study to analyse the sentiments of citizens of India, and other stakeholders and perform analysis with an aim to classify the sentiments as positive and negative categories. Therefore, the objective of this study is to gather the textual data about India's new parliament building from authentic web sources, identify the sentiment of various stakeholders, extracting relevant features from these sentiments and classify these sentiments into positive, and negative categories using machine learning models. The research contributions of the study are as follows:

1. Perform possible exhaustive review of India's new parliament building, its facilities and sentiments of various stakeholders.
2. Attempt to create and annotate the dataset of articles into positive, negative and neutral categories from authentic web sources related to India's new parliament building.
3. Develop machine learning models to automatically classify the sentiments into positive and negative categories.

2 Related Work

In this section, authors discuss the literature from three perspectives. First, the architecture of new parliament building of India, secondly the controversies arrived and thirdly the positive aspect of having a new parliament building.

A. Sculptures and Architectural Designs of India's New Parliament Building

The new building of Indian parliament has three entrances, namely, Gyan dwar, Shakti dwar and Karma dwar depicting knowledge, power and karma gate. Gaj, ashwa, garud, makar, shardula and hams structure the sculptures from Indian temples and mahals. The Foucault pendulum developed by National Council of Science Museum (NCSM) represents India's conception of the universe [5, 10]. The interior of Lok Sabha and Rajya Sabha is based on the theme of peacock and lotus, the national bird and flower of India. Nearly 5,000 works of art, including murals made of metal, stone and wall panels, are shown in the new parliament building. In addition, Sengol is placed in Lok Sabha which is viewed as a representation of justice, power transitions and ethical leadership [16]. Overall the material required for the construction of the new building of the parliament was brought from different states of the country. To some extent, the entire nation came together to build the temple of democracy, exemplifying the true meaning of "Ek Bharat Shreshtha Bharat" [7].

B. Controversies Related to India's New Parliament Building

Since the construction of new parliament building of India began, the various stakeholders have created controversies over its cost and lack of consultation. This section discusses in detail the controversies created by various stakeholders such as environmentalists, conservationists, organizations, politicians and other nations. The environmentalists stated that the project is harmful to the environment and can cause pollution [6]. In addition, they claimed that it is an encroachment of public place by the government. Environmentalists approached the Supreme Court to stop the project till the pandemic is over. However, the supreme court of India upheld the environmental clearance and notification for change in land usage which disappointed the environmentalists [17]. Conservationists fear that revamp of parliament will meddle with the history of current building. However, heritage conservation committee approved the project in the favour of the leading government [18]. In addition, sixty-five organizations also requested to pause the project and use the funds in handling the COVID-19 pandemic [6]. The political parties in opposition boycott the inaugural ceremony and overall construction of new Indian parliament due to reasons such as disqualification, suspension and muting of opposition lawmaking. Also, opposition showed a concern of side-lining head of the state of India for inauguration ceremony. The new parliament's painting in which a map of an "undivided India" is depicted sparked a diplomatic dispute with neighbouring countries and has asked the report on the controversy [5, 19]. In response, the Indian government played down the issue of this controversy by explaining the fact that the Ashokan Empire's expansion is the sole aspect of the artwork that is shown [20].

C. Acquiescence by the Stakeholders

Although there was extensive controversy about the new parliament building, people of India have acknowledged the efforts behind new parliament building. The leader of incumbent government mentioned that the new parliament is the vision of contemporary India. The most popular Indian Actor tweeted saying that the new parliament is the magnificent home for members of parliament [21]. Indian parliamentary stakeholder stated that the "resolve is clear- Akhand Bharat" and that the new parliament building represents the powerful and self-reliant India [6].

Seeing the structure, need, controversies and acquiesces regarding India's new parliament building, authors believed to study the topic extensively.

3 Materials and Methods

In this section, authors discuss the methodology adopted to perform the sentiment analysis of opinions of users on new Indian parliament. The authors have elaborated detailed process of dataset generation, pre-processing, annotation and method of sentiment analysis. Figure 1 displays the process of sentiment analysis and classification with machine learning classifiers.

A. Dataset Collection

The dataset creation process begins with identifying keywords related to parliament. Authors have performed analysis of web URLs related to new Indian parliament to identify keywords which would fetch relevant URLs from the Google search engine. Google is the most popular search engine used by 60% of the internet users to search relevant information [22]. A set of 25 keywords were finalized, listed in Table 1, which helped in fetching relevant web URLs. The queries using keywords were entered into search engine, which resulted in Google displaying the list of URLs related to new Indian parliament. Authors have scrapped these URLs using a crawler

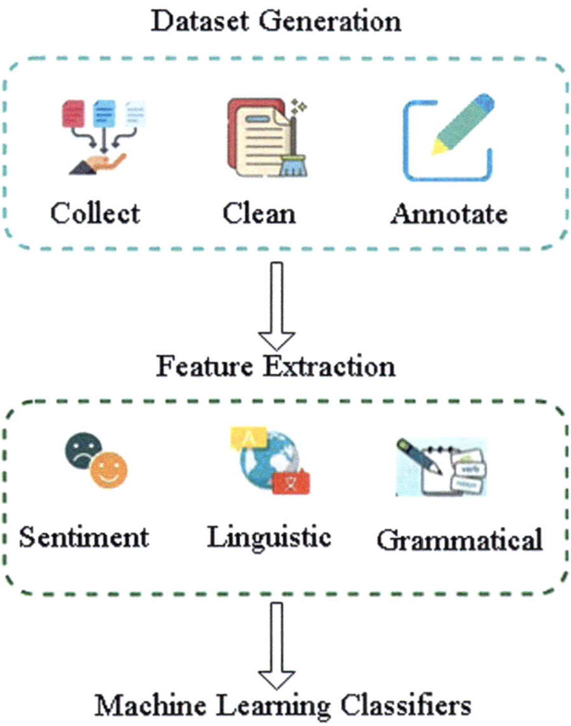

Fig. 1 Proposed methodology of sentiment analysis and classification

module implemented in Python using Beautiful Soup library. The crawler [23, 24] module has the capability to remove duplicate and redundant URLs as well as it also ignores irrelevant URLs. The irrelevancy is dependent on percentage of relevant keywords present in the URL contents. The URL having percentage of relevant keywords greater than the threshold belongs to relevant URLs not otherwise. The crawler was able to scrap 110 such relevant URLs. Authors name this dataset as *New Indian Parliament Dataset (NIPD)*. Further, each URL contents were scrapped paragraph-wise based on the html tags for granular level analysis of sentiments about new Indian parliament. Therefore, NIPD consists of two datasets *New Indian Parliament Dataset-URL (NIPD-U)* and *New Indian Parliament Dataset-Paragraph (NIPD-P)*.

B. Dataset Cleaning

The data cleaning process plays significant role in sentiment analysis. The unnecessary information such as punctuations, one, two or three characters, and duplicate words is removed from the data.

Table 1 List of keywords

Sr. no	Keyword	Sr. no	Keyword
1	Parliament	11	Rafi marg
2	New Indian	12	Replace
3	Indian Parliament	13	Complex
4	New Parliament House	14	Building
5	New Delhi	15	Lok Sabha
6	Sansad Bhavan	16	Rajya Sabha
7	Parliament House	17	Sengol
8	Central Vista	18	Chamber
9	Redevelopment	19	Lower
10	Project	20	Upper

C. Dataset Annotation

The next task in dataset creation process is data annotation. Authors have used SentiWordNet [25] library sentiment library from NLTK in python to compute polarity score URL-wise and paragraph-wise. Based on the polarity score, the target variable is set to positive, negative or neutral. Thus, the *NIPD-U* consists of 129 URLs having 118 positive URLs, and 11 negative URLs. The *NIPD-P* consists of 2033 paragraphs having 1739 positive paragraphs and 294 negative paragraphs. Authors didn't find any URL which could be classified as purely neutral, and thus have not considered for data annotation.

D. Feature Extraction

An important factor in any machine learning-based analysis of textual data is extracting relevant and useful features from the data. In this study, authors have focused on linguistic features, sentimental features and grammatical features. Table displays the list of features with description used in this study to perform automated classification of URLs and paragraphs into positive and negative categories.

E. Machine Learning Techniques

The supervised machine learning algorithms assist in classifying the textual contents into different categories. Therefore, authors have used five machine learning classifiers namely Logistic Regression (LR), Support Vector Machine (SVM), Naive Bayes (NB), Decision Tree (DT) and Random Forest (RF) classifier to perform sentiment classification and evaluated the performance of the model in terms of accuracy, precision, recall and f1-score. In addition, authors have performed analysis of important features which have contributed in classification of URLs and paragraphs in terms of heat map and pair plot (Table 2).

Table 2 Feature list

Sr. no	Feature name	Description
1	Word_Cnt	Counts the total number of words
2	Negative_Count	Count of negative words
3	Positive_Count	Count of positive words
4	Neutral_Count	Count of neutral words
5	Pos_Per	Percentage of positive words
6	Neg_Per	Percentage of negative words
7	Sentence_Polarity_Score	Score of sentences-wise polarity
8	No_of_Verb	Verb count
9	No_of_Noun	Noun count
10	No_of_Adjective	Adjective count
11	No_of_Adverb	Adverb count
12	No_of_Pronoun	Pronoun count
13	No_of_Clean_Word_POS	Number of words with POS tag
14	Total_Sentence	Total number of sentences

4 Experimental Results and Evaluations

This section discusses the results with the experiments carried out using machine learning classifiers. To evaluate the performance of the model, the dataset was divided into train and test set. Table 3 displays the count of URLs and paragraphs in train and test datasets. Figure 2. displays the performance of machine learning classifiers on URL dataset. It can be observed that the NB algorithm shows highest accuracy of 91%, followed by SVM and RF with 89%. LR classifier showed less accuracy of 87%. In terms of precision, RF showed highest precision of 98.6% while DT showed less precision of 93.8%. SVM showed poor performance with recall of 90.3% while RF outperformed others 92.2% of recall. RF outperformed others even with F1-score of 95.4% while SVM showed least F1-score of 94.2%. Figure 3 displays the paragraph-wise analysis of sentiments. It can be observed that the RF model showed highest accuracy of 92.34% while DT showed lowest accuracy of 53.33%. In addition, RF has outperformed other classifiers in terms of precision, recall and F1-score with 98.6%, 92.2%, and 95.4% respectively. Further, LR showed lowest precision of 97.1% while SVM showed lower recall and F1-score of 90.3% and 94.2% respectively. Figure 4 displays the Receiver Operating Curve (ROC) of sentiments based on paragraphs, it can be seen that RF has outperformed other classifiers. Figure 5 displays the pair plot of features showing relationship of important features.

Table 3 Distribution of dataset for train and test

URL	Category	Train	Test	Total
	Positive	27	91	118
	Negative	2	9	11
	Total	**29**	**100**	**129**
Paragraph	Category	Train	Test	Total
	Positive	430	1309	1739
	Negative	77	217	294
	Total	**507**	**1526**	**2033**

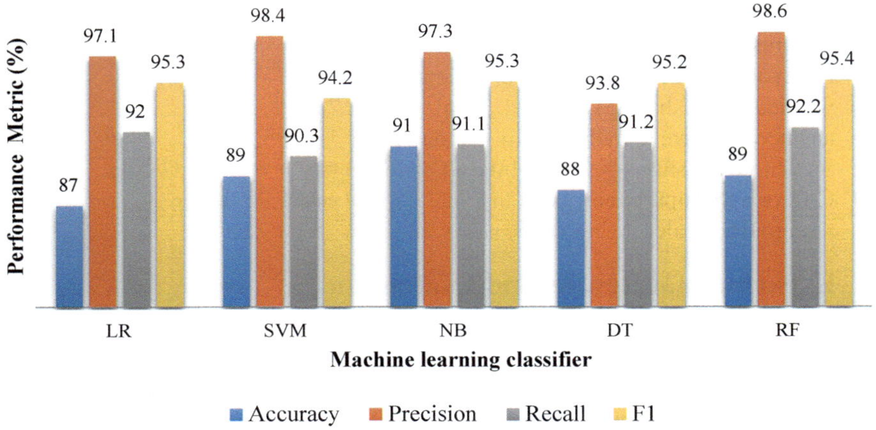

Fig. 2 URL-based sentiment classification

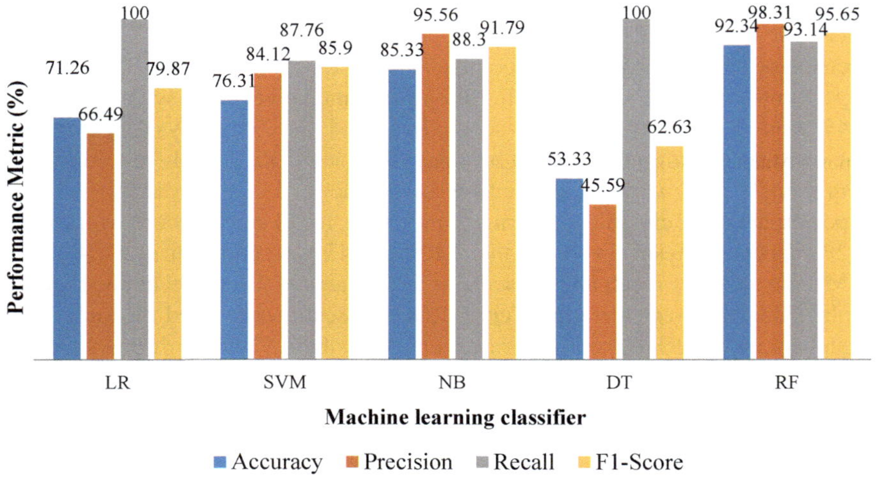

Fig. 3 Paragraph-based sentiment classification

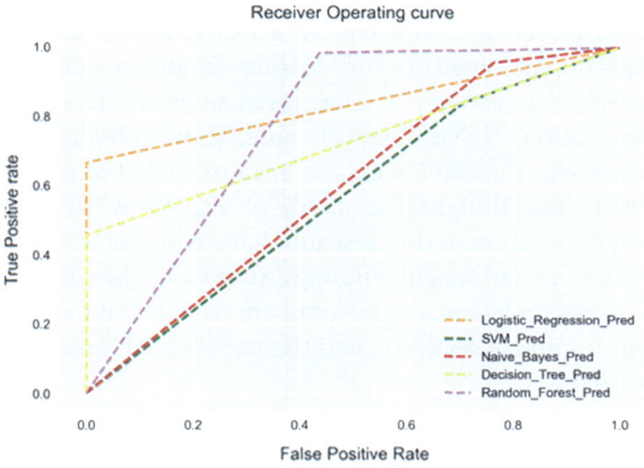

Fig. 4 Paragraph-wise ROC

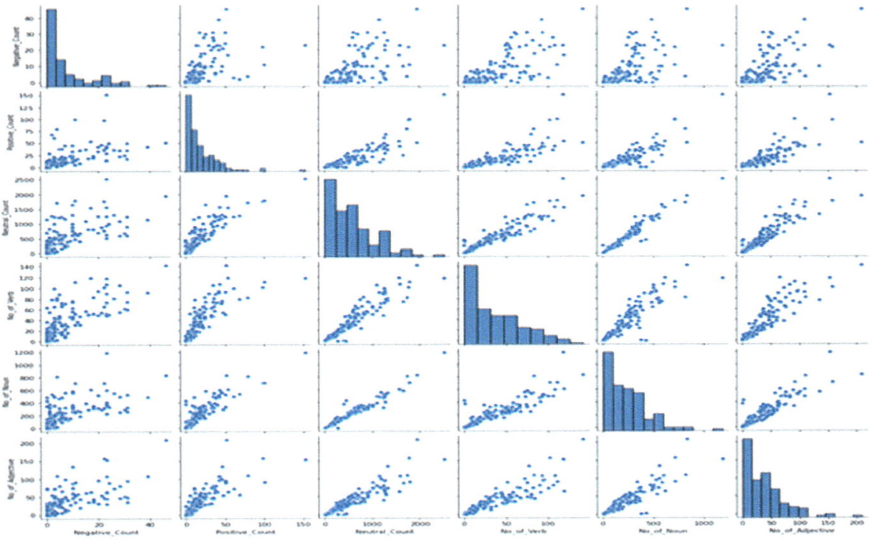

Fig. 5 URL-based feature pair plot

5 Conclusion

In this study, authors have constructed a novel dataset *New Indian Parliament Dataset-URL (NIPD-U)* and *New Indian Parliament Dataset-Paragraph (NIPD-P)* consisting of positive and negative URLs and paragraphs respectively. Further, novel set of features are extracted consisting of sentimental, grammatical and textual feature. Authors have used SentiWordnet library of Python to extract relevant sentimental features and implemented machine learning models to classify the sentimental URLs and paragraphs into positive and negative categories. It was observed that for URL-based classification, Naïve Bayes classifier showed highest accuracy of 91% while logistic regression showed lowest accuracy of 87%. For paragraph-wise analysis, the RF model showed highest accuracy of 92.34% while DT showed lowest accuracy of 53.33%. Authors in this research have considered neutral construction and usage of new Indian parliament building. Authors neither intend to promote nor support and not intend to defend any government or political party. Only sole intention is to perform sentiment analysis, and the represented results are best reported on the considered dataset.

References

1. Singh, M.P.: The decline of the Indian Parliament. India Rev. **14**(3), 352–376 (2015). https://doi.org/10.1080/14736489.2015.1066216
2. "Indian Parliament," Government of India (2023). https://www.india.gov.in/my-government/indian-parliament (Accessed Aug. 29, 2023)
3. Adhya, S., Sanyal, D.K.: What does the Indian Parliament Discuss? An exploratory analysis of the question hour in the Lok Sabha. In: 1st Workshop on Natural Language Processing for Political Sciences, PoliticalNLP 2022 - Proceedings, as part of the 13th Edition of the Language Resources and Evaluation Conference, LREC 2022, pp. 72–78 (2022)
4. Bharatharaj, J., Sasthan Kutty, S.K., Munisamy, A., Krägeloh, C.U.: What do members of parliament in India think of robots? Validation of the Frankenstein syndrome questionnaire and comparison with other population groups. Int. J. Soc. Robot. **14**(9), 2009–2018 (2022). https://doi.org/10.1007/s12369-022-00921-x
5. Wikipedia, New Parliament House, New Delhi, Wikipedia. https://en.wikipedia.org/wiki/New_Parliament_House,_New_Delhi
6. Ritesh, S.: What is 'Central Vista' and why it is being opposed; Zee explains the controversial project, ZEENEWS (2021). https://zeenews.india.com/india/what-is-central-vista-and-why-it-is-being-opposed-zee-explains-the-controversial-project-2361620.html
7. PM Modi inaugurates new Parliament building, The Economic Times (2023). https://economictimes.indiatimes.com/news/india/pm-modi-inaugurates-new-parliament-building/articleshow/100561482.cms?from=mdr
8. Astha, R.: Why India's new parliament building is so controversial, Time 2023. https://time.com/6282819/india-parliament-narendra-modi-controversy/
9. Diapk, D.: New Parliament building will last 150 years, its Houses can seat 150% more MPs, The Times of India (2020). https://timesofindia.indiatimes.com/india/new-parliament-building-will-last-150-yrs-its-houses-can-seat-150-more-mps/articleshow/79671363.cms?from=mdr
10. Door statues, artworks, Foucault's Pendulum: New Parliament's architectural elements capture India's imagination, The Economic Times (2023). https://economictimes.indiatimes.com/news/india/door-statues-artworks-foucaults-pendulum-new-parliaments-architectural-elements-capture-indias-imagination/articleshow/100593455.cms
11. Abdel, H.: New building for the Parliament of India/HCP design, Planning and Management, Archdaily (2023). https://www.archdaily.com/1003508/new-building-for-the-parliament-of-india-hcp-design-planning-and-management (accessed Aug. 29, 2023).
12. New Parliament building for New India, Daily Updates. https://www.drishtiias.com/daily-updates/daily-news-editorials/new-parliament-building-for-new-india

13. Aiyar, Y., Tillin, L.: 'One nation', BJP, and the future of Indian federalism. India Rev. **19**(2), 117–135 (2020). https://doi.org/10.1080/14736489.2020.1744994
14. New Parliament building inauguration updates, The Hindu (2023). https://www.thehindu.com/news/national/new-parliament-building-inauguration-live-prime-minister-narendra-modi-in-delhi-for-ceremony/article66903230.ece
15. Ahmed, H.: The New Parliament building and the democratic future of India, NDTV (2023). https://www.ndtv.com/opinion/the-new-parliament-building-and-the-democratic-future-of-india-4068588
16. What is Sengol? Know its history and significance, Jagran Josh (2023). https://www.jagranjosh.com/general-knowledge/what-is-sengol-1685012740-1
17. Activists 'Disappointed' by Supreme Court's approval to Central Vista Project, The New Indian Express (2021). https://www.newindianexpress.com/nation/2021/jan/05/activists-disappointed-by-supreme-courts-approval-to-central-vista-project-2245891.html
18. Heritage Conservation Committee approves construction of New Parliament Building, NDTV (2021). https://www.ndtv.com/india-news/heritage-conservation-committee-approves-construction-of-new-parliament-building-2350878
19. Nepal Seeks Report on 'Controversial' Mural in India's New Parliament Building, India Today. https://www.indiatoday.in/world/story/nepal-india-new-parliament-building-mural-map-row-2395601-2023-06-20
20. India clarifies on mural in Parliament building that offended Nepal Politicians, Hindustan Times (2023). https://www.hindustantimes.com/india-news/india-downplays-controversial-mural-in-new-parliament-building-depicting-spread-of-ashokan-empire-101685714718698.html
21. What a Magnificent New...': Shah Rukh Khan praises New Parliament building; PM Responds, Business Today (2023). https://www.businesstoday.in/latest/story/what-a-magnificent-new-shah-rukh-khan-praises-new-parliament-building-pm-responds-383123-2023-05-28
22. Barve, Y., Saini, J.R., Pal, K., Kotecha, K.: A novel evolving sentimental bag-of-words approach for feature extraction to detect misinformation. Int. J. Adv. Comput. Sci. Appl. **13**(4), 266–275 (2022). https://doi.org/10.14569/IJACSA.2022.0130431
23. Barve, Y., Saini, J.R., Kotecha, K., Gaikwad, H.: An incremental approach to classify healthcare URLs using a novel 'web document classification model.' Smart Innovat., Syst. Technol. **311**, 653–663 (2023). https://doi.org/10.1007/978-981-19-3571-8_60
24. Barve, Y., Saini, J.R.: Detecting and classifying online health misinformation with 'Content Similarity Measure (CSM)' algorithm: an automated fact-checking-based approach. J. Supercomput. **79**(8), 9127–9156 (2023). https://doi.org/10.1007/s11227-022-05032-y
25. Esuli, A., Sebastiani, F.: SENTIWORDNET: A publicly available lexical resource for opinion mining. In: Proceedings of the 5th International Conference on Language Resources and Evaluation, LREC 2006, European Language Resources Association (ELRA), 2006, pp. 417–422. https://www.scopus.com/inward/record.uri?eid=2-s2.0-85031789656&partnerID=40&md5=90e374da1f67c13ce93da274568ba34b

Adoption of Artificial Intelligence in Business Operations of Technology Firms

Spardha Bisht, Santoshi Sengupta, and Manish Kumar Bisht

Abstract Artificial Intelligence (AI) has brought about significant transformation in the topography of business operations in large technology at a global level. Addressing the growing need of investigating the factors that influence adoption of AI in business processes in technology organizations, this study collects data from 252 middle-level employees about their perception and attitude towards AI in business operations. Statistical analysis of the data shows that while performance expectancy has a significant positive effect on employee attitude and intention to use AI in business operations, effort expectancy does not have any significant effect. Also, it is found that employee attitude mediates the relationship of performance expectancy and effort expectancy with intention to use AI. The study provides us with interesting insights for practitioners about nuances of implementing AI in business operations in large technology organizations.

Keywords Artificial intelligence · Business operations · Technology · Organizations · Adoption

1 Introduction

In the last ten years, advancements in modern technologies such as data analysis, artificial intelligence (AI), and business intelligence applications have significantly eased the development of organizational procedures and outcomes [1]. The rapid

S. Bisht · S. Sengupta (✉)
School of Management, Graphic Era Hill University, Bhimtal Campus, Dehradun, Uttarakhand, India
e-mail: santoshisengupta@gmail.com

S. Bisht
e-mail: spardhabisht9@gmail.com

M. Kumar Bisht
Graphic Era Hill University, Haldwani Campus, Uttarakhand, India
e-mail: manishkumarbisht@geu.ac.in

G. Fortino et al. (eds.), *Proceedings of Third International Conference on Computing and Communication Networks*, Lecture Notes in Networks and Systems 917,
https://doi.org/10.1007/978-981-97-0892-5_49

progress of AI has generated massive volumes of data, which is valuable and plays a notable role in resource allocation [2]. Such humongous amount of data has offered a more effective approach to devising successful marketing and operational strategies through digital analytics [3]. These factors collectively have brought about a substantial improvement in business performance, business operations, employee management, customer relationship management, and service recovery processes [4]. To add to this, the advent of cutting-edge technology has led to higher organizational effectiveness, efficiency, and productivity, as its knowledge base and updated information continually enhance its capabilities. Among these advancements, proponents of AI have particularly envisioned a future in which intelligent computers manage routine tasks presently carried out by humans, thus enabling humans to focus on creative pursuits [5]. AI can identify inefficiencies in business operations and suggest optimal resource allocation to enhance performance. Indeed, AI has become an asset in supporting organizations by providing extensive data and automating various aspects of business processes to accelerate and optimize operations.

1.1 Past Research in AI Adoption in Businesses Operations

The body of scholarly knowledge of AI is at a developing stage where the past work is situated in the multi-disciplinary context of information management, international business, and general management [1]. Prior literature has focused on internationally oriented studies that have provided insights into the relationship between AI and business operations. While symbiosis is at a nascent stage, there are a few significant works that have explored multiple facets of AI such as the journey of AI adoption in organizations [6], impact of AI on business operations [7], applications, challenges, and opportunities in business operations [8].

In the context of technology businesses, it is seen that such organizations benefit from rapid technological advancement, intricate business models, flexible strategies, and a globally dispersed workforce comprising individuals from various linguistic and cultural backgrounds across different time zones [9]. Recent scholarly publications [10, 11] indicate that a growing proportion of businesses worldwide are incorporating AI into their diverse business operations. Notably, Chinese and Indian companies are at the forefront, with nearly 60% of technology professionals stating that their organizations are actively engaged in AI utilization.

Prior investigations have demonstrated that the integration of AI into business processes can evoke employee apprehensions and yield both favorable and unfavorable effects on both employee and business outcomes. These perceived advantages and risks associated with AI adoption within business operations are likely to shape the perspectives of AI consultants toward its adoption, especially given the limited existing research on the determinants of AI adoption. The convergence of AI and business operations has captured the attention of researchers from various perspectives. However, several unresolved inquiries persist concerning the incorporation of AI in business processes, encompassing aspects such as employees' willingness to use AI, its sustained utilization, its achievements, and setbacks, return on investment, and employee reactions to the resultant alterations.

2 Research Objectives and Hypotheses Development

While AI has the potential to revolutionize diverse business operations, a comprehensive understanding of the factors that underpin the successful adoption of AI in business processes is essential. Keeping this in mind, the present study aims to study the relationships of drivers with the attitude of employees and their eventual intention to use AI in business operations.

2.1 Research Objectives

Drawing from the Unified Theory of Acceptance of technology and Usage of Technology (UTAUT) [12], we have identified two drivers—performance expectancy and effort expectancy to check the influence on the employee attitude and intention to use. Thus, the research objectives are as follows:

(i) To investigate the relationships among performance expectancy, effort expectancy, employee attitude, and intention to use AI.
(ii) To investigate the effects of performance expectancy and effort expectancy on intention to use AI.
(iii) To investigate the mediating role of employee attitude in translating performance expectancy and effort expectancy to subsequent intention to use AI.

2.2 Hypotheses Development

Performance expectancy can be defined as the degree to which employees perceive that utilizing AI technology will offer advantages to them and result in enhancements in their overall performance. It plays a crucial role in influencing an individual's intention to use AI in the form of software, digital applications, and chatbots, as it streamlines their tasks and augments their performance impact. Employees might evaluate the performance gains of AI applications and chatbots by comparing the improvements achieved post their usage. If they perceive the value and advancements stemming from AI-based business processes favorably, their inclination to use it will be greater. Similarly, the concept of effort expectancy denotes the ease of utilizing technology and stands as a significant determinant in forecasting the intention to adopt its usage. Earlier studies have indicated that effort expectancy yields a substantial positive impact on the inclination to persistently employ technology. The

Fig. 1 Research model

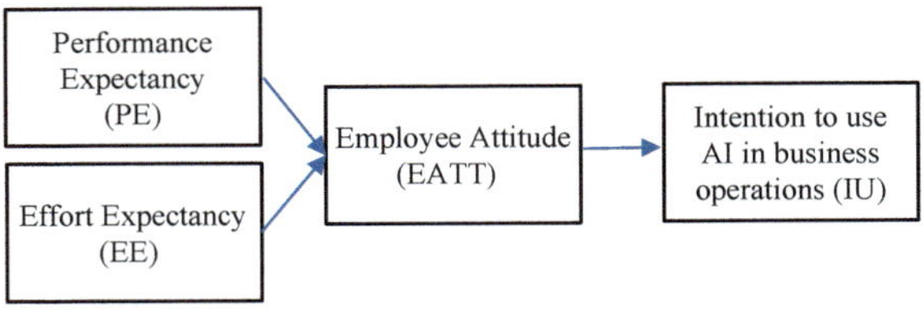

user-friendliness of AI solutions could positively influence their continued utilization, as reduced effort required for app usage generally results in increased adoption rates.

An individual's acceptance of technology is shaped by behavioral, normative, and control beliefs [13]. Behavioral beliefs lead to the formation of attitudes toward a specific behavior, normative beliefs generate a perception of social obligation linked to that behavior, and control beliefs impact the overall perception of engaging in that behavior. The adoption of AI in business operations can result in varied outcomes—some technologies might function effectively and gain swift approval, while others could face setbacks. The success or failure of a technology's performance can influence the attitude held toward that technology. When performance expectations are met, individuals often develop a form of psychological connection that translates into positive attitudes toward the technology, thereby influencing the inclination to use it. Similarly, the effort expectancy can also shape the attitude toward technology usage. This, in turn, may positively affect the intention to use AI in business operations.

Based on the above past research, we hypothesize the following three hypotheses which are also shown in Fig. 1, research model.

H1: There is a positive relationship among performance expectancy, effort expectancy, employee attitude, and intention to use AI in business operations.

H2: Performance Expectancy and effort expectancy have positive effects on intention to use AI in business operations.

H3: Employee attitude mediates the relationship between performance expectancy and effort expectancy with intention to use AI in business operations.

3 Sample

We administered a questionnaire that measured performance expectancy [14], effort expectancy [14], employee attitude [15], and intention to use [14] using MS Forms to 500 employees employed in large technology organizations, of which 252 employees responded and participated in the survey. We gathered information about sex (males = 154 and females = 98), age groups (less than 30 years = 78, 30 to 40 years = 117, More than 40 years = 57), and experience levels (less than 5 years = 109, 5 to

Table 1 Profile of respondents

Characteristic	Details	Number	Percentage (%)
Sex	Males	154	61.11
	Females	98	38.89
Age	Less than 30	78	30.95
	30–40	117	46.43
	More than 40	57	22.62
Experience	Less than 5 years	109	43.25
	5–10 years	83	32.94
	More than 10 years	60	23.81

10 years = 83, and greater than 10 years = 60). The respondent profile is detailed in Table 1.

4 Results and Discussion

We analyzed the relationships among performance expectancy, effort expectancy, employee attitude, and intention to use AI in business operations through Pearson correlations. We found out the effects of performance expectancy and effort expectancy on intention to use AI in business operations using multiple linear regression and tested for the mediating role of attitude in these relationships.

All four variables, viz., performance expectancy, effort expectancy, employee attitude, and intention to use AI in business operations had significant positive correlations. From Table 2, it is observed that all the key study variables have a significant correlation with one other. Cronbach's alpha for all the variables is in the range of 0.61–0.91, thus establishing the reliability of all the scales used for the study. Multiple linear regression results indicated that performance expectancy ($\beta = 0.47$, $t = 8.15$, $p = 0.000$) significantly predicted employees' intention to use AI. We found partial support for hypothesis 2 as effort expectancy did not have a significant impact on employees' intention to use AI. The F-statistic of 39.71 shows that the overall model is significant with an R-square value of 33.7%. In the next step, we tested for the last hypothesis which stated that employee attitude mediates the relationship between performance expectancy and effort expectancy with intention to use AI in business operations. For this we did Hayes and Preacher's (2010) Model 4 approach followed by a bootstrapping procedure in order to test the indirect effects. We found that employee attitude significantly mediated the relationship between performance expectancy and intention to use AI (total effect of PE on EATT: $\beta = 0.497$, $p = 0.000$; direct effect: $\beta = 0.518$, $p = 0.000$). Similarly, it was found that employee attitude had a significant effect on the relationship between effort expectancy and intention to use AI in business operations (total effect of EE on EATT: $\beta = 0.297$, p = 0.000; direct effect: β = 0.268, p = 0.001). Table 3 details the indirect effects. Mediation is significant (support for H2a and H2b) as the bootstrap estimates of the indirect effects fall within the confidence interval which does not include zero.

The research uncovered that the enhancement of performance expectancy positively influences employees' inclination to use AI. This discovery mirrors the outcomes of a previous investigation where a positive correlation was observed between performance gains and employees' utilization of AI tools in conjunction with AI collaboration for performance improvement [7]. However, the impact of effort expectancy on employees' intention to adopt AI was deemed insignificant, aligning with the findings of [8] but contrasting the expectations of the UTAUT framework [12]. This inconsistency in outcomes can be attributed to disparities in technology types, tasks, and user contexts. In our study, the technological context revolves around AI, the task context pertains to AI adoption spanning various processes, and the user context involves business managers in technology organizations. These distinct contexts set our study apart from earlier research [11], where the technological context was focused on IT aiding operational tasks, the task context centered on operational activities, and the user context consisted of operational employees. In addition to it, the study identified the significant role of employee attitude in mediating the connection between performance expectancy and the intention to adopt AI, as well as between effort expectancy and the intention to adopt AI. This outcome lends support to the established constructs within the UTAUT theory when applied to the context of AI adoption across various business processes.

Table 2 Means, SDs, Cronbach's alpha, and Correlations among the variables

Key Variables	Mean	SD	Cronbach's alpha	PE	EE	ATT	IU
Performance Expectancy (PE)	5.57	0.94	0.85	1	0.52**	0.49**	0.49**
Effort Expectancy (EE)	5.83	1.18	0.73		1	0.36*	0.31**
Employee Attitude (EATT)	5.64	1.90	0.65			1	0.27**
Intention to Use (IU)	5.19	1.37	0.87				1

Note n = 252, ** $p < 0.01$, * $p < 0.05$

Table 3 Indirect effects

Mediation	Indirect effect	bootSE	bootLLCI	bootULCI	Decision
H2a: PE → EATT → IU	0.047	0.030	0.005	0.119	Supported
H2b: EE → EATT → IU	0.037	0.029	0.000	0.094	Supported

Note n = 252; Bootstrap sample size 2,000; 95% Confidence Interval; LLCI: Lower limit of confidence interval; ULCI: Upper limit of confidence interval

5 Implications and Conclusion

Our research offers some important implications for business managers. Firstly, technology organizations that aim to integrate AI for operational enhancement must work on employee attitude. Our findings show that performance expectancy and effort expectancy play an important role in influencing employee attitude which further influences the subsequent intention to use AI in business operations. To successfully implement AI, organizations should invest in better infrastructure, leadership support, and continuous learning opportunities, so that the performance benefits should be visible, and usability can be highlighted. Managers should prioritize skill updates to leverage AI's potential for informed workforce analytics. Secondly, technology organizations should develop effective strategies to counteract negative attitudes and behaviors among employees towards AI adoption in business operations as it plays an important role in the final implementation of AI. Companies integrating AI into business should enhance training to foster a comprehensive understanding of the human-AI relationship. Thirdly, our study provides valuable insights for practitioners, guiding them in AI implementation by understanding factors influencing attitudes and intentions. This empowers leaders to employ strategies that enhance collaboration between AI and business automation. Thus, to conclude, the present research contributes to an advanced comprehension of the factors influencing the adoption of AI across diverse business processes within technology organizations. Consequently, our research offers a promising avenue for future inquiries within the emerging domain of AI implementation across organizational processes.

References

1. Vrontis, D., Christofi, M., Pereira, V., Tarba, S., Makrides, A., Trichina, E.: Artificial intelligence, robotics, advanced technologies and human resource management: a systematic review. Int. J. Hum. Resour. Manag. **33**(6), 1237–1266 (2022). https://doi.org/10.1080/09585192.2020.1871398
2. Saboo, A.R.: Using big data to model time-varying effects for marketing resource (RE) allocation. MIS Q. **40**(4), 911–939 (2016). https://doi.org/10.25300/misq/2016/40.4.06
3. Gupta, S., Leszkiewicz, A., Kumar, V., Bijmolt, T., Potapov, D.: Digital analytics: modeling for insights and new methods. J. Interact. Mark. **51**, 26–43 (2020). https://doi.org/10.1016/j.intmar.2020.04.003
4. Gursoy, D., Chi, O.H., Lu, L., Nunkoo, R.: Consumers acceptance of artificially intelligent (AI) device use in service delivery. Int. J. Inf. Manage. **49**, 157–169 (2019). https://doi.org/10.1016/j.ijinfomgt.2019.03.008
5. Jaiswal, A., Arun, C.J., Varma, A.: Rebooting employees: upskilling for artificial intelligence in multinational corporations. Int. J. Hum. Resour. Manag. **33**(6), 1179–1208 (2022). https://doi.org/10.1080/09585192.2021.1891114
6. Altemeyer, B.: Making the business case for AI in HR: two case studies. Strateg. HR Rev. **18**(2), 66–70 (2019). https://doi.org/10.1108/shr-12-2018-0101
7. Tarafdar, M., Beath, C.M., Ross, J.W.: Using AI to enhance business operations. In: How AI Is Transforming the Organization, pp. 67–86 (2020)

8. Dogru, A.K., Keskin, B.B.: AI in operations management: applications, challenges and opportunities. J. Data, Inf. Manag. **2**(2), 67–74 (2020). https://doi.org/10.1007/s42488-020-00023-1
9. Sengupta, S., Patel, P., Zaman, S.: Fostering business growth and innovation through internationalization: the interesting illustration of the Indian IT industry. In: Palgrave Studies in Democracy, Innovation and Entrepreneurship for Growth, pp. 361–387 (2021)
10. Malik, A., Budhwar, P., Mohan, H., Srikanth, N.R.: Employee experience –the missing link for engaging employees: Insights from an MNE's AI-based HR ecosystem. Hum. Resour. Manage. **62**(1), 97–115 (2023). https://doi.org/10.1002/hrm.22133
11. Wang, T., Jung, C.H., Kang, M.H., Chung, Y.S.: Exploring determinants of adoption intentions towards Enterprise 2.0 applications: an empirical study. Behav. Inf. Technol. **33**(10), 1048–1064 (2014). https://doi.org/10.1080/0144929X.2013.781221
12. Venkatesh, V., Thong, J.Y.L., Xu, X.: Unified theory of acceptance and use of technology: A synthesis and the road ahead. J. Assoc. Inf. Syst. **17**(5), 328–376 (2016). https://doi.org/10.17705/1jais.00428
13. Ajzen, I.: The theory of planned behavior. Organ. Behav. Hum. Decis. Process.ocess. **50**(2), 179–211 (1991). https://doi.org/10.1016/0749-5978(91)90020-T
14. Venkatesh, V., Thong, J.Y., Xu, X.: Consumer acceptance and use of information technology: extending the unified theory of acceptance and use of technology. MIS Quart **36**(1), 157–178 (2012)
15. Dwivedi, Y.K., Rana, N.P., Janssen, M., Lal, B., Williams, M.D., Clement, M.: An empirical validation of a unified model of electronic government adoption (UMEGA). Gov Inf Quart **34**(2), 211–230 (2017)

An Artificial Intelligence Approach to Quantifying Exercise Form for Optimal Performance and Injury Prevention

K. R. Sowmia, T. Jayaganeshan, F. Mohammed Abraar Khan, S. Madhesh, and S. Kabilesh

Abstract Human body posture estimation remains one of the most challenging problems despite substantial research and improvements in the fields of artificial intelligence and computer vision. Human pose detection has several uses, including assisted living, video surveillance, biometrics, public security, and at-home health monitoring. Nowadays, teenagers between the ages of sixteen and twenty-five are encouraged to engage in rigorous training because they are interested in maintaining a specific degree of physical fitness in their bodies. If they do not maintain good form or posture, this tough training activity could lead to muscular injuries or any other small or serious problems. They should engage a personal trainer or someone who can help them follow the right methods to sustain and prevent injuries if they want to keep up with all of this. This only applies to those who perform these intense exercises in a gym or other public space; at-home exercisers would not be able to afford personal trainers and instructors. One can utilize an AI-trained model to maintain good body posture for intense workouts. The analysis and evaluation of the participants' exercise movements will be done by the study using a combination of motion sensors, cameras, and machine learning algorithms. The AI-powered system will examine each participant's form, offer real-time feedback, and correct mistakes made while training with various workouts. This AI-trained model serves as a personalized trainer that can be tweaked to the user's specifications. The AI model assigns a preference category to each workout and notifies the user if it detects a posture issue.

Keywords Posture estimation · Biometrics · Media-pipe · Computer vision · Performance

K. R. Sowmia · T. Jayaganeshan · F. Mohammed Abraar Khan · S. Madhesh (✉) · S. Kabilesh
Department of Artificial Intelligence and Machine Learning, Rajalakshmi Engineering College, Chennai, Tamil Nadu, India
e-mail: madheshshanmugam@gmail.com

G. Fortino et al. (eds.), *Proceedings of Third International Conference on Computing and Communication Networks*, Lecture Notes in Networks and Systems 917,
https://doi.org/10.1007/978-981-97-0892-5_50

1 Introduction

Fitness has become one of the most popular fields among youngsters in the modern era. Many people encourage and inspire them to start working out, take excellent care of their bodies, and display their grit and determination in developing their physiques and being stunning [1]. Many people choose fitness as a way to move past trauma, keep their minds stable, and boost their confidence [2]. Therefore, there is a high likelihood that people who are getting into fitness will suffer injuries and relapses as a result of their improper form during their earlier stages of exercise, given the growing popularity of fitness around the world. AI-powered gym form analysis uses cutting-edge algorithms to monitor exercise movements, analyze data from various sensors and cameras, and identify incorrect or potentially harmful forms. It then corrects these issues by sending a warning signal using an in-app buzzer when the user does not maintain proper form or performs an exercise that could lead to issues. By offering a more precise and effective technique to avoid injuries and boost resistance training performance, this technology has the potential to completely transform the fitness sector [3].

The main goal of this study is to determine whether AI-powered gym form analysis may reduce injury risk and boost resistance training performance. The study specifically compares the precision and efficiency of a human feedback system to that of an AI-powered form analysis system [4]. The AI-powered system's usefulness for improving exercise form and preventing injuries will also be evaluated based on the participant's impressions and opinions. Although the application of AI to gym form analysis is still in its infancy, it has already demonstrated promise in terms of enhancing exercise performance and lowering the risk of injury. This study will add to the growing body of knowledge on artificial intelligence's applications in fitness and athletic performance and offer evidence-based justification for the adoption of form analysis tools using artificial intelligence in gyms and training facilities [5]. The results of this study may also guide the creation of more sophisticated and effective AI algorithms for injury prevention and form analysis in the gym [6]. This type of AI-powered technology can be employed in a variety of sports that require intense training, extensive endurance work, etc. As a result, deep learning has considerably enhanced the performance of human posture estimation in recent years [7]. The medi-apipe utilized in this model generates 17 key points that can be used to recognize and categorize the movements of the joints used in exercising.

2 Literature Survey

The program learns to handle many jobs at once, depending on the operator's recognized positions. According to "Real-Time Human Pose Estimation and Control for Dynamic Virtual Characters" (H. Lu, X. Huang, J. Xie, and J. Wang), in order to better control moving virtual characters in the OpenAI Gym environment, this study presents a real-time human position estimation and control system. The method generates realistic movements for the virtual characters by combining deep learning and physics-based modeling [8].

Z. Cao, T. Simon, S.-E. Wei, and Y. Sheikh, "Real-Time Multiple People Pose Estimation," in this research offer a deep learning-based system for doing real-time pose estimation for people [9]. The system is demonstrated to perform better than existing approaches on several datasets, including the CO dataset. X. Zhu, H. Yang, J. Wang, A. Yuille, and Y. Yang, State-of-the-art performance is demonstrated on the COCO dataset using the developed system. Article by A. Toshev and C. Szegedy titled "DeepPose: Human Pose Estimation via Deep Neural Networks" In order to estimate human poses, a deep neural network architecture is proposed in this paper. The architecture is shown to achieve state-of-the-art performances when trained on the MPII Human POSE dataset [10].

Using deep learning approaches, these works showcase the current state of the art in posture estimation and its applications, such as camera relocalization [11]. A method for real-time exercise detection and pose estimation using convolutional neural networks is proposed in "Real-time Exercise Recognition and Pose Estimation with Convolutional Neural Networks" by C. Hsu and J. Huang. High accuracy is demonstrated by testing on a dataset consisting of exercises completed by multiple individuals.

"Pose Estimation for Exercise Recognition Using LSTM-Based Architecture" by S. Chakraborty and S. K. Kopparapu: This paper proposes a pose estimation system for exercise recognition using a long-short-term memory (LSTM)-based architecture. The system is tested on a dataset of exercise videos and is shown to achieve high accuracy in recognizing exercise types [12].

3 System Architecture

Figure 1shows the architecture diagram of machine learning model which inputs the data in the form of parameters, which represent joints. The input data is carried out in the form of a video feed, which captures the motion and movements of the user. The data is processed, and the captured motion video is segregated by frames. The frames are then analyzed by the model, which checks for improper posture. If it detects any error in the user's posture, it sends a warning signal, reminding them that they have been performing the workout in an inaccurate manner. The model also checks for the user's availability in front of the camera to be able to capture their movements. The model also helps the user count their reps in order to keep track of their workouts. Thus, the AI-powered system is used to detect real-time posture estimation to prevent injuries and improve its performance and accuracy.

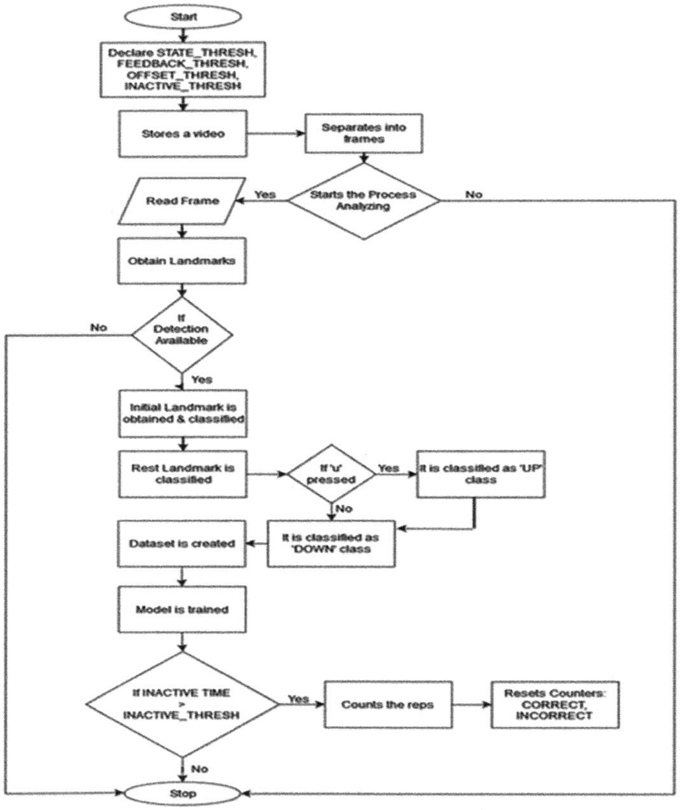

Fig. 1 Architecture diagram

4 System Implementation

4.1 Model Implementation

The above-mentioned figures represent the working condition of the model, resulting in a change in body postures and an increase in rep counts. Figure 2 shows the rep count as UP when the hand movement is above the chest, and Fig. 3 shows the rep count as DOWN when the hand movement is below the chest. Thus, providing a clear idea of how to carry out a workout without any mistakes.

The training methods that serve as the foundation of our working model have the aim of building a reliable AI model for quantifying workout form.

1. Data Collection and Labeling.
2. Data Augmentation.
3. Transfer Learning.

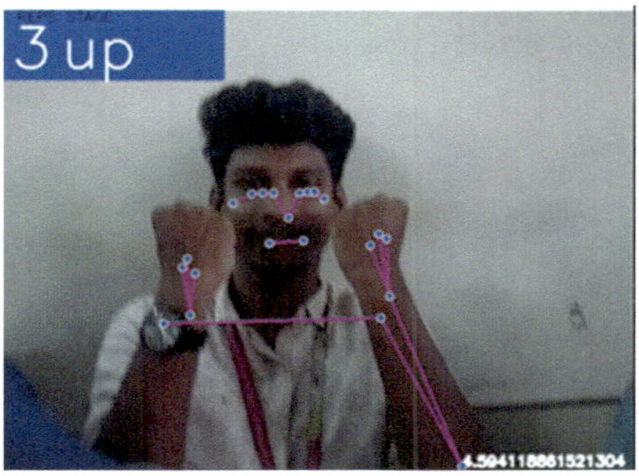

Fig. 2 This represents the working model II

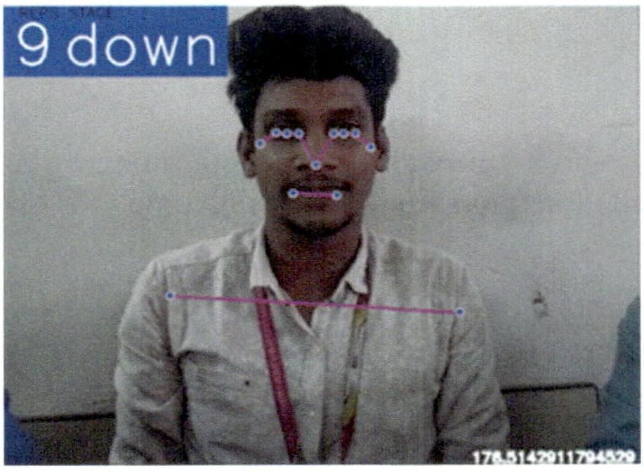

Fig. 3 This Represents the working model I

4. Pose Estimation Model.
5. Model Architecture.
6. Loss Function.
7. Hyperparameter Tuning.
8. Regularization Strategies.
9. Training Pipeline.
10. Assessment Metrics.
11. Early stopping.

The model has undergone a number of algorithmic changes in an effort to increase user flexibility and efficiency. Following testing, the following functioning models are employed.

4.2 *Logistic Regression*

The statistical method of logistic regression is used for binary classification problems where the objective is to predict a binary result (for example, true or false). A logistic

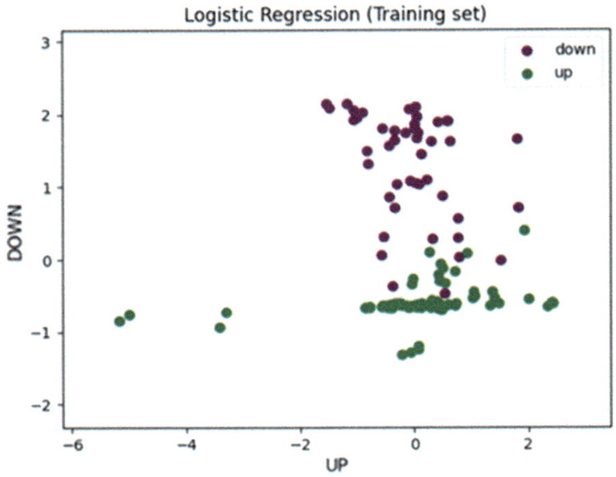

Fig. 4 Represents the scatter plot of logistic regression

function is used to represent the connection between a collection of input factors and the likelihood of a binary outcome [13] (Fig. 4).

4.3 Ridge Classifier, Random Forest, and Gradient-Boosting

A particular kind of linear binary classification model called the Ridge classifier employs L2 regularization to avoid overfitting. The model finds the line based on the input attributes that most effectively divides the data into two classes [14]. A lot of decision trees are built during the training phase of the random forest ensemble learning approach for classification, regression, and other tasks, which then output the class or mean prediction of each individual tree [15]. A series of weak models, often decision trees, are sequentially combined to create a powerful prediction model using the gradient-boosting classifier machine learning approach [16].

The performance metrics for the classification algorithms such as Logistic Regression, Ridge classifier, Random Forest, and gradient boosting classifier models are calculated and evaluated. Performance metrics such as Accuracy, precision, and Recall are calculated for the deep learning algorithms (Table 1).

Table 1 Comparing logistic regression, ridge classifier, random forest, and gradient boosting classifier model with performance metrics

Performance metrics	Logistic regression	Ridge classifier	Random forest	Gradient boosting classifier
Accuracy	95.08	96.72	96.72	93.44
Precision	95.34	97.61	95.45	93.18
Recall	89.47	94.73	89.47	84.21

The data mentioned above was taken from the dataset used for predicting human posture estimation using a supervised machine learning algorithm. The model's first stage is to gather and preprocess the posture information about a human as a CSV file. The dataset is then divided into training data and testing data after the model is pre-processed. The ratio that is being used is 70:30, or 70% of the training data and the remaining 30% of the testing data, which is the ratio that is being provided. To create a precise and effective model, multiple supervised machine learning models are used. When the model is finally assessed, the accuracy found is higher than 96% for the Ridge Classifier algorithm and the Random Forest algorithm, which run perfectly with the training data and the testing data compared to the other two. The model can achieve accuracy levels well above 98% when trained on a sizable training sample. In light of this, the outcome of the human posture estimation model demonstrates how Artificial Intelligence can be applied in the physical sector shortly to improve people's quality of life. These statistics can be used as a preventative measure to raise the standard of living for future generations by lowering their injury rate and improving efficiency while exercising (Fig. 5).

The comparative study of all four models that are implemented in the project is shown in graphical form. The above-mentioned graphical form compares all four working models by taking the average of accuracy, recall, and precision. Thus, proving that the model provides us with an efficiency of more than 92% when trained with three epochs.

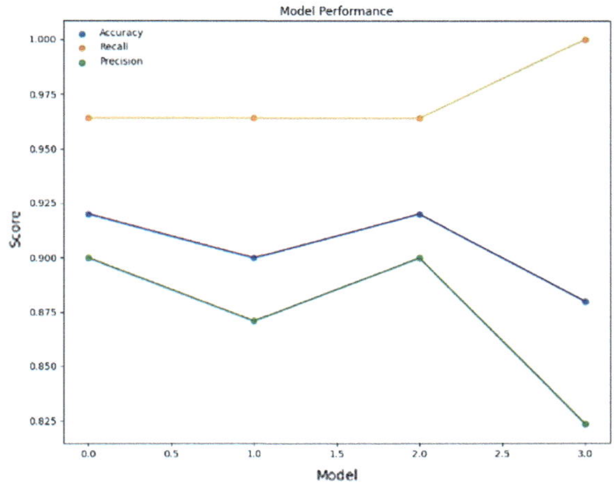

Fig. 5 Represents the line chart of performance metrics

5 Conclusion

A fascinating field of study with numerous potential advantages for people and the fitness industry as a whole is AI-powered gym form analysis for injury prevention and performance enhancement in resistance training. Real-time monitoring and analysis of exercise form using AI algorithms can help identify improper form and offer individualized feedback, reducing the risk of injury and improving performance. With multiple commercial tools on the market, the state of the art in gym form analysis is already producing encouraging results. The incorporation of more sophisticated sensors and cameras, the creation of more complex AI algorithms, and the extension of gym form analysis to other types of exercise are all areas that could use improvement. Despite these difficulties, AI-driven gym form analysis has significant potential advantages. The technology has the potential to increase exercise adherence and performance, lower the chance of injury, and eventually result in a population that is healthier and more active by offering individuals personalized feedback and adjustments. Future research is necessary to fully comprehend the advantages and restrictions of AI-powered gym form analysis. In conclusion, AI-powered gym form analysis has the potential to revolutionize the fitness industry and improve the health and well-being of individuals around the world. While there are still challenges to be overcome, technology is rapidly advancing, and we can expect to see continued progress in the years ahead.

References

1. Ravikumar, S., Vijay, K., Pavithra, S., Prithi, S., Kannan, S.: Prediction of rental demands using different machine learning algorithms. In: 2023 International Conference on Computer Communication and Informatics (ICCCI), Coimbatore, India, pp. 1–6 (2023). https://doi.org/10.1109/ICCCI56745.2023.10128331.MD
2. Hossain, Z., Sohel, F., Shiratuddin, M.F., Laga, H.: A comprehensive survey of deep learning for image captioning. ACM Comput. Surv. **51**(6), 36 (2019). Article 118. https://doi.org/10.1145/3295748
3. Martinez, G.H. et al.: Single-network whole-body pose estimation. In: IEEE/CVF International Conference on Computer Vision (ICCV), Seoul, Korea (South), pp. 6981–6990 (2019). https://doi.org/10.1109/ICCV.2019.00708
4. Tao, Y., Hu, H., Zhou, H.: Integration of vision and inertial sensors for 3D arm motion tracking in home-based rehabilitation. Int. J. Robot. Res. **26**(6):607–624 (2007). https://doi.org/10.1177/0278364907079278
5. Kondragunta, J., Hirtz, G.: Gait parameter estimation of elderly people using 3D human pose estimation in early detection of dementia. In: 2020 42nd Annual International Conference of the IEEE Engineering in Medicine & Biology Society (EMBC), Montreal, QC, Canada, pp. 5798–5801 (2020). https://doi.org/10.1109/EMBC44109.2020.9175766
6. Ge, L., Liang, H., Yuan, J., Thalmann, D.: Real-time 3D hand pose estimation with 3D convolutional neural networks. IEEE Trans Pattern Anal Mach Intell. Apr **41**(4):956–970 (2019). https://doi.org/10.1109/TPAMI.2018.2827052
7. Samuel, P., Jayashree K., Babu R., Vijay K.: Artificial intelligence, machine learning, and IoT architecture to support smart governance. In: Saini, K., Mummoorthy, A., Chandrika, R., Gowri

Ganesh, N. (eds.) AI, IoT, and Blockchain Breakthroughs in E-Governance, pp. 95–113. IGI Global (2023). https://doi.org/10.4018/978-1-6684-7697-0.ch007
8. Lee, S., Lim, Y., Lim K.: Multimodal sensor fusion models for real-time exercise repetition counting with IMU sensors and respiration data. Inf. Fusion **104**, 102153 (2024). ISSN 1566–2535. https://doi.org/10.1016/j.inffus.2023.102153
9. Tang, H., Wang, Q., Chen, H.: Research on 3D Human Pose Estimation Using RGBD Camera, pp. 538–541 (2019). https://doi.org/10.1109/ICEIEC.2019.8784591
10. Jiang, Y., Cao, C., Zhu, X., Ma, Y., Cao, Q.: RGBD-based real-time 3D human pose estimation for fitness assessment. In: 2020 3rd World Conference on Mechanical Engineering and Intelligent Manufacturing (WCMEIM), Shanghai, China, pp. 103–108 (2020). https://doi.org/10.1109/WCMEIM52463.2020.00028
11. Kumar, D., Sinha, A.: Yoga pose detection and classification using deep learning. Int. J. Sci. Res. Comput. Sci. Eng. Inf. Technol. (2020). https://doi.org/10.32628/CSEIT206623
12. Cao, Z., Simon, T., Wei, S.-E., Sheikh Y.: Realtime multi-person 2D pose estimation using part affinity fields. In: 2017 IEEE Conference on Computer Vision and Pattern Recognition (CVPR), Honolulu, HI, USA, pp. 1302–1310 (2017). https://doi.org/10.1109/CVPR.2017.143
13. Buizza, C., Fischer, T., Demiris, Y.: Real-time multi-person pose tracking using data assimilation. In: 2020 IEEE Winter Conference on Applications of Computer Vision (WACV), Snowmass, CO, USA, pp. 438–447 (2020). https://doi.org/10.1109/WACV45572.2020.9093442
14. Kotte, H., Kravcik, M., Duong-Trung, N.: Real-time Posture Correction in Gym Exercises: A Computer Vision-Based Approach for Performance Analysis, Error Classification and Feedback (2023)
15. Hussain, A., Zafar, K., Baig, A., Almakki, R., AlSuwaidan, L., Khan, S.: Sensor-Based Gym Physical Exercise Recognition: Data Acquisition and Experiments (2022). https://doi.org/10.3390/s22072489
16. Faujdar, N., Saraswat, S., Sharma, S.: Human Pose Estimation using Artificial Intelligence with Virtual Gym Tracker, pp. 1–5 (2023). https://doi.org/10.1109/ISCON57294.2023.10112064

Predicting Content Popularity on Social Media: An Analytical Approach Using Regression Modeling

Heba Al-Mamouri and Wadhah R. Baiee

Abstract The significant emergence of the "popularity" phenomena has been fueled by the quick rise of influential social media platforms like Facebook and YouTube as well as the pervasive integration of electronic gadgets into daily life. This popularity essence essentially entails the rapid accrual of substantial views, frequently reaching into the thousands or millions, across videos, posts, and various content types, serving as a tangible reflection of user inclinations. The task of predicting content popularity is a formidable one due to its reliance on an array of factors, encompassing visual and social attributes such as views, likes, comments, as well as variables like publication time, publisher identity, duration, and content specifics. This manuscript presents a comprehensive exploration of this subject matter, delving into recent applications of machine learning techniques for the prediction of content popularity. It underscores the significance of judiciously selecting predictive attributes and appropriately configuring data models to attain accurate prognostications. The research work encompasses an array of regression models harnessed in machine learning, including decision trees, random forests, support vector machines, ridge regression, and both linear and non-linear regression. Diverse classes of attributes employed for popularity prediction are delineated, encompassing text-based features, visual characteristics, metadata with a social dimension, and the fusion of multiple attributes. The paper further outlines the prevalent assessment metrics employed for evaluating regression models, encompassing mean absolute error, mean squared error, and root mean squared error. Also, it includes a table that summarizes the references, models, content types, features, and results of various studies related to popularity prediction.

Keywords Regression · Popularity · Trending · Social media · Video Prediction

H. Al-Mamouri · W. R. Baiee (✉)
Software Department, College of Information Technology, University of Babylon, Babylon, Iraq
e-mail: wadhah.baiee@itnet.uobabylon.edu.iq

G. Fortino et al. (eds.), *Proceedings of Third International Conference on Computing and Communication Networks*, Lecture Notes in Networks and Systems 917,
https://doi.org/10.1007/978-981-97-0892-5_51

1 Introduction

Individuals from diverse corners of the globe have the opportunity to freely share their content and visually document their daily experiences through popular social media platforms such as YouTube, Instagram, and TikTok. This surge in technology adoption and widespread use of electronic devices has facilitated this global connectivity. The potential for such content to gain widespread attention is significant, especially when the content creator possesses substantial real-world influence or recognition [1]. Notably, much of the content that garners widespread attention does so primarily due to this factor. However, the dynamics of popularity are also influenced by an array of additional factors, encompassing the intrinsic value of the content, the overall reputation of the creator, video duration, release timing, content quality, and follower count [2]. The ability to predict which content will achieve viral status and amass a substantial number of views remains a complex endeavor, given that only a fraction of material attains such widespread recognition. This phenomenon has given rise to the term "viral video" [3]. Enterprises capable of forecasting content trends stand to enhance their profits by strategically positioning advertisements [4]. In fact, projections indicate that companies were slated to allocate 41% of their marketing budget to online platforms in 2021 [5], investing in the content that is likely to generate attention, and selecting the most appealing content for their target audience [6]. Studies have focused on various attributes, such as visual, textual, and social features, indicating that the fusion of multiple features, including visual, social, and temporal attributes, is considered effective for estimating popularity. However, predicting popularity remains a challenge, and the definition and processing of predictive features significantly impact the results of predictive models. Extracted features must be converted into categorical or numerical representations for ML algorithms to handle them effectively [7]. Researchers have developed several strategies to enhance the accuracy of predicting popularity. In recent times, Artificial Intelligence (AI) techniques, particularly machine learning (ML) methods, have been highly effective in identifying patterns between content variables and popularity. When dealing with textual content, natural language processing (NLP) can be used. ML aims to create models capable of resolving issues autonomously, often using datasets generated from similar tasks as a reference. A classification task or a regression task can be used to design the popularity forecast task, depending on the intended prediction output. Regression analysis is widely used to address prediction-based issues. The goal of regression analysis is to determine how the dependent and independent variables are related. In contrast to the independent variable, which represents the input to the model or the underlying causes of the outcome, the dependent variable offers the output and outcome of the forecast. Examining the effects of changes in independent variables on the value of the dependent variable can be done with regression models [8].

2 Popularity Prediction in Social Media

A lot of academic research has been done on the prominent issue of predicting popularity. By spotting repeating trends and examining textual and visual elements that could have an impact, it seeks to predict the future popularity of material. Popularity cannot be predicted with certainty by any one criterion.

Video titles that are in line with current trends in other media or in real life, on the other hand, have been found to increase the likelihood that a video would become viral. Additionally, viral material has been linked to videos with shorter titles and lengths [9]. Studying content popularity not only helps understand content consumption patterns but also enables the improvement of services. Applications of these studies include identifying the distinguishing characteristics of viral TikTok videos compared to non-viral ones [10], analyzing popular hashtags, identifying Challenge winners [11], and making recommendations and retweets. Researchers employ various machine learning techniques, popularity measurement methods, and features to ascertain content popularity. These features can be categorized as follows.

2.1 Text Features

Text features in popular content include catchy titles, compelling storytelling, clear and succinct language, emotional appeal, unique perspective or insights, relevant and timely information, and well-structured organization. These features contribute to the content's popularity and success by grabbing audience attention, establishing a relationship, delivering valuable information, and making it easy to consume. To predict popularity, specific elements such as the title, description, and tags of a post are considered as features, utilizing natural language processing techniques [12]. Including more tags, especially trending ones, increases the likelihood of the article appearing in search results [3]. In the case of newspapers, these features can be derived from online descriptions, videos, images, and even user-posted comments on social media platforms [13]. Researchers employ various tools and algorithms to analyze the text and extract multiple features. In this study, GloVe, a pretrained dictionary for language features, has been utilized to process text features like titles, tags, and descriptions [12]. Furthermore, it has been observed that popular videos tend to have shorter names and lengths [9].

2.2 Visual Features

Visual features in popular content refer to specific characteristics or attributes of the visual elements within the content that contribute to its popularity. These features can provide insights into the aesthetics, composition, or visual appeal of the content and how they influence audience engagement and interest, include high-quality visuals, eye-catching thumbnails, creative and engaging visuals, consistent branding and aesthetics, relevant and compelling visuals, emotional impact, and storytelling through visuals. These characteristics are collected from videos and photos using machine learning (ML) approaches and deep learning (for instance, CNN, ANN, etc.) or by manually choosing features from the frames that best describe the content. Trzcinski proposed a regression method to forecast the popularity of content before it is published using visual features, such as deep neural network outputs or scene dynamics metrics [1]. In addition, the researchers are dealing with the issue of predicting an image's or video's popularity by utilizing a number of crucial variables, such as how the semantic and sentiment data extracted from the image affect its popularity and the effect of an image's visual content [14].

2.3 Metadata (Social) Features

Metadata (social) features in popular content refer to specific characteristics or attributes related to the social aspect of the content. These characteristics offer perceptions into how the audience interacts with, shares, and consumes the information on social media sites. They include metrics such as likes, shares, and comments, platform-specific engagement indicators, user-generated content, influencer or celebrity engagement, trending topics or hashtags, virality indicators, and audience demographics and targeting. These features contribute to the content's popularity and success by reflecting audience engagement, participation, and alignment with current trends. In addition, metadata characteristics offered by the application or website where the content was published, such as genre, source, number of views, publication date, duration, comments, and likes, play a role in determining popularity [1]. Additionally, we investigate how social information, such as the popularity of a picture owner, affects popularity [14].

2.4 Multi-Features Fusion

The practice of mixing and integrating various features derived from different sources or modalities to improve the performance of a prediction model or analysis is referred to as multi-feature fusion. By combining numerous features, such as visual, text, social, temporal, or other relevant properties, researchers hope to increase the accuracy and resilience of their models by leveraging the complimentary information supplied by each feature. This fusion process can be carried out at several phases, including feature extraction, feature representation, and model training and prediction. Overall, multi-feature fusion approaches are critical in leveraging the potential of various data sources and modalities, allowing for more complete and effective analysis and modeling in a variety of domains. In this paper, the combination of textual, visual, and metadata features is explored [3]. The integration of text, social, and visual information has been shown to improve the accuracy of popularity prediction. Additionally, social factors have been found to be a more reliable indicator of video popularity compared to visual cues [1].

3 Popularity Measure

Depending on the social media platform, various social metrics are used to determine how popular a piece of content is, such as the number of likes, comments, or views on Facebook, the number of retweets on Twitter, and other metrics [14]. The scientists are thinking the "traditional" metric is views. In some instances, it doesn't accurately reflect the users' and content's shared interests. Since there are many videos that have a huge number of views, but they are not common, but the duration of watching the video or interacting with the video through liking or commenting can give an indication of what attracts the user. Here, we can define popularity is the relationship that exists between a certain item and the user of it [15]. Another metric is the video completion rate. This is a reliable sign that you're producing engaging content for your audience. Many social media algorithms use the video completion rate as a crucial indicator, so it makes sense to concentrate on increasing it. The related works and models for predicting popularity are presented in these papers.

4 Regression Models in Machine Learning

Regression is the process of inferring a relationship from data presented to show features of a data. This relationship can then be used for a variety of calculations, including estimating future values and determining whether or not there is a relationship between different variables [8]. We explain the models used to predict.

4.1 Linear Regression

Linear regression used to determine how two variables (coefficients) a0 and a1 relate to one another in a linear model that forecasts the value of one dependent variable (z) against one independent variable (x) in the form [8, 16].

$$z = \alpha_0 + a_1 x$$

α_0: is the straight line's intersection. (The value of z when it crosses the z-axis)

a_1 is its gradient.

Also known as "simple linear".

For example, the number of views(x) can be used to estimate video popularity(y). The video with highest number of views is considered more popular than the others; in this case, only one variable (the number of views) was used to predict this While when used with multiple variables to estimate it is known as "multiple linear regression" [16] and [17]. Utilize a multiple linear regression model to estimate the popularity of videos based on a variety of variables drawn from enormous amounts of user behavior data [18].

4.2 Non-Linear Regression

Similar to linear regression, which uses a straight-line equation like = (c + mx) to represent association, non-linear regression employs a curve to show association.

4.3 Support Vector Machine

The regression algorithm known as "support vector regression," as its name suggests, enables both linear and non-linear regressions. In contrast to SVM, which is used to predict discrete categorical labels, SVR is a regressor that is used to predict continuous ordered variables. Regression models are used to minimize the error between predicted value and actual value [19]. Support vector regression with Gaussian radial basis was used by the researcher to forecast the popularity of an online video as determined by the quantity of views. This research also shows the value of using visual features to forecast popularity before content is published, such as deep neural network outputs or scene dynamics metrics [1]. They used regression and classification methods to predict the popularity of images and videos on Instagram, and they are getting results for popularity score prediction with a RMSE = 0.002 and popularity class prediction with accuracy of 90.77% [20]. Researchers have discovered that, based on user features and posts, hashtags, image assessment, and user history features, the Support Vector Regression (SVR)'s prediction accuracy could reach up to 73.1%, which is better than in earlier studies on a large dataset [21].

4.4 Ridge Regression

It is a regularized linear regressor in its basic form. The "alpha" parameter of the regularized term controls the regularization of the model, reducing the variance of the predictions. Ridge and lasso methods are used to model regressions with many features. Particularly with lasso regression, it is possible to analyze cases with a million features. L1 and L2 regularization are used in both Lasso and Ridge regression to avoid overfitting [22].

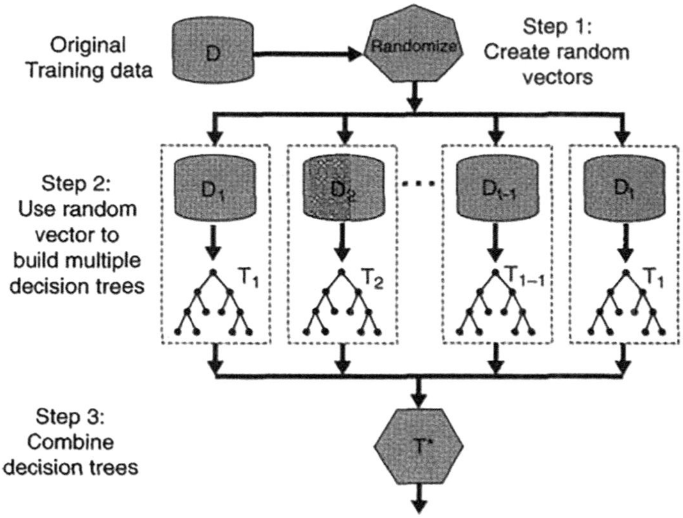

Fig. 1 Random forest

4.5 Random Forest

It is a supervised approach to machine learning that uses classification and regression to create a forest of diverse trees at random. It blends the results of distinct decision trees, each of which is based on the values of a different set of random vectors, in order to obtain predictions. It is a meta predictor that makes use of averaging to improve forecast accuracy while lowering overfitting. This is accomplished by applying various categorized decision trees to various subsamples of the data collection [23]. The classification predictions are combined using a majority voting method, and the regression results are generated using the average estimate of these trees (Fig. 1).

4.6 Decision Tree

CART is a form of supervised machine learning tree that specifies "what is the input and what is the relevant output based on our data. "The central aim of this method is to forecast the value of a specific target variable. The commonly utilized decision tree rule is the conditional statement, often denoted as the "if–then-else" statement. Decision trees are used to address CART classification and regression problems. A binary tree is used to represent the model. Each node in a tree represents a split point on a single input variable (x) and a collection of leaves that make up the tree. (Assuming the variable is numeric). Moving through the binary tree with a single

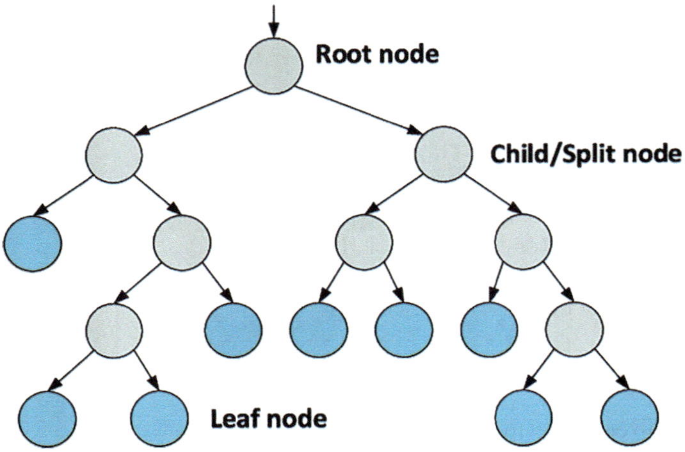

Fig. 2 Decision tree

input record allows CART to make predictions because the leaf nodes of the tree hold an output variable (y) that is used to do so. On the basis of training data, the tree is trained by selecting splits using a greedy algorithm. Additionally, pruning can help trees learn more generally, and stopping factors are used to gage how much a tree has learned [24, 30, 31] (Fig. 2).

The top models and results for each related works are displayed in Table 1.

5 Evaluation Metrics

Evaluation is essential in every model. Evaluation permits us to gage the model's effectiveness and draw parallels with other models. The result of regression is a continuous value. Consequently, it becomes necessary to employ a metric that calculates the disparity between anticipated and real values. There are numerous evaluation metrics available [29].

Mean Absolute Error: It is the average of the difference between the actual values and the predictions.

Mean Squared Error (MSE): It is the most widely used regression measure. The average of the squared difference between the goal value and the values predicted by the regression model is computed.

Root Mean Squared Error (RMSE): It is square (MSE). It corresponds to the average squared difference between the goal value and the value predicted by the regression model.

Table 1 Shows a collection of the related works

References	Model	Content type	Features	Metric	Result
[1]	Vector regression with gaussian radial basis functions	Video	–Visual, social features –Early view counts	Spearmanrank correlation	SRC = 0.94
[25]	LN model	Videos and News	Textual feature	RSE	Graphic
[26]	Support vector regression	Post	–User features –Post features	Accuracy	A = 73.1
[11]	Deep learning model, deep challenger	Video	User past data	f-score	F = 0. 494
[10]	Logistic regression	Video	Social feature (number of followers)	Accuracy	A = 0.93
[27]	Multivariate linear regression model	Video	Early patterns	MRSEs	Outperformed the baseline models
[20]	Polynomial SVR KNN,NB,DT,RFclassification model	Image/ Video	Image/ video context features	Accuracy	A = 90.77
[28]	XGBoost	Video	Video quality, video duration, number of views	Accuracy precision	A = 86% P = 64%

6 Conclusion

In conclusion, the paper highlights the challenges associated with predicting content popularity on social media platforms. It emphasizes the use of machine learning, specifically regression models, as a means to predict trending videos. The importance of selecting appropriate predictive features and understanding video attributes is emphasized. The paper explores different types of features that are utilized in the prediction process. Additionally, it mentions several regression models that can be employed for this purpose. Furthermore, the paper explains evaluation metrics that are commonly used to assess the accuracy of regression models in predicting content popularity.

References

1. Trzcinski, T., Rokita, P.: Predicting popularity of online videos using Support Vector Regression, (2015), https://doi.org/10.1109/TMM.2017.2695439.
2. Lin, H.H., Da Lin, J., Ople, J.J.M., Chen, J.C., Hua, K.L.: Social media popularity prediction based on Multi-Modal Self-Attention mechanisms. IEEE Access **10**, 4448–4455 (2022). https://doi.org/10.1109/ACCESS.2021.3136552.
3. Lv, J., Liu, W., Zhang, M., Gong, H., Wu, B., Ma, H.: Multi-feature fusion for predicting social media popularity. In: MM 2017—Proc. 2017 ACM Multimed. Conf., pp. 1883–1888, (2017). https://doi.org/10.1145/3123266.3127897.
4. Tatar, A., de Amorim, M.D., Fdida, S., Antoniadis, P.: A survey on predicting the popularity of web content. J. Internet Serv. Appl. **5**(1), 1–20 (2014). https://doi.org/10.1186/s13174-014-0008-y.
5. "IAB UK: Digital Ad Spend Boomed 41% In 2021 | The Drum." https://www.thedrum.com/news/2022/04/26/iab-uk-digital-ad-spend-boomed-41-2021 (accessed Nov. 04, 2022).
6. T. Trzcinski, P. Andruszkiewicz, T. Bochenski, and P. Rokita, "Recurrent Neural Networks for Online Video Popularity Prediction," Jul. 2017, https://doi.org/10.1007/978-3-319-60438-1_15.
7. Alasadi, S.A., Bhaya, W.S.: Review of data preprocessing techniques in data mining. J. Eng. Appl. Sci. **12**(16), 4102–4107 (2017). https://doi.org/10.3923/jeasci.2017.4102.4107
8. Humpage, S.: An introduction to regression analysis. Sensors (Peterborough, NH) **17**(9), 68–74 (2000). https://doi.org/10.1002/9781118267912.ch6.
9. Jiang, L., Miao, Y., Yang, Y., Lan, Z., Hauptmann, A.G.: "Viral video style: A closer look at viral videos on YouTube", *ICMR 2014 - Proc.* ACM Int. Conf. Multimed. Retr. **2014**, 193–200 (2014). https://doi.org/10.1145/2578726.2578754.
10. Ling, C., Blackburn, J., De Cristofaro, E., Stringhini, G.: Slapping cats, bopping heads, and oreo shakes: understanding indicators of Virality in TikTok Short Videos. In: ACM International Conference Proceeding Series, pp. 164–173. (2022) https://doi.org/10.1145/3501247.3531551.
11. Ng, L.H.X., Tan, J.Y.H., Tan, D.J.H., Lee, R.K.W.: Will you dance to the challenge?: Predicting user participation of TikTok challenges. In Proceedings of the 2021 IEEE/ACM International Conference on Advances in Social Networks Analysis and Mining, ASONAM 2021. pp. 356–360 (2021) https://doi.org/10.1145/3487351.3488276.
12. Li, Y., Eng, K., Zhang, L.: YouTube Videos Prediction: Will this video be popular ?. Available: http://cs229.stanford.edu/proj2019aut/data/assignment_308832_raw/26647615.pdf (accessed 9 May 2020) (2019).
13. Hamouda, A.E.A.: Sentiment analyzer for arabic comments system, **4**(3), pp. 99–103 (2013).
14. Aloufi, S., Zhu, S., El Saddik, A.: On the prediction of flickr image popularity by analyzing heterogeneous social sensory data. Sensors (Switzerland) **17**(3), 1–27 (2017). https://doi.org/10.3390/s17030631.
15. de Sá, S.L., Rocha, A.A.d.A., Paes, A.: Predicting popularity of video streaming services with representation learning: A survey and a real-world case study. Sensors, **21**(21) (2021), https://doi.org/10.3390/s21217328.
16. Sangwan, N., Bhatnagar, V.: Video popularity prediction based on fuzzy inference system. J. Stat. Manag. Syst. **23**(7), 1173–1185 (2020). https://doi.org/10.1080/09720510.2020.1799577.
17. Pinto, H., Almeida, J.M., Gonçalves, M.A.: Using early view patterns to predict the popularity of YouTube videos. In: WSDM 2013—Proc. 6th ACM Int. Conf. Web Search Data Min., no. February, pp. 365–374 (2013). https://doi.org/10.1145/2433396.2433443.
18. Sibo, Q.I.A.O., Shanchen, P.A.N.G., Min, W.A.N.G., Xue, Z.H.A.I., Feng, D.A.I.: Online video popularity regression prediction model with multichannel dynamic scheduling based on user behavior. Chinese J. Electron. **30**(5), 876–884 (2021). https://doi.org/10.1049/cje.2021.06.010.
19. Mahdevari, S., Shahriar, K., Yagiz, S., Akbarpour Shirazi, M.: A support vector regression model for predicting tunnel boring machine penetration rates. Int. J. Rock Mech. Min. Sci., **72**, pp. 214–229 (2014) https://doi.org/10.1016/j.ijrmms.2014.09.012.

20. Zohourian, A., Sajedi, H., Yavary, A.: A r c h i v e o f S I D Popularity Prediction of Images and Videos on Instagram. Available: www.SID.ir.
21. Purba, K.R., Asirvatham, D., Murugesan, R.K.: Analysis and prediction of instagram users popularity using regression techniques based on metadata, media and hashtags analysis. Eng. Lett. **28**(3), 170–177 (2020).
22. Luo, H., Liu, Y.: A prediction method based on improved ridge regression. In: Proc. IEEE Int. Conf. Softw. Eng. Serv. Sci. ICSESS, **2017-Novem**, pp. 596–599 (2018), https://doi.org/10.1109/ICSESS.2017.8342986.
23. Shreyas, R., Akshata, D.M., Mahanand, B.S., Shagun, B., Abhishek, C.M., "Predicting popularity of online articles using Random Forest regression", Proc.,: 2nd Int. Conf. Cogn. Comput. Inf. Process. CCIP **2016**, 2016 (2016). https://doi.org/10.1109/CCIP.2016.7802890.
24. Brownlee, J.: Master Machine Learning Algorithms. In: Suparyanto dan Rosad 2015, **5**(3), pp. 1–23 (2016).
25. Szabo, G., Huberman, B.A.: Predicting the popularity of online content. Commun. ACM **53**(8), 80–88 (2010). https://doi.org/10.1145/1787234.1787254.
26. Purba, K.R., Asirvatham, D., Murugesan, R.K.: Instagram post popularity trend analysis and prediction using hashtag, image assessment, and user history features. Int. Arab J. Inf. Technol. **18**(1), 85–94 (2021). https://doi.org/10.34028/iajit/18/1/10.
27. Li, C., Liu, J., Ouyang, S.: Characterizing and predicting the popularity of online videos. IEEE Access **4**, 1630–1641 (2016). https://doi.org/10.1109/ACCESS.2016.2552218.
28. Nisa, M.U.N., Mahmood, D., Ahmed, G., Khan, S., Mohammed, M.A., Damaševičius, R.: Optimizing prediction of youtube video popularity using xgboost. Electron., **10**(23) (2021). https://doi.org/10.3390/electronics10232962.
29. Plevris, V., Solorzano, G., Bakas, N., Ben Seghier, M.: Investigation of performance metrics in regression analysis and machine learning-based prediction models. (2022). https://doi.org/10.23967/eccomas.2022.155.
30. Lakshmi, G., Ghonge, M., Obaid, A.J.: Cloud based IoT smart healthcare system for remote patient monitoring. In: EAI Endorsed Transactions on Pervasive Health and Technology. (2021). https://doi.org/10.4108/eai.15-7-2021.170296.
31. Shahab, S., Agarwal P., Mufti, T., Obaid, A.J.: SIoT (Social Internet of Things): A Review. In: Fong, S., Dey, N., Joshi, A. (eds.) ICT analysis and applications. lecture notes in networks and systems, **314**. Springer, Singapore. (2022) https://doi.org/10.1007/978-981-16-5655-2_28.

Hybrid Encryption Algorithm: AES and PRESENT Lightweight with a Chaotic System (PHEPA-AES)

Ibtihal Ali Khanjar, Haider K. Hoomod, and Intisar Abd Yousif

Abstract Cloud computing has gained great popularity recently and has become the focus of attention of companies and institutions, and even exceeded that to reach the level of individuals because of the advantages it enjoys of saving money and managing with a minimum of effort. Despite all the advantages that characterize cloud computing, it remains a crisis of trust between the cloud and its users due to the concern of customers and their fear of leakage and theft of their data, and disclosure by unauthorized persons. Therefore, the best way to secure data on the cloud is to encrypt it before uploading it to the cloud, thus ensuring that this data is not viewed by unauthorized persons or cloud service providers, and even in the event of the collapse of the cloud, this data remains ambiguous. In this paper, more than one encryption algorithm will be combined, a proposed Hybrid encryption algorithm includes new hybrid modified AES and PRESENT (PHEPA-AES). The proposed scheme incorporates the PHEPA-AES hybrid encryption algorithm, which combines the PRESENT-ultra lightweight algorithm, the AES algorithm, and a chaotic map. This combination enhances the efficiency of cryptographic systems while increasing the randomness and security of secret keys. The utilization of the PRESENT algorithm and the chaotic map contributes to the lightweight nature of PHEPA-AES.

Keywords A hybrid · Lightweight · Chaotic · S-box · AES · PRESENT

1 Introduction

There is no doubt that life is changing a lot, especially in the area of information technology IT, and it is becoming more and more important to protect data with security systems. In general, it is challenging to propose a cryptographic technique that is compatible with every possible type of target device.

I. A. Khanjar (✉) · H. K. Hoomod · I. A. Yousif
Education College, Mustansiriyah University, Baghdad, Iraq
e-mail: ibtihalali902@uomustansiriyah.edu.iq

G. Fortino et al. (eds.), *Proceedings of Third International Conference on Computing and Communication Networks*, Lecture Notes in Networks and Systems 917,
https://doi.org/10.1007/978-981-97-0892-5_52

The basic ideas and trends for designing algorithms that can be used in devices with very few resources are different in some ways from how commonly used cryptographic algorithms are designed [1]. Cryptography algorithms can be used to maintain security, but conventional heavyweight algorithms are not appropriate for work-restricted environments [2].

In the lightweight cryptographic (LWC) algorithm to obtain the same level of security, there must be a balance between performance and resources while taking into account how well the algorithm works on low-resource devices [3]. Many block ciphers with lightweight properties have been proposed. Among them PRESENT [4], HIGHT [5], and stream cipher lightweight proposed GRIAN [6] and Trivium [7].

An S-box is one of the most important parts of the lightweight algorithm. An essential component and the algorithm's brain is the S-box. The use of cryptanalysis is used to evaluate the security of any algorithm. The most significant attacks are differential and linear cryptanalysis. These assaults form the foundation for all later assaults.

It is the algorithm's heart and the only nonlinear part. All 4-bit S-box values in LWC block ciphers that employ a 4-bit S-box have been meticulously selected to prevent both linear and differential cryptanalysis [2].

So, we are able to claim that the S-box is the primary factor that determines how secure the lightweight block cipher is. Because of this, the primary focus of a significant number of studies has been on enhancing the S-boxes of producing confusing results. In general, S-boxes come in two types: 4-bit and 8-bit [1].

The rest of the paper is structured as follows: Sect. 2 covers similar works based on cutting-edge approaches. Section 3 provides a summary of the suggested methodology. Section 4 presents the dataset's result analysis and discussion. Finally, Sect. 5 discusses the conclusion.

2 Related Works

Since the 2007 introduction of the PRESENT algorithm, several of literature have focused on the optimization of the PRESENT cipher on a variety of applications and devices.

S. Aldabbagh and I. Al Shaikhli [1] made an improvement in the present algorithm. It selects one dependent S-box by XORing with all elements of the key. Their proposal of a different S-box for each round. Instead of employing a standard S-box, it is applied in each round. The proposed algorithm is more secure than the current algorithm, in addition, each round's PRESENT S-box is 2^8 complicated whereas the proposed algorithm S-box has a round-by-round complexity of 2^{12}.

In [8], the efficacy of AES and PRESENT were compared. Each algorithm was developed on an Android mobile device. The device used in algorithm testing was validated by processing a 20MB file. They measured the average runtime, energy usage, and memory capacity. It was operated at a 25% higher efficiency than the

PRESENT implementations; however, the PRESENT implementation with 80 bits consumed less memory.

In [9], they developed modified hybrid algorithm that is based on the PRESENT algorithm and Salsa20 and that makes use of chaotic maps and various tests were applied to the proposed algorithm to examine its efficiency. The trade-off goal was met by the proposed algorithm between computational speed and ciphertext compliance. In addition to that the 2-D chaotic system has been used in the proposed algorithm to produce pseudo-random keys and increase the complexity of the encryption. The suggested algorithm is a stream cipher combined with a block cipher. The proposed algorithm is suitable and compatible in devices with constrained resources.

In [10], D-Present was developed to improve the PRESENT algorithm with dynamic dependent S-box. The results demonstrated that D-PRESENT is better than PRESENT in terms of performance. The 5.51 percent difference between the computed average avalanche impacts of the two algorithms is a clear indication of this. In addition, the standard deviation of D-PRESENT is less than that of PRESENT, indicating that the hammering distance reported in D PRESENT is closer to its average value than in PRESENT. This indicates that the majority of hammering distance measurements collected in D-PRESENT are close to its mean hammering distance. D PRESENT outscored PRESENT of execution time and throughput, based on the results. This suggests that D-PRESENT can provide increased protection against key-related attacks and lessen the vulnerability to key cryptanalysis. This study confirms prior findings that dynamic key-dependent S-box boosts the algorithm's avalanche impact.

In [11], a new 5_D chaotic system has been proposed for a modified LWC algorithm. New system of chaos has outperformed most of the existing systems, as it was examined by the NIST group for all 15 tests, and through the NIST test, it showed a high degree using complexity and unpredictability for generating the behavior of the chaotic system. Therefore, such a system is more applied in cryptographic systems to generate keys at randomized that are challenging to analyze. This system has been used to generate keys for the PRESENT algorithm that is more applicable in resource constrained environments [23, 24].

3 Methodology

In order to secure data in different kinds of applications, a cryptographic approach must be chosen that makes it obvious how data will be encrypted, what kind of algorithms will be proposed, how the secret keys will be generated and distributed, and other relevant details. It is decided to use symmetric cryptography of type of block cipher in this section.

3.1 AES Algorithm

The NIST The adoption Rijndael as the new In place of the outdated (DES) [12]. The Advanced Encryption Standard (AES) [13] is being implemented. AES algorithm was selected because of the robust security features it offers and the ease with which it's flexible enough to be implemented in either software or hardware. AES algorithm is a block cipher that uses iterative rounds and symmetric keys. The system supports key sizes of 128 bits, 192 bits, and 256 bits, improving cryptographic security but increasing iterative rounds [13].

The AES algorithm has four operations: SubBytes, ShiftRows, MixColumns, and AddRoundKey. TheSubBytes operation is the only one in the AES that can be considered non-linear. Its original function in the encryption process is where each byte is replaced with another according to a lookup table. It does so by computing the multiplicative inverse of GF (2^8), which results in a non-linear mapping being performed between an input of 8 bits and an output of 8 bits [14]. Figure 1 shows the AES round operation and how the state is permuted in the transformation's component.

SubBytes SubBytes is transformation of nonlinear one that can be reversed. It does this by employing 16 sets of identical 256-byte substitution tables, which are referred to as S-boxes. These tables are used to transfer each byte of State to another

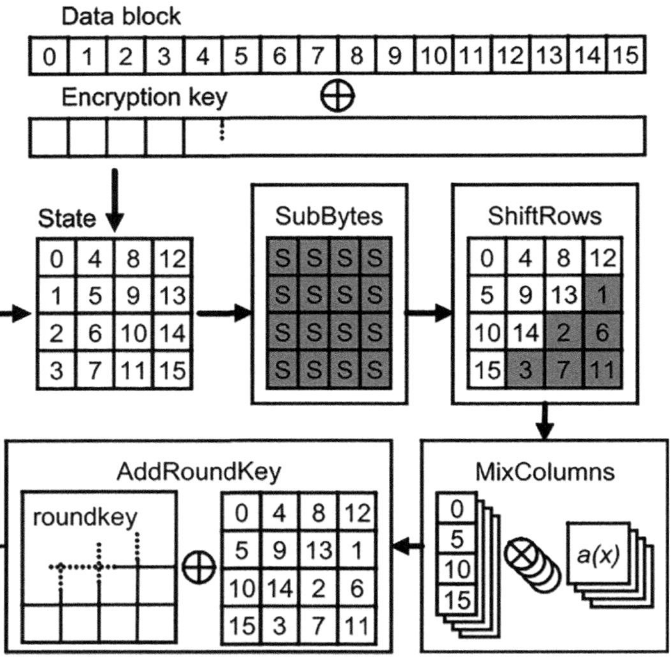

Fig. 1 AES four transformation component

byte in an independent manner. Calculating multiplicative inverses in the Galois Field GF(2^8) and performing an affine transformation [15].

ShiftRows performs a cyclic left shift of one, two, and three bytes, respectively, on the second, third, and fourth rows of State.

Mixcloumn In this stage, each state column is multiplied by the fixed polynomial C(x). In this stage, matrix multiplication is created by multiplying and adding the given data. Here, addition is accomplished with a simple XOR operation, and multiplication is accomplished through an irreducible polynomial [16].

AddRoundkey In this stage, the sub-byte and the state are combined. On the basis of the Rijndael key schedule, the subkey is obtained from the main key for each round. Each subkey occupies the same amount of space as the state. We do the XOR operation with the round's sub-byte key [16].

We will use SubBytes layer and the ShiftRows layer and put them in the PRESENT algorithm to increase the security and complexity of the encryption algorithm because of the strength of S-box of AES algorithm and increase confusion in PRESENT cipher.

3.2 *PRESENT Algorithm*

PRESENT is ultra-lightweight symmetric block cipher whose construction is quite close to that of the popular AES.2007, Bogdanov et al. [4] proposed it, and ISO/IEC 29192-2 [17] adopted it as the standard for lightweight cryptography.

The algorithm was made mostly for small-scale application uses, like RFID tags, sensor nodes, and the IoT. PRESENT is based on (SPN), which has a 64-bit block size and a key size of either 80 bits or 128 bits. Figure 2 illustrates the level layer of the PRESENT algorithm.

In this algorithm, the 128-bit key length will be used to improve security and facilitate comparisons with prior research. A single PRESENT encryption round consists of AddRoundKey layer, a S-Box layer, and a permutation layer (p-layer). Complete encryption is achieved by 31 cycles of the aforementioned processes, with the addition of an extra key after the final round [18].

The PRESENT algorithm converts each of the 31 rounds into plaintext via the bitwise XOR operation. It uses the S-box layer, which XORs each of the four bits with the key bits and the output before passing to entering the P-layer and blending in the 64-bit state. In the final round of the update, the algorithm XORs the bit state with the key. Because of the PRESENT cipher method's simple structure, it can be used with other cryptography algorithms to improve the security of a hybrid algorithm [18]. However, the PRESENT algorithm requires improvement in order to be more efficient in safeguarding resource-constrained environment data [19]. Thus, the proposed research is to create the PRESENT algorithm and improve by using AES algorithm.

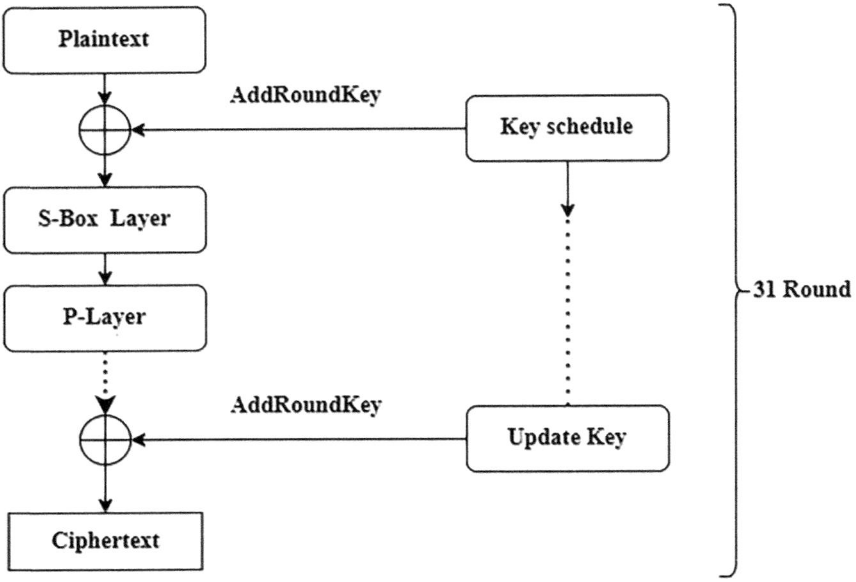

Fig. 2 PRESENT algorithm structure

3.3 Chaotic System

The chaotic map has interesting characteristics like producing a key with a lengthy periodicity, a pseudo randomness number, and sensitivity to change in the system parameters and initial circumstances of chaotic maps are examples of such behaviors. It is an example of chaotic mapping; it is a method that has proven to be effective when applied to the task of encrypting vast amounts of data, including data pertaining to images, audio, and video. The primary objective of the implementation of a chaotic system is order offer encryption techniques with many advantages over classical encryption methods, such as rapid data encryption and high levels of security and acceptable levels of computational overhead and demands on computational capacity, Because of these constraints, researchers have been researching at new chaotic-based data encryption algorithms with digital logics for the purpose of concealing information in quick communication networks that are secure [20, 21].

4 Proposed Encryption Algorithm

We present in this section the hybrid proposed algorithm by combining two parts from AES algorithms (S-box layer and ShiftRow layer) in the PRESENT structure that the key schedule was removed from the PRESENT algorithm and replaced by generated

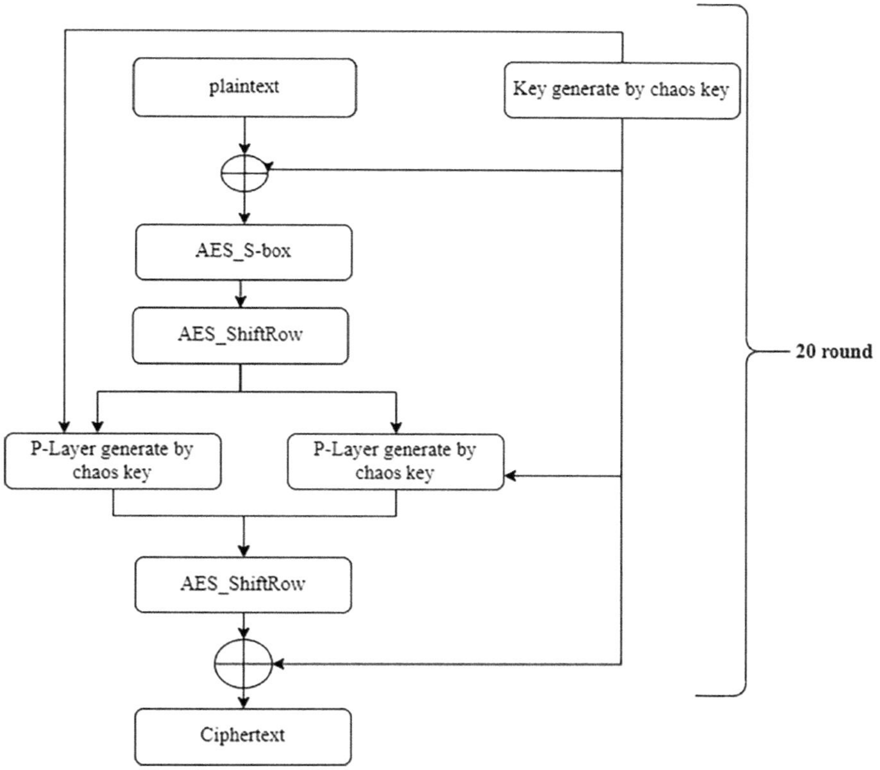

Fig. 3 Diagram of the proposed algorithm

chaos keys. These chaotic keys were employed in order to increase the randomness of the cipher text that was output and provide the encryption process with the best possible strengths. The new proposed algorithm, which is called (PHEPA-AES), could work better had chaos keys. This approach was made to make encryption work well, be resistant to different kinds of attacks, and run quickly enough., and increase the confusion of PRESENT cipher. Chaos key uses random generate the keys in PRESENT to increase the complexity and security of the proposed algorithm. In this proposal, the number of rounds of the PRESENT algorithm was reduced to 20 rounds. Figure 3 illustrates the proposed algorithm.

4.1 The 8-D Chaos Keys Generation

The 8-D messy key generation system is designed and proposed in the key generation process. It is contained in the chaos key generator stage a hybrid chaotic system.

Combining the proposal for a new 4-D chaotic system (named FHII) with the new 4D chaotic system (named as SHII), each with distinct values for the system's beginning conditions and parameters was the method that was utilized to generate the 8-D chaos key values.

The first part of the proposed FHII chaotic system is designed to be more efficient key generator work with four dimensions. The numerical indicator known as the Lyapunov Exponent can be employed to ascertain whether a given system demonstrates chaotic characteristics.

A positive Lyapunov Exponent denotes chaos and positive Exponents of the Lyapunov function greater than one denotes hyper chaos [22].

$$\begin{aligned} x[i+1] &= z[i] - (s \times (b \times x[i] - u \times y[i] - k[i]) + u \times \sin(x[i])) \\ y[i+1] &= r \times (x[i] - \sin(k[i] \times z[i]) + r \times x[i] - b \times k[i] - (r - s \times z[i])) \\ z[i+1] &= (s - \sin(x[i] \times y[i]) - b \times y[i] - r k[i] + u \times x[i]) \\ k[i+1] &= x[i] + (u \times z[i] \times (1 - u \times y[i] \times \sin(1 - x[i]))) \end{aligned} \quad (1)$$

where x, y, z, and k are state variables. $b = [1, 2.985]$, $r = [1.25, 3.91]$, $s = [9.5,10]$, $u = [0,0.005]$ are parameters of this proposed system. We obtain Lyapunov exponents: (0.12887, 0.5214, −1.9998, and 1.00908).

The FHII chaotic system that was proposed is shown in Fig. 4.

The second part of the suggested new 4-D chaos system is the SHII chaos generating. Equation (2) displays the equations and parameters of the SHII chaos system.

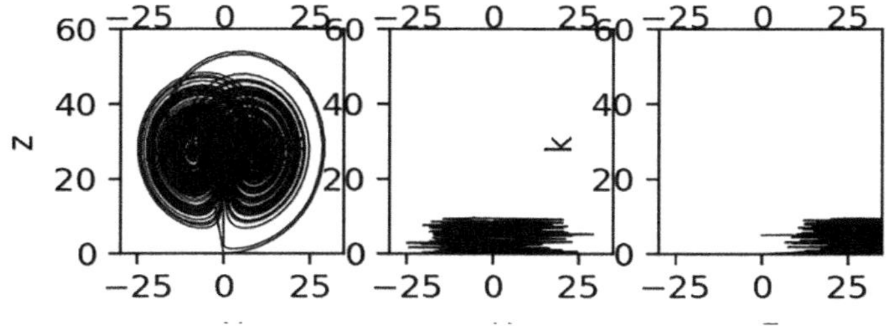

Fig. 4 The 4-D FHII chaotic system

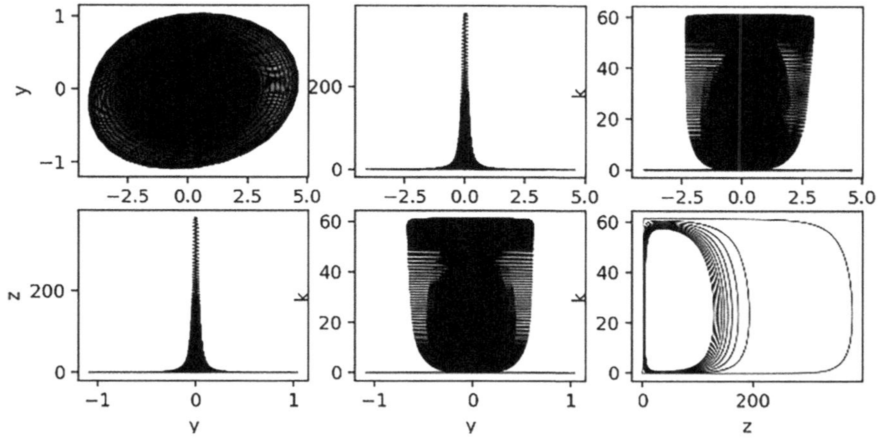

Fig. 5 The 4-D SHII chaotic system results map

$$
\begin{aligned}
x[i+1] &= -(u \times k[i] \times y[i] + u \times y[i] \times (s \times x[i] + \sin(1 - r \times x[i] \times y[i]))) \\
y[i+1] &= (k[i] - b \times z[i] \times \sin(x[i]) - u \times y[i] \times (1 - r \times z[i])) \\
z[i+1] &= u \times \sin(y[i]) \times \begin{pmatrix} r \times \sin(y[i]) - s \times k[i] \times x[i] \\ -u \times x[i] - \sin(b \times y[i] \times (1 - r \times x[i])) \end{pmatrix} \\
k[i+1] &= r \times \sin(z[i]) \times \cos(u \times y[i] \times \sin(1 - k[i] \times y[i]))
\end{aligned} \tag{2}
$$

where x, y, z, and k are state variables. $b = [1.5, 2.75]$, $r = [0.5, 1.5]$, $s = [0.75, 1.10]$, $u = [0, 0.005]$ are parameters of this proposed system. While the $xt(0)$, $yt(0)$, $zt(0)$, and $kt(0)$ are initial values with period from $(-1,1)$, we obtain Lyapunov exponents: (0.3420, 0.8879, 0.2331, and −1.9970). The SHII chaos generator behavior is shown in Fig. 5.

The generation initial values block is used to generate the initial value by taking the most recent two significant numbers (from each K_1 to K_4) that were produced by the modified Lorenz-Logistic chaotic system and then adding those numbers to the starting periods (−1, 1) of the new chaotic system initial conditions. This process is repeated until the desired initial value is produced. For instance, if the final two significant integers of K_1 are equal to 13 and the chosen initial for the first dimension of the chaotic system is equal to 0.300, then the initial that is generated is equal to 0.313.

All of the processes require the use of the created chaos keys (K_1, K_2, K_3, K_4, K_5, K_6, K_7, and K_8), and they will be saved in the file so that their subsequent utilization in the other operations will be simplified.

4.2 Encryption Process of PHEPA-AES

Encryption is defined as the process of transforming plaintext to ciphertext.

Following is a description of the developed algorithm's encryption procedure.

Algorithm 1: encryption process

INPUT: Plaintext(state) and Chaos Key

OUTPUT: Ciphertext

Steps

1. generate chaos key (K20)
2. plaintext
3. for I = 1 to 20 do
4. XOR AddRoundKey (State, chaos Key)
5. AES_S-Box
6. AES_ShiftRow
7. Two P-Layer One of them is the size 64-bit generate by chaos key)
8. AES ShiftRow
9. End for
10. XOR AddRoundKey (update chaos Key)
11. ciphertext
12. End

First, the proposed hybrid encryption algorithm, PHEPA-AES, is comprised of three cryptographic components: the PRESENT-ultra lightweight algorithm, the AES algorithm, and a chaotic map. The combination of these components aims to enhance the efficiency, randomness, and security of cryptographic systems. The utilization of the PRESENT algorithm and the chaotic map contributes to the lightweight nature of PHEPA-AES.

In the encryption process, the input message, which is 64 bits in length, is first padded to a length of 128 bits. Next, 16 bits are shifted to the left based on the first generated key. The resulting 128-bit state is then divided into two 64-bit blocks, each of which is processed by subsequent operations. The plaintext is encrypted using XOR and the chaos key.

Instead of using the S-box approach of the PRESENT algorithm, the binary sequence of the state is replaced with an AES S-box table prior to transmitting it into AES ShiftRow and passing it to the permutation P-layer. One of them is the size 64-bit generate by chaos key, as depicted in the PRESENT algorithm. The entire encryption process is repeated for 20 rounds until the final stage.

To generate the ciphertext, a new key is XOR-encrypted with the previous state of rounds in the final step. The result is a secure and random ciphertext that provides enhanced security for the encrypted data.

The PHEPA-AES hybrid encryption algorithm combines the PRESENT-ultra lightweight algorithm, the AES algorithm, and a chaotic map to improve the efficiency, randomness, and security of cryptographic systems. The encryption process involves padding, bit shifting, XOR, AES S-box table, AES ShiftRow, permutation

P-layer, and a 20-round loop, culminating in the generation of ciphertext through XOR-encryption with the previous round state. This proposed algorithm provides a promising solution to enhancing the security of data transmission in a lightweight and efficient manner.

Algorithm 2: decryption process of PHEPA-AES
INPUT: Ciphertext (state)and Chaos Key
OUTPUT: plaintext
Steps

1. generate chaos key (K20)
2. ciphertext
3. for I = 1 to 20 do
4. XOR AddRoundKey (State, chaos Key)
5. AES InvShiftRow
6. Inv two P-Layer (One of them is the size 64-bit generate by chaos key)
7. AES InvShiftRow
8. AES_InvS-Box
9. End for
10. XOR AddRoundKey (update chaos Key)
11. plaintext
12. End

5 Results and Discussions

Standard tests are used for testing PHEPA-AES encryption to show the encrypting characteristics results such as time tests. Table 1 shows the average of the PHEPA-AES-128bit encryption time with separate iteration rounds (31, 28, 24, and 20).

Figure 6 shows the benchmarking average **PHEPA-AES** encryption time for rounds (20, 24, 28, and 31)

Table 1 Benchmarking performance of the PHEPA-AES average time

Text size (KB)	PHEPA-AES-128 bit time (ms) (31 rounds)	PHEPA-AES-128 bit time (ms) (28 rounds)	PHEPA-AES-128 bit time (ms) (24 rounds)	PHEPA-AES-128 bit time (ms) (20 rounds)
1	0.099	0.087	0.074	0.051
10	2.221	2.078	1.933	0.942
25	9.120	8.952	6.988	5.329
75	44.870	43.231	40.348	38.672
100	82.231	81.443	74.008	67.307
1000	1034.110	1021.707	1000.215	956.529
2000	1887.453	1813.668	1760.588	1245.992
10,000	10,934.626	10,427.554	10,120.884	10,030.649

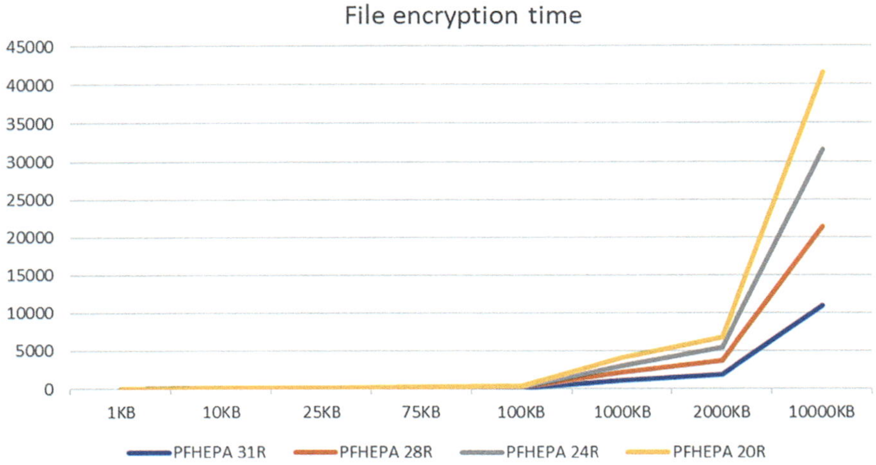

Fig. 6 PHEPA-AES file encryption time with different rounds (20, 24, 28, and 31)

The NIST tests applied to the PHEPA-AES results to ensure the security of proposed the PHEPA-AES and prevent various types of attacks. The PHEPA-AES passed through all NIST tests on each round (20, 24, 28, and 31) due to its functions and chaos as shown in Table 2.

Table 2 NIST experiments results of the PHEPA-AES with different rounds

NIST statistical tests results name	PHEPA-AES			
	Rounds			
	31	28	24	20
Frequency (monobit) test	0.712	0.709	0.699	0.687
Runs test	0.923	0.910	0.903	0.892
Discrete Fourier transform	0.512	0.502	0.498	0.478
Block frequency	0.833	0.831	0.825	0.802
Longest runs test	0.698	0.687	0.676	0.665
Cumulative sums test	0.343	0.330	0.327	0.316
Serial test	0.967	0.959	0.950	0.923
Matrix rank test	0.745	0.741	0.727	0.708
Overlapping template test	0.452	0.448	0.440	0.429
Linear complexity test	0.921	0.918	0.910	0.904
Nonoverlapping template	0.329	0.311	0.304	0.280
Random excursions variant test	0.856	0.854	0.840	0.829
Random excursions test	0.785	0.776	0.757	0.750

6 Conclusion

The work illustrates the efficacy of the suggested algorithm in ensuring data security and privacy in cloud computing. The combination of encryption technologies and the implementation of chaos keys and hybrid encryption algorithms contribute to a secure and efficient system. The results of the testing phase highlight the superior performance of the PHEPA-AES-128 algorithm in terms of speed, making it suitable for high-speed operations and devices.

Designed with a focus on simplicity in its complexity operations, the PHEPA-AES algorithm stands out in its runtime computation. During different encryption rounds, the CPU cycles exhibited an average ranging from 9421 to 10112 cycles. This underscores the algorithm's commitment to achieving efficient and effective computational performance across various scenarios.

References

1. Aldabbagh, S.S.M., Al Shaikhli, I.F.T.: Improving PRESENT lightweight algorithm. In: Proc. - 2013 Int. Conf. Adv. Comput. Sci. Appl. Technol. ACSAT 2013, No. 1, pp. 254–258 (2013). https://doi.org/10.1109/ACSAT.2013.57
2. Isha, Luhach, A.K.: Analysis of Lightweight Cryptographic Solutions for Internet of Things. Indian J. Sci. Technol. **9**(28), 2–8 (2016). https://doi.org/10.17485/ijst/2016/v9i28/98382
3. Singh, S., Sharma, P.K., Moon, S.Y., Park, J.H.: Advanced lightweight encryption algorithms for IoT devices: survey, challenges and solutions. J. Ambient Intell. Humaniz. Comput., 1–18 (2017). https://doi.org/10.1007/s12652-017-0494-4
4. Bogdanov, A., Knudsen, L.R., Leander, G., Paar, C., Poschmann, A.: PRESENT: An Ultra-Lightweight Block Cipher, pp. 450–466 (2007).
5. Hong D. et al.: HIGHT cipher.pdf, pp. 46–59.
6. Hell, M., Johansson, T., Meier, W.: Grain: a stream cipher for constrained environments. Int. J. Wirel. Mob. Comput. **2**(1), 86–93 (2007). https://doi.org/10.1504/IJWMC.2007.013798
7. De Cannière, C.: TRIVIUM: a stream cipher construction inspired by block cipher design principles. In: Lect. Notes Comput. Sci. (including Subser. Lect. Notes Artif. Intell. Lect. Notes Bioinformatics), vol. 4176 LNCS, pp. 171–186 (2006). https://doi.org/10.1007/11836810_13
8. Lara-Niño, C. Andrés, M.S. Miguel, Arturo, D.P.: An evaluation of AES and present ciphers for lightweight cryptography on smartphones. In: 2016 Int. Conf. Electron. Commun. Comput. CONIELECOMP 2016, pp. 87–93 (2016). https://doi.org/10.1109/CONIELECOMP.2016.7438557
9. Kubba, Z.M.J., Hoomod, H.K.: A Hybrid Modified Lightweight Algorithm Combined of Two Cryptography Algorithms PRESENT and Salsa20 Using Chaotic System. In: 1st Int. Sci. Conf. Comput. Appl. Sci. CAS 2019, pp 199–203 (2019)
10. Labio, R.D., Festijo, E.D.: D-present a lightweight block cipher with key dependent Substitution box. In: 2020 Int. Conf. Adv. Comput. Sci. Inf. Syst., pp. 4–9 (2020). https://doi.org/10.1109/ICACSIS51025.2020.9263237
11. Kubba, Z.M.J., Hoomod, H.K.: Modified PRESENT Encryption algorithm based on new 5D Chaotic system. IOP Conf. Ser. Mater. Sci. Eng **928**(3) (2020). https://doi.org/10.1088/1757-899X/928/3/032023
12. Diffie, W., Hellman, M.E.: Exhaustive cryptanalysis of the NBS data encryption standard. Computer (Long. Beach. Calif)., **10**(6), 74–84 (1977). https://doi.org/10.1109/C-M.1977.217750

13. Daemen, J., Rijmen, V.: LNCS 1820 - The Block Cipher Rijndael, no. June (2014). https://doi.org/10.1007/10721064
14. Reyhani-Masoleh, A., Taha, M., Ashmawy, D.: New low-area designs for the AES forward, inverse and combined S-boxes. IEEE Trans. Comput. **69**(12), 1757–1773 (2020). https://doi.org/10.1109/TC.2019.2922601
15. Hämäläinen, P., Alho, T., Hännikäinen, M., Hämäläinen, T.D.: Design and implementation of low-area and low-power AES encryption hardware core. In: Proc. 9th EUROMICRO Conf. Digit. Syst. Des. Archit. Methods Tools, DSD 2006, pp. 577–583 (2006). https://doi.org/10.1109/DSD.2006.40
16. Saifurrab, C., Mirza, S.: AES algorithm using advance key implementation in MATLAB. Int. Res. J. Eng. Technol. **3**(9), 846–850 (2016)
17. Preview, T.S.: International standard ISO/IEC Information technology—security techniques—information security management systems—guidance iTeh standard preview iTeh standard preview, vol. 2017 (2017)
18. Azari, H.D., Joshi, D.P.V.: An efficient implementation of present cipher model with 80 bit and 128 bit key over FPGA based hardware architecture. Int. J. Pure Appl. Math. **119**(14), 1825–1832 (2018). http://www.ijpam.eu
19. Kubba, Z.M.J., Hoomod, H.K.: Developing a lightweight cryptographic algorithm based on DNA computing. In: AIP Conf. Proc., vol. 2290, no. December (2020). https://doi.org/10.1063/5.0027361
20. Ibrahem, M.K.: Perfect Secrecy System Based on Chaotic Key Generator. Iraqi J. Inf. Commun. Technol. **1**(2), 1–12 (2018). https://doi.org/10.31987/ijict.1.2.4
21. Shukla, P.K., Khare, A., Rizvi, M.A., Stalin, S., Kumar, S.: Applied cryptography using chaos function for fast digital logic-based systems in ubiquitous computing. Entropy **17**(3), 1387–1410 (2015). https://doi.org/10.3390/e17031387
22. Mehdi, S.A., Abbood, F.H.: 5D Hyper-Chaotic System Via Improved, no. October 2020 (2018)
23. Shahab S., Agarwal P., Mufti T., Obaid A.J.: SIoT (Social Internet of Things): a review. In: Fong, S., Dey, N., Joshi, A. (eds) ICT Analysis and Applications. Lecture Notes in Networks and Systems, vol. 314. Springer, Singapore (2022). https://doi.org/10.1007/978-981-16-5655-2_28
24. Sharma, A., Kumar, P., Babulal, K.S., Obaid, A.J., Patel, H.: Categorical data clustering using harmony search algorithm for healthcare datasets. Int. J. E-Health Med. Commun. (IJEHMC) **13**(4), 1–15 (2022). https://doi.org/10.4018/IJEHMC.309440

Improving Network Intrusion Detection with Convolutional Neural Networks and Data Balancing Techniques

Yaqot Mohsin Hazzaa and Shahla U. Umar

Abstract In order to maintain network security and detect intrusion attacks, Intrusion Detection Systems (IDS) have become an important technology. Due to the growing usage of the internet, intrusion attacks have become more frequent and can lead to the theft, alteration, or deletion of user data. IDS systems provide an active approach to network security by differentiating between normal and intrusion traffic, which needs to be blocked to protect the network. IDS systems have a wide range of applications in academic and business communities. Researchers have employed deep learning methods for network traffic classification, and conventional neural networks are popular due to their accuracy and versatility in handling various types of data. IoT systems, which are made up of connected devices, are also vulnerable to intrusion attacks, emphasizing the importance of IDS systems. To detect intrusion patterns, researchers have used the NSL-KDD dataset and employed a CNN neural network, which is an imbalanced dataset, meaning that the distribution of classes in the dataset is uneven. The proposed approach includes preprocessing techniques using oversampling by SMOT technique to improve the performance of the CNN-based IDS. The CNN network showed an excellent accuracy rate of 99.87%, and the accuracy rate further increased to 90.12% by incorporating the DLNID model to detect intrusion traffic. This demonstrates the potential of IDS systems and deep learning techniques for maintaining network security and protecting against intrusion attacks.

Keywords Network intrusion detection system (NIDS) · Convolutional neural network (CNN) · Imbalanced data

Y. M. Hazzaa (✉) · S. U. Umar
College of Computer Science and Information Technology, University of Kirkuk, Kirkuk, Iraq
e-mail: yaqothazzaa@gmail.com

Y. M. Hazzaa
Ministry of Youth and Sport, Baghdad, Iraq

G. Fortino et al. (eds.), *Proceedings of Third International Conference on Computing and Communication Networks*, Lecture Notes in Networks and Systems 917,
https://doi.org/10.1007/978-981-97-0892-5_53

1 Introduction

If we want to define the NIDS (Network Intrusion Detection System), it is that system depends on any technique that helps to keep the network secure without allowing for the hackers to log in and steal important information. Some recent studies allow us to use deep learning and data mining for this detection. Figure 1 shows a flowchart diagram of this system. First of all, we need a specific device that can capture and collect data of the network, the main data we need to collect is the packet's headers so there is no need to capture the message content. The captured data is then stored in a file. Before moving to the analyzing step, we must make a filtering one that removes unnecessary data, such like the traffic from trusted sources which is free of intrusion.

The first main step here is feature extraction that includes features like source and destination IP's, source and destination ports too, the flags affected and the protocol used with the number of packets and bytes. Another derived features may contain time window and connection window. This can help to detect fast scanning activities which are close to the intrusion act.

Table 1 shows a global property for the time-window-based features, and Table 2 shows the properties for the connection-window-based features. Sliding window technique is used be TCP protocol; it has the ability to control and manage the flow of the packets between all the hosts in the network. Sliding window protocol gets sure after connection is established of reliability of sequential data packets delivery between hosts. Time-window-based method works with time. So, we can work with continuous packets bits. We use the connection window that works with flows and sends more packets which lets it much faster.

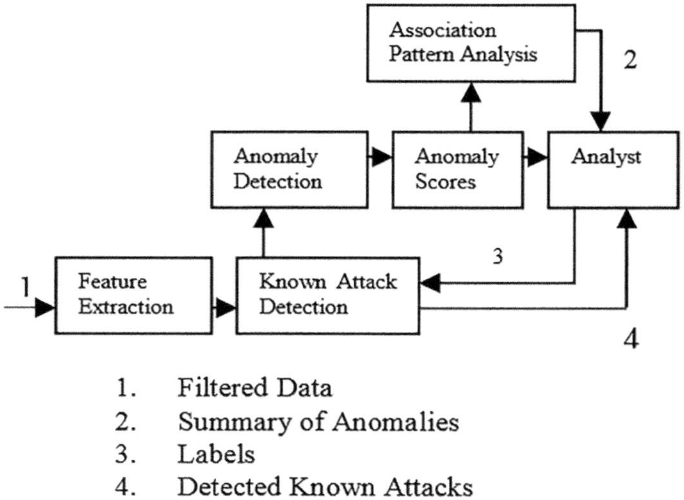

Fig. 1 Flowchart of intrusion detection system

Table 1 Properties for the time-window-based features

Feature	Description
Count-dest	Defines the number of flows that goes to a specific destination IP address in the network in the last T seconds which are coming from the same source
Count-src	Defines the number of flows from one IP address in T seconds that goes to the same destination
Count	The total amount of flows that are coming from the source IP

Table 2 The properties for the connection-window-based features

Count-des-conn	Flows to the specific IP address in the last defines number of flows (usually N) from a specified source
Count-src-conn	The same of Count-des-conn but we have several IP sources to the same destination in N flows
Count-serv-src-conn	Flows to the same specific port in last N flows
Count-serv-dest-conn	Flows to the same source port in last N flows

After this step, we will detect the module of the attack where the signatures might be able to detect. The data resulted (which gives a good view about the system features) is fed to the NIDS that will be used for detection algorithm to isolate the intrusion predicated act, then it gives a score for each network connection. Depending on this score we can define if the connection is an attack or just a usual connection type. This paper is organized as follows: Section two explains related work for sentiment analysis, Section three illustrates the proposed system architecture, Section four portrays the experimental and results, and Section five demonstrates the discussion. Finally, conclusion and future work are presented in Sect. 6.

2 Related Work

Several methods were used for IDS systems, and a large amount of them work on deep learning methods. Here we can define some of these methods that were used.

The first study was proposed by M.Alajanbi et al. where the researchers give a discussion about various methods of intrusion detection [1]. A. Oseni et al. too built a system for intrusion detection to be used in IoT networks that depends on deep learning. Shapley additive explanations were used (also known as SHAP) that give a suitable explain about the network running in the network. Researchers found a method to understand the logical reason of how the deep learning network is making its decision. This study gives nearly 99.15% accuracy and 98.83% F1 score. But this study was focusing on a specific scope for internet of Vehicles when movement is done in fog situation according to Alzubi et al. [2]. Another deep learning system was proposed by M. M. Alani et al. The system was named DeeplIoT. This system was developed to ensure the security for industrial IoT devices. This system was able to reach a 99% accuracy using the proposed dataset of water treatment facilities, nuclear reactors, and power stations. The dataset used by the research was WUSTL-IIOT-2021 dataset [3].

Recurrent neural networks were used too by V. Ravi et al. The researchers developed a new intelligent network to detect the intrusion using features fusion approach. This study approaches a 99% accuracy [4]. The model built by this study was inferior because it was simple and was used by many companies in the scope of IoT. F. Safara et al. proposed a method that depends on association rule of Artificial neural networks (ANN). The researchers study the intrusion types and how to make the detection task in communication networks. This study gives a good improvement in intrusion detection with an accuracy that reaches 98% [5].

Another study is about using deep learning techniques in securing IoT systems. Researchers give a good comparative study of all methods that are able to classify intrusion behavior and how to keep the IoT privacy using deep learning [6].

As a development for rule-based techniques, H. G. An et al. produce a rule dependent network sinkhole detection system [7]. It was an easier system to be developed for intrusion detection with an accuracy of 81.63% with more accuracy about 7% of the study proposed by [8]. H. Asad et al., give another study about rule-based intrusion detection system that may occur in open course networks. The researchers used a dynamic analysis for the system. The study shows the importance of quantitive analysis in specifying the protocol independent intrusion detection system for IoT networks [9].

An investigation of the current state of Intrusion Detection Systems (IDS) and a review of the proposed IDS-based ML algorithms are done in [10]. Also, they evaluate solutions based on architecture, intelligent prediction, and algorithms, and identify challenges and research areas in IoT design.

3 Theoretical Background

3.1 Machine Learning Method

One of the primary approaches for detecting and categorizing intrusion detection system (IDS) methods is through supervised learning. This technique involves using two types of methods. The first is a classification technique that utilizes specific features of the dataset and assigns them to a particular class. For example, there are two classes: "Attack" or "No Attack." Therefore, the goal is to classify all of the available traffic accordingly [11]. This method needs to minimize the false positive states (which are classifying the normal traffic as malicious) and the false negative states (which will see the malicious traffic as a normal one). To increase the available accuracy for this system, we have five types of categories. First one was the inductive rule generation which is a rule learning program; it can generate a rule set [12, 13].

It can define the frequent patterns that can be applied to the network traffic so easily. It can make the result to be understandable to the analyzers. Second category is the genetic algorithm; many searches try to use the genetic algorithm with the IDS system [14]. The Fuzzy Logic is the third category which is a reasoning system that tries to approximate the output predicted value of the system. It depends on fuzzy sets that are defined from the previous data. Then, we have the neural networks; it studies the user behavior in anomaly detection and does not have any need for the explicit user model. First time this technology was used; it was used as a statistical method [15]. The last of five categories is the support vector machine which is a method used in classification and regression, and it is a linear classifier. It separates the data into two classes that can be used in hyperplane [16].

3.2 Feature Selection

One of the machine learning methods where we have subsets of features got from the application of the learning algorithm. This method is infeasible in computation so we need to select specific features, or even if we have a limit number of data samples but with large number of features. Several methods are used here for the data accumulation so we can choose the features that can maximize the accuracy of the system [16].

3.3 Clustering

Clustering is a technique that is used to analyze data in a statistical manner; it is used on a wide range of machine learning and data mining. It is used too in computer vision systems for image and pattern recognition. The main idea here is to group the data that has the same features in one group. Each of these groups is called a cluster, and this grouping method needs to have a specific measurement technique [17]. Clustering is defined as unsupervised learning method. In our case, this method is effective to gather malicious activities together so it will be easy to detect them. The main feature of this technique is that it does not require to label our data to perform the training task [18].

3.4 Statistical Techniques

This method is used when we have information how the relationship between the data points, if we have, then we can perform some mathematical job to get the results that we need. There are three types in this scope which are linear, nonlinear, and decision trees [19], but also we can have more complicated techniques like Markov models and Bayes Estimator. When we want to work with the statistical methods, the task must be done in a specific time like day, week, month…etc. [20, 30].

3.5 Data Mining

There are several methods used in data mining scope [21]; they are different with the functions that they are equipped with, type of the data representation, or the algorithms. The main aim of our task is to make a classification between normal and malicious traffic so we can define if we have an attack or not [22, 23]. There are several types of data mining techniques like rules and decision trees, neural nets, probability models, and any other types of techniques. Some of data mining contributes are [24]:

- It can divide between normal and non-normal activities, so we can run the alarm only when we have a real alarm.
- It can define the bad sensor's data and signatures.
- It can define the patterns that are used in the real time like the IP address and the data that have the same activity.

Any system that uses deep learning needs some important steps to keep sure that the neural network result will be the most accurate available. The main steps required to prepare data for the CNN neural network are data pre-processing, normalization, removing missing or duplicated data, encoding, and data splitting.

3.6 Data Pre-processing

Data pre-processing is the main important task; it is the step where we make the data available in our hands to a form that is compatible with the deep learning module. Here we can be sure that we do not have zero values in the dataset or other missing values which may affect the training task.

In this step, an encoding task is done too; it gives the classifier and the deep learning model to work faster. This pre-processing task allows us to understand the whole model structure so we can understand how to examine it later. There are several additional tasks for pe-processing that includes [25]:

a. **Data Normalization**

First step for data preparation is to give the data a normal distribution form. This task is done prior to the numerical data. The main importance of the normalization is to avoid influence between features that are used so we will get less error when evaluating the whole model. Here we can define the range of the variables that are used, or intervals of it, and the method that we want to use.

In our study, we decide to use the Standard Scalar. This method tries to centralize the data by performing the subtract task from the mean and then scaling it to a constant number. Centralizing method keeps the data (or features) mean to zero and variance will be equal to 1. Scaling is followed by fitting data to make this data compatible with the deep learning model. This is done on the training and testing datasets so the data will be more similar to the ML models [25].

b. **Removing Missing Data**

When we work with the datasets, some of these datasets were created by users and others were collected by sensors systems. So, there might be a missing data or zero values that will affect the training task and decrease the accuracy of the model. All these values must be deleted for several reasons like increasing difficulty of the model, biasing of the estimation values, and wrong estimating values [26].

c. **Label Encoding**

When we work with deep learning models, these models cannot work with direct images or text data so, we need to convert these strings values to a numeric values that are understand by the model. The simplest way is to assign a specific integer value for each string value in the dataset [27].

Figure 2 shows the difference between the data types that we work with like text, array, number, etc. So, these types of data will not be familiar to the computer system (in the same way for the deep learning model) until it is converted to encoded type [6].

d. **Data Splitting**

This is the next step; we must split the dataset into training and testing subsets. There are several types for splitting like 60–40, 70–30, 80–20. We decide to use the type of 80% training set and 20% for system test and evaluation. It is the best used form. If we increase the size of training dataset by more than 80% then the accuracy will increase for a small amount and consume more time and processing power [28].

e. **Classification Using CNN**

This is the last step in our work to define if we have an attack. Neural networks were used on a wide range to work for classification tasks like the human brain task.

Fig. 2 Differences between regular data and computer data

Conventional Neural Network is a type of reasoning system that can predicate an output value from the input features with a high accuracy. There are multiple types for the output. It might be 0 or 1 (yes or no) or a value predicted with the highest accuracy in the model [29].

4 Proposed System

Figure 3 shows the flowchart of the proposed method. It starts by collecting data and then makes a preprocessing step on it. The main aim of the preprocessing is deleting zero values for example. Then we make an encoding task to let the network to work with integers and scaling it around zero values. Data then is split into training dataset to train the network and testing dataset to evaluate the network performance. The network type used was conventional neural network (CNN) that gives the final prediction of the output if it is an attack or not.

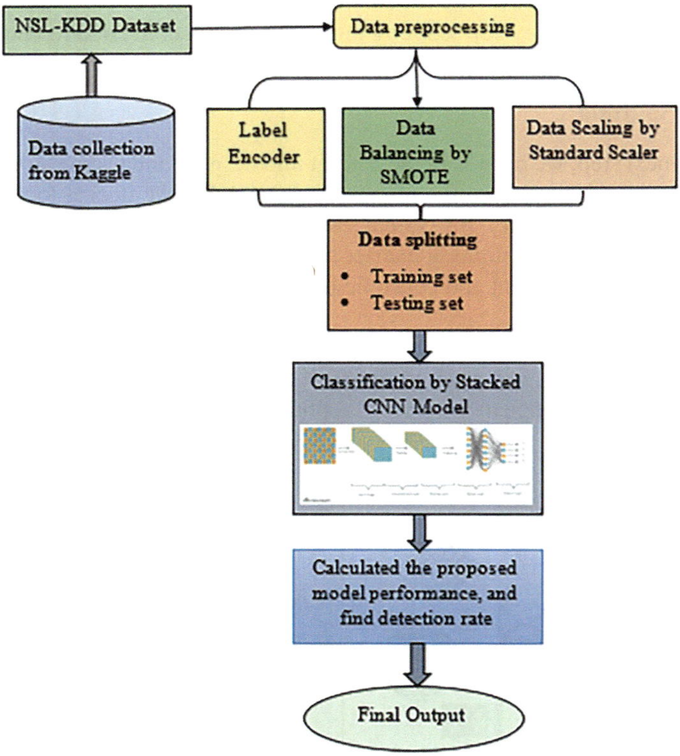

Fig. 3 The proposed system steps

The stacked CNN model is a developed model that makes it to learn how to get the best combination of the resulting prediction of several neural networks. It means we can increase the probability of well-learned networks and decrease it for the less accurate parts. Here, we can use the weighted average ensemble that can improve the performance of the total model. Stacked CNN model consists of two levels. Level 0, where we will enter the training data set and let the models to make their own predictions. In level 1, we will pass the results of level 0 to another neural network or a meta learner to make predictions. It is important that the meta-learner is trained on a separate dataset from the examples used to train the level 0 models to avoid overfitting.

5 Experimental and Results

In the following subsections, the experiments and results have been shown. An experimental setup is introduced in subsection a. The dataset description is demonstrated in subsection b. Lastly, the results and analysis are presented in subsection c.

a. **Experimental Setup**

The experiments were applied on the Windows 64-bits OS. Additionally, Python 3.8.8 has been used. Also, to implement deep learning algorithms TensorFlow and Keras frameworks have been installed.

b. **Dataset**

NSL-KDD: It is a developed edition of KDD-99 that was developed earlier [5]; the main problem was that the KDD dataset does not reflect the real live networks. There is a lack of data sets containing data about IDS on public. But we can say this is the best dataset available for the searchers to use [11, 19].

c. **Results and Analysis**

As mentioned before, the proposed work consists of two machine learning algorithms and three types of deep learning algorithms. Since, the Data is not balanced we did an up-sampling process on the minority classes and then concatenated the up-sampled data frames with the higher-class data frame. For the MNB we create a grid search object using the classifier and the hyperparameter dictionary and set the number of folds for cross-validation to 5. Also, for LR the maximum iteration was set at 500. On the other hand, the deep learning algorithms are applied in 10 epochs with 128 batch sizes. Finally, for each algorithm, the measures of accuracy, precision, recall, F1 score, and execution time performance have been computed as shown in the following Table 3 and Figs. 4 and 5.

Figure 4. shows the accuracy of the model after several Epochs. As we can see that after 30 Epochs the training accuracy was stable. The training accuracy was high too but it vibrates because of the variance of the data used. And Fig. 5 shows a loss-function graph for the proposed model. As we can see from the figure, the model error was minimal in training and testing tasks too.

Table 3 Results of the algorithms

Model	Accuracy	Precision	Recall
Stacked CNN model	0.9978	0.9978	0.9978
DLNID model	0.9012	0.9052	0.9012

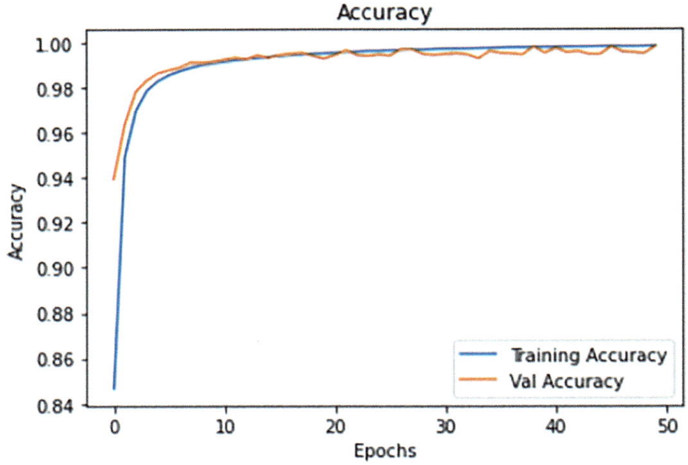

Fig. 4 Accuracy results of the algorithms

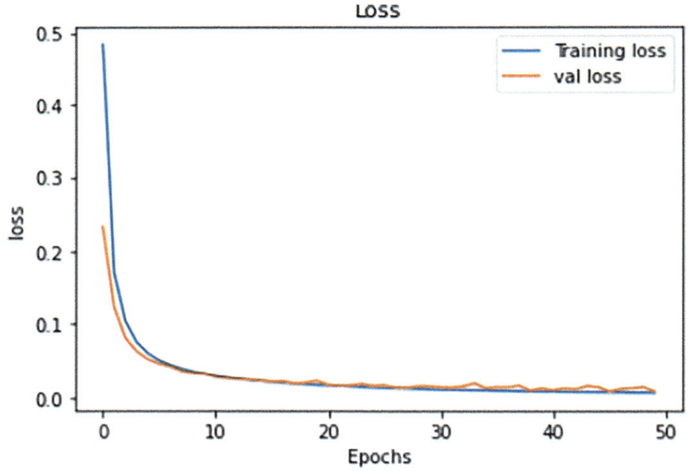

Fig. 5 Execution time of the algorithms

6 Discussion

The results of this study demonstrate the potential of machine learning and deep learning algorithms for sentiment analysis of IDS attacks. The deep learning models, in particular, showed significantly high accuracy compared to the normal classifiers. The Stacked CNN model gives a high accuracy compared to DLNID model. This suggests that the use of more advanced algorithms can improve the performance of sentiment analysis for IDS detection.

However, it is important to note that the accuracy of the models may be affected by various factors, such as the quality and quantity of the data, the preprocessing steps taken, and the specific parameters of the algorithms. Further optimization of the models, using techniques such as swarm optimization or bat optimization, may be necessary to achieve even higher accuracy.

7 Conclusion

The method proposed in our search aims mainly to classify and separate between several intrusion attacks and normal activities using deep learning and mainly conventional neural networks. The dataset was used and got from Kaggle website which is NSL-KDD dataset. The system was able to give the difference between four attack types which are U2R, R2L, DoS, and Probe.

After defining the accuracy of the system, it was 99.78% for the model used and selected. It was a better method that gave better accuracy results than the normal classification methods.

References

1. Alajanbi, M., Mohd Arfian Ismail, R.A.H., Sulaiman, J.: Intrusion detection: a review. Mesopotamian. J. Cybersecur. (2021)
2. Alzubi, O.A., Alzubi, J.A., Alazab, M., Alrabea, A., Awajan, A., Qiqieh, I.: Optimized machine learning-based intrusion detection system for fog and edge computing environment. Electronics (2022)
3. Alani, M.M., Damiani, E., Ghosh, U.: DeepIIoT: An Explainable Deep Learning Based Intrusion Detection System for Industrial IOT. In: *Proceedings of the 2022 IEEE 42nd International Conference on Distributed Computing Systems Workshops (ICDCSW)*, Hong Kong, China, 18–21 July 2022. IEEE, New York, NY, USA (2022)
4. Ravi, V., Chaganti, R., Alazab, M.: Recurrent deep learning-based feature fusion ensemble meta-classifier approach for intelligent network intrusion detection system. Comput. Electr. Eng. (2022)
5. Safara, F., Souri, A., Serrizadeh, M.: Improved intrusion detection method for communication networks using association rule mining and artificial neural networks. IET Commun. (2020)
6. Abdel-Basset, M., Moustafa, N., Hawash, H., Ding, W.: Deep Learning Techniques for IoT Security and Privacy; Springer: New York, NY, USA (2022)

7. An, G.H., Cho, T.H.: Improving sinkhole attack detection rate through knowledge-based specification rule for a sinkhole attack intrusion detection technique of IoT. Int. J. Comput. Netw. Appl. (IJCNA) (2022)
8. Cervantes, C., Poplade, D., Nogueira, M., Santos, A.: Detection of sinkhole attacks for supporting secure routing on 6LoWPAN for Internet of Things. In: *Proceedings of the 2015 IFIP/IEEE International Symposium on Integrated Network Management (IM)*, Ottawa, ON, Canada, 11–15 May 2015; IEEE, New York, NY, USA (2015)
9. Asad, H., Gashi, I.: Dynamical analysis of diversity in rule-based open source network intrusion detection systems. Empir. Softw. Eng. (2022)
10. Najam, C.U., Fakhrudeen, A.M.: On the performance of intrusion detection systems for the internet of things: State-of-the-Art in Research. Int. J. Nonlinear Anal. Appl. **14**(1), 1413–1436 (2023)
11. Shanmugavadivu, R., Dr. Nagarajan, N.: Network intrusion detection system using fuzzy logic. Indian J. Comput. Sci. Eng. (IJCSE) (2011)
12. Almseidin, M., Al-Sawwa, J., Alkasassbeh, M.: Anomaly-based intrusion detection system using fuzzy logic
13. https://arxiv.org/ftp/arxiv/papers/2107/2107.12299.pdf, available online, accessed 2/2/2023
14. Suhaimi, H., Suliman, S.I., Dr. Musirin, I.: Network intrusion detection system by using genetic algorithm. Indones. J. Electr. Eng. Comput. Sci. **16**(3), 1593. https://doi.org/10.11591/ijeecs.v16.i3.pp1593-1599
15. Krishna, A. et al. (2020) Intrusion detection and prevention system using deep learning. In: *Proceedings of the International Conference on Electronics and Sustainable Communication Systems, ICESC 2020*. https://doi.org/10.1109/ICESC48915.2020.9155711
16. Sathish Kumar, P., Raaza, A.: Study and analysis of intrusion detection system using random forest and linear regression. Period. Eng. Nat. Sci. **6**(1), 197–200 (2018), ISSN 2303–4521. http://pen.ius.edu.ba
17. Bharti, K. K., Shukla, Sweta Jain, S.: Intrusion detection using clustering. Int. J. Comput. Commun. Technol. (2010)
18. Damodaram, A., Bhatra Charyulu, N.Ch.: Algorithm for clustering with intrusion detection using modified and hashed K – means algorithms. Adv. Comput. Sci., Eng. Appl., 737–744
19. Ms. Phutane, T., Prof. Pathan, A.: Intrusion detection system using decision tree and Apriori algorithm. Int. J. Comput. Eng. & Technol. (IJCET) **6**(7), 10–19 (2015), Article ID: IJCET_06_07_002. http://www.iaeme.com/IJCET/
20. Pei, J., Upadhyaya Faisal Farooq, S.J., Govindaraju, V.: *Data Mining for Intrusion Detection.* State University of New York at Buffalo
21. Abaimov, S., Bianchi, G.: A survey on the application of deep learning for code injection detection. Array **11**, 100077 (2021). https://doi.org/10.1016/j.array.2021.100077
22. Abdallah, E.E., Eleisah, W., Otoom, A.F.: Intrusion detection systems using supervised machine learning techniques: a survey. Procedia Comput. (2022)
23. Kasongo, S.M.: A deep learning technique for intrusion detection system using a Recurrent Neural Networks based framework. Comput. Commun. **199**, 113–125 (2023). https://doi.org/10.1016/j.comcom.2022.12.010
24. Debar, H.: An introduction to intrusion-detection systems. Télécom SudParis - Institut Mines-Télécom (2009)
25. Yadav, S., Kalpana, R.: Data preprocessing for intrusion detection system using encoding and normalization approaches, 265–269 (2019).https://doi.org/10.1109/ICoAC48765.2019.246851
26. Roy, S., Li, J., Choi, B.J., Bai, Y.: A lightweight supervised intrusion detection mechanism for IoT networks. Futur. Gener. Comput. Syst. **127**, 276–285 (2022)
27. Hancock, J.T., Khoshgoftaar, T.M.: Survey on categorical data for neural networks. J. Big Data **7**(1), 1–41 (2020)
28. Syed, N.F., Ge, M., Baig, Z.: Fog-cloud based intrusion detection system using Recurrent Neural Networks and feature selection for IoT networks. Comput. Netw. **225**, 109662 (2023)

29. Alferaidi, A., Yadav, K., Alharbi, Y., Razmjooy, N., Viriyasitavat, W., Gulati, K., ... , Dhiman, G.: Distributed deep CNN-LSTM model for intrusion detection method in IoT-based vehicles. Math. Probl. Eng. (2022)
30. Iqbal, M.O., Obaid, A.J., Agarwal, P., Mufti, T., Hassan, A.R.: Blockchain technology and decentralized applications using blockchain. In: Tuba, M., Akashe, S., Joshi, A. (eds.) *ICT Infrastructure and Computing. Lecture Notes in Networks and Systems*, vol. 520. Springer, Singapore (2023). https://doi.org/10.1007/978-981-19-5331-6_57

MbAbI: A Benchmark Dataset for Malayalam Text Understanding and Reasoning

K. Reji Rahmath, P. C. Reghu Raj, and P. C. Rafeeque

Abstract This paper proposes a Malayalam Dataset containing 40000 instances, intended to assist the researchers working on Question Answering systems. This data set is the first of its kind. The Facebook bAbI dataset was translated to create this Malayalam bAbI (or, MbAbI) dataset. It comprises 20 different tasks of varying complexity. The tasks include simple supporting facts to complex path finding tasks. We have machine translated these tasks from English to Malayalam and tested the baseline QA models on these datasets. We have obtained state-of-the-art results with MbAbI dataset using deep learning and transformer models. In order to benefit the low-resource Malayalam research community we have made this dataset publicly available.

Keywords Question answering system · Malayalam language · MbAbI dataset · Language understanding · Reading comprehension

1 Introduction

The introduction of deep learning models and language models, especially transformer based models, made a notable improvement in handling Natural Language Processing tasks. The advancement is obvious in various tasks like natural language inference, sentiment analysis, and answering questions, etc. Large number

K. Reji Rahmath (✉)
Government Engineering College, Palakkad, India
e-mail: rahmathrejik@gmail.com

P. C. Reghu Raj
Government Engineering College, Kozhikkode, India

P. C. Rafeeque
Government College of Engineering, Kannur, India
e-mail: rafeeqpc@gcek.ac.in

K. Reji Rahmath · P. C. Reghu Raj · P. C. Rafeeque
APJ Abdul Kalam Technological University, Thiruvananthapuram, Kerala, India

G. Fortino et al. (eds.), *Proceedings of Third International Conference on Computing and Communication Networks*, Lecture Notes in Networks and Systems 917,
https://doi.org/10.1007/978-981-97-0892-5_54

of language models are available with different parameters in various languages like English, Chinese, Korean, and Arabic [8, 10]. However, compared to English language, the development and application of language models in most of the other languages are deficient. This is due to the unavailability of datasets and language specific word embedding models.

Low resource languages like Malayalam are still evolving in comparison to the English language's development of question-answering tasks. This is primarily caused by the lack of Malayalam QA datasets. We are presenting the MbAbI dataset[1] in accordance with Facebook bAbI dataset,[2] which includes 40000 question and answer pairs in Malayalam. There are 20 different jobs in the dataset, varying in complexity. Each task contains 1000 instances. Both the training and test sets of data include 20,000 instances each. MbAbI dataset is made publicly available in both Malayalam and transliterated version[1].

Question answering(QA) is a Language Processing(NLP) task which involves developing models or systems capable of answering questions posed by human beings. QA's objective is to make it possible for machines to comprehend and analyze human language so they can extract relevant information from a given context and provide accurate answers [9].

There are different types of question-answering tasks, and they can vary in complexity [11]. In factoid question answering, the questions usually seek specific factual information. The answer is often a short phrase, word, or number [16]. To answer the proposed question through machine learning approach, the system has to be trained with a lot of example questions and answers.

This paper's primary contributions are as follows:

1. Translating the Facebook bAbI project into Malayalam and producing a benchmark dataset, named MbAbI, with 40000 samples for text comprehension and reasoning.
2. MbAbI dataset contains 20 tasks with varying complexity and each task consists of 1000 training and 1000 testing samples (Malayalam and transliterated Malayalam version).
3. Achieved state-of-the-art results on the dataset by testing baseline QA models using deep learning and transformer algorithms.

Question answering has practical applications in search engines, virtual assistants, customer support, information retrieval, and more. As NLP techniques continue to advance, QA systems are becoming more sophisticated and capable of handling increasingly complex questions and contexts. This paper has been organized as follows: The next section discusses similar datasets with short answers created for reasoning exercises in several languages. The third section describes the dataset duration process in detail, including examples for each task. The next part details the experiments performed with these tasks followed by the report of results obtained.

[1] https://github.com/rahmathrejik/Question-Answering-Malayalam/tree/MbAbI.

[2] https://research.fb.com/downloads/babi/.

2 Literature Review

Question answering datasets are widely employed to train and assess question-answering models. There are a variety of datasets with particular characteristics that are suitable for the specific application or language.

The bAbI dataset is a collection of artificially generated language understanding and reasoning tasks designed to evaluate the capabilities of question answering and reasoning systems. It was introduced by Facebook AI Research(FAIR) in this research paper [18]. The name "bAbI" stands for "Bronze Age Babies", which refers to the idea that these tasks represent the building blocks of human-like language understanding and reasoning. The bAbI dataset includes a number of tasks that involve simple stories or scenarios, along with a series of questions about those stories. The questions require different levels of reasoning, such as deduction, induction, and coreference resolution.

HindiRC, Reading Comprehension dataset in Hindi is a small publicly available dataset for grade-based comprehension [1]. Corresponding to primary school classes, the passages in the dataset have been divided into different classes. The questions have been curated manually and classified according to the grades.

The CNN/DailyMail Corpus introduced by researchers at the University of Oxford is a widely used dataset for tasks related to summarizing text and understanding [5]. This dataset includes human-written summaries for each news piece as well as news articles gathered from the Daily Mail and CNN websites. The collection includes news stories from the Daily Mail and CNN websites. It comprises a variety of writing styles prevalent in news articles and covers a wide range of themes. Each article in the collection has one or more summaries authored by humans. Both abstractive and extractive summaries are possible.

Factoid Question-Answer Corpus is an enormous corpus containing 30 million question answer pairs [14]. It has been generated from Freebase knowledge base using Recurrent Neural Network(RNN). Questions were generated from knowledge base facts. The generated question answer pairs were evaluated both automatically and manually.

A dataset called The Children's Book Test (CBT) was created by Facebook AI Research to assess tale comprehension and reasoning [6]. The CBT dataset consists of a number of questions that are taken directly from picture books for kids. Another new dataset named ASQA contains Answer Summaries for Questions which are Ambiguous [15]. It contains factoid questions which have multiple correct answers for different interpretations.

The study [13] analyzes the recent increase in QA datasets. It lists the formats and domains of available resources as well as the flows in active QA systems. This paper also discusses the resources that are available in languages other than English. KenSwQuAD is a QA dataset for Swahili language [17]. This language is a low resource language predominantly spoken in eastern Africa. 1445 texts were annotated with a total of 7526 QA pairs.

3 Dataset Curation

As part of this work, we have built a Malayalam question-answer dataset(MbAbI), with 20k training instances and 20k test instances. The dataset is in accordance with Facebook's bAbI tasks, which are for now only available in Hindi and English [18]. 20 simulated tasks are included in the dataset to test a system's aptitude for responding to questions involving a variety of textual themes. Their level of difficulty varies.

In the curation process, the dataset in English has been translated to Malayalam using Online Google Translator. Due to the possibility of errors in machine translation, we personally examined the results after completing each task. Because the dataset instance story has short and crisp sentences, the frequency of translation errors was quite low. We translated the test dataset after translating the train dataset. We also transliterated the train and test datasets into Malayalam as part of the translation procedure.

The dataset includes thousand training instances and thousand test instances for each task. The activities are referred to by titles like three supporting facts, indefinite knowledge, path discovery, conjunction, etc. [2, 7]. A series of assertions (sometimes referred to as a tale) about how players move around things, a question, a response, and the pertinent statement numbers are examples in the dataset [4]. A brief summary of each task along with an instance, question, and response as in the training dataset is given below. The left side of the example shows the Malayalam version and right side shows the transliterated version.

- **Task 1: Single supporting fact**: Only one statement from a group of statements has the response to the question. It is a reasonably simple activity. For example,

ഡാനിയേൽ ഇടനാഴിയിലേക്ക് തിരിച്ചുപോ- യി.	Daaniyel itanaazhiyilekku thiric- chupoyi.
സാന്ദ്ര തോട്ടത്തിലേക്ക് മാറി.	Saandra thottatthilekku maari.
ഡാനിയൽ എവിടെയാണ്? ഇടനാഴി	Daaniyal eviteyaan? Itanaazhi

- **Task 2: Two supporting facts**: In order to complete this activity and respond to the prompt, two supporting assertions must be integrated. Compared to a single supporting fact kind, it is a little more complicated. For example,

മേരി പൂന്തോട്ടത്തിലേക്ക് മടങ്ങി.	Meri poonthottatthilekku matangi.
മേരിക്ക് അവിടെ ഫുട്ബോൾ ലഭിച്ചു.	Merikku avite phutbol labhicchu.
ഡാനിയേൽ അടുക്കളയിലേക്ക് പോയി.	Daaniyel atukkalayilekku poyi.
മേരി ഫുട്ബോൾ ഉപേക്ഷിച്ചു.	Meri phutbol upekshicchu.
ഫുട്ബോൾ എവിടെയാണ്? പൂന്തോട്ടം	Phutbol eviteyaan? Poonthottam

- **Task 3: Three supporting facts**: Here, three statements ought to be considered to respond to the questions. It is undoubtedly more challenging to respond to than either of the earlier types. For example,

സാന്ദ്ര പാൽ എടുത്തു.	Saandra paal etutthu.
സാന്ദ്ര ഇടനാഴിയിലേക്ക് പോയി.	Saandra itanaazhiyilekku poyi.
ഡാനിയൽ ഓഫീസിലേക്ക് യാത്ര ചെയ്തു.	Daaniyal opheesilekku yaathra cheythu.
സാന്ദ്ര അടുക്കളയിലേക്ക് മടങ്ങി.	Saandra atukkalayilekku matangi.
അടുക്കളയ്ക്ക് മുമ്പ് പാൽ എവിടെയായിരുന്നു? ഇടനാഴി	Atukkalaykku mumpu paal eviteyaayirunnu? Itanaazhi

- **Task 4: Two argument relations**: In a QA system, it is critical to distinguish subjects and objects. Even if the words are rearranged, this job recognizes the two arguments in the given line, similar to the subject and object. For example,

അടുക്കളയുടെ വടക്ക് ഭാഗത്താണ് ഓഫീസ്.	Atukkalayute vatakku bhaagatthaanu opheesu.
അടുക്കളയുടെ തെക്ക് ഭാഗത്താണ് പൂന്തോട്ടം.	Atukkalayute thekku bhaagatthaanu poonthottam.
അടുക്കളയുടെ വടക്ക് എന്താണ്? ഓഫീസ്	Atukkalayute vatakku enthaan? Opheesu

- **Task 5: Three argument relations**: In contrast to task 4, this task's statement has three arguments. The three arguments in the supplied statement are identified in this task. For example,

ബിൽ ഓഫീസിലേക്ക് യാത്ര ചെയ്തു.	Bil opheesilekku yaathra cheythu.
ബിൽ അവിടെ ഫുട്ബോൾ എടുത്തു.	Bil avite phutbol etutthu.
ബിൽ കിടപ്പുമുറിയിലേക്ക് പോയി.	Bil kitappumuriyilekku poyi.
ബിൽ ഫ്രെഡിന് ഫുട്ബോൾ നൽകി.	Bil phredinu phutbol nalki.
ഫ്രെഡിന് ബിൽ എന്താണ് നൽകിയത്? ഫുട്ബോൾ	Phredinu bil enthaanu nalkiyath? Phutbol

- **Task 6: Yes/No queries**: It is a straightforward activity with a yes-or-no response option, similar to the single supporting fact. For example,

ഡാനിയേൽ ഇടനാഴിയിലേക്ക് തിരിച്ചുപോയി.	Daaniyel itanaazhiyilekku thiricchupoyi.
മേരി കിടപ്പുമുറിയിലേക്ക് പോയി.	Meri kitappumuriyilekku poyi.
ഡാനിയൽ കുളിമുറിയിലുണ്ടോ? ഇല്ല	Daaniyal kulimuriyilundo? Illa
മേരി കിടപ്പുമുറിയിലാണോ? അതെ	meri kitappumuriyilaano? Athe

- **Task 7: Counting**: This evaluates the model's capability to count some items that share a particular characteristic. For example,

ജോൺ ആപ്പിൾ മേരിക്ക് കൈമാറി.	Jon aappil merikku kymaari.
മേരി തോട്ടത്തിലേക്ക് മാറി.	Meri thottatthilekku maari.
മേരി അവിടെ ഫുട്ബോൾ പിടിച്ചു.	Meri avite phutbol piticchu.
ജോൺ അവിടെ പാൽ എടുത്തു.	Jon avite paal etutthu.
മേരി എത്ര വസ്തുക്കൾ വഹിക്കുന്നു? രണ്ട്	Meri ethra vasthukkal vahikkunnu? Randu

- **Task 8: Lists/Sets**: This job aims to provide answers in the form of a list of things. For example,

മേരി ഫുട്ബോൾ എടുത്തു.	Meri phutbol etutthu.
മേരി പാൽ എടുത്തു.	Meri paal etutthu.
മേരി ഫുട്ബോൾ ഉപേക്ഷിച്ചു.	Meri phutbol upekshicchu.
മേരി ആപ്പിൾ എടുത്തു.	Meri aappil etutthu.
മേരി എന്താണ് വഹിക്കുന്നത്?	Meri enthaanu vahikkunnath?
പാൽ, ആപ്പിൾ	Paal, aappil

- **Task 9: Simple negation**: It is a challenging task that requires the model to determine whether the provided line of reasoning is wrong. For example,

സാന്ദ്ര ഓഫീസിലേക്ക് യാത്രയായി.	Saandra opheesilekku yaathrayaayi.
ഫ്രെഡ് ഇപ്പോൾ ഓഫീസിലില്ല.	Phredu ippol opheesililla.
ഫ്രെഡ് ഓഫീസിലാണോ? ഇല്ല	Phredu opheesilaano? Illa
സാന്ദ്ര ഓഫീസിലുണ്ടോ? അതെ	saandra opheesilundo? Athe

- **Task 10: Indefinite knowledge**: This exercise looks for potential outcomes, therefore the response could be one of yes, no, or maybe. For example,

ഫ്രെഡ് സ്കൂളിലോ കിടപ്പുമുറിയിലോ ആണ്.	Phredu skoolilo kitappumuriyilo aanu.
സ്കൂളിൽ ഫ്രെഡ് ഉണ്ടോ? ഒരുപക്ഷേ	Skoolil phredu undo? Orupakshe
ഫ്രെഡ് പാർക്കിൽ ഉണ്ടോ? ഇല്ല	Phredu paarkkil undo? Illa

- **Task 11: Basic coreference**: Anaphora resolution is a function of basic coreference. This examines the system's capacity to identify the single subject that is the closest referent. For example,

ഡാനിയൽ ഓഫീസിലേക്ക് മാറി.	Daaniyal opheesilekku maari.
അതിനുശേഷം അദ്ദേഹം ഇടനാഴിയിലേക്ക് മാറി.	Athinushesham addheham itanaazhiyilekku maari.
ഡാനിയൽ എവിടെയാണ്? ഇടനാഴി	Daaniyal eviteyaan? Itanaazhi

- **Task 12: Conjunction**: This task evaluates the capability of the model to respond to queries involving many actors in a statement. For example,

സാന്ദ്രയും ഡാനിയലും അടുക്കളയിലേക്ക് മാറി.	Saandrayum daaniyalum atukkalayilekku maari.
സാന്ദ്രയും ജോണും പൂന്തോട്ടത്തിലേക്ക് മാറി.	Saandrayum jonum poonthottatthilekku maari.
സാന്ദ്ര എവിടെ? പൂന്തോട്ടം	Saandra evite? Poonthottam

- **Task 13: Compound coreference**: The system must recognize several subjects for this assignment. Contrary to basic coreference, this is more challenging. For example,

മേരിയും ഡാനിയലും ഓഫീസിലേക്ക് യാത്ര ചെയ്തു.	Meriyum daaniyalum opheesilekku yaathra cheythu.
അതിനുശേഷം അവർ ഇടനാഴിയിലേക്ക് മാറി.	Athinushesham avar itanaazhiyilekku maari.
സാന്ദ്രയും മേരിയും വീണ്ടും അടുക്കളയിലേക്ക് പോയി.	Saandrayum meriyum veendum atukkalayilekku poyi.
മേരി എവിടെയാണ്? ഓഫീസ്	Meri eviteyaan? Ophees

- **Task 14: Time reasoning**: The purpose of this procedure is to evaluate the system's ability to recognize time. The sequence of happening of events described in the story will be relevant to the inquiry. For example,

ബിൽ ഇന്നലെ സിനിമക്ക് പോയി.	Bill innale sinimakku poyi.
ജൂലി ഇന്ന് രാവിലെ സ്കൂളിൽ പോയി.	Jooli innu raavile skoolil poyi.
ഫ്രെഡ് ഇന്നലെ പാർക്കിൽ പോയി.	Phredu innale paarkkil poyi.
ഇന്നലെ ജൂലി ഓഫീസിലേക്ക് പോയി.	Innale jooli opheesilekku poyi.
സ്കൂളിന് മുമ്പ് ജൂലി എവിടെയായിരുന്നു? ഓഫീസ്	Skoolinu mumpu jooli eviteyaayirunnu? Opheesu

- **Task 15: Basic deduction**: This job verifies the arguments' fundamental deductive features.

ആടുകൾക്ക് ചെന്നായ്ക്കളെ ഭയമാണ്.	Aatukalkku chennaaykkale bhayamaanu.
ജെസീക്ക ഒരു ചെന്നായയാണ്.	Jeseekka oru chennaayayaanu.
എമിലി ഒരു ആടാണ്.	Emili oru aataanu.
എമിലി എന്തിനെയാണ് ഭയക്കുന്നത്? ചെന്നായ	Emili enthineyaanu bhayakkunnath? Chennaaya

- **Task 16: Basic induction**: In this task, fundamental induction qualities are examined. For example,

ജൂലിയസ് ഒരു ഹംസമാണ്.	Jooliyasu oru hamsamaanu.
ജൂലിയസ് വെളുത്തതാണ്.	Jooliyasu velutthathaanu.
ലില്ലി ഒരു ഹംസമാണ്.	Lilli oru hamsamaanu.
ലില്ലിയുടെ നിറം എന്താണ്? വെള്ള	Lilliyute niram enthaan? Vella

- **Task 17: Positional reasoning**: This analyzes the colored items' relative locations as stated in the statements. For example,

ത്രികോണം പിങ്ക് ചതുരത്തിന് മുകളിലാണ്.	Thrikonam pinku chathuratthinu mukalilaanu.
നീല ചതുരം ത്രികോണത്തിന്റെ ഇടത്താണ്.	Neela chathuram thrikonatthinte itatthaanu.
നീല ചതുരത്തിന്റെ വലത്ത് പിങ്ക് ചതുരം ഉണ്ടോ? അതെ	Neela chathuratthinte valatthu pinku chathuram undo? Athe

- **Task 18: Size reasoning**: In this task of size reasoning, the relative sizes of the objects are examined. For example,

സ്യൂട്ട്കേസിൽ ഫുട്ബോൾ യോജിക്കുന്നു.	Syoottkesil phutbol yojikkunnu.
സ്യൂട്ട്കേസ് അലമാരയിൽ യോജിക്കുന്നു.	Syoottkesu alamaarayil yojikkunnu.
പെട്ടി ഫുട്ബോളിനേക്കാൾ ചെറുതാണ്.	Petti phutbolinekkaal cheruthaanu.
പെട്ടി സ്യൂട്ട്കേസിൽ ഒതുങ്ങുമോ? അതെ	Petti syoottkesil othungumo? Athe
അലമാര പെട്ടിയിൽ ഒതുങ്ങുമോ? ഇല്ല	alamaara pettiyil othungumo? Illa

- **Task 19: Path finding**: This challenge evaluates the model's capacity to identify the route between various sites given a set of assertions that account for various locations. For example,

ഓഫീസിന് വടക്ക് ഭാഗത്താണ് അടുക്കള. കിടപ്പുമുറിയുടെ പടിഞ്ഞാറ് ഭാഗത്താണ് പൂന്തോട്ടം. പൂന്തോട്ടത്തിന്റെ പടിഞ്ഞാറ് ഭാഗത്താണ് ഓഫീസ്. അടുക്കളയിൽ നിന്ന് പൂന്തോട്ടത്തിലേക്ക് എങ്ങനെ പോകും? തെക്ക്, പടിഞ്ഞാറ്	Opheesinu vatakku bhaagatthaanu atukkala. Kitappumuriyute patinjaaru bhaagatthaanu poonthottam. Poonthottatthinte patinjaaru bhaagatthaanu opheesu. Atukkalayil ninnu poonthottatthilekku engane pokum? Thekku, patinjaaru

- **Task 20: Agent's motivations**: This task evaluates how a person reacts to his or her current mood or emotion. For example,

സുമിത്തിന് വിശക്കുന്നു. സുമിത്ത് അടുക്കളയിലേക്ക് പോകുന്നു. സുമിത്ത് അവിടെ ആപ്പിൾ പിടിച്ചു. എന്തുകൊണ്ടാണ് സുമിത്ത് അടുക്കളയിലേക്ക് പോയത്? വിശക്കുന്നു	Sumitthinu vishakkunnu. Sumitthu atukkalayilekku pokunnu. Sumitthu avite aappil piticchu. enthukondaanu sumitthu atukkalayilekku poyath? Vishakkunnu

4 Experiments

We tested the dataset with a variety of models, including LSTM, GRU, and Memory Networks [19]. It was also tested with Transformer models like BERT [3]. 100 samples were picked from the 1000 training samples for testing, whereas 900 were used for training. For each task, the number of test samples was set at 1000. The tests were carried out on a GPU using Tensorflow. For memory network-based approach word vector size was set to 64 and based on convergence 100 was set as epochs. For transformer-based experiment, pre-trained multilingual BERT(mBERT) model was chosen for tokenization and question answering. In this case, batch size was set to 32 and only three epochs were used for training.

5 Results

The results obtained by the implementation of Long Short Term Memory, Gate Recurrent Unit, and Memory network in single and combined ways for this dataset are discussed in detail in [12]. Here in Table 1, we only exhibit the comparatively better results obtained with memory network. From the analysis, it is obvious that Memory network have higher accuracy than the other two models. Out of the 20 tasks

Table 1 Test accuracy (%) for baseline QA models using the MbAbI dataset

Task	Accuracy(%)	
	MemNN	mBERT
Single supporting fact	93	100
Two supporting facts	70	94
Three supporting facts	22	80
Two argument relations	80	95
Three argument relations	96	90
Yes/no questions	60	96
Counting	81	92
Lists/sets	87	96
Simple negation	67	99
Indefinite knowledge	58	85
Basic coreference	75	93
Conjunction	93	100
Compound coreference	84	99
Time reasoning	81	90
Basic deduction	69	100
Basic induction	41	65
Positional reasoning	54	78
Size reasoning	80	91
Path finding	8	40
Agent's motivations	88	98

only a few, like path finding and compound coreference have very low accuracy. Other than these, most of the other tasks performs well.

The performance of QA model with transformer architecture implementation, like BERT, shows better accuracy than the deep learning models. This is due to the fact that language models are able to understand the stories much better than traditional deep learning models and also they can handle long term dependencies.

6 Conclusion and Future Work

The primary component of deep learning models is the dataset. It is possible to conduct experiments using machine learning and language models only when there is enough data for the system to be trained on. Even though the proposed MbAbI dataset consists of questions having short answers, different types of tasks will surely be a benchmark to test the model's capability to do reasoning, logical analysis, induction, deduction, and coreference resolution. The QA research in a language like Malayalam with limited resources will unquestionably benefit from this dataset for reasoning and language understanding. We are working to improve this QA system by adding more examples and challenging questions. For further research in this area, we have to develop another dataset containing complicated questions as well.

References

1. Anuranjana, K., Rao, V. and Mamidi, R.: Hindirc: a dataset for reading comprehension in hindi. In: 0th International Conference on Computational Linguistics and Intelligent Text (2019)
2. Chen, D., Mooney, R.: Learning to interpret natural language navigation instructions from observations. In: Proceedings of the AAAI Conference on Artificial Intelligence, vol. 25 (2011)
3. Devlin, J., Chang, M.W., Lee, K., Toutanova, K.: Bert: pre-training of deep bidirectional transformers for language understanding. arXiv preprint arXiv:1810.04805 (2018)
4. Grigar, D.: Twisty Little Passages: An Approach to Interactive Fiction by Nick Montfort, p. 286. MIT Press, Cambridge (2003). Illus. Trade. ISBN: 0-262-13436-5, 2005
5. Hermann, K.M., Kocisky, T., Grefenstette, E., Espeholt, L., Kay, W., Suleyman, M., Blunsom, P.: Teaching machines to read and comprehend. Adv. Neural Inf. Process. Syst. 28 (2015)
6. Hill, F., Bordes, A., Chopra, S., Weston, J.: The goldilocks principle: reading children's books with explicit memory representations. arXiv preprint arXiv:1511.02301 (2015)
7. Levesque, H., Davis, E. and Morgenstern, L.: The winograd schema challenge. In: Thirteenth International Conference on the Principles of Knowledge Representation and Reasoning (2012)
8. Lim, S., Kim, M., Lee, J.: Korquad1. 0: Korean qa dataset for machine reading comprehension. arXiv preprint arXiv:1909.07005 (2019)
9. Lin, J., Quan, D., Sinha, V., Bakshi, K., Huynh, D., Katz, B., Karger, D.R.: What makes a good answer? the role of context in question answering. In: INTERACT (2003)
10. Mozannar, H., Hajal, K.E., Maamary, E., Hajj, H.: Neural arabic question answering. arXiv preprint arXiv:1906.05394 (2019)
11. Ojokoh, B., Adebisi, E.: A review of question answering systems. J. Web Engin. 717–758 (2018)
12. Reji Rahmath K, PC Reghu Raj, and Rafeeque PC. Malayalam question answering system using deep learning approaches. IETE J. Res. 1–13 (2022)
13. Rogers, A., Gardner, M., Augenstein, I.: Qa dataset explosion: A taxonomy of nlp resources for question answering and reading comprehension. ACM Comput. Surv. **55**(10), 1–45 (2023)
14. Serban, I.V., García-Durán, A., Gulcehre, C., Ahn, S., Chandar, S., Courville, A., Bengio, Y.: Generating factoid questions with recurrent neural networks: The 30m factoid question-answer corpus. arXiv preprint arXiv:1603.06807 (2016)
15. Stelmakh, I., Luan, Y., Dhingra, B. and Chang, M.W.: Asqa: factoid questions meet long-form answers. arXiv preprint arXiv:2204.06092 (2022)
16. Wang, M., et al.: A survey of answer extraction techniques in factoid question answering. Comput. Linguist. **1**(1), 1–14 (2006)
17. Wanjawa, B.W., Wanzare, L.D., Indede, F., McOnyango, O., Muchemi, L., Ombui, E.: Kenswquad-a question answering dataset for swahili low-resource language. ACM Trans. Asian Low-Resour. Lang. Inf. Process. **22**(4), 1–20 (2023)
18. Weston, J., Bordes, A., Chopra, S., Rush, A.M., Van Merriënboer, B., Joulin, A., Mikolov, T.: Towards ai-complete question answering: a set of prerequisite toy tasks. arXiv preprint arXiv:1502.05698 (2015)
19. Yu, Y., Si, X., Hu, C., Zhang, J.: A review of recurrent neural networks: Lstm cells and network architectures. Neural Comput. **31**(7), 1235–1270 (2019)

A Novel Approach to Image Restoration and Image Enhancement

Divya Singh, Bhawna Upadhayay, Pradeep Gupta, and Sonam Gupta

Abstract Image restoration and image enhancement are fundamental tasks in computer vision and image processing. Image restoration aims to recover the original information from a degraded or corrupted image, while image enhancement aims to improve the visual quality and interpretability of an image. This research paper presents a comprehensive theoretical framework that integrates both image restoration and image enhancement techniques. We explore various methods, algorithms, and mathematical models to address the challenges associated with these tasks. The proposed framework leverages both classical and deep learning-based approaches to achieve superior performance in restoring and enhancing images. The experimental results demonstrate the effectiveness and versatility of our proposed approach in a wide range of applications.

Keywords Image restoration · Image enhancement · Weiner filter · Histogram equalization · Spatial filter

1 Introduction

This research paper explores the theoretical foundations and presents an integrated approach to address the challenges in image restoration and enhancement, which are crucial in various fields such as medical imaging, surveillance, remote sensing, and digital photography. The paper highlights the broad applicability of these techniques across various domains, such as medicine, surveillance, satellite imaging, and digital photography. Technological advancements, such as hardware capabilities, computational power, and deep learning techniques, have facilitated the development of sophisticated algorithms and frameworks for image restoration and enhancement. However, there are still challenges to be addressed, such as the need to combine

D. Singh · B. Upadhayay · P. Gupta · S. Gupta (✉)
Ajay Kumar Garg Engineering College, Ghaziabad, Uttar Pradesh, India
e-mail: guptasonam6@gmail.com

G. Fortino et al. (eds.), *Proceedings of Third International Conference on Computing and Communication Networks*, Lecture Notes in Networks and Systems 917,
https://doi.org/10.1007/978-981-97-0892-5_55

restoration and enhancement techniques and the effectiveness of these techniques depending on specific degradation types, image content, and application scenarios.

The research paper reviews and analyzes existing image restoration and enhancement techniques, including classical methods and deep learning-based approaches. It proposes a comprehensive theoretical framework that integrates these techniques, enhancing image quality, detail restoration, noise reduction, and contrast enhancement. The framework's performance is evaluated using datasets and metrics, comparing it to state-of-the-art methods. The paper also demonstrates practical applications in medical imaging, surveillance, satellite imaging, and art restoration.

The research aims to enhance the quality and interpretability of visual data in various applications by proposing an integrated framework that combines the strengths of both techniques. It covers both classical and deep learning-based techniques, considering various degradation types. The proposed framework aims to achieve superior results in image quality, detail restoration, noise reduction, and contrast enhancement.

The structure of this paper is as follows: The literature review and associated activities are covered in Sect. 2, which provides details on how the survey's paper was gathered. Sections 3 and 4 discuss the image restoration and enhancement techniques. The integrated approaches are described in Sect. 5. The experimental evaluation and proposed methodology are given in Sect. 6. Section 7 shows the results. At the end of this study, we have highlighted future research areas with conclusion.

2 Literature Review

The integration of various methodologies and algorithms in image processing aims to enhance image quality and restore damaged ones. This integration serves dual functions: enhancing visual quality and repairing corrupted images. The significance of both enhancement and restoration techniques is evident in the vast landscape of image processing [1–3]. The Deep Texture-Structure Decomposition (DTSD) method and its algorithms have significantly improved image quality by disentangling and processing intrinsic textures and structural elements, thereby enhancing visual appeal and enhancing the overall visual experience [4]. Researchers explore various algorithms and methodologies for improving image quality, spanning underwater imaging, historical photograph restoration, and medical image enhancement across various domains. Researchers are using various techniques, including deep learning models, linear approaches, and non-linear algorithms, to tackle unique challenges in different image categories, aiming to overcome their unique challenges [1, 5–7].

SRGAN, a generative adversarial network for image super-resolution, is a significant advancement in image processing, inferring photo-realistic natural images at a remarkable 4 $\times$ upscaling factor [8]. A groundbreaking discovery has bridged the gap between deep convolutional networks-based learning methods and handcrafted image priors, marking a new era in image processing by allowing randomly-initialized neural networks to function as handcrafted priors [9].

Researchers and practitioners in image processing strive to balance the diverse methods with their respective strengths and weaknesses to find the most suitable technique for a specific task [1–3]. The research community is exploring various strategies, such as MIR nets and normalizing flows, to push the boundaries of image processing and improve image enhancement and restoration. These efforts are documented in related websites and IEEE publications, aiming for greater achievement [7, 10].

An innovative approach in image processing has emerged, offering improved image quality, preservation of contours, and luminance. Researchers continuously innovate and explore new techniques and algorithms to balance merits and demerits for new excellence.

3 Image Restoration Techniques

Image restoration techniques, combining mathematical models and deep learning, aim to improve image quality and interpretability by reversing degradations like blur, noise, and compression artifacts.

3.1 Degradation Models

Image restoration techniques often utilize mathematical models to describe the degradation process, capturing characteristics of image degradations like blur, noise, and compression artifacts. Common degradation models include:

- **Point Spread Function (PSF) Model**: This represents the blurring effect in an image, caused by lens imperfections, motion blur, or out-of-focus capture, and refers to the spread of a point in the original scene.
- **Noise Models**: Understanding noise characteristics is crucial for effective denoising techniques, as it can be modeled using statistical distributions like Gaussian, Poisson, or salt-and-pepper noise models.
- **Compression Artifact Models**: Image compression algorithms introduce artifacts due to lossy compression techniques. Models that capture these artifacts aid in their removal or reduction during the restoration process.

3.2 Classical Restoration Techniques

Classical image restoration techniques involve mathematical operations and algorithms to recover the original information from degraded images. Some common classical restoration techniques include:

- **Deconvolution**: Deconvolution techniques aim to invert the degradation process by estimating the inverse filter or restoring the original image from the blurred image using algorithms like Wiener deconvolution or Richardson-Lucy deconvolution.
- **Image Denoising**: Denoising techniques remove or reduce noise from images. These include spatial filters like Gaussian filtering, median filtering, or bilateral filtering, as well as transform domain methods like wavelet denoising or non-local means denoising.
- **Inpainting**: Inpainting techniques restore missing or damaged portions of an image by inferring plausible information based on the surrounding context. Methods like exemplar-based inpainting or diffusion-based inpainting are commonly used.
- **Super-Resolution**: Super-resolution techniques aim to enhance the resolution and details of low-resolution images. These methods utilize either single-image-based approaches or multiple-image-based approaches to generate a high-resolution output.

3.3 Deep Learning-Based Restoration Methods

Deep learning has revolutionized image restoration by using large-scale datasets and neural network architectures to learn complex mappings between degraded and clean images. Some popular approaches include:

- **Convolutional Neural Networks (CNNs)**: CNN-based methods utilize deep neural networks with convolutional layers to learn feature representations and perform restoration tasks such as denoising, deblurring, or inpainting.
- **Generative Adversarial Networks (GANs)**: GANs consist of a generator network that generates restored images and a discriminator network that discriminates between the generated images and the ground truth clean images. GANs have been successful in tasks like super-resolution and image inpainting.
- **Attention Mechanisms**: Deep learning models employ attention mechanisms to focus on crucial image regions, enabling adaptive resource allocation for restoration, thereby enhancing fine detail restoration and overall performance.

4 Image Enhancement Techniques

Image enhancement techniques improve visual quality, specific features, and information by adjusting contrast, brightness, sharpness, and color balance while preserving details. They can be applied individually or combined for various applications like photography, digital media, and computer vision tasks. Here are some common image enhancement techniques.

4.1 Histogram Equalization

Histogram equalization redistributes the intensity values of an image to enhance its contrast. It stretches the histogram to cover the full dynamic range, effectively spreading out the pixel values and making the image visually more appealing. However, it may lead to unnatural results in certain cases.

4.2 Contrast Stretching

Contrast stretching adjusts the image's contrast by mapping the original intensity range to a desired range. This technique expands the range of pixel values to enhance the differences between dark and bright regions, thereby improving the image's overall contrast.

4.3 Gamma Correction

Gamma correction modifies the relationship between the pixel values and their displayed intensities. It adjusts the gamma parameter to control the brightness and contrast of an image. This technique is particularly useful in compensating for nonlinearities in image acquisition or display devices.

4.4 Spatial Filtering

Spatial filtering techniques apply filters or convolution kernels to the image pixels to enhance specific features or suppress noise. Common filters include Gaussian filters for blurring or smoothing, Laplacian filters for edge enhancement, and sharpening filters to enhance fine details.

4.5 Adaptive Filtering

Adaptive filtering techniques adjust filter parameters based on local image characteristics. These filters dynamically adapt to different regions of the image, preserving details in textured areas while reducing noise or enhancing edges in smoother regions.

4.6 Retinex-Based Methods

Retinex-based methods aim to simulate human perception by separating an image into its illumination and reflectance components. These methods enhance the image by adjusting the illumination and preserving or enhancing the reflectance details, leading to improved visual quality.

4.7 Tone Mapping

Tone mapping techniques are commonly used in high dynamic range (HDR) imaging to display HDR images on devices with limited dynamic range. Tone mapping operators compress the HDR image's tonal range while preserving details and enhancing local contrast, resulting in visually pleasing images.

4.8 Color Correction

Color correction techniques adjust the color balance and color appearance of an image. They can be used to correct color cast, enhance or modify specific color channels, or adjust the overall color balance to achieve desired visual effects.

4.9 Deep Learning-Based Methods

Deep learning techniques, particularly convolutional neural networks (CNNs), have also been applied to image enhancement tasks. These methods learn complex mappings between input images and their enhanced versions, leveraging large-scale datasets to improve visual quality, enhance details, and reduce artifacts.

5 Integration of Image Restoration and Image Enhancement

The integration of image restoration and enhancement techniques in image processing can enhance results, visual quality, and interpretability, making them more visually appealing and informative. This approach is useful in medical imaging, surveillance, digital photography, and multimedia content creation, enhancing the interpretation of images. Here are some approaches for integrating image restoration and image enhancement.

5.1 Joint Restoration and Enhancement Frameworks

Instead of treating restoration and enhancement as separate steps, joint frameworks are developed to simultaneously address both tasks. These frameworks combine the restoration and enhancement objectives into a unified optimization problem, leveraging the interplay between these tasks. By considering restoration and enhancement as complementary processes, joint frameworks can produce visually pleasing and informative results.

5.2 Feedback Mechanisms

Feedback mechanisms can be incorporated into the restoration and enhancement pipeline to iteratively refine the results. Initially, restoration techniques are applied to recover image details, followed by enhancement techniques that adjust the visual appearance. The enhanced image can then be fed back to the restoration stage to further refine the details. This iterative process improves the overall restoration and enhancement quality.

5.3 Adaptive Approaches

Adaptive approaches dynamically adjust the restoration and enhancement parameters based on the image content and degradation characteristics. These approaches analyze the image properties, such as noise level, contrast, or sharpness, and adaptively select appropriate restoration and enhancement techniques to achieve optimal results. Adaptive algorithms can enhance image details while preserving image aesthetics and avoiding over-enhancement.

5.4 Deep Learning-Based Integration

Deep learning techniques have shown promise in integrating image restoration and enhancement tasks. Deep neural networks can learn the mapping between degraded images and their corresponding restored and enhanced versions in an end-to-end manner. These networks can capture complex relationships between restoration and enhancement processes, leading to improved results. Architectures such as autoencoders, encoder-decoder networks, or generative adversarial networks (GANs) have been employed for joint restoration and enhancement tasks.

5.5 Performance Evaluation

The integration of image restoration and enhancement techniques requires appropriate evaluation metrics to assess the effectiveness of the combined approach. In addition to traditional metrics like PSNR and SSIM, perceptual quality metrics that consider human visual perception are essential for evaluating the visual appeal and interpretability of the enhanced images. User studies or subjective assessments can provide valuable insights into the performance of the integrated techniques.

6 Experimental Evaluation and Proposed Model

Experimental evaluation is a crucial method for evaluating the performance and effectiveness of image restoration and enhancement techniques. It involves designing experiments, selecting datasets, choosing metrics, and analyzing results using multiple metrics to assess visual quality, detail preservation, and perceptual appeal. Here are key considerations for conducting experimental evaluation.

6.1 Datasets

Selecting suitable datasets for evaluating image restoration and enhancement techniques is crucial. These should cover real-world scenarios and include ground truth reference images. Common datasets include ImageNet, COCO, BSDS500, DIV2K, and Set5/Set14. Data augmentation techniques and preprocessing steps ensure consistency. Data should be divided into training, validation, and testing sets for unbiased evaluation. Ethical handling and privacy regulations are essential.

6.2 Evaluation Metrics

To evaluate techniques' performance, use metrics that accurately reflect human perception. Traditional metrics like PSNR, SSIM, and MSE may not accurately represent visual quality. Instead, consider perceptual quality metrics like subjective assessments or human visual perception models.

- **Peak Signal-to-Noise Ratio (PSNR)**: PSNR is a metric used to measure the quality of restored or enhanced images by comparing the original reference image to the processed image in terms of signal power and noise level. Higher PSNR values indicate better image quality, but may not always accurately represent subtle visual differences.

- **Structural Similarity Index (SSIM)**: SSIM is a perceptual quality metric that assesses structural similarity between reference and processed images, considering luminance, contrast, and structural information. It ranges from 0 to 1, with 1 indicating perfect similarity, aligning better with human perception than PSNR.
- **Mean Squared Error (MSE)**: MSE is a widely used metric for evaluating image restoration and enhancement techniques, calculating the average squared difference between reference and processed images. Lower values indicate better image quality, but it's sensitive to outliers and may not align well with human perception.

6.3 Comparative Analysis of Restoration Techniques

It involves considering various factors to evaluate their performance and suitability for specific tasks. This helps researchers and practitioners identify the most effective approaches, enabling informed decision-making and the adoption of techniques that best meet the intended applications' requirements.

Here are some key aspects to consider:

- **Restoration Quality**: Evaluate the quality of the restored images using objective metrics such as PSNR, SSIM, MSE, SCS, or VIF. Compare the quantitative results obtained by different restoration techniques to assess their ability to recover image details, reduce noise, remove artifacts, and restore the original image characteristics.
- **Computational Complexity**: The study examines the computational complexity of restoration techniques, considering time and memory requirements. It aims to find techniques that balance efficiency with restoration quality, with faster processing times and lower resource demands being desirable in real-time or resource-constrained applications.
- **Robustness**: Examine the robustness of restoration techniques against different types and levels of image degradation. Evaluate how well the techniques perform when faced with challenging degradation scenarios such as noise, blur, compression artifacts, or mixed degradation types. Robust techniques should produce satisfactory results across a wide range of degradation conditions (Fig. 1).

6.4 Comparative Analysis of Enhancement Techniques

It involves evaluating their performance in improving image quality, enhancing visual appearance, and addressing specific enhancement goals. Here are some factors to consider for a comparative analysis:

- **Enhancement Quality**: The study evaluates the quality of enhanced images through visual inspection and subjective assessments, focusing on brightness, contrast, sharpness, color balance, noise reduction, and dynamic range, and compares different techniques.

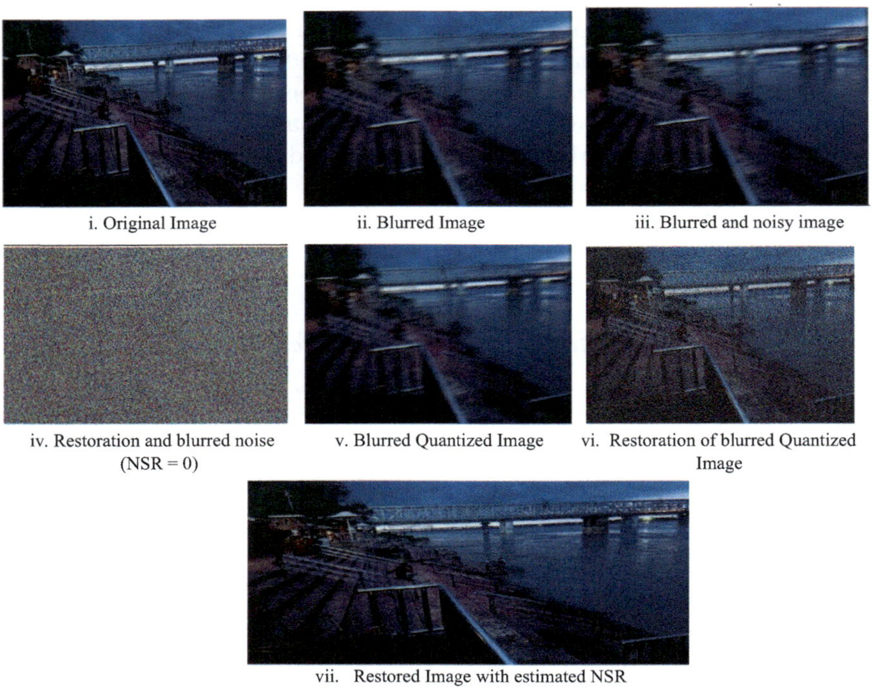

Fig. 1 Image restoration process using weiner filter

- **Enhancement Goals**: Consider the specific enhancement goals of the techniques, such as contrast enhancement, color correction, detail enhancement, noise reduction, or artifact removal. Assess how well the techniques achieve these goals and whether they can be tailored to meet specific requirements in different applications.
- **Computational Efficiency**: Analyze the computational efficiency of enhancement techniques, including the processing time and resource requirements. Consider techniques that offer a good balance between computational complexity and enhancement quality, as faster and more efficient algorithms are desirable for real-time or large-scale processing.
- **Applicability to Different Image Types**: Evaluate how well enhancement techniques perform on different types of images, such as natural scenes, portraits, landscapes, or specific domains like medical or satellite images. Techniques that can adapt to different image characteristics and content while consistently delivering high-quality enhancements are preferable.
- **Artifacts and Side Effects**: Examine the presence of artifacts or side effects introduced by enhancement techniques. Assess whether the techniques introduce unwanted noise, haloing, over-sharpening, or color distortions. Techniques that

minimize or effectively control such artifacts are considered more reliable and suitable for practical use.

- **User Control and Customization**: Consider the level of user control and customization offered by enhancement techniques. Determine whether the techniques provide adjustable parameters or interactive interfaces that allow users to fine-tune the enhancement process according to their preferences or specific requirements.
- **Comparative Studies and Benchmarks**: Refer to comparative studies and benchmark datasets available in the literature. Analyze the results and comparisons provided in research papers, conference proceedings, or publicly available benchmarks to gain insights into the performance of different enhancement techniques and their strengths and limitations (Fig. 2).

6.5 Integrated Approach Evaluation

This paper evaluates an integrated approach combining image restoration and enhancement techniques from various perspectives, including restoration quality, enhancement quality, computational efficiency, subjective evaluation, comparative analysis, and application-specific evaluation, providing insights into the technique's strengths, limitations, and applicability. Here are some factors to consider during the evaluation:

- **Restoration Quality**: Assess the quality of restored images using objective metrics like PSNR, SSIM, MSE, SCS, or VIF, comparing them to standalone techniques, and evaluating the integrated approach's ability to restore details, reduce noise, and remove artifacts.
- **Enhancement Quality**: The integrated approach's enhanced images quality can be assessed through visual inspection and subjective assessments, focusing on improvements in brightness, contrast, sharpness, color balance, and overall appeal.
- **Synergy and Integration**: The integration of restoration and enhancement techniques within a system is crucial for ensuring quality and coherence. The integrated approach combines the strengths of both techniques to achieve superior results compared to using them independently.
- **Computational Efficiency**: Analyze the computational efficiency of the integrated approach. Consider the processing time and resource requirements of the combined restoration and enhancement process. Assess whether the integrated approach is feasible for real-time or near-real-time applications, and whether it can handle large-scale image processing efficiently.
- **Subjective Evaluation**: Subjective evaluations are conducted to evaluate the quality and appeal of images produced by an integrated approach, aiming to gather feedback from users on its effectiveness and user satisfaction.
- **Comparative Analysis**: The study compares the performance of an integrated approach to standalone restoration and enhancement techniques, evaluating their

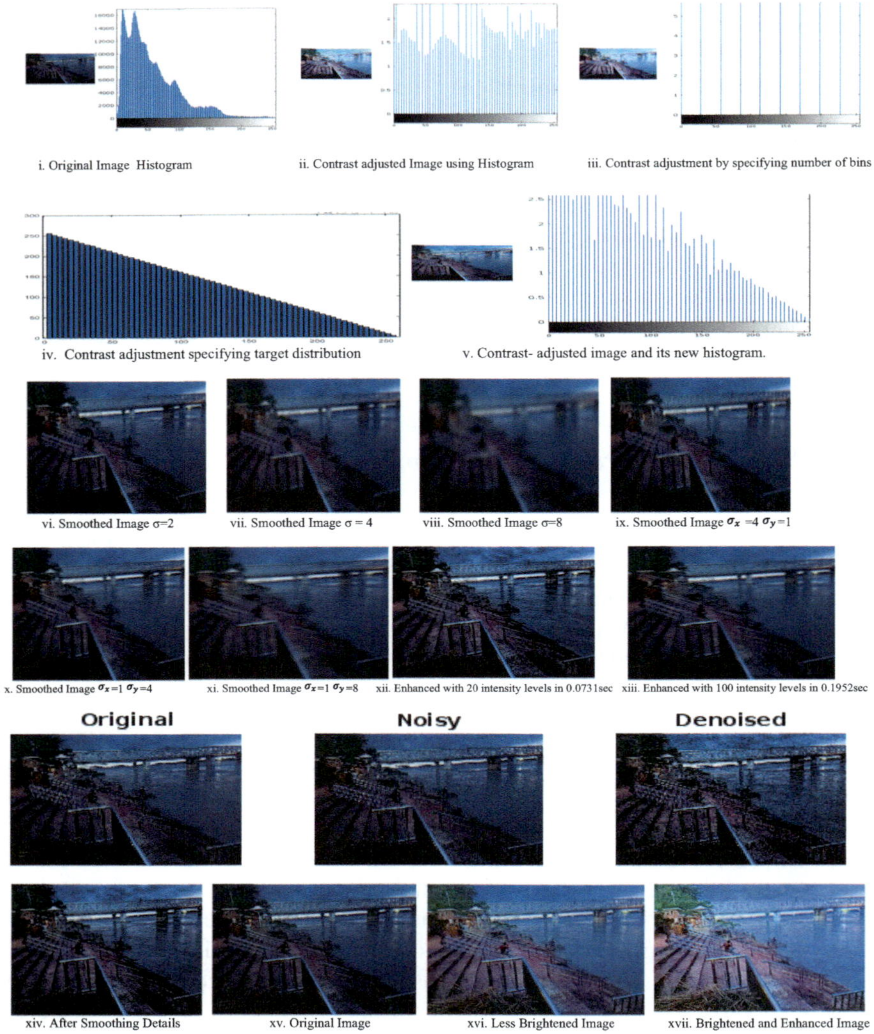

Fig. 2 Image enhancement process using histogram equalization, gaussian and laplacian filter

performance, quality, efficiency, and user satisfaction, and determining significant improvements (Fig. 3).

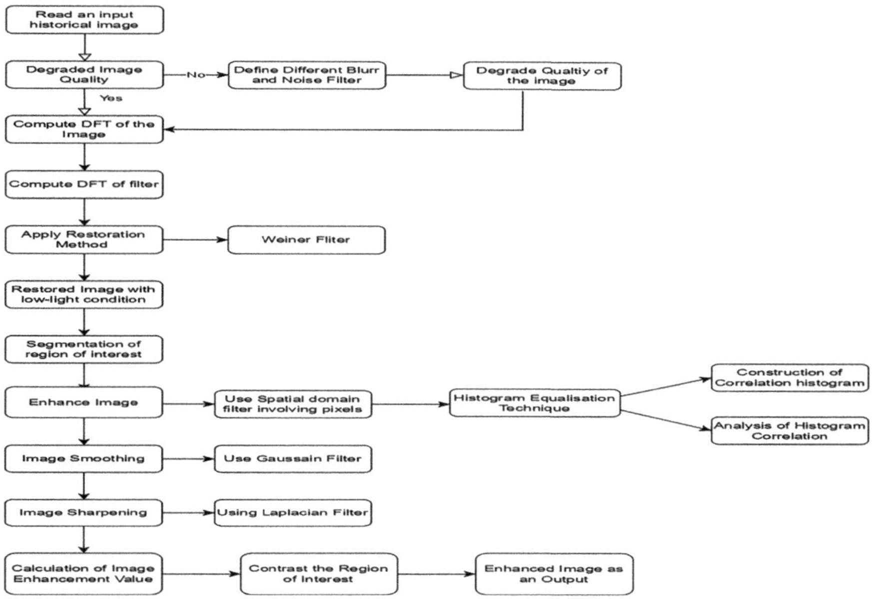

Fig. 3 Proposed architecture for integrated approach

7 Experimental Result

The proposed method involves integrating various filters during repair and improvement processes. The first step involves restoring deteriorated photos using a restoration filter. The inverse Gaussian and Weiner filters were used to determine the extent of image correction. The best filters were tested on historical images. The next stage involves enhancement using a Histogram Equalizer, Gaussian, and Laplacian filter. The quality of the image and restoration can be improved more effectively by integrating multiple methods and algorithms than applying filters separately. This approach is particularly relevant in today's modern era, where demand for picture improvement is increasing.

Results of the Experiment:

The quality of the image is, t_quality = 0.0714.

The noisy image's peak signal-to-noise ratio is 30.1248.

The peak signal-to-noise ratio of the denoised image is 23.2301.

The mean-squared error is 469.4368.

8 Conclusion

Image repair and enhancement techniques have significantly improved the quality, interpretability, and aesthetic appeal of photographs, enhancing their use in multimedia, medical imaging, surveillance, satellite imaging, art restoration, and digital photography. These techniques have improved situational awareness, remote sensing analysis, preservation of cultural artifacts, and diagnostic accuracy. However, challenges like cross-modal restoration, interpretability, adversarial robustness, perceptual quality evaluation, hybrid techniques, and real-world degradations remain. Resolving these issues and exploring new study areas can lead to better solutions, enabling informed decision-making and enhanced photo aesthetics.

Despite advancements in image restoration and enhancement techniques, challenges such as real-world degradations, perceptual quality evaluation, hybrid methodologies, dataset variety, and computational effectiveness remain. Complex degradations like motion blur, sensor noise, and compression artifacts affect real-world photos. Future studies should focus on developing reliable and flexible methods, providing better perceptual quality assessment measures, and utilizing hybrid approaches combining model-based and data-driven deep learning techniques. Expanding dataset variety and increasing computing efficiency are crucial for real-time applications and limited resources. Image restoration techniques are essential in fields like medical imaging, surveillance and security, art restoration and conservation, and digital photography and multimedia.

References

1. Reader, A.J., Corda, G., Mehranian, A., da Costa-Luis, C., Ellis, S., Schnabel, J.A.: Deep learning for pet image reconstruction. IEEE Trans. Radiat. Plasma Med. Sci. (2020)
2. Zhang, L., Qi, Z., Meng, F.: A review on the construction of business intelligence system based on unstructured image data. Procedia Comput. Sci. **199** (2022)
3. Trongtirakul, T., Ladyzhensky, D., Chiracharit, W., Agaian, S.: Non-linear contrast stretching with optimizations. In: Mobile Multimedia/Image Processing, Security, and Applications (2019)
4. Zhao, L., Wang, K., Zhang, J., Wang, A., Bai, H.: Learning deep texture-structure decomposition for low-light image restoration and enhancement. Elsevier (2022)
5. Aslam, M., Naseer, I.: Removal of the noise & blurriness using global & local image enhancement equalization techniques. IJCISV1
6. Wang, Y., Wan, R., Yang, W., Li, H., Chau, L.P., Kot, A.: Low-light image enhancement with normalizing flow. In: Proceedings of the AAAI Conference on Artificial Intelligence
7. Zamir, S.W., Arora, A., Khan, S., Hayat, M., Khan, F.S., Yang, M.-H.: Learning enriched features for real image restoration and enhancement. arXiv:2003.06792v2[cs.CV] (2020)
8. Ledig, C., Theis, L., Huszár, F., Caballero, J., Cunningham, A., Acosta, A., Shi, W.: Photo-realistic single image super-resolution using a generative adversarial network. In: Proceedings of the IEEE Conference on Computer Vision and Pattern Recognition (CVPR) (2017)
9. Ulyanov, D., Vedaldi, A., Lempitsky, V.: Deep image prior. In: Proceedings of the IEEE Conference on Computer Vision and Pattern Recognition (CVPR) (2018)
10. Swarup, J., Sreedevi, I.: DWT based historical image enhancement technique using adaptive gamma correction. J. Algebr. Stat. **13**

Smart Helmet with Crash Detection

K Antony Kumar, S Ravikumar, Angeline Lydia, and Rohit Yadav

Abstract Riding a motorcycle is an exhilarating experience, but it comes with its risks. In 2019, the National Highway Traffic. Safety Administration reported that motorcycle riders were 28 times more likely to die in a crash than passenger car occupants. A smart helmet is a wearable that can be used to enhance the safety of motorcycle riders. It includes features such as crash detection and wear detection, which can prevent the bike from starting unless the rider is wearing a helmet. It is also equipped with accelerometers, gyroscopes which help to detect crashes, and also connected with a GSM module chip which sends a message to important contacts and nearby hospital in case of emergency.

Keywords Crash detection · GSM module · Internet of Things · Smart helmet · Wear detection

1 Introduction

Motorcycle riding is the most popular mode of transportation, around the world. In many nations, the growth of the economy and society is facilitated by transportation made possible by motor vehicles. But it also poses significant risks to riders, who are more vulnerable to accidents and injuries compared to other vehicles on the

K. Antony Kumar (✉) · S. Ravikumar · A. Lydia · R. Yadav
Department of Computer Science and Engineering, Vel Tech Rangarajan Dr. Sagunthala R&D Institute of Science and Technology, Chennai, Tamil Nadu, India
e-mail: antonykumar@veltech.edu.in

S. Ravikumar
e-mail: ravikumars@veltech.edu.in

A. Lydia
e-mail: drangelinelydia@veltech.edu.in

R. Yadav
e-mail: vtu20315@veltech.edu.in

G. Fortino et al. (eds.), *Proceedings of Third International Conference on Computing and Communication Networks*, Lecture Notes in Networks and Systems 917,
https://doi.org/10.1007/978-981-97-0892-5_56

road. Around 1.35 million lives are lost on roadways each year worldwide. This horrifying death toll equals over 3,700 individuals every day who pass away in crashes involving a variety of vehicles, including cars, buses, motorbikes, bicycles, trucks, and pedestrians. According to global report, crash injuries are the number one killer of children and teenagers worldwide and the eighth-leading cause of mortality overall (5–29 years old). The main causes of traffic accidents include poor road conditions, reckless overtaking, driving when exhausted or intoxicated, health problems, cell phone use, and other distractions like eating and drinking while operating a vehicle [1]. Additionally, accidents might happen as a result of other drivers' untimely replies under dangerous circumstances. There were 412,432 traffic incidents reported in all States and Union Territories in 2021. These sad occurrences resulted in 153,972 fatalities and 384,448 injuries. In 2021, there were an average of 12.6% more traffic accidents than there were in 2020. Similar to this, the number of fatalities and injuries from traffic accidents rose by 16.9 and 10.39%, respectively [2].

According to the National Highway Traffic Safety Administration (NHTSA), motorcycle riders have a startling 28 times greater chance of dying in a collision per mile traveled than passengers in passenger automobiles. In addition, motorcycle riders are more likely to suffer from head and neck injuries in a crash, which can lead to fatal or permanent disability. The World Health Organization's research shows that appropriately donning a helmet can lower the risk of fatal injuries by 42% and the risk of brain injuries by 69%. A total of 46,593 people died in 2021 as a result of not wearing a helmet, of whom 13,716 (29.5%) were passengers and 32,877 (70.6%) were drivers. The significant loss of life and associated financial losses are mostly preventable and call for immediate action to be taken for their effective prevention. Helmets are the most effective safety measure for riders which reduces the risk of a head injury by 69 and the risk of death by 37 in the event of a crash, according to the NHTSA. However, not all riders wear helmets, and even those who do may not wear them properly or consistently. This is why smart helmets with crash detection and wear detection features are necessary.

Smart helmets that incorporate advanced sensors and systems to enhance the safety of riders. They are equipped with features such as crash detection and wear detection, which can prevent accidents from happening in the first place and also provide critical support in the event of a crash. The crash detection feature uses sensors like accelerometers and gyroscopes to detect if the rider has been involved in an accident this can include detecting sudden changes in speed and orientation. If a crash is detected the helmet can automatically send an alert to emergency services, providing the rider's location and other relevant information. The reaction time of emergency services might be significantly sped up as a result, perhaps saving the rider's life.

According to Road Accidents in India 2021 report, there were an average of 12.6% more traffic accidents than there were in 2020. Similar to this, the number of fatalities and injuries from traffic accidents rose by 16.9 and 10.39%, respectively as shown in Table 1.

A total of 46,593 people died in 2021 as a result of not wearing a helmet, of whom 13,716 (29.5%) were passengers and 32,877 (70.6%) were drivers as given

Table 1 Number of accidents, fatalities, and injured people for last 5 years

Year	Accidents	% Variation from the earlier time period	Fatalities	% Variation from the earlier time period	Persons injured	% Variation from the earlier time period
2017	4,64,910	−3.28	1,47,913	−1.9	4,70,975	−4.78
2018	4,67,044	0.46	1,51,417	2.37	4,69,418	−0.33
2019	4,49,002	−3.86	1,51,113	−0.2	4,51,361	−3.85
2020	3,66,138	−18.46	1,31,714	−12.84	3,48,279	−22.84
2021	4,12,432	12.64	1,53,972	16.9	3,84,448	10.39

Table 2 Number of people died and injured due to non-use of helmets in 2021

Particulars	Without helmet	
	Killed	Injured
Drivers	32,877	57,264
Passenger	13,716	36,499
Total	46,593	93,763

in Table 2. According to 2021 road accident data, the percentage proportion of each type of vehicle in total accidents is often close to the percentage share of deaths and injuries.

1.1 Objectives

The proposed approach intends to create a device that improves driver safety, with the primary goal of reducing the fatality rate connected to two-wheeler accidents. The apparatus also incorporates an alcohol sensor to assess whether the user has ingested alcohol.

1.2 Scope

The proposed strategy seeks to reduce the severity of injuries caused by motorbike accidents by actively preventing them. The solution attempts to improve the general safety and security of motorbike trips by including a smart helmet.

1.3 Motivation

This study's inspiration comes from its potential to transform motorcycle safety and save lives. It advances the use of technology to address an urgent public health issue, makes roadways safer for motorbike riders, and establishes a model for future developments in wearable safety technology.

In Sect. 2 of this paper, related work has been discussed along with its benefits and drawbacks. In Sect. 3, the technique for the smart helmet and accident detection system is presented, along with an explanation of the operation of key system components. Section 4 presents the hardware description of the proposed system. Section 5 describes about the Testing and validation of the proposed work. The conclusion is provided in Sect. 6 at the end.

2 Related Works

Motorcycles are a popular mode of transportation all over the world, but they come with a lot of risks because their riders are more susceptible to accidents and injuries than occupants of other vehicles [3]. According to the National Highway Traffic Safety Administration (NHTSA), people driving motorbikes had a higher chance of dying in collisions than people driving passenger automobiles [4]. Head and neck injuries are particularly common among motorcycle riders involved in crashes, which can lead to serious, permanent disabilities or even death [5]. Wearing a helmet is a crucial safety precaution that can dramatically reduce the incidence of fatalities and brain injuries in motorcycle accidents, according to the NHTSA [6]. Even individuals who occasionally use helmets may not do it appropriately or consistently [7]. Motorcycle riders do not always wear them. Smart helmets with cutting-edge sensors and systems have been created to improve rider safety in response to this problem [8]. With technologies like collision detection and wear detection, these intelligent helmets can help avoid accidents and offer assistance in the case of one [9]. The crash detection function makes use of accelerometers and gyroscopes as sensors to identify abrupt changes in direction or speed that may be caused by an accident. When an accident is detected, the smart helmet may instantly notify emergency services and provide them with the rider's position and other relevant information, thereby reducing response times and saving lives. Smart helmets can detect crashes as well as whether the rider consistently and correctly wears the helmet. In order to prevent accidents, the wear detection feature, which checks if the rider is wearing the helmet before starting the motorbike ride, is essential.

An RF transmitter and RF receiver make up the smart helmet that is suggested in [3]. The helmet's transmitter and receiver are attached, with the bike's ignition switch serving as the receiver. The bike won't start unless and until the driver wears a helmet, which eventually aids in averting any fatal injuries in the event of an accident. The bike will stop instantly if the rider takes his or her helmet off while operating it.

The client–server architecture used by the system described in paper [10] includes local hospitals, regional transport offices (RTO), and vehicles. serving as clients, and the driver's phone serving as the server. When a vehicle collides, the client's smartphone takes a picture of the situation and sends it to the server. This procedure, among other things, determines the seriousness of the event and evaluates whether essential safety measures have been taken.

The suggested technology consists of a receiver built into the motorcycle and a transmitter built into the helmet. This set up is intended to determine whether the driver is using the car while not drinking any alcohol and wearing a helmet [11].

In order to determine if the rider is wearing the helmet or not, a developed system makes use of an infrared (IR) sensor mounted on the helmet. The IR sensor sends a signal to a NodeMCU device that is connected to the motorcycle's battery and ignition system. The NodeMCU blocks the ignition system if the helmet is not being worn, preventing the motorbike from starting until the rider dons the helmet. The system also includes a crash detection sensor that may notify the NodeMCU in the case of a collision. The NodeMCU can then carry out a range of actions, including shutting off the ignition to prevent more accidents or alerting a selected contact of an emergency [12]. The system is designed to be low-power and efficient and employs only the necessary components to minimize size and weight. For the purpose of locating the rider and informing emergency contacts of his or her position in the event of an accident, the system may additionally incorporate a triple-axis accelerometer and a GSM module [13]. A 20-s timeout allows the rider to stop system-generated misleading notifications. By making sure the rider is wearing a helmet before starting the motorbike and offering a quick and efficient means to warn emergency services and contacts in the case of an accident, the system is a crucial safety element that can help lower the risk of accidents.

3 Proposed Modeling

The device consists of a helmet equipped with an IR sensor that can detect whether or not the rider is donning a helmet. The sensor alerts a NodeMCLU device that is linked to the bike's battery and ignition system when it determines that the helmet is being worn. The NodeMCLU receives a signal from the helmet and uses it to identify whether the helmet is being worn. The NodeMCU will disable the ignition system if the helmet is not being worn, which will prevent the bike from starting. Before starting the bike, this is done to make sure the rider is wearing a helmet, which can assist prevent accidents and injuries. The bike also has a crash detection sensor in addition to the helmet with the IR sensor. Typically, this sensor is fixed to the bike's frame or another place where it can sense forces from impacts. The NodeMCU device receives a signal from the sensor if it notices a collision. Depending on the individual implementation, the NodeMCIJ has a range of options when it gets a crash signal from the bike's sensor. It may, for instance, automatically inform emergency services or a specified contact to a situation. Additionally, it could turn off the motorcycle's

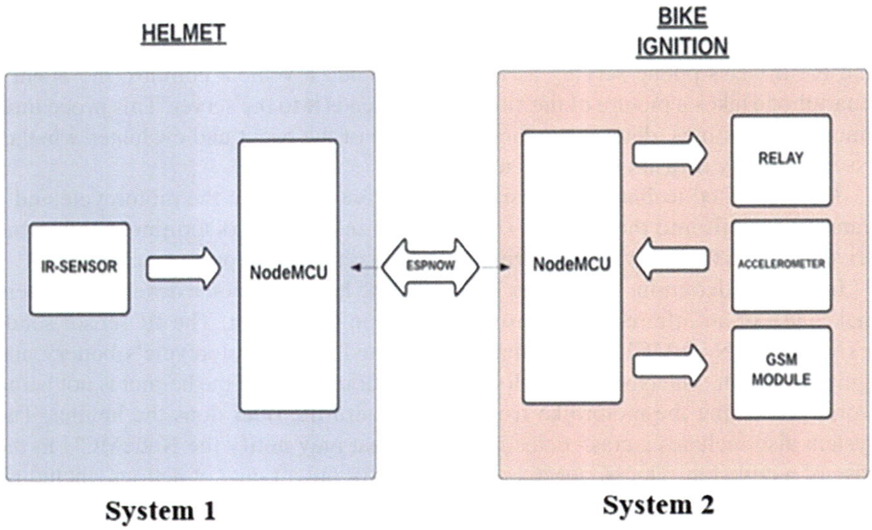

Fig. 1 System architecture

ignition system to stop future harm or mishaps. Overall, by making sure riders are wearing helmets before starting the bike and acting accordingly in the case of a mishap, this system adds an extra layer of protection for them. As demonstrated in Fig. 1, the system uses the NodeMCU and an IR sensor to determine if the cyclist is wearing a helmet or not. The first system's NodeMCU, which is in charge of processing the incoming signals and delivering the data to the second system, is coupled to the IR sensor, which is positioned on the helmet. The esp-now protocol, used by the NodeMCU, offers a dependable and secure wireless connection with little battery usage.

A NodeMCU in the second system analyzes the signal from the NodeMCU in the first system to determine whether or not the helmet is being worn. The second method activates a relay when the helmet is donned, completing the circuit of the bike's battery and ignition system and enabling the bike to start. If the helmet is not noticed, though. The relay failed to turn on, preventing the bike from starting.

In the event of an accident, the second system is also equipped with a triple-axis accelerometer and a GSM module for the purpose of locating and alerting crucial contacts and services [14]. The cyclist may have been engaged in an accident if the triple-axis accelerometer notices any abrupt changes in motion or contact. The rider's position and any extra details are subsequently sent through notification via the GSM module to a specified list of emergency contacts, such as family members or medical services.

The rider has a 20-s timeout that enables them to cancel the notice if it was accidentally activated in order to prevent false notifications. The system uses only the components that are required in order to minimize the size and weight of the apparatus while still being low-power and efficient. This technology, which makes sure the user is wearing a helmet before starting the bike, is a crucial safety component that can significantly lower the risk of accidents. In the case of an accident, it also offers a rapid and efficient means to notify contacts and emergency personnel, possibly saving lives.

4 Hardware Description

4.1 ESP32 Node MCU

The ESP32 Node Micro Controller Unit's on-chip board configuration is shown in Fig. 2. NodeMCU ESP32 is a low-cost and compact development board based on the ESP32 microcontroller. It is appropriate for a range of Internet of Things projects since it supports both Wi-Fi and Bluetooth communication. A USB-to-serial converter included in the board makes it simple to code using the Arduino IDE. It has a built-in 4 MB flash memory for program and data storage, and supports external SPI flash memory up to 16 MB. A voltage regulator on the board enables it to be supplied by a 3.3 V battery or a 5 V USB power supply. NodeMCU ESP32 has a variety of software libraries available, making it easy to implement complex features such as HTTP and MQTT protocols, and OTA (over-the-air) firmware updates.

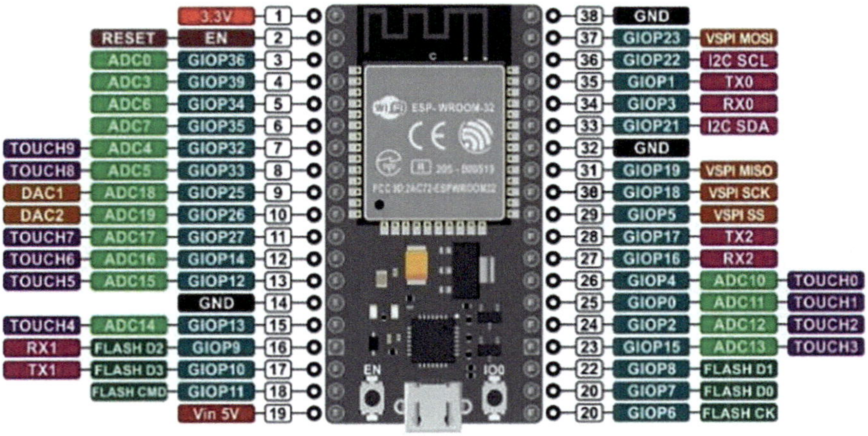

Fig. 2 On chip board configuration

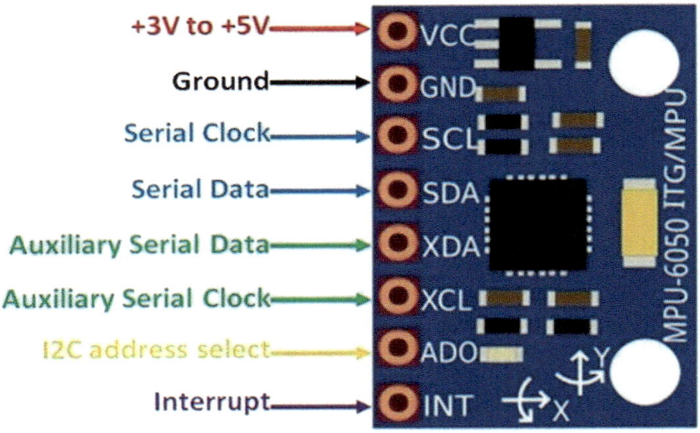

Fig. 3 MPU6050 sensor

4.2 *MPU6050 Sensor*

The MPU6050 is a 6-axis sensor that can measure acceleration in three axes (x, y, and z) and rotational motion (gyroscope) in three axes (x, y, and z) as shown in Fig. 3. The sensor is designed to operate on very little power and can operate in a low-power mode at 3.9 A.

The sensor can be configured to operate in a variety of different modes, with different sample rates, filtering options, and other settings that can be adjusted to optimize performance for specific applications. The MPU6050 includes an on-chip temperature sensor that can be used to measure the ambient temperature.

The algorithm creates a methodical procedure to guarantee a motorcycle's safe functioning. The first step is to look for any evidence of damage or malfunction in the bike's engine. To allow qualified people to ride, a valid driver's license must be verified. The next step is to confirm the usage of a helmet, a crucial safety necessity. The rider's suitability for safe operation is also assessed by the algorithm. Additionally, it scans for nearby vehicles and reacts to crashes appropriately by turning off the engine and sending a notification with GPS locations. The algorithm places a strong emphasis on safety, adherence to driving laws, and prompt reactions to possible threats.

Algorithm

Step 1: Check the bike's engine status
Step 2: Validate the driving license
Step 3: validate the helmet status
Step 4: Turn on RF transmitter
Step 5: Check the status of Driver
    5.1 Case 1: Status of Alcoholic
    5.2 Case 2: If the driver is drowsy
    5.3 Case 3: If the traffic signal is red
        Then Engine=OFF
            Sent alert message
Step 6: IF any two vehicles are close to each other
            Buzzer=ON
Step 7: IF vehicle clash with another vehicle
        Then Engine=OFF
            Sent alert message with GPS location

5 Testing

The motorcycle rider's safety is given top priority by the intelligent helmet, which also takes precautions to lessen the possibility of mishaps. In the event that any safety procedures are broken, the rider is not permitted to start the motorcycle as shown in Fig. 4.

The relay module is activated when the helmet is on, starting the ignition system and allowing the bike to start. An accelerometer detects changes in forces acting on it when acceleration changes. During a crash, sudden changes in acceleration occur, resulting in a significant change in the magnitude as shown in Fig. 5. The accelerometer detects this sudden change and triggers an alert or safety system to prevent or reduce damage or injuries. When a crash is detected, an IFTTT webhook is

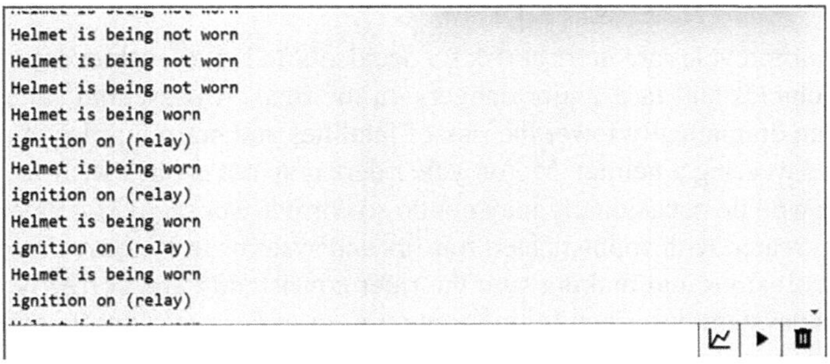

Fig. 4 Validation of helmet wearing

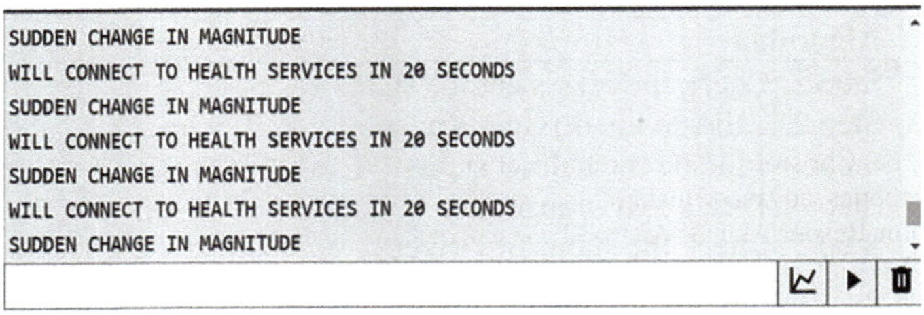

Fig. 5 Validation of magnitude

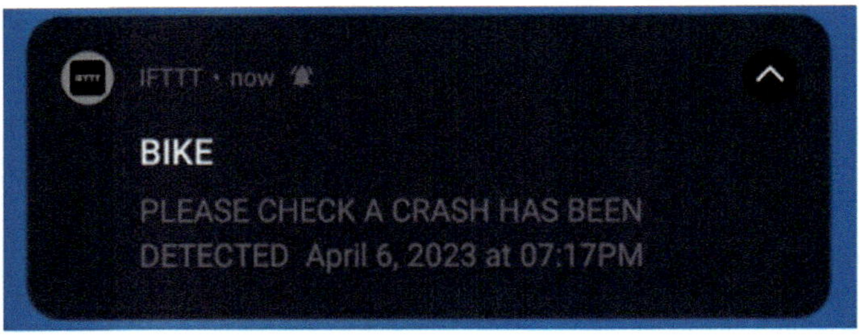

Fig. 6 Alert message for crash detection

triggered, which sends a notification to the user through the selected method such as SMS, email, or push notification as shown in Fig. 6. The Sim800L module calls the specified number when a crash is detected, allowing for quick response to potential accidents. The proposed approach offers enhanced openness and straightforward traceability. The accuracy of accident detection has been nearly perfect and close to 100% throughout the testing period. The accident detection system is quite precise and operates in automated recording mode.

6 Conclusion

Riders on motorcycles are more at risk for accidents and injuries than drivers of other types of vehicles and face major dangers on the road. An essential safety precaution that can dramatically lower the risk of fatalities and brain injuries in motorbike accidents is wearing a helmet. Motorcycle riders may not always wear helmets, and even those who do occasionally may not do so correctly or regularly. Smart helmets have been created with sophisticated sensors and systems to increase rider safety by detecting collisions and making sure the rider consistently and correctly wears the helmet. The system described in this context is an important safety feature that can help reduce the risk of accidents by providing a quick and effective way to alert emergency services and contacts in the event of an accident. With the increasing adoption of technology in the automotive industry, the development of smart helmets could revolutionize motorcycle safety and save countless lives in the future. Spreading awareness on the aspects of road accidents and road safety requirements must become a key for safer roads and safer communities, once and for all. Battery backup is one of the limitations of this model.

References

1. Kumar, K.K., Durai, S., Vadivel, M.T., Kumar, K.A.: Smart traffic system using raspberry pi by applying dynamic color changer algorithm. In: 2017 IEEE International Conference on Smart Technologies and Management for Computing, Communication, Controls, Energy and Materials (ICSTM), pp. 146–150. IEEE (2017)
2. Road Accidents ın India 2021, Ministry of Road Transport and Highways, Report.
3. Agarwal, N., Singh, A.K., Singh, P.P., Sahani, R.: Smart helmet. Int. Res. J. Eng. Technol. **2**(02), 3 (2015)
4. Jesudoss, A., Vybhavi, R., Anusha, B.: Design of smart helmet for accident avoidance. In: 2019 International Conference on Communication and Signal Processing (ICCSP), pp. 0774–0778. IEEE (2019)
5. Nanda, S., Joshi, H., Khairnar, S.: An IOT based smart system for accident prevention and detection. In: 2018 Fourth İnternational Conference on Computing Communication Control and Automation (ICCUBEA), pp. 1–6. IEEE (2018)
6. Rahman, M.A., Ahsanuzzaman, S.M., Rahman, I., Ahmed, T., Ahsan, A.: Iot based smart helmet and accident identification system. In: 2020 IEEE Region 10 Symposium (TENSYMP), pp. 14–17. IEEE (2020)
7. Selvathi, D., Pavithra, P., Preethi, T.: Intelligent transportation system for accident prevention and detection. In: 2017 International Conference on Intelligent Computing and Control Systems (ICICCS), pp. 442–446. IEEE (2017)
8. Desai, N., Kulkarni, P., Joy, L., Raut, P.: IoT based post crash assistance system. In: 2020 4th International Conference on Trends in Electronics and Informatics (ICOEI)(48184), pp. 734–739. IEEE (2020)
9. Gour, M.S., Kumara, P., Manjunatha, S., Kumar, S., Chetan, H.: Arduino based smart and intelligent helmet system for two-wheelers. In: 2020 IEEE International Conference on Distributed Computing, VLSI, Electrical Circuits and Robotics (DISCOVER), pp. 236–240. IEEE (2020)
10. Liao, C., Shou, G., Liu, Y., Hu, Y., Guo, Z.: Intelligent traffic accident detection system based on mobile edge computing. In: 2017 3rd IEEE International Conference on Computer and Communications (ICCC), pp. 2110–2115. IEEE (2017)
11. Vijayan, S., Govind, V.T., Mathews, M., Surendran, S., Sabah, M.: Alcohol detection using smart helmet system. IJETCSE **8**(1) (2014)
12. Kumar, K.A., Belinda, M.C.M., Lydia, A., Ravikumar, S.: Secure data transmission through steganography with ECC. In: 2022 International Conference on Advances in Computing, Communication and Applied Informatics (ACCAI), pp. 1–7. IEEE (2022)
13. Saranya, R., Kumar, A., Rajendran, P., Kumar, N., Keerthana, P.: Smart device to check harmful chemicals in fruits and vegetables. In: 2023 7th International Conference on Computing Methodologies and Communication (ICCMC), pp. 866–871. IEEE (2023)
14. Vivek, J., Kumar, K.A., Sahtihi, R., Sravani, R.S., Sravani, V.: Automated extraterritorial averting system. J. Comput. Theor. Nanosci. **15**(11–12), 3538–3541 (2018)

Digital Payment Systems and Financial Inclusion: Examine How Digital Payment Systems, Such as Mobile Wallets and Digital Currencies, Can Improve Financial Inclusion by Providing Access to Banking Services for the Unbanked and Underbanked Population

Mukul Gupta, Deepa Gupta, and Priti Rai

Abstract For economic development and poverty alleviation, financial inclusion is essential. This study's overarching goal is to assess the prevalence of financial exclusion across various demographic subsets in both urban and rural settings, and to explore how the widespread use of digital payment systems might influence those figures. Two hundred people were randomly selected to participate in a cross-sectional survey. Questions about financial exclusion and the use of digital payment systems were included in the survey framework used to collect the data. Financial inclusion was shown to be higher in urban regions across all age groups compared to rural areas, although this discrepancy persisted even after controlling for other factors. Individuals with a high adoption of digital payment systems were shown to have a higher likelihood of being fully banked, while those with a moderate adoption level had a lower likelihood. These results highlight the necessity for tailored initiatives to promote financial inclusion among various demographic groups and shed light on the significance of digital payment systems in closing the financial inclusion gap.

Keywords Digital payment systems · Digital financial inclusion · Mobile wallets · Digital currencies · Banking services · Unbanked population · Underbanked population

M. Gupta (✉) · D. Gupta · P. Rai
GL Bajaj Institute of Management, Greater Noida, India
e-mail: prof.dr.mukulgupta@gmail.com

G. Fortino et al. (eds.), *Proceedings of Third International Conference on Computing and Communication Networks*, Lecture Notes in Networks and Systems 917,
https://doi.org/10.1007/978-981-97-0892-5_57

1 Introduction

1.1 Background of the Study

As a result of these rapid technical developments, the financial services industry is undergoing a dramatic transformation, and digital payment systems are at the centre of this revolution. The widespread adoption of digital payment systems, such as mobile money and cryptocurrency, might dramatically expand access to banking services [1]. This revolutionary shift is motivated by a common goal: reaching the unbanked and underbanked people of the world with the benefits of banking and financial services. Due to their low socioeconomic status, these communities have a difficult time gaining access to basic banking services including savings accounts, loans, and insurance. In this investigation, we explore the world of digital payment systems, analysing their potential to help close the wealth gap by making traditional banking services more available, affordable, and secure for underserved communities [2]. We'll take a look at how the rise of digital payment methods is altering the economy and whether or not that shift can lead to more people worldwide having access to formal banking services [3].

1.2 The Evolution and Impact of Digital Payments in Modern Society

Payment done via digital means is said to be digital. A digital payment is made totally digitally between the giver and the receiver. Therefore, the name implies that monetary transactions occur digitally. In the case of online transactions, no actual money changes hands. All electronic money transfers happen in cyberspace [4]. It's a convenient and speedy method of sending money.

Digital payments are having an effect on our daily lives and introducing useful and novel services as a result of recent technological developments. The RBI Bulletin reports tremendous growth in the use of electronic payments in India [5]. Cash Karo India found that electronic wallets are the preferred way of payment among its consumers. The haphazard growth in smartphone ownership has enabled the widespread adoption of mobile payments [6].

1.3 Digital Financial Inclusion: A Global Effort to Combat Poverty

Digital financial inclusion is one of the most talked-about areas of international development right now. Companies and governments from all across the world have recently begun to take notice. A project to expand access to banking services in developing and emerging economies has been spearheaded by the G-20 and the World Bank since 2010, with the goal of reducing poverty in these regions. Even though over two billion people lack access to basic banking and credit facilities, the vast majority of financially excluded people hold a cell phone as an asset [7]. Since the cost of living while poor is high, most transactions between the poor are conducted in hard currency. Making ends meet is difficult for those making less than $2 a day due to high fees associated with moving funds, a lack of access to credit, and the fact that natural disasters and illnesses can further impoverish those without savings or insurance. With the goal of increasing access to low-cost, high-quality financial services such as bill pay, wire transfers, local and international money transfers, insurance, credit, and savings [8]. It's crucial that people adopt technological innovations like smartphones and internet cafes. To quote the Centre for Global Development Access Project: "access to technology to, including use of, basic financial institutions by the excluded and disadvantaged population" [9] is what the CGAP defines as "digital financial inclusion."

2 Related Works

In [10] financial inclusion could be facilitated by the use of digital gadgets, as was pointed out. This approach uses digital tools and technologies to increase the accessibility of banking services, particularly in emerging countries and other economies on the rise, where a large population segment now lacks this basic necessity. Digitalization is having a significant impact, as evidenced by Rizzo's study, narrowing the gap between those who are excluded from and those who have access to financial services. More people and areas are now able to join the formal financial system because of the proliferation of digital banking, which is good for the economy as a whole and can help spur growth and stability. Research like that conducted by Rizzo is essential if we are to gain a deeper grasp of the consequences and promise of digital finance to expand access to banking services around the world as it continues to rapidly develop.

In [11] the card is still widely used for online transactions, even though many other payment methods have been on the market for almost two decades. The lack of privacy protection, user agency, and transaction oversight is a major drawback of such a system. According to the authors, making a virtual card, and particularly a credit card, would provide users more agencies over their online activities and ensure their privacy when making purchases in secret.

In [12] it was emphasised how important it is to ensure that people from all walks of life, but especially those who are socially and economically marginalised, have unfettered access to financial services. In this context, "financial inclusion" refers to the practise of expanding access to credit for people who have been underserved by conventional financial institutions in the past. It represents a dedication to reducing economic inequality, with special attention paid to uplifting the most disadvantaged members of society. The studies conducted by Malladi, Soni, and Srinivasan highlight the significance of accessible credit in supporting economic growth and stability among groups that have been historically excluded from mainstream financial systems. From this vantage point, it's clear that more study and policymaking are needed to advance the concepts of financial inclusion and guarantee that all people in a society have fair and equal access to financial resources.

In [13] the Indian e-payment system was broken down. Electronic payment systems have gained widespread use over the past decade, yet many people in the country still don't utilise them properly out of laziness or distrust. The government recognises the need to combat these challenges by promoting the system via television and online education initiatives.

In [14] in this paper, we outlined the fundamental features that characterise the current financial landscape in the field of digital financial products. These capabilities span the gamut of digital advancements made possible by internet services, from electronic payment systems to remote account management and electronic document handling. The work by Naumenkova, Mishchenko, and Dorofeiev illuminates the revolutionary effects of digitalization in the financial industry, where tech-driven solutions have altered the ways in which individuals and businesses interact with their money. Convenience, efficiency, security, and accessibility in financial services are all affected by the opportunities and threats presented by this dynamic digital ecosystem. Studying digital financial products and services is crucial to understanding and managing the changing environment of modern finance as financial institutions and fintech companies innovate and adapt to this digital paradigm.

In [15] the safety of online payment methods was investigated. The widespread adoption of electronic payment systems like debit cards, credit cards, online banking, mobile banking, etc. has led to quicker and more comfortable transactions. However, some people are worried about the implications of these changes for national security. Researchers argued that by boosting security knowledge among e-users and bolstering security at each stage of e-payment, risks may be mitigated and user confidence could be boosted (Table 1).

3 Research Methodology

The goal of this study was to determine how widespread financial exclusion is; hence a cross-sectional survey was used to gather data. The study asked people in both urban and rural settings to ensure an accurate cross-section of the population. Financial exclusion was measured across demographic groups, and the effect of digital payment systems, especially mobile wallets, on financial inclusion was investigated. This research was conducted entirely through quantitative methods.

Table 1 Summary of the research on digital payment systems and financial inclusion

Author(s)	Title	Publication	Results
El Haddad, G., Hage, H., and Aïmeur, E.	E-payment plan: a conditional multipayment scheme based on user personalization and plan agreement	International Conference on E-Technologies, 2017	In order to facilitate electronic payments, the study introduces a conditional multipayment mechanism that makes use of user customisation and plan agreement. The material given does not include any specific outcomes from the study
Malladi, C. M., Soni, R. K., and Srinivasan, S.	Digital financial inclusion: next frontiers—challenges and opportunities	CSI Transactions on ICT, 2021	The problems and opportunities associated with digital financial inclusion are discussed in the article. The information given does not indicate particular outcomes
Gupta, D.	Digital payments in India: a conceptual study	Studies in Indian Place Names, 2020	The research on digital payments in India is conceptualised in the article. The information at hand does not specify the study's precise findings
Naumenkova, S., Mishchenko, S., and Dorofeiev, D.	Digital financial inclusion: evidence from Ukraine	Investment Management & Financial Innovations, 2019	The study offers data on Ukraine's use of digital finance. The information is not specific enough to summarise findings or results
Saxena, S., Vyas, S., Kumar, B. S., and Gupta, S.	Survey on online electronic payments security	2019 Amity International Conference on Artificial Intelligence, 2019	The study examines the safety of electronic payments made online. The information given does not include the detailed survey results

3.1 Sampling and Sample Size

My study's sample population would include people from both urban and rural settings, and I'd be interested in gauging the effects of digital payment methods and financial exclusion among the former. In this study, 200 participants were selected using stratified sampling.

3.2 *Tools Used for Data Collection*

We created a detailed questionnaire for primary data collection (Data Collection). Ask me whatever you like, but make sure it relates to my area of study (i.e. mobile wallets, digital payment systems, and financial inclusion). Two hundred people were chosen at random from both urban and rural locations, and we asked them to fill out this survey.

3.3 *Data Set*

The study involved 200 participants selected through stratified sampling. Data collection was done using a 15-question questionnaire, with 7 questions related to demographics and 10 questions concerning financial inclusion and digital payment adoption. The Likert Scale (ranging from 1—Strongly Agree to 5—Strongly Disagree) was employed for questionnaire responses to capture nuanced opinions. Statistical analysis using SPSS was carried out to analyse the collected data comprehensively.

3.4 *Tools Used for Data Analysis*

Statistical packages (like SPSS) were used to analyse the survey data. Means, proportions, and standard deviations were employed for descriptive purposes, while Inferential Statistics (t-test and Logistics Regression) were used to compare the prevalence of financial exclusion and the use of digital payment systems across demographic groups (Fig. 1).

4 Data Analysis and Results

An overview of a population's socioeconomic position and demographics is provided below (Table 2 and Fig. 2). Fifty per cent of the population is comprised of people aged 25–34, with a further quarter aged 18–24, 15% aged 35–44, and 10% aged 44 and up. There are roughly as many males as females (60–40%). There are 50% middle-income earners (those making between Rs 26,000 and Rs 35,000) and 50% high-income earners (those making over Rs 36,000) in the sample. People with college degrees make up the bulk (60%), followed by those with high school diplomas

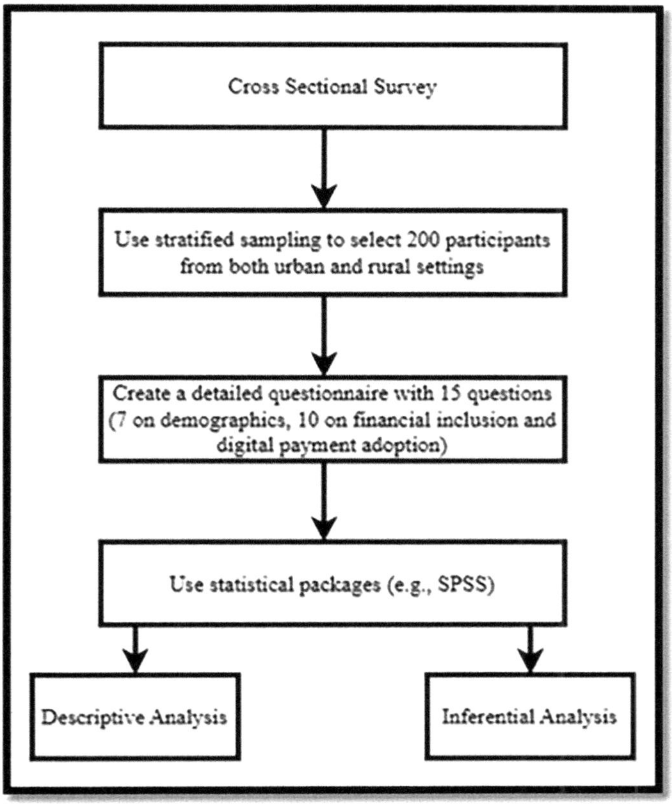

Fig. 1 Flowchart of research methodology

(30%), and then "Others" (10%). The sample is composed of 50% each of the people living in rural and urban areas. There are 50% working adults, 30% students, and 20% "Others" in this sample. Fewer than 5% are unemployed. 45% of the population is considered "Fully Banked," 40% is "Underbanked," and 15% is "Unbanked" in terms of their financial inclusion status.

The average disparity in financial inclusion between city and country dwellers, broken down by age group, is presented in Fig. 3 and Table 3. The data shows a mean difference of 30 percentage points between urban and rural locations, and this gap widens with age (from 18 to 24 years old, 25 to 34 years old, and 35 to 44 years old and older than 44 years old). This suggests that there is a clear 30-percentage-point advantage for the urban population over the rural population in terms of financial inclusion across all age groups.

Table 4 and Fig. 4 displays the findings of separate samples t-tests used to determine whether or not the differences in means presented in Table 2 are statistically significant. For those in the 18–24 age brackets, the t-statistic is 30.18 with 49 degrees

Table 2 Demographic and financial inclusion profile

Categories	Sub-categories	Frequency	Percentage
Age group	18–24	50	25
	25–34	100	50
	35–44	30	15
	Above 44	20	10
Gender	Male	120	60
	Female	80	40
Income level (per month)	Rs 25,000	100	50
	Rs 26,000–Rs 35,000	80	40
	Above Rs 36,000	20	10
Education level	High school	60	30
	Graduate	120	60
	Others	20	10
Residence	Rural	100	50
	Urban	100	50
Occupation	Employed	100	50
	Unemployed	10	5
	Student	60	30
	Others	30	15
Financial inclusion status	Fully banked	90	45
	Underbanked	80	40
	Unbanked	30	15

of freedom, yielding a *p* 0.001. Since the *p*-value is less than 0.005, we conclude that the null hypothesis must be false. This indicates that there is a statistically significant difference between financial inclusion rates in urban and rural areas among people aged 18–24. The *t*-statistic for those in the age bracket 25–34 is also extremely significant, coming in at 35.53 with 99 degrees of freedom and a *p*-value of 0.001. Again, the *p*-value is smaller than the cutoff, suggesting that the difference in mean is significant even at this younger age. With 29 degrees of freedom, the *t*-statistic for those between the ages of 35 and 44 is 28.13, yielding a *p*-value of 0.002. Again, the *p*-value is lower than the threshold for statistical significance. The *t*-statistic for those above the age of 44 is 32.61 with 19 degrees of freedom, yielding a significance level of 0.001. Mean differences are statistically significant, as the *p*-value is smaller than the significance level across all age groups.

Since the discrepancy between urban and rural financial inclusion rates persists across all age groups, we may conclude that it is not a result of coincidence.

Individuals' levels of financial inclusion (Table 5, Full Banked, Underbanked, and Unbanked) and usage of digital payment methods (Table 5, High, Moderate, and Low) are cross-tabulated (Fig. 5).

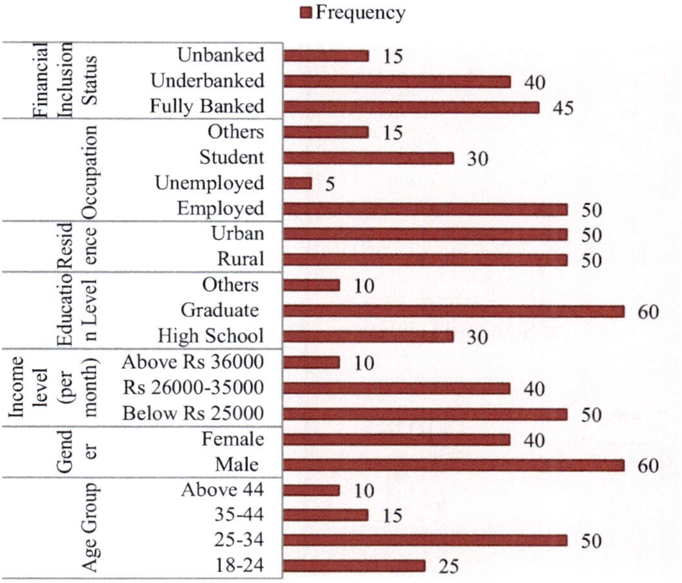

Fig. 2 Demographic and financial inclusion profile

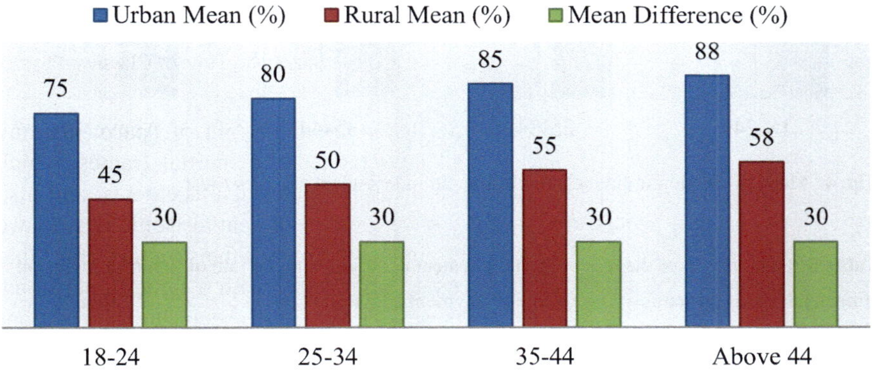

Fig. 3 Mean difference in financial inclusion between urban and rural areas by age group

Table 3 Mean difference in financial inclusion between urban and rural areas by age group

Age group	Urban mean (%)	Rural mean (%)	Mean difference (%)
18–24	75	45	30
25–34	80	50	30
35–44	85	55	30
Above 44	88	58	30

Table 4 Financial inclusion mean difference independent samples *t*-test

Age group	Sample size	Mean difference	Standard deviation	*t*-statistics	Degree of freedom	Critical *t*-value	*P*-value
18–24	50	30	0.993	30.18	49	2.0096	0.001
25–34	100	30	0.845	35.53	99	1.9236	0.001
35–44	30	30	1.066	28.13	29	2.0152	0.002
Above 44	20	30	0.922	32.61	19	2.0963	0.001

Significance value is 0.005

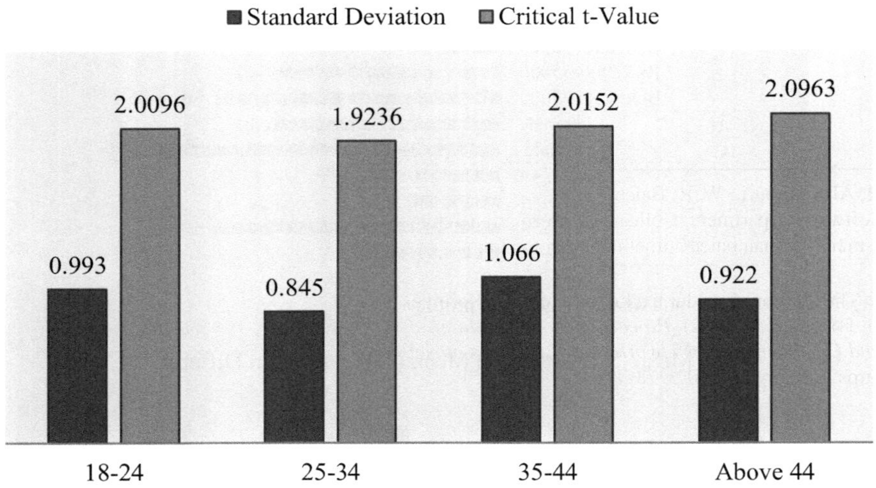

Fig. 4 Mean distinction in financial inclusion: an independent samples *t*-test

Table 5 Comparison of the rate of digital payment adoption with the rate of financial inclusion

Financial inclusion status	Digital payment adoption			Total
	High	Moderate	Low	
Fully banked	45	40	5	90
Underbanked	30	40	10	80
Unbanked	10	15	5	30

The findings of a logistic regression analysis are shown in Table 6 and Fig. 6, with respondents' familiarity with digital payment methods serving as the independent variable and respondents' degree of financial inclusion serving as the dependent variable.

When compared to the baseline group (Fully Banked), "Digital Payment Adoption (High)" shows the impact of a "High" level of adoption of digital payment

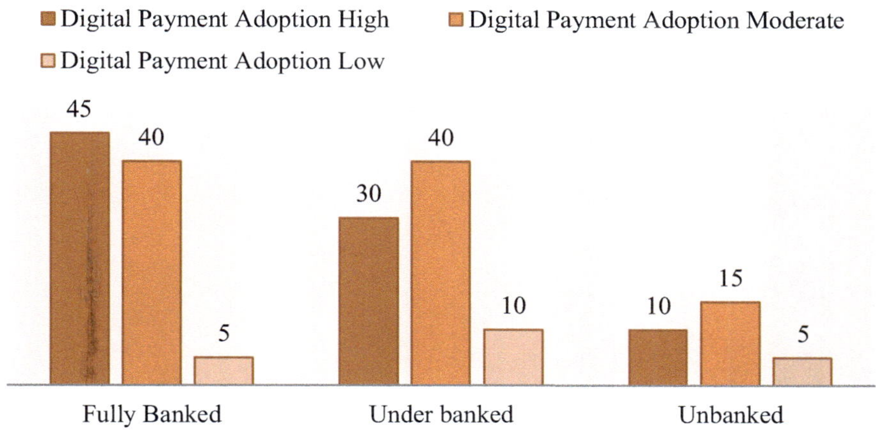

Fig. 5 Comparison of the rate of digital payment adoption with the rate of financial inclusion

Table 6 Logistic regression results for financial inclusion status and digital payment adoption

Independent variable	Coefficient	Std. error	Wald-statistic	*p*-value	Odds ratio (OR)	95% CI for OR
Constant	−0.7121	2.0142	−0.3526	0.002	0.4152	(0.0239, 9.8697)
Digital payment adoption (high)	0.4012	1.0236	0.3663	0.001	1.5263	(0.0489, 14.6454)
Digital payment adoption (moderate)	−1.2363	1.0523	−1.2531	0.003	0.2536	(0.0228, 3.2753)
Digital payment adoption (low)	0.4125	1.1263	−1.5236	0.004	1.4123	(0.0215, 4.2363)

methods. A coefficient of 0.4012, a standard error of 1.0236, and a significance level of 0.001 are associated with it. In other words, the probability of being "Fully Banked" increases by a factor of 1.5263 for those with a "High" level of digital payment adoption compared to those with a "Low" level of adoption. The impact of a "Moderate" degree of digital payment adoption relative to the reference category is depicted by the "Digital Payment Adoption (Moderate)" effect. A *p*-value of 0.003 is associated with a coefficient of −1.2363 and a standard error of 1.0523. Individuals who are "Moderate" adopters of digital payment methods are 0.2536 times less likely to be "Fully Banked" than those who are "Low" adopters of these methods. The result of having a "Low" degree of digital payment adoption relative to the

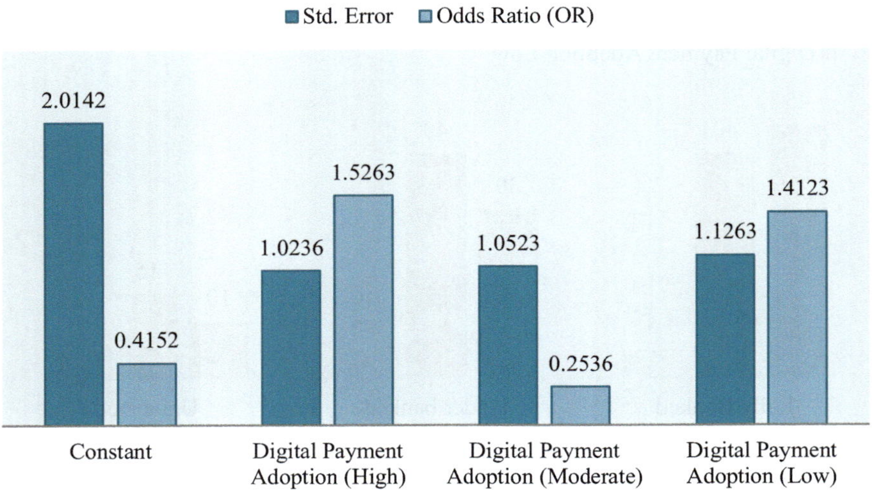

Fig. 6 Logistic regression results for financial inclusion status and digital payment adoption

reference category is depicted by "Digital Payment Adoption (Low)". The *p*-value is 0.004, the standard error is 1.1263, and the coefficient is 0.4125. Individuals with a "Low" degree of adoption of digital payment technologies are 1.423% more likely to be "Fully Banked" than those with a "Low" level of adoption.

This logistic regression study sheds light on the correlation between a person's level of digital payment acceptance and their financial inclusion. The direction and importance of these associations can be deduced from the coefficients, *p*-values, and odds ratios. A "High" level of digital payment usage is connected with a higher likelihood of being "Fully Banked," whereas a "Moderate" level is associated with a lower likelihood compared to a "Low" level. These results can be used to inform plans to expand access to financial services among different demographics.

5 Conclusions and Future Scope

Understanding the complex dynamics of financial inclusion and the game-changing potential of digital payment systems is now much easier thanks to this study. The need of implementing targeted interventions to narrow the glaring financial inclusion gap between urban and rural areas is highlighted by the fact that this gap persists across age groups. Notably, our research showed that the use of digital payment systems plays a crucial role in promoting financial inclusion, with those who use them frequently being more likely to be completely banked. Further research is warranted, however, because of the unexpected discovery that those with moderate adoption are less likely to obtain full banking status. It's obvious that a cookie-cutter approach to financial inclusion won't cut it. To fully capitalise on digital payments' potential to promote financial inclusion, specific approaches should be developed in light of relevant demographic factors and patterns of adoption. To make financial inclusion a reality for all people in society as technology develops, governments, financial institutions, and other stakeholders must be flexible and creative. Deeper research into the societal and economic effects of these technologies, evaluations of their scalability, security, and usability in a variety of settings, and examinations of regulatory frameworks to encourage their responsible adoption can all be expected in the future.

Conflict of Interest All authors declare that there is no conflict of interest regarding the publication of this research paper.

References

1. Aeron, P., Jain, S., Kumar, A.: Revisiting trust toward E-retailers among Indian online consumers. J. Internet Commer. **18**(1), 45–72 (2019)
2. Khandker, V., Joshi, K.P.: Price determination for 4G service using price sensitivity model in India. J. Revenue Pricing Manag. **18**(2), 93–99 (2019)
3. Shaikh, N., Shah, N.D.: Role of government and RBI in financial inclusion—an ecosystem of development. Stud. Indian Place Names **40**(8), 261–271 (2020)
4. Agrawal, C.A., Kavishwar, S.: Digital financial inclusion of women in India. Our Herit. **68**(9), 723–735 (2020)
5. Omotubora, A., Basu, S.: Regulation for E-payment systems: analytical approaches beyond private ordering. J. Afr. Law **62**(2), 281–313 (2018)
6. Chen, A.N., Griffin, K., Zeltmann, S., Ota, M., Ozeki, R.: Demographic background, perceptions, and E-Payment usage among young Japanese. In: Institute for Global Business Research Conference Proceedings, Nashville, TN, USA, 133 (2018)
7. Badovinac, N., Simic, D.: E-Payment systems using multi-card smartcard. In: Advances in Operational Research in the Balkans, pp. 237–249. Springer, Cham (2020)
8. Ahuja, K.: Challenges for Indian EPayment system. Mmu J. Manag. Pract. **10**(1), 27–29 (2019)
9. Fatonah, S., Yulandari, A., Wibowo, F.W.: A review of E-payment system in E-commerce. In: Journal of Physics: Conference Series, vol. 1140, no. 1, pp. 1–7. IOP Publishing (2018)
10. Rizzo, M.: Digital Finance: Empowering the Poor Via New Technologies. The Word Bank, Washington DC (2014)
11. El Haddad, G., Hage, H., Aïmeur, E.: E-payment plan: a conditional multipayment scheme based on user personalization and plan agreement. In: International Conference on ETechnologies, pp. 285–299. Springer, Cham (2017)
12. Malladi, C.M., Soni, R.K., Srinivasan, S.: Digital financial inclusion: next frontiers—challenges and opportunities. CSI Trans. ICT **9**(2), 127–134 (2021)
13. Gupta, D.: Digital payments in India: a conceptual study. Stud. Indian Place Names **40**(3), 5906–5914 (2020)
14. Naumenkova, S., Mishchenko, S., Dorofeiev, D.: Digital financial inclusion: evidence from Ukraine. Invest. Manag. Financ. Innov. **16**(3), 194–205 (2019)
15. Saxena, S., Vyas, S., Kumar, B.S., Gupta, S.: Survey on online electronic payments security. In: 2019 Amity International Conference on Artificial Intelligence (AICAI), pp. 756–751. IEEE (2019)

Enhanced Prediction of Breast Cancer Using Machine Learning Ensemble Models and Techniques

E. Chandralekha, S Ravikumar, K Antony Kumar, and M. J. Carmel Mary Belinda

Abstract Breast cancer was acknowledged as one of the world's most formidable diseases. The key to improving patient outcomes and general wellbeing had been to make an accurate and timely diagnosis. This research used Machine Learning methods to investigate several ensemble approaches to breast cancer diagnosis prediction. This research made extensive use of the Breast Cancer Wisconsin dataset. The research aimed to provide a comprehensive evaluation of the predictive powers of a variety of five ensemble models, including Random Forest, Gradient Boosting, AdaBoost, Bagging, and Extra Trees. The approach included several criteria for assessment, including accuracy, precision, recall, and F1-score. Further analysis was done with the use of the ROC curve, the precision-recall curve, and other statistical tools. It's important to note that among all the models tested, AdaBoost performed the best.

Keywords Machine learning · Ensemble techniques · Random forest · AdaBoost · Bagging · Extra trees · Breast cancer

1 Introduction

A lot of people get breast cancer, which starts in the flesh of the breast. This cancer mostly affects women, but men can acquire it. First sign of illness is uncontrolled proliferation of abnormal breast tissue cells. Once advanced, these malignant cells can infect surrounding tissues and spread through the blood and lymphatic systems.

E. Chandralekha (✉) · S. Ravikumar · K. Antony Kumar
Vel Tech Rangarajan Dr. Sagunthala R&D Institute of Science and Technology, Chennai, Tamilnadu, India
e-mail: clchandralekha08@gmail.com

S. Ravikumar
e-mail: ravikumars@veltech.edu.in

M. J. Carmel Mary Belinda
Saveetha School of Engineering, SIMATS, Chennai, Tamilnadu, India

G. Fortino et al. (eds.), *Proceedings of Third International Conference on Computing and Communication Networks*, Lecture Notes in Networks and Systems 917,
https://doi.org/10.1007/978-981-97-0892-5_58

Gender (women are more likely to develop breast cancer), age (risk rises with age), a personal or family history of the disease, certain inherited gene mutations (such as BRCA1 and BRCA2), hormonal factors (early menstruation or late menopause), and environmental factors are important breast cancer facts. The two most frequent breast cancer subtypes are invasive ductal carcinoma and invasive lobular carcinoma. Different cells give rise to and develop various variations. Classifying breast tumors by hormone receptor status (estrogen or progesterone receptors) and HER2/neu status is important for treatment. Many convincing arguments explain why breast cancer affects public health so much. Since it is the most frequent cancer in women worldwide, it is crucial. To make matters worse, breast cancer kills more women than any other malignancy. A breast cancer diagnosis also affects patients' families emotionally, physically, and socially, in addition to their physical condition. Correct breast cancer diagnosis affects several patient care aspects. First, it detects breast cancer early, which improves treatment and survival rates since less intensive medicines work better. Second, an accurate diagnosis is needed to design a treatment plan that accounts for the patient's age, gender, and other factors to maximize the likelihood of a successful outcome, whether that plan involves surgery, radiation therapy, chemotherapy, hormone therapy, targeted therapies, or a combination of these. Thirdly, diagnosis helps patients and healthcare providers make treatment options and understand the prognosis by staging the cancer and predicting its outcome. Finally, proper diagnosis is essential for post-therapy monitoring to assess the patient's response to treatment and detect any cancer recurrence.

Breast cancer is a prevalent malignant disease that affects many people. Breast cancer is caused by abnormal cell growth. This sickness can impact individuals, families, and communities. Medical experts, researchers, and volunteers must address the issue. Breast cancer starts in chest mammary lobules [1, 2]. Early detection is crucial because the illness may start without symptoms. Although the illness is more common in women, men might be affected occasionally. Different types of breast cancer have different characteristics. These differences affect therapy. Recently, doctors have improved breast cancer detection and treatment [3, 4]. Mammograms and MRIs detect issues early [5]. To choose the best medicines, they examine cancer cell genes. This has helped more breast cancer patients recover and live better. Machine learning has influenced breast cancer, a common malignancy. This technology is crucial to addressing this global health crisis. Early disease detection and treatment benefit individuals, families, and communities worldwide. Machine learning analyzes breast cancer data to help clinicians and researchers identify cancer tendencies [6, 7]. Machine learning uses computers to learn from breast images, biopsies, and gene information to discover minor issues clinicians may overlook. This helps doctors detect cancer earlier, improve treatments, and extend patient lifespans. Computers can classify malignancies by genetics and assist clinicians choose the best treatments [8, 9]. This unique treatment helps each person recover faster and with fewer side effects.

This research examined how detailed breast cancer diagnosis affects patient outcomes. Early detection enhances treatment efficacy and patient survival, and precise diagnosis influences treatment planning, prognostic information, and post-treatment monitoring. The study examined these key components to demonstrate the clinical importance of correct breast cancer detection. The study has five sections: introduction, literature review, methodology, suggested work, experiment findings, and conclusion.

2 Previous Work

This section includes a comprehensive literature review that sheds light on the current study and places it in the larger research context.

To diagnose breast cancer, two research used various machine learning datasets and methods. Naji et al. [10] used SVM, RF, LR, DT, and K-NN in their initial investigation, with SVM showing the most promise in terms of efficiency, accuracy, and area under the curve (AUC). Bhise et al. [11] employed RFE for feature selection and CNN as the classifier in their second investigation. Using activation functions like ReLu, they found that SVM, KNN, LR, and NB classifiers provided the most accurate probabilistic predictions.

ABUS was created by Wang et al. [12] as an innovative method of breast screening that makes 3D breast imaging. To speed up image evaluations while keeping responsiveness high and decreasing false positives, an innovative 3D CNN was developed for automated cancer detection. Using a threshold loss mechanism and dense deep supervision, this technique successfully differentiated between normal and cancerous tissue. An automated system was introduced by Naseem et al. [13] to detect breast tumor using ensembles of some classifiers. They explored a range of machine learning algorithms, including the adaptation of Artificial Neural Networks (ANN) for this application. The study involved a comparison of ensemble models and an assessment of the influence of balanced class weight, yielding a notable accuracy rate of 98.83%. This system's potential impact on the medical sector and research community is significant, as it outperformed established methods [14].

Current methods for breast cancer detection come with their own set of limitations. These include the risk of missing cancers in their early stages due to limited sensitivity, the issue of false positives causing unwarranted anxiety, the dependency on specialized experts which can lead to inconsistencies, and the use of invasive procedures that bring discomfort and potential hazards for patients [3]. Zheng et al. [15] presented the DLA-EABA method, which employs cutting-edge computational techniques, for detecting breast cancer. They used deep learning and CNNs to classify breast masses in MRI, ultrasound, tomosynthesis, and mammography diagnostic tasks. Classification and error estimates were executed in fully connected and softmax layers, while convolutional, LSTM, and Max-pooling layers were also utilized. The research focused on merging machine learning approaches with feature selection and extraction to find the best methodology. Compared to other systems, it has 97.2% accuracy, 98.3% sensitivity, and 96.5% specificity.

Haq et al. [16] selected relevant features from their dataset using supervised (Relief algorithm) and unsupervised (Autoencoder, PCA algorithms) methods. To identify breast cancer accurately, a support vector machine was trained and tested using these features. They used k-fold cross-validation to test and optimize the model. Relief algorithm-selected features performed better in detection. Their approach was so successful that it outperformed existing methods in the literature by a margin of 99.91%. Their method is a serious candidate for breast cancer diagnosis in healthcare systems.

Breast cancer detection has improved with machine learning. These innovations use deep learning and machine learning techniques to improve diagnosis accuracy. However, these advances face obstacles. Training them requires large datasets and may make sophisticated model interpretation challenging. Integrating machine learning algorithms into healthcare settings raises ethical and practical issues. To use machine learning for breast cancer diagnosis, certain barriers must be overcome.

3 Proposed Work

This subsection expands on the suggested method for screening the Wisconsin cohort for breast cancer. The strategic development of an effective breast cancer identification approach was significantly influenced by the utilization of this dataset, marking a notable contribution to the advancement of knowledge and management of this crucial medical issue. Figure 1 shows the work flow of this research work.

3.1 Data Acquisition and Data Cleaning

Machine learning and medical research rely on the Breast Cancer Wisconsin (Diagnostic) dataset (WBCD). It is commonly used to diagnose breast cancer and originated in a clinical trial at the University of Wisconsin Hospitals in Madison, USA. Each of 569 breast biopsies in this dataset has 30 cell nuclei characteristics. These criteria include Mean radius, texture, perimeter, area, smoothness, compactness, concavity, concave spots, symmetry, fractal dimension, Radius mistake, Error in texture Perimeter, Area, Smoothness, Compactness, Concavity, Concave points, Symmetry, Fractal dimension errors Worst radius, texture, perimeter, area, smoothness, compactness, concavity, concave points, symmetry, and fractal dimension. The WBCD dataset is essential for medical researchers and data analysts studying breast cancer and its detection since it covers these cell nuclei properties.

Missing values were handled by inserting the average value for each column into the relevant cells. This action was done to ensure the integrity of the data set. The dataset was separated into two parts: 'X,' which held the features required for prediction, and 'y,' which stood in for the target variable. This segmentation is a critical stage in the development of forecasting models. Additional checks were

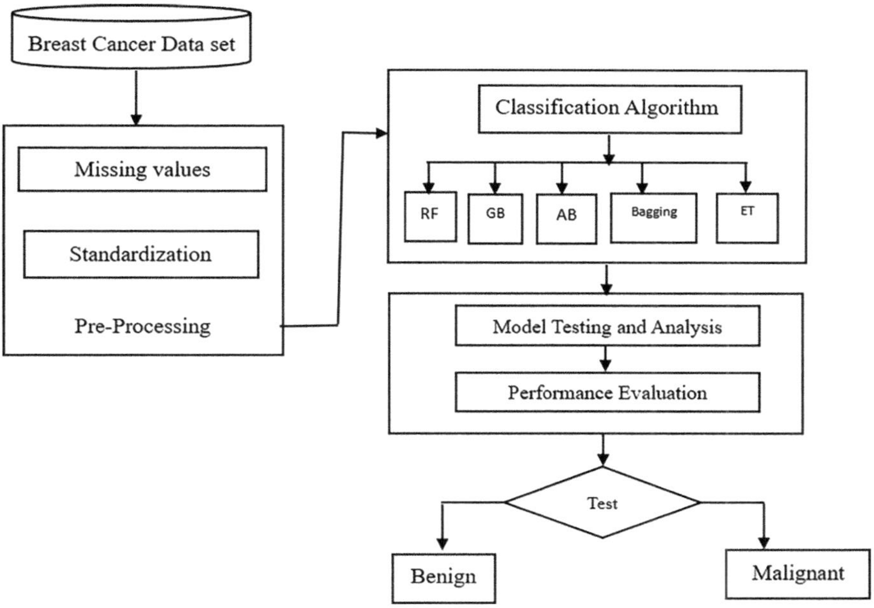

Fig. 1 Work flow of the proposed work

performed to identify and, if necessary, fill in missing values in the target variables 'y_train' and 'y_test' with zeros, ensuring completeness of the data. This was done to make sure all of the numbers added up correctly. The characteristics were then standardized using the 'StandardScaler' method. This ensured that all features were normalized, which facilitated fair comparisons during training and predictions and ultimately boosted our models' performance.

3.2 *Ensemble Models*

Ensemble learning combines many smaller models into one larger model to create a more resilient and accurate prediction model. This method improved the model's performance by predicting novel outcomes more accurately and reducing overfitting to the training dataset. Random Forest, Gradient Boosting, AdaBoost, Bagging, and Extra Trees were the main ensemble models. These models improved interoperability and prediction using multiple methods.

Random Forest. Random Forest method was used to train a forest of decision trees to predict breast cancer outcomes during the testing phase. Predictions were derived by combining the outputs of these trees. Each tree's attributes and data samples were selected at random. This step was taken to address the problem of overfitting, which occurs when a model becomes very specific to the training data and has trouble generalizing to new, unseen data. The model's accuracy is improved by averaging each tree's predictions. Randomly selecting features and samples has increased tree variation in the ensemble, improving model robustness and effectiveness. The equation served as a representation of this phenomenon. This was represented by the Eq. (1).

$$\text{Prediction}_{RF} = \frac{1}{N}\sum_{i=1}^{N}\text{Prediction}_{\text{Tree}_i} \tag{1}$$

Gradient Boosting. In breast cancer analysis, Gradient Boosting created prediction models. This strategy trained many weak models, frequently using decision trees, to repair prior errors. Each model's faults were used to train the next, boosting the model's capacity to differentiate benign and malignant tumors using tiny needle aspirates. Equation (2) shows how this method improved the model's breast cancer diagnosis, notably in difficult cases and with crucial variables.

$$\text{Final Prediction}_{GB} = P_{WM1} + \alpha_1 \times PE_{WM1} + P_{WM2} + \alpha_2 \times PE_{WM2} + \cdots \tag{2}$$

where, α_1, α_2, ... represent the weights assigned to the prediction errors of each weak model.

AdaBoost. In the arena of breast tumor classification, AdaBoost was used to create a series of weak classifiers, which were less accurate decision trees. These classifiers were trained to focus more on samples that were previously misclassified in each round of training. This iterative process aimed to gradually enhance the model's accuracy by addressing challenging cases. The final classification was determined by combining the predictions of these weak classifiers with different weights. Emphasis was placed on misclassified samples by assigning them higher weights. This approach helped the model better handle intricate patterns within the breast cancer dataset, which could lead to improved predictive performance. In terms of mathematical representation, the AdaBoost ensemble's prediction $H(x)$ for a specific sample x was given by the Eq. (3).

$$H(x) = \text{sign}\left(\sum_{t=1}^{T}\alpha_t h_t(x)\right) \tag{3}$$

Here, T represented the number of weak classifiers, α_t denoted the weight associated with the prediction $h_t(x)$ of the tth classifier, and sign(·) was the sign function used to assign a class label based on the weighted sum. The weights α_t were determined based on the performance of the weak classifier in each iteration, with higher weights assigned to misclassified samples. This iterative process aimed to improve the ensemble's accuracy over multiple rounds.

Bagging. In breast cancer detection, Bagging (Bootstrap Aggregating) was used to train several instances of the same classifier on different parts of the dataset created through bootstrapping. Each classifier instance learned from a slightly modified version of the original data. This method aimed to lower the chances of overfitting and variance that could arise with individual classifiers. Ultimately, the predictions of all these classifiers were combined, often through majority voting, to arrive at a final decision. In mathematical terms, the prediction $H(x)$ made by the Bagging ensemble for a specific sample x was expressed in Eq. (4):

$$H(x) = \text{mode}(\{h_1(x), h_1(x), \ldots, h_T(x)\}) \tag{4}$$

In this equation, T represented the number of individual classifiers in the ensemble, $h_t(x)$ was the prediction made by the tth classifier, and mode indicated the most frequently occurring prediction among these individual classifiers.

Extra Trees. Extra Trees, also known as Extremely Randomized Trees, is a machine learning technique that, like Random Forests, can be applied to the diagnosis of breast cancer. Nonetheless, it introduces even more uncertainty by picking feature split points at random, without taking optimal thresholds into account. The model is more flexible and may produce more reliable results as a result of the additional randomness. Equation (5) describes the Extra Trees prediction formula.

$$H(x) = \text{sign}\left(\sum_{i=1}^{N} h_i(x)\right) \tag{5}$$

Here, N is the total number of decision trees used to make a prediction, and $h_i(x)$ is the output of the ith tree. A class label is determined by adding up these guesses using the sign function.

3.3 *Model Training and Evaluation*

In this research, several ensemble ML classifiers were learned and tested. Models included Random Forest, Gradient Boosting, AdaBoost, Bagging, and Extra Trees. These models were trained and evaluated using this study's dataset. Each ensemble model was trained with the training data. Models were applied to testing data to make predictions after training.

Breast cancer predictive models must be assessed for accuracy, precision, recall, and F1-score. The evaluation metrics are accuracy, precision, recall, and F1 score. Model accuracy is the percentage of correct predictions. Accurate breast cancer diagnosis requires fewer patient classification errors. The model's cancer predictions are measured by "precision." Precision reduces false alarms and patient stress. Sensitivity recall: Recall tests the model's cancer detection accuracy. When the model has good recall, it doesn't miss many real cases, speeding diagnosis and treatment. A model's accuracy and quantity of correct predictions determine its F1-score. It helps find a breast cancer model with the right sensitivity and specificity.

Breast cancer detection requires balancing cancer sensitivity and false alarm specificity. These indicators can show if the model strikes this balance, improving patient care and outcomes. ROC and precision-recall curves can evaluate classification models like breast cancer diagnostic prediction models. Precision-recall curves help with imbalanced datasets or rare classes like cancer cases, but ROC curves show how well a classifier distinguishes between classes across multiple decision thresholds. These curves help assess model sensitivity, specificity, and precision. They help choose appropriate criteria and models to ensure the model's performance meets intended results, such as a more accurate and quick breast cancer diagnosis. Evaluation accuracy was assessed to standardize evaluation. A detailed categorization report included precision, recall, F1-score, and support. Additionally, a confusion matrix was created to graphically show accurate, wrong, and unexpected categorization findings. ROC and accuracy-Recall curves assessed model sensitivity, specificity, and accuracy. This study's evaluation method revealed how well the models distinguished benign and malignant breast cancers.

4 Experimental Results

Breast cancer detection models are thoroughly assessed in the experimental data section. This experiment used accuracy, precision, recall, F1-score, area under the receiver operating characteristic curve, and precision-recall curve. This study tests the models' capacity to identify benign from malignant tumors, focusing on breast cancer diagnosis.

A model's capacity to discriminate two groups is quantified by its AUC-ROC value. The strengths and weaknesses of the models can be better understood with the use of this data. The ROC curves of the various models—Random Forest, Gradient Boosting, AdaBoost, Bagging, and Extra Trees—are depicted in Figs. 2, 4, 6, 8, and 10. The Precision-Recall Curve is depicted in Figs. 3, 5, 7, 9, 11 for the various models (Random Forest, Gradient Boosting, AdaBoost, Bagging, Extra Trees). AdaBoost excelled above all others. Table 1 illustrates Random Forest, Gradient Boosting, AdaBoost, Bagging, and Extra Trees recall, precision, and F1 score. AdaBoost has the highest precision, recall, and F1 score. These measurements show its extraordinary ability to reduce false positives and pinpoint true positives. Figure 12 shows AdaBoost's enhanced F1 score, precision, and recall. As other models performed similarly, their competency is comparable. Random Forest, Gradient Boosting, AdaBoost, Bagging, and Extra Trees accuracy results were also shown in Table 2. Graphically, Fig. 13 showed these accuracy findings. These tables and graphs thoroughly evaluate model performance, demonstrating AdaBoost's superiority across metrics.

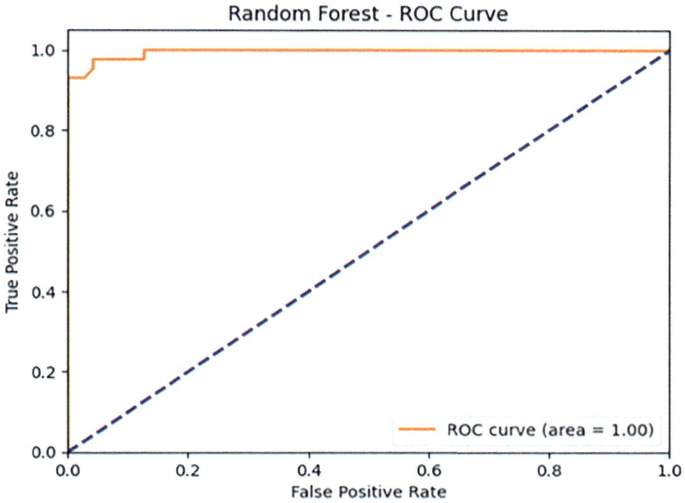

Fig. 2 Random forest ROC curve

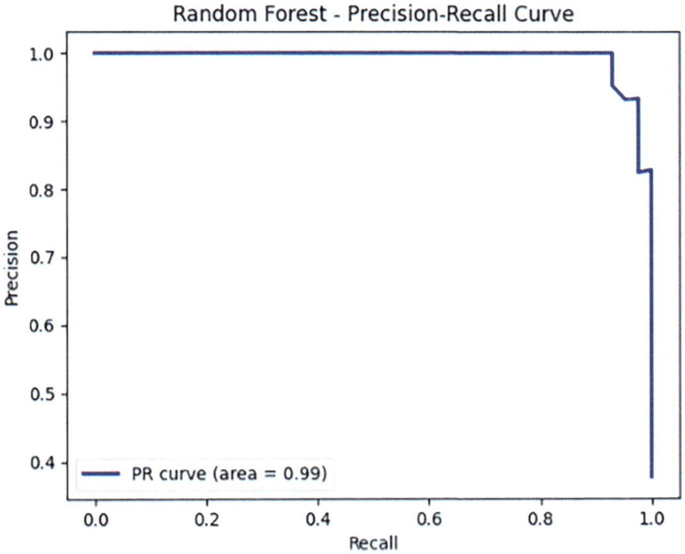

Fig. 3 Random forest precision-recall curve

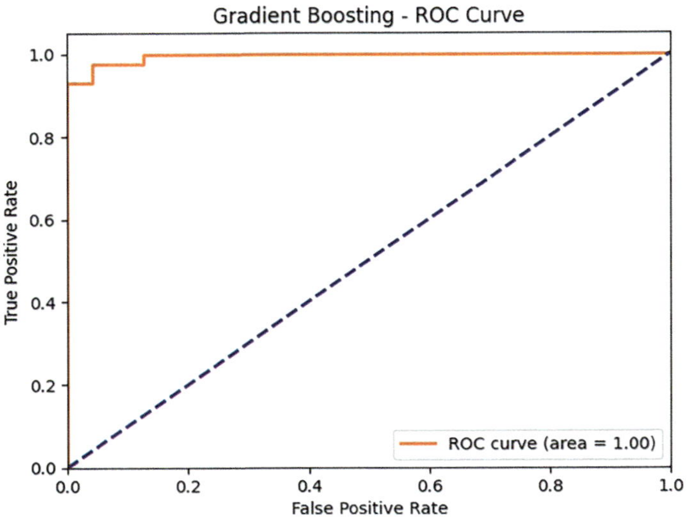

Fig. 4 ROC curve for gradient boosting

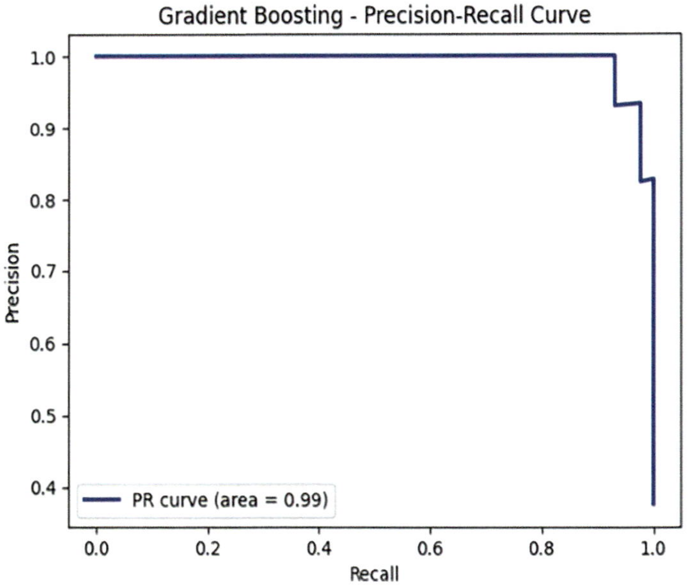

Fig. 5 GBoosting precision-recall curve

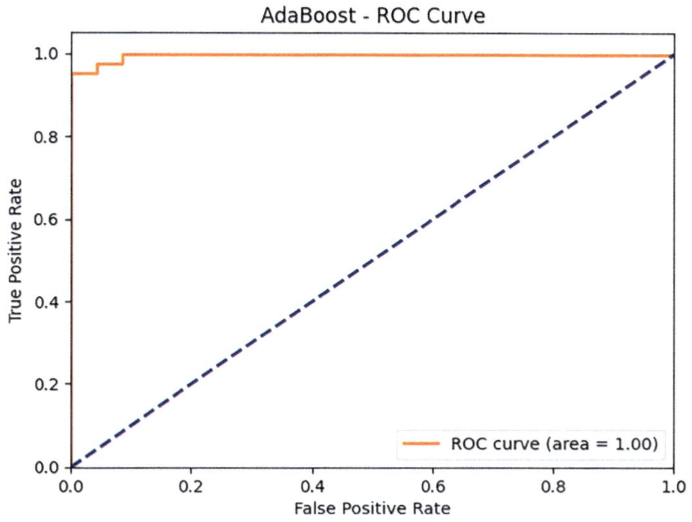

Fig. 6 AdaBoost ROC curve

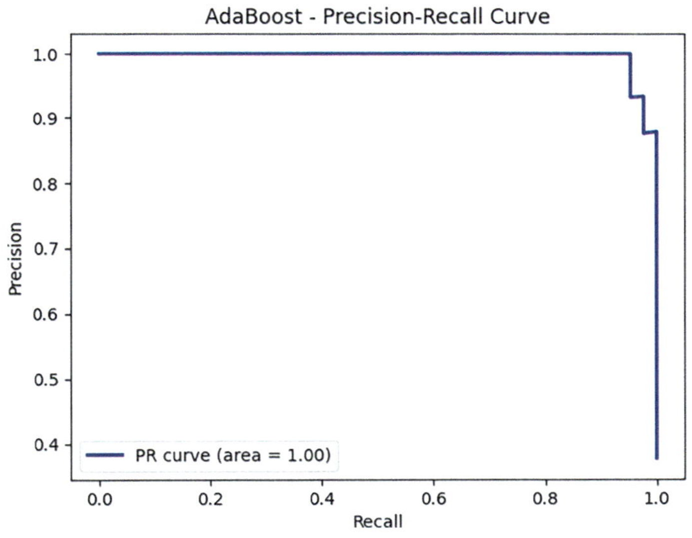

Fig. 7 AdaBoost precision-recall curve

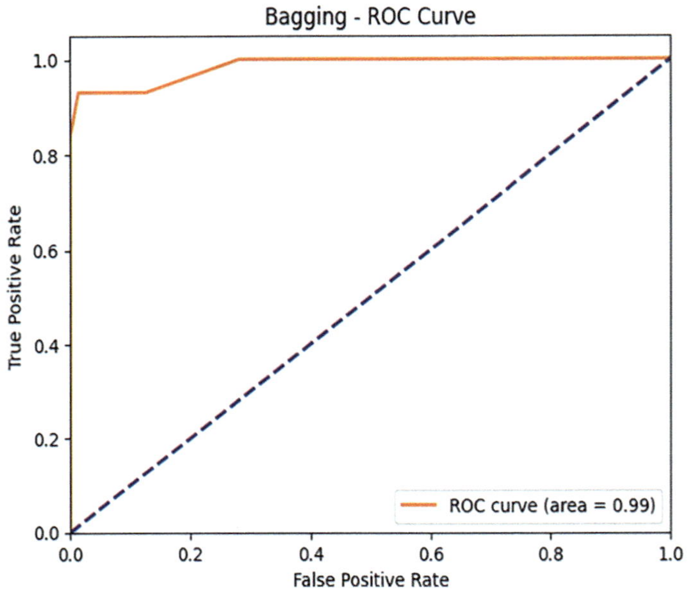

Fig. 8 Bagging ROC curve

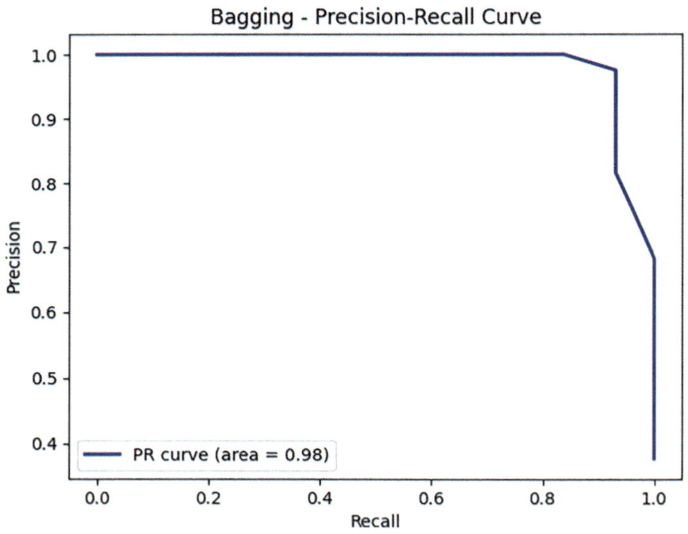

Fig. 9 Bagging precision-recall curve

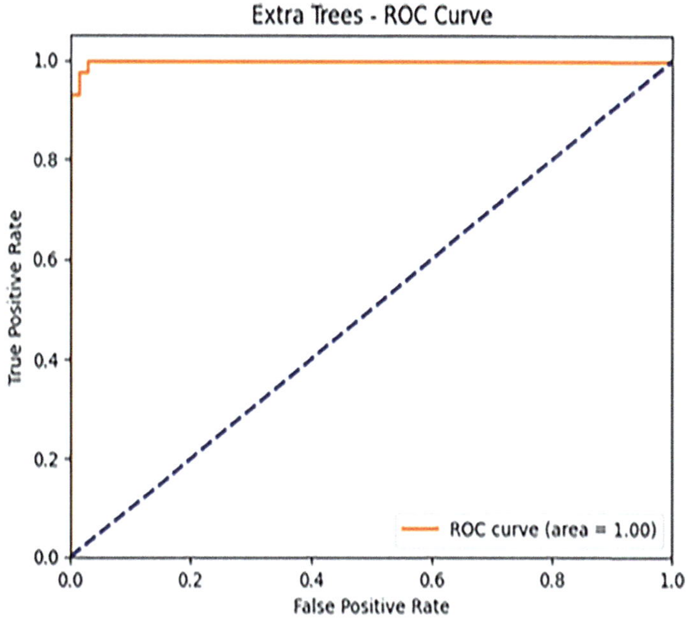

Fig. 10 Extra trees ROC curve

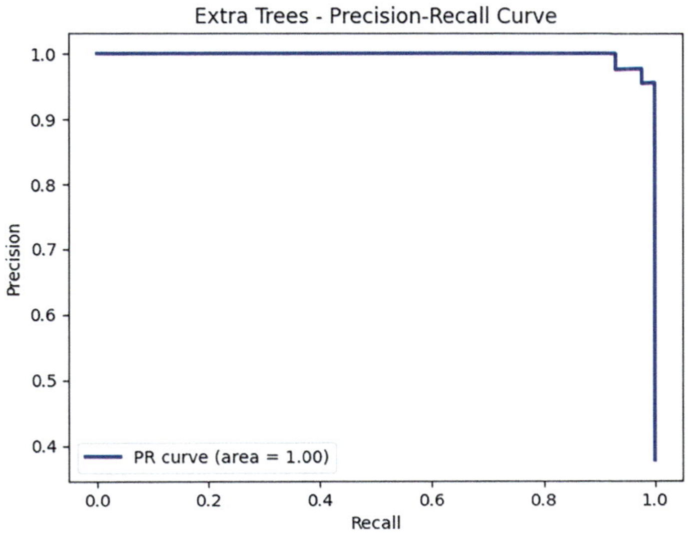

Fig. 11 Extra trees precision-recall curve

Table 1 Precision, recall, and F1 score of models

Model	Precision	Recall	F1 score
Random forest	0.96	0.99	0.97
Gradient boosting	0.96	0.97	0.97
AdaBoost	0.97	0.99	0.98
Bagging	0.96	0.97	0.97
Extra trees	0.96	0.99	0.97

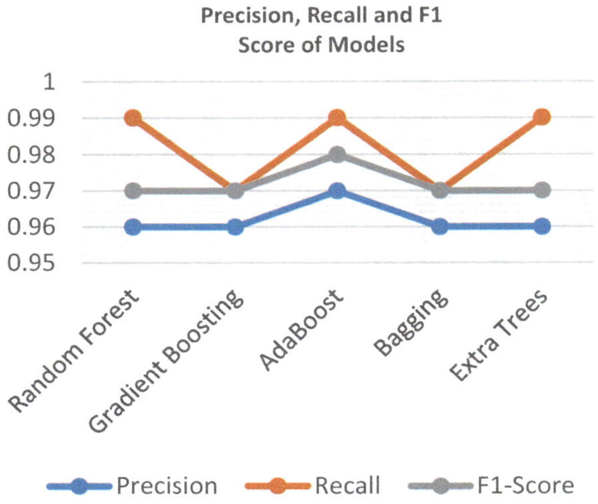

Fig. 12 Precision, recall, and F1 score of all models

Table 2 Accuracy of all models

Model	Accuracy
Random forest	96
Gradient boosting	96
AdaBoost	97
Bagging	96
Extra trees	96

5 Conclusion

In concluding the study, the importance of precise model selection and thorough feature analysis was underscored in the comprehensive exploration of breast cancer detection models. The ensemble models—Random Forest, Gradient Boosting, AdaBoost, Bagging, and Extra Trees—exhibited varying performance across diverse evaluation metrics. AdaBoost consistently emerged as a standout performer, excelling in key metrics like precision, recall, F1 score, and accuracy, highlighting its ability to minimize false positives and maximize true positives, crucial

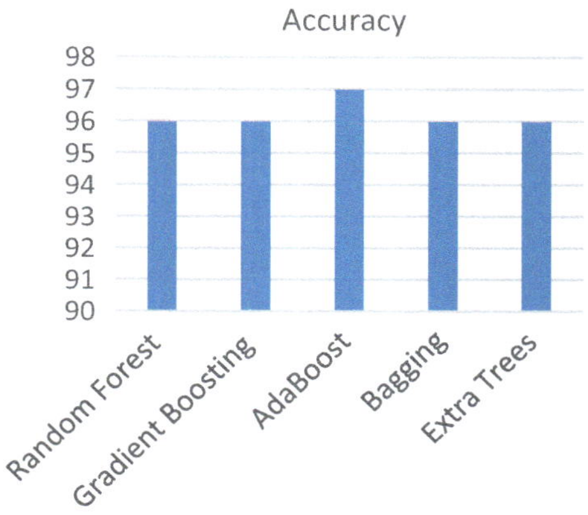

Fig. 13 Accuracy of models

in medical diagnoses. In-depth representations like the precision-recall curve and AUC-ROC helped us understand model performance and make informed decisions.

The proposed strategy was validated and its potential to improve early diagnosis was reinforced by the successful combination of machine learning techniques, rigorous evaluation measures, and intuitive visualizations in breast cancer detection. These results emphasized the significance of cutting-edge technologies in improving medical diagnostics, which may eventually result in better quality healthcare. Although AdaBoost showed the most consistent improvement across all models in our investigation, it is important to remember that even the weaker models have value and can be improved upon for use in certain therapeutic settings. This study laid the groundwork for future inquiry and development of methods for detecting breast cancer, making significant progress in a crucial area of study. The research contains some important caveats that need to be considered. It is heavily dependent on one dataset, which may not be representative of all healthcare settings. There is a possibility that the selected machine learning models will not perform as well in all healthcare settings, and that the data used will contain inaccuracies. Also, the possibility of overfitting is not thoroughly investigated, and the study does not offer in-depth guidance for fine-tuning the model. There are also no in-depth reviews of the research's clinical validity, ethics, or potential consequences. Addressing these caveats and discussing potential future research avenues would strengthen the research's relevance to the field of breast cancer detection and therapy.

References

1. Roy, S., Jeyabalan, J., Gochhait, S., Sugumaran, P., Gromiha, M.M.: Metadata analysis to get insight into drug resistant ovarian cancer. Ingénierie des Systèmes d'Information **28**(2) (2023)
2. Geller, J.: Data mining: practical machine learning tools and techniques with Java implementations. SIGMOD Record **31**(1), 77 (2002)

3. Islam, M.M., Haque, M.R., Iqbal, H., Hasan, M.M., Hasan, M., Kabir, M.N.: Breast cancer prediction: a comparative study using machine learning techniques. SN Comput. Sci. **1**, 1–14 (2020)
4. Assegie, T.A.: An optimized K-Nearest Neighbor based breast cancer detection. J. Robot. Control (JRC) **2**(3), 115–118 (2021)
5. Mangukiya, M., Vaghani, A., Savani, M.: Breast cancer detection with machine learning. Int. J. Res. Appl. Sci. Eng. Technol. **10**(2), 141–145 (2022)
6. Ortega, J.H.J.C., Resureccion, M.R., Natividad, L.R.Q., Bantug, E.T., Lagman, A.C., Lopez, S.R.: An analysis of classification of breast cancer dataset using J48 algorithm. Int. J. Adv. Trends Comput. Sci. Eng. **9** (2020)
7. Dubey, A.K., Gupta, U., Jain, S.: Analysis of k-means clustering approach on the breast cancer Wisconsin dataset. Int. J. Comput. Assist. Radiol. Surg. **11**, 2033–2047 (2016)
8. Fatima, N., Liu, L., Hong, S., Ahmed, H.: Prediction of breast cancer, comparative review of machine learning techniques, and their analysis. IEEE Access **8**, 150360–150376 (2020)
9. Manikandan, J., Devakadacham, S.R., Shanthalakshmi, M., Raj, Y.A., Vijay, K.: An efficient technique for the better recognition of oral cancer using support vector machine. In: 2023 7th International Conference on Intelligent Computing and Control Systems (ICICCS), pp. 1252–1257 (2023)
10. Naji, M.A., El Filali, S., Aarika, K., Benlahmar, E.H., Abdelouhahid, R.A., Debauche, O.: Machine learning algorithms for breast cancer prediction and diagnosis. Procedia Comput. Sci. **191**, 487–492 (2021)
11. Bhise, S., Gadekar, S., Gaur, A.S., Bepari, S., Deepmala Kale, D.S.A.: Breast cancer detection using machine learning techniques. Int. J. Eng. Res. Technol. **10**(7), 2278–0181 (2021)
12. Wang, Y., Wang, N., Xu, M., Yu, J., Qin, C., Luo, X., Ni, D.: Deeply-supervised networks with threshold loss for cancer detection in automated breast ultrasound. IEEE Trans. Med. Imaging **39**(4), 866–876 (2019)
13. Naseem, U., Rashid, J., Ali, L., Kim, J., Haq, Q.E.U., Awan, M.J., Imran, M.: An automatic detection of breast cancer diagnosis and prognosis based on machine learning using ensemble of classifiers. IEEE Access **10**, 78242–78252 (2022)
14. Mohammad, W.T., Teete, R., Al-Aaraj, H., Rubbai, Y.S.Y., Arabyat, M.M.: Diagnosis of breast cancer pathology on the Wisconsin dataset with the help of data mining classification and clustering techniques. Appl. Bionics Biomech. (2022)
15. Zheng, J., Lin, D., Gao, Z., Wang, S., He, M., Fan, J.: Deep learning assisted efficient AdaBoost algorithm for breast cancer detection and early diagnosis. IEEE Access **8**, 96946–96954 (2020)
16. Haq, A.U., Li, J.P., Saboor, A., Khan, J., Wali, S., Ahmad, S., Ali, A., Khan, G.A., Zhou, W.: Detection of breast cancer through clinical data using supervised and unsupervised feature selection techniques. IEEE Access **9**, 22090–22105 (2021)

Enhancing Adaptive E-Learning with Generative AI: Expanding the Horizon Beyond Recommendation Systems

Venkata Bhanu Prasad Tolety and Venkateswara Prasad Evani

Abstract In this paper, we introduce a pioneering approach to e-learning recommendation systems by harnessing the capabilities of Generative AI. Traditional recommendation methods have been trained on small-scale special-purpose datasets and they often fall short when deployed in e-learning applications, due to their inability to capture evolving user preferences and content diversity. In this work, we instead turn to Generative AI systems such as large language models (LLMs), which are extremely powerful general-purpose systems trained over Internet-scale data. We design specific prompting strategies that enable our Generative AI-based recommendation system to dynamically adapt to learners' evolving preferences as they progress through course content, transcending the limitations of state-of-the-art systems, which still rely heavily on collaborative and content-based filtering. Beyond mere content suggestions, our system engages learners in real-time, providing personalized explanations and practice materials, and even generating bespoke learning resources. Experimental validation on real-world data demonstrates remarkable improvements in user engagement and learning outcomes compared to conventional systems. This Generative AI-driven adaptive e-learning approach not only enhances recommendations but also redefines the e-learning experience itself, marking a significant leap forward in creating a more adaptive, engaging, and enriching educational environment.

Keywords Generative AI · Adaptive E-Learning · Recommendation systems · Prompt engineering · User engagement · Personalization · Learning outcomes · Course content · Real-time interaction · Educational innovation

---

V. B. P. Tolety (✉)
Scholar, Jawaharlal Nehru Technological University, Kakinada, India
e-mail: tvbphyd@gmail.com

V. P. Evani
GVP College of Engineering for Women, Visakhapatnam, India

G. Fortino et al. (eds.), *Proceedings of Third International Conference on Computing and Communication Networks*, Lecture Notes in Networks and Systems 917,
https://doi.org/10.1007/978-981-97-0892-5_59

1 Introduction

In the rapidly evolving landscape of digital education, the quest for personalized and adaptive learning experiences has been at the forefront of research and innovation. As technology continues to revolutionize the way we access and consume educational content, the role of artificial intelligence (AI) in enhancing the e-learning journey becomes increasingly profound. Building upon the foundation laid by our previous research endeavors, which explored the efficacy of graph neural networks and hybrid content-collaborative filtering recommendation systems, we embark on a new expedition in the realm of adaptive e-learning.

This research conference paper, titled "Enhancing Adaptive E-Learning with Generative AI," marks the continuation of our journey to revolutionize the e-learning landscape. While our earlier works focused on recommendation systems, this paper represents a paradigm shift. We delve into the exciting world of Generative AI, pushing the boundaries of what's possible in adaptive e-learning beyond the realms of conventional recommendation systems. This research endeavor embarks on a new and exciting trajectory, one that explores the fusion of Generative Artificial Intelligence (AI), particularly Large Language Models (LLMs) like ChatGPT, with e-learning. Our objective: to redefine the boundaries of adaptive e-learning by harnessing the creative, dynamic, and human-like text generation capabilities of LLMs.

The foundation of this study lies in the innovative application of Generative AI, specifically utilizing advanced language models, such as ChatGPT, for prompt engineering in e-learning. Our exploration spans various dimensions of this transformative approach, from leveraging vast datasets to augmenting the instructional content to match individual learner needs. To provide a comprehensive understanding, we draw insights from an extensive dataset, encompassing critical fields like course titles, URLs, pricing information, subscription data, user reviews, instructional levels, content details, publication timestamps, and more.

Our approach goes beyond the conventional structured dataset fields and dives deep into the world of natural language interactions between learners and virtual companions powered by generative AI. Through innovative prompt engineering techniques, we aim to construct dynamic, conversational interactions that cater to individual learner needs, bridging the gap between humans and machines in the educational context.

We are combining Generative AI with e-learning to create adaptive learning experiences. This means not just recommending content but generating custom materials in real-time. Our goal is to enhance education by making it more dynamic and engaging.

In this paper, we will explore our research's methodology, results, and implications. We are shaping the future of e-learning, where education isn't just recommended; it's dynamically generated for each learner. Our aim is to advance adaptive e-learning using Generative AI. Generative AI uses AI to create new content, such as text, images, and more. It's powered by large language models that can handle various tasks like new content creation, summarization, Q&A, classification, code generation, and more. These models analyze human-generated data patterns to produce fresh content, typically using supervised learning alongside labels.

Why is this relevant at this moment? Despite the longstanding presence of AI and machine learning, their significance has gained newfound prominence. Generative AI models have reached a pivotal juncture in terms of precision and capabilities, unlocking possibilities that were previously beyond reach, even as recently as a year ago. Moreover, they have become economically viable and user-friendly, accommodating individuals without a technical background. These models and tools are easily accessible, with many being open source and customizable. While they do demand considerable GPU power, the cloud now provides this resource, making them more accessible than ever before.

Generative AI versus Traditional Systems:

1. Recommendation Approach: Traditional systems use existing data to suggest content, while Generative AI creates entirely new content, enabling dynamic customization.
2. Enhanced Personalization: Generative AI tailors content not only to preferences but also to specific learning needs and styles.
3. Interactive Engagement: Generative AI models like ChatGPT engage in real-time conversations, adapting responses based on interactions, a feature lacking in traditional systems.
4. Adaptability: Generative AI evolves with user interactions, whereas traditional systems often rely on fixed algorithms.
5. Content Creation: Generative AI can generate fresh content, including text, images, and code, for customized learning materials and more.

2 Related Work

The quest for personalized and adaptive e-learning experiences has spurred significant research and innovation in the field of educational technology. As we explore the landscape of related work, it becomes evident that while recommendation systems have been instrumental in shaping e-learning, the integration of generative AI stands at the forefront of transformative possibilities [1].

1. Traditional Recommendation Systems: Traditional recommendation systems, like collaborative filtering and content-based filtering, have long powered personalized content delivery in e-learning by suggesting course materials based on user behavior and preferences. [1]. However, the inherent limitations of these methods have prompted the search for more dynamic and adaptive approaches.
2. Adaptive Learning: The concept of adaptive learning has been a driving force in e-learning research. Venkata Bhanu Prasad Tolety and Evani Venkateswara Prasad introduced a hybrid recommendation system that adapted to changing user profiles and content sources [1]. Their model demonstrated a significant

improvement over traditional filtering techniques, emphasizing the importance of adaptability in modern e-learning.

3. Personalization: Personalization has been a focal point in e-learning research, with several studies emphasizing the importance of tailoring content to individual learner profiles. This includes considering learning styles, knowledge levels, and preferences. The work of Sunil and M. N. Doja, for instance, introduced a recommendation system that factored in learning styles and knowledge levels to guide learners toward courses that align with their specific needs and learning preferences [2].
4. Hybrid Recommender Systems: The concept of hybrid recommender systems has gained prominence, aiming to combine the strengths of various recommendation techniques. Bhaskaran and Marappan explored strategies for improving hybrid recommenders, introducing modifications like the Anarchic Society Optimization, and enhanced sequential pattern extraction [3]. These enhancements aimed to provide more accurate and precise recommendations, reflecting the growing demand for superior recommendation quality in e-learning.
5. Content-Based Filtering: Content-based filtering approaches have gained traction, primarily by modeling and analyzing the content itself. Tahir et al. proposed a dynamic recommendation system for learning objects (DRFLO) that leveraged automatic feature extraction and classification to provide contextually relevant learning objects [4]. This approach marked a departure from traditional recommendation systems by considering not only user behavior but also the content's inherent attributes.
6. Personalized Content Creation: The work of Murad and Linda explored the creation of personalized e-learning experiences through recommendation systems that integrated student profiling, knowledge estimation, assessment, and feedback [5]. This approach aimed to enhance student learning by recommending online video-based learning materials that aligned with each student's unique profile and knowledge level.
7. Recommendations using Graph Neural Networks: Venkata Bhanu Prasad Tolety and Evani Venkateswara Prasad introduced a novel recommendation system for e-learning platforms, which utilizes graph neural networks (GNNs) for user and content profiling, outperforms various content-based and collaborative filtering baselines in e-learning [6]
8. Context-Aware Recommendations for Museums - Trichopoulos et al., explored personalized recommendations using large language models, enhancing visitor experiences in museums [7]
9. Impact of Generative AI and Language Models—Wu et al., examined the widespread influence of generative AI and categorized language models, emphasizing the importance of personalized explanations [8]
10. Contributions of GPT in Language Generation—Zheng et al., highlighted GPT's significant contributions to coherent language generation, focusing on contextual relevance [9]

11. Potential of Generative LLMs in Recommendations—Ji et al., investigated the capability of generative LLMs in understanding user preferences and enhancing recommendation accuracy, with insights from Wang et al., 2023's comprehensive survey [10]

While the previous research contributions have significantly advanced the field of adaptive e-learning, our current paper on "Enhancing Adaptive E-Learning with Generative AI" takes a pioneering step beyond traditional recommendation systems. By leveraging the creative potential of generative AI models, our research explores the uncharted territory of dynamically generating customized learning materials, interactive dialogues, and adaptive content creation [11]. In doing so, we aim to push the boundaries of adaptive e-learning into a new era, where learners are not only recommended content but actively engage in meaningful, personalized educational experiences.

3 Overview of the Proposed Approach

Before examining the details of our proposed approach, it is essential to set the stage for our innovative exploration. Considering the rich landscape of related work, which has underscored the pivotal role of recommendation systems and adaptive learning in e-education, our paper takes a pioneering leap into the realm of generative AI. The proposed approach represents an evolution beyond traditional recommendation systems. We harness the transformative power of generative AI models to redefine the dynamics of adaptive e-learning. Through dynamic content generation, personalized interactions, and adaptability, we aim to revolutionize the e-learning experience. Our approach aspires to empower learners by not only recommending content but by actively engaging them in meaningful, personalized educational journeys.

We provide a comprehensive overview of our approach, detailing its key components and potential impact on the future of adaptive e-learning. In the modern age of advanced language models and AI-driven applications, LangChain emerges as a versatile framework centered around Large Language Models (LLMs). This framework, poised to reshape the landscape of AI applications, offers a unique approach to harnessing the power of LLMs for diverse purposes, including chatbots, Generative Question-Answering (GQA), summarization, and beyond.

At its core, LangChain introduces the concept of "chaining" together various components to unlock more sophisticated use cases powered by LLMs. These components form intricate chains, combining the following elements:

1. Prompt Templates: These predefined structures aid various interactions, like chatbot conversations and Q&A sessions, providing a structured starting point.
2. Large Language Models (LLMs): Utilizing advanced models like GPT-3.5 and BLOOM, LLMs power LangChain to understand and generate human-like text.

3. Agents: These intelligent intermediaries bridge user queries and data interpretation. They come equipped with tools like web search and calculators for complex queries.
4. Memory: LangChain's memory, both short-term and long-term, enables agents to maintain context and information across interactions, resulting in more coherent and informed responses.

The LangChain framework, described in greater detail in subsequent sections of this paper, empowers developers and users to create dynamic and adaptive AI-driven solutions. It bridges the gap between unstructured data and human-like interaction, making it a powerful tool for diverse applications, from educational platforms to customer support.

In our exploration of LangChain, we focus on agents. These key components bridge user queries and LLMs, breaking down complex questions for comprehensive responses. For instance, when presented with a question like "What is the distance between Earth and the Moon?" the agent systematically identifies the entities involved, conducts research to retrieve relevant astronomical data, and performs calculations, all to provide an accurate answer. Furthermore, we delve into the practical application of LangChain's agents for querying structured data, showcasing their utility in academic research contexts. These specialized agents facilitate the extraction of insights from datasets, making them invaluable tools for researchers in various fields.

To enable these capabilities, LangChain integrates advanced language models like those from OpenAI, converging cutting-edge technologies. Users can access LangChain by configuring the OpenAI API key, enabling AI-driven data analysis and interactions. In this section, we have laid the groundwork for our research, highlighting LangChain, agents, and OpenAI's pivotal roles in dynamic content generation, personalized interactions, and adaptability in e-learning. LangChain, a versatile framework, seamlessly integrates advanced language models, ushering in a new era of AI-driven solutions. Agents bridge the gap between user queries and language models. Integration with OpenAI through API keys signifies a convergence of state-of-the-art technologies. In the forthcoming section, we delve into the specifics of our proposed approach.

4 Proposed Approach

In our experiment, we set up a controlled environment within a Google Colab Notebook, employing Python code alongside four fundamental components: LangChain, LangChain Agents, OpenAI, and the OpenAI API Key. Additionally, we enriched our experimentation with a meticulously selected course dataset sourced from a leading e-learning platform. This dataset provided us with indispensable insights into various course attributes, including 'Index,' 'Title,' 'Summary,' 'Enrollment,' 'Rating,' 'Reviews,' 'Link,' 'Difficulty Level,' 'Price,' and 'Duration (in hours).'

Our approach consisted of the following key stages:

1. Environment Setup: We initiated our experiment by setting up the requisite environment, seamlessly installing LangChain and OpenAI within the Google Colab Notebook. To unlock OpenAI's capabilities, we utilized an API Key.
2. Dataset Acquisition: Our carefully curated dataset, meticulously sourced from a prominent e-learning platform, found its place within our Google Colab Notebook. Leveraging the Python library pandas, we executed the essential task of importing this dataset, setting the stage for our comprehensive analysis.
3. Exploratory Data Analysis (EDA): An initial exploration of the dataset played a pivotal role in enhancing our understanding of its underlying structure and contents, furnishing us with invaluable insights into the available data.
4. LangChain Integration: With seamless precision, we integrated LangChain Agents, skillfully engineered to engage with CSV database files. These agents assumed a central role in our experiment, facilitating the querying and extraction of insightful information from the dataset.
5. Agent Initialization: A LangChain Agent was meticulously crafted and configured, with considerations for the OpenAI model, the dataset's path, and verbose mode for comprehensive feedback.
6. Running LangChain Agent: With the LangChain Agent at our disposal, we embarked on a journey of communication with the dataset using natural language queries in English. Through an array of diverse queries, we demonstrated the extraordinary capabilities of generative AI in extracting profound insights from structured data (Fig. 1).

Our results underscore the effectiveness of this approach in querying and deriving valuable information from a structured dataset. By leveraging the combined capabilities of LangChain, OpenAI, and LangChain Agents, we have unveiled a powerful methodology poised to shape the future of data-driven applications within the ever-evolving landscape of e-learning.

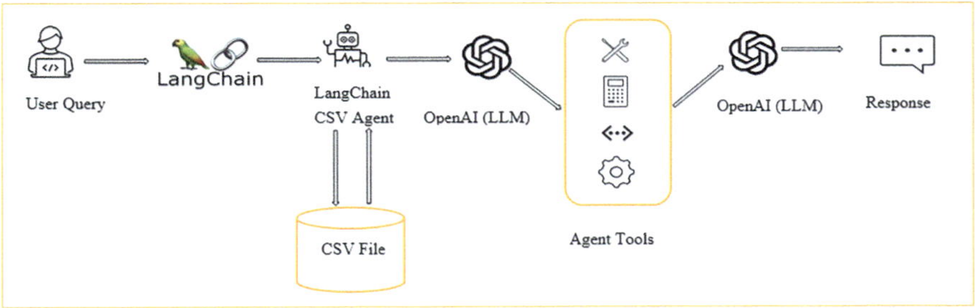

Fig. 1 Proposed approach

5 Experiments

Datasets: Our dataset originates from a prominent online learning platform renowned for its vast repository of 130,000 expert-led video courses. This dataset consists of 42,432 lectures, across nine distinct categories, with our specific focus on the technology-related dataset. In this specialization, we zeroed in on a dataset comprising 9,964 meticulously curated records, each delivering crucial insights into various facets, including 'Index,' 'Title,' 'Summary,' 'Enrollment,' 'Rating,' 'Reviews,' 'Link,' 'Difficulty Level,' 'Price,' and 'Duration (in hours).' This dataset is unique because it's the largest of its kind, focusing on how people access e-learning content. This distinctive attribute renders it supremely aligned with the objectives of our research endeavors.

5.1 Results

In this section, we present the outcomes of our exploration into the vast dataset of e-learning courses, employing a combination of LangChain, LangChain Agents, and OpenAI's generative capabilities. Our approach successfully addressed a spectrum of queries, showcasing the potential of generative AI for enriching e-learning platforms.

Course Summaries: We initiated our experiment by creating concise summaries for beginner-level courses, providing learners with valuable insights into each course's topic and expected learning outcomes.

```
query=("Generate a concise summary for a beginner-level course from the dataset. Include details about the topic and what learners can expect to learn")
agent.run(query)

'The 2020 Complete Python Bootcamp: From Zero to Hero in Python is a beginner-level course with the most enrollment (1,100,093). Learners can expect to learn Python from
the basics to creating their own applications and games.'
```

Summarizing courses at beginner level with rating 4.7 and above.

```
query=("Summarize in detail about courses with a rating above 4.7 and categorized at Beginner level.")
agent.run(query)

'There are 111 courses with a rating above 4.7 and categorized at Beginner level. These courses cover a wide range of topics, fro
m iOS app development to modern JavaScript, microservices with Node JS and React, server side rendering with React and Redux, gam
e development with GameMaker Studio 2, game audio, fundraising, concurrency in Go (Golang), and Unity game development. The cours
es range in price from 54 to 497, and in duration from 17 to 60 hours.'
```

Tailored Recommendations: For busy professionals seeking quick skill acquisition, our approach recommended five courses with durations of fewer than 25 h, aligning with their hectic schedules.

```
query = ("Recommend 5 courses that have a duration of fewer than 25 hours and is suitable for busy professionals looking to acquire new skills quickly")
agent.run(query)

'The 5 recommended courses are Machine Learning A-Z™: Hands-On Python & R In Data Science, React - The Complete Guide (incl Hooks, React Router, Redux), Modern React wit
h Redux [2020 Update], Build Responsive Real World Websites with HTML5 and CSS3, and The Complete Web Developer in 2020: Zero to Mastery.'
```

Recommend 5 top rated advanced courses in programming for experienced learners.

```
query=("Recommend 5 top rated advanced courses in programming for experienced learners.")
agent.run(query)

'The 5 top rated advanced courses in programming for experienced learners are Automate the Boring Stuff with Python Programming,
The Complete Node.js Developer Course (3rd Edition), The Ultimate MySQL Bootcamp: Go from SQL Beginner to Expert, Docker Mastery:
with Kubernetes +Swarm from a Docker Captain, and R Programming A-Z™: R For Data Science With Real Exercises.'
```

Quality Control: We assisted the learning platform in refining its course catalog by identifying poorly rated courses. We meticulously listed the ten lowest-rated courses, facilitating streamlined catalog management.

```
query = "We want to refine/discard poorly rated coursess from the catalogue and list bottom most 10 courses along with their ratings which have very poor ratings"
agent.run(query)

'The bottom 10 courses with their ratings are: Python Flask Web Development: REST API, Postman & HTTP Requests (Rating: 1.3), AWS Certification Course (Rating: 1.4), Rea
ct Course For Complete Beginner (Rating: 1.4), Spring Framework, Hibernate & Java: Oracle, Database & Web App (Rating: 1.4), Machine Learning for ABSOLUTE beginners! [Ap
ril 2020] (Rating: 1.5), Certification Course For Jaava Learne (Rating: 1.5), Projects in Android (Rating: 1.6), Learn Basics Of Angular From Scratch (Rating: 1.6), Java
Script Tutorial (Rating: 1.6).'
```

Analyze and provide detailed analysis for making quality courses.

```
query=("Analyze and provide detailed analysis for making quality courses")
agent.run(query)

'Courses with a rating of 4.5 or higher have an average enrollment of 12,425, average reviews of 2,242, average price of $266, and average duration of 35 hours. This sug
gests that courses with higher ratings tend to have higher enrollments, more reviews, higher prices, and longer durations.'
```

Popularity Analysis: By examining courses with high enrollments, we uncovered the factors contributing to their popularity. This analysis shed light on enrollment trends and the elements influencing learners' choices.

```
query = ("Analyze courses with higher enrollments, presenting the enrollment numbers and elucidating the reasons contributing to their increased popularity.")
agent.run(query)

'The course with the highest enrollment is 2020 Complete Python Bootcamp: From Zero to Hero in Python, with 1,100,093 enrollments. This is likely due to its beginner-fri
endly nature, low price, and short duration.'
```

Topic-based Recommendations: Exploring specific topics like Python, DevOps, and JavaScript, we provided personalized course recommendations, along with detailed titles and summaries, catering to learners' diverse interests.

```
query = ("Provide course recommendations from the dataset based on specific topics or keywords")
query= (query + "such as Python, DevOps, and JavaScript, including their titles and brief summaries.")
agent.run(query)

'The following courses are recommended based on the topics or keywords Python, DevOps, and JavaScript: \n\n1. 2020 Complete Python Bootcamp: From Zero to Hero in Python
- Learn Python like a Professional! Start from the basics and go all the way to creating your own applications and games!\n2. The Web Developer Bootcamp - The only cours
e you need to learn web development - HTML, CSS, JS, Node, and More!\n3. Machine Learning A-Z™: Hands-On Python & R In Data Science - Learn to create Machine Learning Al
gorithms in Python and R from two Data Science experts. Code templates included.\n4. The Complete JavaScript Course 2020: Build Real Projects! - Master JavaScript with t
he most complete course on the market! Projects, challenges, quizzes and more!\n5. Automate the Boring Stuff with Python Programming - A practical programming course for
office workers, academics, and administrators who want to improve their productivity.\n6. Java Programming Masterclass for Software Developers - Learn Java...'
```

Course Metrics Visualization: Here is Scatter plot analysis depicting the connection between course prices and the number of reviews, further enhancing insights into course performance.

```
query=("Produce a scatter plot that showcases the relationship between course prices (X-axis) ")
query=(query+ "and the number of reviews (Y-axis), with points grouped by course durations")
agent(query);
```

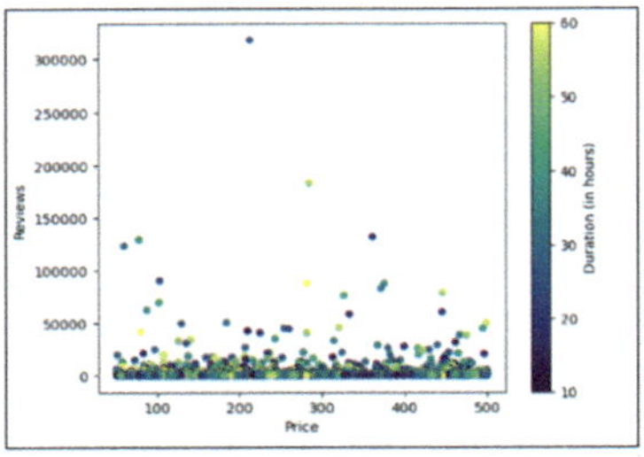

Using a bubble chart, we masterfully illustrated the interplay between course prices, ratings, and enrollments, providing a holistic view of course dynamics.

```
query=("Generate a bubble chart that visualizes the relationship between course prices (X-axis), ratings (Y-axis), and enrollments (size of bubbles)")
agent(query);
```

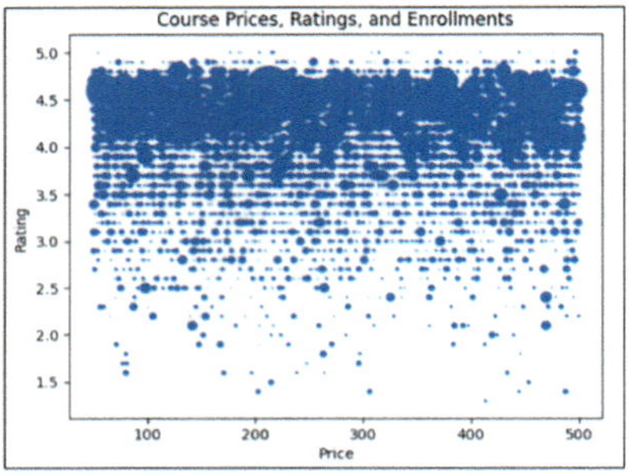

The results demonstrate the versatility and effectiveness of our generative AI approach in enhancing e-learning platforms with tailored content, improved course quality, and valuable insights into user preferences and engagement pattern.

6 Limitations

In this paper, we have unveiled a promising approach with diverse potential applications in e-learning domain. However, it's crucial to acknowledge some of the inherent limitations. Firstly, the efficacy of our approach depends on both data quality and quantity, as well as the performance of generative AI models. Variability in these factors can impact outcomes significantly. Secondly, the utilization of generative AI introduces ethical concerns, especially regarding potential biases in generated content. Moreover, the computational resources required for generative models can limit accessibility in resource-constrained environments. Furthermore, many generative models lack inherent support for continuous learning and adaptation, potentially hindering responsiveness to evolving educational contexts. Addressing these constraints will be pivotal for the comprehensive integration of generative AI in adaptive e-learning.

7 Conclusion

This paper has undertaken a pioneering journey to enhance adaptive e-learning through the integration of generative AI, transcending the limitations of traditional recommendation systems. Building upon prior research, where the constraints of conventional content and collaborative filtering techniques for e-learning were revealed, this study introduces an innovative approach. By harnessing the creative potential of generative AI models, along with the versatile LangChain framework, intelligent agents, and OpenAI, significant improvements in recommendation accuracy and effectiveness have been demonstrated. This novel approach not only addresses the challenges posed by dynamic user preferences but also fosters personalized learning experiences tailored to individual needs, pace, and style. With notable advancements over established content and collaborative filtering methods, our research lays the foundation for e-learning practitioners to shape the future of more precise and adaptive recommender systems, ushering in a new era of education that goes beyond traditional recommendations.

References

1. Venkata Bhanu Prasad Tolety, Evani Venkateswara Prasad: Hybrid content and collaborative filtering-based recommendation system for e-learning platforms. Bull. Electr. Eng. Inform. (BEEI), **11**(3), (2022).
2. Sunil, Doja, M.N.: An improved recommender system for E-Learning environments to enhance learning capabilities of learners. In: P. K. Singh et al. (Eds.): Proceedings of ICETIT 2019, LNEE 605, pp. 604–612, (2020).
3. Bhaskaran, S., Marappan, R.: Design and analysis of an efficient machine learning based hybrid recommendation system with enhanced density-based spatial clustering for digital e-learning applications. Complex Intell. Syst. (2021). https://doi.org/10.1007/s40747-021-00509-4
4. Tahir, S., Hafeez, Y., Abbas, M.A., et al.: Smart learning objects retrieval for E-Learning with contextual recommendation based on collaborative filtering. Educ. Inf. Technol. (2022). https://doi.org/10.1007/s10639-022-10966-0
5. Murad, H., Yang, L.: Personalized E-Learning recommender system using multimedia data. Int. J. Adv. Comput. Sci. Appl.Comput. Sci. Appl. **9**(9), 565–567 (2018)
6. Venkata Bhanu Prasad Tolety, Evani Venkateswara Prasad: Graph neural networks for E-Learning recommendation systems. Int. J. Recent. Innov. Trends Comput. Commun. (IJRITCC), **11**(9s), pp. 43–50, (2023), https://doi.org/10.17762/ijritcc.v11i9s.7395
7. Trichopoulos, G., Konstantakis, M., Alexandridis, G., Caridakis, G.: September 10). Large language models as recommendation systems in museums (2023). https://doi.org/10.3390/electronics12183829
8. Wu, L., Zheng, Z., Qiu, Z., Wang, H., Gu, H., Shen, T J., Qin, C., Zhu, C., Zhu, H., Liu, Q., Xiong, H., & Chen, E.:. A survey on large language models for recommendation. (2023). https://doi.org/10.48550/arXiv.2305.19860
9. Zheng, Z., Qiu, Z., Hu, X., Wu, L., Zhu, H., Xiong, H.: Generative job recommendations with large language model. (2023). https://doi.org/10.48550/arXiv.2307.02157.
10. Ji, J., Li, Z., Xu, S., Hua, W., Ge, Y., Tan, J., & Zhang, Y.: GenRec: Large language model for generative recommendation. (2023). https://arxiv.org/abs/2307.00457.

11. Recommendation as instruction following: a large language model empowered recommendation approach Junjie Zhang1, Ruobing Xie2, Yupeng Hou1, Way, arXiv:2305.07001v1 [cs.IR] (2023).
12. Hyung Won Chung, Le Hou, Shayne Longpre, Barret Zoph, Yi Tay, William Fedus, Eric Li, Xuezhi Wang, Mostafa Dehghani, Siddhartha Brahma, et al. 2022. Scaling instruction-finetuned language models. arXiv preprint arXiv:2210.11416 (2022)
13. Shijie Geng, Shuchang Liu, Zuohui Fu, Yingqiang Ge, Yongfeng Zhang.: Recommendation as language processing (rlp): A unified pretrain, personalized prompt & predict paradigm (p5). In RecSys (2022)
14. Sunhao Dai, Ninglu Shao, Haiyuan Zhao, Weijie Yu, Zihua Si, Chen Xu, Zhongxiang Sun, Xiao Zhang, and Jun Xu. 2023. Uncovering ChatGPT's Capabilities inRecommender Systems. arXiv preprint arXiv:2305.02182 (2023)
15. GPT4Rec: A Generative Framework for Personalized Recommendation and User Interests Interpretation Jinming Li1, ,Wentao Zhang2,Tian Wang2,Guanglei Xiong2, Alan Lu2, Gerard Medioni2 SIGIR eCom'23, July 27, 2023, Taipei, Taiwan.
16. AI-assisted Learning with ChatGPT and Large Language Models: Implications for Higher Education. Laato, Samuli &Morschheuser, Benedikt & Hamari, Juho & Björne, Jari. (2023).
17. ChatGPT and large language models in academia: opportunities and challenges, Jesse G. Meyer,corresponding author1 Ryan J. Urbanowicz,1 Patrick C. N. Martin,1 Karen O'Connor,2 Ruowang Li,1 Pei-Chen Peng,1 Tiffani J. Bright,1 Nicholas Tatonetti,1 Kyoung Jae Won,1 Graciela Gonzalez-Hernandez,1 and Jason H. Moorcorresponding author1, 2023 Jul 13. https://doi.org/10.1186/s13040-023-00339-9.
18. Education in the Era of Generative Artificial Intelligence (AI): Understanding the Potential Benefits of ChatGPT in Promoting Teaching and Learning, David BAİDOO-ANU Leticia OWUSU ANSAH, **7**:1, 52–62 (2023).
19. Pavlik, J.V.: Collaborating with ChatGPT: considering the implications of generative artificial intelligence for journalism and media education. J.Ism & Mass Commun. Educ., **0**(0). https://doi.org/10.1177/10776958221149577.
20. Wang, W., Chen, Y., Heffernan, N.: A generative model-based tutoring system for math word problems. (2020). arXiv preprint arXiv:2010.04.
21. Zhai, X.: ChatGPT User experience: Implications for education. (2022). Available at SSRN: https://ssrn.com/abstract=4312418 or https://doi.org/10.2139/ssrn.4312418.
22. Generative Recommendation: Towards Next-generation Recommender Paradigm, Wenjie Wang, Xinyu Lin, Fuli Feng, Xiangnan He, Tat-Seng Chua, arXiv:2304.03516, https://doi.org/10.48550/arXiv.2304.03516
23. Wenqiang Lei, Xiangnan He, Yisong Miao, Qingyun Wu, Richang Hong, MinYen Kan, Tat-Seng Chua.: Estimation-action-reflection: Towards deep interaction between conversational and recommender systems. In WSDM. ACM, 304–312 (2020).
24. Chenhao Hu, Shuhua Huang, Yansen Zhang, and Yubao Liu.: Learning to infer user implicit preference in conversational recommendation. In SIGIR. ACM, 256–266 (2022).
25. Bommasani, Rishi, Drew A Hudson, Ehsan Adeli, Russ Altman, Simran Arora, Sydney von Arx, Michael S Bernstein, Jeannette Bohg, Antoine Bosselut, Emma Brunskill et al.: On the opportunities and risks of foundation models. (2021). arXiv preprint arXiv:2108.07258.
26. Hill,J., Randolph Ford, W., &Farreras, I. G.: Real conversations with artificial intelligence: A comparison between human–human online conversations and human–chatbot conversations. Computers in Human Behavior, **49**, 245–250 (2015). https://doi.org/10.1016/j.chb.2015.02.026.
27. Generative AI at Work, Erik Brynjolfsson, Danielle Li, Lindsey R. Raymond, Working Paper 31161, http://www.nber.org/papers/w31161.
28. Bubeck, Sebastien, Varun Chandrasekaran, Ronen Eldan, Johannes Gehrke, Eric Horvitz, Ece Kamar, Peter Lee, Yin Tat Lee, Yuanzhi Li, Scott Lundberg et al.: Sparks of artificial general intelligence: Early experiments with gpt-4, (2023). arXiv preprint arXiv:2303.12712.
29. Brown, Tom B., Benjamin Mann, Nick Ryder, Melanie Subbiah, Jared Kaplan, Prafulla Dhariwal, Arvind Neelakantan, Pranav Shyam, Girish Sastry, Amanda Askell, Sandhini

Agarwal, Ariel Herbert-Voss, Gretchen Krueger, Tom Henighan, Rewon Child, Aditya Ramesh, Daniel M. Ziegler, Jeffrey Wu, Clemens Winter, Christopher Hesse, Mark Chen, Eric Sigler, Mateusz Litwin, Scott Gray, Benjamin Chess, Jack Clark, Christopher Berner, Sam McCandlish, Alec Radford, Ilya Sutskever, Dario Amodei: Language models are Few-Shot learners. (2020). arXiv:2005.14165 [cs].
30. Reconceptualizing ChatGPT and generative AI as a student-driven innovation in higher education Author links open overlay panelYun Dai a, Ang Liu b, Cher Ping Lim c, https://doi.org/10.1016/j.procir.2023.05.002.

Venkata Bhanu Prasad Tolety is presently engaged in his doctoral studies at JNTU Kakinada. Prior to this, he attained his bachelor's degree (B. Tech) in computer science and information technology from JNT University, Hyderabad, India. Subsequently, he completed his master's degree in computer science engineering from the same institu-tion. Currently, he holds a professional position at Accenture and is an active member of both IEEE and CSI. Additionally, he possesses certifications in Microsoft Azure Cloud and Google Cloud. For contact, you can reach him via email at tvbphyd@gmail.com.

E.V. Prasad a professor at GVP College of Engineering for Women in Visakhapatnam, has 37 years of academic experience. He holds a Ph.D. in Computer Science Engineering from the University of Roorkee, an ME in Control Systems from Madras University, and a BE in Electronics and Communications Engineering from SV University, Tirupati. Dr. E.V. Prasad has guided 11 Ph.D. students, co-authored six books, and authored 102 research publications. He has international experience, including visits to Malaysia and Sweden. He received the State Award for Meritorious Teachers in 2008 and is recognized as the 'best teacher' by CSE students. Contact him via email at profevprasad@yahoo.com.

A Fusion Framework of Pre-trained Deep Learning Models for Oral Squamous Cell Carcinoma Classification

Muhammad Attique Khan, Momina Mir, Muhammad Sami Ullah, Ameer Hamza, Kiran Jabeen, and Deepak Gupta

Abstract One of the most serious illnesses in the world is oral cancer, and prompt, effective treatment can significantly increase patient survival and cure rates. The manual diagnosis from Histopathologic Images (HI) is time-consuming and requires an expert doctor. To reduce the burden on a doctor, a computerized technique is helpful for the classification of medical image analysis. In this work, we proposed a deep learning framework for classifying Oral Squamous Cell Carcinoma (OSCC) from HI. Two pre-trained deep learning architectures, MobileNet-V2 and DarkNet-19, have been fine-tuned in the proposed framework. Both architectures have been selected based on recent performance and fewer parameters. Both models were trained using the transfer learning concept and extracted features from the global average pooling layers. A Chaotic Crow Search optimization algorithm has been employed, and the best features have been selected. The selected features are finally classified using machine learning classifiers. A publicly available dataset was utilized for experimental purposes and obtained the highest accuracy of 92%. Compared with some state-of-the-art techniques, the proposed framework shows improved accuracy.

Keywords Oral cancer · Deep learning · Transfer learning · Features selection · Classification

1 Introduction

One of the most common fatal illnesses, oral cancer, has long been a major global public health concern. A subtype of head and neck cancers, 475,000 new patients are reported worldwide with oral cancer every year. The survival rate of late-stage

M. A. Khan (✉) · M. Mir · M. S. Ullah · A. Hamza · K. Jabeen
Department of Computer Science, HITEC University, Taxila 47080, Pakistan
e-mail: attique.khan@ieee.org

D. Gupta
Department of Computer Science and Engineering, National Institute of Technology Arunachal Pradesh, Jote, Arunachal Pradesh 710003, India

G. Fortino et al. (eds.), *Proceedings of Third International Conference on Computing and Communication Networks*, Lecture Notes in Networks and Systems 917,
https://doi.org/10.1007/978-981-97-0892-5_60

illness is less than 20%, while that of early-stage illness is over 80% [1]. In recent years, oral cancer incidence rates have rapidly increased in places like South-Central Asia, Melanesia, and Central and Eastern Europe due to escalating tobacco and alcohol usage. Even New Zealand, Australia, and North America have seen some of its spread, albeit not significantly [2].

Oral cancer is the second most frequent type of cancer in Pakistan, and the incidence of cancer is rising there, according to GLOBOCAN [3]. Oral squamous cell carcinoma (OSCC), minor salivary gland carcinomas, verrucous carcinoma, mucosal melanoma, and lymphoma are the most prevalent forms of oral cancer. Lesions, referred to as oral, possibly malignant disorders, are typically the source of OSCC [4]. Malignant squamous cells are larger than normal cells and differ significantly in shape from one another at the histopathological level [5].

Deep learning has been extensively applied in medical image analysis and has become a potent technique for resolving image categorization issues [6]. Convolutional neural networks (CNNs) can automatically classify various cancer lesions, oral cancer [1], breast cancer [7], skin cancer [8], lung cancer [9], and a named a few more [10, 11]. In deep learning, the key problems the computer vision researcher faces are selecting deep models for training and selecting the best features. Several pre-trained models exist, including VGG16, AlexNet, DarkNet, and MobileNet. In addition, the researchers used several customized models for the classification tasks. Transfer learning is usually employed for classification tasks; however, few researchers focused on the training from scratch.

Several computerized techniques have recently been introduced for classifying OSCC and normal images. Figueroa et al. [12] used the gradient-weighted class activation mapping developed by Selvaraju et al. to incorporate interpretability and explainability into Convolutional Neural Networks (CNNs). Images from bespoke mobile oral screening devices were trained using a two-stage method that used data augmentation techniques and Li et al.'s guided attention inference network (GAIN). By using attention maps as trustworthy priors, GAIN's architecture—which includes classification, attention mining, and bounding box streams—optimized CNN classification and segmentation accuracy. The resulting attention maps gave the network information about how it makes decisions, helping it to concentrate on precise lesion spots and ignore unimportant areas. The study showed how well this method worked to classify oral lesions as malignant or potentially malignant in a way that was easier to understand and dependable. Using patient hyperspectral pictures, Jeyaraj et al. [13]created a deep learning method for the automatic, computer-aided identification of oral cancer. During training on 100 picture datasets, this partitioned deep Convolutional Neural Network (CNN) was able to distinguish between malignant tumors and benign tissues with a classification accuracy of 91.4%, sensitivity of 0.94, and specificity of 0.91. Using 500 training patterns, a classification accuracy of 94.5% was obtained for malignant tumors compared to normal tissues. By contrasting these findings with those of conventional medical image classification algorithms, the study demonstrated how much the suggested method improved the accuracy of oral cancer diagnosis.

Al Duhayyim et al. [14] introduced a unique CAD model named CADOC-SFOFC (Computer Aided Diagnosis for OC using Sailfish Optimization with Fusion-based classification). The model utilized a fusion-based feature extraction process incorporating VGGNet16 and Residual Network (ResNet) models. Feature vectors were fused and input into an extreme learning machine (ELM) model for classification, with the Sailfish Optimization (SFO) algorithm improving parameter selection for enhanced performance. Experimental analysis of the Kaggle dataset demonstrated the superior performance of the CADOC-SFOFC model compared to similar methods, achieving a maximum accuracy of 98.11%. The results indicated that CADOC-SFOFC has substantial potential as a cost-effective and non-invasive tool for oral cancer detection and classification. In their discussion of the oral cancer pandemic, Welikala et al. [15] emphasized the disease's major impact, especially in low- and middle-income nations, where it was the cause of 177,384 fatalities in 2018. The goal of the MeMoSA project was to automate the recognition of oral lesions for prompt and economical diagnosis. The study used deep neural networks—ResNet-101 for image classification and Faster R-CNN for object detection—as well as a novel technique to aggregate annotations from many clinicians to develop automated systems that could identify intricate patterns. ResNet-101 had excellent F1 ratings of 87.07% for lesion recognition and 78.30% for identifying images that needed to be referred to in the first evaluation. Faster R-CNN object detection achieved an F1 score of 41.18% for lesions that needed to be referred. An intelligent multisampling tensor model was presented by Huang et al. [16] for the classification of cysts and mouth cancer from magnetic resonance imaging (MRI). To be more precise, the method first used four straightforward sampling processes to encode the input images, which helped the model learn a variety of global, influential, correlated, and regional properties. These derived representations were then combined using a representation fusion technique. The representations were then contracted using a sequence of matrix product states, which mapped them into a high-dimensional space for classifier operations. Oral MRI cancer data were used to assess the suggested method, and the experimental findings showed that the strategy could produce competitive classification results. Joshi et al. [17] aimed to reduce overfitting by employing a Convolutional Neural Network (CNN) to generate the required feature map with dimension reduction. The methodology involved utilizing Enhanced CNN and autoencoder techniques to enhance the efficiency of the feature extraction process and compress information. Pooling and deconvolution were applied to minimize the difference between input and output data. The results showed varying accuracy and processing time values across different sample groups of Confocal Laser Endomicroscopy (CLE) images. Based on the results, the proposed solution demonstrated superiority over the current system, improving classification accuracy by an average of 5–5.5% and reducing processing time by 20–30 ms on average.

However, the MobileNet and DarkNet architectures have some impressive performance in terms of accuracy and less computational time due to fewer parameters. We fine-tuned these models to classify normal and OSCC Histopathologic images in this work. Our major contributions of this work are as follows:

- Fine-tuned MobileNet-V2 and DarkNet-19 pre-trained models based on the additional layers and fixed hyperparameters setting using literature knowledge.
- The best features have been selected using a Chaotic Crow Search Optimization algorithm.
- Best features are classified using machine learning classifiers and compared the performance with several state-of-the-art techniques.

2 Proposed Methodology

The proposed oral squamous cell carcinoma classification framework has been presented in this section using a fusion of pre-trained model information and optimal feature selection. Figure 1 shows the proposed framework that consists of some important middle steps. The original dataset images were processed initially and passed to the two fine-tuned deep learning models, MobileNet-V2 and DarkNet-19. Both models are trained with manual initialization of hyperparameters and performed training. The training was performed using the transfer learning concept. After that, deep features are extracted, and the best features are selected using the Chaos Crow Search Optimization algorithm. The best-selected features are finally classified using machine learning classifiers.

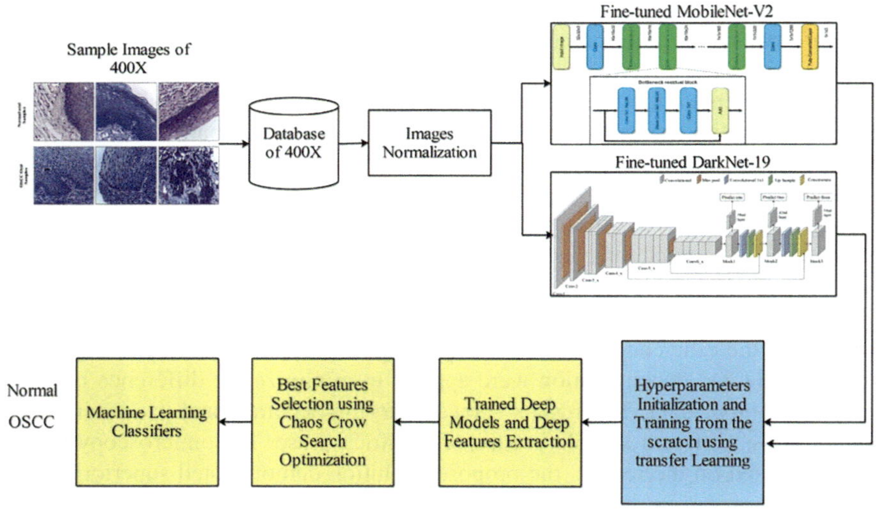

Fig. 1 Proposed framework of oral squamous cell carcinoma classification

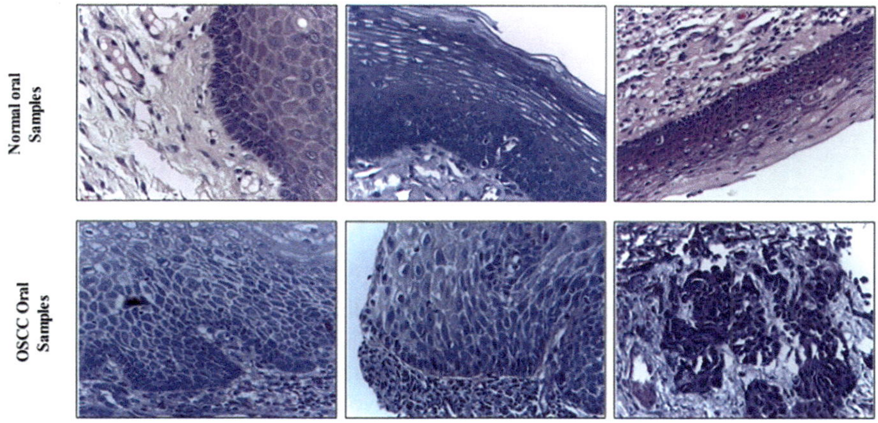

Fig. 2 Sample images of normal and oral squamous cell carcinoma (OSCC)

2.1 Dataset Description

A useful tool for investigating the histology of oral cancer is the dataset used in this work [18]. An openly accessible dataset has been employed to assess the suggested framework. The dataset has two types of resolution images: 100× and 400×.

400× Resolution Set: The second dataset comprises 400× resolution images that are separated into images showing abnormal and normal oral cancer. There are 696 images in this set; 495 of them show OSCC, while 201 show normal oral cavity tissue.

Image Acquisition Details: The Leica DM 750 microscope and the ICC50 HD camera were used to capture the images for the dataset. The high degree of resolution offered by the microscope and camera combination allowed for a thorough examination of the histology of mouth cancer. The dataset's images were all 2048 by 1536 pixels, which improved the amount of detail that could be collected. A few sample images of the selected dataset have been shown in Fig. 2.

2.2 Deep Learning Features

Fine-Tuned DarkNet-19: In 2016, the Darknet-19 idea was introduced. It is a fundamental part of the DarkNet architecture [19]. Because of its remarkable effectiveness and performance in a variety of tasks, such as object detection and classification, it has gained widespread recognition, especially in the field of computer vision [20]. A specific DarkNet framework called Darknet-19 was developed to offer a quick

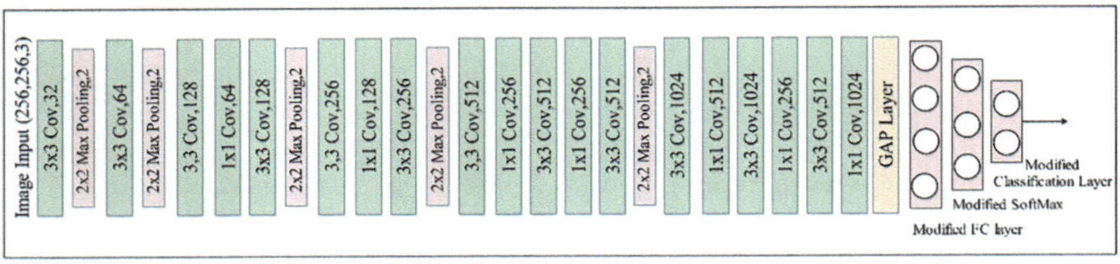

Fig. 3 Architecture of fined-tuned darknet19 model

and effective way to handle computer vision tasks such as real-time object detection. A suitable balance between computing performance and precision is the goal of this network's architecture. In order to efficiently capture detailed information in visual input, the architecture makes use of convolutional layers with modest filter sizes, such as 3 × 3 and 1 × 1 conv [21]. Darknet-19 benefits from the optimization process by using shortcut connections to improve information transfer between many tiers [22]. Darknet-19 is well-known for its ability to recognize objects in real-time. It does this by using global average pooling instead of fully linked layers, which helps to reduce the number of parameters in the model. The system's design also includes using the leaky rectified linear unit (Leaky ReLU) activation function to induce non-linearity [23]. Darknet-19 is notable for its capacity to achieve high performance in object recognition tasks while having a small size, making it appropriate for settings and applications with limited resources and requiring real-time processing. The architecture of modified darknet19 model is presented in Fig. 3.

Fine-tuned MobileNet-V2: MobileNet-V2 architecture is primarily designed for areas with limited resources and mobility [24]. The primary rationale behind this architecture type is its ability to minimize memory consumption and computational costs and design compliance with a greater emphasis on mobile applications [25]. It is significant to remember that the categorization layer of the design above originally contained roughly a thousand nodes. The main novelty of MobileNet-V2 is the use of inverted residuals, where the high-dimensional input is first subjected to a lightweight depth-wise separable convolution, and then a linear bottleneck layer is applied to capture intricate patterns [26]. This design option is ideal for deployment on mobile and edge devices since it lowers the computational cost and number of parameters. To maintain information flow throughout the network, MobileNet-V2 now includes the concept of linear bottlenecks [27]. The worldwide depth-wise separable convolution, which enables better parameter and computing resource usage, significantly improves the architecture's efficiency. With a decent mix of efficiency and accuracy, MobileNet-V2 has succeeded in several computer vision tasks [28]. The number of parameters in MobileNet-V2 is 3.5 M. The fact that it is so widely used proves how well it works to enable deep learning applications on devices with limited resources, making artificial intelligence technologies more widely available [29]. The architecture of the modified MobileNet-V2 model is presented in Fig. 4.

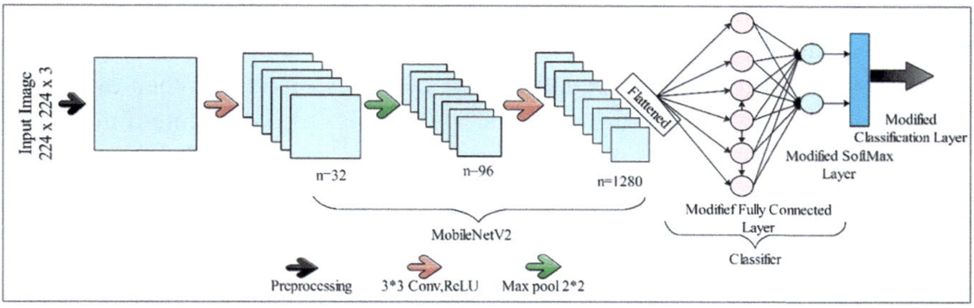

Fig. 4 Architecture of fined-tuned MobileNet-V2 model

Feature Extraction and Fusion: Both models were initially trained on a dataset of 1,000 unique classes (ImageNet). We improved the FC Layer by including a new FC Layer that encompasses selected dataset classes. The modified model was trained using deep transfer knowledge. The hyperparameters used for training are the learning rate, number of epochs, mini-batch size, and optimizer. The specific values allocated to these hyperparameters are 0.0001, 20, 12, and SGDM, respectively. After training, both trained models were utilized for feature extraction. The features were extracted from both models' global average pool activation. The dimensions of extracted features from both models were $N \times 1024$ and $N \times 1280$, respectively. These extracted features are later fused using a simple serial-based approach []. Later, a feature selection technique was implemented for the best feature selection.

2.3 *Chaotic Crow Search Features Selection*

Crow birds maintain complex social structures and interactions among them [30]. The crow search algorithm is an evolutionary and population-based algorithm that maps social interaction and conduction of crows to optimization of features in the computing domain.

In this algorithm, solutions are denoted by N, which represents the population of crows. The problem dimension is denoted by k. At a specific iteration e, the location of each crow is represented by a. A vector denotes the location concerning each iteration as follows:

$$g_a^e = \left[g_{a1}^e, g_{a2}^e, \ldots, g_{ak}^e\right] \text{ for } i = 1,2,3,\ldots,N \tag{1}$$

In dimension k, g_a^e denotes a possible solution for crow at location a at a specific iteration.

Suppose in a specific scenario crow a wants to steal from crow b in a specific scenario. Two possible situations may arise here; these are as follows:

1. Crow *b* was unable to know that crow *a* is following her, then crow *a* will get to know about food store of crow *b*, and after this crow *a* will update her current position as per given below equation.

$$g_a^{(e+1)} = q_a^{(e)} + u_a * ht_b^{(e)} * (m_b^{(e)} - q_b^{(e)}) \tag{2}$$

where length of flight is indicated by *ht* and u_i is a random number belonging to a range $\varepsilon = [0, 1]$ [0, 1].
2. Crow *b* knows that crow *a* is following her in order to get information about her food source. In this scenario, crow *a* is fooled by crow *b* by her random moves.

   Both of these cases can be combined together and mathematically can be represented as

$$g_a^{(e+1)} = \begin{cases} g_a^{(t)} + u_i * ht_b^{(e)} * \left(m_b^{(e)} - q_a^{(e)}\right), & u_b \geq AP_a^e \\ \text{Choose a random position,} & \text{otherwise} \end{cases} \tag{3}$$

where u_a and u_b are random numbers belonging to a range $\varepsilon = [0, 1]$. The probability of awareness for crow *b* is denoted by AP_a^e at a specific iteration *e*. Crows' ability to search is influenced by the value of *ht*; large values of *ht* support global search, while low levels support local search [31]. A fitness function assesses the whole population of crows. Each crow updates its position according to the fitness value provided by a fitness function. The feasibility of each new position is verified.

The memory of each crow is updated using the equation given below:

$$m_a^{(e+1)} = \begin{cases} g_a^{(e+1)} \mathit{iff} \left(g_a^{(e+1)}\right) \text{ is better than } f\left(g_a^{(e)}\right) \\ m_a^{(e)}, \quad \text{otherwise} \end{cases} \tag{4}$$

The exploration and exploitation of initial CSA is increased by the introduction of a new time varying function. The solution set is converted to a binary form, where the solution can only exist in the binary form {0, 1}. Agents from continuous space are transformed to binary space using the following equation.

$$g_a^{(e+1)} = \begin{cases} 1 \text{ if}\left(S(g_a^{(e+1)})\right) \geq r\,\text{and}() \\ 0 \text{ otherwise} \end{cases} \tag{5}$$

where

$$s = \frac{1}{1 + \text{euler}(e)^{10\left(g_a^{(e+1)} - 0.5\right)}}$$

One description of chaos is a phenomenon. Any modification to its initial condition can potentially cause non-linear changes in subsequent behavior.

Chaotic variables are employed in CCSA in place of random variables to alter the location of the crow.

In the above equation, the chaotic map value at *j*th iteration is represented by C_j and C_z represents the value of a chaotic map that is obtained at *z*th iteration. The selected features of this equation are finally passed to classifiers for the final results.

3 Results and Discussion

In this section, the suggested framework's experimental procedure has been covered. A publicly accessible dataset with two classes—normal and OSCC—has been used to assess the suggested framework. With a cross-validation value of 10, the training and testing ratio was 50:50. The following neural network based classifiers are used to categorize features: narrow neural network (N3), medium neural network (MN2), wide neural network (WN2), bi-layered neural network (Bi-Layered N2), and tri-layered neural network (Tri-layered N2). Recall, precision, F1-core, accuracy, duration, and other performance indicators are used to gauge each classifier's performance. Using MATLAB R2023a and a personal computer equipped with a Core i7 13th generation CPU, 12 GB GPU, and 16 GB of RAM, the full experiment was conducted.

3.1 Results

According to the results, MobileNet-V2 achieved an accuracy of 86.3% and an error rate of 13.7% for the narrow neural network classifier in 41.543 s. In contrast, DarkNet-19 achieved an improved accuracy of 89.2%, a lower error rate of 13.7% to 10.8%, and a shorter execution time of 38.994 s. The combination of both models produced an improved accuracy and error rate of 90.1% and 9.9%, respectively. But there is a significant increase in execution time. With a time of 43.343 s, MobileNet-V2 achieved an accuracy of 88.1% and an error rate of 11.9% in a medium neural network classifier. In contrast, DarkNet-19 achieved an improved accuracy of 90.0%, a reduction in error rate from 11.9% to 10%, and a time reduction of 39.091 s in execution. The combination of both models produced an improved accuracy and error rate of 90.7% and 9.3%, respectively. Nevertheless, the execution time has significantly increased to 54.401. In the wide neural network classifier, MobileNet-V2 achieved an accuracy of 90.6% and an error rate of 9.4% in 41.023 s. In contrast, DarkNet-19 achieved an accuracy of 89.5% and an error rate of 10.5%, but the execution time was also shortened at 37.192 s. The combination of both models produced an improved accuracy and error rate of 91.4% and 8.6%, respectively. Nevertheless, there is a significant increase in execution time—50.113.

Table 1 Classification results of normal and OSCC using the middle steps of the proposed framework

Classifier	Features extraction methods			Performance measures		
	MobileNet-V2	DarkNet-19	Fusion	accuracy (%)	Error rate (%)	Time (s)
N^3	✓			86.3	13.7	41.543
		✓		89.2	10.8	38.994
			✓	90.1	9.9	51.554
MN^2	✓			88.1	11.9	43.343
		✓		90.0	10	39.091
			✓	90.7	9.3	54.401
Wide N^2	✓			90.6	9.4	41.023
		✓		89.5	10.5	37.192
			✓	91.4	8.6	50.113
Bi-layered N^2	✓			89.4	10.6	46.004
		✓		88.6	11.4	41.884
			✓	90.5	9.5	59.027
Tri-layered N^2	✓			87.4	12.6	44.376
		✓		88.8	11.2	40.900
			✓	90.2	9.8	56.115

In a bi-layered neural network classifier, MobileNet-V2 achieved an accuracy of 89.4% and an error rate of 10.6% in 46.004 s. In contrast, DarkNet-19 achieved an accuracy of 88.6% but an increased error rate of 10.6% to 11.4%, along with a reduced execution time of 41.884 s. The combination of both models produced an improved accuracy and error rate of 90.5% and 9.5%, respectively. But the execution time has significantly increased to 59.027. MobileNet-V2 achieved an accuracy of 87.4% and an error rate of 12.6% in the Tri-layered neural network classifier in 44.376 s. In contrast, DarkNet-19 achieved an improved accuracy of 88.8%, a reduction in error rate from 12.6% to 11.2%, and a reduction in execution time of 40.900 s. The combination of both models produced an improved accuracy and error rate of 90.2% and 9.8%, respectively. But the execution time has significantly increased to 56.115. The accuracy was enhanced overall by the fusion procedure, although the calculation time was increased (Table 1).

Finally, the proposed framework's complete results are presented in Table 2. It is evident from the results that wide neural network classifier obtained higest accuracy of 92.9% with sensitivity, precision, and F1 score values are 92.84%, 92.70, and 92.17%, respectively. The performance of this classifier can be further confirmed by a confusion matrix, as shown in Fig. 5. The diagonal values in this figure show the correct predicted values. In addition, the rest of the classifiers obtained accuracy of 91.62, 92.50, 91.90, and 91.64%, respectively. The minimum execution time is 7.6697 (s) for Medium NN. Comparing the results of this table with Table 1, it is

Table 2 Classification results of oral cavity cancer using proposed framework of 400× images

Classifier	Sensitivity rate (%)	Precision rate (%)	F1 score (%)	FPR	AUC	Accuracy (%)	Time (s)
N^3	91.6	91.25	91.25	0.080	0.96	91.62	15.157
MN^2	92.62	92.44	92.14	0.07	0.97	92.50	**7.6697**
Wide N^2	92.84	92.70	92.17	0.07	0.97	**92.90**	11.897
Bi-layered N^2	91.74	91.50	91.16	0.08	0.95	91.64	16.235
Tri-layered N^2	91.50	91.52	91.15	0.08	0.94	91.62	23.57

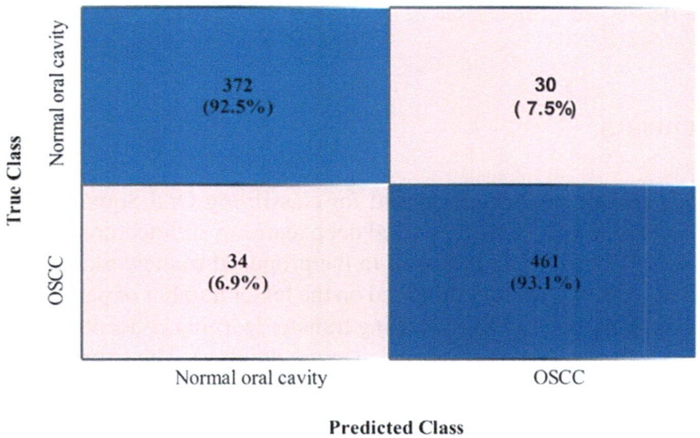

Fig. 5 Confusion matrix of Wide N^2 using the proposed framework

shown that the proposed framework obtained accuracy is improved and time is significantly reduced. Moreover, Table 3 shows the comparison of the proposed framework accuracy with the combination of several pre-trained CNN architectures and Chaotic Crow Search optimization. In this table, it is analyzed that the combination of MobileNet-V2 and DarkNet-19 is better than all other combinations.

Table 3 Comparison of proposed framework with other combinations of networks

Method	Dataset	Accuracy (%)	Time (s)
VGG16 + AlexNet + Chaotic Crow Search	Histopathologic oral cancer 400×	88.40	27.883
VGG19 + AlexNet + Chaotic Crow Search	Histopathologic oral cancer 400×	88.94	31.554
ResNet50 + AlexNet + Chaotic Crow Search	Histopathologic oral cancer 400×	90.32	25.774
Resnet-101 + AlexNet + Chaotic Crow Search	Histopathologic oral cancer 400×	91.4	22.826
ResNet-101 + MobileNet-V2 + Chaotic Crow Search	Histopathologic oral cancer 400×	91.8	20.993
Proposed = ResNet-101 + MobileNet-V2 + Chaotic Crow Search	Histopathologic oral cancer 400×	92.90	11.897

4 Conclusion

A fusion framework has been proposed for classifying Oral Squamous Cell Carcinoma (OSCC) from HI. Two pre-trained deep learning architectures, MobileNet-V2 and DarkNet-19, have been fine-tuned in the proposed framework. The selection of pre-trained models was performed based on the lesser number of parameters and top-5 accuracy. Both models are trained using transfer learning concepts and features are extracted from the deeper layer (global average pooling). The extracted features are fused using a serial approach and performed classification. The results of this step improved the accuracy than fine-tuned MobileNet-V2 and DarkNet-19; however, the computational time was also increased. Therefore, we implemented a Chaotic Crow Search optimization algorithm and selected the best features. The selected features are classified using machine learning classifiers. Comparison is conducted with recent techniques and shows improved accuracy and lessen the overall computational time. In future, the attention mechanism has been added in the pre-trained models for better accuracy and computational time. In addition, an improved whale optimization technique will be implemented for the best feature selection.

References

1. Rahman, A.-U., Alqahtani, A., Aldhafferi, N., Nasir, M.U., Khan, M.F., Khan, M.A., et al.: Histopathologic oral cancer prediction using oral squamous cell carcinoma biopsy empowered with transfer learning. Sensors **22**, 3833 (2022)
2. Ananthakrishnan, B., Shaik, A., Kumar, S., Narendran, S., Mattu, K., Kavitha, M.S.: Automated detection and classification of oral squamous cell carcinoma using deep neural networks. Diagnostics **13**, 918 (2023)

3. Ahmad, M., Irfan, M.A., Sadique, U., Haq, I. u., Jan, A., Khattak, M.I., et al.: Multi-method analysis of histopathological image for early diagnosis of oral squamous cell carcinoma using deep learning and hybrid techniques. Cancers **15**, 5247 (2023)
4. Liyanage, V., Tao, M., Park, J.S., Wang, K.N., Azimi, S.: Malignant and non-malignant oral lesions classification and diagnosis with deep neural networks. J. Dent. **137**, 104657 (2023)
5. Begum, S.H., Vidyullatha, P.: Deep learning model for automatic detection of oral squamous cell carcinoma (OSCC) using Histopathological Images. Int. J. Comput. Digit. Syst. **13** (2023)
6. Mostafa, R.R., Zhang, Y.-D., Baili, J., Alhaisoni, M., Tariq, U., et al.: Deep-Net: fine-tuned deep neural network multi-features fusion for brain tumor recognition. Comput., Mater. & Contin. **76** (2023)
7. Wang, J., Wang, S., Zhang, Y.: SNSVM: SqueezeNet-guided SVM for breast cancer diagnosis. Comput., Mater. & Contin. **76** (2023)
8. Dillshad, V., Nazir, M., Saidani, O., Alturki, N., Kadry, S.: D2LFS2Net: Multi-class skin lesion diagnosis using deep learning and variance-controlled Marine Predator optimisation: an application for precision medicine. CAAI Trans. Intell. Technol. (2023)
9. Ijaz, M., Ashraf, I., Zahid, U., Yasin, A., Ali, S., Attique Khan, M., et al.: A decision support system for lung colon cancer classification using fusion of deep neural networks and normal distribution based gray wolf optimization. In: ACM Transactions on Asian and Low-Resource Language Information Processing (2023)
10. Ali, M.U., Zafar, A., Tanveer, J., Kim, S.H., Alsulami, M.M., et al.: Deep learning network selection and optimized information fusion for enhanced COVID-19 detection. Int. J. Imaging Syst. Technol.
11. Khan, A., Alhaisoni, M., Alqahtani, A., Alsubai, S., Alharbi, M., et al.: Multimodal brain tumor detection and classification using deep saliency map and improved dragonfly optimization algorithm. Int. J. Imaging Syst. Technol. **33**, 572–587 (2023)
12. Figueroa, K.C., Song, B., Sunny, S., Li, S., Gurushanth, K., Mendonca, P., et al.: Interpretable deep learning approach for oral cancer classification using guided attention inference network. J. Biomed. Opt. **27**, 015001–015001 (2022)
13. Jeyaraj, P.R., Samuel Nadar, E.R.: Computer-assisted medical image classification for early diagnosis of oral cancer employing deep learning algorithm. J. Cancer Res. Clin. Oncol. **145**, 829–837 (2019)
14. Al Duhayyim, M., Malibari, A., Dhahbi, S., Nour, M.K., Al-Turaiki, I., Obayya, M.I., et al.: Sailfish optimization with deep learning based oral cancer classification model. Comput. Syst. Sci. Eng., **45**, 753–767 (2023)
15. Welikala, R.A., Remagnino, P., Lim, J.H., Chan, C.S., Rajendran, S., Kallarakkal, T.G., et al.: Automated detection and classification of oral lesions using deep learning for early detection of oral cancer. IEEE Access **8**, 132677–132693 (2020)
16. Huang, C., Zhang, G., Chen, S., de Albuquerque, V.H.C.: An intelligent multisampling tensor model for oral cancer classification. IEEE Trans. Industr. Inf. **18**, 7853–7861 (2022)
17. Joshi, P., Alsadoon, O.H., Alsadoon, A., AlSallami, N., Rashid, T.A., Prasad, P., et al.: Deep learning for size and microscope feature extraction and classification in Oral Cancer: enhanced convolution neural network. Multimed. Tools Appl. **82**, 6197–6220 (2023)
18. Rahman, T.Y., Mahanta, L.B., Das, A.K., Sarma, J.D.: Histopathological imaging database for oral cancer analysis. Data Brief **29**, 105114 (2020)
19. Redmon, J., Farhadi, A.: YOLO9000: better, faster, stronger. In: Proceedings of the IEEE Conference on Computer Vision and Pattern Recognition, pp. 7263–7271 (2017)
20. Al-Haija, Q.A., Smadi, M., Al-Bataineh, O.M.: Identifying phasic dopamine releases using DarkNet-19 convolutional neural network. In: 2021 IEEE International IoT, Electronics and Mechatronics Conference (IEMTRONICS), pp. 1–5 (2021)
21. Mahum, R., Azim, T.: Proteins Classification Using An Improve Darknet-53 Deep Learning Model. In: 2022 Mohammad Ali Jinnah University International Conference on Computing (MAJICC), pp. 1–6 (2022)
22. Choudhry, Z.A., Shahid, H., Aziz, S., Naqvi, S.Z.H., Khan, M.U.: DarkNet-19 Based Intelligent Diagnostic System for Ocular Diseases. Iran. J. Sci. Technol., Trans. Electr. Eng. **46**, 959–970 (2022)

23. Sudha, D., BS, S.: An automated system of identifying kidney stones using Darknet-19, a deep learning model in MATLAB. J. Coast. Life Med. **11**, 1578–1585 (2023)
24. Hamid, Y., Wani, S., Soomro, A.B., Alwan, A.A., Gulzar, Y.: Smart seed classification system based on MobileNetV2 architecture. In: 2022 2nd International Conference on Computing and Information Technology (ICCIT), pp. 217–222 (2022)
25. Sandler, M., Howard, A., Zhu, M., Zhmoginov, A., Chen, L.-C.: Mobilenetv2: Inverted residuals and linear bottlenecks. In: Proceedings of the IEEE Conference on Computer Vision and Pattern Recognition, pp. 4510–4520 (2018)
26. Akay, M., Du, Y., Sershen, C.L., Wu, M., Chen, T.Y., Assassi, S., et al.: Deep learning classification of systemic sclerosis skin using the MobileNetV2 model. IEEE Open J. Eng. Med. Biol. **2**, 104–110 (2021)
27. Knapheide, J., Stabernack, B., Kuhnke, M.: A high throughput MobileNetV2 FPGA implementation based on a flexible architecture for depthwise separable convolution. In: 2020 30th International Conference on Field-Programmable Logic and Applications (FPL), pp. 277–283 (2020)
28. Dong, K., Zhou, C., Ruan, Y., Li, Y.: MobileNetV2 model for image classification. In: 2020 2nd International Conference on Information Technology and Computer Application (ITCA), pp 476–480 (2020)
29. Sheikh, S., Qidwai, U.: Using MobileNetV2 to classify the severity of diabetic retinopathy. Int. J. Simul. Syst. Sci. Technol **21** (2020)
30. Takeda, K., Takahashi, N., Izawa, E.-I.: Social encounters produce different autonomic response between dominants and subordinates in crows. R. Soc. Open Sci. **9**, 220972 (2022)
31. Adamu, A., Abdullahi, M., Junaidu, S.B., Hassan, I.H.: An hybrid particle swarm optimization with crow search algorithm for feature selection. Mach. Learn. Appl. **6**, 100108 (2021)

Author Index

G. Fortino et al. (eds.), *Proceedings of Third International Conference on Computing and Communication Networks*, Lecture Notes in Networks and Systems 917,
https://doi.org/10.1007/978-981-97-0892-5

GPSR Compliance

The European Union's (EU) General Product Safety Regulation (GPSR) is a set of rules that requires consumer products to be safe and our obligations to ensure this.

If you have any concerns about our products, you can contact us on ProductSafety@springernature.com

In case Publisher is established outside the EU, the EU authorized representative is:

Springer Nature Customer Service Center GmbH
Europaplatz 3
69115 Heidelberg, Germany

Batch number: 10726030

Printed by Printforce, the Netherlands